Proceedings of the Workshop on

Limit Equilibrium, Plasticity and Generalized Stress-Strain In Geotechnical Engineering

McGill University
May 28-30, 1980

Sponsored by:
National Science Foundation, U.S.A.
National Science and Engineering Research Council, Canada
McGill University
Univ. of Colorado

Raymond K. Yong and Hon-Yim Ko, Workshop Co-Chairmen

Published by the
American Society of Civil Engineers
345 East 47th Street
New York, New York 10017

```
TA
710
.A1
W67
1980
```

Copyright © 1981 by the American Society of Civil Engineers,
All Rights Reserved.
Library of Congress Catalog Card No. 81-69233
ISBN 0-87262-282-7
Manufactured in the United States of America

TABLE OF CONTENTS

Part 1

1. Introduction and Organization of Workshop 1

2. Pre-Workshop Information Package 13

3. Challenge from Industry and Academia 60

4. (A) Data to be Predicted

   (B) Predictions and Comparisons

   (i) Strain Predictions Using Hyperbolic Stress-Relations, by J. M. Duncan 245

   (ii) Evaluation of Predictions, by E. Kavazanjian, J. K. Mitchell and R. Bonaparte 254

   (iii) Nonlinear Hyperelastic (Green) Constitutive Models for Soils: Predictions and Comparisons, by A. F. Saleeb and W. F. Chen . 265

   (iv) Prediction of Soil Behavior by Endochronic Theory, by A. M. Ansal, R. J. Krizek and Z. P. Bazant 286

   (v) Plasticity Models for Soils-Comparison and Discussion, by E. Mizuno and W. F. Chen 328

   (vi) Predictions of Stress-Strain Behavior for Ottawa Sand, by P. V. Lade 352

   (vii) Comparison of Cap Model Predictions with Laboratory Data for Soils, by G. Y. Baladi and I. S. Sandler 364

   (viii) Prediction of the Response of the Natural Clays X and Y Using the Bounding Surface Model, by Y. F. Dafalias, L. R. Herrmann and J. S. DeNatale 402

   (ix) Predictions and Comparison, by K. Akai, T. Adachi and F. Oka 416

5. Reports from Group Chairmen

   Group 1—J. T. Christian 93

   Group 2—H. B. Poorooshasb and E. T. Selig 102

   Group 3—W. D. L. Finn 132

   Group 4—R. F. Scott 151

6. Synthesis and Recommendations 166

7. Post Workshop Comparisons 176

Part 2
1. Position Papers by Predictors
    - (i) Hyperbolic Stress-Strain Relationships 44
      by J. M. Duncan

    - (ii) Stress-Deformation Predictions Using a General Phenomenological Model 46
      by E. Kavazanjian, J. K. Mitchell and R. Bonaparte

    - (iii) Nonlinear Hyperelastic (Green) Constitutive Models for Soils: Theory and Calibration 49
      by A. F. Saleeb and W. F. Chen

    - (iv) Critical Appraisal of Endochronic Theory for Soils 52
      by Z. P. Bazant, A. M. Ansal and R. J. Krizek

    - (v) Plasticity Models for Soils 55
      by E. Mizuno and W. F. Chen

    - (vi) A Critical State Model for Predicting the Behavior of Clays ... 59
      by C. P. Wroth and G. T. Houlsby

    - (vii) Elasto-Plastic Stress-Strain Model for Sand 62
      by P. V. Lade

    - (viii) Examples of the Use of the Cap Model for Simulating the Stress-Strain Behavior of Soils 64
      by G. Y. Baladi and I. S. Sandler

    - (ix) Description of Natural Clay Behavior by a Simple Bounding Surface Plasticity Formulation 71
      by Y. F. Dafalias, L. R. Herrmann and J. Scott DeNatale

    - (x) Constitutive Theory for Soil 74
      by J. H. Prevost

    - (xi) Constitutive Models for Clays and Sands 81
      by K. Akai, T. Adachi and F. Oka

    - (xii) A Constitutive Law for Sands and Clays—Position, Prediction and Evaluation 83
      by D. Kolymbas and G. Gudehus

2. List of Workshop Participants 86

Part 3
Subject Index 86

Author Index 87

NSF/NSERC NORTH AMERICAN WORKSHOP ON
GENERALIZED STRESS-STRAIN AND PLASTICITY THEORIES FOR SOILS

INTRODUCTION AND ORGANIZATION OF WORKSHOP
by Raymond N. Yong and Hon-Yim Ko

Introduction

The following set of Proceedings provides a detailed presentation of the material prepared for the WORKSHOP, the deliberations of the various discussion groups, and the conclusions derived from the concensus input and discussions from the various participants in the WORKSHOP. This North American WORKSHOP on Generalized Stress-Strain and Plasticity Theories for Soils, funded by the National Science Foundation, U.S.A. and the National Sciences and Engineering Research Council of Canada, was held at McGill University in the period of May 28 through 30, 1980.

The reasons for the organization and the operation of the WORKSHOP can be traced to several identifiable needs:

(1) The need to study the problems associated with the proper development of generalized stress-strain relations and plasticity principles for accurate modelling of the rheological performance of soils.

(2) The requirement for a critical examination and discussion of the adequacy of presently available test procedures, techniques, and analytical tools for the study of generalized stress-strain and soil plasticity.

(3) The need to evaluate and determine the conditions and requirements for constitutive formulations of soils to meet the design/analytical demands from geotechnical engineering practice.

(4) The need to provide the necessary recommendations from the deliberations of (1), (2) and (3), for the researchers to establish research priorities and needs in their further studies of the problem, and for the practicing profession to guide them into proper usage of present theories.

Purpose and Objectives

Responding to the above needs, the objectives for the WORKSHOP focused directly on those needs and provided the purpose of the WORKSHOP - i.e. to bring together the more informed active researchers, in the discussions, evaluations, deliberations, of not only the state-of-the art for the various aspects of the problems of modelling, testing and evaluation, but also to exmaine the deficiencies, the merits and requirements for successful solution of the problem of specification of proper and pertinent soil constitutive relationships and yield/failure criteria.

Because of the wide scope and range of material to be covered, the Specific Objectives of the WORKSHOP dealt directly with the development and use of the various models for prediction of stress-strain performance and yield. The area of soil testing-- i.e., instruments, techniques, procedures and methods for data reduction and analysis-- was deliberately assigned for

assigned for examination, study and discussion in the ASTM
Symposium on Shear Strength of Soils which was scheduled for
June 25, 1980.* In concert with the above, the application of
stress-strain and yield theories to Geotechnical Engineering
Practice was assigned to ASCE as a One-Day Symposium--scheduled
for October 28, 1980 in Hollywood, Florida.**

Implementation of the WORKSHOP Plan

To fulfill the objectives of the WORKSHOP, the work/study
program focused around predictive models and formulations
presently available for stress-strain and the plastic yielding
of soils, and the requirements and procedures needed to provide
a more accurate modelling of the stress-strain and yield
performance of soils. One of the main concerns of the WORKSHOP
was with one's capability or ability to accurately or properly
formulate stress-strain relations and yield theories applicable
to soils subjected to various stress paths and stress histories.

To provide the focus for the WORKSHOP, several "predictive"
models reported in recent literature which were currently in use
--and also in the stage of advanced development--were selected
for examination and study. These ranged from the
empirical/phenomenological to the various forms of cap and state
boundary models. A number of key proponents of these models,

---
*The Proceedings of the ASTM Symposium will appear as STP-740.
**The Proceedings of the ASCE Symposium will appear as
"Applications of Plasticity and Generalized Stress-Strain
Relationships in Geotechnical Engineering.

identified as Predictors, were asked to use their "models" to formulate stress-strain relationships and plastic yielding criteria based upon soil information and task data supplied to them some few months before the WORKSHOP. Using their developed relationships, the Predictors were then required to apply them to predict soil behavior (for the same soils tested) for certain specified conditions and situations.

It is useful to observe that in selecting the kinds of soil and test data to be provided as a basic data set to the Predictor, the Organizers of the WORKSHOP were motivated by several overriding concerns:

1. Test data reduced to a simple format for general application--to allow for a broad spectrum of user application.

2. A simple range of soils--sand, remoulded clay and a natural soil.

3. Simple loading paths for calibration purposes from which the capability of the models were to be tested for application or prediction of performance of these same soils subject to more complex loading patterns.

4. Capability of models for use in general as opposed to specialized situations--both in terms of calibration data requirements and applications.

The Predictors were:

Professors K. Akai and T. Adachi, Kyoto University

Dr. G.Y. Baladi, U.S. Army Corps of Engineers and Dr. I. Sandler, Weidlinger Associates

Professor Z.P. Bazant, Northwestern University and Dr. A. Ansal, Istanbul Technical University

Professor W.F. Chen, Purdue University

Professor J.M. Duncan, University of California, Berkeley

Professor Y.F. Dafalias and Professor L.R. Herrmann, University of California at Davis

Professor G. Gudehus, Karlsruhe University

Professor P.V. Lade, University of California, Los Angeles

Professor J.K. Mitchell, University of California, Berkeley and Professor E. Kavazanjian, Stanford University

Professor J.H. Prevost, Princeton University

Professor C.P. Wroth, Oxford University

The actual conduct of the WORKSHOP can be seen in the program details given in the Information Package presented in Section 2. The three distinct parts of the deliberations consisted of:

1. Presentations, discussions, and comparisons of the various constitutive relationships proposed by the various Predictors, in Individual Working Groups. Four Working Groups consisting of 17 to 20 participants per Group were formed to allow for closer interaction between the participants during the deliberations. The various models used to arrive at the stress-strain relations and yield or failure of the soils evaluated by all the Predictors in each Group were examined and comparisons made with actual performance of the soils. The Working Group Chairmen, Secretaries and the participants in each group were:

Chairman:	J.T. Christian	H. Poorooshasb	W.D.L. Finn	R.F. Scott
Secretary:	F.C. Townsend	E.T. Selig	A.S. Saada	H.W. Olsen
	Z.P. Bazant	G. Gudehus	H.E. Read	A. Ansal
	J.H. Prevost	G.Y. Baladi	S. Koh	H. Ludwig
	K.R. Demars	R.J. Krizek	J.M. Duncan	Y.F. Dafalias
	I. Sandler	P.V. Lade	T. Adachi	I. Vardoulakis
	W.F. Chen	J.K. Mitchell	C.P. Wroth	E. Kavazanjian
	B.O. Hardin	G.W. Clough	C.C. Ladd	S.G. Wright
	R.D. Holtz	S. Sture	H.E. Wahls	K.Y. Lo
	L. Domaschuk	F. Tavenas	A.M. Hanna	W. Marcuson
	M.M. Baligh	T.C. Kenney	K.R. Rowe	S.J. Poulos
	R.P. Khera	R.L. McNeill	H.M. Horn	C.S. Chang
	L. Hansen	R.S. Ladd	K.H. Stokoe	G.T. Houlsby
	R. Singh	R.C. Kirby	V. Drnevich	L.M. Kraft
	K.T. Law	M.M. Tabba	R.E. Brown	V. Silvestri
	M. Soulie	I. Lam	P. Mayne	S. Leroueil
	J.H.A. Crooks			

2. Presentation of the various models for stress-strain and yielding by the Predictors to all participants in the Plenary Sessions. Examination of new concepts and theories and the requirements for modelling of actual test performance of the soils were detailed. The review and discussion by participants of model requirements addressed the adequacies and deficiencies of present models and theories, and the kinds of models needed to solve the problems faced in actual practice.

3. A listing of the priorities and needs for further research into the problems for establishment of more accurate models for stress-strain and yielding.

Organization and Presentation of WORKSHOP Proceedings

To provide for a more continuous presentation of the material covered before, during and after the WORKSHOP, this set of Proceedings is divided into two distinct parts. Part 1 contains the details concerned with the central issues of the WORKSHOP while Part 2 provides the background position papers from the Predictors, and a listing of all participants.

The seven (7) sections in Part 1 deal with the following items:

Section 1 - Introduction and Organization of Workshop

The background development of the WORKSHOP is given together with the purpose and objectives of the WORKSHOP. The other items in this section deal with the implementation of the objectives, the selection of test data and models, and the use of four working discussion groups, and the reporting procedure for the Proceedings.

Section 2 - Information Package

The information package supplied to all participants before the WORKSHOP contains the basic data set initially sent to all Predictors some months before the WORKSHOP. By including the basic data set in the information package, it was assumed that some participants might also wish to test their own pet theories [or other available theories], and to use their own evaluations as a basis for discussion in the WORKSHOP.

The Challenge Paper contained in the information package was intended to serve as a vehicle to provide all the participants with a capsule view of the PROBLEM as presently understood from a general understanding of the state-of-the art. To complete the information package, the agenda and selected background references were also provided in order to allow all participants to better prepare themselves for the WORKSHOP.

Section 3 - Challenges from Industry and Academia

A separate challenge, each from industry and academia, for predictive requirements and objectives is presented in this section. These challenges were given in the course of the WORKSHOP in the Plenary Sessions.

Section 4 - Predictions and Comparisons

The data and test results to be predicted by the Predictors are presented in this section. These are followed by the actual predictions made by the individual Predictors themselves. The ordering of the predictions begin with the empirical/ phenomenological models and end with the more sophisticated forms of plasticity and hypoelasticity models. The predictions and evaluations have been prepared by the individual Predictors.

Section 5 - Reports from Group Chairmen

The chairmen from each of the four working groups have provided their synthesis and evluation of the discussions, predictions and comparisons made in the working sessions. Recommendations and conclusions drawn by the chairmen in their individual reports pertain particularly to those models considered in each separate group. Note that not all models were presented in each group, since individual Predictors could not be present in each of all the concurrent working sessions.

Section 6 - Synthesis and Recommendations

The various comments given by the Group chairmen and several other pertinent observations are synthesized in this section, the outcome of which has allowed for a set of recommendations to be proposed and developed.

Section 7 - Post WORKSHOP Comparisons

Following the WORKSHOP, all Predictors were requested to provide the Organizers with a digitized output of their predictions in order that comparisons could be made on a more uniform basis using a simple variance technique. Most of the Predictors were able to comply with this request. The detailed graphs presented in this section provide a quantitative comparison of the predictive capability of those models tested, using the actual test data given in Section 4 and the digitized values provided by the Predictors.

With the presentation of material in this section, Part 1 can be considered as complete in that the main purpose and objectives of the WORKSHOP have been fully addressed.

Part 2 of the Proceedings is provided to allow the individual Predictors the opportunity to show the basic development and structure of their own models, and also to show how the models were calibrated and applied. The observations and comments made by the individual Predictors are indeed most pertinent and valuable.

The concluding material in Part 2 contains the list of all participants in the WORKSHOP.

Concluding Remarks

The success of the WORKSHOP is a testimony to not only the preparatory work done by the Task Committee associated with the Grantees funded by NSF and NSERC, but also to the unstinting cooperation and enthusiastic participation by the Predictors, Chairmen and Secretaries of the Working Groups. The input, intense give and take by all participants heightened the interaction between all participants and predictors. All these combined to provide for a very stimulating WORKSHOP. The Task Committee associated with the Grantees, H.Y. Ko for the NSF Grant and R.N. Yong for the NSERC Grant, were:

> Professor A.S. Saada, Case Western Reserve University
> Dr. H.W. Olsen, U.S.G.S.
> Professor E.T. Selig, University of Massachusetts
> Professor K.Y. Lo, University of Western Ontario
> Professor F.C. Townsend, University of Florida
> Professor R.D. Japp, McGill University

Particular attention and acknowledgement should also be given to the laboratories which supplied the data for the Predictors--University of Colorado, Case Western Reserve University and McGill University.

We also wish to record the assistance provided by Professor S. Sture, of the University of Colorado, and by Mrs. M.L. Powell of the Geotechnical Research Centre, McGill University, and the postgraduate students D.R. Elmonayeri, S. O. Li, J. C. Mould,

K. S. Ng, N. Sciadas, K. H. S. Siu, K. M. Tang, K. Y. Tang and G. Wong from Mc Gill University and the University of Colorado.

Finally, we note once again that the WORKSHOP would not have been made possible without the financial support provided by the National Science Foundation (U.S.A.) and the Natural Sciences and Engineering Research Council (Canada).

2. PRE-WORKSHOP INFORMATION PACKAGE

About six months before the Workshop, a package of data was sent to the Predictors who had shown an interest in participating in the prediction exercise at the Workshop. Data sent included physical properteis of four soils, identified as clays x and y, Kaolin and Ottawa sand, and stress-strain data from each soil along certain loading paths. The stress paths along which the behavior of each soil was to be predicted were also specified.

A challenge paper by R.N. Yong and H.-Y. Ko was included in the package, which was later sent to all invited Workshop participants, in the hope that they would arrive at the Workshop fully prepared to contribute to the discussions.

This Section contains the data package and the challenge paper. It is hoped that the data herein presented would be useful to the future development of any new or refinement of existing constitutive models.

(INFORMATION PACKAGE FOR)

WORKSHOP

on

LIMIT EQUILIBRIUM, PLASTICITY
AND GENERALIZED STRESS-STRAIN
IN GEOTECHNICAL ENGINEERING

McGILL UNIVERSITY
MAY 28-30, 1980

SPONSORED BY:

National Science Foundation, U.S.A.
Natural Sciences and Engineering Research Council, Canada
McGill University
University of Colorado

WORKSHOP CO-CHAIRMEN:

Raymond N. Yong
Hon-Yim Ko

McGill University
GEOTECHNICAL RESEARCH CENTRE

DIRECTOR
DR. RAYMOND N. YONG
WILLIAM SCOTT PROFESSOR OF
CIVIL ENGINEERING AND APPLIED MECHANICS

NSF/NSERC NORTH AMERICAN WORKSHOP ON
Plasticity Theories and Generalized Stress-Strain Modelling of Soils

McGill University
Montreal

28-30 May 1980

TENTATIVE SCHEDULE

ALL meetings will be held on Campus.

HOUSING for participants will be provided at the University Residences of McGill University.

PARTICIPANTS should expect to arrive on TUESDAY, 27 May 1980

Postal Address: 817 SHERBROOKE ST. W., MONTREAL, PQ. CANADA H3A 2K6

Tel. (514) 392-4751   Télex 05-268510

Montreal Arrival Information for NSF/NSERC Workshop on Plasticity

1. PARTICIPANTS should plan to arrive on Tuesday, May 27, 1980.

2. AIRPORT in Montreal is located in *Dorval* —

    Upon arrival, please take ground transportation (AIRPORT BUS) to the end of line for *Downtown* - i.e. *City* airport bus.

    This brings you to Sheraton-Mt. Royal Hotel on Peel St.

    Use map below to get to the University Residences at the top of University Street - on your right - for your lodgings. Note this is an 8-block walk. Taxis are available at the Sheraton-Mt. Royal Hotel, if desired.

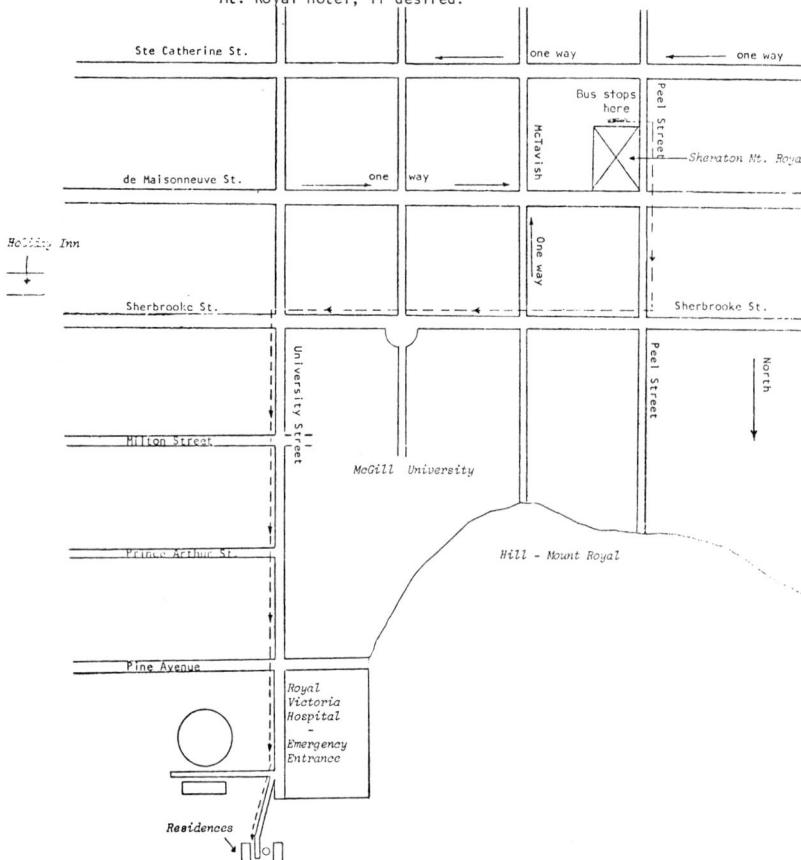

PRE-WORKSHOP PACKAGE

Tuesday, 27 May 1980

   8.00 pm - 10.00 pm - R E G I S T R A T I O N

Wednesday, 28 May

   8.30 am - 8.45 am - WELCOME and INTRODUCTION to WORKSHOP
                                    Reasons for Specific Objectives

   8.45 am - 9.05 am - *Challenge Paper* - ANALYTICAL AND PHYSICAL MODELS
   9.05 am - 9.25 am - Presentation of Data Base and Charge to *Predictors*
   9.25 am - 10.20 am - Individual Predictor Presentations

   10.20 am - 10.40 am - *Coffee Break*

   10.40 am - 12.30 pm - Continuation of Individual Predictor Presentations

   12.30 pm - 2.00 pm - *Group Lunch*

   2.00 pm - 3.45 pm - Continuation of Individual Predictor Presentations

   3.45 pm - 4.10 pm - *Coffee Break*

   4.10 pm - 5.30 pm - (a) Evaluation of Predictions
                                    (b) First Charge to Working Discussion Groups -
                                          CRITIQUE OF PREDICTION MODELS - INCLUDING TEST REQUIREMENTS

   6.15 pm - 7.30 pm - *Group Dinner*

   7.30 pm - 10.00 pm - Working Discussion Groups - deliberations.

Thursday, 29 May

   8.30 am - 10.00 am - "First" Reports from Individual Working Discussion Groups
   10.00 am - 10.30 am - *Challenge Paper:* USER NEEDS, ACADEMIC VIEWPOINT

   10.30 am - 10.50 am - *Coffee Break*

   10.50 am - 11.20 am - *Challenge Paper:* USER NEEDS, INDUSTRY VIEWPOINT
   11.20 am - 12.15 pm - Discussion
   12.15 pm - 12.30 pm - Second Charge to Working Discussion Groups
                                    EVALUATION OF MODELS IN RELATION TO THEIR APPLICABILITY IN ENGINEERING PRACTICE

   12.30 pm - 2.00 pm - *Group Lunch*

   2.00 pm - 5.00 pm - Deliberations - Individual Working Discussion Groups
   6.00 pm   - *No host cocktails*

Friday, 30 May

   8.30 am - 10.00 am - "Second" Reports from Individual Working Discussion Groups
   10.00 am - 10.20 am - Third Charge to Working Discussion Groups -
                                    RESEARCH NEEDS AND GUIDES TO PRACTICE

   10.20 am - 10.40 am - *Coffee Break*

   10.40 am - 12.30 pm - Final Deliberations of Working Discussion Groups

   12.30 pm - 1.30 pm - *Group Lunch*

   1.30 pm - 2.50 pm - Final Report from Individual Working Discussion Groups

   2.50 pm - 3.10 pm - *Coffee Break*

   3.10 pm - 4.30 pm - Formulate and Present WORKSHOP RECOMMENDATIONS

   4.30 pm   - * * * END * * *

McGill University
GEOTECHNICAL RESEARCH CENTRE

DIRECTOR
DR. RAYMOND N. YONG
WILLIAM SCOTT PROFESSOR OF
CIVIL ENGINEERING AND APPLIED MECHANICS

January 17, 1980

To: All Predictors

Professor K. Akai	Kyoto University, Japan
Dr. G.Y.Baladi/Dr. I.S.Sandler	U.S.Army, WES/Weidlinger Assocs.
Professor Z.P. Bazant/Dr. A.Ansal	Northwestern University, Evanston/ Istanbul Technical University, Turkey
Professor W.F.Chen	Purdue University, West Lafayette
Professor J.M.Duncan	University of California, Berkeley
Professor Dr.-Ing. G.Gudehus/ Dr. L. Vardoulakis	University of Karlsruhe, W.Germany/ University of Minnesota, Minneapolis
Professor L.R.Herrman	University of California, Davis
Professor P.Lade	University of California, Los Angeles
Professor J.K.Mitchell/ Professor E.Kavazanjian	University of California, Berkeley/ Stanford University
Professor J.H.Prevost	Princeton University, Princeton
Professor K.C.Valanis	University of Cincinnati
Professor P.Wroth	Oxford University, England

Dear Colleagues:

The enclosed three sets of data represent the test measurements taken from three separate situations of testing. In each instance, you will be introduced to the method or methods used for data reduction and the nomenclature. As you will observe, all the measurements have been reduced into a format which gives the information in terms of the principal stresses and principal strains. The tests are either drained or undrained tests, and the effective stresses are so noted.

We have screened a whole host of data from very many laboratories and reputable researchers and have now assured ourselves that the data you have represents quality information from very reputable research laboratories. All efforts have been expended in ensuring that problems with regard to boundaries, distribution of stresses and strains, etc. have been met as well as can be expected in any good reputable research laboratory.

The test data you have in the enclosed have been obtained for tests on:

(a) two natural soil samples,

(b) a laboratory-prepared kaolinite sample, and

(c) sand samples.

2...

Postal Address: 817 SHERBROOKE ST. W., MONTREAL, PQ. CANADA H3A 2K6   Tel. (514) 392-4751   Telex 05-268510

To: All Predictors   January 17, 1980

In the case of the natural soil samples, we should note that the soil is not necessarily totally homogeneous, although attempts have been made to ensure that the soil samples obtained from the block samples themselves are at least as homogeneous as they might be. The tests are essentially true triaxial tests or as close to a true triaxial test as one would find. We understand that many of you have reservations in regard to uncontrolled boundary friction problems etc., etc. but ask that you fully appreciate that one part of our Workshop problem is indeed to focus on the quality requirements and kinds of data generation required for modelling procedures. Our idealistic objective is to reach the stage where stress-strain modelling could work with good quality data obtained from established reputable research laboratories where stress loading conditions approach those that assumed as field conditions. Hence the samples tested were subject to stress controlled loading as opposed to strain controlled loading conditions. The other conditions for sample preparation and testing are noted on the three separate situations posed with the data.

The questions asked are also given on the information sheet accompanying the data. They vary for the three separate situations shown. We realise that you will have some questions concerning the information given to you and would ask that you contact me - not only for questions to be answered, but also for whatever feedback you might think would be useful to us. We have tried to reduce the data to the simplest form for use and think that what you now have will be suitable for most, if not all of you. We appreciate that some of you might have special demands, but hope that these demands can also be met with the method of data presentation given herein. Please note that all the results have been reported to the last significant figure noted in the data.

We shall look forward to hearing from you in the near future. We will be sending to you some information at a later time on the common format for presentation of your predictions. We hope to be able to ask you to present your information in graphical form on standard plots which will allow us to compare easily between all predictors.

I thank you very much for your kind cooperation. If you could acknowledge receipt of this package for our files, this would be greatly appreciated.

Yours sincerely,

Encs.
RNY:P

A - PROBLEM DEVELOPMENT FOR PREDICTION

Natural Clays

Appended are two sets of experimental data from consolidated, drained triaxial tests conducted on two natural soils designated as Clay X and Clay Y.

For both soils the triaxial tests were conducted on prismatic samples trimmed from block samples. All samples were 100% saturated.

In the experimental data that follows, the subscript 1 refers to the principal stress axis which was always maintained in the vertical direction. (i.e. ε_1 is the strain in the vertical direction, and σ_1 is the stress in the vertical or *axial* direction. Directions 2 and 3 are constant.

As is the convention in soil mechanics, the positive sign has been used to denote compression or contraction. All applied stresses were of a compressive type. A positive ε_1 denotes axial compression. ε_2 and ε_3 may be positive or negative depending upon whether the sample contracts or expands in the lateral directions. (i.e. ε_2 and ε_3 positive = compressive strains, and ε_2 and ε_3 negative = expansion.)

During the testing the measured quantities were *axial deformation*, *lateral deformation* in the minor direction, and *volume change*. Thus the strain in the *intermediate direction*, (ε_2) is a derived (calculated) quantity rather than a direct measurement.

In the tests reported, and in those to be predicted, the stress ratio m was kept constant throughout the test. The stress ratio m is defined as:

$$m = \frac{\sigma_2 - \sigma_3}{\sigma_1 - \sigma_3}$$

The stress ratio m = 0 denotes the conventional triaxial test situation in which $\sigma_2 = \sigma_3$. The condition m = 1 is met when $\sigma_1 = \sigma_2$. This latter condition should not be confused with the conventional *extension* test since for the tests reported herein, σ_1 is still in the vertical direction.

CLAY X

For Clay X, an additional condition is imposed, namely that of a constant mean normal stress. This condition is achieved by an appropriate reduction of the minor stress as indicated by the results presented.

Given (Constant Mean Stress) *Principal Stress-Strain Data*

1) $\sigma_c = 10$psi   m = 0, m = 1
2) $\sigma_c = 20$psi   m = 0, m = 1
3) $\sigma_c = 30$psi   m = 0, m = 1
4) $\sigma_c = 30$psi   m = 0

Predict (Constant Mean Stress) *Stress-Strain Curves for*

$\sigma_c = 10$psi   m = 0.25, 0.50, 0.75
$\sigma_c = 20$psi   m = 0.25, 0.50, 0.75
$\sigma_c = 30$psi   m = 0.25, 0.50, 0.75
$\sigma_c = 40$psi   m = 0

CLAY Y

For Clay Y, the tests were of the conventional type in which the minor principal stress was kept constant, and the stress ratio maintained by adjustments in σ_1 and σ_2.

Given *Principal Stress-Strain Data*

1) $\sigma_c = 2.50$psi   m = 0, m = 1
2) $\sigma_c = 5.0$psi   m = 0, m = 1
3) $\sigma_c = 10.0$psi   m = 0, m = 1

Predict *Stress-Strain Curves for*

$\sigma_c = 2.50$ psi m = .25, .50, .75
$\sigma_c = 5.0$ psi m = .25, .50, .75
$\sigma_c = 10.0$ psi m = .25, .50, .75

A-1

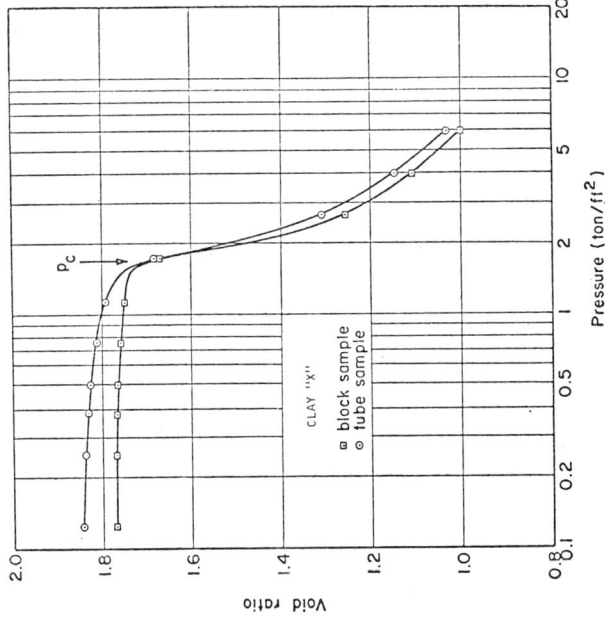

PHYSICAL PROPERTIES OF CLAY "X"

SOIL	PROPERTIES
Natural water content (%)	66 ± 2
Liquid Limit (%)	48
Plastic Limit (%)	28
Plasticity Index (%)	20
Liquidity Index	1.8 ± 2.0
Clay content (%)	79
Silt content (%)	20
Sand content (%)	1
Activity	0.25
Sensitivity by Laboratory Vane	50
Specific Gravity	2.80

CLAY "X"

TRIAXIAL TEST RESULTS
CONSTANT MEAN STRESS TESTS

1. Consolidation Pressure $\sigma_c = 10.0$ psi

σ_1'	σ_2'	σ_3'	ϵ_1	ϵ_2	ϵ_3	$\frac{dV}{V}$
psi	psi	psi	%	%	%	%

Stress Ratio m = 0:

10.00	10.00	10.00	0.00	0.00	0.00	0.00
11.00	9.50	9.50	0.14	-0.17	0.03	0.00
12.00	9.00	9.00	0.28	-0.35	0.05	-0.02
13.00	8.50	8.50	0.42	-0.43	0.06	-0.05
14.00	8.00	8.00	0.56	-0.65	0.04	-0.05
15.00	7.50	7.50	0.70	-0.78	0.03	-0.05
16.00	7.00	7.00	0.88	-0.77	-0.13	-0.02
17.00	6.50	6.50	1.10	-0.80	-0.31	-0.01
18.00	6.00	6.00	1.30	-0.94	-0.40	-0.04
19.00	5.50	5.50	1.93	-1.45	-0.34	0.14
20.00	5.00	5.00	–	–	–	–

Stress Ratio m = 1.0:

10.00	10.00	10.00	0.00	0.00	0.00	0.00
10.50	10.50	9.00	0.16	-0.13	-0.02	0.01
11.00	11.00	8.00	0.33	-0.28	-0.04	0.01
11.50	11.50	7.00	0.49	-0.40	-0.08	0.01
12.00	12.00	6.00	0.65	-0.49	-0.12	0.04
12.50	12.50	5.00	0.80	-0.51	-0.18	0.11
13.00	13.00	4.00	0.90	-0.53	-0.25	0.12
13.50	13.50	3.00	–	–	–	–

Consolidation Pressure $\sigma_c = 20.0$ psi

σ_1'	σ_2'	σ_3'	ϵ_1	ϵ_2	ϵ_3	$\frac{dV}{V}$
psi	psi	psi	%	%	%	%

Stress Ratio m = 0:

20.00	20.00	20.00	0.00	0.00	0.00	0.00
21.33	19.33	19.33	0.28	-0.16	0.00	0.12
22.67	18.67	18.67	0.49	-0.26	0.00	0.23
24.00	18.00	18.00	0.94	-0.17	0.00	0.77
25.33	17.33	17.33	1.41	-0.17	-0.09	1.15
26.67	16.67	16.67	2.27	0.07	-0.32	2.02
28.00	16.00	16.00	3.68	0.74	-1.05	3.37
29.33	15.33	15.33	5.81	1.71	-2.34	5.18
30.67	14.67	14.67	10.38	3.37	-4.63	9.12
32.00	14.00	14.00	12.32	2.78	-5.03	10.07
33.33	13.33	13.33	–	–	–	–

Stress Ratio m = 1:

20.00	20.00	20.00	0.00	0.00	0.00	0.00
20.70	20.70	18.60	0.25	-0.09	0.18	0.34
21.40	21.40	17.20	0.50	-0.06	0.18	0.62
22.08	22.10	15.80	0.75	0.72	-0.01	1.46
22.76	22.80	14.40	1.15	1.92	-0.38	2.69
23.46	23.50	13.00	1.74	3.21	-1.02	3.93
24.17	24.20	11.60	2.81	5.29	-2.65	5.45
24.89	24.90	10.20	4.84	9.93	-6.73	8.04
25.60	25.60	8.80	–	–	–	–

CLAY "X"

PHYSICAL PROPERTIES OF CLAY "Y"

SOIL	PROPERTIES
Natural Water Content (%)	64 ± 0.5
Apparent Preconsolidation Pressure	1.6 TSF
Liquid Limit (%)	46
Plastic Limit (%)	28
Liquidity Index	1.8
Clay content (%)	79
Silt content (%)	20
Sensitivity	80
Specific Gravity	2.80

3. Consolidation Pressure $\sigma_c = 30.0$ psi

σ_1'	σ_2'	σ_3'	ϵ_1	ϵ_2	ϵ_3	$\frac{dV}{V}$
psi	psi	psi	%	%	%	%

Stress Ratio m = 0:

30.00	30.00	30.00	0.00	0.00	0.00	0.00
32.00	29.00	29.00	0.32	0.06	0.00	0.38
34.00	28.00	28.00	0.58	0.20	0.00	0.78
36.00	27.00	27.00	1.15	-0.23	0.00	0.92
38.00	26.00	26.00	1.97	-0.51	0.00	1.46
40.00	25.00	25.00	3.18	-0.90	-0.16	2.12
42.00	24.00	24.00	4.71	-0.77	-0.80	3.14
44.00	23.00	23.00	6.32	-0.99	-1.13	4.20
46.00	22.00	22.00	8.03	-1.47	-1.45	5.11
48.00	21.00	21.00	10.67	1.06	-4.34	7.39
50.00	20.00	20.00	13.29	1.04	-5.31	9.02
52.00	19.00	19.00	15.70	1.99	-7.89	9.80
54.00	18.00	18.00	–	–	–	–

Stress Ratio m = 1:

30.00	30.00	30.00	0.00	0.00	0.00	0.00
31.00	31.00	28.00	0.31	0.54	-0.18	0.67
32.00	32.00	26.00	0.70	1.27	-0.55	1.42
33.00	33.00	24.00	1.20	2.10	-1.00	2.30
34.00	34.00	22.00	1.79	3.10	-1.72	3.17
35.00	35.00	20.00	2.54	4.61	-3.30	3.85
36.00	36.00	18.00	3.24	6.36	-4.98	4.62
37.00	37.00	16.00	3.76	8.52	-6.83	5.45
38.00	38.00	14.00	5.44	12.07	-11.22	6.29
39.00	39.00	12.00	–	–	–	–

STRESS STRAIN FOR SOILS

CLAY "Y"

TRIAXIAL TEST RESULTS

Consolidation Pressure $\sigma_c = 2.50$ psi

σ_1' psi	σ_2' psi	σ_3' psi	ε_1 %	ε_2 %	ε_3 %	$\frac{dV}{V}$ %
Stress Ratio m = 0:						
2.50	2.50	2.50	0.000	0.000	0.000	0.000
4.17	2.50	2.50	0.063	-0.013	-0.013	0.037
5.83	2.50	2.50	0.163	-0.059	-0.059	0.045
7.48	2.50	2.50	0.313	-0.092	-0.092	0.129
9.15	2.50	2.50	0.413	-0.107	-0.107	0.199
10.77	2.50	2.50	0.685	-0.217	-0.217	0.251
12.42	2.50	2.50	0.838	-0.222	-0.222	0.394
14.04	2.50	2.50	1.063	-0.322	-0.322	0.419
15.75	2.50	2.50	1.462	-0.415	-0.415	0.632
Stress Ratio m = 1.0:						
2.50	2.50	2.50	0.000	0.000	0.000	0.000
5.83	5.83	2.50	0.150	0.139	-0.125	0.164
9.15	9.17	2.50	0.300	0.170	-0.233	0.237
12.45	12.50	2.50	0.475	0.338	-0.358	0.455
15.75	15.83	2.50	0.625	0.317	-0.465	0.477

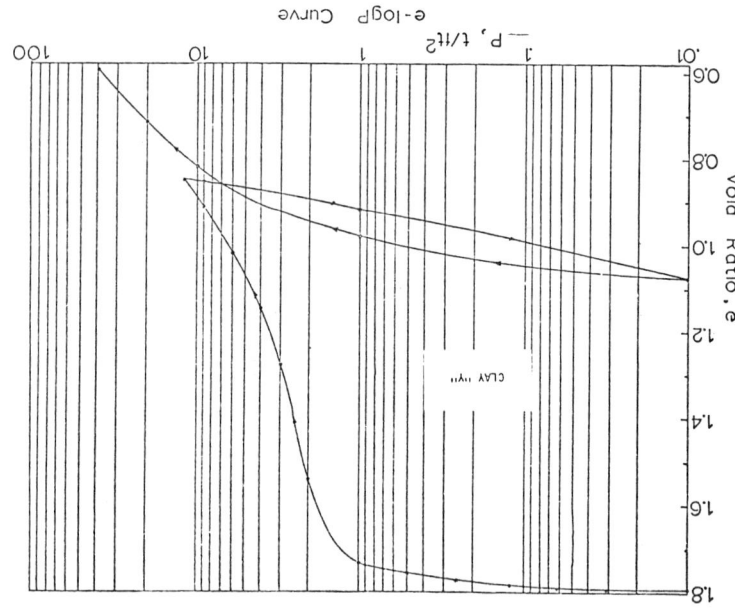

e-logP Curve

CLAY "Y"

P, t/ft² vs Void Ratio, e

CLAY "Y"

2. Consolidation Pressure $\sigma_c' = 5.00$ psi

σ_1'	σ_2'	σ_3'	ε_1	ε_2	ε_3	$\frac{dV}{V}$
psi	psi	psi	%	%	%	%

Stress Ratio m = 0:

5.00	5.00	5.00	0.000	0.000	0.000	0.000
8.33	5.00	5.00	0.100	-0.059	0.000	0.041
11.65	5.00	5.00	0.300	-0.037	-0.100	0.163
14.94	5.00	5.00	0.650	-0.193	-0.198	0.259
18.17	5.00	5.00	1.350	-0.461	-0.300	0.589
19.36	5.00	5.00	4.470	-2.485	-1.090	0.895

Stress Ratio m = 1.0:

5.00	5.00	5.00	0.000	0.000	0.000	0.000
8.33	8.33	5.00	0.200	0.182	0.000	0.382
11.66	11.67	5.00	0.400	0.376	0.000	0.776
14.95	15.00	5.00	0.600	0.452	-0.088	0.964
18.21	18.33	5.00	0.850	0.350	-0.255	0.945

CLAY "Y"

3. Consolidation Pressure $\sigma_c' = 10.00$ psi

σ_1'	σ_2'	σ_3'	ε_1	ε_2	ε_3	$\frac{dV}{V}$
psi	psi	psi	%	%	%	%

Stress Ratio m = 0:

10.00	10.00	10.00	0.000	0.000	0.000	0.000
13.33	10.00	10.00	0.125	-0.039	-0.085	0.001
16.65	10.00	10.00	0.375	-0.229	-0.113	0.033
19.92	10.00	10.00	0.775	-0.439	-0.163	0.173
22.96	10.00	10.00	2.750	-1.532	-0.533	0.685
25.15	10.00	10.00	9.960	-7.147	-0.913	1.900
26.60	10.00	10.00	10.420	–	–	–

Stress Ratio m = 1.0:

10.00	10.00	10.00	0.000	0.000	0.000	0.000
13.33	13.30	10.00	0.075	0.000	-0.040	0.035
16.66	16.67	10.00	0.200	0.335	-0.085	0.450
19.97	20.00	10.00	0.450	1.476	-0.323	1.603
23.20	23.33	10.00	0.800	1.800	-0.660	1.940
26.45	26.67	10.00	1.250	3.394	-1.773	2.871
27.65	28.33	10.00	3.900	–	-3.723	2.900

B - PROBLEM DEVELOPMENT FOR PREDICTION

Laboratory Prepared Kaolinite Clay

THE following data is available for a laboratory prepared kaolinite clay:

1. Consolidation and rebound
2. Water content during undrained testing (give on every test). Average = 38.8%)
3. Liquid and plastic limit (W_L = 62.5%, W_p = 39%)
4. Specific gravity - 2.61
5. K_o = 0.48

All the samples were K_o consolidated with a cell pressure of 58psi and a backpressure of 18psi (and whatever additional axial load is necessary to maintain K_o condition). They were then left to rebound to a cell pressure of 58psi and a back pressure of 18psi (in other words the excess axial load necessary for K_o consolidation was released). Loads were then applied to the specimens in an undrained condition, and the pore water pressure measured during deformation up to failure. All the tests were controlled true stress tests and started from an all around pressure of 58psi and a pore pressure of 18psi. Note that positive strains represent compressive strains and that negative pore pressures are taken with a base value of 18psi back pressure as zero p.p. Thus a -1psi pore pressure is 17psi residual p.p.

You are given the results of four basic tests numbered 1, 4, 10 and 13.

Test 1 – A compression test in which the axial stress is increased while cell pressure is kept constant.

Test 4 – A compression test with a constant mean normal stress in which the axial stress is increased and the lateral stresses decreased such that the sum of the three principal stresses remains constant during the test.

Test 10 – An extension test in which the axial stress is decreased while cell pressure is kept constant.

Test 13 – An extension test with a constant mean normal stress in which the axial stress is decreased and the lateral stress increased such that the sum of the three principal stresses remains constant during the test.

QUESTIONS TO BE ANSWERED

Given the total principal stresses for nine other tests, predict the effective principal stresses and the principal strains:

1. Page B-3 gives the results of the consolidation tests.
2. Pages B-4 to B-7 give the results of the four basic tests.
3. Pages B-8 to B-16 give the data of the nine tests whose results are to be predicted.
4. Page B-17 gives the total stress paths of the four basic tests in two different representations.
5. Page B-18 gives the total stress paths of the nine tests whose results are to be predicted.
6. On pages B-17 and B-18 the number of each test is shown on its corresponding total stress path.

TEST No. 1
CONVENTIONAL TRIAXIAL COMPRESSION TEST

Axial Strain	Pore Pressure Developed	Effective Principal Stresses		
		σ'_1	σ'_2	σ'_3
.0000	.00	40.05	40.05	40.05
.0022	6.70	53.90	33.30	33.30
.0043	9.40	59.10	30.60	30.60
.0065	10.50	62.90	29.50	29.50
.0087	11.60	65.90	28.40	28.40
.0109	12.40	69.90	27.60	27.60
.0131	12.90	72.40	27.10	27.10
.0153	13.30	74.70	26.70	26.70
.0175	13.40	76.30	26.60	26.60
.0197	13.40	77.80	26.60	26.60
.0219	13.60	78.80	26.40	26.40
.0241	13.80	79.20	26.20	26.20
.0263	13.90	80.00	26.10	26.10
.0285	14.00	80.50	26.00	26.00
.0308	14.00	81.00	26.00	26.00
.0352	14.10	81.60	25.90	25.90
.0397	14.30	82.00	25.70	25.70
.0442	14.40	82.30	25.60	25.60
.0488	14.60	82.40	25.40	25.40
.0533	14.80	82.50	25.20	25.20

Water Content: 38.5

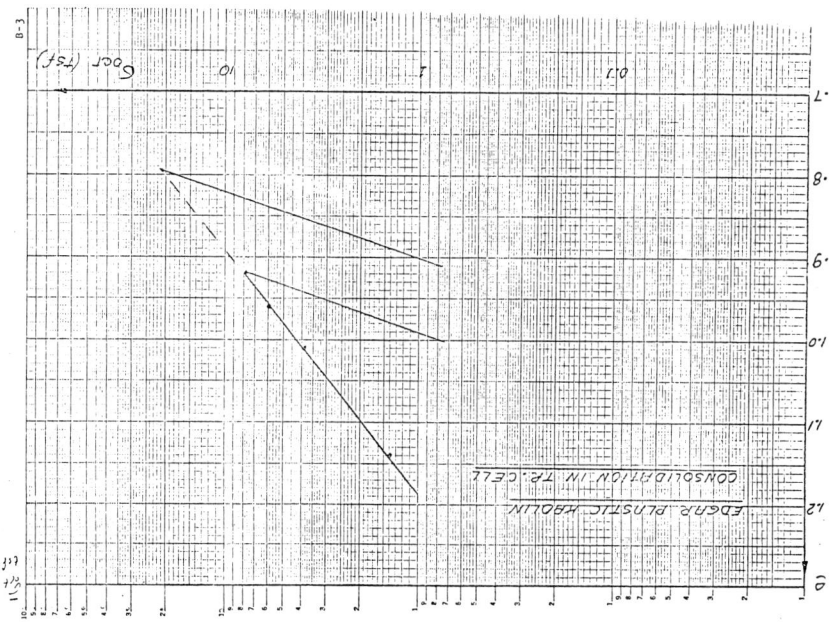

STRESS STRAIN FOR SOILS

TEST No. 4
COMPRESSION TEST - CONSTANT MEAN NORMAL STRESS

Axial Strain	Pore Pressure Developed	Effective Principal Stresses		
		σ_1'	σ_2'	σ_3'
.0000	.00	40.00	40.00	40.00
.0022	1.20	55.67	30.37	30.37
.0043	1.20	59.47	28.47	28.47
.0065	.60	64.07	27.07	27.07
.0087	.00	67.80	26.10	26.10
.0109	-1.00	71.20	25.90	25.90
.0131	-1.40	73.40	25.40	25.40
.0153	-1.90	75.30	25.20	25.20
.0175	-2.30	76.50	25.20	25.20
.0219	-2.80	78.40	25.00	25.00
.0263	-3.00	79.13	24.93	24.93
.0307	-3.20	79.87	24.87	24.87
.0352	-3.30	80.43	24.73	24.73
.0397	-3.40	80.80	24.70	24.70
.0442	-3.50	81.10	24.70	24.70
.0487	-3.50	81.37	24.57	24.57

Water Content = 39.1

Test No. 10
EXTENSION TEST - CONSTANT MEAN NORMAL STRESS

Axial Strain	Pore Pressure Developed	Effective Principal Stresses		
		σ_1'	σ_2'	σ_3'
.0000	.00	40.00	40.00	40.00
-.0043	1.40	43.27	43.27	29.27
-.0087	3.00	42.83	42.83	25.33
-.0130	4.20	42.47	42.47	22.47
-.0173	5.00	42.30	42.30	20.40
-.0215	6.00	41.83	41.83	18.33
-.0250	6.60	41.73	41.73	16.73
-.0300	7.00	41.77	41.77	15.47
-.0342	7.50	41.70	41.70	14.10
-.0426	7.90	42.03	42.03	12.23
-.0509	8.00	42.57	42.57	10.87
-.0592	8.00	43.07	43.07	9.87
-.0674	7.90	43.73	43.73	8.83
-.0745	7.70	44.47	44.47	7.97
-.0836	7.30	45.43	45.43	7.23
-.0915	7.00	46.23	46.23	6.53
-.0995	6.70	46.73	46.73	6.43
-.1073	6.60	47.10	47.10	6.00
-.1151	6.60	47.30	47.30	5.60
-.1229	6.60	47.40	47.40	5.40
-.1306	6.60	47.47	47.47	5.27

Water Content = 39.1

TEST No. 13

CONVENTIONAL TRIAXIAL EXTENSION TEST

Axial Strain	Pore Pressure Developed	Effective Principal Stresses		
		σ'_1	σ'_2	σ'_3
.0000	.00	40.00	40.00	40.00
-.0022	-1.10	41.10	41.10	29.10
-.0043	-1.00	41.00	41.00	27.20
-.0087	-.80	40.80	40.80	24.30
-.0130	-.30	40.30	40.30	21.50
-.0173	.00	40.00	40.00	19.60
-.0215	.30	39.70	39.70	17.50
-.0258	.40	39.60	39.60	15.90
-.0342	.40	39.60	39.60	14.10
-.0426	.20	39.80	39.80	12.20
-.0510	-.20	40.20	40.20	11.10
-.0592	-.80	40.80	40.80	10.10
-.0674	-1.40	41.40	41.40	9.40
-.0755	-2.00	42.00	42.00	8.70
-.0836	-2.70	42.70	42.70	8.10
-.0916	-3.40	43.40	43.40	7.40
-.0995	-4.00	44.00	44.00	7.00
-.1074	-4.70	44.70	44.70	6.70
-.1152	-4.30	44.30	44.30	5.30
-.1230	-4.20	44.20	44.20	4.40

Water Content = 38.9

TEST No. 2

Water content: 38.8%

TOTAL PRINCIPAL STRESSES			PORE PRESSURE	ANGLE OF MAJOR PRINCIPAL STRESS AXES TO VERTICAL IS 15°		
				PRINCIPAL STRAINS		
σ_1	σ_2	σ_3		ϵ_1	ϵ_2	ϵ_3
58.00	58.00	58.00	18.10	0	0	0
78.90	56.00	56.50				
86.22	58.00	55.98				
92.69	58.00	55.51				
96.78	58.00	55.22				
100.01	58.00	54.99				
102.28	58.00	54.82				
104.32	58.00	54.68				
105.72	58.00	54.58				
107.88	58.00	54.42				
109.60	58.00	54.30				
110.79	58.00	54.21				
111.76	58.00	54.15				
112.40	58.00	54.10				
112.94	58.00	54.06				
113.48	58.00	54.02				
113.91	58.00	53.99				
114.13	58.00	53.97				
114.34	58.00	53.96				

FOR this test the major principal stress is inclined at 15° to the vertical axis of the sample as shown in the sketch below. The intermediate principal stress (σ_2) is horizontal in all tests.

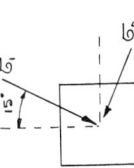

B-9

TEST No. 3

Water content : 38.8%

ANGLE OF MAJOR PRINCIPAL STRESS AXES TO VERTICAL IS 37.5°

TOTAL PRINCIPAL STRESSES			PORE PRESSURE	PRINCIPAL STRAINS		
σ_1	σ_2	σ_3		ϵ_1	ϵ_2	ϵ_3
58.00	58.00	58.00	18.00	0	0	0
74.18	58.00	51.82				
81.14	58.00	49.16				
88.10	58.00	46.50				
91.98	58.00	45.02				
94.08	58.00	44.22				
96.19	58.00	43.41				
97.64	58.00	42.86				
98.45	58.00	42.55				
99.58	58.00	42.12				
100.39	58.00	41.81				
101.20	58.00	41.50				
101.85	58.00	41.25				
102.17	58.00	41.13				
102.33	58.00	41.07				

TEST No. 5

Water Content : 39.7%

ANGLE OF MAJOR PRINCIPAL STRESS AXES TO VERTICAL IS 15°

TOTAL PRINCIPAL STRESSES			PORE PRESSURE	PRINCIPAL STRAINS		
σ_1	σ_2	σ_3		ϵ_1	ϵ_2	ϵ_3
58.00	58.00	58.00	18.00	0	0	0
72.36	51.57	50.08				
73.62	51.00	49.38				
77.34	49.33	47.32				
79.87	48.20	45.93				
81.58	47.43	44.98				
82.92	46.83	44.25				
83.74	46.47	43.79				
84.41	46.17	43.42				
84.86	45.97	43.18				
85.67	45.60	42.73				
86.27	45.33	42.40				
86.72	45.13	42.15				
87.01	45.00	41.99				
87.31	44.87	41.82				
87.46	44.80	41.74				
87.53	44.77	41.70				

TEST No. 6

Water content : 39.4%

ANGLE OF MAJOR PRINCIPAL STRESS AXES TO VERTICAL IS 31.75°

TOTAL PRINCIPAL STRESSES			PORE PRESSURE	PRINCIPAL STRAINS		
σ_1	σ_2	σ_3		ε_1	ε_2	ε_3
58.00	58.00	58.00	18.00	0	0	0
70.33	54.80	48.87				
74.70	53.67	45.63				
77.40	52.97	43.63				
79.45	52.43	42.11				
80.74	52.10	41.16				
81.51	51.90	40.59				
82.79	51.57	39.64				
83.82	51.30	38.88				
84.34	51.17	38.50				
84.98	51.00	38.02				
85.36	50.90	37.74				
85.88	50.77	37.36				
86.26	50.67	37.07				

TEST No. 7

Water content : 38.8%

ANGLE OF MAJOR PRINCIPAL STRESS AXES TO VERTICAL IS 45°

TOTAL PRINCIPAL STRESSES			PORE PRESSURE	PRINCIPAL STRAINS		
σ_1	σ_2	σ_3		ε_1	ε_2	ε_3
58.00	58.00	58.00	18.00	0	0	0
64.10	58.00	51.90				
68.10	58.00	47.90				
70.10	58.00	45.90				
72.10	58.00	43.90				
74.10	58.00	41.90				
77.80	58.00	38.20				
79.40	58.00	36.60				
81.00	58.00	35.00				
81.80	58.00	34.20				
82.50	58.00	33.50				
83.40	58.00	32.60				
84.10	58.00	31.90				
84.20	58.00	31.80				
84.40	58.00	31.60				

TEST No. 8

Water content : 39.4%

ANGLE OF MAJOR PRINCIPAL STRESS AXES TO VERTICAL IS 58.25°

TOTAL PRINCIPAL STRESSES			PORE PRESSURE	PRINCIPAL STRAINS		
σ_1	σ_2	σ_3		ε_1	ε_2	ε_3
58.00	58.00	58.00	18.10	0	0	0
63.71	60.00	50.29				
64.47	60.27	49.26				
65.61	60.67	47.72				
66.47	60.97	46.57				
67.32	61.27	45.41				
68.56	61.70	43.74				
69.51	62.03	42.46				
70.37	62.33	41.30				
71.03	62.57	40.40				
71.70	62.80	39.50				
72.46	63.07	38.47				
73.03	63.27	37.70				
73.41	63.40	37.19				
73.70	63.50	36.80				
74.08	63.63	36.29				

TEST No. 9

Water content : 38.8%

ANGLE OF MAJOR PRINCIPAL STRESS AXES TO VERTICAL IS 75°

TOTAL PRINCIPAL STRESSES			PORE PRESSURE	PRINCIPAL STRAINS		
σ_1	σ_2	σ_3		ε_1	ε_2	ε_3
58.00	58.00	58.00	18.00	0	0	0
62.72	61.83	49.44				
63.87	62.77	47.36				
64.90	63.60	45.50				
65.68	64.23	44.09				
66.29	64.73	42.97				
66.87	65.20	41.93				
67.36	65.60	41.04				
67.85	66.00	40.15				
68.59	66.60	38.81				
69.37	67.23	37.39				
69.99	67.73	36.28				
70.65	68.27	35.09				
71.18	68.70	34.12				
71.75	69.17	33.08				
72.08	69.43	32.48				
72.37	69.67	31.96				
72.58	69.83	31.59				

TEST No. 11

Water content : 38.8%

ANGLE OF MAJOR PRINCIPAL STRESS AXES TO VERTICAL IS 58.25°

TOTAL PRINCIPAL STRESSES			PORE PRESSURE	PRINCIPAL STRAINS		
σ_1	σ_2	σ_3		ϵ_1	ϵ_2	ϵ_3
58.00	58.00	58.00	18.00	0	0	0
62.94	58.00	45.06				
63.87	58.00	42.63				
64.43	58.00	41.17				
65.29	58.00	38.91				
65.97	58.00	37.13				
66.65	58.00	35.35				
67.21	58.00	33.89				
67.76	58.00	32.44				
68.20	58.00	31.30				
68.51	58.00	30.49				
68.94	58.00	29.36				
69.37	58.00	28.23				
70.30	58.00	25.80				
69.93	58.00	26.77				

TEST No. 12

Water content : 38.7%

ANGLE OF MAJOR PRINCIPAL STRESS AXES TO VERTICAL IS 75°

TOTAL PRINCIPAL STRESSES			PORE PRESSURE	PRINCIPAL STRAINS		
σ_1	σ_2	σ_3		ϵ_1	ϵ_2	ϵ_3
58.00	58.00	58.00	17.80	0	0	0
58.83	58.00	46.47				
59.07	58.00	43.13				
59.26	58.00	40.44				
59.39	58.00	38.61				
59.54	58.00	36.56				
59.65	58.00	35.38				
59.62	58.00	33.98				
59.72	58.00	32.68				
59.82	58.00	30.64				
59.96	58.00	28.91				
60.09	58.00	26.97				
60.23	58.00	25.47				
60.33	58.00	23.96				
60.44	58.00	22.77				
60.53	58.00	21.91				
60.59	58.00	21.37				
60.63	58.00	20.94				
60.66	58.00	20.51				
60.69	58.00					

STRESS STRAIN FOR SOILS

* Same scale for obscissa & ordinate
* Same numbers on the two Graphs correspond to same test

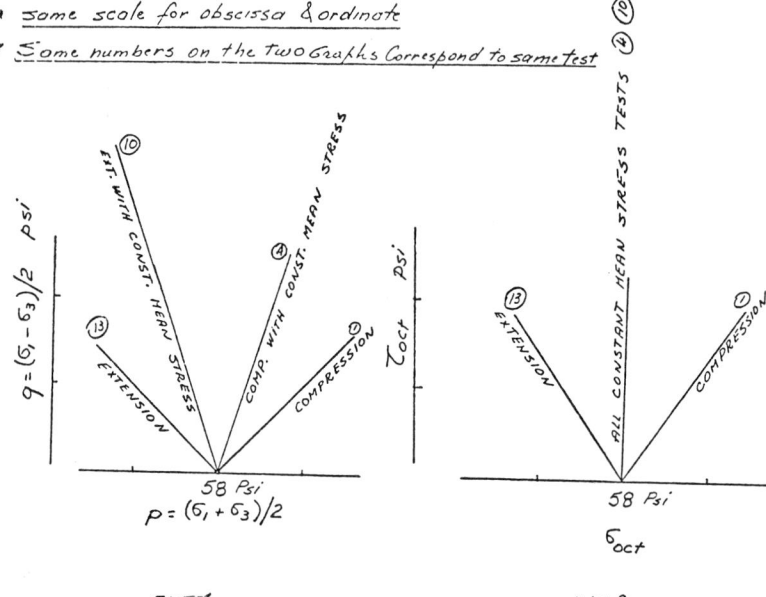

FIG. 1   FIG. 2

* Same Scale for obscissa & ordinate
** Same numbers on the two Graphs correspond to Same test

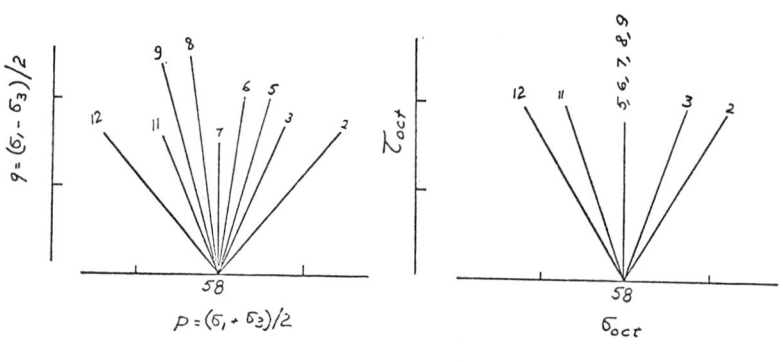

FIG. 3   FIG. 4.

PREDICTION PROBLEM NO. 3

An Ottawa sand, whose gradation is given in Fig. 1, was compacted by aerial pluviation to 1.73 gm/c.c. (relative density of 87%) and tested in the dry state three-dimensionally for its stress-strain properties. The stress paths shown in Fig. 2 were used to generate the data base shown in Table 1 for formulating constitutive models. Using the model(s), it is required to make predictions for the stress-strain response of the same soil when tested in the same test equipment along stress paths different from those used in the model formulation. The stress paths for prediction are shown in Fig. 3, and the predicted strain response is to be entered in Table 2.

The test equipment used is a flexible, fluid cushion, cubical device with stress control. The vertical axis, z, and the horizontal axes, x and y, are principal stress axes.

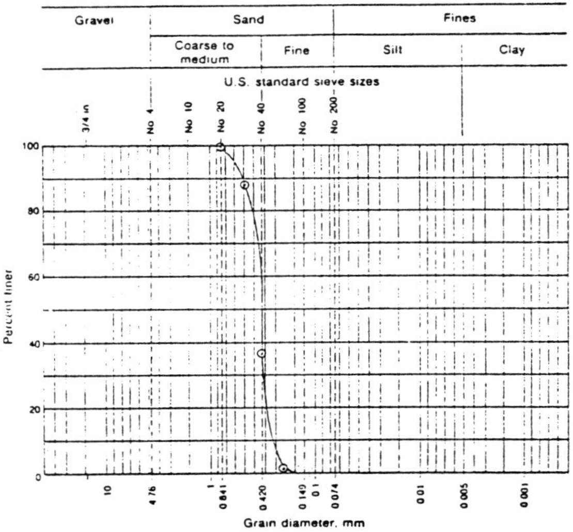

FIG. 1 GRAIN SIZE DISTRIBUTION FOR OTTAWA SAND

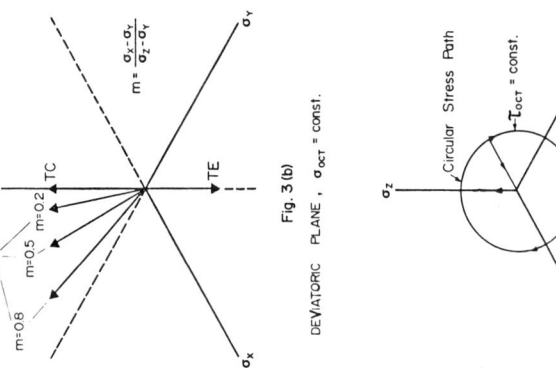

Fig. 3 (b)

DEVIATORIC PLANE, σ_{oct} = const.

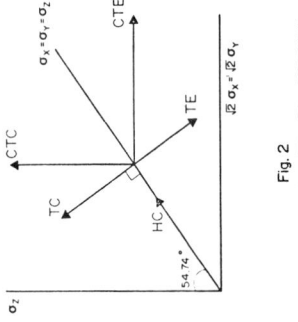

Fig. 2

STRESS PATHS USED TO GENERATE DATA BASE FOR MODELING.

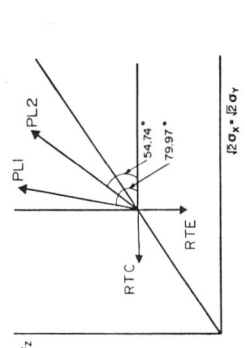

Fig. 3 (c)

DEVIATORIC PLANE, σ_{oct} = const.

Fig. 3 (a)

Fig. 3 STRESS PATHS FOR PREDICTIONS

C-5 OTTAWA SAND

HYDROSTATIC TEST (HC) RESULTS

σ, psi	ε_{avg}, %
4 (datum)	0
5	.002
6	.0044
7	.0066
8	.009
9	.011
10	.0135
8	.012
6	.009
4	.005

TABLE 1 (Next 11 pages)

C-7 OTTAWA SAND

CONVENTIONAL TRIAXIAL COMPRESSION TEST (CTC) RESULTS
Constant Lateral Stress Tests
Lateral Stress = 10 psi; Stress Ratio, m = 0

σ_x	σ_y	σ_z	ε_x, %	ε_y, %	ε_z, %
10	10	10	0	0	0
10	10	12	0	0	+ .011
10	10	13	- .001	0	+ .020
10	10	14	- .001	0	+ .031
10	10	15	- .003	0	+ .045
10	10	16	- .004	0	+ .058
10	10	17	- .006	- .002	+ .072
10	10	18	- .008	- .005	+ .087
10	10	19	- .011	- .009	+ .103
10	10	20	- .017	- .012	+ .123
10	10	18	- .017	- .017	+ .116
10	10	15	- .019	- .017	+ .106
10	10	12	- .016	- .016	+ .092
10	10	10	- .015	- .014	+ .077
10	10	12	- .016	- .014	+ .084
10	10	15	- .016	- .015	+ .095
10	10	18	- .017	- .016	+ .109
10	10	20	- .020	- .020	+ .123
10	10	22	- .037	- .031	+ .164
10	10	24	- .053	- .050	+ .212
10	10	26	- .079	- .070	+ .256
10	10	28	- .107	- .099	+ .310
10	10	30	- .157	- .141	+ .379
10	10	32	- .197	- .186	+ .455
10	10	34	- .242	- .230	+ .542
10	10	36	- .289	- .287	+ .635
10	10	38	- .347	- .352	+ .732
10	10	40	- .478	- .493	+ .926
10	10	42	- .629	- .653	+1.126
10	10	44	- .783	- .819	+1.334
10	10	25	- .776	- .812	+1.271
10	10	15	- .705	- .724	+1.112
10	10	10	- .575	- .589	+ .907
10	10	15	- .527	- .535	+ .842
10	10	25	- .526	- .537	+ .872
10	10	35	- .550	- .556	+ .967
10	10	44	- .610	- .628	+1.150
10	10	45	- .741	- .775	+1.417
10	10	46	- .763	- .795	+1.462
10	10	47	- .789	- .817	+1.504
10	10	48	- .831	- .860	+1.562
10	10	49	- .887	- .916	+1.632
10	10	50	- .947	- .978	+1.707
10	10	50	-1.026	-1.063	+1.801
10	10	51	-1.063	-1.104	+1.852
10	10	52	-1.157	-1.205	+1.957
10	10	60	-13.929	-15.070	+13.364

OTTAWA SAND

HYDROSTATIC TEST (HC) RESULTS

σ, psi	ε, %
5	0
10	.015
15	.030
20	.039
25	.048
30	.057
35	.066
40	.076

OTTAWA SAND

CONVENTIONAL TRIAXIAL EXTENSION TEST (CTE) RESULTS

Stress Ratio, m = 1

σ_x, psi	σ_y, psi	σ_z, psi	ϵ_x, %	ϵ_y, %	ϵ_z, %
10	10	10	0	0	0
12	12	10	.020	.020	- .010
14	14	10	.054	.054	- .031
16	16	10	.097	.094	- .065
18	18	10	.160	.154	- .126
20	20	10	.224	.216	- .197
25	25	10	.387	.380	- .438
30	30	10	.599	.599	- .835
25	25	10	.596	.588	- .825
20	20	10	.582	.572	- .804
15	15	10	.551	.544	- .757
10	10	10	.491	.488	- .651
20	20	10	.533	.531	- .683
25	25	10	.565	.561	- .714
30	30	10	.613	.612	- .783
35	35	10	.793	.786	-1.158
40	40	10	1.083	1.057	-1.807
45	45	10	1.411	1.391	-2.707
50	50	10	1.791	1.765	-3.926
55	55	10	2.213	2.225	-5.428
50	50	10	2.210	2.223	-5.422
40	40	10	2.175	2.189	-5.386
30	30	10	2.121	2.133	-5.297
20	20	10	2.008	2.035	-5.047
10	10	10	1.705	1.766	-4.014

CONVENTIONAL TRIAXIAL COMPRESSION TEST (CTC) RESULTS

Lateral Stress = 5 psi
Stress Ratio, m = 0

σ_x	σ_y	σ_z	ϵ_x, %	ϵ_y, %	ϵ_z, %
5	5	5	0	0	+ .009
5	5	6	0	0	+ .025
5	5	7	0	0	+ .042
5	5	8	0	0	+ .063
5	5	9	.001	- .002	+ .088
5	5	10	- .003	- .008	+ .115
5	5	11	- .007	- .016	+ .148
5	5	12	- .014	- .030	+ .183
5	5	13	- .026	- .045	+ .225
5	5	14	- .033	- .060	+ .264
5	5	15	- .055	- .060	+ .253
5	5	12	- .055	- .055	+ .240
5	5	10	- .052	- .039	+ .215
5	5	7.5	- .038	- .041	+ .176
5	5	5	- .040	- .042	+ .192
5	5	7.5	- .041	- .050	+ .212
5	5	10	- .043	- .062	+ .233
5	5	12	- .049	- .073	+ .258
5	5	14	- .058	- .088	+ .278
5	5	15	- .076	- .114	+ .311
5	5	16	- .106	- .148	+ .369
5	5	17	- .149	- .188	+ .432
5	5	18	- .191	- .245	+ .505
5	5	19	- .246	- .318	+ .584
5	5	20	- .320	- .387	+ .690
5	5	21	- .392	- .475	+ .783
5	5	22	- .483	- .730	+ .895
5	5	23	- .742	-1.084	+1.198
5	5	25	-1.098	-1.542	+1.590
5	5	27	-1.542	-2.312	+2.066
5	5	29	-2.321	-3.754	+2.846
5	5	31	-3.820	-15.077	+4.206
5	5	33	-3.820		
5	5	34	-13.796	-15.077	+18.205

OTTAWA SAND

C-11

OTTAWA SAND

TRIAXIAL COMPRESSION TEST (TC) RESULTS

Constant Mean Stress Test

Mean Stress = 5 psi

Stress Ratio, m = 0

σ_x	σ_y	σ_z	ϵ_x, %	ϵ_y, %	ϵ_z, %
5.0	5.0	5.0	0	0	0
4.5	4.5	6.0	−.001	0	+.015
4.0	4.0	7.0	−.005	0	+.050
4.0	3.5	8.0	−.015	−.007	+.103
3.5	3.0	9.0	−.054	−.047	+.207
3.0	2.5	10.0	−.161	−.157	+.394
2.5	1.5	12.0	−1.332	−1.367	+1.652
1.5	3.0	9.0	−1.278	−1.320	+1.614
3.0	4.0	7.0	−1.137	−1.169	+1.524
4.0	5.0	5.0	−.861	−.886	+1.307
5.0	4.0	7.0	−.862	−.887	+1.334
4.0	3.0	9.0	−.888	−.909	+1.395
3.0	2.0	11.0	−1.063	−1.084	+1.591
2.0	1.5	12.0	−1.562	−1.593	+2.010
1.5	1.25	12.5	−13.844	−14.891	+17.692

TRIAXIAL COMPRESSION TEST (TC) RESULTS

Constant Mean Stress Test

Mean Stress = 10 psi

Stress Ratio, m = 0

σ_x	σ_y	σ_z	ϵ_x, %	ϵ_y, %	ϵ_z, %
10	10	10	0	0	0
9	9	12	−.001	0	+.026
8	8	14	−.015	−.007	+.074
7	7	16	−.036	−.033	+.163
6	6	18	−.255	−.253	+.538
8	8	14	−.247	−.240	+.512
9	9	12	−.208	−.204	+.468
10	10	10	−.173	−.176	+.432
9	9	12	−.177	−.176	+.443
8	8	14	−.181	−.182	+.459
6	6	18	−.224	−.218	+.528
5.5	5.5	19	−.362	−.356	+.722
5	5	20	−.560	−.547	+.981
4.5	4.5	21	−1.220	−1.209	+1.776
4.0	4.0	22	−2.091	−2.088	+2.720
3.5	3.5	23	−3.239	−3.257	+3.915

OTTAWA SAND

C-13

OTTAWA SAND

TRIAXIAL COMPRESSION TEST (TC) RESULTS

Constant Mean Stress Test

Mean Stress = 20 psi

Stress Ratio, m = 0

σ_x, psi	σ_y, psi	σ_z, psi	ε_x, %	ε_y, %	ε_z, %
20	20	20	0	0	0
19	19	22	.001	.001	.017
18	18	24	0	.003	.041
17	17	26	-.016	.013	.073
16	16	28	-.047	.030	.117
15	15	30	-.082	.052	.168
14	14	32	-.119	.087	.236
13	13	34	-.171	.146	.335
12	12	36	-.266	.241	.477
14	14	32	-.262	.239	.468
14	14	32	-.245	.224	.442
16	16	28	-.217	.202	.407
18	18	24	-.175	.168	.351
20	20	20	-.183	.171	.364
18	18	24	-.193	.180	.386
16	16	28	-.221	.196	.418
14	14	32	-.240	.208	.442
13	13	34	-.265	.229	.487
12	12	36	-.376	.349	.659
11	11	38	-.687	.652	1.042
10	10	40	-1.184	1.122	1.613
9	9	42	-2.394	2.258	2.795
8	8	44	-14.066	15.469	17.714
7.5	7.5	45			

TRIAXIAL EXTENSION TEST (TE) RESULTS

Constant Mean Stress Test

Mean Stress = 5 psi

Stress Ratio, m = 1

σ_x	σ_y	σ_z	ε_x, %	ε_y, %	ε_z, %
5	5	5	0	0	0
5.5	5.5	4	+.013	+.011	-.011
6	6	3	+.055	+.049	-.072
6.25	6.25	2.5	+.107	+.100	-.162
6.5	6.5	2.0	+.188	+.178	-.320
6	6	3	+.188	+.173	-.305
5	5	5	+.174	+.169	-.232
6	6	3	+.180	+.174	-.258
6.5	6.5	1.5	+.204	+.194	-.321
6.75	6.75	1.0	+.352	+.334	-.680
7	7	1.0	+.660	+.602	-1.597
7.25	7.25	.5	+3.824	+6.017	-14.304

OTTAWA SAND

TRIAXIAL EXTENSION TEST (TE) RESULTS

Constant Mean Stress Test

Mean Stress = 10 psi
Stress Ratio, m = 1

σ_x	σ_y	σ_z	ε_x, %	ε_y, %	ε_z, %
10.0	10.0	10.0	0	0	0
11.0	8.0	11.0	+.016	-.003	+.012
11.0	7.0	11.5	+.026	-.012	+.021
11.5	6.0	11.5	+.045	-.043	+.039
12.0	5.0	12.0	+.085	-.110	+.068
12.5	5.0	12.5	+.206	-.227	+.117
13.0	4.0	13.0	+.325	-.444	+.188
13.5	3.0	13.5	+.325	-.421	+.184
12.5	5.0	12.5	+.324	-.371	+.174
11.5	7.0	11.5	+.305	-.288	+.156
10.5	9.0	10.5	+.292	-.244	+.142
10.0	10.0	10.0	+.292	-.247	+.148
11.0	8.0	11.0	+.297	-.268	+.159
12.0	6.0	12.0	+.318	-.319	+.176
13.0	4.0	13.0	+.347	-.380	+.194
13.5	3.0	13.5	+.498	-.803	+.287
14.0	2.0	14.0	+2.638	-12.106	+.837
14.5	1.0	14.5	+2.679	-12.450	+.831
13.0	4.0	13.0	+2.539	-11.065	+.768
11.0	8.0	11.0	+2.230	-10.756	+.707
10.0	10.0	10.0			

TRIAXIAL EXTENSION TEST (TE) RESULTS

Constant Mean Stress Test

Mean Stress = 20 psi
Stress Ratio, m = 1

σ_x	σ_y	σ_z	ε_x, %	ε_y, %	ε_z, %
20	20	20	+.009	+.011	-.007
21	21	18	+.025	+.025	-.025
22	22	16	+.050	+.053	-.068
22	23	14	+.092	+.092	-.154
23	24	12	+.168	+.160	-.319
24	25	10	+.162	+.159	-.297
25	23	14	+.133	+.148	-.252
23	21	18	+.120	+.135	-.221
21	20	20	+.126	+.141	-.236
20	22	16	+.146	+.154	-.270
22	24	12	+.156	+.159	-.286
24	24.5	11	+.168	+.166	-.312
24.5	25	10	+.212	+.211	-.418
25	25.5	9	+.274	+.292	-.600
25.5	26	8	+.361	+.403	-.897
26	26.5	7	+.478	+.517	-1.328
26.5	27	6	+.656	+.693	-2.093
27	27.5	5	+1.060	+1.063	-4.186
27.5	28	4	+2.368	+2.966	-12.947
28	28.5	3.5			
28.5					

OTTAWA SAND

SIMPLE SHEAR TEST (SS) WITH CONSTANT MEAN STRESS AT 10 PSI

Stress Ratio, m = 0.2

Principal Stresses			Principal Strains		
σ_x	σ_y	σ_z	ε_x, %	ε_y, %	ε_z, %
10	10	10			
8	9	13			
7	8.5	14.5			
6	8	16			
5	7.5	17.5			
4	7	19			
6	8	16			
8	9	13			
10	10	10			
8	9	13			
6	8	16			
5	7.5	17.5			
4	7	19			
3.5	6.75	19.75			
3	6.5	20.5			
2.5	6.25	21.25			
2	6	22			
1.5	5.75	22.75			

TABLE 2 (Next 9 pages)

OTTAWA SAND

SIMPLE SHEAR TEST (SS) WITH CONSTANT MEAN STRESS AT 5 PSI

Stress Ratio, m = 0.5

Principal Stresses			Principal Strains		
σ_x	σ_y	σ_z	ϵ_x, %	ϵ_y, %	ϵ_z, %
5	5	5			
4	5	6			
3	5	7			
2.5	5	7.5			
2	5	8			
3	5	7			
4	5	6			
5	5	5			
4	5	6			
3	5	7			
2	5	8			
1.5	5	8.5			
1.0	5	9			
.75	5	9.25			
.50	5	9.5			
.25	5	9.75			

SIMPLE SHEAR TEST (SS) WITH CONSTANT MEAN STRESS AT 20 PSI

Stress Ratio, m = 0.5

Principal Stresses			Principal Strains		
σ_x	σ_y	σ_z	ϵ_x, %	ϵ_y, %	ϵ_z, %
20	20	20			
18	20	22			
16	20	24			
14	20	26			
12	20	28			
11	20	29			
10	20	30			
12	20	28			
14	20	26			
16	20	24			
18	20	22			
20	20	20			
18	20	22			
14	20	26			
12	20	28			
11	20	29			
10	20	30			
9	20	31			
8	20	32			
7	20	33			
5	20	35			
4	20	36			
3	20	37			
2	20	38			
1	20	39			
0	20	40			

OTTAWA SAND

SIMPLE SHEAR TEST (SS) WITH CONSTANT MEAN STRESS AT 10 PSI

Stress Ratio, $m = 0.8$

Principal Stresses			Principal Strains		
σ_x	σ_y	σ_z	ε_x, %	ε_y, %	ε_z, %
10	10	10			
8.5	10.5	11.0			
7.0	11.0	12.0			
5.5	11.5	13.0			
4.0	12.0	14.0			
7.0	11.0	12.0			
10.0	10.0	10.0			
8.5	10.5	11.0			
7.0	11.0	12.0			
5.5	11.5	13.0			
4.0	12.0	14.0			
3.25	12.25	14.5			
2.5	12.50	15.0			
1.75	12.75	15.5			
1.0	13.0	16.0			
0.0	13.333	16.667			

OTTAWA SAND

REDUCED TRIAXIAL COMPRESSION TEST (RTC)

Stress Ratio, $m = 0$

Principal Stresses			Principal Strains		
σ_x	σ_y	σ_z	ε_x, %	ε_y, %	ε_z, %
20	20	20			
19	19	20			
18	18	20			
17	17	20			
16	16	20			
15	15	20			
17	17	20			
19	19	20			
20	20	20			
19	19	20			
17	17	20			
15	15	20			
14	14	20			
13	13	20			
12	12	20			
11	11	20			
10	10	20			
9	9	20			
8	8	20			
7	7	20			
6	6	20			
5	5	20			
4	4	20			
3.5	3.5	20			
3	3	20			
2.5	2.5	20			
2	2	20			

OTTAWA SAND

REDUCED TRIAXIAL EXTENSION TEST (RTE)

Stress Ratio, m = 1.0

Principal Stresses			Principal Strains		
σ_x	σ_y	σ_z	ε_x, %	ε_y, %	ε_z, %
20	20	20			
20	20	18			
20	20	16			
20	20	14			
20	20	12			
20	20	9			
20	20	7			
20	20	10			
20	20	14			
20	20	18			
20	20	20			
20	20	18			
20	20	14			
20	20	10			
20	20	8			
20	20	7			
20	20	6			
20	20	5			
20	20	4			
20	20	3			
20	20	2.5			
20	20	2			
20	20	1.5			

PROPORTIONAL LOADING TEST (PL1)

Stress Ratio, m = 0

Principal Stresses			Principal Strains		
σ_x	σ_y	σ_z	ε_x, %	ε_y, %	ε_z, %
10	10	10			
10.5	10.5	14			
11	11	18			
11.5	11.5	22			
12	12	26			
12.5	12.5	30			
13	13	34			
13.5	13.5	38			
14	14	42			
14.5	14.5	46			
15	15	50			
15.5	15.5	54			
16	16	58			
16.5	16.5	62			
17	17	66			
17.5	17.5	70			
18	18	74			
18.5	18.5	78			
19	19	82			
19.5	19.5	86			
20	20	90			
20.5	20.5	94			
21	21	98			
21.5	21.5	102			
22	22	106			
22.5	22.5	110			

OTTAWA SAND

PROPORTIONAL LOADING TEST (PL2)

Stress Ratio, m = 0

Principal Stresses			Principal Strains		
σ_x	σ_y	σ_z	ϵ_x, %	ϵ_y, %	ϵ_z, %
10	10	10			
11	11	12			
12	12	14			
13	13	16			
15	15	20			
20	20	30			
25	25	40			
30	30	50			
35	35	60			
40	40	70			
45	45	80			
50	50	90			
55	55	100			
45	45	80			
35	35	60			
25	25	40			
15	15	20			
10	10	10			

CIRCULAR STRESS PATH (CSP) WITH CONSTANT MEAN STRESS AT 10 PSI

Principal Stresses			Principal Strains		
σ_x	σ_y	σ_z	ϵ_x, %	ϵ_y, %	ϵ_z, %
10	10	10			
9	9	12			
8	8	14			
7	7	16			
6.07	8.04	15.89			
5.18	9.31	15.51			
4.49	10.69	14.81			
4.11	11.96	13.93			
4	13.0	13.0			
4.11	13.93	11.96			
4.49	14.81	10.69			
5.18	15.51	9.31			
6.07	15.89	8.04			
7.00	16.0	7.0			
8.04	15.89	6.07			
9.31	15.51	5.18			
10.69	14.81	4.49			
11.96	13.93	4.11			
13.00	13.0	4.0			
13.93	11.96	4.11			
14.81	10.69	4.49			
15.51	9.31	5.18			
15.89	8.04	6.07			
16	7.0	7.00			
15.89	6.07	8.04			
15.51	5.18	9.31			
14.81	4.49	10.69			
13.93	4.11	11.96			
13.0	4.0	13.0			
11.96	4.11	13.93			
10.69	4.49	14.81			
9.31	5.18	15.51			
8.04	6.07	15.89			
7.00	7.0	16.0			
6.07	8.04	15.89			
5.18	9.31	15.51			
4.49	10.69	14.81			
4.11	11.96	13.93			
4	13.0	13.0			
6	12	12			
8	11	11			
10	10	10			

SOIL CONSTITUTIVE RELATIONSHIPS AND MODELLING OF SOIL BEHAVIOUR

by

R. N. Yong and H. Y. Ko

Challenge Paper

NSF/NSERC

North American Workshop
on
Plasticity Theories and Generalized Stress-Strain Modelling of Soils

Montreal

28-30 May 1980

SOIL CONSTITUTIVE RELATIONSHIPS AND MODELLING OF SOIL BEHAVIOUR

R. N. Yong[1] and H. Y. Ko[2]

INTRODUCTION

The increasing demands seen today for a better capability in design and analysis in geotechnical engineering practice, are reflected in corresponding requirements for more accurate predictions of the performance of soils under load. Responding to these needs and requirements are several recently developed methods of analysis and models of soil behaviour - as can be seen in the published literature.[*] It is noted however that whilst some of these models are still in the ongoing stage of development and hence cannot be fully evaluated, there appears to be differing opinions and judgements concerning the reliability and capability of many of the presently available theories and models to properly predict constitutive relationships consistent with actual physical performance.

With a better understanding of soil properties and soil behaviour - through many recent studies - and especially with the availability of sophisticated numerical techniques and computer based computational tools, it is evident that the quality of the set of material inputs should at least keep pace with the high level of computational capability. It is commonly believed that the

---

[1] William Scott Professor of Civil Engineering and Applied Mechanics, and Director, Geotechnical Research Centre, McGill University, Montreal, Quebec, Canada.

[2] Professor of Civil Engineering and Environmental Engineering, University of Colorado, Boulder, U.S.A.

[*] Note that because of the great wealth of available reports and published information on methods, procedures, theories, and models for prediction of stress-strain and yield/failure, it is impractical to identify them in the text. Representative items can be seen in the *Reading List* provided with the Workshop package.

problem of modelling actual soil behaviour is indeed too difficult to accomplish faithfully over the entire range of stress-strain performance - even for a limited range of typical soils. On the other hand, it is contended that the problem of fully utilizing the available models lies with the practitioners and users who have failed to completely understand the limits of application and indeed the requirements of the models, theories and methods used to provide the predictions desired.

It is because of

(a) the existence of some apparent areas of disagreement between the performance of analytically predicted models and actual physical performances,

(b) insufficient comprehension and understanding of both physical and analytical models, and

(c) perhaps some inadequacy of some of the computational models used,

that the NSF/NSERC Workshop has been organized to focus on the development and validation of predictive models designed to meet the present day requirements of geotechnical engineering. This Workshop forms the first part of a three-part program of assessment and study of the State-of-the-Art in analytical modelling of soil behaviour and application to geotechnical problems. It will be followed by a one-day ASTM Symposium[*] which deals with soil strength testing methods and analysis of data, and a one-day ASCE Symposium (2 sessions)[**] dealing with the application of current understanding of limit equilibrium and plasticity theories to geotechnical engineering problems. The thrust of the studies and applications will be to static and

---

[*] ASTM Symposium - June 25, 1980, Chicago, Illinois.

[**] ASCE 2-Session Series - October 1980 Annual Meeting of ASCE, Hollywood, Florida.

quasi-static situations.

In order to focus on the applicability of various constitutive models, a prediction exercise was conceived for this Workshop. Several sets of laboratory test data were gathered on the stress-strain behaviour of different types of soils and supplied to the proponents of the models. These proponents were identified as *Predictors*. The Predictors were requested to consider the data as being representative of the kinds of detailed test information available from good test laboratories, and to make use of these data in formulating their models quantitatively. In addition, the Predictors were to utilize the models in predicting the response of the same soils under general three-dimensional stress paths different from those given in the original data base. Four types of soils were used in this prediction; two were undisturbed natural clays, one was a remoulded clay, and one was an Ottawa sand. It is felt that by requiring the proposed constitutive models to be exercised on these widely different soils to predict responses under generalized load paths, it would be possible to identify the range of applicability of each model.

PROBLEM DEVELOPMENT

Analyses of soils subject to complex stress states have traditionally been conducted either as stability problems or problems associated with load-deformation response. By and large, stability problems are examined with the aid of limit equilibrium principles whilst load-deformation performances are evaluated through the use of analytical models incorporating the stress-strain performance of the soil. It is interesting to recall that the slip line studies in soil mechanics by Coulomb, Rankine and Levy predate those of Lüders and Hartman for soft steel. The many studies on stability using this auspicious beginning have led to clarifications and refinements in the application of limit equilibrium principles to the solution of stability problems. It should be noted that the general framework of the stability analyses do not as a rule consider the stress-strain properties of the material below the yield/failure "point".

In the application of plasticity theories to the analysis of stability problems, there is a limited range of models chosen to represent the performance of the material - e.g. rigid-perfectly plastic, elastic-perfectly plastic, work-hardening, etc. If it is desired to establish limiting loads, the procedures used require one to define an admissible stress field in addition to an appropriate yield criterion. We note that the stress fields derived therefrom using the method of characteristics as a solution technique for certain kinds of boundary value problems for certain types of soils have been amply verified through physical experiments by many authors. By and large, the soils tested have been subjected to certain ideal conditions and circumstances.

On the assumption that the material behaves as a perfectly plastic material, the normality condition can be invoked and correspondence between the stress and velocity characteristics can be directly established. Thus when the stress characteristics match those velocity characteristics which satisfy the boundary displacement conditions for the problem under consideration, one can be assured of both static and kinematic admissibility in the solution of the problem. Note the importance therefore of an appropriate and valid yield criterion, and also the stress-strain specification of the material following "yield". Obviously, with the specification of perfectly plastic behaviour of the soil beyond yield, the assumption of the coincidence of the stress and velocity characteristics greatly simplifies the manipulative procedures enroute to obtaining an exact solution. The consequences arising from the preceding bear directly on the nature and location of the slip surfaces. If ideal behaviour is really obtained, there is no ambiguity between real and analytic slip surfaces. However, the evidence indicates that ideal behaviour is the exception - as opposed to the general rule. The soil conditions in actual situations such as overconsolidation effects, anisotropy, dilatancy, bonding, etc. tend to render actual soil behaviour under load away from ideal behaviour requirements.

With the availability of high powered machine computational tools, more sophisticated analytical modelling and numerical techniques can be used to treat the same boundary value problems - permitting thereby a more realistic appreciation of soil behaviour under stress. We note that a form of phenomenological laws covering deformation after "yield" should also consider the magnitude of the total and incremental strains and strain-rates obtained. The basic assumptions generally invoked in the phenomenological approach need further detailed study since neither deformation theories nor associated flow laws based on the concept of plastic potential appear to be totally consistent with available test results. Thus, if the normality criterion is not necessarily applicable to soils, the restriction of coincidence of the stress and velocity characteristics is no longer required. The more recent studies in this area have highlighted some of the difficulties - and successes - with the development and use of the associated field technique for the solution of boundary value problems. It is noted that the velocity field by and large can only be calculated from information taken from the stress field. Thus if the stress field is smaller than the velocity field, it is not immediately clear how the velocity field can be successfully determined for that region outside the stress field. This also generates the continuing debate as to whether the slip surface is coincident with the stress or velocity characteristics. Arguments in support of either position can be mounted - as seen in the published literature.

The preceding discussion is seen to rely on the use of a time-independent elastoplastic model characterization which utilizes a yield function, plastic flow rules, and perhaps a hardening rule. As noted above, idealization of material characteristics to provide for conformity to associated flow rule requirements might serve to simplify calculations, but does not necessarily represent true material behaviour, and hence the solutions obtained are not necessarily valid. However, in working with the uncoupled stress and velocity fields where association is sought through stress-strain information, the burden of responsibility in specification of appropriate stress-strain laws becomes apparent and significant.

As stated earlier, ultimate "limiting" performance in load carrying capability is generally considered as a stability problem - separate from the so-called standard examination of constitutive (i.e. stress-strain) relationships. It should however be noted that the general concept of constitutive relationships covers the total area of concern, and that the limiting state of performance is indeed a more than valid concern of the constitutive equation. In addition, if continued material performance is to be examined, the behaviour of the material under load after the "limiting", "yield", etc. state needs to be further defined and specified. Separation of studies of constitutive relationships into those regions fictitiously identified as "elastic" and "plastic" perhaps does not serve to promote a better understanding of the use of proper analytical models for the treatment of the total load-deformation problem.

It is apparent that if stress-strain performance is to be traced from sub-yield through yield and post-yield, the constitutive relationship must present a continuously changing material property. For a multi-phase material such as soil, since volume change under load is conditioned by past history general compositional characteristics and local environmental conditions, the establishment of a constitutive equation which will properly-reflect the soil performance over the entire range of deformation becomes indeed difficult. The coupling of stress-strain behaviour in sub-yield with that at yield and post-yield into one coherent framework for expression in the analytic form requires one to incorporate the role of pore-water pressure (for a fully saturated soil) in addition to the complexities introduced by experimental test systems which do not necessarily provide the kinds of information needed to allow for proper generalization into the analytic form for successful application to the field problem. The problem of course becomes exceedingly complex when partial saturation is encountered, not only because of the analytic requirements caused by the need to model a third phase, but also because of the many difficulties in formulating and actual implementation of the kinds of test instrumentation and system.

THE PROBLEM

The available evidence shows that by and large, the yield/failure theories and constitutive relationships that have been proposed on the basis of various extensive laboratory studies have directed their attention and actual physical experience to certain well documented soils. In many cases, these have been laboratory prepared soils, whilst in many other cases, selected soils with limited performance characteristics have been chosen. The relationships developed therefrom have been obviously conditioned to respond to the soils tested as well as for the particular test system constraints.

and therefore the parameters used and material properties sensed have been chosen to fit the test circumstance. Extension and projection into a more general framework for wider use and application do not necessarily appear to be sufficiently well-founded.

Four aspects of the problem can be identified as:

1. Soil compositional features:

    Mineralogy, bonding, structure, history, unsaturation/saturation etc. The problems of soil type, local interactions and environmental influence, overconsolidation, pore-water and pore-air considerations, sensitivity, anisotropy, homogeneity, etc. can be cited as problems which need to be examined in conjunction with the feature identified as *initial conditions*.

2. Laboratory testing:

    Test system constraints, simulation of actual field test load imposition in laboratory test, measurement of pertinent strength properties, and ability of test technique to discriminate and isolate the various compositional characteristic influences on response performance.

    Is the test technique too specialized or too complicated? Are the pieces of information obtained from the test system necessarily restricted in use to the "predictive" theory specially developed for the soil tested?

    Is the test sound?

    Cost realities of testing?

3. Field situation:

    Identification of field constraints, boundary conditions, initial conditions, etc. Reconciliation between field variability and limited sampling and testing.

    Is the constitutive relationship only applicable to the laboratory situation?

    Are transfer functions required?

4. Analytical modelling:

    For a successful development of a generalized constitutive relationship, the concerns expressed in 1 through 3 need to be properly incorporated into the "equation".

    Can this be done? Or should one concentrate in development of of specialized relationships applicable over specified regions of interest and limited to particular situations and soils?

Can one afford the model? Will it work in the computer solution technique devised for treatment of the overall geotechnical problem?

CONCLUDING REMARKS

There has been no attempt to set the historical background of the problem under consideration; nor has there been any attempt to develop the many models presently available for application to the problem of specification of yielding and stress-strain performance. There has been no prior judgement regarding the merits or disadvantages of any of the theories now appearing in print. It could well be that some of these can fully answer the concerns expressed in the four-fold problem statement. If such be the case, we can indeed look forward to an illuminating experience in the Workshop.

READING LIST OF SOME PERTINENT PAPERS FOR
NSF/NSERC WORKSHOP ON PLASTICITY

1. AKAI, K., ADACHI, T. and ANDO, N. (1975) *Existence of a unique stress-strain-time relation of clays.* Soils and Found., Vol. 15, No.1, pp. 1-16.

2. AKAI, K., ADACHI, T. and FUJIMOTO, K. (1977) *Constitutive equations for geomechanical materials based on elasto-viscoplasticity.* Proc., Spec. Session 9, Int. Conf. ISMFE, Tokyo - Const. Eq. of Soils.

3. BALADI, G. Y. and ROHANI, B. (1979) *An elastic-plastic constitutive model for saturated sand subjected to monotonic and/or cyclic loadings.* 3rd Int. Conf., Numerical Methods in Geomechanics, pp. 389-404.

4. BAZANT, Z. P., ANSAL, A. M. and KRIZEK, R. J. (1979) *Viscoplasticity of transversely isotropic clays.* Journal of the Engineering Mechanics Division, Proc. ASCE, Vol. 105. pp.

5. BAZANT, Z. P. (1978) *Endochronic inelasticity and incremental plasticity.* Int. Journal of Solids and Struct., Vol. 14, pp. 691-714.

6. BAZANT, Z. P. and BHAT, P. D. (1976) *Endochronic theory of inelasticity and failure of concrete.* Journal Engineering Mechanics Division Proc. ASCE, Vo.1 102, pp. 701-721.

7. CHEN, W. F. (1975) <u>Limit Analysis and Soil Plasticity</u>. Elsevier, Amsterdam.

8. DAFALIAS, Y. F. and POPOV, E. P. (1975) *A model of non-linearly hardening materials for complex loading.* Acta Mechanica, Vol. 21, pp. 173-192.

9. DRUCKER, D. C., GIBSON, R. E. and HANKEL, D. J. (1957) *Soil mechanics and work-hardening theories of plasticity.* Trans. ASCE Vol. 122, pp. 338-346.

10. DUNCAN, J. M., BYRNE, P., WONG, K. S. and MABRY, P. (1979) *Strength, stress-strain and bulk modulus parameters for finite element analyses of stresses and movements in soil masses.* Geotechnical Engineering Report, University of California, Berkeley.

11. DUNCAN, J. M. and CHANG, C. T. (1970) *Non-linear analysis of stress and strain in soils.* Jour. SMFD, ASCE, Vol. 96, No. SM5 September.

12. GUDEHUS, G. and KOLYMBAS, D. (1979) *A constitutive law of the rate type for soils.* Proc. 3rd Int. Conf. Num. Meth. Geomech., Aachen. Vol.1 Editor: Wittke, Balkema. pp. 319-329.

13. GUDEHUS, G. (1980) *A comparison of some constitutive laws for soils under radially symmetric loading and unloading.* Proc. 3rd Int.Conf. Num. Meth. Geomech., Aachen. Vol. 4. Editor: Wittke, Balkema.

14. GUDEHUS, G. (1977) *Some interactions of finite element methods and geomechanics.* A Survey: Chapter 1 of <u>Finite Elements for Geomechanics</u> Wiley.

15. HARDIN, B. O. (1978) *The nature of stress-strain behaviour of soils.* Proc. Spec. Conf. Earthquake Engr. and Soil Dynamics. ASCE. Vol. 1, pp.3-90.

16. LADE, P. V. (1977) *Elastic-plastic stress-strain theory for cohesionless soil with curved yield surfaces.* Int. Journal of Solids Struct., Vol. 13, pp. 1019-1035.

17. PREVOST, J. H. (1978) *Plasticity theory for stress-strain behaviour.* Jour. EMD, ASCE, EM5, pp. 1177-1194.

18. POPPER, Karl R.   *Logic of Scientific Discovery.*

19. SANDLER, I. S., DiMAGGIO, F. L. and BALADI, G. Y. (1976) *Generalized cap model for geological materials,* Jour. GED, ASCE, pp. 683-699.

20. SCHOFIELD, A. N. and WROTH, P. (1968) <u>*Critical state soil mechanism.*</u> McGraw-Hill, N.Y.

21. VALANIS, K. C. 91975) *On the foundations of the endochronic theory of viscoplasticity.* Archives of Mechanics, Vol. 27, No. 5, pp. 857-868.

22. WROTH, P. (1973) *A brief review of the application of plasticity to soil mechanics.* Proc. Symp. on the Role of Plasticity in Soil Mechanics, Cambridge, pp. 1-11.

23. ZIENKIEWICZ, O. C., HUMPHESON, C. and LEWIS, R. W. (1975) *Associated and non-associated viscoplasticity and plasticity in soil mechanics.* Geotechnique, Vol. 25, No. 4, pp. 671-689.

* * * * * * * * * *

a. *Constitutive Equations of Soils,* Proceedings, Speciality Session No. 9 , Int. Conf. ISMFE, Tokyo.(1977) S. Murayamo and A. N. Schofield, editors.

b. *Role of Plasticity in Soil Mechanics,* Proc. of Symposium, Cambridge, 1973. A. C. Palmer, editor.

c. *Roscoe Memorial Symposium: Stress-Strain Behaviour of Soils.* Henley-on-Thames, Foulis, U.S. 1972.   editor: R.G.H.Parry.

3. CHALLENGES FROM INDUSTRY AND ACADEMIA

The two challenge papers, from industry and academic users' points of view, are given in this Section. These were given during the Plenary Sessions at the Workshop, with the purpose of focusing the discussions at the separate Working Groups on the needs from the users in professional practice as well as in academic research.

USER NEEDS: A VIEW FROM INDUSTRY

By

John T. Christian*

The user of a proposed constitutive model has essentially the same requirements whether his purpose is academic instruction, research, or practical application to an engineering problem. Therefore, most of Professor Selig's remarks apply with equal force to the situation of the practicing engineer, for both the teacher and the practitioner need to translate the abstract, theoretical model into forms simple enough to be understood and used by less sophisticated engineers under less than ideal circumstances.

One way to express the needs of the user is to develop a catalogue of requirements. Ideally, the developers of constitutive models would retire to the laboratory and office, returning in due time with fully developed models that can be applied without further modification. The real world is never so simple. There must be feedback from the users to describe not only in what situations the models will be used but how well the models function when they are applied. Nonetheless, there have been several attempts to list research needs. For example, one of the working groups at a recent NSF/NBS Workshop[1] developed the following statement of needs:

---

*Senior Consulting Engineer, Stone & Webster Engineering Corporation, Boston, Massachusetts, U.S.A.

Any stress-strain relations developed for soils under dynamic excitations should comply with the basic laws of mechanics. Such relations should be capable of handling arbitrarily varying cyclic loadings in three dimensions (even if they are to be used in conjunction with two-dimensional codes), but they should be simple from a computational point of view. Furthermore, it should be possible to determine the coefficients of the model in terms of a small number of parameters obtained from standard tests. Some of the experimentally observed soil characteristics that should be considered when developing the stress-strain relations are:

(a) Soils behave generally as nonlinear, hysteretic materials for the range of frequencies of interest during earthquakes.

(b) The stiffness and strength of soils depend on effective stresses.

(c) There is a strong coupling between cyclic shear stresses and volumetric strains. This coupling is small or nonexistent at small shear strains, but it is a major factor in soil behavior at moderate and large cyclic shear strains.

(d) The stress-strain behavior and strength of soil can change during cyclic loading.

This statement is biased somewhat because the workshop was concerned with problems of earthquake engineering and soil dynamics, and several additional requirements could have been added to the list. For example, a generalized constitutive model should be able to deal correctly with unloading and reloading of the soil, and cyclic loading should also be treated properly. The model should be capable of dealing with fully three-dimensional effects. It should work for loading paths different from those in the tests from which the model was developed or in which the values of the parameters were measured. It should be able to handle viscous as well as plastic effects. Generation of excess pore pressures and their dissipation by consolidation should be included. The model should be capable of describing unstable materials, and, for general application to practical geotechnical problems, it should include the effects of partial saturation.

All of the above are desirable in a constitutive model, but no model that is now available or contemplated is able to deal with all of them. One of the purposes of this symposium is to investigate how well many of the available models do deal with some of these problems. As has already been stated in this symposium, a useful model in engineering mechanics is not necessarily one that includes all aspects of material behavior. Instead of trying to find the most complicated model, the user should try to find the simplest one that adequately represents those aspects of his problem that are significant. Thus, it is ultimately fruitless to try to develop a set of guidelines for the most general, all-encompassing constitutive model for soil.

This user would urge that the developers of constitutive models and the participants in this conference follow four general guidelines. First, it should be the objective of every developer of a constitutive model to make the model understandable to potential users. The temptation to express the model in complicated mathematical notation and arcane terminology is almost overwhelming. While this may occasionally impress one's peers, it raises an impenetrable barrier to understanding by the practicing engineer. What the user cannot understand, he will not use. The failure to describe new constitutive models in terms that are sufficiently simple, clear, and unambiguous is probably the greatest single barrier to the application of such models to practical problems.

Second, the developer of the model should indicate how to use the model and how to measure the values of the parameters that must be input to it. One of the reasons for the widespread use of the well-known hyperbolic stress-strain relations is that procedures for obtaining the necessary input values have been explained in the published literature. Further, those parameters are related to data that can be obtained readily from conventional tests on soils.

Third, developers of models should encourage critical experiments to establish the range of validity of the models. One of the very best checks of models in geotechnical engineering, as well as in other branches of engineering and science, is to develop predictions based on the newly developed model and to compare those predictions with experiments carried out independently. When sucessful, such critical experiments are vastly more useful than any number of back-calculations that attempt to verify the model from experimental data that were already available before the model was developed.

Fourth, the practical use of most models requires that they be simplified. This is due in part to the difficulty of obtaining accurate data under field conditions and to the uncertainty in describing actual soil profiles. Therefore, it would be very helpful if developers of constitutive models would indicate where the models can be simplified further and what the range of validity might be for the simpler models. For example, even though it is well known that the behavior of soil under dynamic and cyclic loads is quite complicated, most dynamic analyses are carried out using linearly elastic soil models in which the fundamental parameter describing the behavior of the soil models is the shear modulus. In many cases of dynamic soil-structure interaction for earthquakes, this is the only type of model that is used.

The deliberations of this workshop will be very heavily concerned with the relations between theory and laboratory experiment. This is a proper interplay of theory and experiment and should lead to several valuable insights into the behavior of the several models to be investigated. However, the practical user must also bear in mind that the ultimate application of these models is in the field, and there are substantial differences between laboratory behavior and field behavior of soils. As mentioned above, the major parameter used to describe the dynamic behavior of soils in earthquake engineering is the shear modulus. This happens to be a quantity that is accurately measured in the field by such procedures as cross-hole shear wave velocity measurements and in the laboratory by experimental equipment such as the resonant column device. In fact, this may be one of the very few constitutive parameters that is measured as accurately in the laboratory as in the field. When values of the shear modulus measured in the laboratory on undisturbed samples of clay are compared to values obtained from field measurements, a substantial discrepancy is usually encountered. It has been discovered that the values of shear modulus measured in the laboratory increase dramatically as a result of secondary compression and that, when this effect is used to correct the laboratory values, they can be made to agree quite closely with the field data[2,3]. How much correction and what sort of correction is needed for other constitutive parameters is not known, but this example should provide a warning that, in addition to laboratory verification of constitutive models, field verification is important.

This workshop is most unusual for a variety of reasons and the organizers ought to be commended for assembling it. The financial support by both the United States and Canadian governments and the cooperation between the researchers in both countries have been exemplary. A large number of workers have assembled both from the research laboratories and from the practicing community. However, the greatest praise must go to the developers of the models, who have been willing to make critical predictions of how well their models would perform in describing the behavior of soil under unusual testing conditions. Such intellectual courage is rare and the predictors are to be commended for exhibiting it. In the last analysis, open predictions together with the resulting discussions will lead to an improvement in the understanding of soil behavior that will be of benefit to the researchers as well as the practitioners. That alone is sufficient benefit to the user community to warrant holding this workshop.

References

1. Report of Panel No. 3: Analytical Procedures and Mathematical Modeling, Research Needs and Priorities for Geotechnical Earthquake Engineering Applications, NSF/NBS Workshop at University of Texas, Austin, Texas, K. L. Lee, W. F. Marcuson III, K. H. Stokoe, and F. Y. Yokel (eds.), June 2 and 3, 1977, p. 35.

2. Anderson, Donald G., and Woods, Richard D., "Time - Dependent Increase in Shear Modulus of Clay," Journal of the Geotechnical Engineering Division, ASCE, Vol. 102, No. GT5, May 1976, pp. 525-537.

3. Trudeau, Paul J., Whitman, Robert V., and Christian, John T., "Shear Wave Velocity and Modulus of a Marine Clay," Journal, Boston Society of Civil Engineers, Vol. 61, No. 1, January 1974, pp. 12-25.

"User Needs - Academic Viewpoint"

Challenge Paper

by

Ernest T. Selig[1]

Introduction

The purpose of this paper is to help provide a focus for the workshop discussions on soil constitutive models. In preparing these remarks I had to ask myself the question, "What does 'user needs' from the academic viewpoint mean and how do these needs differ from the industry viewpoint?" Of course, there must be some overlap between academic and industry user needs because many university faculty are known to have a consulting practice. The distinction is primarily in teaching and research. An understanding of the existing models and their capabilities is needed in order to teach this subject; an assessment of the present state-of-the-art is needed to determine research needs.

The following discussion will present my views about the goals of this workshop and give suggestions about how they can be approached. For the most part, I believe that these views are equally appropriate for academic and industry users because the proposed steps form the basis for meeting both sets of needs.

Workshop Goals

The following are specific objectives proposed for the workshop:
1. To understand the concept behind each model.
2. To determine the required parameters for each model and how to obtain them.
3. To determine the advantages and limitations of each model.
4. To assess the application possibilities of the models.
5. To establish needs for further research.

---

[1] Professor of Civil Engineering, University of Massachusetts, Amherst, MA

The first day of the workshop should concentrate on items 1 and 2. The second day will emphasize items 3 and 4. The third day should address item 5. Among the main purposes of using the prediction approach to this workshop are to provide a common basis for comparison of the various models, and to permit quantitative evaluations by having specific numerical examples of application.

Guidelines for Model Description

The following are the proposed minimum requirements for describing each model to provide the information needed to assess and compare the models:

1. Important features to specify.
    a) Whether the soil is represented as 1, 2, or 3 phase system.
    b) Nonlinear and inelastic features.
    c) Ability to model dilatency.
    d) Total or effective stress approach.
    e) Representation of stress-induced or fabric anisotropy.
    f) Degree to which stress history is considered by model.
    g) Type of kinematic hardening.
    h) Limitations on soil type or conditions which can be represented.
2. Describe the parameters incorporated in the model and indicate the preferred tests to determine their numerical values.
3. Indicate the tests believed to be the most appropriate for validating a model and assessing its suitability.

As a minimum, the workshop groups are challenged to obtain the above information for each model.

Assessment of Model Application

To assess the applicability of the models to geotechnical engineering problems, the following questions are posed as a guide to the evaluation:

1. How well do the models represent actual soil behavior, and furthermore how well do they need to?
2. Is the value of the models primarily in aiding understanding soil behavior or in problem solving for design or analysis?

3. Assuming that finite element computer methods are needed to use the models:
   a) At what level of problem is this justified?
   b) What are relative complexities or computer costs in using the various models?
   c) Is analysis a more suitable application than design?
   d) Are deformation problems more important applications than stability problems?
4. Considering the realities of field soil variability and uncertainty of field conditions, is it primarily precision that is being improved in using the models rather than accuracy? Are the models useful even when available soil data are inadequate for determining the values of the model parameters?
5. What have been the successful applications of the models, and, considering 1 to 4 above, what are the recommended applications? Have the models been used by other than their developers, and to what extent is adequate documentation of the model available?

Concluding Comments

The stated objectives of the workshop should always be kept in mind during the deliberations. The first task should be to obtain, for each model, the information listed in the guidelines for model description. This can be done as part of the prediction presentations. Then the applicability of the models should be discussed using the posed questions as a guide. When this point in the workshop program is reached, the research needs should become evident.

Obtaining even preliminary information and answers for all of the above items is a challenge, but it is a challenge which must be met if the state-of-the-art is to be properly assessed. Otherwise the subject of soil stress-strain modeling will remain a subject of interest to a few theoreticians and of limited value in geotechnical engineering practice.

4. PREDICTIONS AND COMPARISONS

In this Section, the test data which were to be predicted were first presented in the same format as those given in Pre-Workshop Data Package in Section 2. These data were shown in the Workshop and the Predictors were given the opportunity to present their predictions during the individual working group sessions.

The predictions had been submitted in writing before the test data were made known. After the Workshop, the Predictors were given the opportunity to add their own comparisons to their written contributions. Their prediction and comparison papers are included in this Section, except for the following three cases.

Professor C.P. Wroth presented his predictions in Discussion Group 2 at the Workshop. However, he did not submit a written documentation for the predictions and comparisons. The reader is referred to the Chairman's report from Group 2 for a discussion of Professor Wroth's predictions.

Both Professor J.H. Prevost and Professor G. Gudehus chose to combine their predictions and comparisons with their position papers, which appear in Part 2 of these Proceedings.

Test No. 2

Water Content: 38.8% KAOLINITE

ANGLE OF MAJOR PRINCIPAL STRESS AXES TO VERTICAL IS 15°

TOTAL PRINCIPAL STRESSES			PORE PRESSURE	PRINCIPAL STRAINS		
σ_1	σ_2	σ_3		ε_1	ε_2	ε_3
58.00	58.00	58.00	18.1	.0	.0	.0
78.90	58.00	56.50	23.5	.0024	-.0011	-.0013
86.22	58.00	55.98	26.0	.0047	-.0022	-.0026
92.69	58.00	55.51	27.8	.0074	-.0033	-.0042
96.78	58.00	55.22	28.9	.0101	-.0044	-.0058
100.01	58.00	54.99	29.6	.0128	-.0055	-.0074
102.28	58.00	54.82	30.1	.0155	-.0065	-.0090
104.32	58.00	54.68	30.6	.0182	-.0076	-.0106
105.72	58.00	54.58	30.9	.0208	-.0088	-.0120
107.88	58.00	54.42	31.4	.0261	-.0110	-.0151
109.60	58.00	54.30	31.8	.0310	-.0132	-.0179
110.79	58.00	54.21	32.2	.0365	-.0154	-.0211
111.76	58.00	54.15	32.6	.0421	-.0176	-.0244
112.40	58.00	54.10	32.8	.0473	-.0199	-.0274
112.94	58.00	54.06	33.0	.0525	-.0222	-.0303
113.48	58.00	54.02	33.1	.0575	-.0244	-.0331
113.91	58.00	53.99	33.3	.0626	-.0267	-.0359
114.13	58.00	53.97	33.4	.0679	-.0290	-.0389
114.34	58.00	53.96	33.5	.0726	-.0313	-.0413

FOR this test the major principal stress is inclined at 15° to the vertical axis of the sample as shown in the sketch below. The intermediate principal stress (σ_2) is horizontal in all tests.

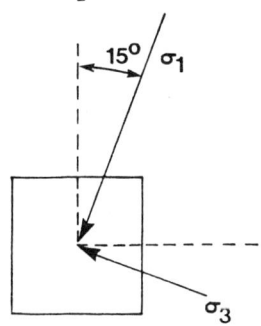

Test No. 3

Water Content: 38.8% KAOLINITE

ANGLE OF MAJOR PRINCIPAL STRESS AXES TO VERTICAL IS 37.5°						
TOTAL PRINCIPAL STRESSES			PORE PRESSURE	PRINCIPAL STRAINS		
σ_1	σ_2	σ_3		ε_1	ε_2	ε_3
58.00	58.00	58.00	18.0	.0	.0	.0
74.18	58.00	51.82	22.2	.0025	-.0011	-.0014
81.14	58.00	49.16	25.0	.0070	-.0022	-.0048
88.10	58.00	46.50	26.3	.0120	-.0033	-.0087
91.98	58.00	45.02	27.5	.0181	-.0043	-.0137
94.08	58.00	44.22	28.0	.0231	-.0054	-.0176
96.19	58.00	43.41	28.3	.0279	-.0065	-.0213
97.64	58.00	42.86	28.7	.0340	-.0076	-.0263
98.45	58.00	42.55	28.9	.0395	-.0087	-.0308
99.58	58.00	42.12	29.0	.0446	-.0098	-.0348
100.39	58.00	41.81	29.1	.0499	-.0109	-.0389
101.20	58.00	41.50	29.2	.0554	-.0120	-.0433
101.85	58.00	41.25	29.4	.0609	-.0132	-.0478
102.17	58.00	41.13	29.4	.0656	-.0143	-.0513
102.33	58.00	41.07	29.4	.0695	-.0154	-.0541

Test No. 5

Water Content: 39.7%

ANGLE OF MAJOR PRINCIPAL STRESS AXES TO VERTICAL IS 15°						
TOTAL PRINCIPAL STRESSES			PORE PRESSURE	PRINCIPAL STRAINS		
σ_1	σ_2	σ_3		ε_1	ε_2	ε_3
58.00	58.00	58.00	18.0	.0	.0	.0
72.36	51.57	50.08	20.0	.0019	-.0005	-.0013
73.62	51.00	49.38	20.1	.0031	-.0011	-.0021
77.34	49.33	47.32	20.5	.0054	-.0022	-.0032
79.87	48.20	45.93	20.5	.0083	-.0033	-.0050
81.58	47.43	44.98	20.6	.0110	-.0044	-.0067
82.92	46.83	44.25	20.6	.0138	-.0055	-.0083
83.74	46.47	43.79	20.6	.0163	-.0066	-.0097
84.41	46.17	43.42	20.6	.0192	-.0077	-.0115
84.86	45.97	43.18	20.6	.0218	-.0088	-.0130
85.67	45.60	42.73	20.7	.0273	-.0110	-.0162
86.27	45.33	42.4	20.9	.0326	-.0133	-.0193
86.72	45.13	41.15	20.9	.0376	-.0155	-.0221
87.01	45.00	41.99	21.1	.0432	-.0178	-.0254
87.31	44.87	41.82	21.2	.0488	-.0200	-.0288
87.46	44.80	41.74	21.5	.0552	-.0223	-.0329
87.53	44.77	41.70	21.7	.0634	-.0246	-.0388

Test No. 6

Water Content: 39.4% KAOLINITE

ANGLE OF MAJOR PRINCIPAL STRESS AXES TO VERTICAL IS 31.75°						
TOTAL PRINCIPAL STRESSES			PORE PRESSURE	PRINCIPAL STRAINS		
σ_1	σ_2	σ_3		ε_1	ε_2	ε_3
58.00	58.00	58.00	18.0	.0	.0	.0
70.33	54.80	48.87	20.0	.0016	-.0005	-.0011
74.70	53.67	45.63	20.7	.0045	-.0011	-.0034
77.40	52.97	43.63	21.0	.0074	-.0016	-.0058
79.45	52.43	42.11	21.6	.0122	-.0022	-.0100
80.74	52.10	41.16	21.8	.0162	-.0027	-.0135
81.51	51.90	40.59	22.0	.0196	-.0033	-.0163
82.79	51.57	39.64	22.4	.0259	-.0044	-.0216
83.82	51.30	38.88	22.8	.0334	-.0055	-.0279
84.34	51.17	38.50	23.0	.0387	-.0066	-.0321
84.98	51.00	38.02	23.2	.0453	-.0077	-.0377
85.36	50.90	37.74	23.2	.0496	-.0088	-.0409
85.88	50.77	37.36	23.4	.0619	-.0110	-.0509
86.26	50.67	37.07	24.0	.0770	-.0132	-.0638

Test No. 7

Water Content: 38.8% KAOLINITE

ANGLE OF MAJOR PRINCIPAL STRESS AXES TO VERTICAL IS 45°						
TOTAL PRINCIPAL STRESSES			PORE PRESSURE	PRINCIPAL STRAINS		
σ_1	σ_2	σ_3		ε_1	ε_2	ε_3
58.00	58.00	58.00	18.0	.0	.0	.0
64.10	58.00	51.90	19.0	.0011	.0	-.0011
68.10	58.00	47.90	20.3	.0018	.0	-.0018
70.10	58.00	45.90	21.0	.0023	.0	-.0023
72.10	58.00	43.90	21.9	.0066	.0	-.0066
74.10	58.00	41.90	22.6	.0094	.0	-.0094
77.80	58.00	38.20	24.0	.0188	.0011	-.0199
79.40	58.00	36.60	24.4	.0261	.0022	-.0283
81.00	58.00	35.00	24.5	.0350	.0027	-.0377
81.80	58.00	34.20	24.5	.0416	.0032	-.0448
82.50	58.00	33.50	24.5	.0490	.0038	-.0528
83.40	58.00	32.60	24.4	.0600	.0043	-.0644
84.10	58.00	31.90	24.2	.0758	.0032	-.0790
84.20	58.00	31.80	24.2	.0861	.0022	-.0883
84.40	58.00	31.60	24.4	.1086	.0	-.1086

Test No. 8

Water Content: 39.4%
KAOLINITE

ANGLE OF MAJOR PRINCIPAL STRESS AXES TO VERTICAL IS 58.25°						
TOTAL PRINCIPAL STRESSES			PORE PRESSURE	PRINCIPAL STRAINS		
σ_1	σ_2	σ_3		ε_1	ε_2	ε_3
58.00	58.00	58.00	18.1	.0	.0	.0
63.71	60.00	50.29	19.4	.0011	.0005	-.0016
64.47	60.27	49.26	19.9	.0018	.0011	-.0029
65.61	60.67	47.72	20.8	.0032	.0022	-.0054
66.47	60.97	46.57	21.6	.0052	.0033	-.0085
67.32	61.27	45.41	22.4	.0072	.0043	-.0115
68.56	61.70	43.74	23.6	.0107	.0065	-.0172
69.51	62.03	42.46	24.2	.0143	.0086	-.0229
70.37	62.33	41.30	25.0	.0188	.0108	-.0296
71.03	62.57	40.40	25.4	.0223	.0129	-.0352
71.70	62.80	39.50	25.8	.0266	.0150	-.0417
72.46	63.07	38.47	26.2	.0311	.0172	-.0483
73.03	63.27	37.70	26.4	.0363	.0193	-.0555
73.41	63.40	37.19	26.4	.0410	.0214	-.0623
73.70	63.50	36.80	26.4	.0456	.0234	-.0691
74.08	63.63	36.29	26.4	.0505	.0255	-.0760

Test No. 9

Water Content: 38.8%

ANGLE OF MAJOR PRINCIPAL STRESS AXES TO VERTICAL IS 75°						
TOTAL PRINCIPAL STRESSES			PORE PRESSURE	PRINCIPAL STRAINS		
σ_1	σ_2	σ_3		ε_1	ε_2	ε_3
58.00	58.00	58.00	18.0	.0	.0	.0
62.72	61.83	49.44	19.8	.0024	.0022	-.0046
63.87	62.77	47.36	21.3	.0048	.0043	-.0091
64.90	63.60	45.50	22.7	.0072	.0065	-.0137
65.68	64.23	44.09	23.6	.0095	.0086	-.0182
66.29	64.73	42.97	24.5	.0121	.0108	-.0228
66.87	65.20	41.93	25.0	.0146	.0129	-.0274
67.36	65.60	41.04	25.4	.0169	.0150	-.0319
67.85	66.00	40.15	25.7	.0193	.0171	-.0364
68.59	66.60	38.81	26.0	.0244	.0213	-.0457
69.37	67.23	37.39	26.0	.0295	.0255	-.0550
69.99	67.73	36.28	25.9	.0345	.0296	-.0641
70.65	68.27	35.09	25.5	.0397	.0337	-.0734
71.18	68.70	34.12	25.1	.0448	.0378	-.0825
71.75	69.17	33.08	24.8	.0499	.0418	-.0917
72.08	69.43	32.48	24.6	.0522	.0438	-.0960
72.37	69.67	31.96	24.3	.0547	.0458	-.1005
72.58	69.83	31.59	24.2	.0578	.0478	-.1055

Test No. 11

Water Content: 38.8%   KAOLINITE

ANGLE OF MAJOR PRINCIPAL STRESS AXES TO VERTICAL IS 58.25°						
TOTAL PRINCIPAL STRESSES			PORE PRESSURE	PRINCIPAL STRAINS		
σ_1	σ_2	σ_3		ε_1	ε_2	ε_3
58.00	58.00	58.00	18.0	.0	.0	.0
62.94	58.00	45.06	18.0	.0022	.0011	-.0033
63.87	58.00	42.63	18.5	.0042	.0022	-.0063
64.43	58.00	41.17	18.8	.0060	.0032	-.0092
65.29	58.00	38.91	19.1	.0079	.0043	-.0122
65.97	58.00	37.13	19.5	.0114	.0065	-.0179
66.65	58.00	35.35	19.8	.0154	.0086	-.0240
67.21	58.00	33.89	19.9	.0197	.0108	-.0304
67.76	58.00	32.44	19.95	.0242	.0129	-.0371
68.20	58.00	31.30	19.8	.0285	.0150	-.0435
68.51	58.00	30.49	19.6	.0324	.0171	-.0495
68.94	58.00	29.36	19.2	.0373	.0192	-.0565
69.37	58.00	28.23	18.7	.0432	.0213	-.0645
70.30	58.00	25.80	18.3	.0478	.0234	-.0712
69.93	58.00	26.77	18.0	.0526	.0244	-.0770

Test No. 12

Water Content: 38.7%

ANGLE OF MAJOR PRINCIPAL STRESS AXES TO VERTICAL IS 75°						
TOTAL PRINCIPAL STRESSES			PORE PRESSURE	PRINCIPAL STRAINS		
σ_1	σ_2	σ_3		ε_1	ε_2	ε_3
58.00	58.00	58.00	17.8	.0	.0	.0
58.83	58.00	46.47	17.1	.0024	.0022	-.0045
59.07	58.00	43.13	17.7	.0048	.0043	-.0091
59.26	58.00	40.44	18.3	.0072	.0065	-.0137
59.39	58.00	38.61	18.9	.0096	.0086	-.0182
59.54	58.00	36.56	19.1	.0120	.0108	-.0227
59.62	58.00	35.38	19.3	.0143	.0129	-.0272
59.72	58.00	33.98	19.4	.0166	.0150	-.0316
59.82	58.00	32.68	19.4	.0189	.0171	-.0360
59.96	58.00	30.64	19.0	.0239	.0213	-.0453
60.09	58.00	28.91	18.5	.0290	.0255	-.0545
60.23	58.00	26.97	18.0	.0339	.0296	-.0635
60.33	58.00	25.47	17.0	.0391	.0337	-.0728
60.44	58.00	23.96	16.4	.0442	.0378	-.0819
60.53	58.00	22.77	15.8	.0493	.0418	-.0911
60.59	58.00	21.91	15.5	.0521	.0438	-.0959
60.63	58.00	21.37	15.2	.0550	.0458	-.1008
60.66	58.00	20.94	15.0	.0575	.0478	-.1052
60.69	58.00	20.51	14.9	.0604	.0497	-.1101

CLAY "X"

Consolidation Pressure, σ_c = 10.0 psi

σ_1 (psi)	σ_2 (psi)	σ_3 (psi)	ε_1 (%)	ε_2 (%)	ε_3 (%)	$\frac{dV}{V}$ (%)
Stress Ratio m = 1/4						
10.00	10.00	10.00	0.00	0.00	0.00	0.00
11.17	9.66	9.17	0.09	-0.49	0.00	-0.40
12.33	9.33	8.33	0.16	-0.89	0.00	-0.73
12.39	9.00	7.50	0.23	-1.13	0.00	-0.90
14.66	8.66	6.67	0.33	-1.23	-0.05	-0.95
15.83	8.33	5.83	0.41	-1.32	-0.07	-0.98
17.00	8.00	5.00	0.56	-1.34	-0.13	-0.90
18.17	7.66	4.17	0.81	-1.51	-0.20	-0.90
19.33	7.33	3.33	-	-	-	-
Stress Ratio m = 1/2						
10.00	10.00	10.00	0.00	0.00	0.00	0.00
11.00	10.00	9.00	0.20	-0.10	-0.10	0.00
12.00	10.00	8.00	0.43	-0.25	-0.18	0.00
13.00	10.00	7.00	0.63	-0.38	-0.25	0.00
14.00	10.00	6.00	0.83	-0.52	-0.31	0.00
15.00	10.00	5.00	1.03	-0.63	-0.40	0.00
16.00	10.00	4.00	1.25	-0.45	-0.80	0.00
17.00	10.00	3.00	1.45	-0.15	-1.30	0.00
17.50	10.00	2.50	-	-	-	-
Stress Ratio m = 3/4						
10.00	10.00	10.00	0.00	0.00	0.00	0.00
10.83	10.33	8.83	0.12	-	-	0.00
11.67	10.67	7.67	0.24	-	-	0.00
12.50	11.00	6.50	0.32	-	-	0.15
13.33	11.33	5.33	0.42	-	-	0.53
14.16	11.66	4.16	0.52	-	-	0.88
15.00	12.00	3.00	0.62	-	-	1.35
15.41	12.16	2.41	-	-	-	-

CLAY "X"

Consolidation Pressure, σ_c = 20.0 psi

σ_1 (psi)	σ_2 (psi)	σ_3 (psi)	ε_1 (%)	ε_2 (%)	ε_3 (%)	$\frac{dV}{V}$ (%)
\multicolumn{7}{c}{Stress Ratio m = 1/4}
20.00	20.00	20.00	0.00	0.00	0.00	0.00
21.20	19.60	19.20	0.40	0.35	-0.07	0.67
22.40	19.20	18.40	0.81	0.35	-0.11	1.05
23.60	18.80	17.80	1.16	0.31	-0.22	1.25
24.80	18.40	16.80	1.72	0.82	-0.34	2.15
25.80	18.40	15.80	2.29	1.11	-0.63	2.78
27.00	18.00	15.00	3.00	1.43	-1.09	3.34
28.16	17.67	14.17	4.22	2.79	-1.98	5.02
29.33	17.33	13.33	5.63	3.17	-2.90	5.89
30.50	17.00	12.50	8.72	4.95	-5.59	8.07
31.67	16.66	11.67	11.79	6.14	-7.98	9.94
32.84	16.33	10.84	-	-	-	-
\multicolumn{7}{c}{Stress Ratio m = 1/2}
20.00	20.00	20.00	0.00	0.00	0.00	0.00
21.00	20.00	19.00	0.35	-0.28	0.06	0.14
22.00	20.00	18.00	0.72	-0.31	0.05	0.45
22.29	20.00	17.00	0.94	-0.31	-0.01	0.61
24.00	20.00	16.00	1.55	-0.45	-0.13	0.98
25.00	20.00	15.00	1.99	-0.31	-0.35	1.33
26.00	20.00	14.00	2.97	0.50	-0.81	2.67
27.00	20.00	13.00	4.15	1.32	-1.45	4.02
28.00	20.00	12.00	5.60	2.54	-2.64	5.50
29.00	20.00	11.00	7.87	4.07	-5.22	6.72
30.00	20.00	10.00	-	-	-	-

CLAY "X"

Consolidation Pressure, σ_c = 30.0 psi

σ_1 (psi)	σ_2 (psi)	σ_3 (psi)	ε_1 (%)	ε_2 (%)	ε_3 (%)	$\frac{dV}{V}$ (%)
\multicolumn{7}{c}{Stress Ratio m = 1/4}

σ_1 (psi)	σ_2 (psi)	σ_3 (psi)	ε_1 (%)	ε_2 (%)	ε_3 (%)	$\frac{dV}{V}$ (%)
30.00	30.00	30.00	0.00	0.00	0.00	0.00
31.75	29.50	28.75	0.28	0.04	0.00	0.33
33.50	29.00	27.50	0.67	0.25	-0.10	0.82
35.25	28.50	26.25	1.16	0.18	-0.14	1.20
37.00	28.00	25.00	1.70	0.17	-0.29	1.59
38.75	27.50	23.75	2.40	0.70	-0.63	2.47
40.50	27.00	22.50	3.49	0.70	-0.97	3.23
42.25	26.50	21.25	4.47	1.19	-1.38	4.28
44.00	26.00	20.00	5.65	1.79	-1.73	5.70
45.75	25.50	18.75	7.43	2.08	-2.67	6.85
47.50	25.00	17.50	9.47	2.94	-3.69	8.72
49.25	24.50	16.25	10.45	3.55	-4.57	9.43
51.00	24.00	15.00	-	-	-	-

Stress Ratio m = 1/2

σ_1 (psi)	σ_2 (psi)	σ_3 (psi)	ε_1 (%)	ε_2 (%)	ε_3 (%)	$\frac{dV}{V}$ (%)
30.00	30.00	30.00	0.00	0.00	0.00	0.00
31.25	30.00	28.75	0.31	-0.11	0.00	0.20
32.50	30.00	27.50	0.62	-0.14	-0.06	0.42
33.75	30.00	26.25	0.94	0.14	-0.30	0.88
35.00	30.00	25.00	1.30	0.41	-0.53	1.20
36.25	30.00	23.75	1.77	0.88	-0.83	1.82
37.50	30.00	22.50	2.34	1.51	-1.33	2.51
39.00	30.00	21.00	3.20	1.67	-1.71	3.16
40.50	30.00	19.50	4.41	1.89	-2.49	3.84
42.00	30.00	18.00	8.80	2.92	-4.65	5.07
43.50	30.00	16.50	10.50	4.87	-8.67	6.71
45.00	30.00	15.00	-	-	-	-

Stress Ratio m = 3/4

σ_1 (psi)	σ_2 (psi)	σ_3 (psi)	ε_1 (%)	ε_2 (%)	ε_3 (%)	$\frac{dV}{V}$ (%)
30.00	30.00	30.00	0.00	0.00	0.00	0.00
31.25	30.50	28.25	0.41	0.14	0.00	0.36
32.50	31.00	26.50	0.88	-0.14	0.00	0.74
33.75	31.50	24.75	1.35	0.10	-0.37	1.07
35.00	32.00	23.00	1.74	0.46	-0.53	1.67
36.25	32.50	21.25	2.01	0.77	-0.87	1.90
37.50	33.00	19.50	2.62	1.00	-1.36	2.26
38.75	33.50	17.75	3.58	2.22	-2.81	2.99
40.00	34.00	16.00	5.39	5.24	-6.53	4.10
41.25	34.50	14.25	-	-	-	-

CLAY "X"

Consolidation Pressure, σ_c = 40.0 psi

σ_1 (psi)	σ_2 (psi)	σ_3 (psi)	ε_1 (%)	ε_2 (%)	ε_3 (%)	$\frac{dV}{V}$ (%)
Stress Ratio m = 0						
40.00	40.00	40.00	0.00	0.00	0.00	0.00
41.82	39.00	39.00	0.29	-0.36	0.00	-0.07
43.61	38.00	38.00	0.74	-0.70	0.00	0.05
45.41	37.00	37.00	1.41	-1.17	0.00	0.24
47.30	36.00	36.00	2.19	-0.96	0.00	1.23
49.10	35.00	35.00	2.89	-0.61	0.00	2.20
50.92	34.00	34.00	3.49	-0.35	0.00	3.14
52.70	33.00	33.00	4.03	-0.89	0.00	3.14
54.50	32.00	32.00	4.91	-1.06	0.00	3.85
56.40	31.00	31.00	5.68	-1.13	0.00	4.55
58.20	30.00	30.00	6.87	-1.37	0.00	5.50
60.00	29.00	29.00	8.51	-1.63	-0.50	6.38
61.80	28.00	28.00	10.27	-2.07	-1.43	6.71
63.60	27.00	27.00	12.92	-1.96	-2.95	8.01
65.30	26.00	26.00	15.70	-1.11	-4.65	9.95
67.10	25.00	25.00	18.50	-0.30	-6.50	11.70
72.00	24.00	24.00	-	-	-	-

CLAY "Y"

Consolidation Pressure, σ_c = 2.5 psi

σ_1 (psi)	σ_2 (psi)	σ_3 (psi)	$\frac{\Delta V}{V}$ (%)	ε_1 (%)	ε_2 (%)	ε_3 (%)
Stress Ratio m = 1/4						
5.82	3.33	2.50	+0.161	+0.275	+0.058	-0.163
9.14	4.17	2.50	+0.302	+0.600	+0.189	-0.488
12.40	5.00	2.50	+0.399	+1.000	+0.011	-0.613
15.60	5.83	2.50	+0.918	+1.750	+0.081	-0.913
17.20	6.25	2.50	+0.922	+1.900	-0.031	-0.948
Stress Ratio m = 1/2						
5.82	4.17	2.50	+0.354	+0.275	+0.092	-0.013
9.15	5.83	2.50	+0.612	+0.675	-0.033	-0.030
12.40	7.50	2.50	+0.926	+1.150	-0.137	-0.088
15.53	9.17	2.50	+1.715	+2.325	+1.828	-2.438
16.94	10.00	2.50	+1.786	+3.875	+2.024	-4.113
Stress Ratio m = 3/4						
5.83	5.00	2.50	+0.044	+0.125	-0.019	-0.063
9.15	7.50	2.50	-0.139	+0.250	-0.189	-0.200
12.45	10.00	2.50	-0.179	+0.450	-0.367	-0.263
15.76	12.50	2.50	-0.035	+0.675	-0.285	-0.425
17.34	13.75	2.50	+0.238	+1.200	+1.825	-2.789
18.90	15.00	2.50	+0.298	+1.500	+1.936	-3.138

CLAY "Y"

Consolidation Pressure, σ_c = 5.0 psi

σ_1 (psi)	σ_2 (psi)	σ_3 (psi)	$\frac{\Delta V}{V}$ (%)	ε_1 (%)	ε_2 (%)	ε_3 (%)
\multicolumn{7}{c}{Stress Ratio m = 1/4}
8.33	5.83	5.00	-0.016	+0.200	-0.151	-0.065
11.66	6.67	5.00	-0.033	+0.375	-0.313	-0.098
14.95	7.50	5.00	-0.032	+0.575	-0.469	-0.138
18.24	8.33	5.00	-0.023	+0.750	-0.558	-0.170
21.45	9.17	5.00	+0.079	+1.500	-1.199	-0.223
\multicolumn{7}{c}{Stress Ratio m = 1/2}
8.33	6.67	5.00	+0.264	+0.200	+0.112	-0.048
11.65	8.33	5.00	+0.424	+0.350	+0.139	-0.065
14.92	10.00	5.00	+0.704	+0.812	+0.062	-0.170
18.10	11.67	5.00	+0.742	+1.700	-0.751	-0.208
19.60	12.50	5.00	+1.104	+2.750	-1.239	-0.408
21.00	13.33	5.00	+1.790	+4.200	+0.073	-2.483
\multicolumn{7}{c}{Stress Ratio m = 3/4}
8.33	7.50	5.00	0.0	+0.250	-0.213	-0.038
11.66	10.00	5.00	+0.086	+0.400	-0.226	-0.088
14.95	12.50	5.00	+0.211	+0.625	-0.277	-0.137
18.24	15.00	5.00	+0.480	+0.850	-0.083	-0.288
21.50	17.50	5.00	+0.522	+1.075	-0.078	-0.475

CLAY "Y"

Consolidation Pressure, σ_c = 10.0 psi

σ_1 (psi)	σ_2 (psi)	σ_3 (psi)	$\frac{\Delta V}{V}$ (%)	ε_1 (%)	ε_2 (%)	ε_3 (%)
\multicolumn{7}{c}{Stress Ratio m = 1/4}
13.33	10.83	10.00	+0.001	+0.025	+0.013	-0.038
16.66	11.67	10.00	+0.037	+0.200	-0.063	-0.100
19.94	12.50	10.00	+0.101	+0.600	-0.349	-0.150
23.20	13.33	10.00	+0.256	+1.125	-0.569	-0.200
26.32	14.17	10.00	+0.732	+1.900	-0.856	-0.312
27.90	14.58	10.00	+0.854	+2.300	-1.046	-0.400
28.46	15.00	10.00	+2.590	+8.350	-1.735	-4.025
29.63	15.42	10.00	+3.110	+10.425	-1.715	-5.600
\multicolumn{7}{c}{Stress Ratio m = 1/2}
13.33	11.67	10.00	+0.003	+0.075	-0.039	-0.033
16.66	13.33	10.00	+0.028	+0.225	-0.147	-0.050
19.95	15.00	10.00	+0.067	+0.500	-0.378	-0.055
23.20	16.67	10.00	+0.151	+0.913	-0.649	-0.113
26.38	18.33	10.00	+1.146	+1.850	-0.279	-0.425
27.87	19.17	10.00	+1.234	+2.550	-0.741	-0.575
29.27	20.00	10.00	+1.522	+3.775	-1.103	-1.150
\multicolumn{7}{c}{Stress Ratio m = 3/4}
13.33	12.50	10.00	+0.001	+0.075	-0.049	-0.025
16.66	15.00	10.00	+0.033	+0.250	-0.142	-0.075
19.95	17.50	10.00	+0.295	+0.500	-0.079	-0.125
23.15	20.00	10.00	+1.310	+1.300	+0.523	-0.513
25.91	22.50	10.00	+2.170	+4.650	+0.820	-3.300
27.37	23.75	10.00	+2.372	+5.425	+0.897	-3.950
28.50	25.00	10.00	+2.810	+8.000	+0.110	-5.300
30.00	26.25	10.00	+2.850	+8.150	+0.113	-5.413

OTTAWA SAND
SIMPLE SHEAR TEST
σ_{oct} = 10 psi
   m = 0.2

σ_x	σ_y	σ_z	ε_x (%)	ε_y (%)	ε_z (%)
10	10	10	0.0	0.0	0.0
8	9	13	-0.011	-0.008	0.061
7	8.5	14.5	-0.028	-0.013	0.106
6	8	16	-0.070	-0.025	0.189
5	7.5	17.5	-0.175	-0.041	0.317
4	7	19	-0.387	-0.097	0.531
6	8	16	-0.381	-0.097	0.518
8	9	13	-0.334	-0.086	0.486
10	10	10	-0.234	-0.055	0.405
8	9	13	-0.241	-0.056	0.431
6	8	16	-0.270	-0.063	0.467
5	7.5	17.5	-0.296	-0.068	0.498
4	7	19	-0.405	-0.087	0.592
3.5	6.75	19.75	-0.529	-0.116	0.699
3	6.5	20.5	-0.868	-0.220	0.957
2.5	6.25	21.25	-1.493	-0.393	1.369
?	6.	22	-3.442	-0.798	2.311
1.5	5.75	22.75	-13.686	-12.843	18.374

OTTAWA SAND
SIMPLE SHEAR TEST
σ_{oct} = 5 psi
  m = 0.5

σ_x	σ_y	σ_z	ε_x (%)	ε_y (%)	ε_z (%)
5.0	5.0	5.0	0.0	0.0	0.0
4.0	5.0	6.0	-0.003	0.003	0.013
3.0	5.0	7.0	-0.018	0.005	0.037
2.5	5.0	7.5	-0.034	0.007	0.059
2.0	5.0	8.0	-0.085	0.011	0.103
3.0	5.0	7.0	-0.081	0.012	0.101
4.0	5.0	6.0	-0.063	0.013	0.090
5.0	5.0	5.0	-0.039	0.014	0.070
4.0	5.0	6.0	-0.042	0.015	0.078
3.0	5.0	7.0	-0.050	0.016	0.090
2.0	5.0	8.0	-0.079	0.017	0.113
1.5	5.0	8.5	-0.194	0.027	0.184
1.0	5.0	9.0	-0.483	0.052	0.330
0.75	5.0	9.25	-0.971	0.090	0.528
0.50	5.0	9.5	-1.334	0.115	0.668
0.25	5.0	9.75	-12.071	1.384	4.224

OTTAWA SAND
SIMPLE SHEAR TEST
σ_{oct} = 20 psi
  m = 0.5

σ_x	σ_y	σ_z	ε_x (%)	ε_y (%)	ε_z (%)
20	20	20	0.0	0.0	0.0
18	20	22	0.001	0.009	0.020
16	20	24	-0.003	0.020	0.042
14	20	26	-0.026	0.030	0.073
12	20	28	-0.062	0.037	0.115
11	20	29	-0.089	0.045	0.147
10	20	30	-0.124	0.089	0.187
12	20	28	-0.119	0.089	0.182
14	20	26	-0.111	0.090	0.175
16	20	24	-0.096	0.091	0.165
18	20	22	-0.074	0.092	0.152
20	20	20	-0.052	0.093	0.134
18	20	22	-0.059	0.093	0.139
14	20	26	-0.066	0.094	0.148
12	20	28	-0.091	0.095	0.173
11	20	29	-0.101	0.098	0.180
10	20	30	-0.119	0.103	0.195
9	20	31	-0.153	0.109	0.224
8	20	32	-0.248	0.151	0.295
7	20	33	-0.402	0.198	0.383
5	20	35	-1.118	0.490	0.730
4	20	36	-1.867	0.612	0.910
3	20	37	-5.437	0.835	2.103
2	20	38	-5.437	1.133	2.646
1	20	39	-5.439	1.271	2.750
0	20	40	-5.694	1.389	2.831

OTTAWA SAND
SIMPLE SHEAR TEST
$\sigma_{oct} = 10$ psi
$m = 0.8$

σ_x	σ_y	σ_z	ε_x (%)	ε_y (%)	ε_z (%)
10.0	10.0	10.0	0.0	0.0	0.0
8.5	10.5	11.0	-0.006	0.011	0.016
7.0	11.0	12.0	-0.020	0.028	0.049
5.5	11.5	13.0	-0.104	0.059	0.115
4.0	12.0	14.0	-0.286	0.127	0.216
7.0	11.0	12.0	-0.243	0.125	0.200
10.0	10.0	10.0	-0.154	0.121	0.173
8.5	10.5	11.0	-0.159	0.124	0.180
7.0	11.0	12.0	-0.164	0.125	0.189
5.5	11.5	13.0	-0.185	0.127	0.199
4.0	12.0	14.0	-0.237	0.139	0.222
3.25	12.25	14.5	-0.399	0.187	0.292
2.5	12.50	15.0	-0.715	0.273	0.406
1.75	12.75	15.5	-1.565	0.511	0.670
1.0	13.0	16.0	-6.710	1.061	1.569
0.0	13.33	16.67	-11.156	1.438	1.699

OTTAWA SAND

REDUCED TRIAXIAL COMPRESSION TEST

m = 0

σ_x	σ_y	σ_z	ε_x (%)	ε_y (%)	ε_z (%)
20	20	20	0.0	0.0	0.0
19	19	20	0.0	0.001	0.003
18	18	20	0.002	0.004	0.008
17	17	20	0.003	0.006	0.013
16	16	20	0.003	0.007	0.016
15	15	20	0.002	0.007	0.019
17	17	20	0.004	0.007	0.019
19	19	20	0.007	0.010	0.018
20	20	20	0.010	0.013	0.018
19	19	20	0.010	0.013	0.019
17	17	20	0.007	0.013	0.020
15	15	20	0.004	0.012	0.022
14	14	20	0.001	0.011	0.027
13	13	20	-0.002	0.007	0.032
12	12	20	-0.007	0.003	0.042
11	11	20	-0.014	0.005	0.053
10	10	20	-0.024	0.012	0.068
9	9	20	-0.038	0.024	0.088
8	8	20	-0.055	0.041	0.116
7	7	20	-0.087	0.075	0.168
6	6	20	-0.146	0.136	0.251
5	5	20	-0.304	0.305	0.490
4	4	20	-0.608	0.611	0.843
3.5	3.5	20	-1.034	-1.032	1.278
3	3	20	-1.617	-1.608	1.820
2.5	2.5	20	-3.033	-3.044	3.053
2	2	20	-13.986	-14.795	18.148

OTTAWA SAND

REDUCED TRIAXIAL EXTENSION TEST

m = 1.0

σ_x	σ_y	σ_z	ε_x (%)	ε_y (%)	ε_z (%)
20	20	20	0.0	0.0	0.0
20	20	18	0.005	0.005	-0.008
20	20	16	0.009	0.011	-0.023
20	20	14	0.016	0.016	-0.046
20	20	12	0.028	0.026	-0.087
20	20	9	0.079	0.072	-0.248
20	20	7	0.149	0.148	-0.490
20	20	10	0.150	0.152	-0.474
20	20	14	0.147	0.154	-0.431
20	20	18	0.138	0.146	-0.373
20	20	20	0.132	0.139	-0.344
20	20	18	0.131	0.139	-0.349
20	20	14	0.134	0.142	-0.372
20	20	10	0.141	0.150	-0.414
20	20	8	0.150	0.154	-0.452
20	20	7	0.163	0.161	-0.499
20	20	6	0.219	0.207	-0.693
20	20	5	0.322	0.314	-1.094
20	20	4	0.473	0.478	-1.750
20	20	3	0.818	0.797	-3.646
20	20	2.5	1.290	1.162	-7.971
20	20	2	1.655	1.554	-10.184
20	20	1.5	2.548	2.847	-16.379

OTTAWA SAND

PROPORTIONAL LOADING TEST (PLI)

m = 0

σ_x	σ_y	σ_z	ε_x (%)	ε_y (%)	ε_z (%)
10	10	10	0.0	0.0	0.0
10.5	10.5	14	0.003	-0.001	0.043
11	11	18	0.005	-0.008	0.094
11.5	11.5	22	-0.001	-0.023	0.151
12	12	26	-0.012	-0.042	0.208
12.5	12.5	30	-0.027	-0.063	0.266
13	13	34	-0.055	-0.094	0.340
13.5	13.5	38	-0.084	-0.125	0.412
14	14	42	-0.127	-0.170	0.515
14.5	14.5	46	-0.161	-0.221	0.668
15	15	50	-0.182	-0.249	0.744
15.5	15.5	54	-0.233	-0.302	0.893
16	16	58	-0.273	-0.351	1.022
16.5	16.5	62	-0.312	-0.396	1.157
17	17	66	-0.357	-0.446	1.291
17.5	17.5	70	-0.420	-0.510	1.448
18	18	74	-0.490	-0.584	1.597
18.5	18.5	78	-0.559	-0.654	1.722
19	19	82	-0.649	-0.745	1.857
19.5	19.5	86	-0.788	-0.891	2.054
20	20	90	-0.962	-1.071	2.287
20.5	20.5	94	-1.146	-1.260	2.518
21	21	98	-1.364	-1.495	2.786
21.5	21.5	102	-1.686	-1.815	3.125
22	22	106	-2.110	-2.230	3.534
22.5	22.5	110	-13.330	-14.289	11.397

OTTAWA SAND

PROPORTIONAL LOADING (PL2)

m = 0

σ_x	σ_y	σ_z	ε_x (%)	ε_y (%)	ε_z (%)
10	10	10	0.0	0.0	0.0
11	11	12	0.005	0.006	0.015
12	12	14	0.010	0.011	0.032
13	13	16	0.014	0.013	0.048
15	15	20	0.022	0.017	0.076
20	20	30	0.039	0.021	0.142
25	25	40	0.049	0.027	0.191
30	30	50	0.057	0.034	0.239
35	35	60	0.068	0.051	0.294
40	40	70	0.077	0.060	0.342
45	45	80	0.099	0.077	0.417
50	50	90	0.134	0.138	0.542
55	55	100	0.151	0.173	0.631
45	45	80	0.143	0.171	0.602
35	35	60	0.128	0.161	0.557
25	25	40	0.105	0.139	0.480
15	15	20	0.084	0.109	0.332
10	10	10	0.070	0.067	0.202

OTTAWA SAND

CIRCULAR STRESS PATH

$\sigma_{oct} = 10$ psi

σ_x	σ_y	σ_z	ε_x (%)	ε_y (%)	ε_z (%)
10.0	10.0	10.0	0.0	0.0	0.0
9.0	9.0	12.0	-0.002	-0.005	0.022
8.0	8.0	14.0	-0.014	-0.024	0.068
7.0	7.0	16.0	-0.050	-0.074	0.187
6.07	8.04	15.89	-0.066	-0.065	0.210
5.18	9.31	15.51	-0.095	-0.033	0.235
4.49	10.69	14.81	-0.141	0.055	0.262
4.11	11.96	13.93	-0.177	0.124	0.269
4.0	13.0	13.0	-0.201	0.171	0.266
4.11	13.93	11.96	-0.218	0.215	0.255
4.49	14.81	10.69	-0.222	0.261	0.234
5.18	15.51	9.31	-0.213	0.319	0.191
6.07	15.89	8.04	-0.198	0.372	0.138
7.00	16.0	7.0	-0.176	0.433	0.071
8.04	15.89	6.07	-0.143	0.522	-0.033
9.31	15.51	5.18	-0.079	0.623	-0.165
10.69	14.81	4.49	0.012	0.687	-0.314
11.96	13.93	4.11	0.118	0.721	-0.448
13.00	13.0	4.0	0.213	0.719	-0.490
13.93	11.96	4.11	0.293	0.624	-0.535
14.81	10.69	4.49	0.335	0.584	-0.543
15.51	9.31	5.18	0.392	0.512	-0.532
15.89	8.04	6.07	0.462	0.435	-0.509
16.0	7.0	7.0	0.519	0.337	-0.472
15.89	6.07	8.04	0.581	0.219	-0.415
15.51	5.18	9.31	0.699	0.116	-0.311
14.81	4.49	10.69	0.737	0.050	-0.245
13.93	4.11	11.96	0.735	-0.037	-0.161
13.0	4.0	13.0	0.694	-0.073	-0.102
11.96	4.11	13.93	0.661	-0.098	-0.049
10.69	4.49	14.81	0.627	-0.101	-0.006
9.31	5.18	15.51	0.566	-0.096	0.066
8.04	6.07	15.89	0.508	-0.080	0.100
7.00	7.00	16.0	0.326	-0.042	0.165
6.07	8.04	15.89	0.268	0.027	0.200
5.18	9.31	15.51	0.169	0.106	0.250
4.49	10.69	14.81	0.109	0.202	0.263
4.11	11.96	13.93	0.046	0.280	0.271
4.0	13.0	13.0	0.022	0.339	0.257
6.0	12.0	12.0	0.041	0.344	0.248
8.0	11.0	11.0	0.086	0.330	0.234
10.0	10.0	10.0	0.134	0.278	0.207

REPORT ON WORKING GROUP ONE

Workshop on Limit Equilibrium Plasticity,
and Generalized Stress-Strain in
Geotechnical Engineering, Montreal, May 28-30, 1980

By

John T. Christian*

Working Group One discussed the models and predictions developed by J. H. Prevost, Z. P. Bazant, I. S. Sandler, and W. F. Chen. Since W. F. Chen presented results for two models, five models were considered in all. Membership in the working group varied somewhat over the three days of the workshop as people moved from one working group to another. F. C. Townsend was the committee member assigned permanently to the working group. The author served as chairman.

The five models discussed by the working group spanned a broad range of constitutive relations for soils. Prevost's model is based on incremental plasticity theory with anisotropic kinematic hardening for deviatoric behavior and isotropic consolidation effects. Sandler's model, which is one of a family of models developed jointly with G. Y. Baladi, is based on the capped model but modified to include kinematic hardening. It was generally agreed that these two models are now quite similar even though

---

*Senior Consulting Engineer, Stone & Webster Engineering Corporation, Boston, Massachusetts

they start from different conceptual points of view. Bazant uses the endochronic model. Chen proposed one model based on fitting stress-strain curves according to hyperelastic theory. His second model used shear and bulk moduli modified to account for stress levels, which are limited by the strength of the soil.

The models and the corresponding predictions are described in the accompanying papers prepared by each of the predictors. These go into more detail than was possible in a working group; in particular, the descriptions of the predictions are much more comprehensive than those presented at the working group sessions. Therefore, the following paragraphs concentrate on observations, discussions, and conclusions instead of presenting the details of the models. The reader is referred to the individual papers for further treatment of any model.

Comparison of Predictions with Observations

Each of the four predictors compared predicted and observed behavior for some of the soils and some of the tests. The discussion was limited to those results actually presented to the working group; in some cases the prediction papers describe additional results.

Sandler predicted the response of the remolded kaolinite only. The pore pressure predictions were in fair agreement with experimental values. The stress paths were well predicted with some relatively minor discrepancy between the modeled and actual paths. The predicted stress-strain curves were fairly good. Baladi presented results from a companion study to Working Group Two.

Bazant predicted the response of clays X and Y only. The stress-strain curves were quite good for both clays. However, the predictions of volumetric strain were seriously in error, in some instances having the wrong sign. Bazant indicated that computer problems precluded predictions for the other soils.

Prevost predicted the behavior of the sand, except for the circular stress path, and the kaolin clay. For the sand, the initial portions of the Z direction strains were very good, with some deviations between observation and prediction at large strains. The predicted X and Y strains differed slightly from experimental results. For the kaolin, the initial behavior of the strains was quite good, but there was some discrepancy between prediction and experiment at larger strains.

Chen predicted the behavior of all four soils using two models. The deformation model based on hyperelastic theory predicted strains that were fairly accurate, but on the large side, for the initial portions of the tests. However, at larger strains the predictions deviated badly from experiment because the model did not limit the allowable stresses to be within the strength of the soil. The model using variable bulk and shear moduli predicted the experimental behavior much better than did the hyperelastic model.

In this group, only Chen attempted to predict the behavior during the test with the circular stress path. His results were in poor agreement with experiment. Some of the other predictors stated frankly that they doubted their models would work for this case; others simply ignored it. From these results and those reported by other working groups, it is doubtful that the present state of knowledge of constitutive relations allows a dependable prediction to be made for this type of stress path.

It should be noted that, for the most part, the strains involved in these predictions are small so that considerable experimental error can be expected. The agreement, or the lack of it, between prediction and experiment should be interpreted with that in mind.

Input Parameters and Discussion of Models

Prevost described his model clearly and in great detail. His procedure for obtaining input parameters is to conduct triaxial compression and extension tests at various levels of initial stress and void ratio. The stress-strain data are plotted as smooth curves, from which several data points are obtained for each curve. Values of eight parameters are required throughout the stress-strain relation. The data from the smoothed test curves are fed into a computer program that calculates the eight parameters. This program will be made available in the proceedings of the symposium.

Typically, Prevost has found that seven to ten failure surfaces are adequate for a prediction. The positions of the same surface in extension and compression are identified as occurring where the slopes of the stress-strain curve are the same in extension and compression. Prevost recommended predicting the behavior of the soil in a direct simple shear test and comparing this with the experimental results of such a test. This procedure, though imperfect, provides a good test of the selection of input parameters. About two days were required to make the predictions for the symposium.

Bazant explained that his model is not simple and that experience is needed to use it correctly. The model was originally developed for concrete and has only recently been adapted to cyclic simple shear. Approximately one week was needed to obtain the parameters required for this symposium and some results were not obtained because of computer problems.

The endochronic model uses families of parameters for functions that describe the material behavior. The choice of which functions to use depends on judgment. The parameters are adjusted until a satisfactory fit to the test data is obtained. The reaction of the working group was that use of the endochronic model involves essentially curve fitting between experimental data and selected functions. The selection of the functions covers a broad range, which is a strength of the approach. However, the physical and analytical implications of the choices of basic functions and parameters are obscure, and the working group felt that this fact undermined confidence in its use. There was general agreement that the developers of the model needed to do much work to understand the model and to explain it clearly to the profession.

Chen's models were generally understood by the working group, so the discussion concentrated on how he had obtained his values for input data. It developed that the second order hyperelastic stress-strain relation had been fitted through only three points on each experimental curve. It was suggested that a best fit through many more points would have given better results. The lack of a failure cut-off on the hyperelastic model has already been noted.

Sandler's discussion was unavoidably brief, but Baladi's presentations to Working Group Two are on essentially the same model. The isotropic consolidation portion of the triaxial shear test was used to locate the position of the elliptical cap in the Sandler-Baladi model. The movement of the cap as a function of deviatoric stress and strain is determined from the shearing portion of the test. A total of nine constants is required. Sandler pointed out that the model is really a family of models. Selections among different types of behavior, ultimate failure envelopes, anisotropy, and the like, are made on the basis of the observed behavior of the soil. The specific values of the parameters are then chosen so as best to fit the data.

Research Needs

The working group established the following list of questions and subjects for further investigation:

1. What do we expect from these models? That is, do the repeatability and accuracy of the data from soil tests warrant the sophistication of some of the models? Further, are some of the observed deviations between prediction and experiment within the limit of experimental accuracy? It makes little sense to require a model to make predictions that cannot be observed or to develop one that requires input that cannot be measured.

2. What are the effects of proportional loading, of the choice of stress path, and of rotation of the principal axes of stress on strain? These have been and are still vexing questions for theorists and experimenters alike.

3. How can these models be implemented in Finite Element or Finite Difference computer programs? How does the implementation affect the model, and vice versa?

4. How can predictions from boundary value problems be best tested? Models, centrifuge devices, and field tests are all candidates.

5. What are the effects of sample disturbance on the values to be used in a constitutive model and how can these effects be

corrected? This is a major question for the effective use of simple models, and it is reasonable to expect that the effects of sample disturbance will be more severe for complicated models and less easily corrected.

6. What improvements can be made in quality testing?

7. Can in situ tests be used or developed for use with sophisticated constitutive models? Most current in situ tests provide data that are effectively used in relatively simple models only.

8. How can the constitutive models be improved? Such efforts should consider the shapes and locations of ultimate surfaces, hardening parameters, non-associated flow parameters, behavior at small values of strain, and description of kinematic hardening.

There was a consensus on the need for research in the first seven areas. In contrast, a significant minority felt that there was much less need for research on the eighth question, and the opinion was expressed that the sophistication of the models may have outrun the data on which they are based, the users' understanding of the models, their implementation in analytical procedures and computer programs, and their testing in laboratory and field.

Final Comments

The author of these notes came away from the workshop with two strong impressions. First, the researchers in the field of constitutive relations have developed a variety of sophisticated models to explain the behavior of soil. They have presented the profession with a substantial job of assimilation, comparison, and understanding. This will require a significant effort in the coming years.

Second, accurate predictions were made with very simple and very complicated models; inaccurate predictions were made with equally simple or complicated models. The accuracy of the predictions depended more on the predictor's experience with the behavior of soils and less on the sophistication of his model. This is likely to be true not only in a symposium such as this, but also in a case of practical design.

The workshop was made possible by the efforts of many people, but it depended ultimately on the willingness of the predictors to expose the performance of their models to public examination. This required much hard work on their part, but, more important, it required patience and, on occasion, good humor. The participants in the symposium and the profession should be grateful.

Professor F. C. Townsend, as the committee member assigned to the working group, kept notes of the proceedings and provided much other assistance. This help is gratefully acknowledged.

Group 2 Report

on

Workshop on Plasticity Theories and
Generalized Stress-Strain Modelling of Soils

by H. B. Poorooshasb[1] and E. T. Selig[2], M.ASCE

Introduction

A brief, simple account of the four models submitted and discussed by Group 2 will be presented first. This we feel is appropriate, since frequently during our meetings, the participants indicated their difficulties in following the salient points of the various models being discussed. The main sources of these difficulties are two. First, in formulating their model for soil behavior, the respective authors use tensor notation, a notation which not all engineers are familiar with or can employ easily. Second, the authors assume that the interested readers have a full and working knowledge of classical plasticity, an assumption which is not always justified.

Various terms such as "theoretical," "empirical," "analytical" and "phenomenological" are used to describe the models. These terms tend to be misused or given a variety of meanings in describing different models. This terminology problem has caused some confusion. Based on a review of the models, it is our opinion that none of them are theoretical models. Instead, they are analytical models fit empirically to experimental data. In this sense, they all may also be called phenomenological models.

In what follows, the behavior of an element of soil subjected to a very simple stress state will first be examined. Then the more dominant aspects of the models will be discussed. The terms used will be defined as we understand them. The purpose of this presentation is to provide the necessary background for evaluating the models, since few people understand many of the models. We hope that this review is consistent with the intentions of the authors in our

---

[1]Professor of Civil Engineering, Concordia University, Montreal, Quebec.
[2]Professor of Civil Engineering, University of Massachusetts, Amherst, Mass.

Basic Concepts

The soil element shown in Fig. 1a is subjected to a simple state of stress defined by normal effective stress σ and shearing stress τ. Thus, the state of stress acting on the element can be represented in a two-dimensional space (coordinate system) of τ and σ, which is the stress space. Note that the effect of the stress component perpendicular to the plane of the sheet of paper shown in Fig. 1 is ignored in what follows. If the successive states of stress that the element follows are plotted in the stress space (τ,σ), a curve such as OCA is obtained. This is the stress path. The strain space and the strain path are similarly defined: if the successive states of strain followed by the soil element are plotted in the strain space (ε,v), where ε is the shearing strain and v the total volumetric strain (Fig. 1a), then the strain path O'C'A' is obtained (Fig. 1d).

Now let the state of stress (τ,σ) of Fig. 1a change by a small increment $(d\tau, d\sigma)$. The stress path would move from A to B in the stress space of Fig. 1b and the strain path from point A' to B' (Fig. 1d). Since both AB and A'B' are of magnitude and direction, they can be represented by incremental vectors. This is the definition of the stress increment vector (AB Fig. 1b) and strain increment vector (A'B' Fig. 1d). Often, and for convenience, stress and strain spaces are superimposed in the sense that the strain increment vector A'B' resulting from incremental stress change AB is indicated by vector $\vec{d\varepsilon}$ (equal and parallel to A'B') positioned at point A in the stress space, Fig. 1e.

Let the state of stress in Fig. 2 at point A be changed to any point on the circumference of the circle with a small radius (i.e., change from A to B or from A to C) and be subsequently returned to its original position A. Two things may happen: i) the element may regain its original shape and volume, or ii) the element may not regain its original shape and/or volume. In the first instance (case i.), the soil is said to have behaved elastically. In the second case (ii.), where irrecoverable straining (whether volumetric or shearing or both) of the element has taken place, the element is said to have yielded (or undergone plastic straining), and the total strain increment vector for the loading portion of the incremental path (i.e., moving from A to B) is written in the form

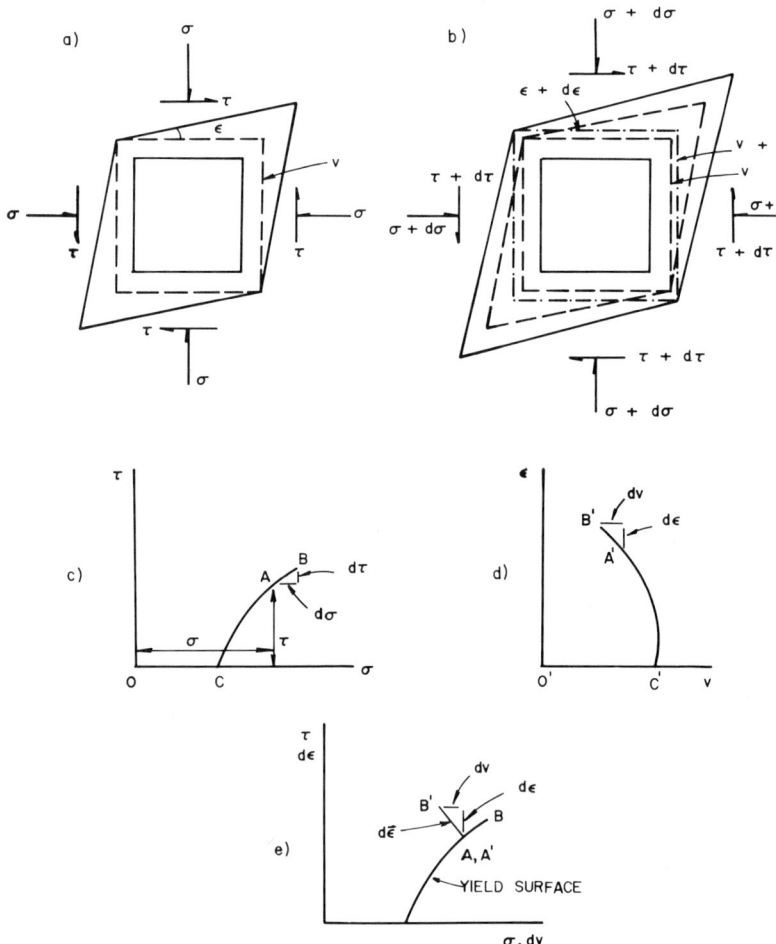

Fig. 1. Stress and Strain Representation

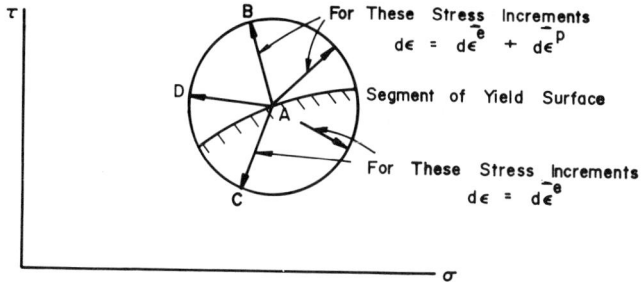

Fig. 2. Segment of Yield Surface

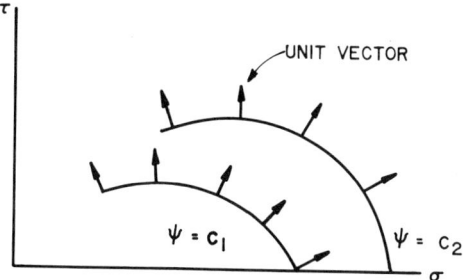

Fig. 3. Family of Plastic Potential Curves (ψ = constant)

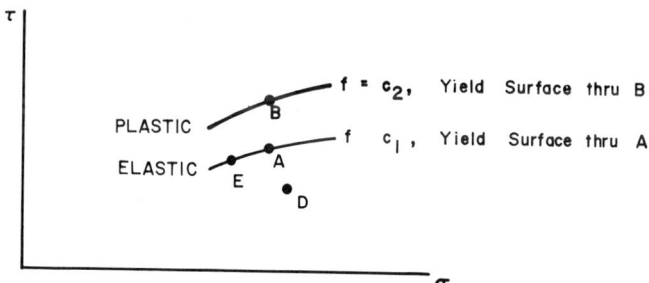

Fig. 4. Family of Yield Surfaces (f = constant)

$$d\vec{\varepsilon} = d\vec{\varepsilon}^e + d\vec{\varepsilon}^p \quad , \tag{1}$$

where $d\vec{\varepsilon}$ is the total strain increment vector, $d\vec{\varepsilon}^e$ is the elastic (recoverable) strain increment, and $d\vec{\varepsilon}^p$ is the plastic (irrecoverable) strain increment vector. The borderline separating those stress increment vectors which produce irrecoverable strains and those for which only recoverable strains are encountered, if indeed it exists, constitutes a segment of the yield surface passing through A. Stated otherwise, the yield surface, if it can be located at least locally, separates the domain for which only recoverable (elastic) strains are encountered from the domain for which both elastic and irrecoverable (plastic) strains are encountered.

In the early stages of application of theory of plasticity to soils, it was observed that when plastic strains were encountered (i.e., the soil element was yielding), then at least for cohesionless granular media the <u>direction</u> of the plastic strain increment vector defined by $d\vec{\varepsilon}^p = d\vec{\varepsilon} - d\vec{\varepsilon}^e$ was, to a very close approximation, independent of the direction of the stress increment vector. That is, while for the incremental stress path AB and AD, different magnitudes of plastic strain increments were measured, their direction in the strain space remained essentially unchanged. This led to the introduction and use of the concept of plastic potential, which had been employed previously and for many years in the classical theory of plasticity, to indicate the direction of the plastic strain increment vector.

The argument was as follows. If, at any point in the stress space, the direction of the plastic strain increment vector was constant, then at each point, a unit vector could be assigned to represent this direction (Fig. 3). Also, these directions (collection of vectors) could be analytically represented as the gradient of scalar function ψ. Obviously, the ψ = constant curves would be perpendicular to the strain increment vectors by the very definition of gradient. A very familiar situation to the geotechnical engineer is the steady state flow of fluids through isotropic soils. Here, the vectors representing the velocity of flow are proportional to gradient of loss of energy h, i.e., $v_x = -k(\partial h/\partial x)$, etc. It is known from elementary soil mechanics that the set of vectors v (components v_x, v_y, v_z) are perpendicular to contours of equal (constant) head.

In terms of stress components (τ,σ) used in our simple example, the plastic strain increment components are defined by

$$d\varepsilon^p = d\lambda \, \frac{\partial \psi}{\partial \tau} \quad , \tag{2}$$

and

$$dv^p = d\lambda \, \frac{\partial \psi}{\partial \sigma} \quad , \tag{3}$$

where $d\lambda$ is an incremental scalar factor of proportionality. The magnitude of $d\lambda$ depends on the "extent" of yielding of the element.

If the stress path in Fig. 4 moves from point A, at which point the yield locus is defined by the equality $f(\tau,\sigma) = c_1$ (c_1 = a constant), to point B, for which the yield locus is defined by $f(\tau,\sigma) = c_2$ (c_2 also = a constant > c_1), then $d\lambda$ is assumed to be dependent on $c_2 - c_1$. If the stress point A moves to D (a point below the $f = c_1$ locus), the soil is said to be unloading and $d\vec{\varepsilon}^p = 0$. If the stress point moves to another neighboring point on the yield surface (say point E), then $d\lambda = 0$ and $d\vec{\varepsilon}^p$ would also be zero. To express these conditions analytically, use may be made of singularity brackets <>, in which

$<-|x|> = 0$,

$<0> = 0$, and

$<+|x|> = x$.

Thus, Equations 2 and 3 are restated in the form

$$d\varepsilon^p = <d\lambda> \frac{\partial \psi}{\partial \tau} \quad , \tag{4}$$

and

$$dv^p = <d\lambda> \frac{\partial \psi}{\partial \sigma} \quad . \tag{5}$$

Now making use of the assumption that $d\lambda$ is proportioned to df, where

$$df = c_2 - c_1 \quad , \tag{6}$$

in which c_2 is incrementally larger than c_1, these equations may further be modified to read

$$d\varepsilon^p = <h(\frac{\partial f}{\partial \tau} d\tau + \frac{\partial f}{\partial \sigma} d\sigma)> \frac{\partial \psi}{\partial \tau} \quad , \tag{7}$$

and

$$dv^p = <h(\frac{\partial f}{\partial \tau} d\tau + \frac{\partial f}{\partial \sigma} d\sigma)> \frac{\partial \psi}{\partial \sigma} \quad , \tag{8}$$

where h is a proportionality factor ($d\lambda = hdf$), and

$$df = \frac{\partial f}{\partial \tau} d\tau + \frac{\partial f}{\partial \sigma} d\sigma \quad .$$

If the family of f = constant and ψ = constant surfaces coincide, then the plastic strain increment vectors would be perpendicular to the yield (f = constant) surfaces, since they are perpendicular to ψ = constant surfaces. This condition is referred to as the normality condition. It simply means that the plastic strain increment vector at a given point in the stress space is normal to the segment of the yield surface passing through this point. When normality exists, $\psi = f$ and

$$d\varepsilon^p = \langle h(\frac{\partial f}{\partial \tau} d\tau + \frac{\partial f}{\partial \sigma} d\sigma)\rangle \frac{\partial f}{\partial \tau} \quad , \text{ and} \tag{9}$$

$$dv^p = \langle h(\frac{\partial f}{\partial \tau} d\tau + \frac{\partial f}{\partial \sigma} d\sigma)\rangle \frac{\partial f}{\partial \sigma} \quad . \tag{10}$$

The $f = \psi$ condition is of central importance in theory of plasticity and further remarks are appropriate. To ensure a unique and stable solution to boundary value problems, two postulates are made regarding the behavior of ideal plastic solids. These postulates are sufficient, but not necessary. The first deals with infinitesimal increment of work and second the complementary work in a stress cycle.

According to the first postulate, any infinitesimal increment of plastic work is non-negative. In terms of stress and strain parameters adopted above, this first postulate statement would read

$$d\sigma dv^p + d\tau d\varepsilon^p \geq 0 \quad . \tag{11}$$

Now consider Fig. 5a, which shows a portion of the yield surface, the strain increment vector $\vec{d\varepsilon^p}$ and two stress increments AB and AC, the magnitude of which are so chosen as to produce the same $\vec{d\varepsilon^p}$. For incremental stress path $\vec{AB}$

$$d\sigma dv^p + d\tau d\varepsilon^p = \vec{AB} \cdot \vec{d\varepsilon^p} \quad , \tag{12}$$

and is positive, thus satisfying the inequality. However, for path AC

$$d\sigma dv^p + d\tau d\varepsilon^p = \vec{AC} \cdot \vec{d\varepsilon^p} \tag{13}$$

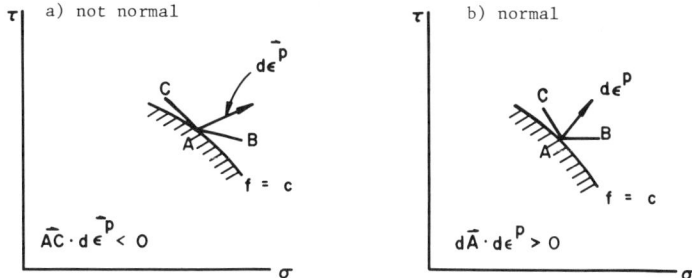

Fig. 5. Plastic Strain Increment in Relation to the Yield Surface

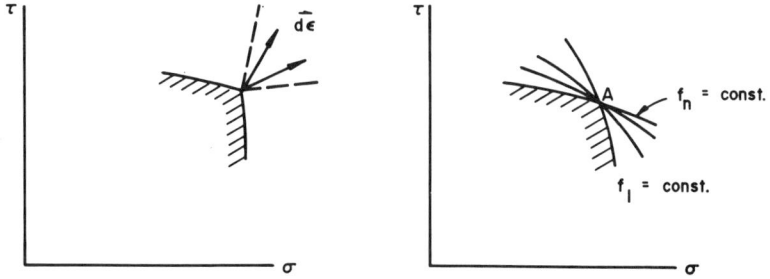

Fig. 6. Yield Surface with a Corner

Fig. 7. Family of n Yield Surfaces Passing Through Point A

is negative (since the angle between AC and $d\vec{\varepsilon}^p$ is obtuse), and the above inequality is not satisfied. Thus, if the portion of yield surface under consideration is smooth (i.e., does not have a corner), the only situation which would satisfy the first postulate is when $d\vec{\varepsilon}^p$ is normal to the yield surface, as shown in Fig. 5b.

If the yield surface contains a corner (Fig. 6), then the plastic strain increment vector must lie between adjacent normals to the two portions of the yield surface. In this event, it appears that the strain increment vectors are not uniquely determined. To overcome this, let it be assumed that n yield surfaces pass through a typical stress point such as A in Fig. 7. Then the plastic strain increment vector may be assumed to be composed of the contribution from the individual yield surfaces, i.e.

$$d\varepsilon^p = \langle h_1(\frac{\partial f_1}{\partial \sigma} d\sigma + \frac{\partial f_1}{\partial \tau} d\tau) \rangle \frac{\partial \Psi_1}{\partial \tau} +$$
$$\langle h_2(\frac{\partial f_2}{\partial \sigma} d\sigma + \frac{\partial f_2}{\partial \tau} d\tau) \rangle \frac{\partial \Psi_2}{\partial \tau} +$$
$$\ldots\ldots\ldots + \quad (14)$$
$$\langle h_n(\frac{\partial f_n}{\partial \sigma} d\sigma + \frac{\partial f_n}{\partial \tau} d\tau) \rangle \frac{\partial \Psi_n}{\partial \tau} ,$$

and

$$dv^p = \langle h_1(\frac{\partial f_1}{\partial \sigma} d\sigma + \frac{\partial f_1}{\partial \tau} d\tau) \rangle \frac{\partial \Psi_1}{\partial \sigma} +$$
$$\langle h_2(\frac{\partial f_2}{\partial \sigma} d\sigma + \frac{\partial f_2}{\partial \tau} d\tau) \rangle \frac{\partial \Psi_2}{\partial \sigma} +$$
$$\ldots\ldots\ldots + \quad (15)$$
$$\langle h_n(\frac{\partial f_n}{\partial \sigma} d\sigma + \frac{\partial f_n}{\partial \tau} d\tau) \rangle \frac{\partial \Psi_n}{\partial \sigma} ,$$

where h_1, h_2,....h_n are constants of proportionality associated with f_1, f_2, f_3,....f_n, respectively. If normality is assumed (i.e., the first postulate adopted), then ψ_1, ψ_2, ...ψ_n in the above expressions may be replaced by f_1, f_2, ...f_n.

To our knowledge, the first model advocating intersecting yield surfaces was proposed by Prassad, who studied under the supervision of Professor Lelievr at the University of Waterloo. According to this model, proposed for normally consolidated clays, the two yield surfaces were assumed to be i) a line parallel to the σ axis, and ii) a curved surface (Fig. 8). The region in the immediate vicinity of point A is divided into four zones by these two curves. In zone (i), only elastic strains would be encountered. In zone (ii), only plastic volumetric strains would occur, i.e., if a stress increment starting from A moved to a neighboring point in zone (ii), only volumetric strains would occur. In zone (iii), both shearing and volumetric strains would be present. Finally, in zone (iv), only shearing strains would occur. In Fig. 8 are also shown the new yield loci resulting from movement of A to a neighboring point in the four zones mentioned above.

To conclude this section, two important properties of yield loci, which are direct results of assuming normality, will be mentioned. First, the yield loci must be convex. Otherwise, stress increments may be found such that the infinitesimal increment of work performed may be negative, thus violating the first postulate. For precisely the same reason, a yield locus, if it moves, must move in the same direction as that indicated by the outward normal to the yield surface passing through the given stress point A.

With the above introductory remarks, we are now in a position to give a brief account of the four models discussed by our group.

Models Proposed by Lade and Baladi

Since the Lade and Baladi models are in certain aspects similar, they will be discussed together.

Figure 9a shows a model proposed at the early stages of the application of theory of plasticity to cohesionless media. It first assumes that yield loci exist and are in the form of straight line passing through origin ($\tau = 0$, $\sigma = 0$). Hence, $f(\tau,\sigma) = \tau/\sigma$ = constant would represent the family of yield loci. It also assumes the plastic potential function ψ exists, and that for a given cohesionless medium at a given void ratio, the ψ = constant curves plotted in the stress space are in the form of similar almost elliptical curves. Here, $\psi \neq f$ and the normality condition is not assumed. One of the main shortcomings of this model was that for any proportional loading scheme (for which τ/σ remains constant), the model would predict only elastic behavior and no plastic strains. To remove this limitation in a subsequent, modified version, the yield surface was assumed to be in the form $f(\tau,\sigma) = (\tau/\sigma) + m \ln \sigma$ = constant (Fig. 9b). This removed the difficulty with proportional loading, but then introduced another inconsistency. If a soil were consolidated isotropically ($\tau = 0$) to a point A, and then unloaded to point B and sheared under a σ = constant condition, according to the model it

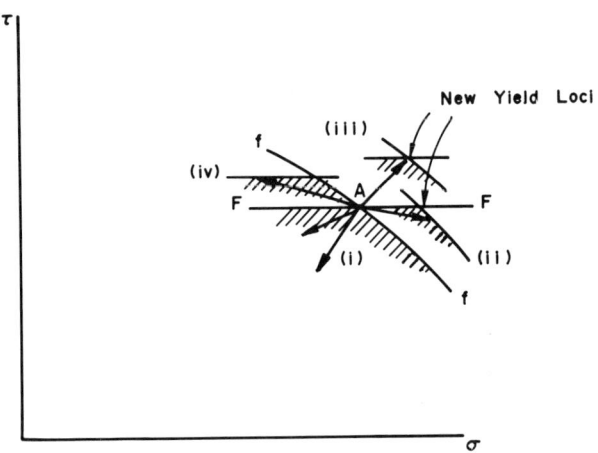

Fig. 8. Prassad's Model for Normally Consolidated Clays

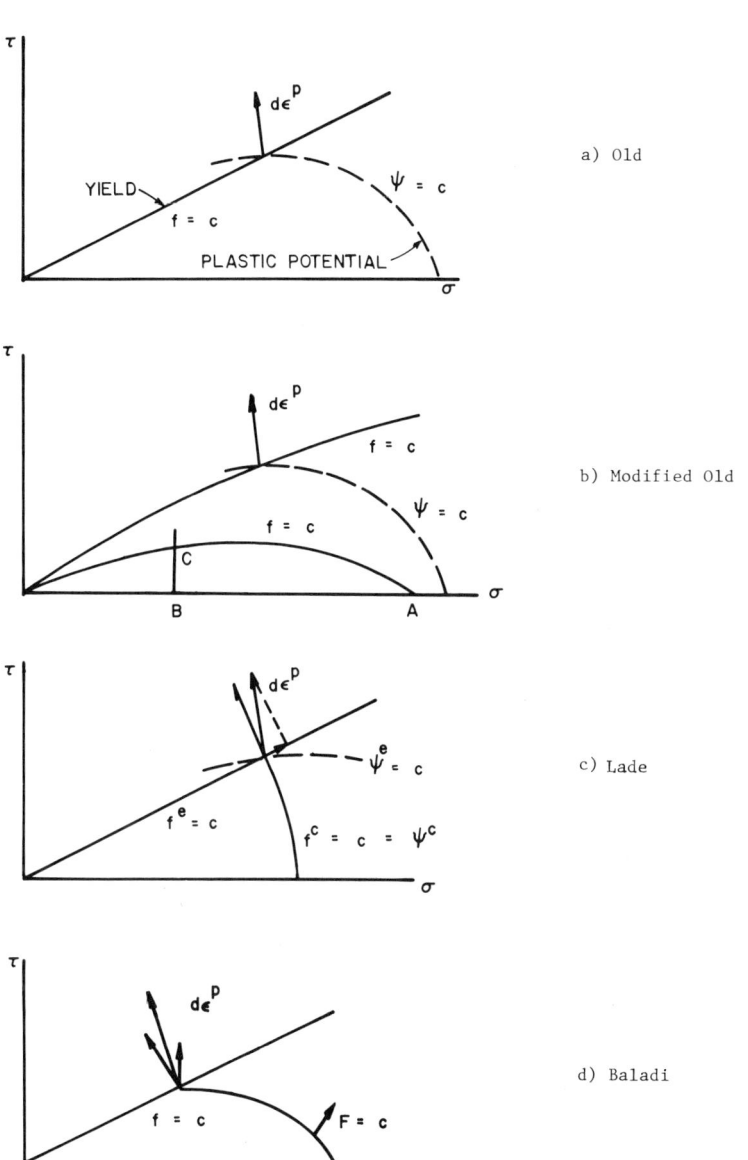

Fig. 9. Description of Lade and Baladi Models

would not yield until stress state C was achieved. However, both models are very simple in nature and so their application to solution of boundary value problems would, comparatively, be simple.

The model proposed by Lade introduces the concept of "spherical cap" yield surface and is shown in Fig. 9c. Here, the strain increment vector is assumed to be the sum of an elastic component $d\varepsilon^e$, a plastic collapse component $d\varepsilon_c^p$, due to a "contribution" from f^c (Fig. 9c), and a plastic expansion component $d\varepsilon_e^p$, due to a "contribution" from f^e, which are radial lines passing through the origin. Thus, Lade's equations for the plastic component of ε are

$$d\varepsilon^p = d\varepsilon_e^p + d\varepsilon_c^p =$$

$$\langle h_c (\frac{\partial F^c}{\partial \sigma} d\sigma + \frac{\partial F^c}{\partial \tau} d\tau) \rangle \frac{\partial F^c}{\partial \tau} + \qquad (16)$$

$$\langle h_e (\frac{\partial F^e}{\partial \sigma} d\sigma + \frac{\partial F^e}{\partial \tau} d\tau) \rangle \frac{d\psi^e}{\partial \tau} ,$$

and

$$dv^p = dv_c^p + dv_c^p =$$

$$\langle h_c (\frac{\partial F^c}{\partial \sigma} d\sigma + \frac{\partial F^e}{\partial \tau} d\tau) \rangle \frac{dF^c}{\partial \sigma} + \qquad (17)$$

$$\langle h_e (\frac{\partial F^e}{\partial \sigma} d\sigma + \frac{\partial F^e}{\partial \tau} d\tau) \frac{d\psi^e}{d\sigma} .$$

It is worth noting that, while for collapse yielding Lade assumes the normality condition $F^c = \psi^c$, he does not make the same assumptions for expansive yielding, i.e., $F^e \neq \psi^e$. It may also be noted that since the magnitude of the vector $\vec{d\varepsilon_c^p}$ is usually quite small compared to $d\varepsilon_e^p$, then Lade's model does not appear to be vastly different from the old model. This is not true, however. Lade's model can account for many observations that the old model could not, for example yielding during proportional loading, and undrained soil behavior. It appears to be a good model and is likely to be used extensively in the future.

The model presented by Baladi and his colleagues is shown in Fig. 9d. Here, the yield loci are assumed to consist of a single "limit" line $f(\tau,\sigma)$ = constant and a curved surface $F(\tau,\sigma,k)$ = constant, where k is in turn a function of v^p, the plastic volumetric strain. Note that the $f(\tau,\sigma)$ = c surface is fixed in (τ,σ) stress space; only the $F(\tau,\sigma,k)$ = c surface moves, depending on the value of k. Thus the concept of intersecting yield loci is utilized when the stress point touches the $f(\tau,\sigma) = c$ surface. The incremental stress strain loci proposed by Baladi are

$$d\varepsilon^p = \langle d\lambda \rangle \frac{\partial F}{\partial \tau} \quad , \tag{18}$$

and

$$dv^p = \langle d\lambda \rangle \frac{\partial F}{\partial \sigma} \quad , \tag{19}$$

in general, i.e., when the stress point is not at the limit state. The coefficient $d\lambda$ is obtained as follows, assuming for simplicity in illustration that the behavior is isotropic, linear elastic. The total strain increments in a loading process are

$$d\varepsilon = \frac{d\tau}{G} + d\lambda \frac{\partial F}{\partial \tau} \quad , \tag{20}$$

and

$$d\sigma = \frac{d\sigma}{k} + d\lambda \frac{\partial F}{\partial \sigma} \quad , \tag{21}$$

where k is the modified bulk modulus in two dimensions, i.e., for purely elastic behavior it is assumed that $\Delta\varepsilon^e = \Delta\sigma/k$. Rearranging terms in the above equation and multiplying the first equation by $\frac{\partial F}{\partial \sigma}$ results in

$$(Gd\varepsilon - Gd\lambda\frac{\partial F}{\partial \tau}) \frac{\partial F}{\partial \tau} = \frac{\partial F}{\partial \tau} d\tau \quad , \tag{22a}$$

$$(kd\sigma - kd\lambda\frac{\partial F}{\partial \sigma}) \frac{\partial F}{\partial \sigma} = \frac{\partial F}{\partial \sigma} d\sigma \quad , \tag{22b}$$

which, upon addition, leads to

$$(Gd\varepsilon - Gd\lambda\frac{\partial F}{\partial \tau}) \frac{\partial F}{\partial \tau} + (kd\sigma - kd\lambda\frac{\partial F}{\partial \sigma}) \frac{\partial F}{\partial \sigma} = \frac{\partial F}{\partial \tau} + \frac{\partial F}{\partial \sigma} d\sigma \quad . \tag{23}$$

But by assumption, $F[\tau,\sigma,k(v^p)] = 0$,

$$\frac{\partial F}{\partial \tau} d\tau + \frac{\partial F}{\partial \sigma} d\sigma = -\frac{\partial F}{\partial k} \cdot \frac{\partial k}{\partial \sigma^p} \cdot d\sigma^p$$

$$= -\frac{\partial F}{\partial k} \cdot \frac{\partial k}{\partial v^p} \{d\lambda \frac{\partial F}{\partial \sigma}\} \quad . \tag{24}$$

Eliminating $\frac{\partial F}{\partial \tau} d\tau + \frac{\partial F}{\partial \sigma} d\sigma$ between the last two equations and solving for dλ results in

$$d\lambda = \frac{(\frac{\partial F}{\partial \tau})Gd\varepsilon + (\frac{\partial F}{\partial \sigma})kdv}{G(\frac{\partial F}{\partial \tau})^2 + k(\frac{\partial F}{\partial \sigma})^2 - \frac{\partial F}{\partial k} \cdot \frac{\partial k}{\partial vp} \cdot \frac{\partial F}{\partial \sigma}} \quad . \tag{25}$$

The first fundamental difference between Lade's formulation and Baladi's is that in Lade's model, both yield loci move with the stress point (in a loading process), while in Baladi's model, the yield surface f = c is a limit surface and hence fixed in space. Lade's model is a "non-associated" rule, in that $d\varepsilon_e^p$ is not normal to the corresponding yield loci, while in Baladi's mode normality is assumed for both yield surfaces. The area enclosed by the F = c, f = c curves represents elastic behavior in the Baladi model, which differs fr the behavior of real soil.

There is no doubt that the introduction of intersecting yield surfaces re moved many previously unresolved difficulties. For example, both models of Lade and Baladi can account for yielding along proportional loading, which the simple model (the old model) could not. And indeed, if no restriction is put on the form (shape) of the yield loci, complicated behavior can be accounted for. In fact, Baladi stated during the course of one of our meetings that, with judicial choice of his yield functions, he could apply it to describe the behavior of "....all soils, even rock."

Modelling Difficulties

There are still certain unresolved questions about soil behavior that effect the validity of the proposed models. One is the shape and "uniqueness" o the yield surface in the general space of stress, as distinct from the rather simple state of stress (τ,σ) used in the models. To illustrate this point, re erence is made to Fig. 10, which shows the results obtained by Tong, another graduate student at the University of Waterloo who worked under joint supervision of Professors Lelievre and Holubec. The tests were performed on hollow cylindrical samples of Ottawa sand. The stress space shown in Fig. 10 is slightly more complicated than the one used so far in this paper. Its axes represent the three principal stress components used in cylindrical coordinate These are the axial (σ_{zz}), radial (σ_{rr}) and tangential $(\sigma_{\theta\theta})$ components, eac divided by the mean normal stress p equal to $1/3(\sigma_{rr} + \sigma_{\theta\theta} + \sigma_{zz})$. In Fig. 10 is shown an element first subjected to a stress level of σ_{zz}/p, then unloaded along the $\bar{\sigma}_{zz}$ axis ($\sigma_{rr} = \sigma_{\theta\theta}$, σ_{zz} increasing), and subsequently reloaded alon a radial path until flow was achieved. The initiation of yielding (the point

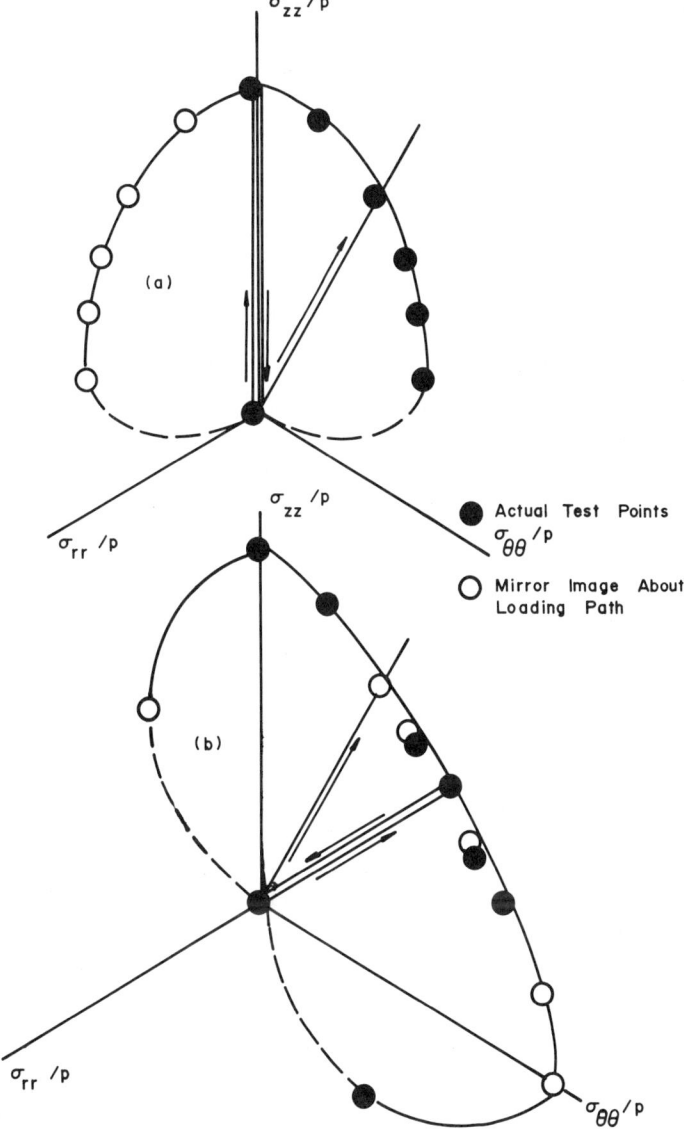

Fig. 10. Test Results by Tong on Hollow Cylindrical Samples of Ottawa Sand

after which irrecoverable axial strains were observed) was recorded and is shown by the appropriate point in the diagram. In a second series of tests, the same procedure was followed but the sample was initially subjected to a ($\sigma_{\iota\iota} = \sigma_{zz}$, $\sigma_{\theta\theta}$ decreasing) loading path and then unloaded. The results of these tests are shown in Fig. 10b.

Certain features of these test results by Tong are worthy of note. First, both yield surfaces contain the origin. Thus, a sample subjected to, for example, a compression test, then unloaded and subsequently subjected to an extension test, would start yielding as soon as the stress path crosses the p axis (origin). In a cyclic test, crossing of the p axis appears to wipe out the memory of the previous loading history of the sample completely. Second, near the origin, both yield surfaces appear to be concave in defiance of the first postulate of the ideal plastic material. Third, the shape of the two yield loci are so remarkably different that it raises the question of the uniqueness of a yield locus. This question is: given that a sample loaded to a point A would yield at point B upon unloading and subsequent reloading, would a similar sample loaded to point B yield at point A upon unloading and reloading along a path leading to A? Present results indicate that this may not, in fact, be the case.

Further insight may be gained by introducing an additional concept, which was first proposed by Phillips and Sierakosuski.* Consider a sample subjected to an axial stress σ_A and let the associated axial plastic strain be ε^p. Figure 11a shows the behavior of an ideal elastic-plastic material: upon unloading from point A, the material would behave elastically ($\Delta\varepsilon^p = 0$), and upon reloading from B to A (=C), the material would still behave elastically ($\Delta\varepsilon^p = 0$). It is only for stress levels higher than σ_A that plastic strains are encountered. This is the behavior of an ideal elastic-plastic material, a behavior which has been assumed in all the definitions and arguments presented in this paper.

In Fig. 11b, a close approximation to the stress strain curve for a "real" material subjected to a similar loading process as discussed above is shown. Upon unloading to point B, the material behaves elastically (unless the level of σ_B is reduced below a certain level, say σ_y); reloading from B to C is still elastic ($\Delta\varepsilon^p = 0$), but for stress levels higher than σ_c (where $\sigma_c < \sigma_A$), plastic

---

*Phillips, A. and Sierakosuski, R.L., "On the Concept of the Yield Surface," Acta Mechanica 1965, pp. 29-35.

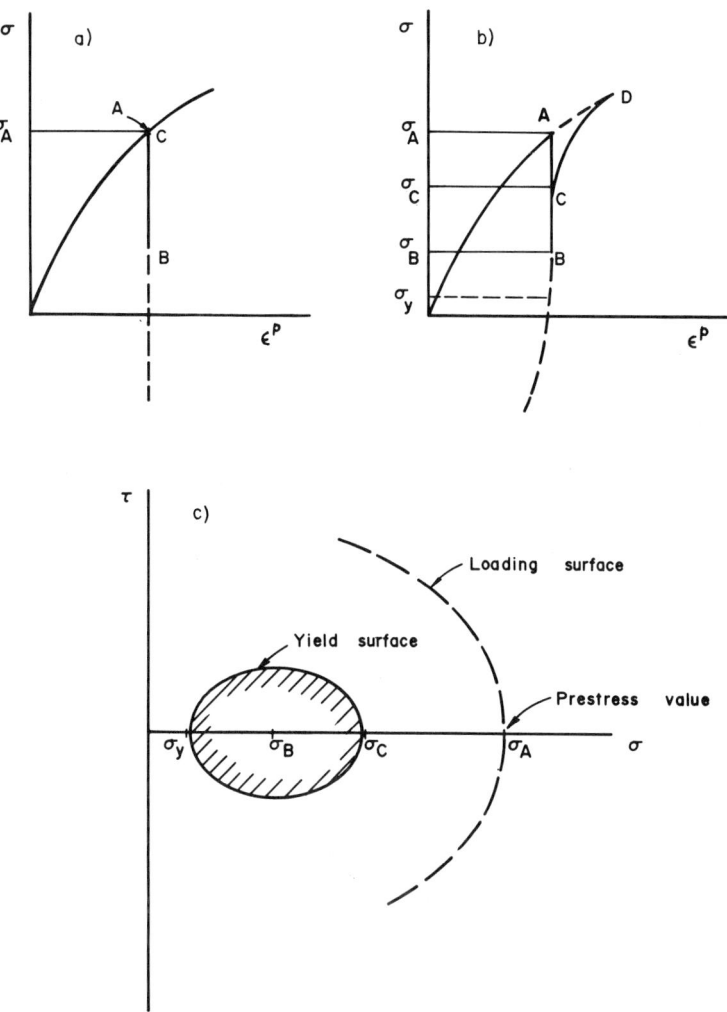

Fig. 11. Illustration of Elastic-Plastic Behavior

strains are experienced ($\Delta\varepsilon^p \neq 0$).

Let point σ_A in the (τ,σ) stress plane represent the state of stress σ_A (Fig. 11c). For the simple uniaxial stress test conducted above, it may be said that if stress moves to levels higher than σ_A (i.e., loading continues), the sample will experience plastic strain, and if stress moves to levels lower than σ_A (i.e., the sample is unloaded), only elastic strains are encountered. Therefore, there should exist some locus in the stress space which borders the "only-elastic" and the "elastic-plastic" zones. This locus is the so-called "loading surface." The yield surface corresponding to the pre-stress state σ_A before unloading would pass through stress points σ_c and σ_y (compare Figs. 11c and 11b). An infinite number of loading surfaces exist between σ_c and σ_A, since for every intermediate point between σ_c and σ_A, a new loading surface may be plotted, but no two loading surfaces may intersect. For ideal materials, the "loading surfaces" and the "yield surfaces" coincide.

The yield surfaces shown in Fig. 10 are plotted in the three-dimensional stress space (σ_{rr}, $\sigma_{\theta\theta}$, σ_{zz}). Comparable results for a soil sample plotted in the (τ,σ) stress plane are not available. For this reason, it was decided to include Figs. 12 and 13, showing the results of tests carried out on Titanium 50A, discussed at the workshop session by Professor Ellyin. To appreciate Figs. 12 and 13, the cross-hatched yield locus of Fig. 12 will be discussed. Here, the sample was subjected to a pre-stress value of $\tau = 0$, $\sigma_A = 272$ MP_a. Hence, the loading surface passes through this point and is indicated by the broken curve in Fig. 12. The sample is then unloaded along the $\tau = 0$ axis to some σ_B level smaller than σ_A and subsequently sheared. The initiation of yielding was recorded and noted on the corresponding stress path in the (τ,σ) plane. The test was repeated until sufficient points were obtained to enable the drawing of the smooth curve bounding the shaded zone.

Titanium is a fairly well-behaved alloy. Its specimen can easily be machined to precise dimensions and the testing conditions are exact and quite controllable. Yet its deformation pattern does not appear to be very simple, as the jungle of yield loci of Figs. 12 and 13 indicate. We can expect soil behavior to be much more complicated.

Model Proposed by Kavazanjian, Mitchell and Bonaparte

The model proposed by Kavazanjian, Mitchell and Bonaparte does not follow the lines of classical plasticity as explained above. Perhaps for this reason, the authors have chosen to refer to their model as a "phenomenological model."

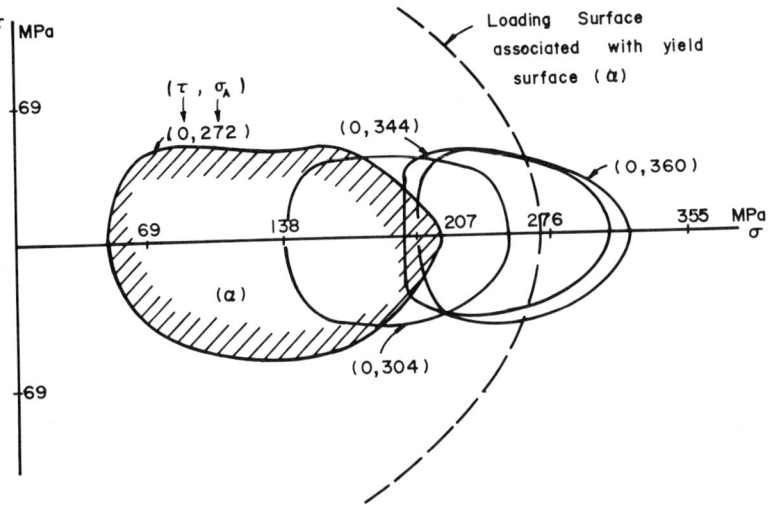

Fig. 12. Illustration of Yield Surfaces for Titanium Initially Loaded under Hydrostatic Stress

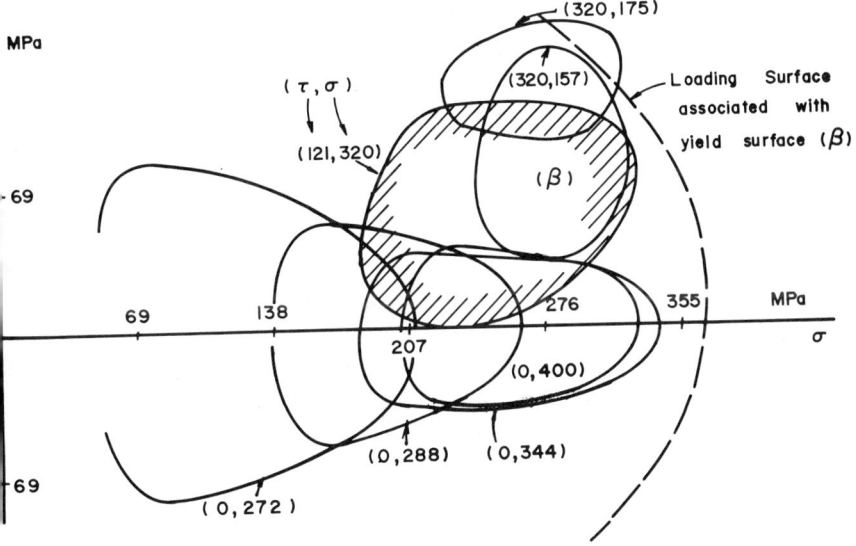

Fig. 13. Illustration of Yield Surfaces for Titanium Initially Loaded in Shear and Compression

Fundamentally speaking, however, all of the models can be considered phenomenological.

The time effects of the KMB model were not considered in this workshop, so these will not be discussed. The model basically assumes that

$$\vec{\varepsilon} = \vec{\varepsilon}_\tau + \vec{\varepsilon}_v \quad , \tag{26}$$

where $\vec{\varepsilon}$ is the total strain vector (as distinct from incremental strain vector $d\vec{\varepsilon}$), $\vec{\varepsilon}_v$ is the total strain due to change of volume, and $\vec{\varepsilon}_\tau$ is the total strain which is obtained from exerting the shear stress component τ.

If the contours of equal water content for a normally consolidated clay are plotted in the (τ,σ) stress space, where σ is the <u>effective</u> stress and τ the shearing stress, a diagram is obtained which is generally known as the Rendulic diagram, after the person who first proposed it in 1936 (Fig. 14). This diagram indicates that w is a function of τ and σ. Now since a change in volume is directly related to a change in water content, then obviously

$$v \sim \Delta W_A^B = w(\tau_B, \sigma_B) - w(\tau_A, \sigma_A) \quad , \tag{27}$$

i.e., v is independent of stress path followed. Thus, a sample starting from A and arriving at point B would undergo the same volumetric strain, regardless of the shape of the stress path taken, provided, of course, that the stress path does not include unloading processes.

To calculate ε_τ, the authors use the normalized stress-strain relationship in the form

$$\varepsilon_\tau = \varepsilon(\overline{D}) = f(\tau/\tau_f) \quad , \tag{28}$$

where τ_f is the maximum value of τ the sample can withstand under the existing value of σ. Since a functional dependence between ε_τ and $\overline{D}$ is assumed, it appears that the authors are postulating the following: a sample starting from point A and arriving at point B would experience the same change of volume and the same change of shape (associated with shearing strains), regardless of the stress path followed. However, this assumption is not a true representation of the behavior of most soils. For example, take two samples, both initially at a water content w_A, and at the stress state $\tau = 0$, $\sigma = \sigma_A$. Let the first sample be sheared along the undrained path denoted by w_A = constant and then sheared

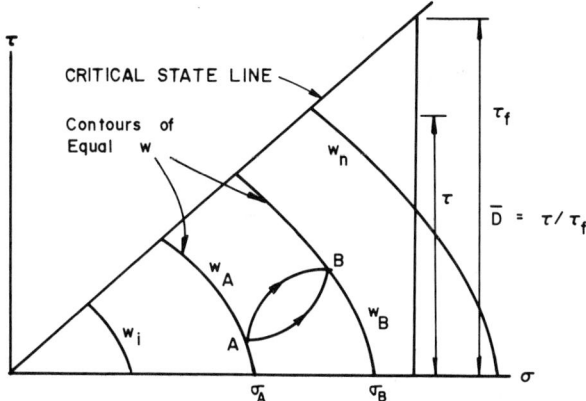

Fig. 14. Rendulic Diagram used in Kavazanjian, Mitchell and Bonaparte Model

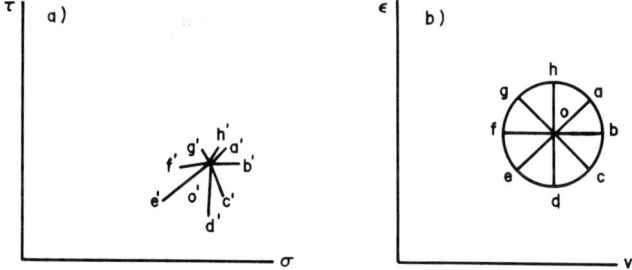

Fig. 15. Illustration of Gudehus Model

along the critical state line. Let the second sample be consolidated under $\tau = 0$ condition until σ_B is reached, and then tested undrained until the critical state line is reached. The first sample would experience infinite strain, while the strain required for the second sample would be finite.

Model Proposed by Gudehus

In the model proposed by Gudehus and his colleagues, the stress increment is expressed as a function of the current state of stress and the total strain increment. In terms of stress and strain variables used in this paper, the proposed relationships appear to be expressible in the form

$$\dot{\tau} = a_2 \tau + 2a_3 \tau\sigma + a_4 \dot{\varepsilon} + a_5 \dot{\varepsilon}\dot{\sigma} + \qquad (29)$$

$$a_6 (\sigma\dot{\varepsilon} + \frac{1}{2} \tau\dot{v}) + a_7 (\tau^2 \dot{\varepsilon} + \sigma\tau\dot{v} + \sigma^2 \dot{\varepsilon})$$

$$+ a_8 (\frac{1}{4} \tau\dot{v}^2 + \tau\sigma\dot{v} + \sigma^2 \dot{\varepsilon}) + a_9 (\frac{1}{2} \dot{v}\dot{\varepsilon}\sigma^2)$$

$$+ \frac{1}{2} \sigma^2 \dot{\varepsilon}\dot{v} + \sigma\tau\dot{\varepsilon}^2 + \tau\sigma\frac{1}{4}\dot{v}^2 + \dot{\varepsilon}\dot{v}\tau^2 + \dot{\varepsilon}^2\tau\sigma + \frac{1}{4} \tau\sigma\dot{v}^2 \quad ,$$

and a similar expression for $\dot{\sigma}$. A dot above a symbol indicates the rate of change with time.

To gain some insight, let the state of strain of the element shown in Fig. 1 be represented by a point σ' in the stress space, and the corresponding state of stress by point σ in the strain space (Figs. 15a and 15b). Now let a unit strain increment of arbitrary direction be imposed on the sample (oa, ob, oc, etc. in Fig. 15a). The corresponding stress increment in the stress space will be o'a', o'b', o'c', etc., as shown in Fig. 15a. Both magnitude and direction of these stress increments would depend not only on the strain increments oa, ob, etc., but also on the state of stress o'. The model uses a piecewise linear technique in constructing the stress path, which the element must follow for a specified strain path. The equations proposed by Gudehus are such that some of the known stress-strain properties of the soils can be accounted for, for example, the behavior under proportional loading and under similar loading paths. There are quite a few aspects, however, that the model cannot accommodate. One is the strain softening behavior. In a shear test on a dense sand sample with constant normal stress (say constant σ test), the stress-strain curve attains a peak value (corresponding to ϕ_{max}), after which, if straining is continued, the soil undergoes strain softening. Gudehus' model apparently

cannot account for this. Referring to triaxial testing, Gudehus states "....
the peak almost disappears if the test is carried out in such a manner that
bifurcation is avoided." While some truth exists in this statement, it is very
doubtful that the large difference between ϕ_{max} and $\phi_{residual}$ observed in soils
such as dense sands and overconsolidated clays can be attributed to sample inhomogeneity brought about by testing conditions alone.

Models Compared to Predictions

In the first two workshop sessions of Group 2, the predictors described how they obtained their model parameters from the data provided. Then the predictions were compared to the actual test measurements that were provided to the participants, including the predictors, for the first time at these sessions. A detailed account of these models and the predictions using them is given in papers submitted afterwards by the predictors, and reproduced elsewhere in these workshop proceedings. A brief summary will be given here to indicate how well a particular model appeared to predict a particular test. Conclusions drawn from these comparisons must take into consideration the fact that good agreement between prediction and measurement does not necessarily mean a theoretically correct model, and vice versa.

Baladi predicted the behavior of sand, clay X and clay Y. For the cases involving the simple loading processes, the predictions were quite good. Lade only predicted the sand behavior. Again, for the simple loading processes, his results were quite close to the experimental observations. However, for the circular loading path test on the sand, both Baladi's and Lade's predictions were completely different from the measured values. Difficulty in defining the appropriate yield surfaces is probably the reason for these discrepancies.

Mitchell and his colleagues satisfactorily predicted clay X behavior.

The success of the model proposed by Gudehus could not be judged. His results were not provided in the requested form which permitted a comparison with the measured results. To the extent that a judgment could be made, the predictions did not seem to be very close to the test results. Gudehus argued that the data provided to him included many inconsistencies. However, he was given the same data as that used by the other predictors, who did obtain some successful predictions.

Features of Models

In the third group session, an effort was devoted to developing a list of features that could be used to classify and compare the models. Although it was recognized that what the desired features are in any case depends upon the application, nevertheless, the features that can be used to describe any model should be identifiable without regard to applications. What may change in each application is the importance of each feature.

The resulting list of features given in Table 1 was prepared without regard to applications. The relationship of each of the four models to the features is also given in Table 1. These ratings were given by the predictors, but subject to challenge by the other participants if it was possible to make a judgment, and if this judgment was in disagreement with the predictor's position. Most of the features will be self-explanatory. However, comments on some are needed. "History" is considered different from stress path dependency in that it lumps all the past events experienced by a soil sample before it is subjected to a given loading process. Such events may leave the sample initially overconsolidated and with induced anisotropy, for example. The "yes" response from the predictors means that the model is sensitive to such effects and can account for them. The item "coaxiality," as suggested by Gudehus, refers to whether during rotation of principal stress axes, there is coincidence of stress rate increment vector and strain rate increment vector, or strain rate increment vector and stress increment vector (not its rate). A "yes" response means that the proposed model can account for the effects of rotation of principal axes.

Behavior Complications

Some of the complications of real soil behavior were discussed in order to gain a perspective on model requirements and expected accuracy. This discussion began with the presentation by Ellyin of experiments on samples of titanium. These experiments have been described earlier in this paper. Several participants indicated that they had observed similar behavior for soil. A point needing emphasis is that these complications relate to yielding at stress conditions that do not involve failure. The conclusion from this discussion was that none of the models can predict such complicated behavior. The question then posed was what are the limitations on the stress paths that can be reasonably represented by the models? Some participants suggested that we are limited to monotonic loading with, at most, small reversals. However, others pointed out that large reversals are often present in problems of interest. Thus a model that is only suitable for monotonic loading would be of

Table 1. Model Features

	Kavazanjian-Mitchell	Baladi	Lade	Gudehus
1. Failure criteria exist	Yes (Duncan-Chang)	Yes (any)	Yes	Yes
2. No. of phases (solid-liquid-gas)	2	2	2	2
3. Volume change representation				
Dilatancy	?	Yes	Yes	Yes
Contractancy	Yes	Yes	Yes	Yes
4. Inelastic	Yes	Yes	Yes	No
5. Tensile behavior	No	Yes	No	No
6. Strain softening	No	Yes	Yes	?
7. Strain hardening	Yes	Yes	Yes	?
8. Total or effective stress representation	Both	Both	Both	Both
9. Pore pressure represented	Yes	Yes	Yes	Yes
10. History	Yes	Yes	Yes	Yes
11. Loading path dependency	?	Yes	Yes	Yes
12. Time effects			No	
Creep	Yes	No		Yes
Consolidation	Yes	No		Yes
Strain ratio	No	Yes		Yes
Load rate	Yes	Yes		Yes
13. Coaxiality	No	Yes	Yes	Yes
14. Anisotropy	Yes	Yes	No	Only stress-induced
15. Soil type	To OCR < 2 NC "clay"	All	All	Sand, NC clay
16. Stability for:				
Stress control	?	Yes	Yes	Yes
Strain control	?	?	?	?

limited value.

The need for both loading and unloading capability in a model for even simple situations was illustrated by an example given by Poorooshasb. He considered the case of a footing resting on the level surface of a deep stratum of sand. Prior to loading, the state of stress is represented by a vertical stress equal to the unit weight times the depth, and a horizontal stress equal to K_o times the vertical stress. Upon loading by the foundation, the soil elements immediately below the footing would load and yield. This situation is similar to the triaxial compression test. Away from the footing, the stress path followed by the element would fall below the K_o line and, presumably, unload. Some mechanism must be built into the soil model to account for both types of behavior, if the foundation response in this problem is to be correctly predicted. The alternative is to "calibrate" the model for a particular response such as vertical footing settlement, with the realization that the model may not be correctly representing other response.

A variety of opinions were presented on these issues, without resolution. There was general consensus that the models were reasonably good at handling loading problems, but limited in handling unloading problems, much less unloading with subsequent reloading. However, the degree of error in unloading situations could not be determined.

The following was suggested as a summary of the session:

1. Any model should be capable of accounting for large reversals of loading.

2. For certain types of boundary value problems, the results of models which have been formulated for monotonic loading only are quite satisfactory, from a theoretician's point of view.

3. In other cases, these simple models for monotonic loading may not be theoretically correct, but from an engineer's viewpoint may be quite satisfactory.

4. Many cases exist, however, for which the simple models discussed in our group cannot even be used to generate the required governing equations.

Model Assessment

The participants, particularly the predictors, were asked to indicate what tests are needed to check the models. Gudehus indicated the need to test the effects of loading history on behavior at a given stress state. Lade suggested that a very critical test would be one in which shear tractions are applied on the sample boundaries such as represented by a torsional shear or simple shear test. The importance of this type of test is the effect of reorientation of the principal stress axes. Lade indicated that basically elastic models will not do well in such cases, whereas plastic models should be quite satisfactory. Baladi emphasized the importance of doing replicate tests to determine experimental data scatter, in order that the models are not improperly judged. Clough indicated the need to consider a wider variety of soil conditions, such as partially saturated soils and cemented sands.

The predictors were then asked to indicate the minimum tests needed to obtain parameters for their models. For example, can the parameters be obtained from a standard penetration test or cone test in the field? Is a conventional triaxial test sufficient for more detailed parameter determination, or should a cubical triaxial test be used?

For sand, Baladi requested one loading-unloading isotropic consolidation test, plus two triaxial compression tests including unloading, as a minimum. However, if extension stress states are to be predicted, Baladi also would prefer to have triaxial extension test results. In addition, he indicated that to the extent that the stress states other than the axis symmetric affect soil behavior, he could not represent it by using parameters from the specified tests.

Gudehus requested one triaxial compression test on a sample with lubricated ends and a height less than its diameter, plus a K_o test as a minimum. In response to questions, he agreed that the effect of confining pressure would not be given by these tests, and so would have to be assumed.

Lade desires as a minimum an isotropic compression test and three triaxial compression tests with unloading. For sand and overconsolidated clay, the tests should be drained with volume change measurement. For normally consolidated clay, the tests should be undrained with pore pressure measurement.

Kavazanjian and Mitchell require an isotropic compression test and a triaxial compression test, preferably undrained.

Concern was expressed that none of the predictors were satisfied with field measurements to get their parameters. This is a severe limitation in practice, because the requested lab test results will not usually be available.

In a related discussion, the predictors were asked whether they used a computer to get their results. Mitchell and Baladi indicated that they used a pocket calculator. Lade used a pocket calculator to obtain his parameters. However, he used a computer to make his predictions because a program was available, not because it was necessary. Gudehus used a computer to obtain the results for the circular stress path, but used a programmable pocket calculator for the rest of his predictions.

Selig pointed out that all of the predictions and test results represent a single soil element, whereas applications using the finite element method involve additional factors such as numerical stability, and ease of incorporating the soil model. The real proof of the suitability of the models must involve case study predictions. The participants were asked to suggest minimum requirements for such tests. One suggestion of a simple example was expansion of a spherical cavity in soil. The use of centrifuge model tests was suggested. Several persons expressed the opinion that obtaining reliable experimental results for boundary value problems was difficult and that failure of the analytical methods to predict the experimental results would not necessarily invalidate the analytical models.

The predictors were next asked to indicate what further model development is needed. Overconsolidated soils and anisotropic properties were among the examples cited that are not well handled by most available models. It was also evident that work is needed on representing the effects of rotation of principal stress axes. The need for improvement in kinematic hardening models was indicated to handle large stress reversals. Gudehus believes new state parameters are needed to supplement the usual stress and density parameters. However, the view was expressed by several persons that what is needed most is experience with the models, rather than further development of the models.

This led to the final topic of discussion, which concerned application experience. The predictors were asked for what types of problems have their models been used, and have they been used to any significant extent by anyone other than the developers.

Baladi reported that he has used his model to solve complicated dynamic problems and that the cost has not been excessive. Because some of the parameters were not available and had to be estimated, in some cases a variation of parameters approach was used with the model to estimate the consequences of this uncertainty on the results. Baladi indicated that his model is fully documented in the literature and that others have used it.

Gudehus reported that his model is being used for shock problems. Application experience is limited, he indicated, but the model is documented and has been used by others. Unfortunately, military restrictions have prevented the publication of this experience.

Lade indicated that his model has been documented, that application experience has been obtained and that others have used his model.

The Kavazanjian-Mitchell model is new and so is only partially documented. Application to problems is just beginning.

Summary

This workshop report first summarizes basic concepts of plasticity models, including a definition of yield surface and plastic potential. The main features of the models by Lade, Baladi, Gudehus and Kavazanjian and Mitchell were then described. Test results for sand and for titanium metal were presented to illustrate the complex nature of material yielding and show why accurate modeling of soil behavior is not expected in general.

A comparison of the predictions to test results showed good agreement for simple, monotonic loading stress paths, but poor agreement for the severe case of the circular stress path.

A list of features that can be used to classify the models was prepared by the workshop group and is presented in this paper.

In general, the predictors preferred conventional triaxial and K_o tests for obtaining their model parameters. None indicated the capability of using test data to estimate the required parameters.

Further work is needed to model overconsolidated soils and anisotropic properties. However, experience in application of the models in finite element programs was considered a higher priority than further model development.

The workshop was a very valuable exchange of information on soil models. A good perspective on the state-of-the-art was obtained, and all participants should have gained a better understanding of the capabilities of the various models.

EVALUATION OF CRITICAL STATE AND HYPERBOLIC
STRESS-STRAIN MODELS
WORKING GROUP THREE
by

W.D. Liam Finn, M.ASCE*

INTRODUCTION

The workshop on "Plasticity Theories and Generalized Stress-Strain Modelling of Soils" held at McGill University in Montreal, May 28-30, 1980 is one of the more significant events in the history of the development of comprehensive constitutive relations for soils. For the first time an opportunity was provided to test the predictive capability of a wide variety of constitutive laws on the same data under independently controlled conditions. The discussions of the predictions and the spirited probing of the foundations and applications of the constitutive relations by the participants in the workshop resulted in a clearer understanding of the limitations of all existing constitutive relations and the substantial inherent difficulties in developing a truly general constitutive relationship for soils.

Predictors from Canada, England, Germany, Japan, Turkey and the United States participated in the workshop. Engineers experienced in the development and application of stress-strain relationships for soils were also invited to participate in the evaluation of the predictions and the general discussions on constitutive relations. To insure detailed study of all the predictions and careful critical assessment of the stress-strain models participants were divided into groups, each under a chairman. For the duration of the workshop each group discussed the constitutive relations assigned to them and evaluated the stress-strain and porewater pressure predictions made using these relations. At regular intervals during the workshop plenary sessions were held in which the various groups reported on their

*Univ. of British Columbia, Vancouver, B.C., CANADA

discussions.

I chaired the group who were assigned the task of assessing the predictions and theories of Wroth and Houlsby (8), Akai, Adachi and Oka (1) and Duncan (5). Wroth and Houlsby based their prediction on the Cambridge Critical State Theory (3). The theory used by Akai et al is an extension of the Critical State Theory to include such time dependent phenomena as creep, stress relaxation, and secondary consolidation. Duncan's predictions were based on the hyperbolic stress-strain relationship developed by Duncan and Chang (7).

The critical state and hyperbolic stress-strain models are long established in geotechnical research and engineering practice and the group members were familiar with them. Discussion of these theories was at a sophisticated level and dealt with basic assumptions, inherent limitations, optimum selection of parameters, exploration of difficulties in predictions and possible extensions of the theories. The Akai et al model though ultimately based on elements of critical state theory was new to all group members and, in fact, had been developed specifically for the workshop. Much of the discussion on this theory was devoted to acquainting the group with the elements of the theory and the operation of the model. The group therefore did not arrive at any firm conclusions about the appropriateness or utility of the Akai model.

The group also discussed the adequacy of the data base, the characteristics of the loading paths chosen for the prediction exercises, the general requirements for a stress-strain model and the needs and problems associated with applying stress-strain models in engineering practice. The position papers of the predictors deal with many of these points in considerable detail and reflect quite well many of the points that were considered during the group discussions.

In what follows I endeavour to present only the essence of the group discussions. My presentation is based on tapes of the discussions, summary notes taken at the time, and a recent review of the position papers (8,2,6) and prediction papers (8,1,5) submitted by the predictors.

THE DATA BASE

A comprehensive and representative data base is a fundamental requirement for testing validly the predictive capability of constitutive relations. It is imperative that errors in predictions which may be due to deficiencies of the data base should not be casually attributed to deficiencies in the constitutive relations. Therefore, before discussing the quality of the predictions I will attempt to assess the adequacy of the data base. Data were provided for four soils; two natural highly sensitive clays X and Y, a kaolinite, Edgar Plastic Kaolin, K_0-consolidated in the laboratory and an Ottawa sand (9).

In the case of the natural clays X and Y, the experimental data were obtained from consolidated, drained triaxial tests on fully saturated prismatic samples in a true triaxial apparatus. The notation used to describe the stresses in the data base is somewhat unconventional; σ_1 is the axial (vertical) stress irrespective of whether it is the major principal stress or not and the orthogonal directions 2 and 3 remain constant. The notation for principal strains is to be similarly interpreted. This notation is very convenient for the presentation of data but some participants found it somewhat confusing if not ambiguous.

The prismatic test samples were trimmed from block samples. There is no guarantee that the block samples were homogeneous, although attempts were made to ensure that each test sample itself was as homogeneous as possible. The monitored quantities were axial deformation, ε_1, lateral deformation, ε_3

and the volumetric strain, $\Delta v/v$. The lateral strain, ε_2, given in the data set is a derived quantity. Had ε_2 been also measured the relation

$$\varepsilon_1 + \varepsilon_2 + \varepsilon_3 = \varepsilon_{v\text{(volumetric strain)}} \qquad (1)$$

would have provided a useful check on the consistency of the data.

The lateral strains are not equal, $\varepsilon_2 \neq \varepsilon_3$, during either the compression or extension tests with the exception of the compression test on clay Y at a consolidation pressure $\sigma_c' = 2.5$ psi. From the given simple stress paths in the data base it is not possible to determine whether this results from anisotropy, inhomogeneity or both. During our discussions the opinion emerged that the samples were anisotropic and that the axes of anisotropy may not be similarly oriented in all specimens. To establish the existence of anisotropy the behaviour of the soil must be probed by applying <u>the same probing stress system</u> at different directions to some reference axes embedded in the material. The strain response to these probes will help to define principal axes of anisotropy if they exist.

All the tests on clay X were carried out at <u>constant mean normal stress</u>. This condition has interesting implications for theories which separate shear and hydrostatic behaviour and attribute volume changes to changes in effective hydrostatic stress only. Tests on clay Y did not maintain constant mean normal stress. The lateral stress σ_3 was kept constant and compression or extension failure induced by changing the values of σ_1 and σ_2.

It should be stressed that the extension tests in the data base for clays X and Y do not follow the usual deformation modes in conventional laboratory triaxial equipment. In the conventional triaxial test the axial direction is the direction of the major principal stress in compression tests and of the minor principal stress in extension tests. In the data base, the axial direction is the direction of the major principal stress in the compression tests but in the extension tests the direction of extension and of the minor principal stress is at right angles to the axial direction in the compression test. In the octahedral plane the stress path in the extension test ($\sigma_1 = \sigma_2$) plots at $60°$ to the stress path during compression ($\sigma_2 = \sigma_3$) instead of the usual $180°$. This means that the kind of extension experienced during rebound caused by excavation, for example, is not modelled by the data. Furthermore, in the case of clay X, both the compression and extension occur without any hydrostatic loading or unloading because of the constraint of constant mean normal stress. Therefore, apart from any concerns about anisotropy, it would not seem that a comprehensive stress-strain model could be fully calibrated for clay X. Predictions therefore would be, in theory, possible only for stress paths that did not invoke any features of response not present in the data base. Therefore, the stress paths for which predictions are required are all subjected to the constraint of constant mean normal stress. In the case of clay Y for which the mean normal stress is not kept constant, there is no hydrostatic unloading in the extension tests.

Edgar Plastic Kaolin

A more varied data base was provided for the Edgar Plastic Kaolin consisting of consolidation and rebound data and the results of compression and extension tests at constant mean normal stress and constant cell pressure. In the compression tests the stress along the axis of consolidation is increased while the stresses in the orthogonal directions are maintained equal; in the extension tests the axial stress is decreased. Therefore, the data represents the deformation modes in conventional triaxial equipment and allows the representation of what is commonly understood by compressive and extensive behaviour of clays. In the deviatoric stress plane the compression and extension paths for the kaolin plot at $180°$ to each other and therefore may be expected to bound a large range in behavioural response.

Because of the K_0-consolidation the kaolinite is expected to show significantly different behaviour in conventional extension and compression modes and to be anisotropic with rotational symmetry about the vertical axis. However, for this clay as for clay X and clay Y, not enough information is given to characterize the anisotropy.

Ottawa Sand

The sand was compacted by aerial pluviation to 1.73 gm/cc corresponding to a relative density, D_r = 87%. It was tested dry in a flexible fluid cushion cubical device with strain control. Test data were provided from hydrostatic compression tests, conventional compression and triaxial extension tests and triaxial compression and extension tests at constant mean normal stress. Data were not provided to establish the degree of anisotropy of the sand. Examination of the strain data suggests that the sand may be reasonably isotropic about the vertical axis.

In the case of Ottawa sand the conventional modes of deformation during compression and extension tests were followed but, in the case of extension tests, no hydrostatic unloading occurred. Failure was induced in the conventional triaxial extension tests by increasing σ_x and σ_y to induce failure while keeping σ_z constant. No isotropic rebound data was provided.

General Assessment

The test data was very carefully selected and represents high quality information from very reputable research laboratories. Nevertheless, some scatter is to be expected resulting from variations in the degree of homogeneity and anisotropy of the samples especially in the case of natural clays. It would have been useful to have measures of the scatter of data from allegedly similar samples. Without this measure it is sometimes difficult to interpret the differences between experimental and predicted results. This is particularly true for the stress paths for clays X and Y which all lie in a 60^0 sector of the octahedral plane.

Test data was not provided that would allow the complete characterization of the anisotropy of the soils. Anisotropy appears to be significant in the natural clays and the kaolin. The predictions for stress paths at various angles to the stress paths used for calibrating the models cannot, therefore, include a full measure of the effects of anisotropy. The quality of the prediction will depend, apart from the overall capability of the model, on the relative importance of anisotropy in comparison with other factors. Data is not provided on extension behaviour in the usual sense of the stress path in extension making 180° with the axial compression path in the octahedral plane for clays X and Y nor is data provided for extension with hydrostatic unloading for clays X,Y and the Ottawa sand.

Perhaps the main effect of some of the limitations cited above is that they may have handicapped some of the more comprehensive theories that could have incorporated the missing information. The predictions of the simpler theories which ignore anisotropy or differences in compression or extension behaviour, for example, were not affected by the missing information. Despite some of these limitations, the test data provided for the calibration and evaluation of the stress-strain models is the finest collection of general data yet made available for such an exercise.

STRESS PATHS AND DEFORMATION FIELDS

The stress paths both for model calibration and response predictions, in the case of clays X and Y, were specified by the parameter m defined by

$$m = \frac{\sigma_2 - \sigma_3}{\sigma_1 - \sigma_2} \tag{2}$$

Therefore, compression tests were defined by m=0 and extension tests by m=1.

For the natural clays the stress-strain curves were to be predicted for
3 other values of m between 0 and 1 at various initial effective confining
pressures σ_c'. As pointed out earlier, all these paths lie in a $60°$ sector of
the octahedral shear plane. This limited separation of the stress paths
would not be expected to result in radically different stress-strain response
for samples with the same initial consolidation stresses. The possible
limits are shown in Fig. 9 of Ref. 4 for clay X. This narrow range would
probably be representative of the scatter that might be expected in data on
natural clays at intermediate values of m. Therefore, for this clay it may
be difficult to interpret the reasons for moderate divergences between predictions and experimental data. Divergence may be due to scatter in the experimental data, incomplete representation of anisotropy because of deficiencies
in the data for calibration or limitations in the stress-strain model.

In the case of clay Y, the divergence between the stress-strain
curves for m=0 and m=1 at a given initial consolidation stress at low strains
is negligible but at large strains the divergence is greater than for clay X
(Fig. 16, Ref. 4). It would be expected for these clays that any model well
calibrated to a stress-strain curve interpolated between those for m=0 and
m=1 would give a very good representation of the required stress-strain
responses at stress levels less than about 70% of the failure strength.
Only after pronounced yielding occurred would such a simple approach prove
inadequate for clay Y.

In the case of the Edgar Plastic Kaolin, the principal stresses
are defined by the conventional terminology that σ_1 is the major principal
stress. The principal stresses are applied to the sample so that σ_2 always
remains horizontal and the angle that σ_1 makes with the vertical axis of the
sample is specified, i.e., the orientation with respect to the axis of consolidation is defined. It is not specified how these stresses were applied.

One reasonable assumption is that a cubical test sample was cut from consolidated blocks with sides oriented along the principal axes and that the sample was then tested in the true triaxial apparatus. If this is done then the principal strains which are determined from boundary displacements of the equipment are forced to be coincident with the principal stresses.

In the case of Ottawa sand, simple shear states are defined by $m=0.2$, 0.5 and 0.8. These states should not be confused with the deformation modes imposed in the more conventional simple shear apparatus in which deformation is induced by the application of a shear stress. For isotropic materials the distinction may be irrelevant but for anisotropic materials the different modes of _imposed_ deformations may have a bearing on the measured strain field.

ASSESSMENT OF PREDICTIONS AND MODELS

The position papers (8,2,6) and the papers comparing predictions with experimental data (8,1,5) should be consulted for a detailed explanation of the stress-strain models and an accurate picture of the quality of the predictions made using them. It is not useful or necessary to repeat much of that information here. What is offered in this assessment is a distillation of the extensive analysis of the predictions by our discussion group and our attempts to relate differences between predictions and experimental data to problems with the data base and the assumptions and structure of the stress-strain models. Extensive agreement was reached by the discussion group on most of the important topics discussed. However, the available documentation required some personal interpretation and judgement.

Critical State Theory

Wroth and Houlsby (8) used the Modified Cam-Clay Theory (3) to make predictions for clays X and Y and the Edgar Plastic Kaolin. They did not offer predictions for Ottawa sand because, as they put it, "the Modified Cam-Clay Model is known not to be suitable for modelling sand". The predictors introduce a minor alteration to the Cam-Clay Model. The consolidation and critical state lines are now considered to be straight lines in $\ln V - \ln p'$ space instead of the $V - \ln p'$ space in which V is the specific volume. This change results in a slight simplification of the mathematics and ensures that the bulk modulus is truly proportional to pressure.

The procedure for determining the yield surface and the critical state line is demonstrated for clay X and illustrates the judgement and understanding of test mechanics that is fundamental to calibrating a stress-strain model. The tests on clay X were of the load control type. Therefore, each load increment was applied under essentially undrained conditions and the total load maintained constant until equilibrium was reached under drained conditions. The effective stress path, the strains developed and the final failure load all depend on the sizes of the load increments. The failure load will be less than that achieved under a slow strain controlled test. The critical state model (Fig. 5, Ref. 8) is very effective in illustrating these differences in behaviour.

The uncertainty regarding the proper failure load requires some judgement in selecting the slope of the critical state line and the height of the elliptical yield surface which must cut the critical state line at $p'_c/2$. Fig. 9 of Ref. (8) indicates the kind of compromises that are made to obtain an adequate fit to the calibration data. The critical state line and the yield surface are determined primarily by compression data. The extension side in Fig. 9 is a simple isotropic generalization obtained by rotating the yield surface about the p' axis.

The Cam-Clay Model is based on data from triaxial compression tests on isotropic clays. Therefore, at its present state of development it cannot model anisotropic behaviour, the responses of soils that are significantly affected by variations in the intermediate principal stress or materials with pronounced differences in responses to conventional compression and extension stress paths. The compression and extension data on clay X supplied for model calibration bound only a $60°$ sector in the deviatoric plane and all the stress paths for which predictions are required lie in this sector. Therefore, if the calibration data is reasonably well modelled one would expect the predicted stress-strain response to be reasonably good.

Wroth and Houlsby (8) show very good approximations to selected calibration data but do not show the predicted stress-strain response for the stress paths specified by the various values of m. They state that they used the parameters derived from the extension and compression tests to model the tests at intermediate m values and that they found the effects of values of m to be small within the specified range of $0 \leq m \leq 1$. However, no comparisons of predicted and experimental values are given for clay X or clay Y except for $\sigma_c' = 40$ psi, m=0.

The K_0-consolidated kaolin is not only anisotropic but exhibits distinctly different behaviour in conventional compression and extension modes. The Critical State Model is not expected to be effective in modelling this range of behaviour. The expectation is confirmed by comparing the performance of the model in matching the calibration data. The match in compression is rather good but very poor in extension (Fig. 16, Ref. 8). This result was expected by Wroth and Houlsby who carried out the calibration merely to demonstrate this point. No other predictions were carried out using the Critical State Model.

The Cam-Clay Model has been under research and development for about 25 years. Within recent years it has been incorporated into finite element analyses and is finding increasing use in certain applications offshore. In view of this history, it is somewhat disappointing that its predictive capability is not more clearly illustrated by giving the predictions for the stress paths specified for the predictors. This is particularly true in the case of the K_0-consolidated kaolin which would be representative of many clays in nature and typical of some of the applications of the Cam-Clay Model in the field. Many of the specified stress paths would violate the assumptions of the Cam-Clay Model but it would have been interesting to see how sensitive the model might be to deviations from ideal conditions. One of the attractions of the Cam-Clay Model is its relative simplicity. This simplicity is bought at the price of restrictive assumptions, chiefly isotropy. If the simplicity is to be maintained in applications it is imperative to document the effect of deviations from the assumptions.

An alternative approach is to extend the model to more general conditions. In the group discussion various suggestions were made. The deviator stress, $q = \sigma_1 - \sigma_3$, could be generalized to the octahedral shear stress, the yield surface could be rotated about the K_0-consolidation line instead of the p' axis; different semi-ellipses could be used in compression and extension. Other ideas are given by Wroth and Houlsby in their position paper (8), which should also be consulted for a detailed discussion of some other limitations of the theory.

Our group came to be conclusion that the Cam-Clay Model is one of the best models available for making qualitative predictions without computations about the nature of the response of clays to loads. It is quite effective in demonstrating the different responses of heavily and lightly overconsolidated clays, the probable range in pore pressures, the different effects of different stress paths on volume change, effective stresses and shear strains. It seems an excellent structure for integrating many aspects of test data and relating the fundamental parameters controlling soil behaviour under load.

Hyperbolic Stress-Strain Model

Duncan (5,6) based his predictions of stress-strain response on the hyperbolic stress-strain model developed by Duncan and Chang (7). This model is based on incrementally linear elasticity. The original model used as elastic constants, the tangent values of Young's modulus, E_t, to the hyperbolic stress-strain curve and constant values of Poisson's ratio. A more recent version of the model using a bulk modulus dependent on the confining stress instead of the constant Poisson's ratio was used in making the predictions. This change increases the accuracy of the model.

The model has been applied extensively in engineering practice for the analyses of dams, excavations and various types of soil-structure interaction. Therefore, like the Cam-Clay Model, its strengths and limitations are well understood. The position paper by Duncan (6) gives a very lucid explanation of the theory and sets out very clearly its limitations. Many of these points were discussed in detail by our group and Duncan's presentation of the limitations of this model reflects closely the consensus of the group.

The parameters of the hyperbolic model are obtained from conventional triaxial compression and extension tests only and the soil is assumed to be isotropic. Therefore, in theory, it is inapplicable to most of the stress paths for which predictions were required. Duncan, however, applied the model to these stresses and his approach to getting as good a fit as possible is worth studying for its practical applications.

First, he takes explicit account by judgement of the scatter in test data although no measure of such scatter is provided. The parameters of the model are determined using data from the conventional stress paths. Using these parameters, the stress-strain response in the non-conventional paths were determined and compared with the experimental data. On the basis of these responses final adjustments were made in the parameters and estimates of uncertainty were established which were applied to all predictions. For clays X and Y, since the stress paths for prediction were defined by values of m and the calibration data was given for m=0 and m=1, interpolation of the model parameters for intermediate values of m was a reasonable approach for the stress paths specified for the prediction exercises.

At normal working stress levels, i.e., below 60%-70% of the failure stress, the predictions of the hyperbolic model are in many instances quite good. At larger stresses and higher strains the predictions are less satisfactory and in some instances the range of uncertainty is too large to be very meaningful. Since the method uses elastic analysis, shearing deformations do not result in volumetric strains and so such strains are only predicted for hydrostatic loading and unloading. Thus, no volume changes are predicted for the constant mean normal stress tests.

The response to drained and undrained loading is determined by determining the parameters of the model under the appropriate conditions. In general, the success of the hyperbolic model rests on duplicating the field conditions as closely as possible in the laboratory tests to determine the model parameters. These parameters are not fundamental soil properties but correlation co-efficients applicable for a limited range of conditions. Obviously, the more extensive and thorough the testing the better the correlations that will be achieved.

Akai, Adachi and Oka Model

This is the most comprehensive of the three models assigned to our group for discussion. Basically, it is a critical state model extended to include time effects. The model was developed especially for the symposium. Therefore, even the developers have little experience with its use. Discussion of this model was devoted mainly to developing an understanding of its structure and the group did not reach the point of offering constructive criticism of it. The lower part of the stress-strain curves were predicted fairly well for clays X and Y but the prediction at larger strains greatly underestimated the shear resistance of the clays. The volume changes except for a few stress paths for clay Y were greatly underestimated.

The deviator stress vs. axial strain curves for the kaolin were probably the best predictions of the model but any plots involving the octahedral shear stress were generally poor even though the model was developed in terms of the second invariant of the deviatoric stress.

The predictions for the Ottawa sand, especially those derived using averaged soil properties, are better for the higher values of the stress parameter m. In general, the predictions are qualitatively correct but it is difficult to reach an objective conclusion of the overall effectiveness of the model for sand. It will be recalled that Wroth and Houlsby expressed some reservations about the applicability of a critical state model to sand.

The prediction exercises were the first application of this new theory. At such an early stage in its development and verification it is not possible to offer a critical assessment of the theory. It is obviously very comprehensive, not only because the critical state model is generalized by using the second invariant of the deviator stress rather than the usual deviator $q = \sigma_1 - \sigma_3$, but because it attempts to incorporate rate effects. This added potential would be extremely useful in many applications particularly with regard to offshore construction in very soft clays. Therefore, future developments of this theory will be awaited with interest.

General Comments on Workshop

The workshop was highly successful in generating a clearer and more widely shared understanding about the requirements of a comprehensive stress-strain model and the kind and quality of data needed to verify its capability. The data base for calibrating models must be comprehensive enough for the calibration of sophisticated stress-strain models to their full potential and the data for testing predictive capability must involve stress paths that will clearly discriminate between restricted and general stress-strain models. Some of this data was lacking for the workshop so that most of the sophisticated models operated in a restricted mode, for example, assuming isotropic behaviour even for the K_0-consolidated kaolinite.

Although sophisticated equipment is necessary to provide the variety of stress paths required for testing stress-strain models it would be useful also to provide duplicate test data for the simpler stress paths such as conventional compression and extension data from conventional triaxial test equipment. Also, a measure of the scatter in the test data is required to provide a criterion for judging the divergence of predictions from test data.

The procedure adopted at the workshop of splitting the members into groups for the discussion of particular predictions and theories resulted in the group members becoming very knowledgeable about the particular theories assigned to them and led to very thorough appraisal of both theories and predictions. On the other hand, it denied the group members the opportunity for any meaningful disucssion of other predictions and theories and precluded general theoretical discussions about the assumptions and foundations of the various theories. Perhaps one of the three days might have been usefully devoted to discussions of selected general questions which developed during the group discussions.

One of the invaluable benefits of the workshop was the opportunity to cross-examine each predictor on the foundations of his theory and how he went about making his predictions, about the role that judgement played in achieving his results, and his approach to handling stress paths not lying within the strict theoretical bounds of his theory. In general, the time and the opportunity to pursue those troublesome points usually glossed over in publications or never referred to at all was one of the great benefits of the workshop.

The workshop did not crown a king of stress-strain relations. Despite very different underlying assumptions a number of theories gave predictions of comparable quality. The lack of discrimination between sophisticated theories may result from some of the deficiencies in the test data referred to above or may reflect the lack of basic physical correspondence between models and reality, resulting in models that are highly dependent on direct correlation of model parameters rather than on the measurement of physical properties of the soils.

The workshop on stress-strain relations focussing on testing the predictive capability of a wide variety of stress-strain model was a unique and highly successful event. Its impact on the development of stress-strain relations for soils will be far reaching not only through its direct influence on those who attended but also through the impact of the publications of the proceedings. The success of this event should provide the necessary encouragement for a second workshop of this kind at an appropriate time.

ACKNOWLEDGEMENTS

I wish to express my appreciation to the members of our discussion group for a stimulating workshop. On behalf of the group, I express our appreciation to Professors Adachi, Duncan and Wroth for lucid and entertaining presentations and for their good-humoured submission to probing and pointed questions. I am very grateful to Professor Harvey Wahls who kept careful notes of the proceedings and to Professor Saada of the Organizing Committee for his considerable help with discussions. I express my appreciation to the Organizing Committee for their invitation to act as chairman of a discussion group and for their heroic efforts to ensure a successful workshop.

REFERENCES

1. Akai, K., Adachi, T. and Oka, F. (1980), "Predictions and Comparison", Proceedings, NSF/NSERC North American Workshop on Plasticity and Generalized Stress-Strain in Geotechnical Engineering, McGill University, Montreal, May 28-30, Vol. 1, pp.

2. Akai, K., Adachi, T. and Oka, F. (1980), "Constitutive Models for Clays and Sands", Proceedings, NSF/NSERC North American Workshop on Plasticity and Generalized Stress-Strain in Geotechnical Engineering, McGill University, Montreal, May 28-30, Vol. 2, pp.

3. Burland, J.B. (1967), "Deformation of Soft Clay", Ph.D. Thesis, University of Cambridge, Cambridge, U.K.

4. Dafalias, Y.F., Herrman, L.R. and Denatali, J.S. (1980), "Prediction of the Response of the Natural Clays X and Y using the Boundary Surface Model", Proceedings, NSF/NSERC North American Workshop on Plasticity and Generalized Stress-Strain in Geotechnical Engineering, McGill University, Montreal, May 28-30, Vol. 1, pp.

5. Duncan, J.M. (1980), "Strains Predicted using Hyperbolic Stress-Strain Relationships", Proceedings, NSF/NSERC North American Workshop on Plasticity and Generalized Stress-Strain in Geotechnical Engineering, McGill University, Montreal, May 28-30, Vol. 1, pp.

6. Duncan, J.M. (1980), "Hyperbolic Stress-Strain Relationships", Proceedings, NSF/NSERC North American Workshop on Plasticity and Generalized Stress-Strain in Geotechnical Engineering, McGill University, Montreal, May 28-30, Vol. 2, pp.

7. Duncan, J.M. and Chang, Y-Y. (1970), "Nonlinear Analysis of Stress and Strain in Soils", Journal of the Soil Mechanics and Foundations Division, ASCE, Vol. 96, No. SM5, September.

8. Wroth, C.P. and Houlsby, G.T. (1980), "A Critical State Model for Predicting the Behaviour of Clays", Proceedings, NSF/NSERC North American Workshop on Plasticity and Generalized Stress-Strain in Geotechnical Engineering, McGill University, Montreal, May 28-30, Vol. 1, pp.

9. Yong, R.N. and Ko, Hon-Yim (1980), "Information Package for Workshop", Proceedings, NSF/NSERC North American Workshop on Plasticity and Generalized Stress-Strain in Geotechnical Engineering, McGill University, Montreal, May 28-30.

NOTATIONS

D_r	=	relative density
E_t	=	tangent modulus
K_0	=	σ'/σ_3' for 1-dimensional consolidation
m	=	parameter for specification of stress path
p'	=	mean normal effective stress
q	=	deviator stress
V	=	specific volume
X,Y	=	designations for two natural clays
$\sigma_1, \sigma_2, \sigma_3$	=	principal stresses
σ_c'	=	consolidation stress
$\varepsilon_1, \varepsilon_2, \varepsilon_3$	=	principal strains
ε_v	=	volumetric strain

NORTH AMERICAN WORKSHOP ON PLASTICITY
Working Group 4 - Results of Session Discussions
R. F. Scott, Chairman

I shall address my remarks in two areas: First, the results of the general discussions that took place among the group on three separate occasions and second conclusions in summary form resulting from the group discussions. A point to be made is that the group generally agreed that the conference was an excellent idea and that a great deal of credit is due to the organizers for both the initial idea and the way in which the material was prepared and presented to the predictors and discussors. I think that it is obvious that this conference would not have been thought of ten years ago and probably would not have been possible as little as five years ago. It has been apparent for a long time that the completely empirical techniques of soil mechanics could not continue as they were in the face of modern computational power and that it was desirable to upgrade the theories of constitutive behavior of soils in order to account as realistically as possible for soil behavior. This is also necessary from the point of view of supplying the computer with the necessary basis for calculations.

In the VIIth International Soil Mechanics and Foundation Engineering Conference at Mexico City in 1969 there was a specialty session on stress-strain relations in soils. It is noteworthy that at that time this was the phrase that was selected to describe the field of endeavor discussed in this conference today. In those days the stress-strain relations referred principally to the single dimension of axial stress and corresponding axial strain in a triaxial test. Behavior of material in three dimensions was only a thought in the minds of a few investigators at that stage. To the majority of soil engineers the words "constitutive relations" or "state variables" were unknown phrases. By the time of this meeting in Montreal it was possible to assemble a large and knowledgeable group both of experimenters and attendees who all understood what the problems were and what the sessions were intended to accomplish. Not only was there a general audience present who were prepared to argue and discuss some of the points made by the experimenters but it proved possible to assemble for the sake of the meeting no less than twelve different theoretical derivations of constitutive relations for clays and sands. It might even have been possible to accumulate more than that number but even so in the period that we are discussing this represents a remarkable advance in the state of the art.

This is the first meeting of its kind at which the experimenters were asked not only to describe their theories but to examine the prediction of tests with which they had not been supplied. For a conference the whole idea of supplying investigators with one set of data and asking them to use their theories to predict the response of tests kept hidden from them at the time of their calculations is novel. As a result of this meeting a number of changes have suggested themselves but with the inclusions of alterations to meet minor criticisms it is hoped that a meeting like this can become a regular forum for the testing and evaluation of new and improved constitutive theories. The suggestions themselves will be considered later.

The working groups consisted of four parties each of which included a number of experimenters ("predictors"). In group 4, discussed here, there were represented five of the constitutive relation theories which were being examined. The proponents consisted of Dr. Ansal, representing the "endochronic" theory of Valanis as modified to soils by Bazant and Ansal, Dr. Dafalias who came to discuss the theory which he and Herrmann had developed and referred to by them as the "bounding surface" theory, Dr. Kavazanjian who concerned himself with the model developed by himself and Mitchell, Dr. Vardoulakis who discussed the constitutive relation for sand proposed by Gudehus and his co-workers over a number of years, and Mr. Houlsby who was concerned with some of the aspects of the "camclay" model developed by Schofield, Wroth and co-workers over the last ten or twelve years. The last is perhaps the oldest of the constitutive models intended to represent a range of clay behavior.

The above investigators were all associated with the formal groups who had been invited to participate in this conference. However, in addition several other people had attempted to utilize one or more of the existing theories in the literature in their own attempts to model and simulate the test data given out by the organizers. This is a particularly valuable exercise since there can be a wide difference in the application of a theory by the people who invented it, and its use by an independent investigator whose knowledge of the relationships is confined to reading the papers written by the inventors. We shall discuss this area more later. Two of these studies were presented in group 4, one by Dr. Chang of the University of Massachusetts who had attempted to use both the camclay model and the model proposed by Lade. The other attempt involved the model (named "IMP" here) proposed for undrained clay behavior by Prevost based on the work by Mroz and Iwan. It was implemented by Scott and Bardet in a limited study of some of the test material behavior. Additional material related to constitutive relations but not directly involved with any of the theories was discussed by Dr. Leroueil of Laval University; he presented a number of results of experiments on clays which bore directly on the conclusions of the various experimenters.

Not all of the experimenters had examined all of the available models for the conference and so the order in which the discussions took place was established to give some continuity to the sessions. As can be seen in the preliminary material of the conference, the organizers had given the "predictors" data on the behavior of three clays and a sand named successively clay x, clay y, kaolin and sand. The endochronic theory (Ansal) and the bounding surface theory (Dafalias) had both been employed in an evaluation of the behavior of clay x and clay y, while the theory developed by Mitchell and Kavazanjian had been used for clay x and kaolin predictions only. The work of Gudehus and his co-workers which was discussed by Vardoulakis was applied to kaolin and sand and finally the camclay model had, as its name suggests, been applied to the three clays, clay x, clay y and kaolin, and this material was presented by Houlsby. Since it was some advantage to the audience to have continuity in the presentations, it was decided to begin with clay x and clay y as treated by Ansal and Dafalias and, since clay x had also been examined by Kavazanjian, the next logical presentation was by him on clay x and kaolin, having dropped clay y out of the discussion by this stage. Then, Vardoulakis took up the Gudehus work on kaolin and sand, omitting both clay x and clay y. The principal discussion of the camclay model was given by Wroth in group 3 and only some aspects of it were presented by Houlsby last in order of our agenda.

Different considerations were raised in the three separate discussions held by group 4. These feelings were distilled by the Chairman and Reporter and presented in general sessions for contemplation by the whole group. The three sessions addressed the following topics in turn: Session 1 - Here the predictors were presented for the first time with the results of the measured tests whose behavior they had to predict. They had not seen these results before and so the predictions were made blind as they should be. As might be expected, this caused a number of problems. The intention had been to have the predictors prepare their predictions of stress paths, stress-strain behavior, etc., on plots which would be compatible with transparencies of the test results handed out to the Chairman of each of the group sessions. Not all of the predictors had made their plots to the correct scale or had even prepared correct plots according to the conference instructions. This meant that in a number of instances comparisons were difficult to achieve on the spot and it was not easy to form appraisals of the goodness of fit of a number of the experimental results. However, this did not prevent wide-ranging and fruitful discussions from taking place on the technique of prediction and comparison. The topics for the first session then were: how were the coefficients of the particular model obtained; which of the tests supplied by the organizers were used in arriving at these coefficients, and which were discarded; how many of the various coefficients were used in the various tests, and what sort of sensitivity could be ascribed to them. Finally, after this general description of the application of a particular technique to the test results, the actual comparisons between predicted and measured data were made and the predictor gave a brief discussion of his opinions on how well the predictions came out. This was necessarily brief since the predictors had only a few minutes to compare their calculations with the measured values.

This first session was held in the evening of the first day of the conference during which all of the predictors had given short talks on the basic theory of their methods. As a consequence, the audience had been subjected to twelve different theories in the course of a rather long day session. There were a number of questions and some confusion regarding some of the methods and so in the evening I asked each of the predictors to give a brief summary of his theory for clarity before the group before addressing the topics of the session. This made for a long evening. It is not the intention of this summary of the work of group 4 to describe in detail the correlations obtained by the different experimenters since these are described in the papers which they have submitted since the conference was held. However, a brief summary of the main points of discussion is in order.

Ansal ("endochronic")

This theory had been applied to the x and y clays only. They had also worked on the kaolin and sand but there had not been enough time prior to the conference to finish the calculations to a state in which the results could be presented. One of the principal questions brought up by Ansal which was a topic of much of the subsequent discussion in other sessions related to the anisotropy of the clays tested. It was apparent from the test results that the clays were anisotropic but some inconsistencies were observed. This meant that the investigators had to make assumptions with respect to the isotropy and the direction of testing of the clays with respect to the axes of isotropy. Once these assumptions had been made, a mathematical optimization technique was used to obtain the coefficients required in the theory to fit the results. In a general subjective sense, it was apparent that the predictions of this theory in terms of stress correlations and volume change predictions were not very good but the volume changes were in the right directions. Ansal had to change two of his coefficients to account for different confining pressures in the tests.

Dafalias ("bounding surface")

After a description of the model in brief, Dafalias indicated that they had employed a sequence of three experiments, one normally consolidated, one slightly overconsolidated and one highly overconsolidated to arrive at the model coefficients required. He had assumed that the behavior was isotropic for the application of his model. He commented that the test data on clay x were obviously poor and inconsistent as indicated in the data that he had been given. In a subjective summary of the results which he gave, it appeared that his predictions for clay x were fairly good on the whole and for clay y somewhat better, both in terms of stress-strain behavior and volumetric performance. He felt that the poorer fit for clay x was a consequence of the anomalous behavior observed in some of the tests presented. In general the volume change fits were much better than the stress predictions. He had employed one set of parameters to describe all overconsolidation ratios for the clays.

Kavazanjian (Mitchell-Kavazanjian)

He had also noticed anomalies in the test results, particularly the lack of isotropy. In his predictions the clay x behavior was represented quite well; specifically the volume change predictions were remarkably good. These predictions applied to the clay x at all of the consolidation pressures and at the different m values. In the case of kaolin the predictions were good for both tests 2 and 12. Some of the predictions were very good indeed.

Vardoulakis (Gudehus and co-workers)

He presented no comparison of the test results since this was being done elsewhere by Gudehus but instead described some of the features of the model and the way in which the coefficients were developed. He had not examined the test results in detail.

Houlsby (camclay)

Again the question about the anisotropy of clay x from the test results was raised. Then the comparison of predicted and experimental behavior was examined. The results were not very satisfactory for clay x although the behaviorial trends were all clearly there. This led to a discussion of brittle and plastic behavior of clays.

In the general discussion centering about the individual presentations a number of points came up for criticism and comment and these will be given in detail at the end of this contribution. The second session was devoted to presentations by several researchers, including Leroueil. He discussed some data that he had obtained illustrating that all clays were anisotropic and that any theory in his opinion which did not include anisotropy would fail to predict the performance of clay. He also pointed out that there were questions relating to the overconsolidation of a clay. A clay may begin a test overconsolidated but end up in a normally consolidated state, and this raised questions about the use of various constitutive relations to predict the material behavior. In his opinion, it was still too early to apply constitutive relations to real soils and a lot more work had to be done before successful applications could be anticipated.

Following this Vardoulakis continued the presentation of some of the work that he had done after the development of the Gudehus constitutive relation theory. He had directed his efforts towards a study of homogeneity in tests and the development of nonhomogeneity during the course of tests on sand. In these tests shear bands develop typically and result in unstable behavior or peaking of the stress-strain curves. He wanted to see if the shear banding could be prevented by better boundary conditions in the tests. In conclusion he felt that it was not useful to try to model tests with peaking behavior since he concluded that it was mostly a function of the test conditions. This meant that the modeling of that behavior and its inclusion in a constitutive relation with appropriate coefficients in a study of a boundary value problem would be completely erroneous.

Following these two discussions, Chang presented the results of his independent study of the camclay model as applied to the behavior of clay y. In the experiments the bullet-shaped yield surface according to the original camclay model was employed rather than the elliptical shape which is now currently in vogue. It turned out that the bullet shape gave too high shear strengths and volumetric strains. He pointed out that the model was originally evolved for very soft clay and clay y is probably too overconsolidated for a fair representation. The original camclay model had also been used to predict the kaolinite clay behavior. There were a number of differences between the prediction and the actual behavior; once again the shear strength was very different.

In response to questions, Chang indicated that the calculations and modeling for camclay had all been done by hand using a calculator and that the development of the fitting coefficients took one or two hours and the actual calculation of the material stress paths, etc., took another two hours or so by calculator.

Chang had also carried out a comparison on the behavior of the sand by using Lade's model. There were some interesting differences in the results of the fitting procedure used by Chang and by Lade in a separate session. Chang discovered that in fitting the test results he had used an elastic coefficient which had turned out much smaller than the value that Lade had used. The difference was ascribed to Lade's experience in using his own model for fitting test data whereas this was the first time that Chang had done it. Chang pointed out that it took a student about a month to perform the calculations, starting from Lade's papers on the subject, of which about 50% of time was reading and 50% was calculating the coefficients. These were all done by hand using a calculator. Each stress path took approximately two or three hours to calculate, again by hand. Of course, if a computer program was written, the actual calculations would proceed much faster.

Following Chang, Scott described the results of calculations which Bardet had made in order to attempt the fitting of the kaolinite clay. The model used here is the one proposed by Prevost (we have suggested the name IMP for it) with some modifications made by Bardet to automate the fitting procedure since it is subject to a considerable amount of personal judgement if it is done by hand. Bardet had no previous experience in this kind of fitting. The comparison of the predictions with the test behavior was quite good in most of the cases for different m values. In the calculations a computer program was used and the fitting was achieved at the cost of a few dollars for ten circles.

The last of the three group sessions was devoted to the research needs of the various constitutive models and some guides to their employment in practice. At first the questions were addressed: what tests were required for the use of the models, what tests would each inventor ideally like to have in order to improve his model, and how these experiments could be achieved in practice for the prediction of field behavior. The other part of the final session related to the practical applicability of each of the models. How could a model be implemented on a computer program, for example in finite element form? How difficult was it to do this? How much storage space was required for typical problems and what were the differences in running time among the various relationships proposed? Again, the considerations of the different predictors are presented in summary form.

Ansal made the point that the endochronic theory, although complex, needs little machine storage since the entire previous history of the element does not need to be remembered, whereas for example in the model proposed by Prevost a great deal of information needs to be recalled and stored by the machine for each stage in a stress path. Up to this meeting the endochronic model had been used mostly to model different normal types of clays and sands. They had not worked before with sensitive or overconsolidated clays as had been the experience with this study. Modifications will be required to take into account overconsolidated and sensitive behavior. They typically use five parameters and employ three tests in an optimization technique in order to obtain the coefficients of the five parameters. They considered that they obtained a great deal more data for this conference than they would typically expect to receive in practice. In fact, they had done some simulations of the dynamic behavior of a small dam for the Waterways Experiment Station and had received much less data than that associated with the conference. They would prefer to have cyclic tests added to the material given to them for prediction and, of course, would prefer to see not one test for each stress path but a variety of tests because of their optimizing procedure. Ansal noted that in the course of the conference they discovered that the way they determined the coefficients of the model is not a suitable technique for practical purposes. They intend to simplify the material functions and try to correlate the various parameters with specific tests. They need to develop a finite element model to try to correlate with field problems to see how successfully that can be implemented. The theory, although complicated conceptually, is simple in application. They tried to keep the various constants in terms of single numbers so that they are kept constant all the time to model the behavior of different clays.

Dafalias would like to have better cyclic loading tests and tests with different loading paths as well as heavily overconsolidated soil tests for clays in order to achieve better fits with the bounding surface model. Once again because of the choice of parameters involved in his model, the computer storage requirements were quite small and he felt that it would not be a difficult job to implement a finite element program. Although the theory refers to eight parameters, he generally works with basically six of them, four of them being the usual ones associated with the camclay model. The model is particularly well adapted to studies of cyclic loading with pore pressure generation. Dafalias has not so far had much experience in trying to model sands with his bounding surface model. They also at present still have trouble with predicting pore pressures and plan to make some improvements in that area.

Kavazanjian pointed out that their model is more a behavioral one than a mathematical model. All of the calculations for this conference were done by hand and some of their results were based on general material behavior as indicated by index properties. Their general philosophy is to use any information about the soil they can get to help guide them in the selection of the coefficients. Again, as was the case with other models, their primary intention had not been to develop one for the specific purposes of this conference but their original intention had been to look at time-dependent behavior as it related to the "stand up" time of cuts, and creep rupture under embankments or in excavations. Originally, it was intended for soft clays only and any other materials where time-dependent behavior was important. In general, in overconsolidated clay creep rupture is not important and so they were not concerned with it.

In each of these individual group sessions there were a variety of contributions from the floor which emphasized and clarified the speakers' comments. Frequently the questions resulted in considerable amplification and explanation of various difficult points in the different theories. In conclusion I would like to try to summarize the various substantial points that came out throughout the whole conference as they appeared to me. I will point to these areas which need further development or in which more or different work is indicated.

CONCLUSIONS AND RECOMMENDATIONS

First of all a certain number of general questions were raised at all of the sessions and this has some impact on future conferences of this type.

1. How were the tests done that were supplied to the predictors? What were the kinds of clays used? Where did they come from and what were their orientations in the tests? The same information is also required for the tests on which comparisons are to be based. A number of people made the point that in a real case in practice all of this information would be available for a prediction or for a modeling of a real boundary value problem.

2. There were problems with test inconsistencies in terms of volume changes and stress-strain paths. This meant that some tests were rejected by some of the experimenters and could not therefore be used in developing their coefficients.

3. The clay x clearly indicated anisotropic behavior but the extent of anisotropy was not clear from the test results. The behavior in two lateral directions were different and this led some of the predictors to conclude that the test samples had been cut at some angle to the direction of prevailing anisotropy. Since this was unknown it was impossible to include it in the theoretical developments.

4. This led to a great many questions on how the tests were performed and the apparatus in which the tests were carried out. Since it appeared part way through the conference that some of the tests had been done in hollow cylinder form, others in cubical form, a number of criticisms arose addressed to those specific tests as arbiters of material behavior. There are known problems, both mechanical and mathematical, associated with these tests and a number of experimenters felt that they were not reliable as a basis for calculation.

5. A consequence of this was a strongly expressed need to get uniform test methods set up. It might be necessary to develop new tests or to work more on existing methods in order to obtain the variety of information needed by some of the models. Clearly, in a number of cases models can utilize more information than can be obtained from standard triaxial tests and the more complex tests are necessary.

6. Associated with this was some criticism of the plotting of the test results and there was feeling that different ways of plotting them should be examined in an attempt to find the best way of presentating results for clarification of the performance of different constitutive law predictions.

7. In a separate area there were a number of comments, mostly from the audience, with respect to the constitutive relations themselves. This gave rise to many questions about the physical meaning of the various constitutive relations. There was a strong need expressed by various participants that constitutive relations to be useful and effective should be as closely tied as possible to the physical properties and behavior of the material. In this way numerical constants and the parameters would be directly related to aspects of clay or sand behavior for which one had a physical intuition.

8. An associated point is the number of parameters involved in a theory. Various audience members clearly felt that the fewer parameters and the fewer the number of variations required in those, the more applicable or useful a particular theory could be. This would also be directly related to the attractiveness of the theory to practitioners. There is obviously a tradeoff involved here in that a theory with only three or four constants cannot be expected completely to represent the behavior of sands and clays, normal and overconsolidated, anisotropic as well as isotropic, and in conditions of drained and undrained behavior. On the other hand, a number of constants of the order of fifteen or twenty and, in particular, constants that varied with the state of consolidation of the material or with the overconsolidation ratio and which had to be changed continuously during a test rendered a theory basically unattractive for everyday use. It is also worth pointing out that the present feeling of most of the participants is that comprehensive constitutive theories would never be extensively used in practice but would be mostly employed in the case of expensive, large or unusual structures to clarify difficult aspects of the behavior for engineering purposes.

9. These expressions of opinion were also directly associated with a desire to have the assumptions required for the various models clearly set out. Because of the time involved and the complexity of the theories expressed, the different predictors or experimenters involved in this conference could not always clearly set out the basic underlying assumptions and restrictions in their models. This made it difficult for other participants to assess readily the advantages and drawbacks of the different proposals.

10. This led to an expression of interest in obtaining or having set out a table of comparative data for the different models

in terms of: different parameters involved, the materials the model was applicable to, the tests required for determination of the coefficients, and possibly an estimate in the time involved in their determination, or in the running time of a standard computer program or standard test configuration using the model. It is to be hoped that this information will be made available at some time in the future preferably by an independent investigator who understands the ramifications of the different constitutive relations proposed.

11. Finally, along the lines of these general remarks, it is apparent that the subjective impression of the goodness of fit of the different models with the test data is inadequate for a realistic assessment. Instead, some kind of measure of fit is required. In the subjective assessments given above, in which a model was described to fit a particular test fairly well or not well, it becomes very difficult to assess the differences between different competing models. It is suggested, for instance, that for a standardized number of test points each model be assessed on the sum of the squares of the differences between the predicted points and the test values. However, this alone might be insufficient to characterize the behavior since its adequacy depends very strongly on what the user wishes to employ the model for. If he is interested only in the behavior one-third of the way towards failure where elastic behavior might predominate, then the initial slope of the model is of prime concern to him. On the other hand, if he is interested in employing the model to predict the failure of a footing or a foundation, then perhaps the thing of most interest is in how well the model anticipates failure conditions in the ground, and so the yield stress prediction is of the most importance. For future meetings, some attention needs to be paid to these points so that practitioners who do not have the time to study all of the models can pick out those which best fit their needs in particular cases without experimenters' bias being injected into the decision process.

Other general comments:

1. It was obvious from many comments by the inventors that they were employing their models in this conference for the prediction of the behavior of soils for which the model had not been originally intended. For example, the model was designed for insensitive, normally consolidated or remolded clays and they were required to give predictions for over-consolidated, sensitive soils. Once again, it would be nice to have a small table showing the limitations of the different constitutive models so that these possibly misleading

applications would not unwittingly be invoked by a practitioner.

2. It is very obvious that anisotropy plays a large part in the behavior of soils and it is desirable that it be incorporated in any constitutive relation.

3. Some feeling surfaced that we are not yet ready to model constitutive behavior, but I think the whole tenor of this conference rejects that notion.

4. There is a good deal of clay behavior and probably that of sand also which is not yet clearly understood. If it is not understood, it is impossible to model it correctly.

5. The number of inconsistencies in the tests indicated that each stress path in the test data supplied should be the result of the average of a number of individual tests. This would avoid the chance of the predictor fitting his coefficients to a test which was erroneous or was performed on an unrepresentative sample. Some more consideration has to be given to tests which exhibit peaks in the stress-strain relationships. It is still not clear that these are genuine material element behaviors, since they are generally associated with the development of shear bands. Obviously, it is desirable to develop constitutive relations which when applied to a particular boundary value problem, for example, the triaxial test, permit the development of shear bands in the model.

6. It would be desirable to develop constitutive relations in which the coefficients are not subject to the judgement of the individual user. It is apparent that in a number of the models presently used the original inventor is much more skillful at applying the model than a person reading the relevant papers. Experience seems to play a large part. It should be possible to code the fitting procedures so that the judgement of the user does not play a part. This is a controversial point, since in all cases in practice the judgement of the practitioner of all kinds of material properties not directly relevant to the test will probably play a part in his assessment of material parameters. Since some of the test calculations were done both by inventors and by users, it appeared that even with the same data base and with the same fitting procedures different people get somewhat different results for the values of the coefficients.

7. It would be nice to have some indication from the inventors of which problems they think their constitutive relations are most applicable to. For example, a particular method might give a good result in the calculation of settlements of footings or foundations, whereas another technique might

best lend itself to a calculation of the stability of an embankment.

8. The more complicated the soil behavior, it almost goes without saying, the more tests are required to describe it.

9. A corollary to 8) is that the more you ask for in a constitutive equation, the more complicated it gets and the more parameters you have to adduce. This might be a law.

10. Some of the tests required for modeling some of the constitutive equations are perhaps not obvious. For example, it seems to be desirable to employ cyclic loading or at least loading/unloading stages of a test in order to establish the features of some constitutive relations even though they may be used to predict monotonic loading behavior subsequently.

11. There was a general feeling that the closer the model is to established physical properties of a soil, the more attractive it is to practitioners.

12. Although a number of models endeavored to characterize undrained behavior of soils, it should be pointed out that in real life there is no such thing as an undrained soil in a field problem. In all circumstances in a field problem there is always going to be redistribution of pore pressures and some drainage effects. This will occur even when there is no significant displacement occurring at the boundaries. An exception is possibly the behavior of very heavily overconsolidated stiff clays.

13. There was a general feeling that the search for the "ideal" ("10") model will continue for some time. The models presented all had defects of one kind or another.

14. There exists a real need for pure tests done by uncommitted investigators as precisely as possible to characterize material behavior. The behavior that the soil exhibits in the laboratory still may not represent that of the same material in the field, because of laboratory test boundary conditions.

15. It is necessary ultimately to compare the predictions of models based on the various constitutive relations with the performance of soil in field tests. There is a great need for believable, correctly-performed, accurate field tests on soils as homogeneously situated as possible. Since this is expensive and difficult to accomplish, one possible solution may be the utilization of centrifuge model tests as a substitute for field standard tests.

16. It also seems likely that the tests which are done in laboratories based on traditional soil tests are not large enough to represent adequately the material behavior for constitutive relation tests. Some consideration should be given to the establishment of much larger tests.

17. In many real life circumstances for which predictions of stress and deformation are required, the information that is required by constitutive relations will always be lacking. Routine boring logs, blow counts on the soil, and a number of index tests on jar samples are all that is available. It is necessary to develop techniques of making such limited information available for use in appropriate, perhaps limited constitutive relations for predictions in such circumstances.

18. It would seem, therefore, that there is a place in the soil engineering world for a large range of constitutive relations extending from those at a lower end which accept simple results including index properties of the material, or some limited field test, to those at the higher end for which all sorts of complicated tests, perhaps of a kind not yet developed, will be required.

All these considerations give rise to the following list of recommendations which may be controversial but which in my opinion are required for the further development of the area of constitutive relations.

A. Better laboratory tests are required. The boundary problem still remains with us and efforts should be made to eliminate it. It seems likely that the better tests should also be larger tests so that statistical scatter in small sample testing be avoided.

B. A data bank of suitable approved, filtered and checked tests is urgently required. At the present moment, experimenters must rely for their data on tests done by other experimenters in different laboratories and on a wide range of non-standard apparatus. Each of the tests that has been done has usually been performed by an experimenter anxious to prove his own techniques, or his own constitutive relations. This leads to a question of bias in different experiments, which places a difficult task on the modeler trying to assess the usefulness and validity of the experimental results. I think it is no longer possible for all developers of theory to carry out and follow through on their own tests. It should be possible for these people to have a set of correct and consistent data to draw upon. For example, many of the tests utilized have been those performed at Imperial College, London, or at Cambridge University or at MIT in past years. It would be very desirable if one group or agency collected these together, sorted them out, found where there were gaps in the data, and performed new or different tests to fill in these gaps. Such a process would be difficult and complicated but ultimately very rewarding. I would suggest that it be undertaken by a group such as those at MIT, for example.

C. It seems desirable that efforts continue in the development of constitutive relations, particularly in the direction of tieing them to physical reality for the different materials involved. In each case there is a desirability for the constitutive relation to be as economical as possible in its use and development of parameters. Dafalias describes this very neatly as the development of "parsimonious" theories.

D. A great deal of effort has been expended in this conference in comparing predictions with test results carried out on triaxial or hollow cylinder or cubical samples. The behavior of soil in the laboratory is one thing and in the field may be quite another. Another exercise that is required would be the comparison of different constitutive relations, based upon measured soil properties, with the performance of a field boundary value problem. For this a properly carried through series of field experiments is necessary, as necessary as properly carried out laboratory tests. At present, these are not available at all and no effort is being devoted to performing them. If at all possible, this situation should be remedied and an organizing agency set up to provide a uniform and consistent series of field experiments covering the whole span of soil interests. Although this goal is ambitious and unlikely to be achieved, some limited program of field tests is very desirable and should be studied. A possible substitute for field tests would be a controlled series of boundary value problems carried out experimentally in a centrifuge as large as possible. These could serve as controls for the evaluation of constitutive relations applied to boundary value problems.

E. There is an educational need to persuade industry to upgrade into use of more sophisticated constitutive models such as the ones that have been discussed here and to educate it into more aspects of solid mechanics that are needed to analyze and utilize these models. This kind of information must be disseminated to undergraduate and graduate students at universities, since the traditional soil mechanics training is not adequate for this purpose.

F. An effort such as this conference should be repeated at regular intervals perhaps of the order of two or three years.

6. SYNTHESIS AND RECOMMENDATIONS

by Raymond N. Yong and Hon-Yim Ko

General Remarks

In the previous sections, the predictions of response of the "test soils" subjected to simple and moderately complex loading paths, using the various models, have been shown and discussed by the various chairmen of the working groups. It should be noted that all Predictors were given the opportunity to test their respective models and to discuss the more specific requirements and capabilities of their models—both in the Plenary Sessions and particularly in the Working Groups. The discussions and comments arising therefrom have been very well documented and summarized by the Working Group Chairmen.

Not all the models were applied for prediction of the performance of all soils tested. In part, this was due to the fact that some models were specifically designed or developed for cohesive or cohesionless soils and therefore could not be easily or usefully used for the "other" type of soil. In some instances, models developed for general application to one kind of soil have been "adapted" for use in the "other" kind of soil. The specific soil performance predictions made by the various Predictors for this documentation, for the tested soils, are shown in Table 1:

SYNTHESIS AND RECOMMENDATIONS

Table 1. Soil Performance Predictions

	Predictions Made for Soils			
Predictor	Kaolinite	Sand	Clay X	Clay Y
Duncan	x	x	x	x
Kavazanjian/Mitchell	x		x	
Saleeb/Chen	x	x	x	x
Bazant/Ansal				
Mizuno/Chen	x	x	x	x
Wroth*				
Lade		x		
Baladi	x	x	x	x
Dafalias/Herrmann			x	x
Prevost	x	x		
Akai/Adachi	x	x	x	x
Gudehus	x	x		

It is important to observe that the "failure" of any particular predictive model to accurately model the performance of a given test soil may or may not be directly attributed to the inadequacy of the predictive model itself. The lack or insufficiency of direct specific test input information can be cited as one of the recurring items in the various discussions pertaining to model requirements. On the other hand, it should also be noted that the "success" of a predictive model to model

---

*No written documentation was provided by this Predictor, although predictions were made at the Workshop. See chairman's report from Discussion Group 2.

the test soil performance--where some other predictive models have failed to do so because of data insufficiency--should not necessarily be attributed to the "genius" of the predictive model. The direct quotes from the Working Group Chairmen are in order:

1. From Finn - "The WORKSHOP did not crown a king of stress-strain relations. Despite very different underlying assumptions, a number of theories gave predictions of comparable quality. The lack of discrimination between sophisticated theories may result from some of the deficiencies in the test data referred to above, or may reflect the lack of physical correspondence between models and reality, resulting in models that are highly dependent on direct correlation of model parameters rather than on the measurement of physical properties of the soils."

2. From Poorooshasb and Selig - "Conclusions drawn from these comparisons must take into consideration the fact that good agreement between prediction and measurement does not necessarily mean a theoretically correct model, and vice versa."

In addition to the "correctness" of a predictive model,--for modelling of a particular test soil in a given loading situation --concern was expressed throughout the WORKSHOP on the fact that the clay soils tested exhibited "anisotropic behavior". The questions and discussions arising therefrom can be reduced to the form of two simple questions:

1. Can we trust the data? -- and the test techniques?

2. Can models developed for isotropic soils be adequately adapted to account for the apparent anisotropic quality of both material and test information?

While the above questions are indeed pertinent, one should perhaps ask the question, "Are _natural_ soils really isotropic?" If the answer is "No", it then becomes clear that not only should predictive models be developed to account for such happenstance, but that material characterization and test techniques themselves should also pay sufficient attention to such a phenomenon. The discussions and concerns for material anisotropy characterization and testing have been addressed in the ASTM Symposium on Shear Testing of Soils, STP-74 (1981), and will not be further developed herein.

In hind sight, it would appear that the charge for prediction of performance of the test soils should have included predictions of soil response to more complex stress path load/unload situations--akin to field problems. It is noteworthy that only a small number of attempts was made to predict the circular stress path performance of the sand tested. Perhaps too much was expected from any predictive model. While it can be argued that--to cite Scott--"The more complicated the soil behavior, it almsot goes without saying, the more tests are required to describe it," one could ask "How many tests? What kinds of tests? How sophisticated?" It is clear that the discussion arising therefrom will be never-ending. Perhaps the more relevant point for consideration is the degree of "accuracy" expected or required from any modelling technique utilized. The answer will eventually have to come from the application of these constitutive models to the solution of field problems, in which case the degree of accuracy required will depend on the type of structure as well as on its importance. Questions of this type were left out of the WORKSHOP and were handled by the ASCE Symposium on Applications of Plasticity and Generalized Stress-Strain Theories in Geotechnical Engineering.

In this WORKSHOP, we have seen the application of predictive models ranging from the behavioral type--e.g. empirical/phenomenological--to the more sophisticated cap and hypoelastic models. Casting aside comparisons made solely on the basis of quality of comparative predictions, and focusing on input requirements for the respective models and facility in handling of the computations required, it would be useful for the reader to seek to balance these three aspects of the problem of predicting soil response performance. The one significant point in the prediction of performance of soils that stands out is the importance of a prior knowledge of regional geology (for the natural soils tested) and history of performance of similar soils. In that regard, behavioral models can be seen to be most attractive. To what extent these models can or need to be extended for complex load/unload situation will remain in the domain of the model developers themselves or others seeking to "improve" the state-of-the-art.

The more comprehensive mathematical/analytical models can indeed be most powerful and useful. The results of the limited exercise conducted in the WORKSHOP can be misleading since not all the models were fully tested to their capabilities. It is clear, however, that much work remains to be done in the case of modelling of complex soil behavior--especially for natural soils subject to nonuniform complex stress situations. The need forcomparable development of material test input must also be stressed if these models are to be successfully utilized.

Summary and Recommendations

From the several very pertinent points and observations made during the course of the WORKSHOP, three distinct and identifiable areas of the problem can be identified.

1. <u>Material Property Input</u> - relating to the methods, equipment, technique, etc. used to provide the various kinds of material property input for application to the modelling procedures;

2. <u>Model Limitations, Simplifications and Capabilities</u> - ranging from requirements for isotropic behavior and simple load conditions to accountability or lack of physical correspondence.

3. <u>Application to Field Loading Situations</u> - relating to homogeneous and uniform stress situations to complex and load/unload situations with coupled shear applications.

The significant points for consideration can be summarized as follows:

A. Standard and well-conditioned physical tests for the production of calibrated input for the models. These must be well-defined tests with no significant influences from boundary testing effects.

B. There is a need to ensure that the data base available for calibrating the models for prediction of stree-strain behavior or of yield, to be comprehensive so that the various kinds of models with their numerous requirements can be satisfactorily calibrated. The tests should include simple and complex load/unload tests with shear application. It would be useful to provide a data bank which would include properly derived material property values that can be used by various investigators and researchers involved in development of analytical models--to test their models and also to improve the capability of their models for prediction of soil performance. In this regard, it would be helpful if studies could be generated which would provide for this kind of data and test results could be made available to the PROFESSION.

C. It would be useful to separate characterisztion testing from validation testing. Studies would be needed to identify what critical tests and measurements are needed and validation test procedures. The distinction between the characterization tests and validation tests lie in the criticality of the tests and measurements.

D. In the testing of the soil materials to provide for characterization or validation, material property accountability is a very large factor. In future tests it is indeed important and necessary to account for anisotropy, volume change characteristics, stress and strain path history, and soil uniformity. Also included in the requirement to identify the orientation of the principal stress axis in regard to the inherent fabric of the soil itself. Studies are needed which would begin to provide all these items.

E. Future research should be directed towards the characterization and quantification of soil anisotropy. Present procedures for quantification of anisotropy through the use of actual test loading methods are not satisfactory since the derived performance of the soil in response to tests along various imposed principal stress directions cannot separate stress-induced anisotropy from inherent material anisotropy. It is clear that stress-induced anisotropy can be in actual fact altered or preconditioned by initial inherent material anisotropy. Research studies should be initiated to quantify the initial inherent material anisotropy as a routine procedure, somewhat along the lines of present tests for such items as consistency limits.

F. Physical testing of soils should identify all the above items and also the kinds of stress and strain paths utilized. In that regard, there is need to provide for a standard type of test procedure, technique, etc. and also a standard uniform method or methods of representation of test results to satisfy all the above requirements.

G. In regard to modelling procedures and models, it is apparent that there is a need to ensure that the models are fundamentally sound and selective. In addition, the models must provide for computational stability and facility. Research is needed to determine whether the present phenomenological models now available are sufficient for the wide range of problems encountered. The various requirements as stated in the above points perhaps could be incorporated and factored into these phenomenological models.

H. In regard to the various forms of plasticity, cap, bounding and hypoelastic models, there is a great need to identify the physical relationships more accurately in these models. It is apparent that the more sophisticated these models are, the more parameters and items that need to be measured in any kind of test procedure. Research is required to fully determine the necessity for the wide range of requirements for the more sophisticated analytical/mathematical models. How well do these models address the problem? How well can they predict?

In summary, it is clear that there are several very significant areas for research and study needed in the next many years. These relate to:

1. Physical testing for determination of material properties--including equipment, instrumentation, standarization of methods of tests and presentation of

test results and generation of data and input for both calibration and validation requirements.

2. Characterization of soil material for testing--to include effects due to load and unload fabric anisotropy quantification and other items related to composition of the material.

3. Analytical models--to include the examination of whether the models are specialized for a particular test system, or whether they can indeed be used as a general modelling procedure for analysis and application of test results derived from various kinds of tests. It is clear that analytical models would indeed be more useful if they were freed from any kind of restrictive requirements for material test input not ordinarily met with the standard test techniques, and especially if they can allow for input from various kinds of tests not restricted to a research laboratory. Model development, therefore, should identify and indeed seek to develop procedures to allow for the above.

7. POST WORKSHOP COMPARISONS

by H.Y. Ko, J.C. Mould and R.N. Yong

After the WORKSHOP, the Predictors were asked to provide their predictions in a digitized form so that the comparison with the laboratory test measurements could be evaluated in a quantitative manner. In the following table, the Predictors who responded to this request are identified with the digitized data they supplied.

	Data Supplied For			
Predictors	Kaolin	Ottawa Sand	Clay X	Clay Y
Duncan	x	x	x	x
Mitchell	x		x	
Chen (hyperelasticity)	x	x	x	x
Chen (plasticity)		x	x	x
Baladi - Sandler		x	x	x
Defalias - Herrmann		x	x	x
Prevost	x	x		
Gudehus	x	x		

The following evaluation was made of the comparisons. Referring to Fig. 1 as an example, the predicted strains were compared with the measured strains by plotting the octahedral shear stress τ_{oct} versus the normal strains ε_x, ε_y, and ε_z. A quantitative measure of the difference was picked to be the area between the predicted and measured curves. This area was computed by the trapezoidal rule at each shear stress level for which a predicted strain was given by the Predictor. The area, which has the units of psi-percent, is positive if the magnitude of the predicted strain exceeded that of the measured value, whether the strains were compressional or extensional after the initial consolidation of hydrostatic loading. For tests where there was unloading and reloading, this definition of positive area continues to apply, with the cumulative strains calculated at each step of loading throughout the tests. The final cut-off points in the area calculations were set by the limits of the final experimental data point in both the stress level and the measured strain, as illustrated for a monotonic loading test in Fig. 1.

For each test evaluated, up to four stress level versus area curves were computed and plotted, for the normal strains ε_x, ε_y, ε_z and the volume strain ε_v which is the sum of the three normal strains. These curves are labeled X, Y, Z and T, respectively. A typical set of curves is given in Fig. 2, at the top of which are given the name of the Predictor (G.Y. Baladi), soil type (clay x), and the test identification (PC = σ_c = 20 psi, m = 0.5). Since all subsequent curves have the same horizontal and vertical axes as the ones being shown in Fig. 2, they will not carry labels. However, the scale for each curve may differ, since attempts to make maximum use of limited space called for varying the scale in plotting these curves. Thus, all curves have the scales identified with a number at the last tic mark on the axes. The units are psi for the vertical axis and psi-percent for the horizontal axis. Care should therefore be exercised in comparing different curves, while attention should first be paid to the scales of plotting for the curves being compared.

The comparison curves provided in this section are first grouped according to the soil, in the order of kaolin, Ottawa sand, clay X, and clay Y, and for each soil, in the order of the Predictors listed in the last Table. In the case of the predictions made by Duncan for clays X and Y, an upper bound (UB) and a lower bound (LB) prediction were made, as explained in his paper earlier. Thus, two sets of comparisons are made for his predictions.

While all Predictors were requested to submit their results in digitized form for this comparison, only those listed above responded.

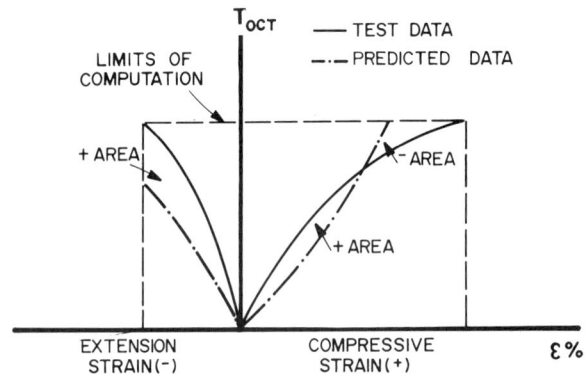

FIG. I   SIGN CONVENTION FOR COMPARISON

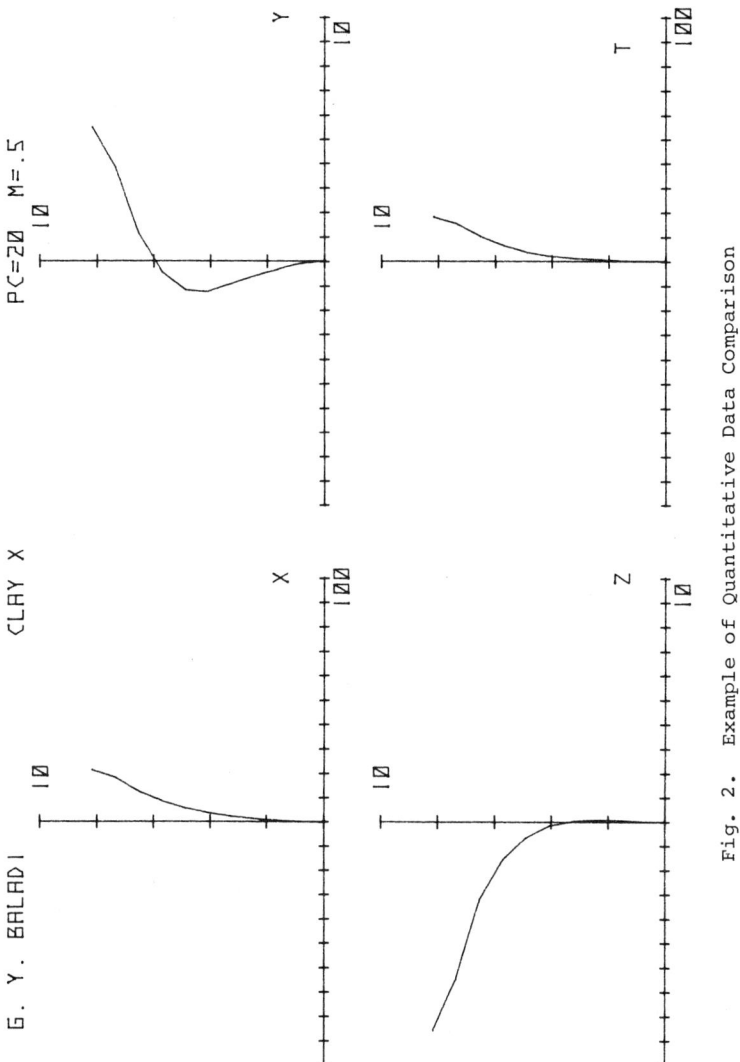

Fig. 2. Example of Quantitative Data Comparison

STRESS STRAIN FOR SOILS

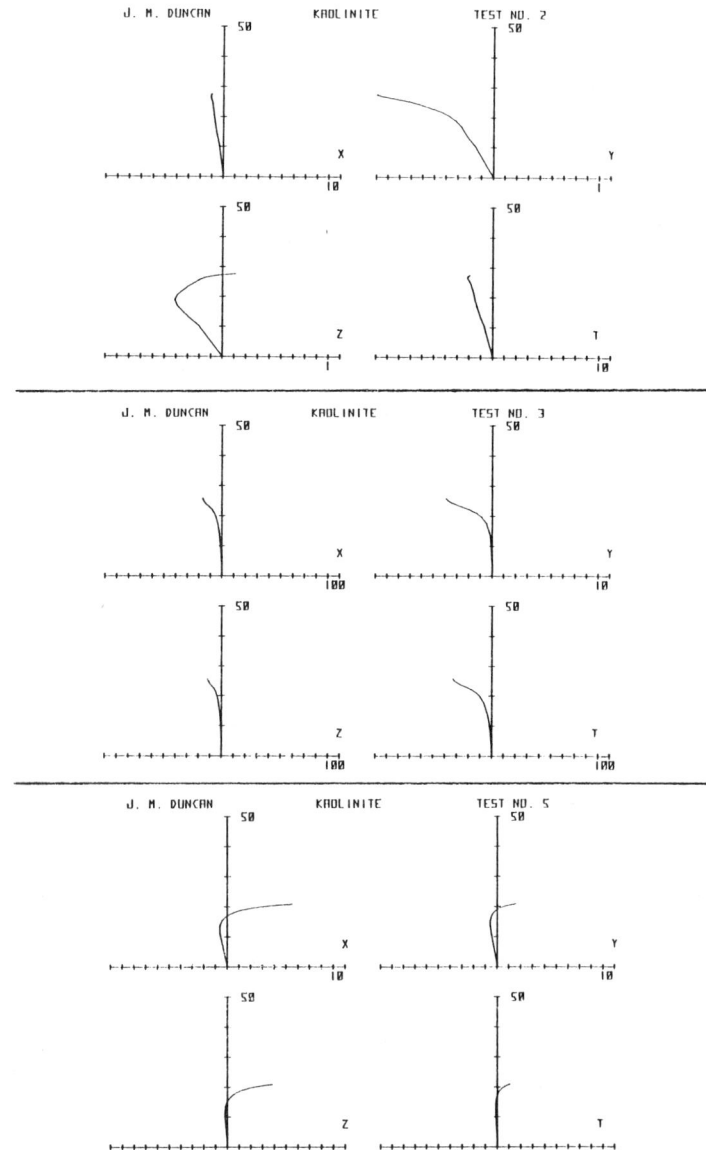

POST WORKSHOP COMPARISONS 181

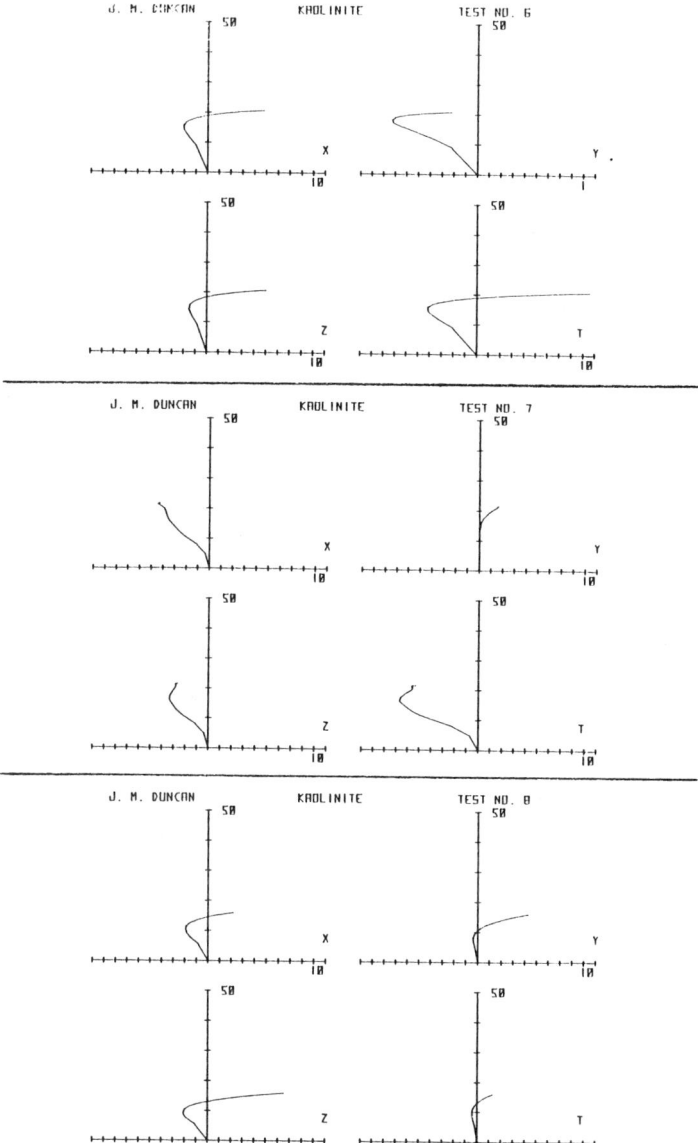

182 STRESS STRAIN FOR SOILS

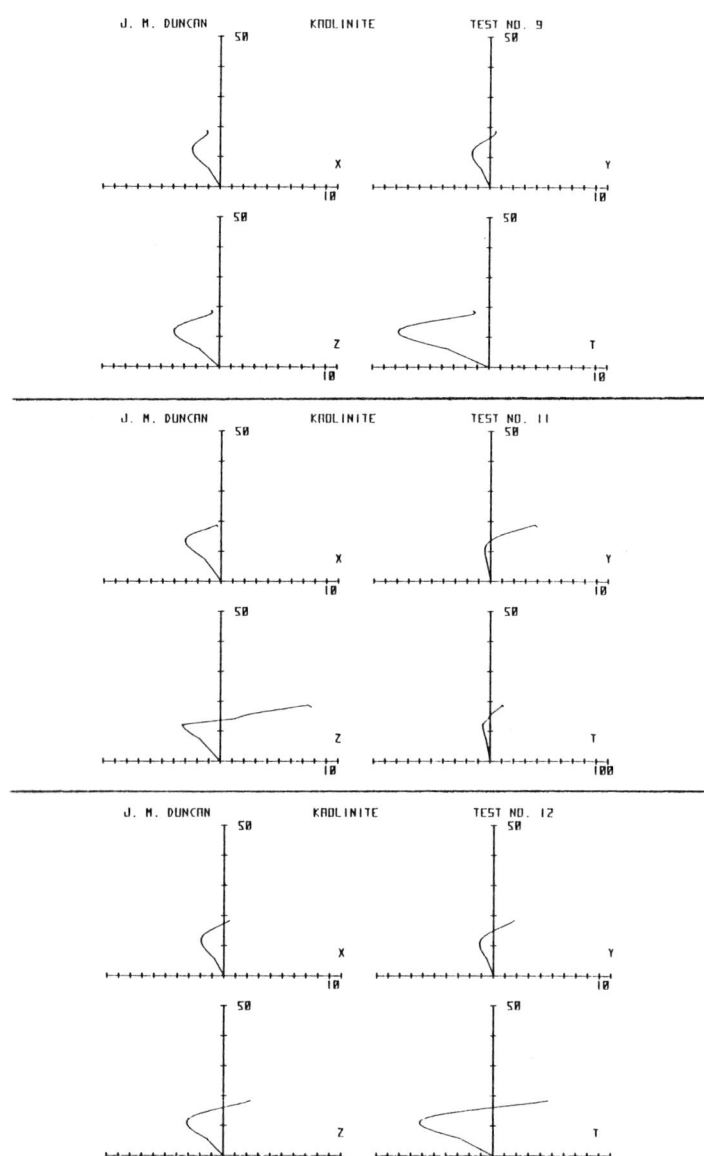

KAVAZANJIAN/MITCHELL/BONAPARTE   KAOLINITE   TEST NO. 2

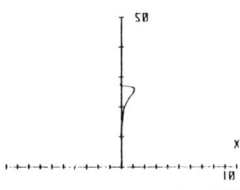

KAVAZANJIAN/MITCHELL/BONAPARTE   KAOLINITE   TEST NO. 3

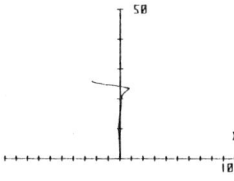

KAVAZANJIAN/MITCHELL/BONAPARTE   KAOLINITE   TEST NO. 5

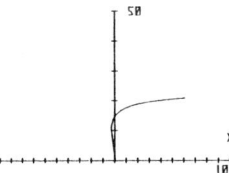

KAVAZANJIAN/MITCHELL/BONAPARTE   KAOLINITE   TEST NO. 6

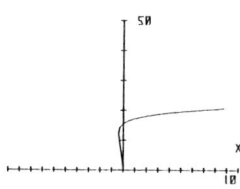

KAVAZANJIAN/MITCHELL/BONAPARTE   KAOLINITE   TEST NO. 7

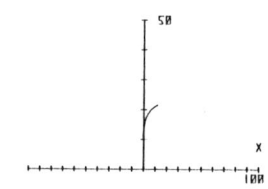

KAVAZANJIAN/MITCHELL/BONAPARTE   KAOLINITE   TEST NO. 8

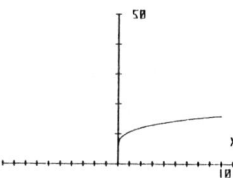

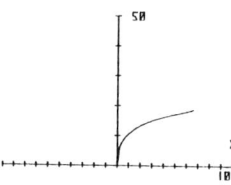

KAVAZANJIAN/MITCHELL/BONAPARTE   KAOLINITE   TEST NO. 9

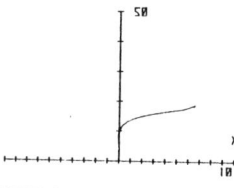

KAVAZANJIAN/MITCHELL/BONAPARTE   KAOLINITE   TEST NO. 11

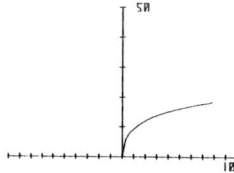

KAVAZANJIAN/MITCHELL/BONAPARTE   KAOLINITE   TEST NO. 12

186 STRESS STRAIN FOR SOILS

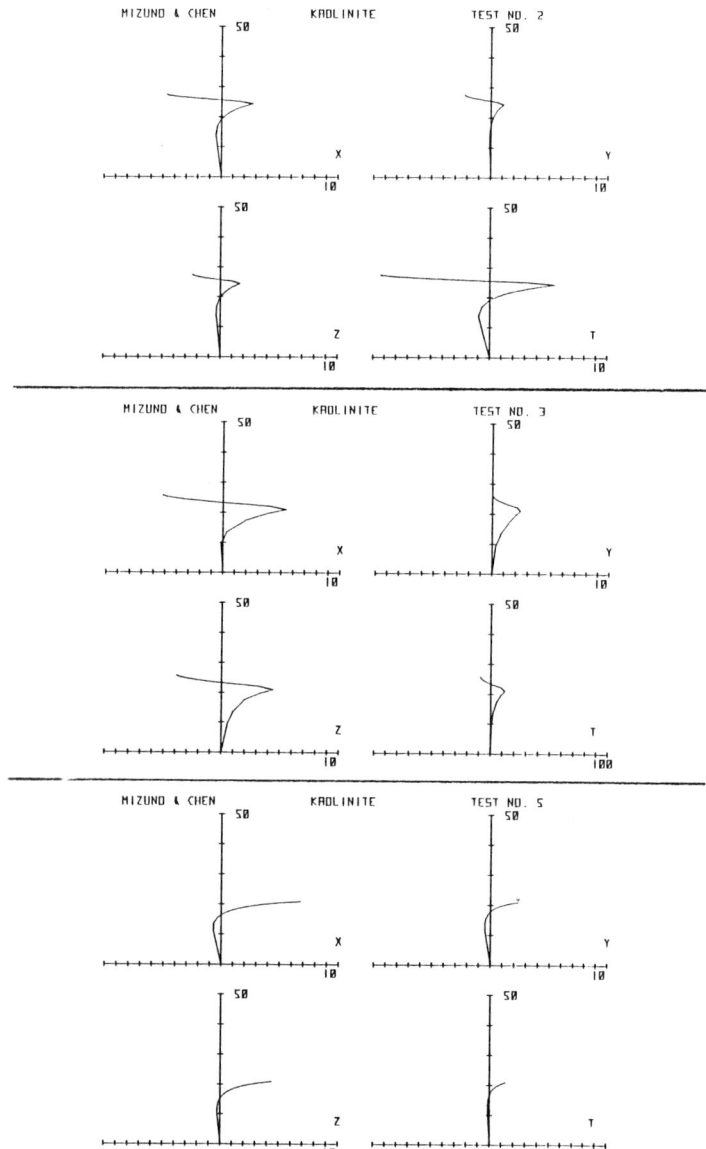

POST WORKSHOP COMPARISONS 187

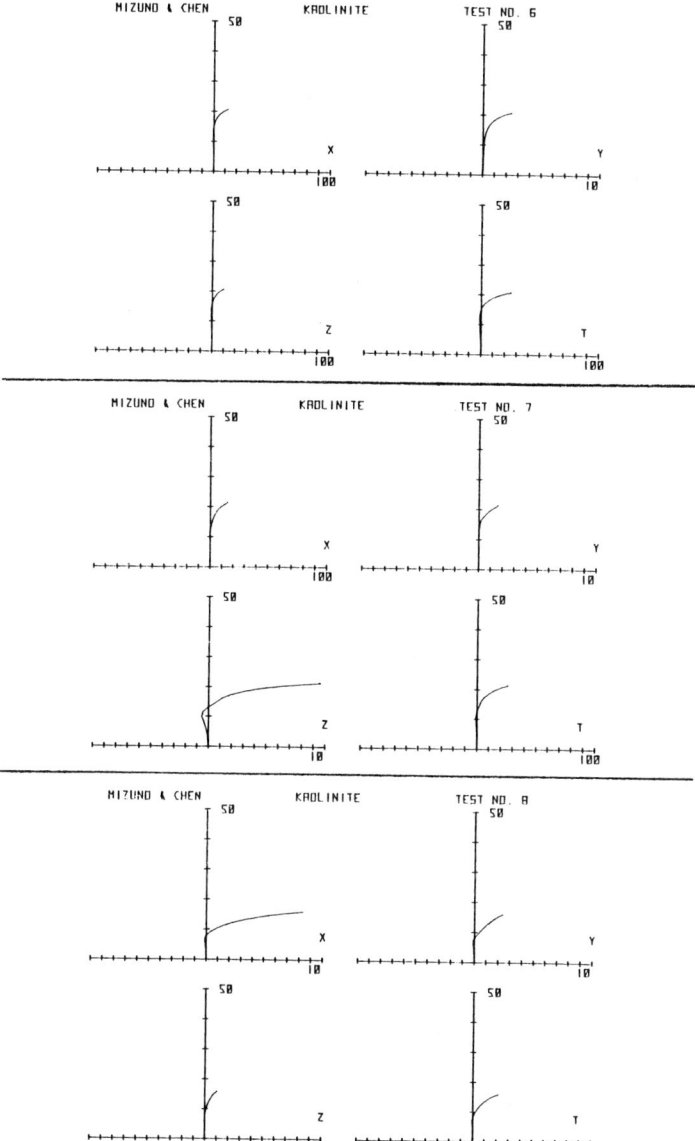

STRESS STRAIN FOR SOILS

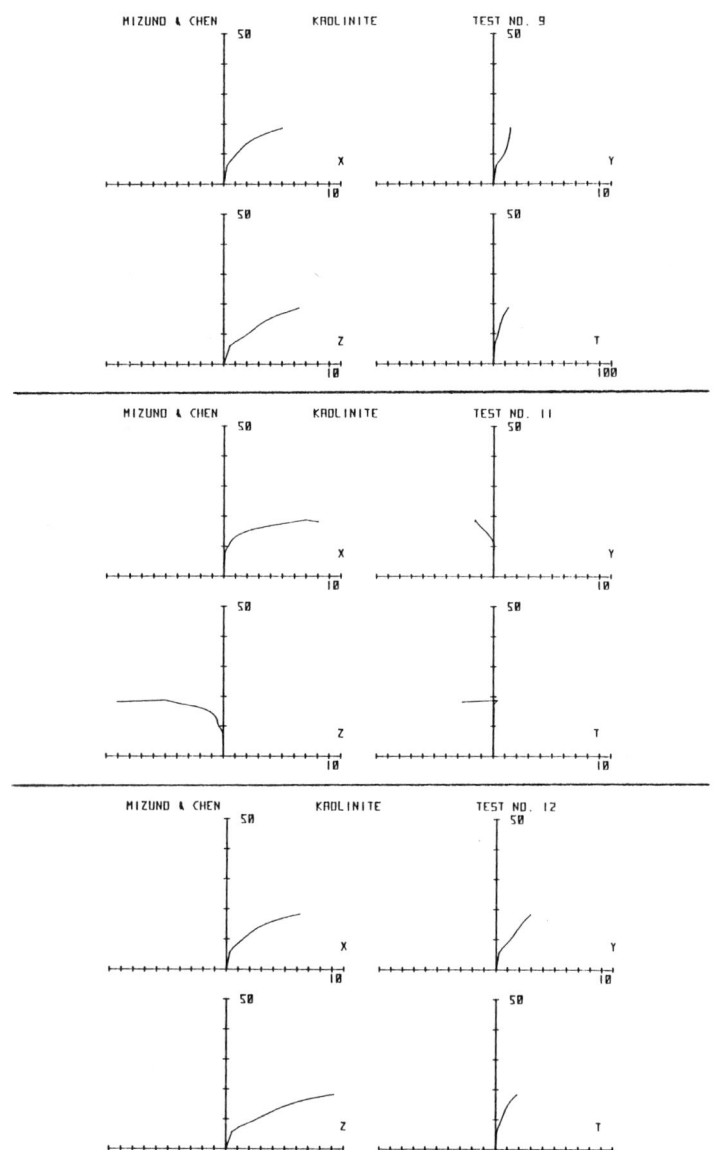

POST WORKSHOP COMPARISONS

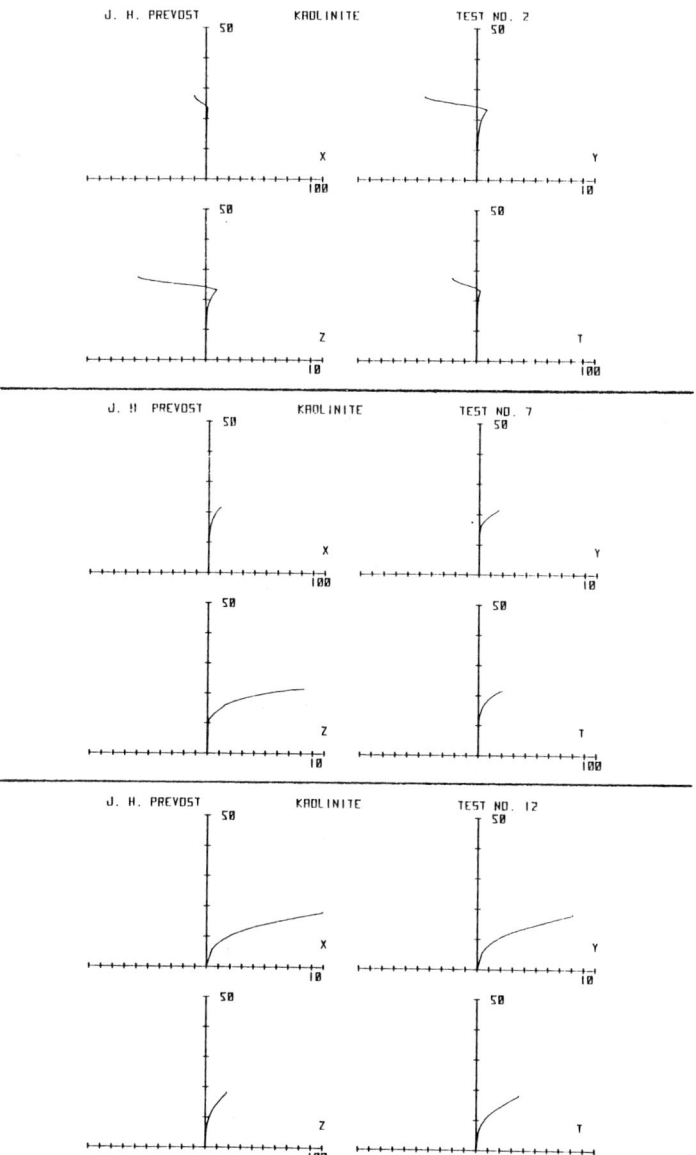

190 STRESS STRAIN FOR SOILS

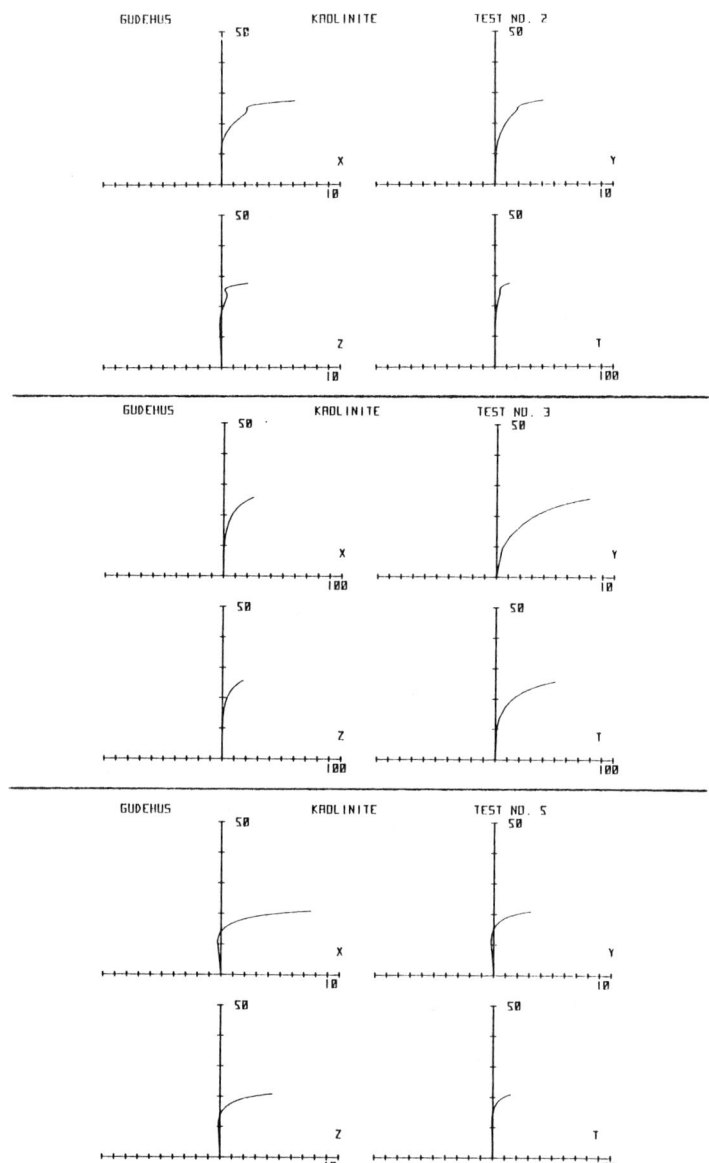

POST WORKSHOP COMPARISONS

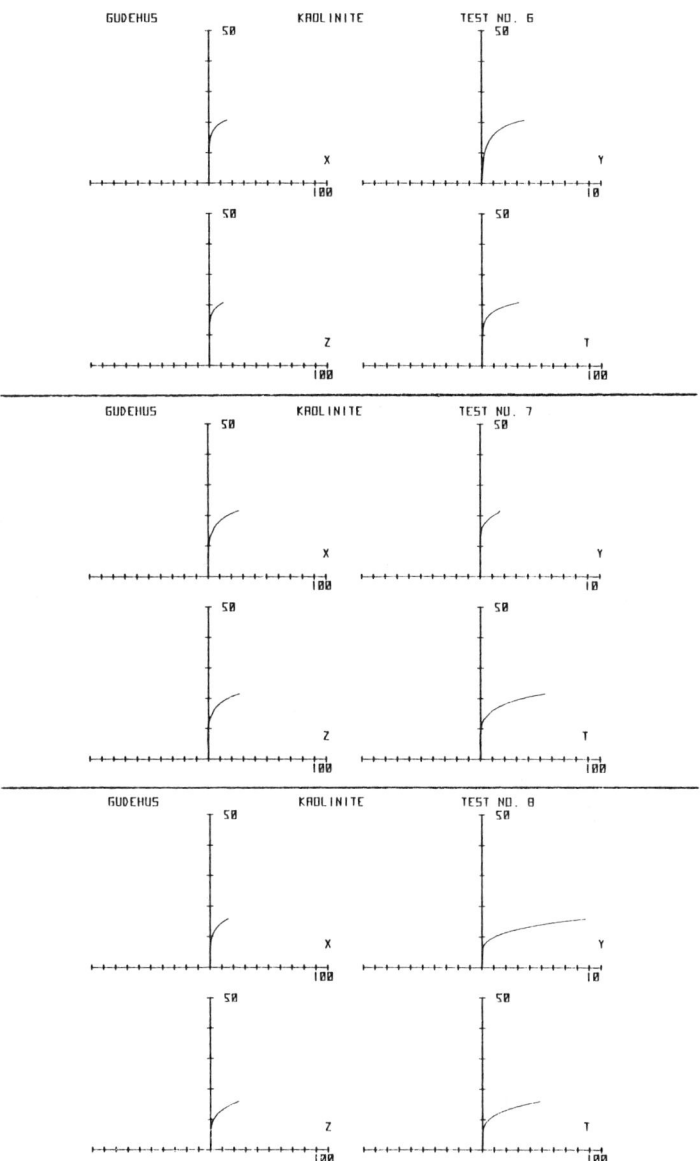

192 STRESS STRAIN FOR SOILS

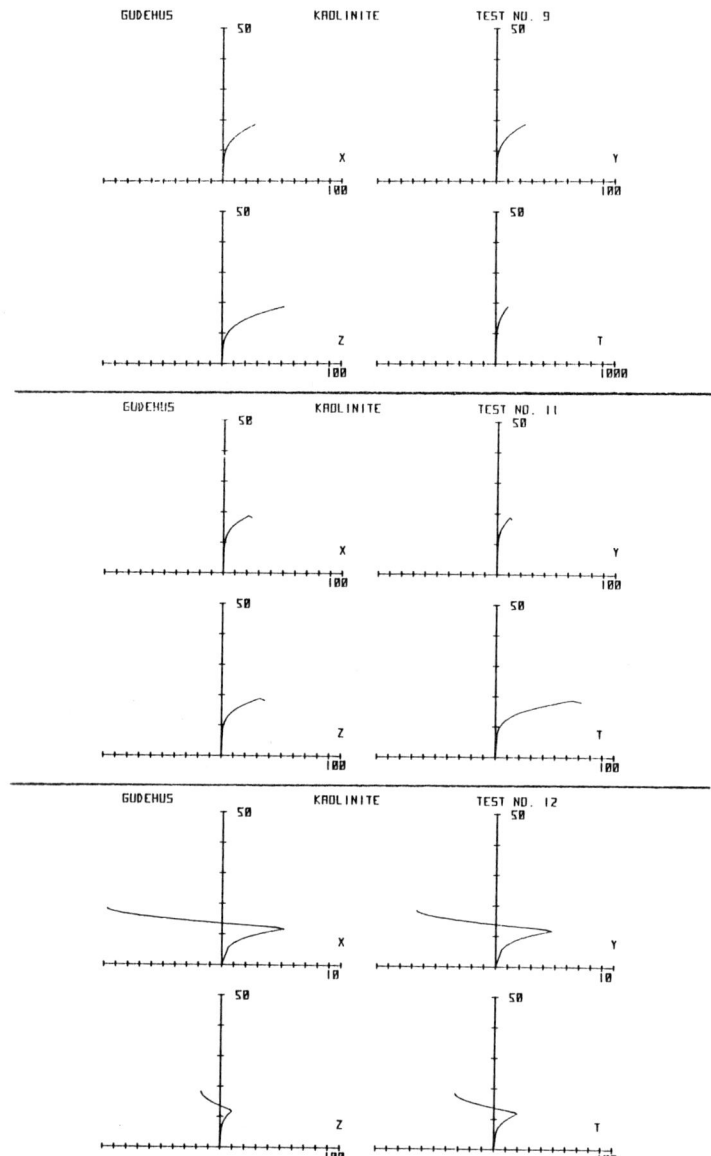

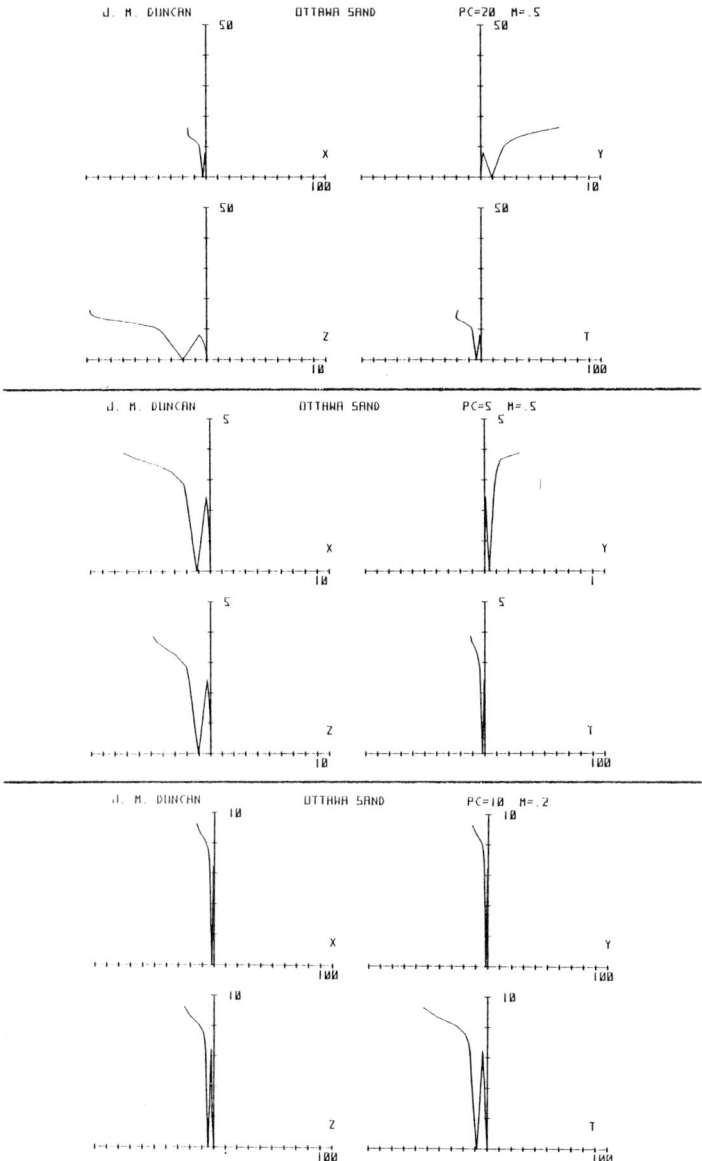

194 STRESS STRAIN FOR SOILS

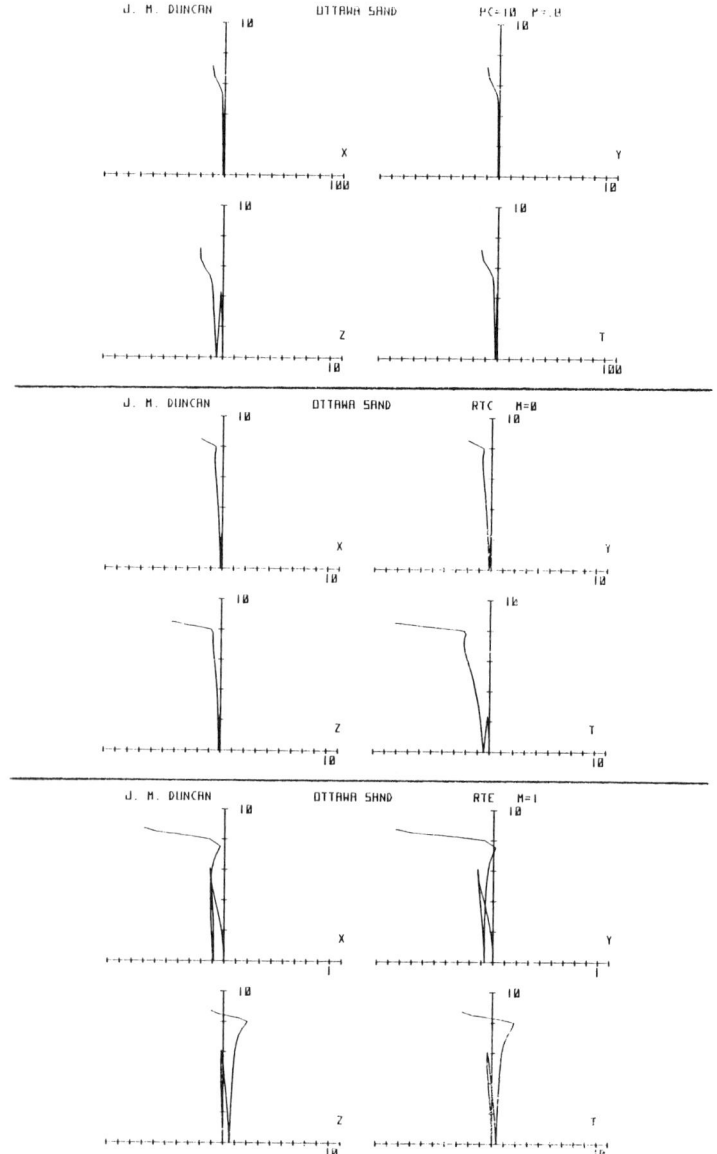

POST WORKSHOP COMPARISONS 195

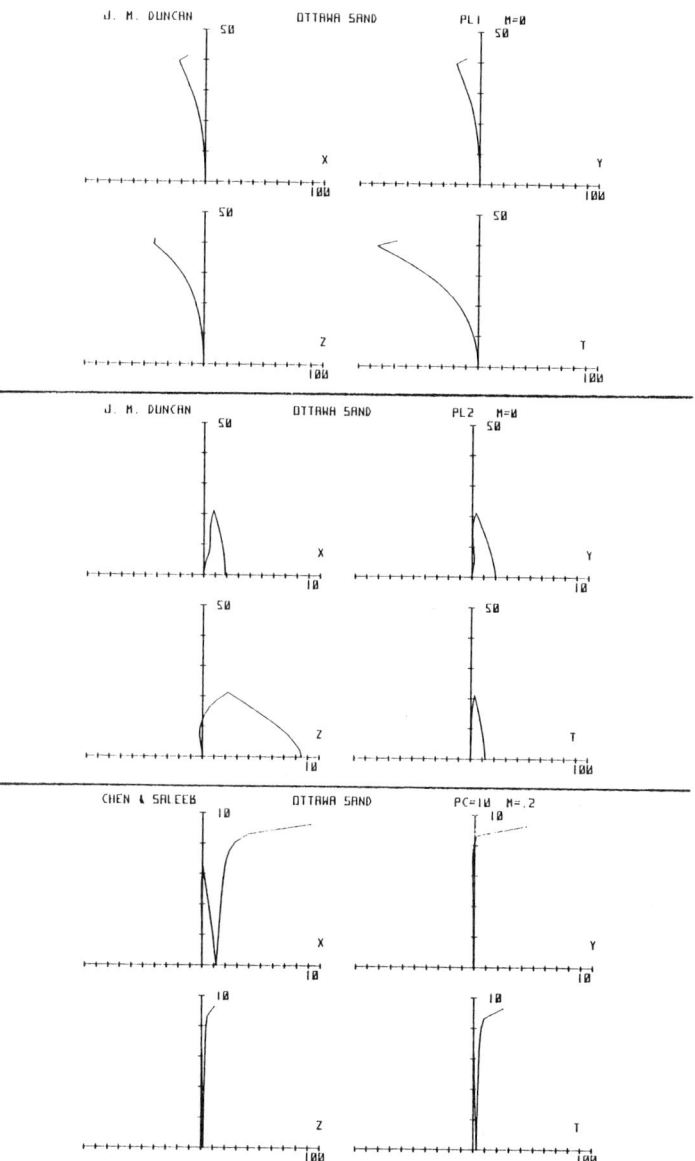

196 STRESS STRAIN FOR SOILS

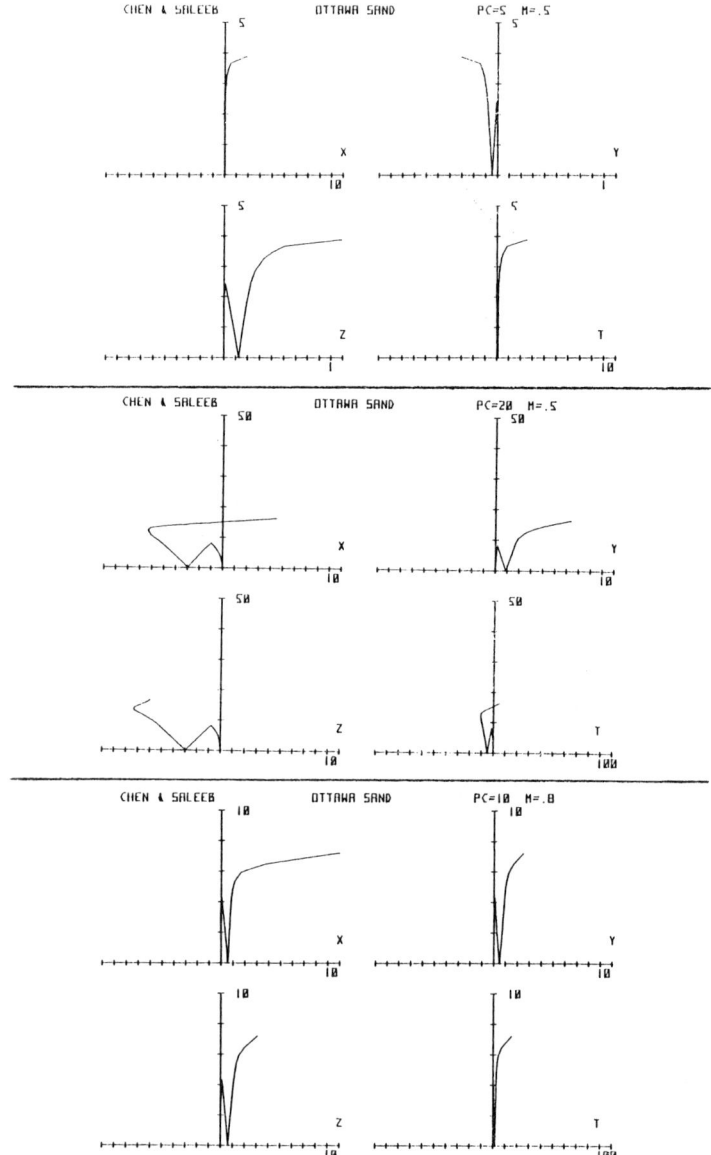

POST WORKSHOP COMPARISONS

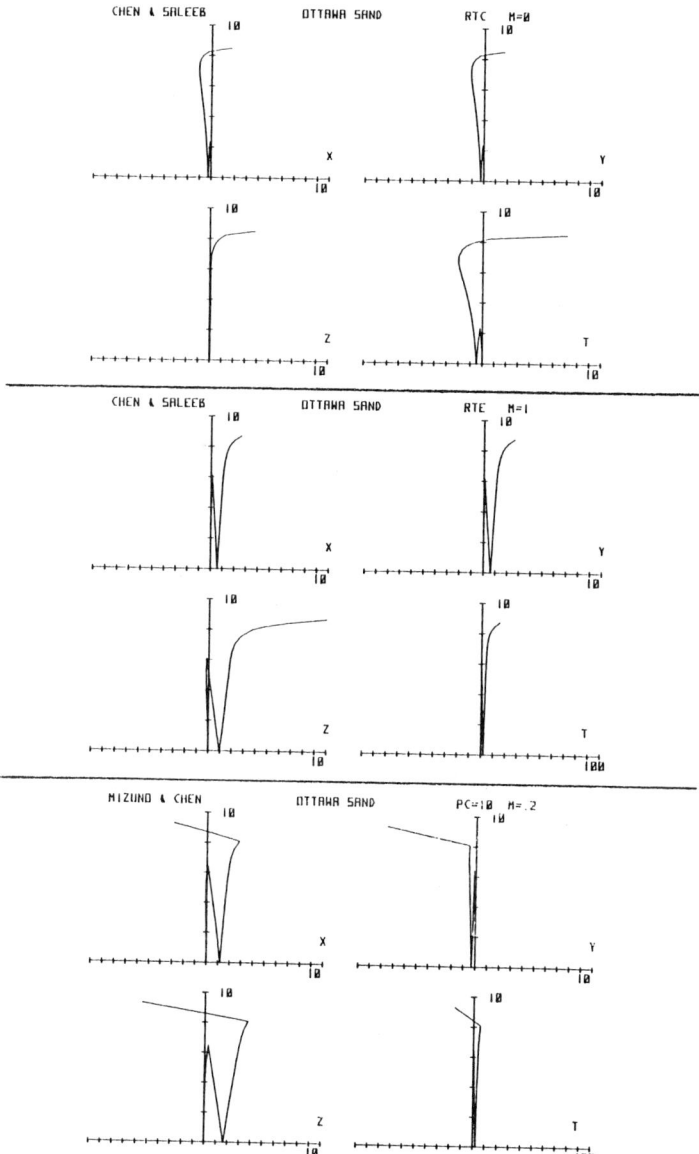

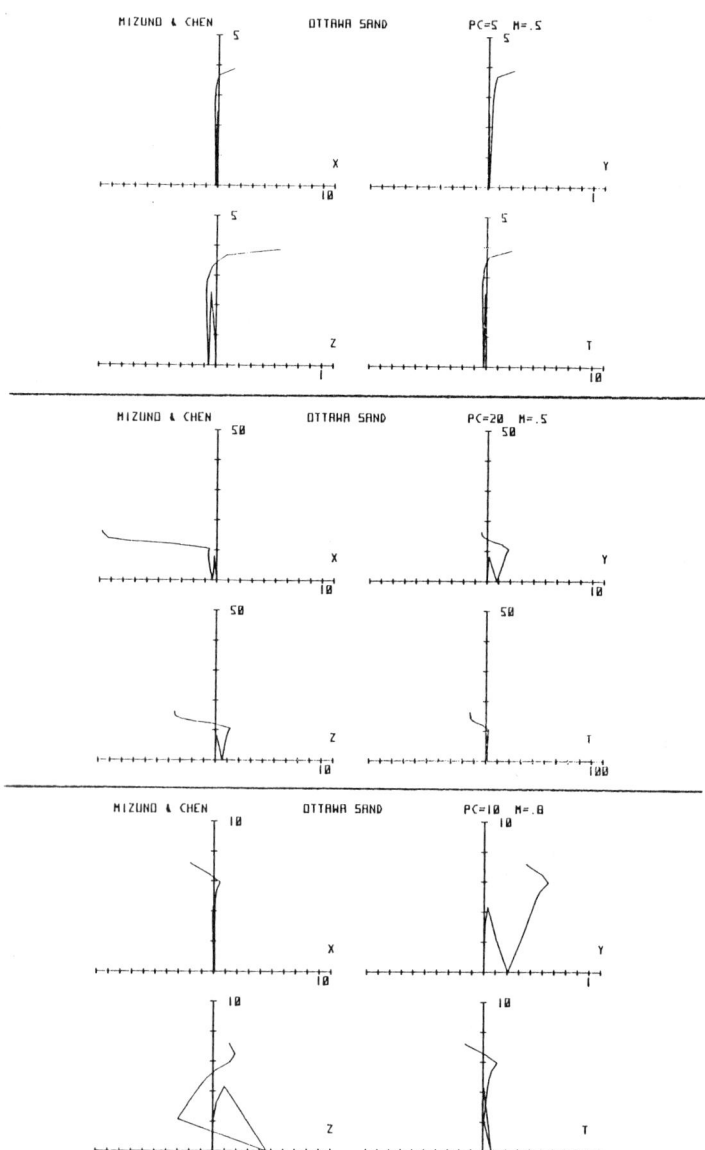

POST WORKSHOP COMPARISONS

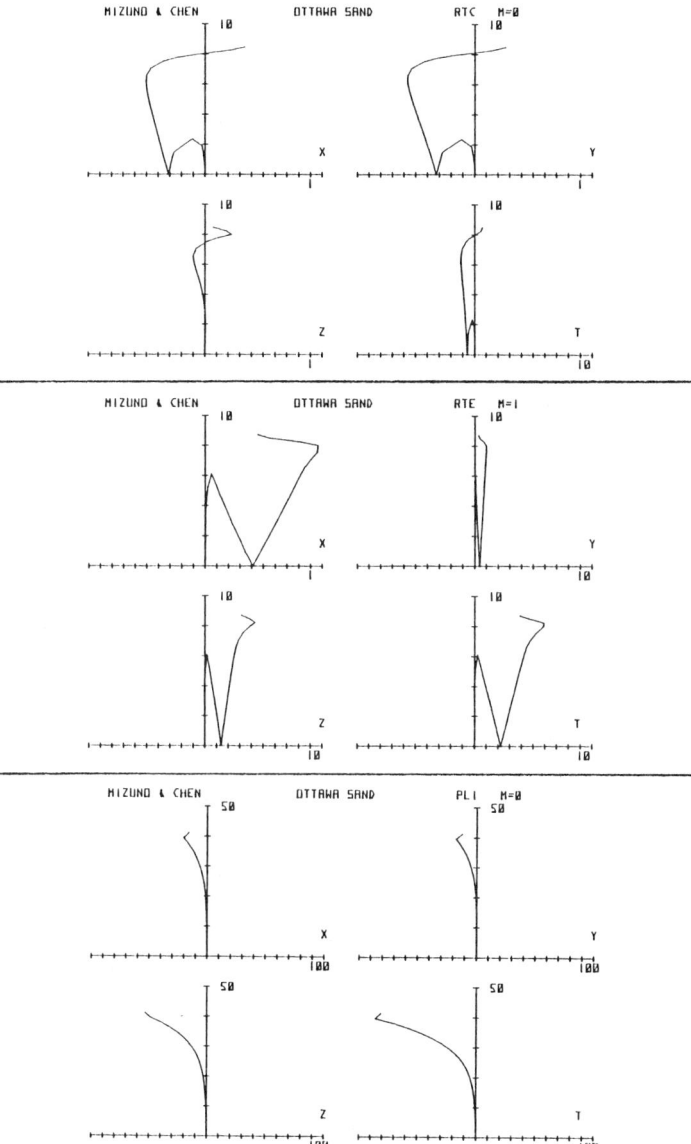

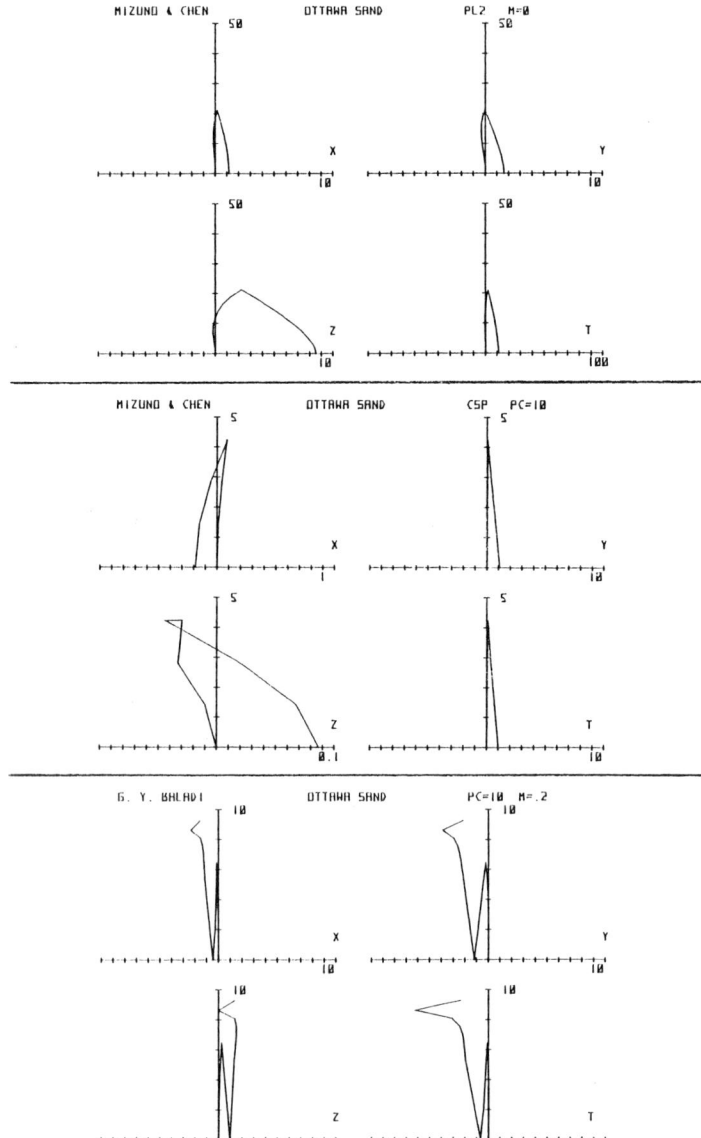

POST WORKSHOP COMPARISONS

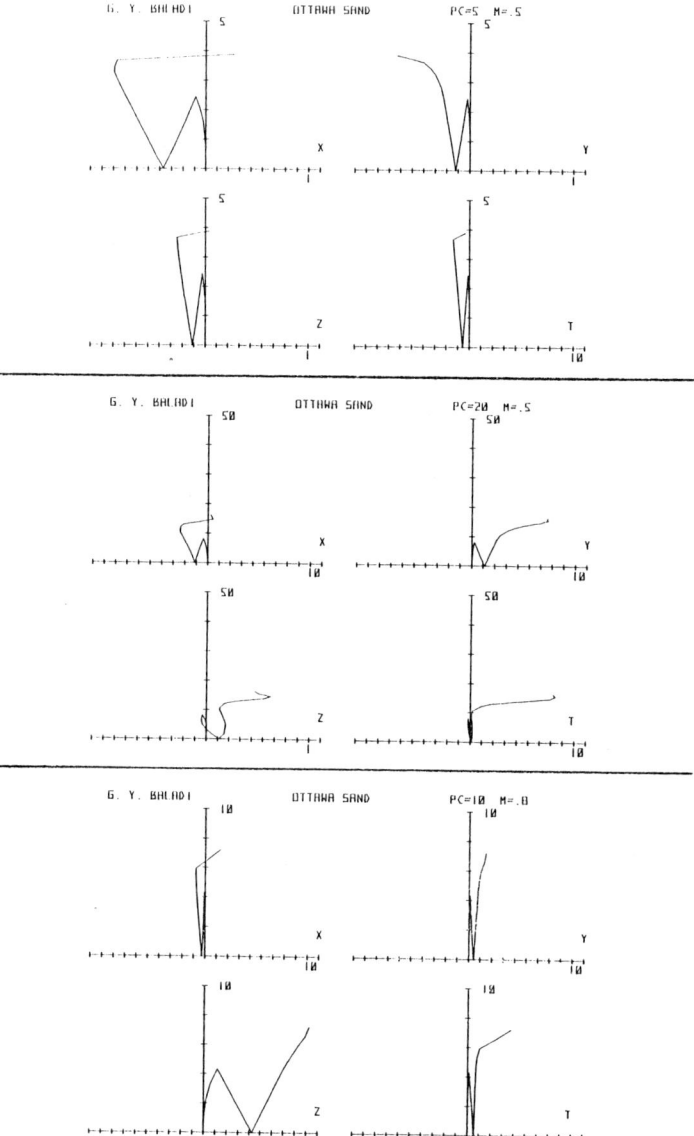

202 STRESS STRAIN FOR SOILS

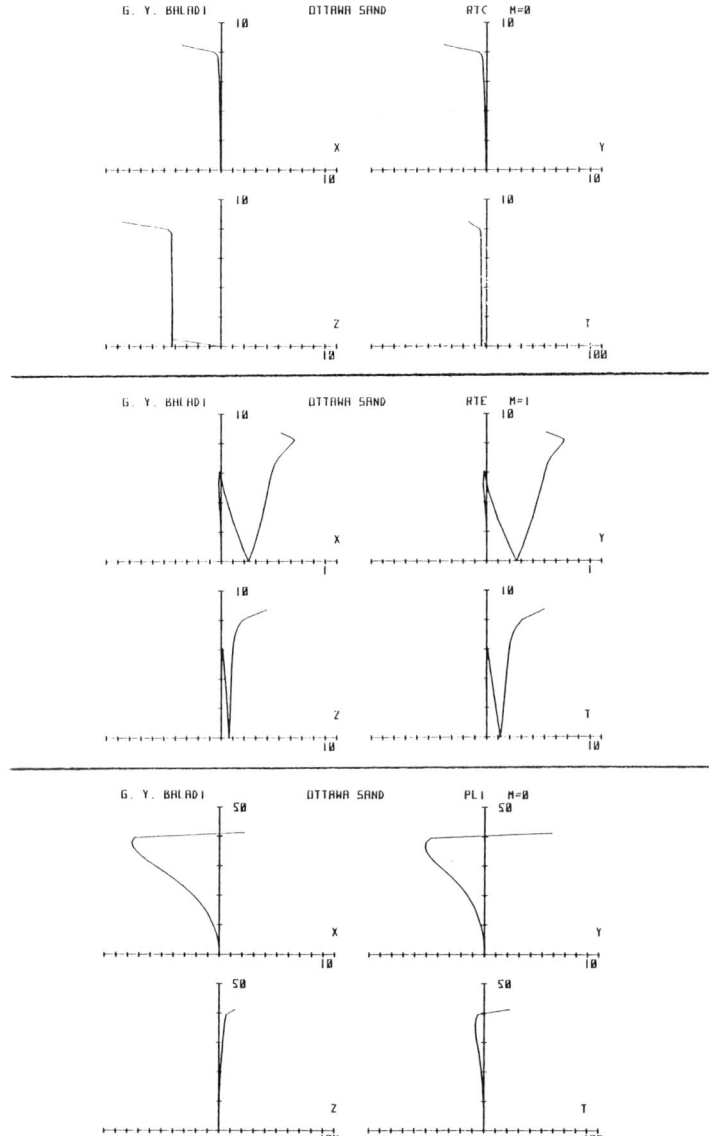

POST WORKSHOP COMPARISONS

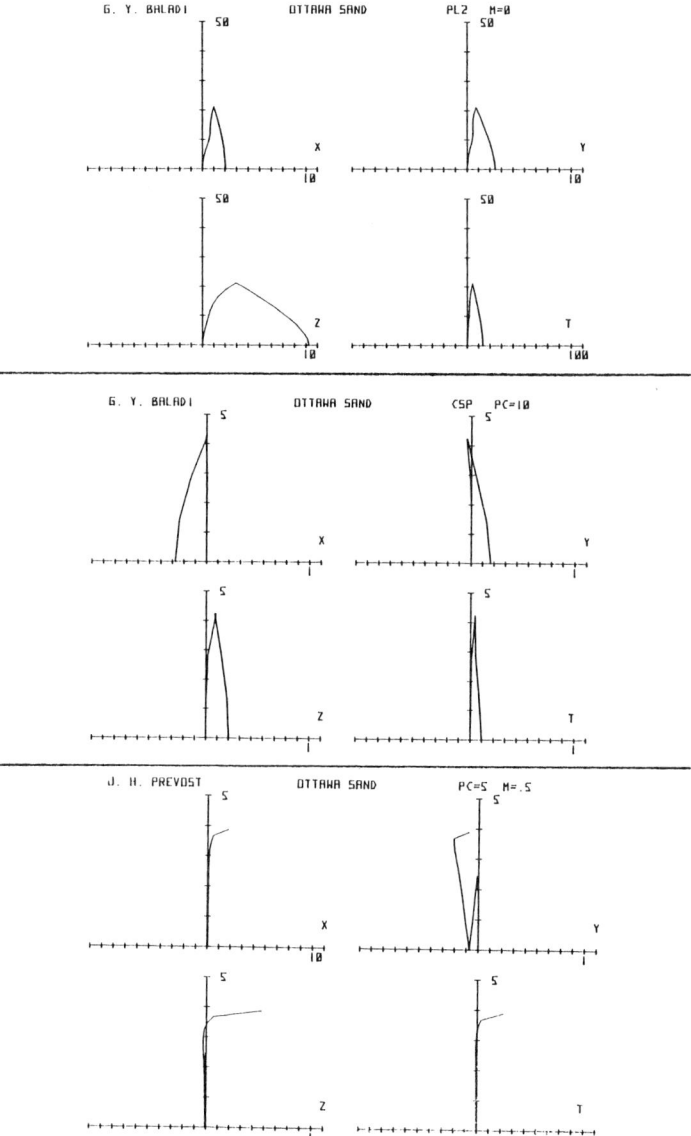

STRESS STRAIN FOR SOILS

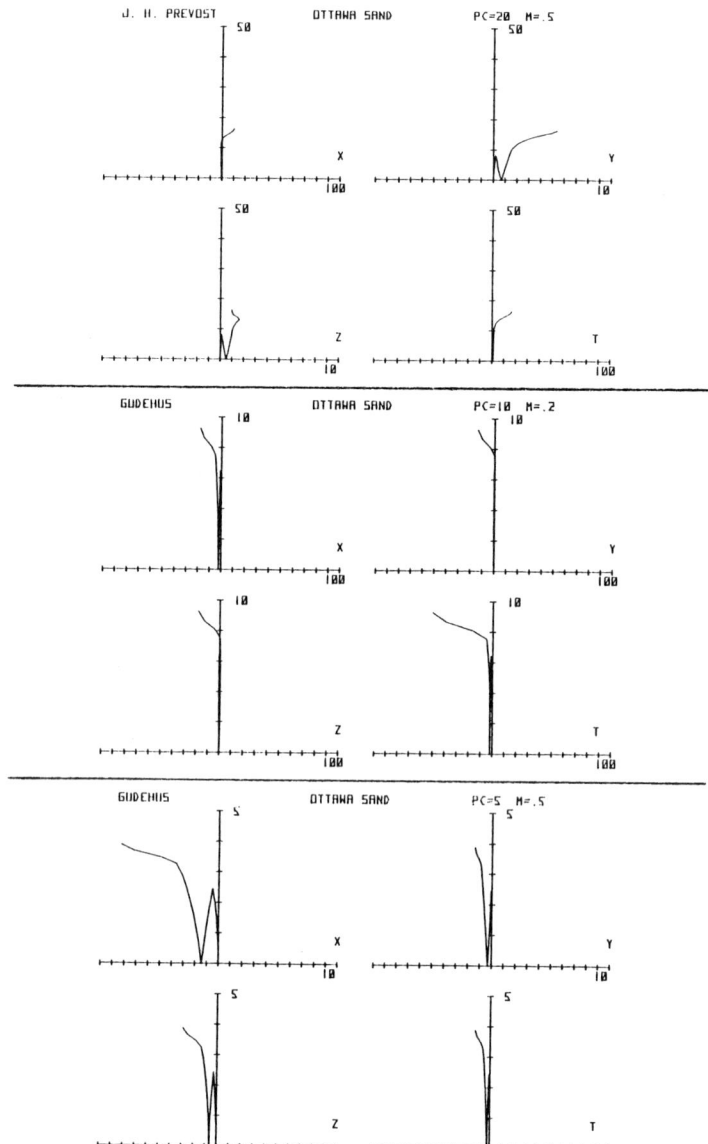

POST WORKSHOP COMPARISONS

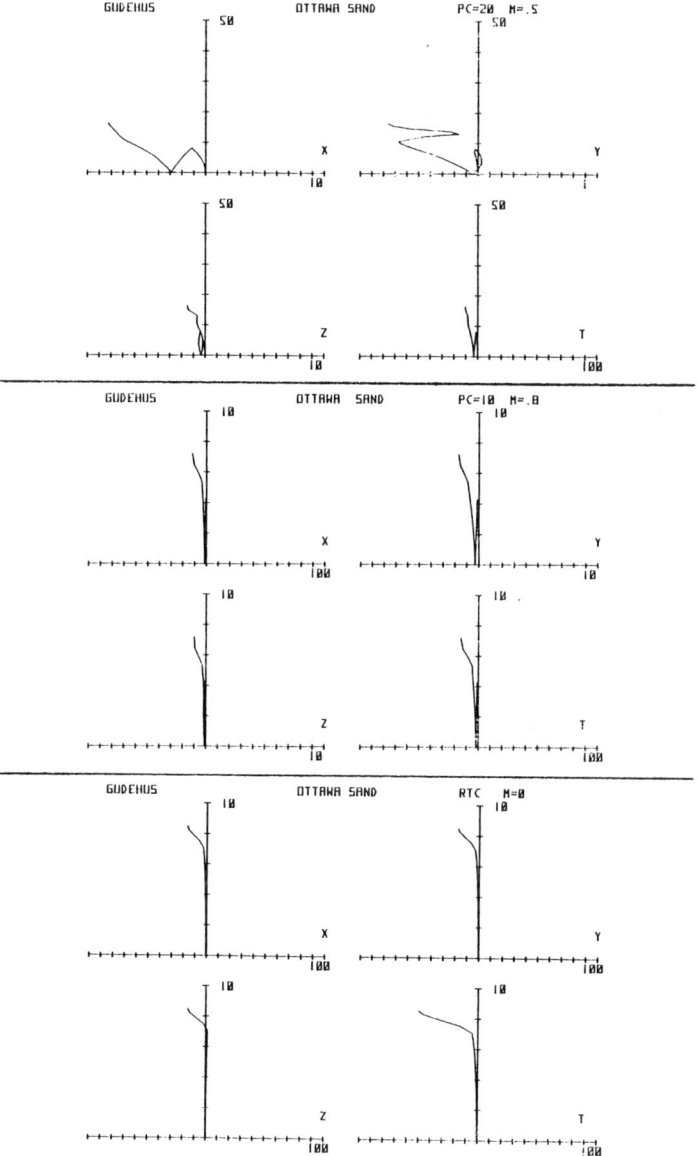

206 STRESS STRAIN FOR SOILS

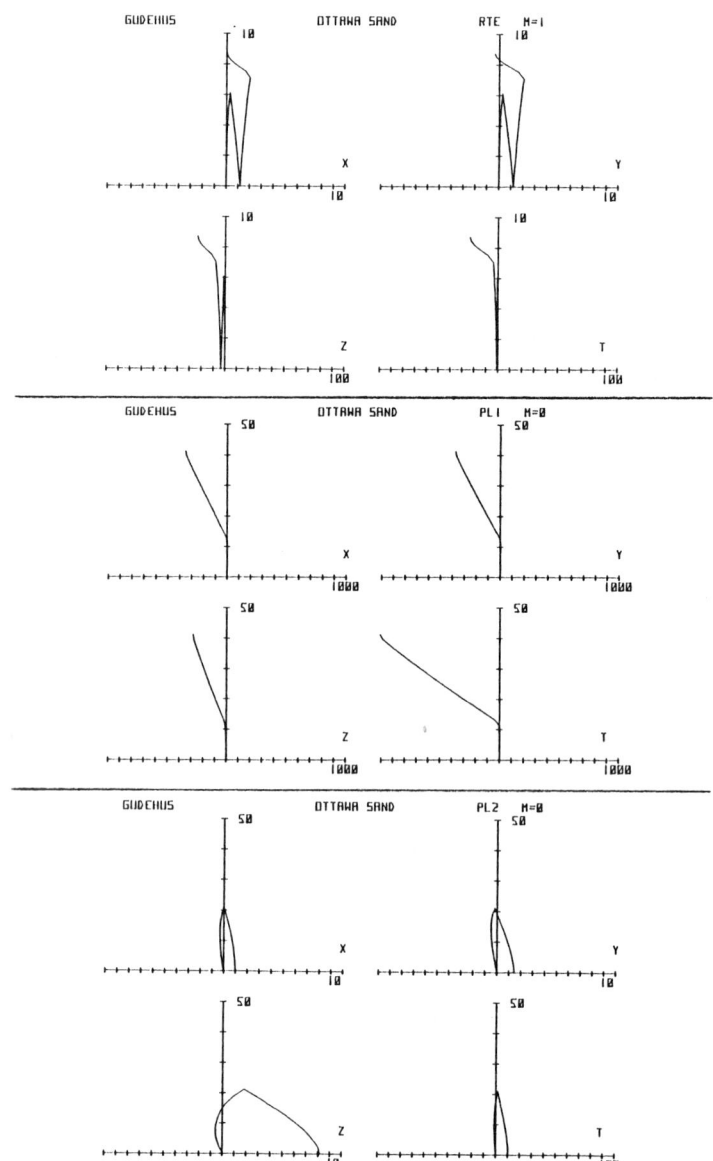

POST WORKSHOP COMPARISONS

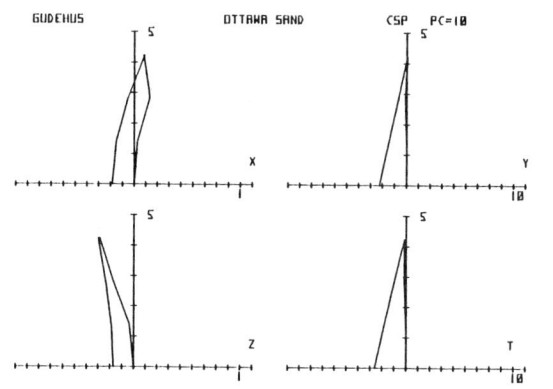

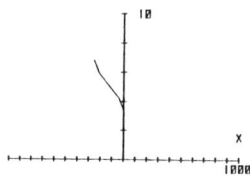

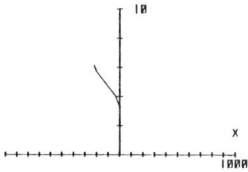

J. M. DUNCAN   CLAY X (LB)   PC=10 M=.75

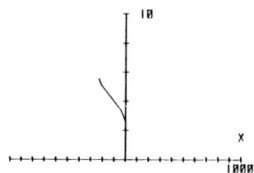

J. M. DUNCAN   CLAY X (UB)   PC=20 M=.25

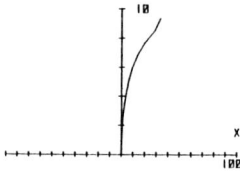

J. M. DUNCAN   CLAY X (LB)   PC=20 M=.5

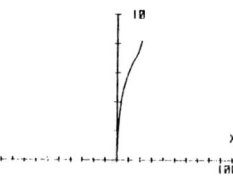

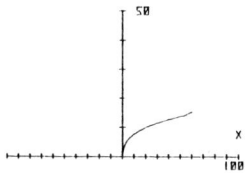

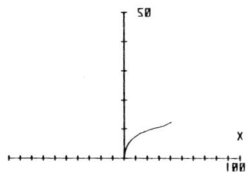

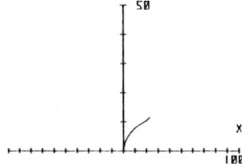

210 STRESS STRAIN FOR SOILS

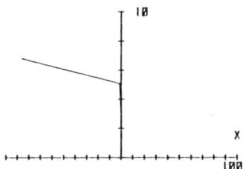

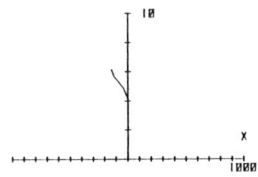

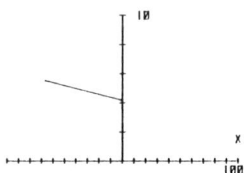

POST WORKSHOP COMPARISONS

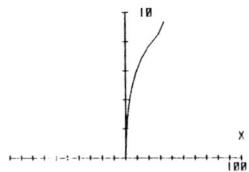

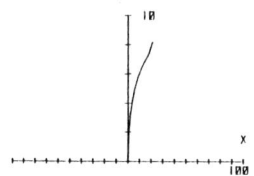

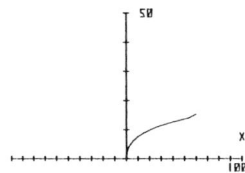

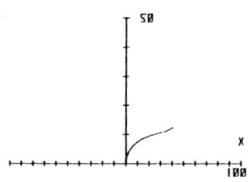

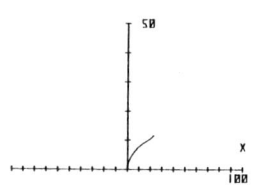

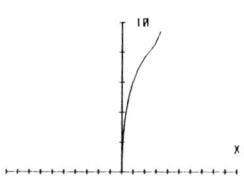

KAVAZANJIAN/MITCHELL/BONAPARTE   CLAY X   PC=20 M=.5

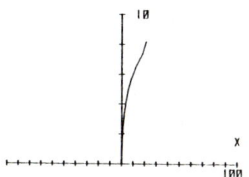

KAVAZANJIAN/MITCHELL/BONAPARTE   CLAY X   PC=30 M=.25

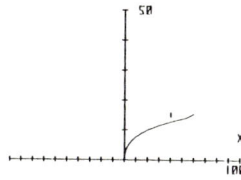

KAVAZANJIAN/MITCHELL/BONAPARTE   CLAY X   PC=30 M=.5

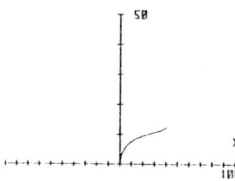

214 STRESS STRAIN FOR SOILS

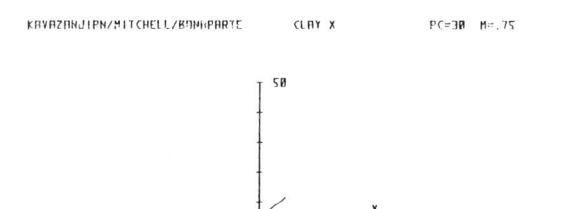

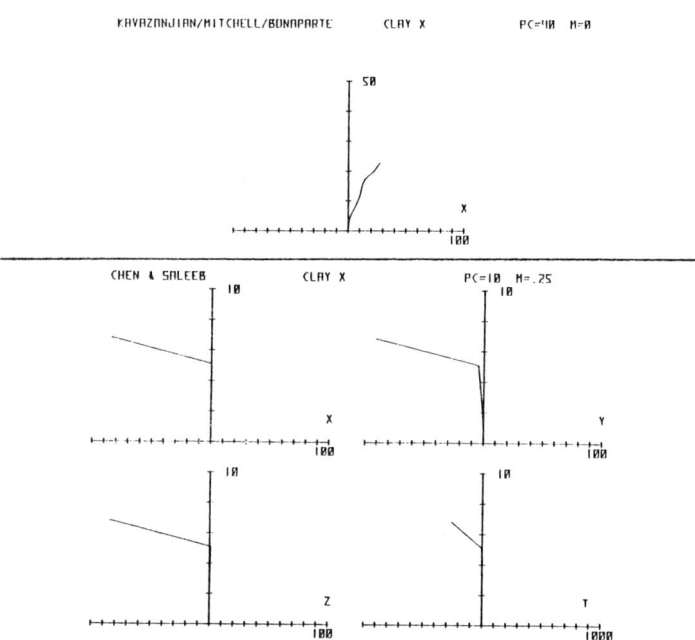

POST WORKSHOP COMPARISONS

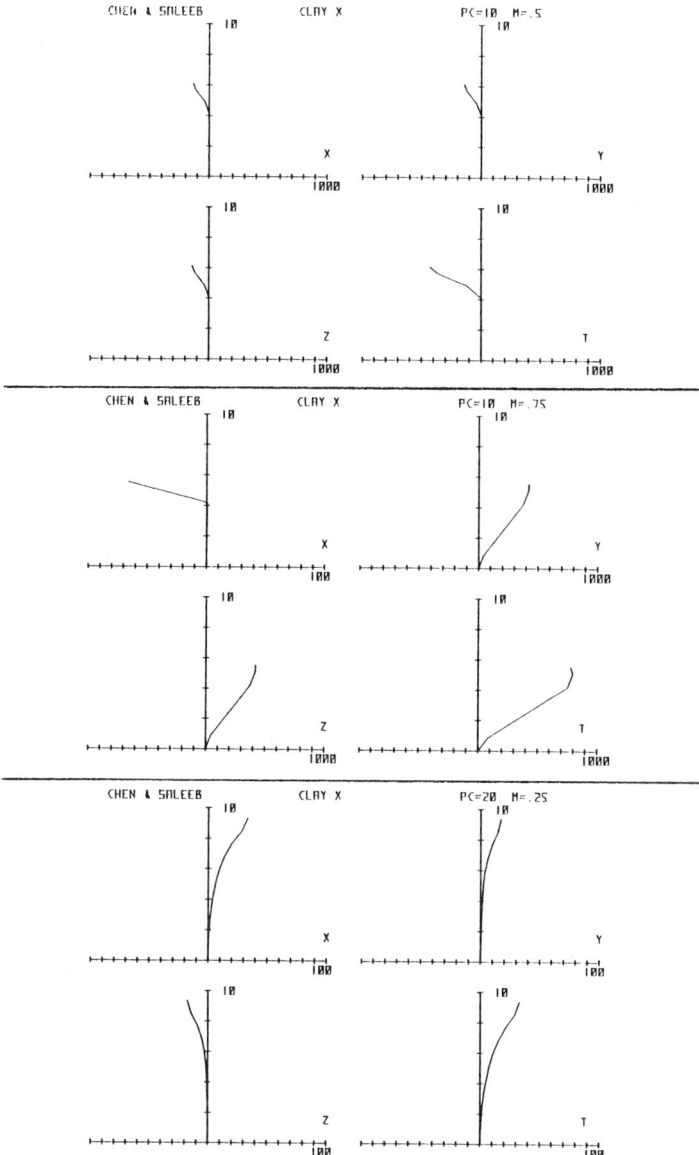

216 STRESS STRAIN FOR SOILS

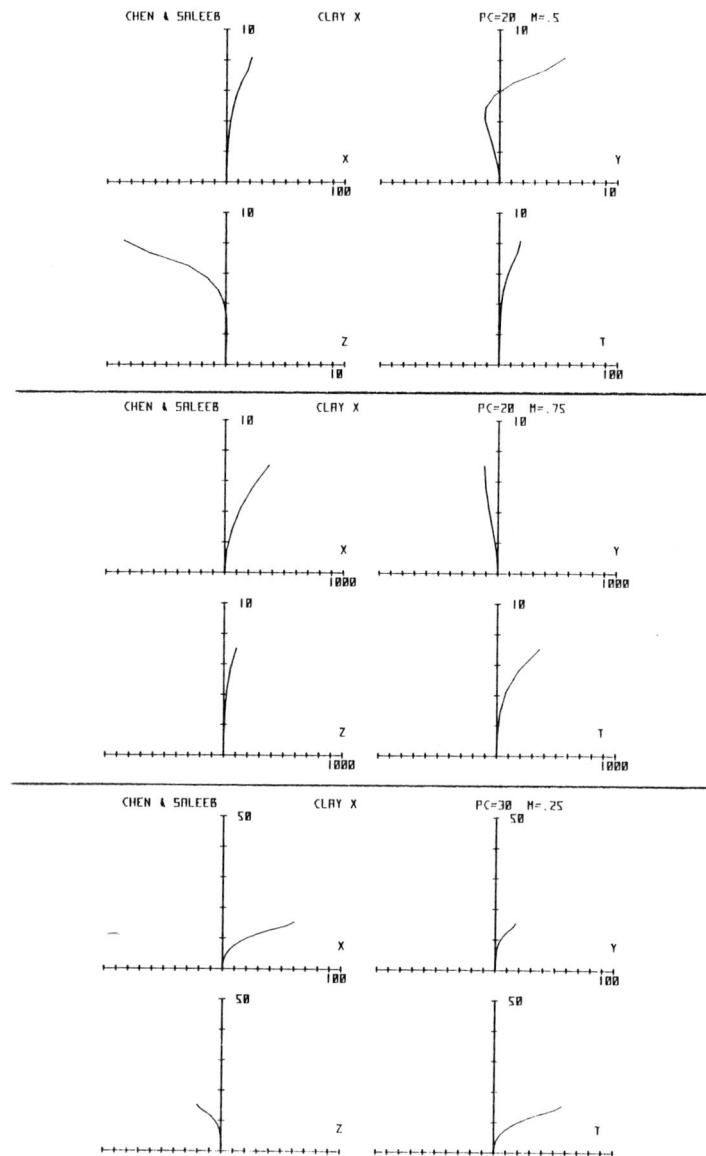

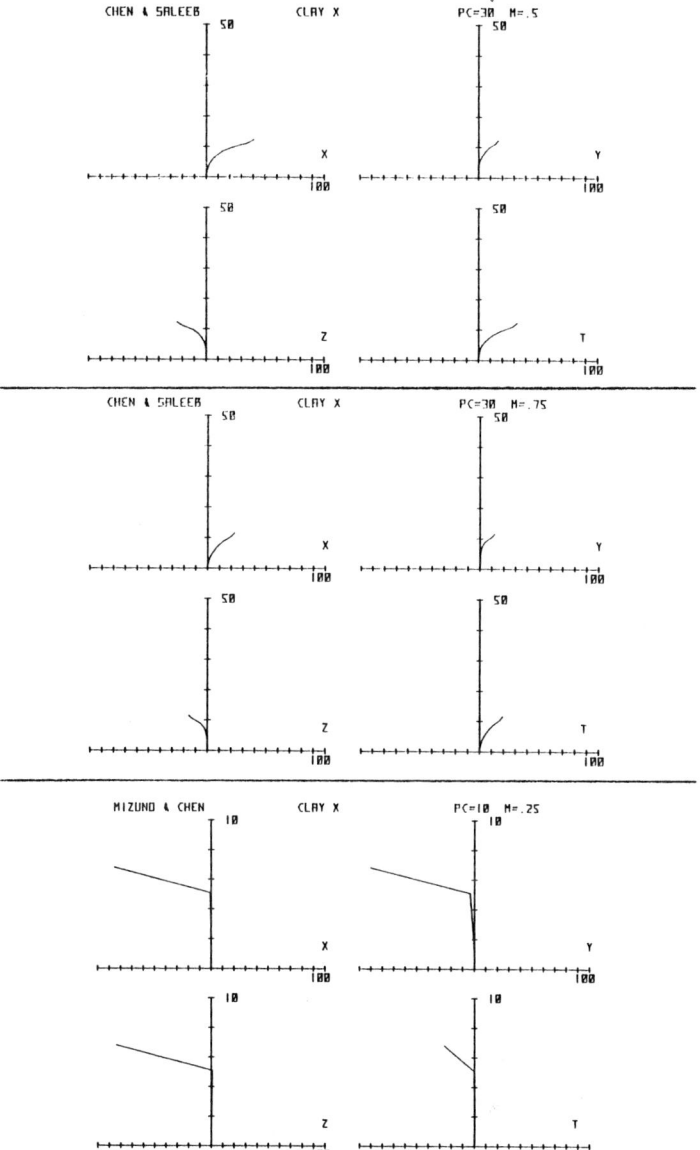

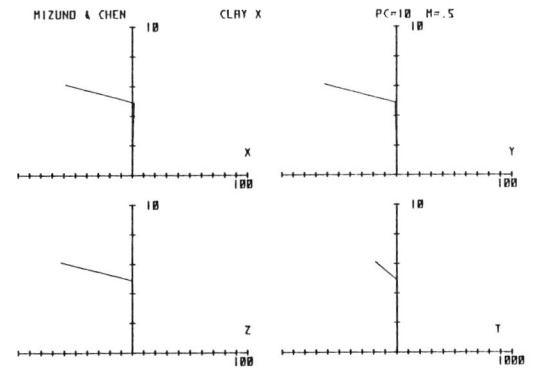
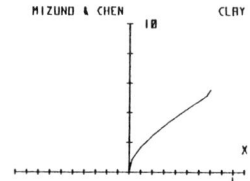
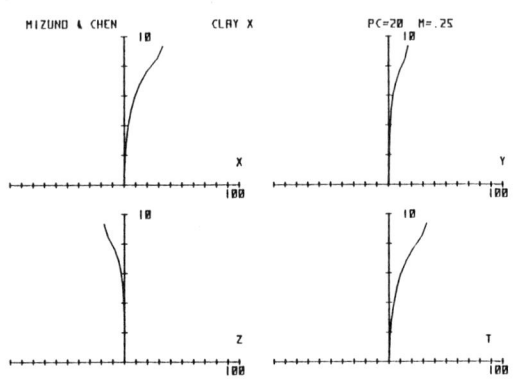

POST WORKSHOP COMPARISONS

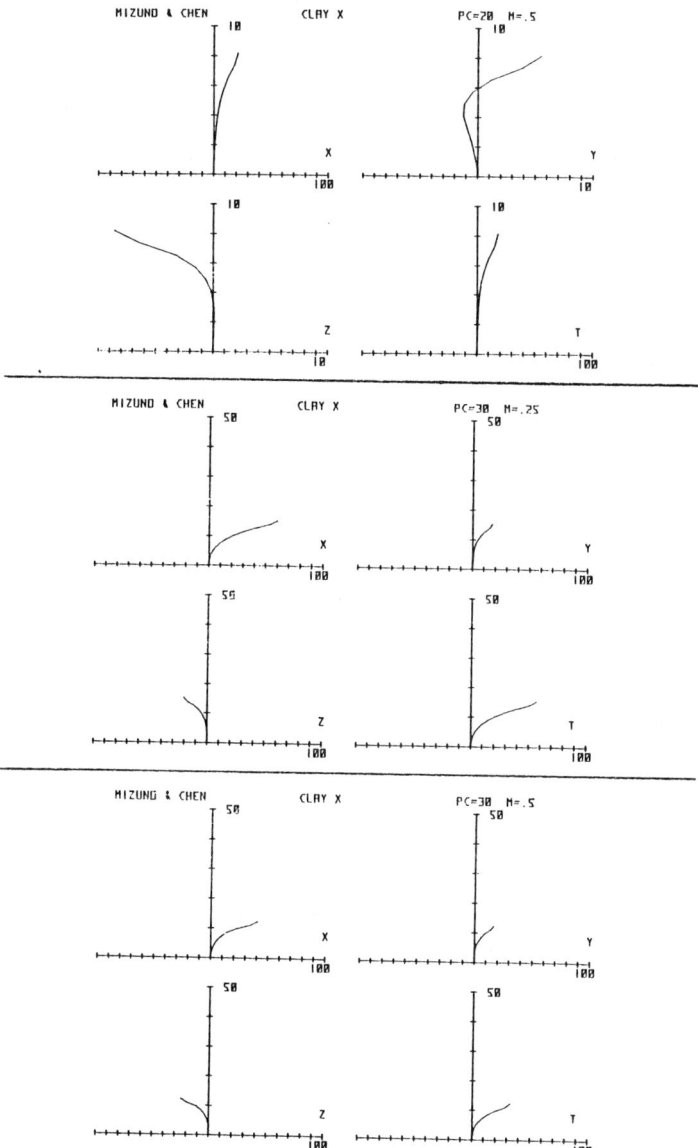

STRESS STRAIN FOR SOILS

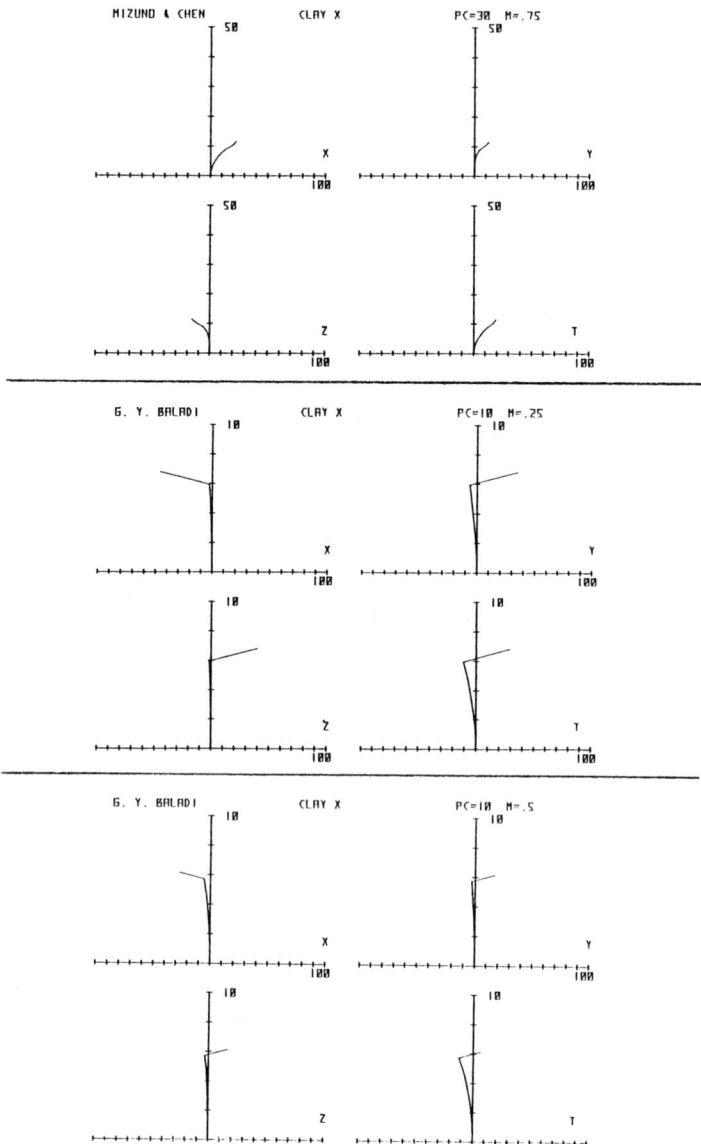

G. Y. BALADI
CLAY X
PC=10 M=.75
10
X
100

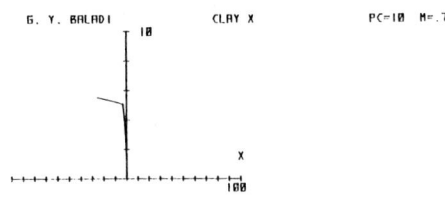

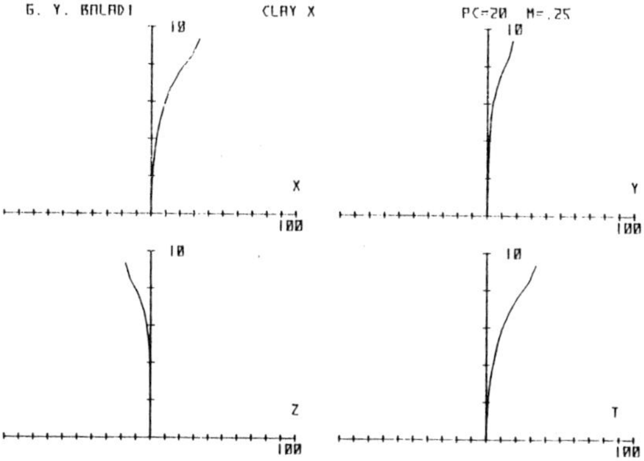
G. Y. BALADI
CLAY X
PC=20 M=.25
10
X
100
10
Y
100
10
Z
100
10
T
100

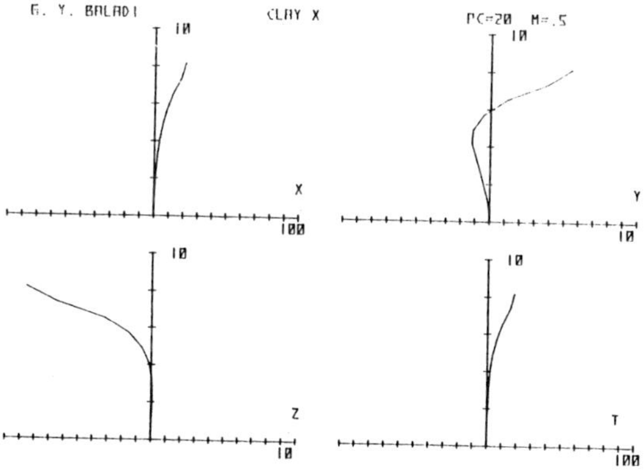
G. Y. BALADI
CLAY X
PC=20 M=.5
10
X
100
10
Y
10
10
Z
10
10
T
100

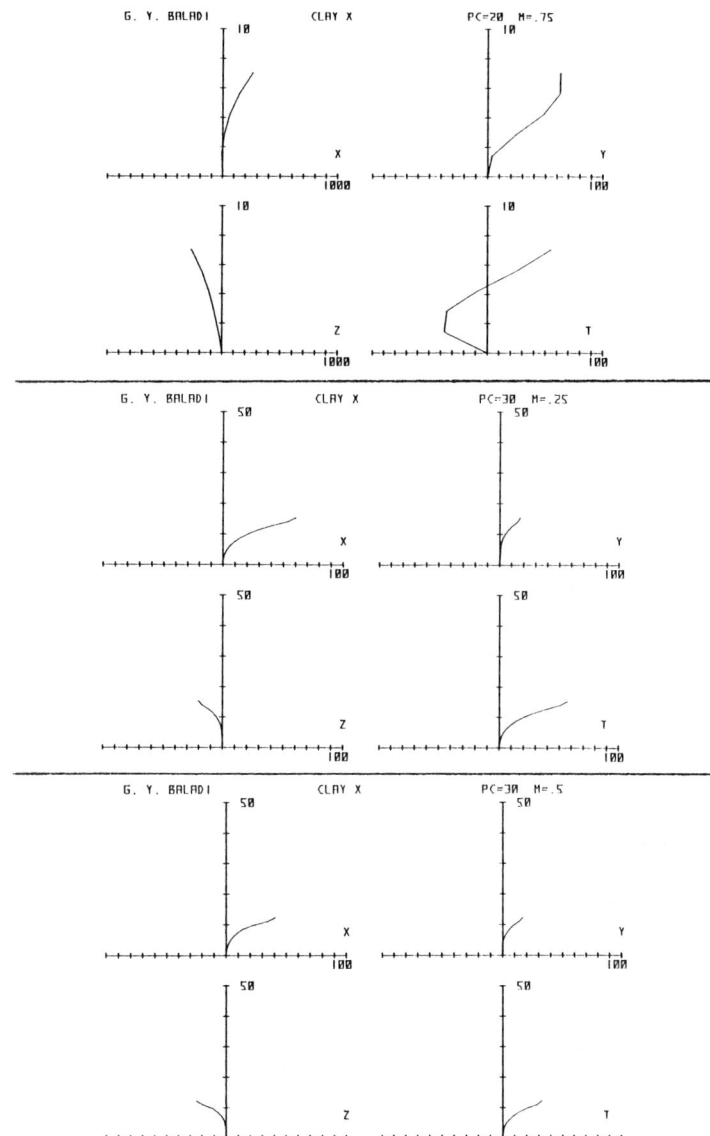

POST WORKSHOP COMPARISONS

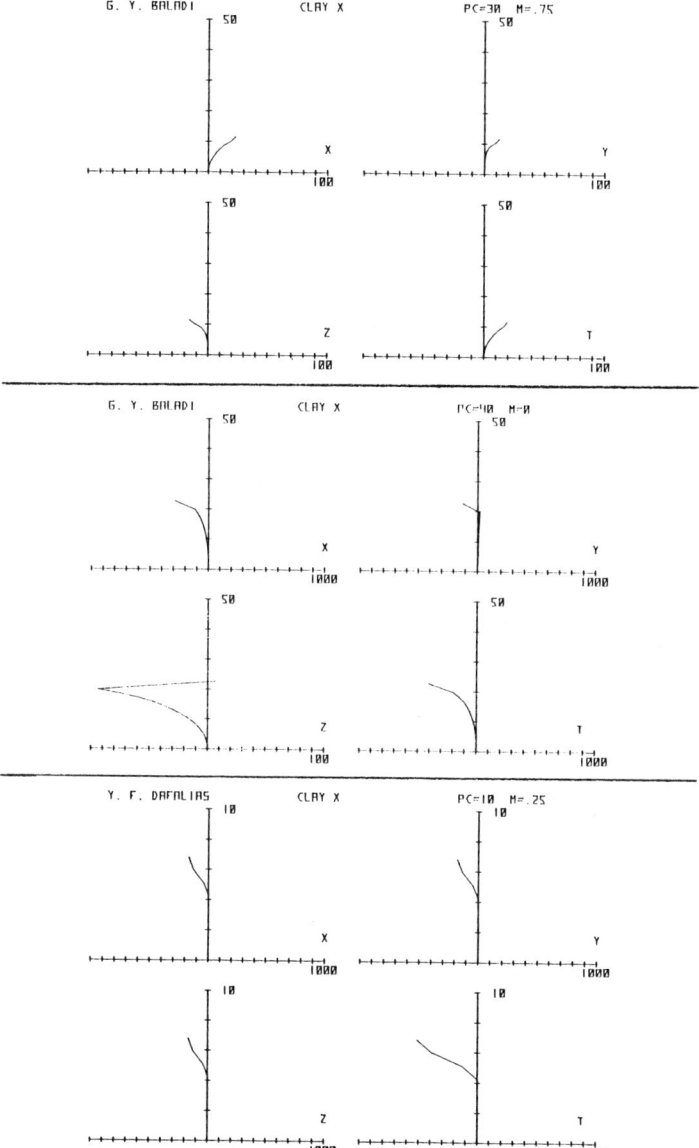

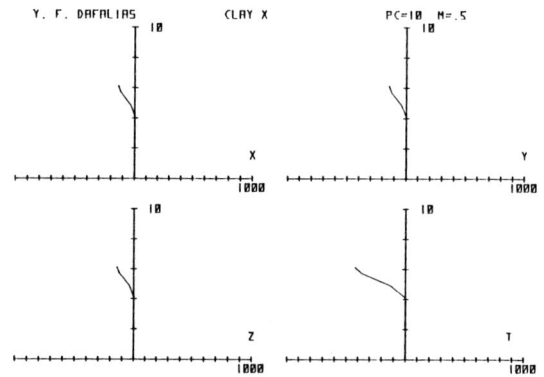

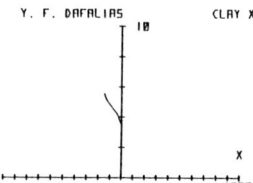

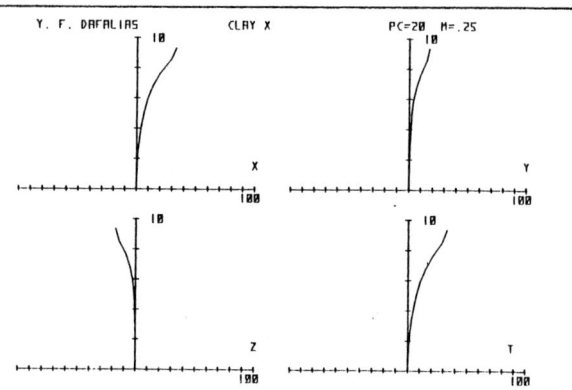

POST WORKSHOP COMPARISONS

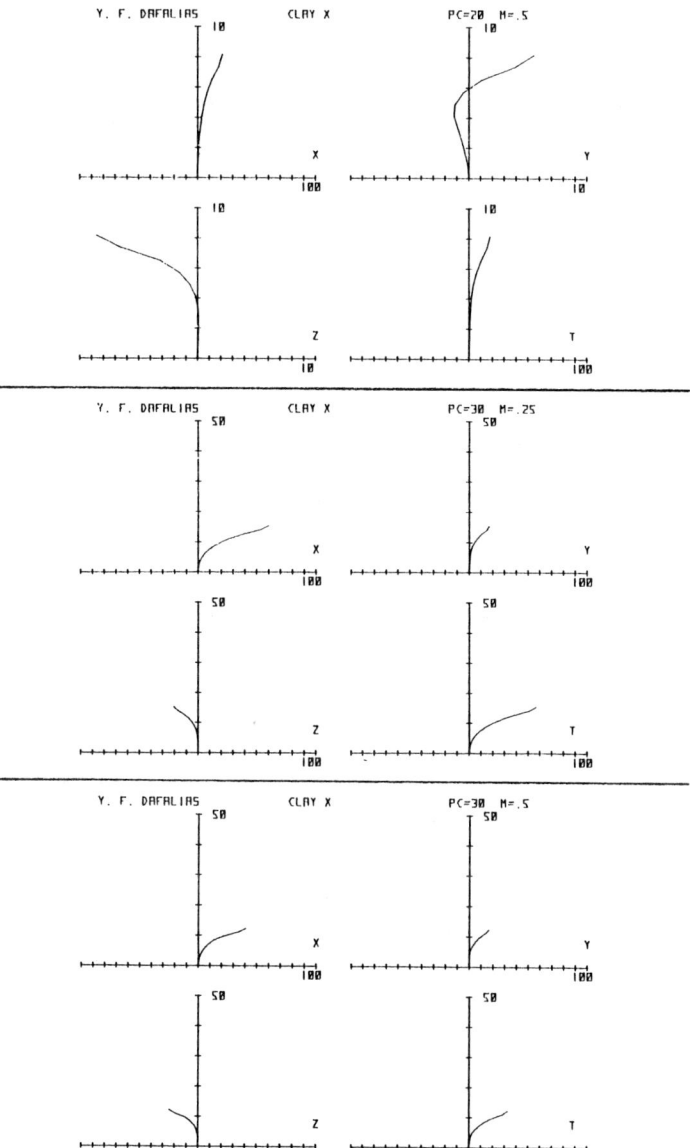

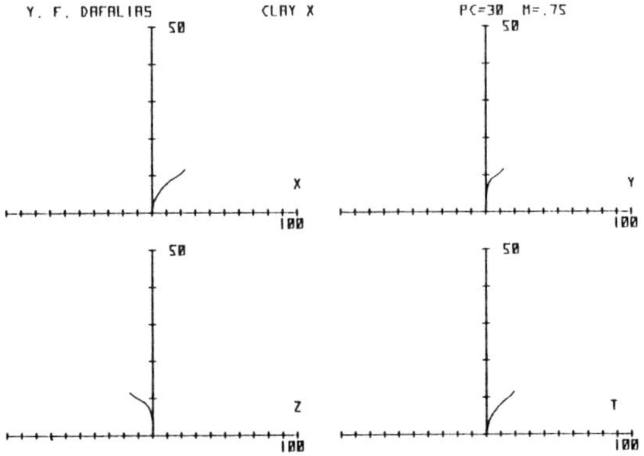

STRESS STRAIN FOR SOILS

Y. F. DAFALIAS   CLAY X   PC=30 M=.75

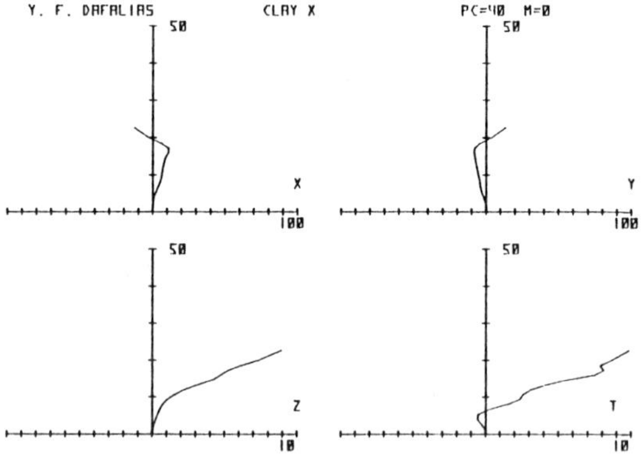

Y. F. DAFALIAS   CLAY X   PC=40 M=0

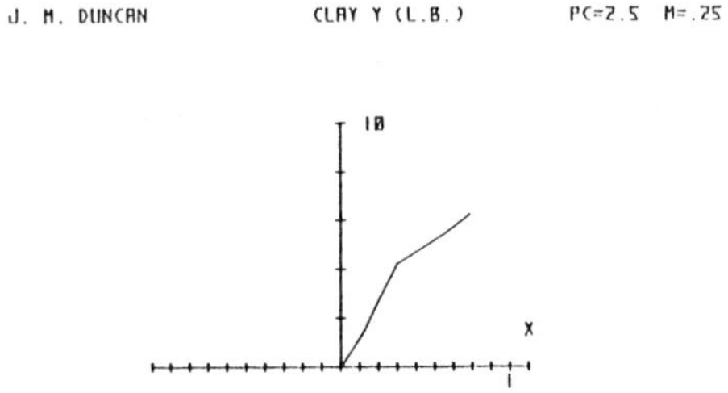

J. M. DUNCAN   CLAY Y (L.B.)   PC=2.5 M=.25

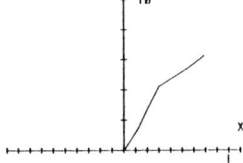

POST WORKSHOP COMPARISONS 227

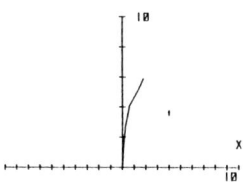

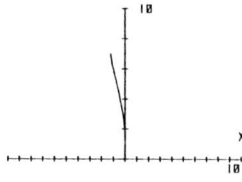

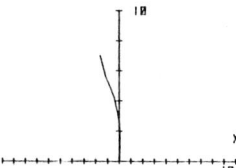

228

J. M. DUNCAN   CLAY Y (L.B.)   PC=5 M=.5

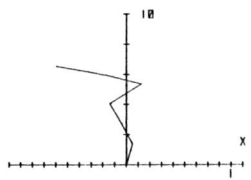

J. M. DUNCAN   CLAY Y (L.B.)   PC=5 M=.75

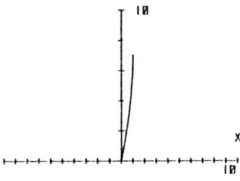

J. M. DUNCAN   CLAY Y (L.B.)   PC=10 M=.25

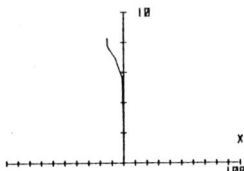

POST WORKSHOP COMPARISONS

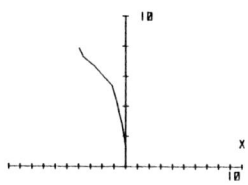

J. M. DUNCAN   CLAY Y (L.B.)   PC=10 M=.5

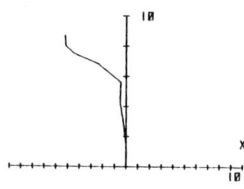

J. M. DUNCAN   CLAY Y (L.B.)   PC=10 M=.75

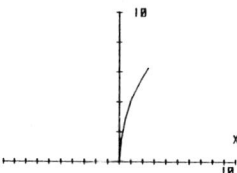

J. M. DUNCAN   CLAY Y (U.B.)   PC=2.5 M=.25

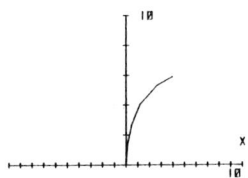

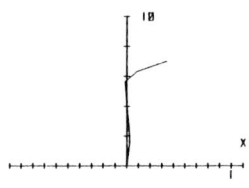

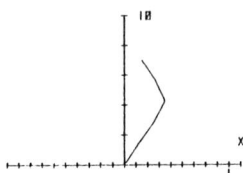

J. M. DUNCAN   CLAY Y (U.B.)   PC=5 M=.5

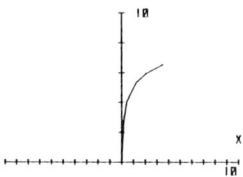

J. M. DUNCAN   CLAY Y (U.B.)   PC=5 M=.75

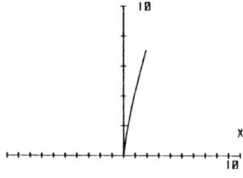

J. M. DUNCAN   CLAY Y (U.B.)   PC=10 M=.25

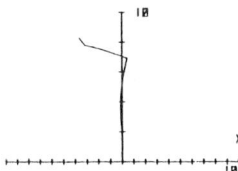

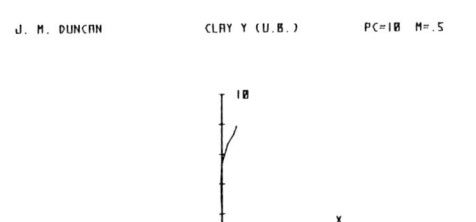

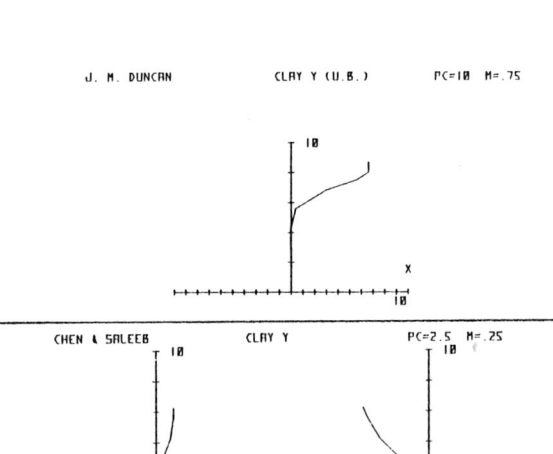

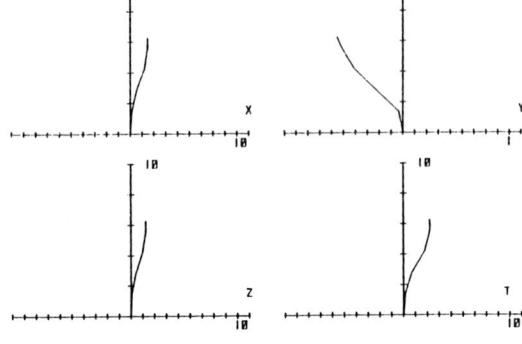

POST WORKSHOP COMPARISONS

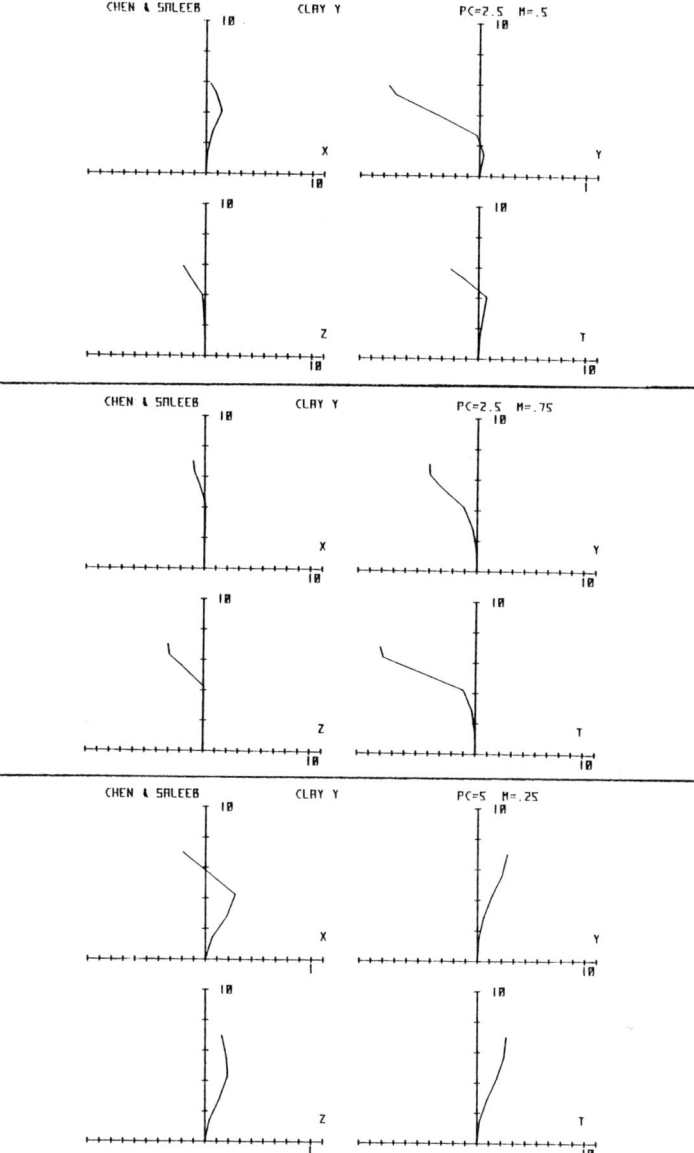

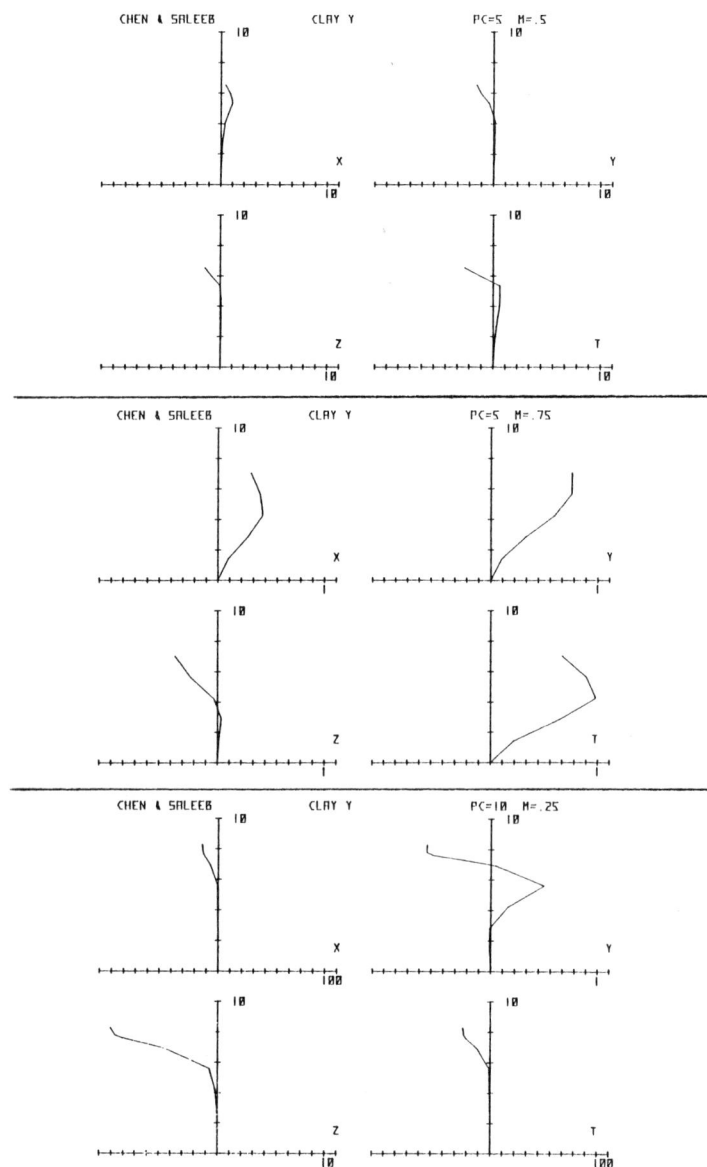

POST WORKSHOP COMPARISONS

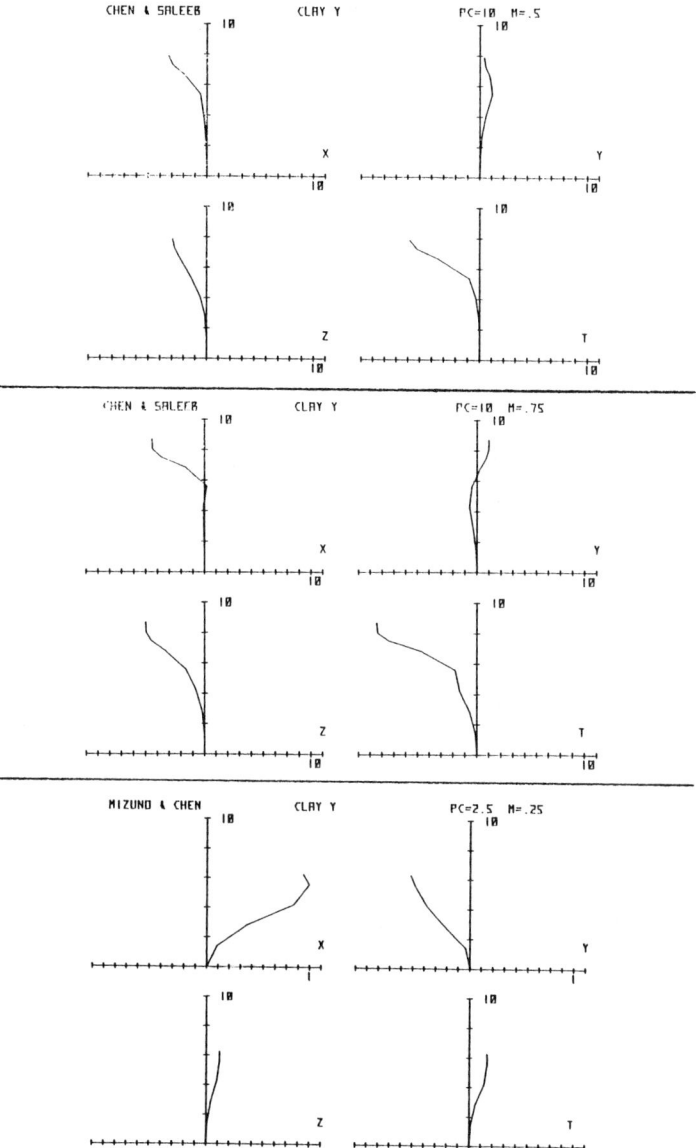

STRESS STRAIN FOR SOILS

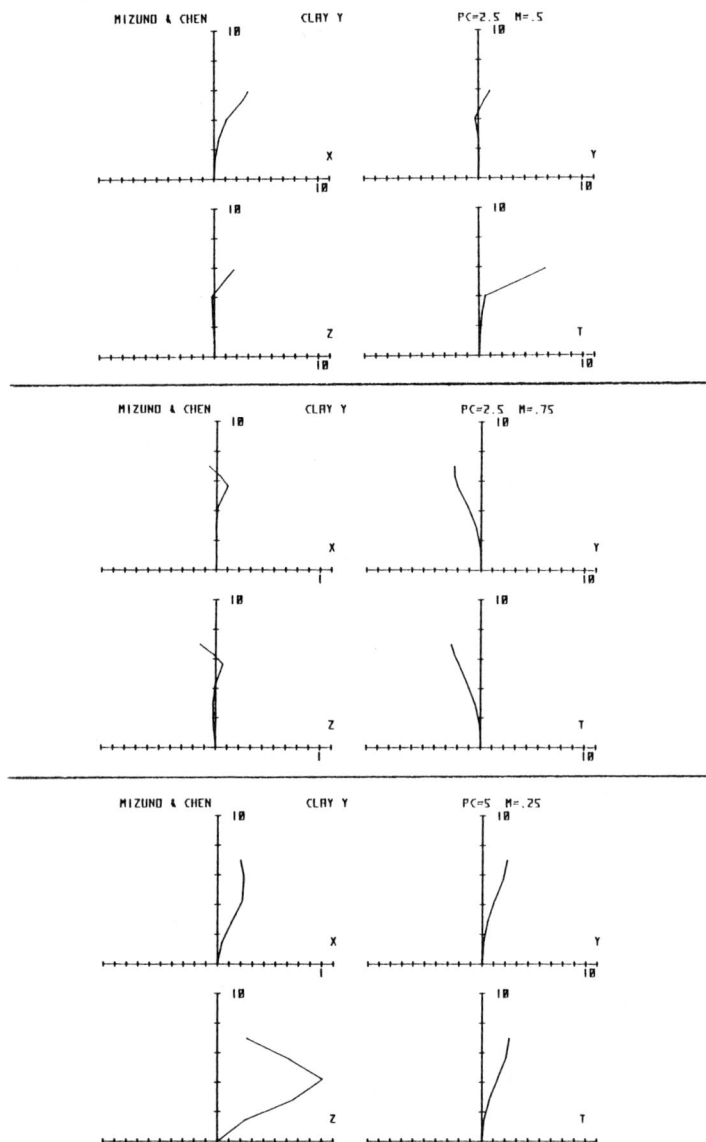

POST WORKSHOP COMPARISONS

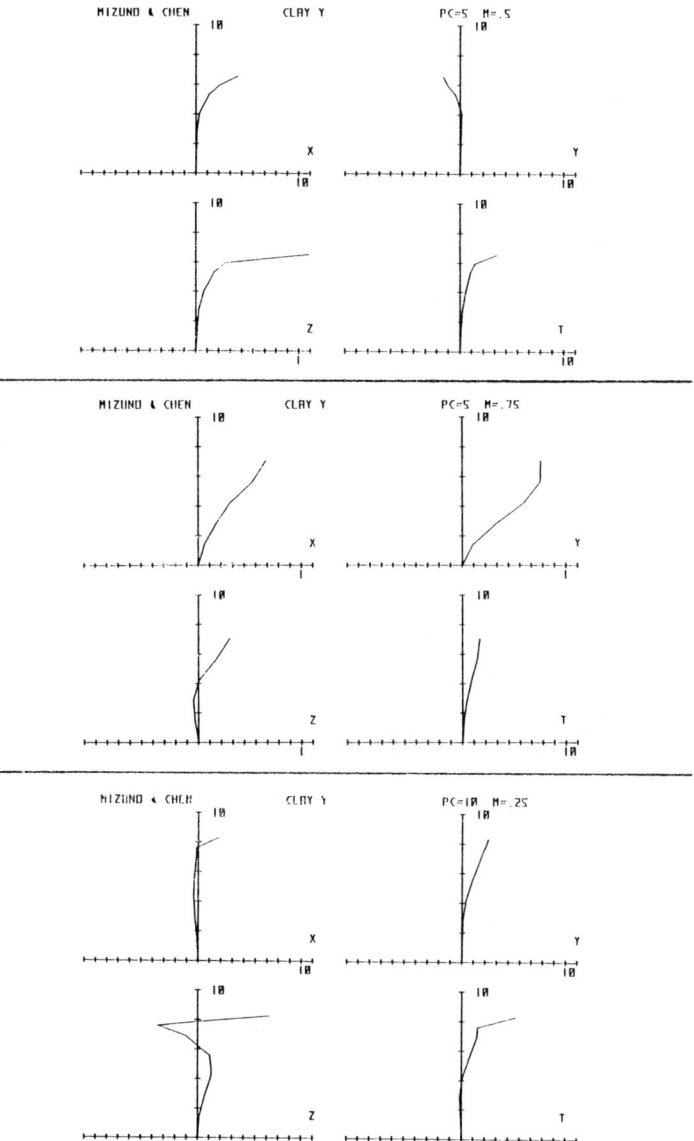

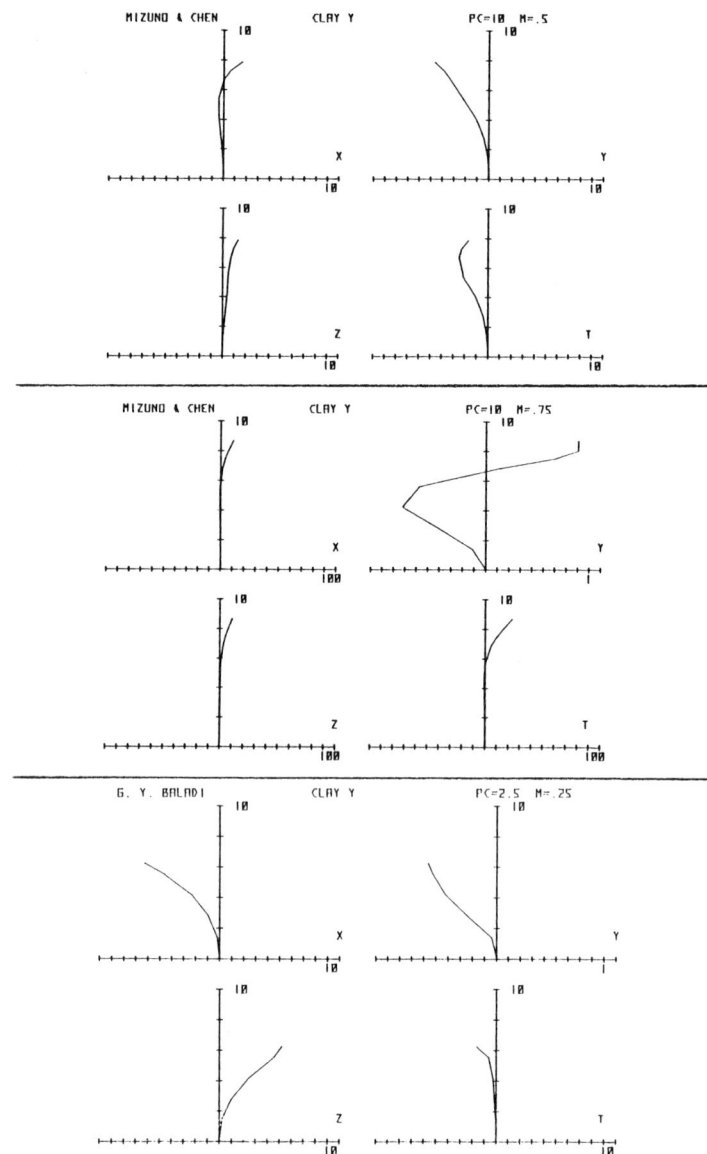

POST WORKSHOP COMPARISONS

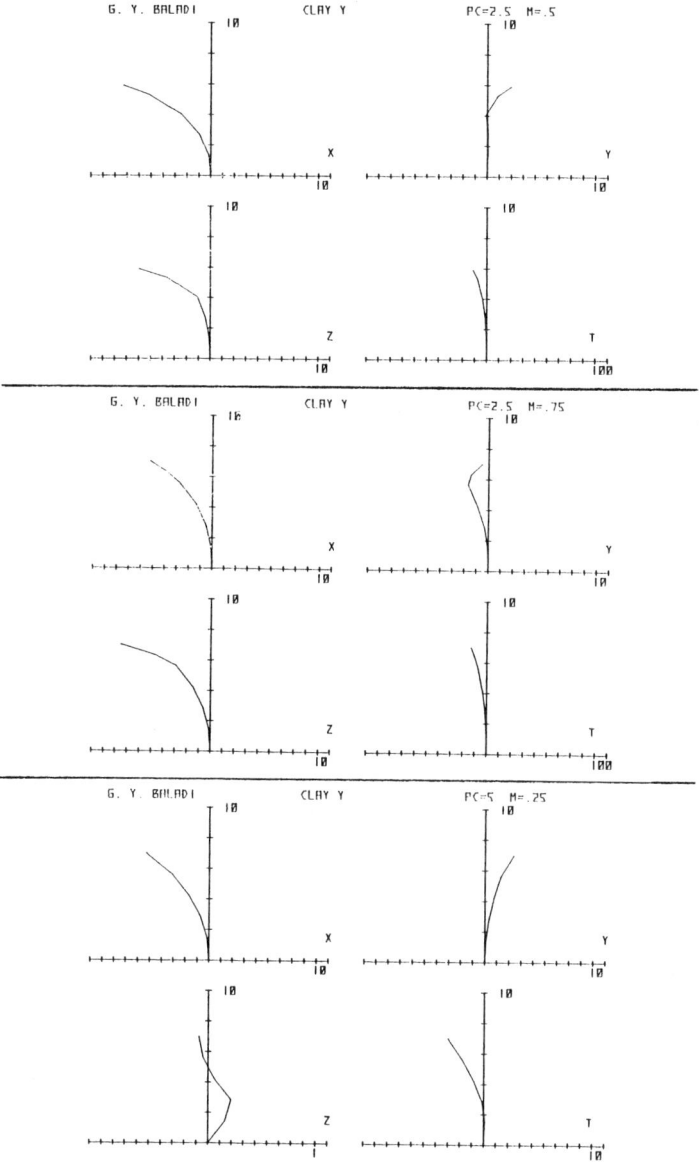

STRESS STRAIN FOR SOILS

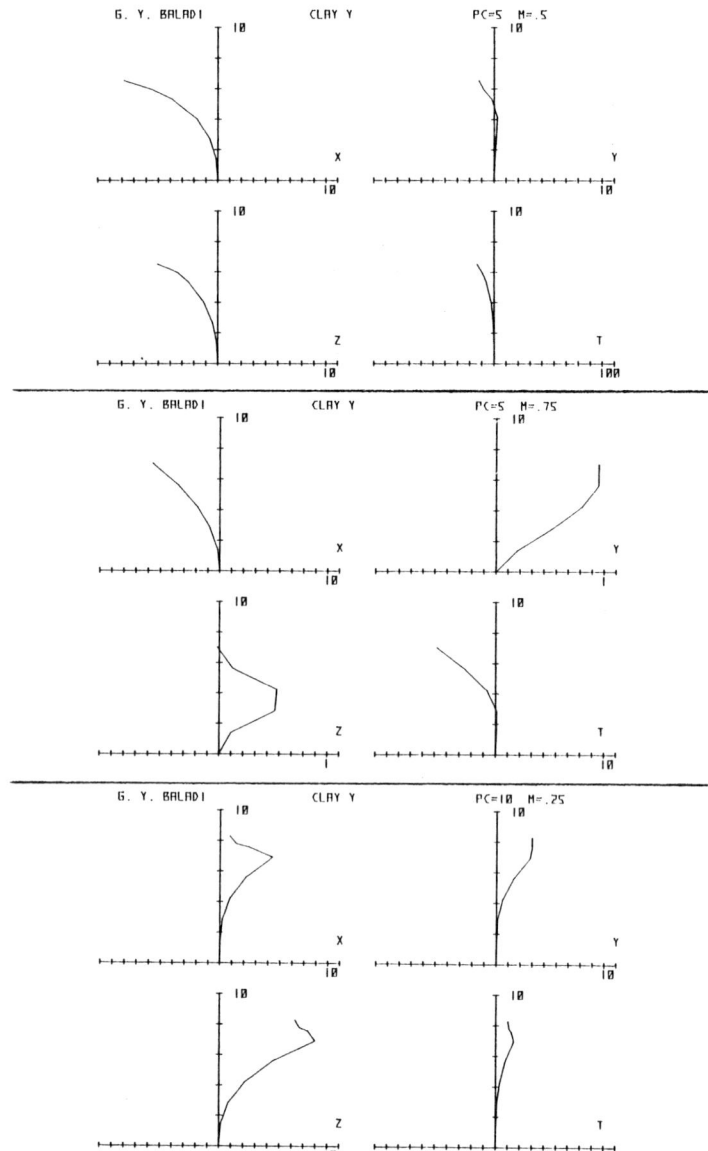

POST WORKSHOP COMPARISONS

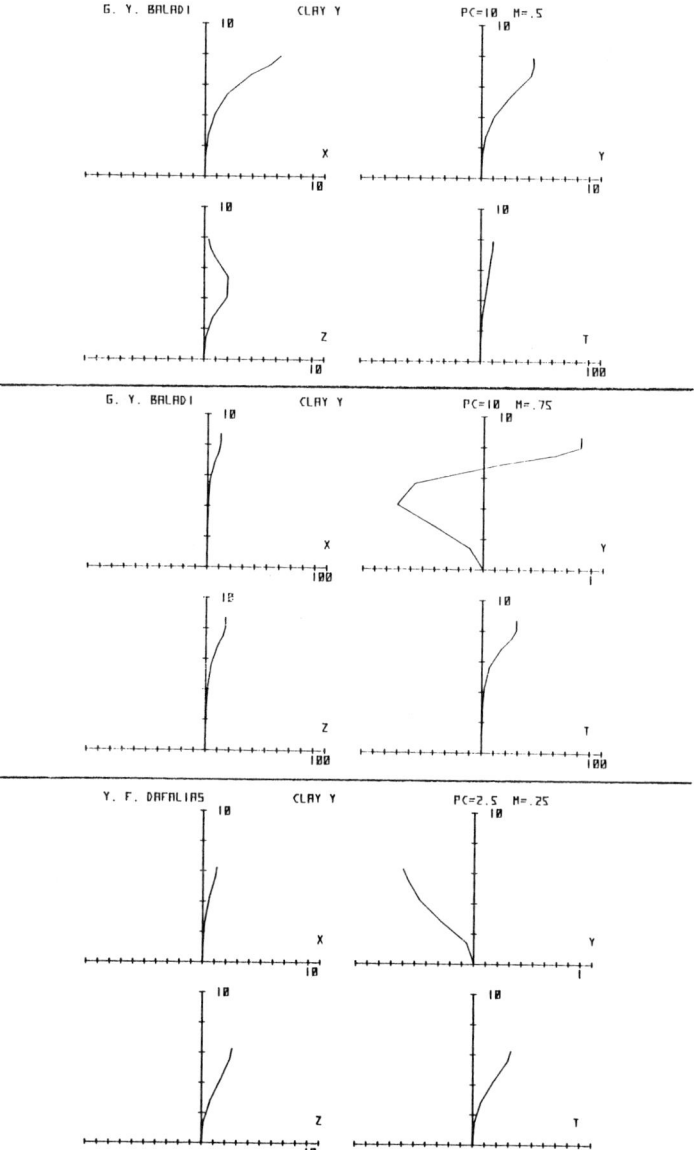

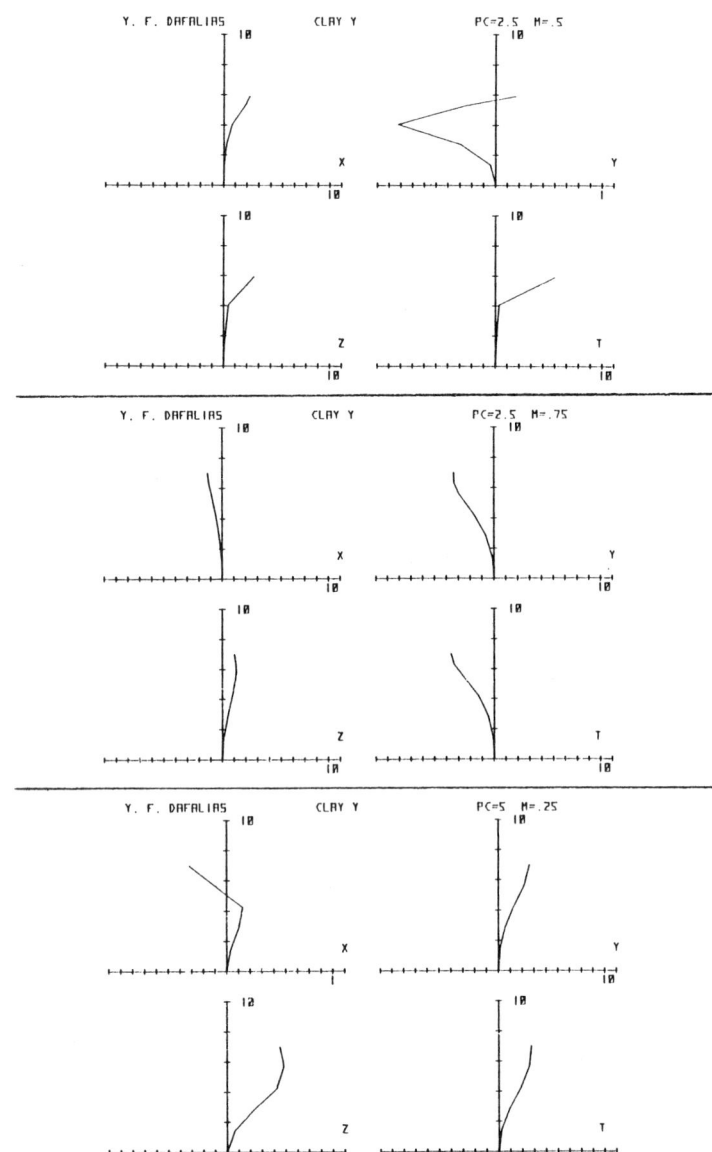

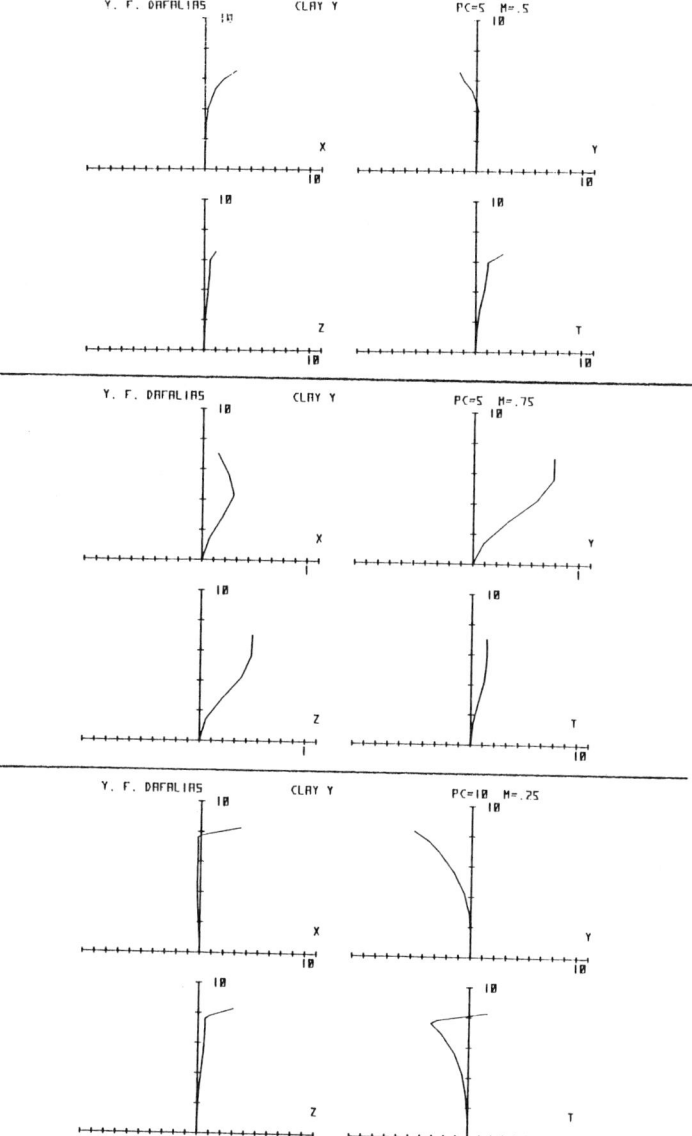

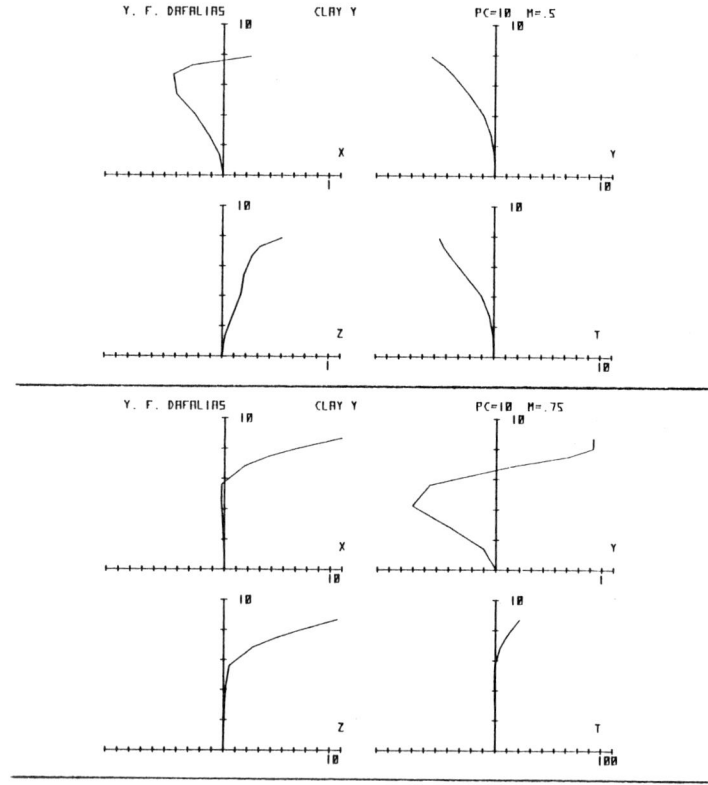

STRAINS PREDICTED USING

HYPERBOLIC STRESS-STRAIN RELATIONSHIPS

by James M. Duncan,[1] M. ASCE

INTRODUCTION

The hyperbolic model was used to predict strains for all four of the soils considered in the Workshop exercise. The predictions and comparisons with the measured values are discussed in the following paragraphs.

CLAY X

The predicted and measured strains for Clay X are shown in Fig. 1. As for all of the soils, ranges of values of strain were predicted for each test based on the estimated uncertainties in the test data and the approximations in the model. The predicted ranges were slightly different for different values of m: The values of strain (ε_1) predicted for m = 3/4 were about 10% smaller than those for m = 1/4. These differences are small compared with the ranges predicted for each value of m, and the full ranges for all values of m are shown together in Fig. 1, with the measured values.

The tests on Clay X were all performed at constant mean normal stress. For this condition the hyperbolic model (which is elastic in nature) predicts zero volume change.

---

[1] Prof. of Civ. Engrg., Univ. of California, Berkeley, CA

CLAY Y

The predicted and measured strains for Clay Y are shown in Fig. 2. As in the case of Clay X, the predicted ranges were only slightly different for the different values of m, and all are shown together in Fig. 2. The predicted strains and strength values were the same for all values of σ_c' (2.5, 5.0, and 10.0 psi).

REMOULDED KAOLINITE

The predicted and measured strains for the remoulded kaolinite are shown in Fig. 3. The calculations were performed using total stresses, and no pore pressures were calculated. Therefore no comparisons of pore pressures or effective stress paths can be made.

OTTAWA SAND

The predicted and measured strains for Ottawa sand are shown in Figs. 4, 5, and 6. Simple shear test results are shown in Fig. 4. The predicted strains for m = 0.5 are smaller than those measured, undoubtedly because the triaxial value of ϕ was used for all values of m, whereas the actual value of ϕ is larger for m = 0.5 (a condition near plane strain) than for m = 0 (triaxial compression) or m = 1.0 (triaxial extension).

Results for proportional loading tests and for reduced triaxial compression and extension tests are shown in Fig. 5. Results for the circular stress path test are shown in Fig. 6.

CONCLUSION

The comparisons shown in Figs. 1 through 6 speak for themselves. Conclusions as to whether the predicted values of strain are in good or poor agreement with the measured values will depend on which test conditions and which types of results are considered most important. These conclusions are left to the judgment of the reader.

Perhaps the two most outstanding characteristics of the hyperbolic model are its simplicity and its wide applicability. The fact that only conventional test results are required for its application, and the fact that it can be used to model any type of soil under either drained or undrained conditions, make the hyperbolic model well-suited for application to a variety of practical problems.

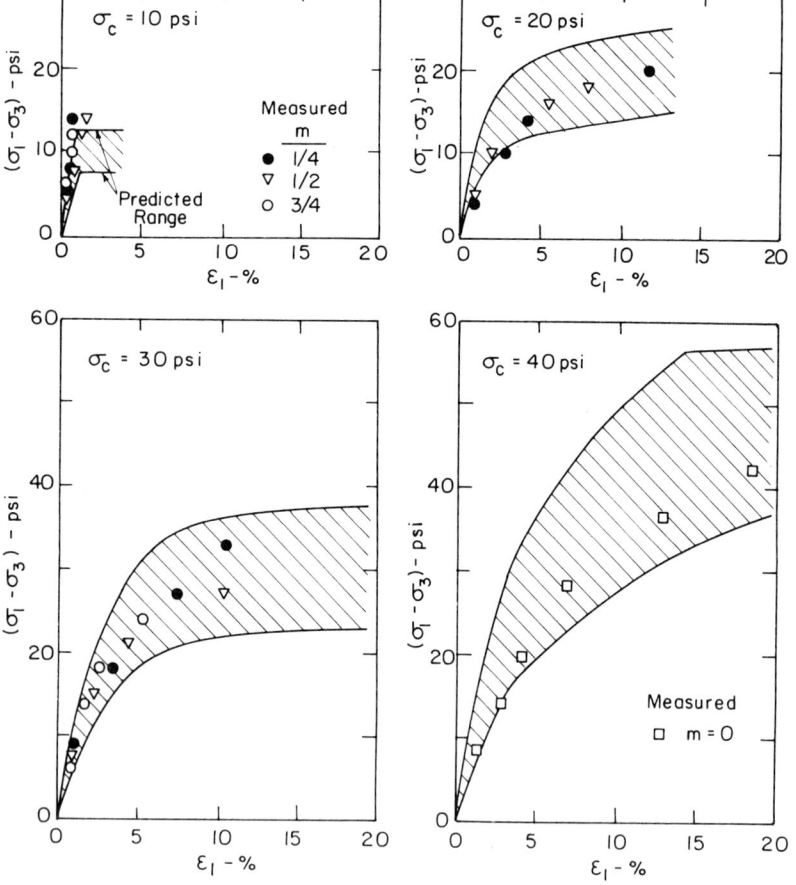

Fig. 1 CLAY X - HYPERBOLIC MODEL - COMPARISON OF PREDICTIONS AND MEASUREMENTS

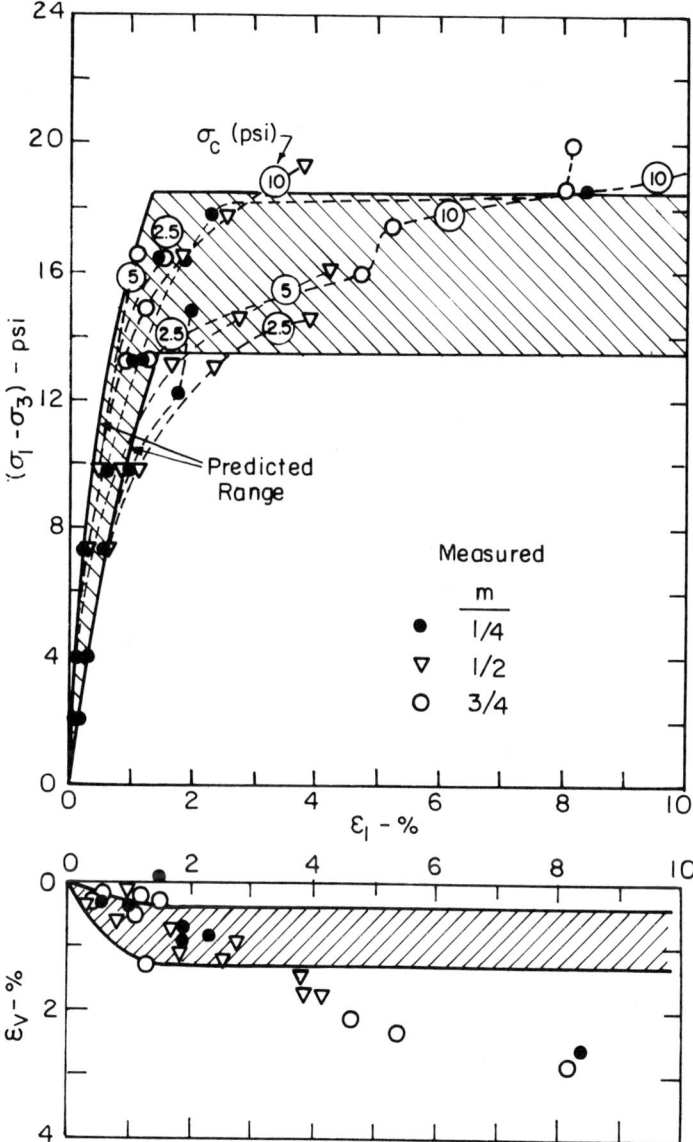

Fig. 2 CLAY Y - HYPERBOLIC MODEL - COMPARISON OF PREDICTIONS AND MEASUREMENTS

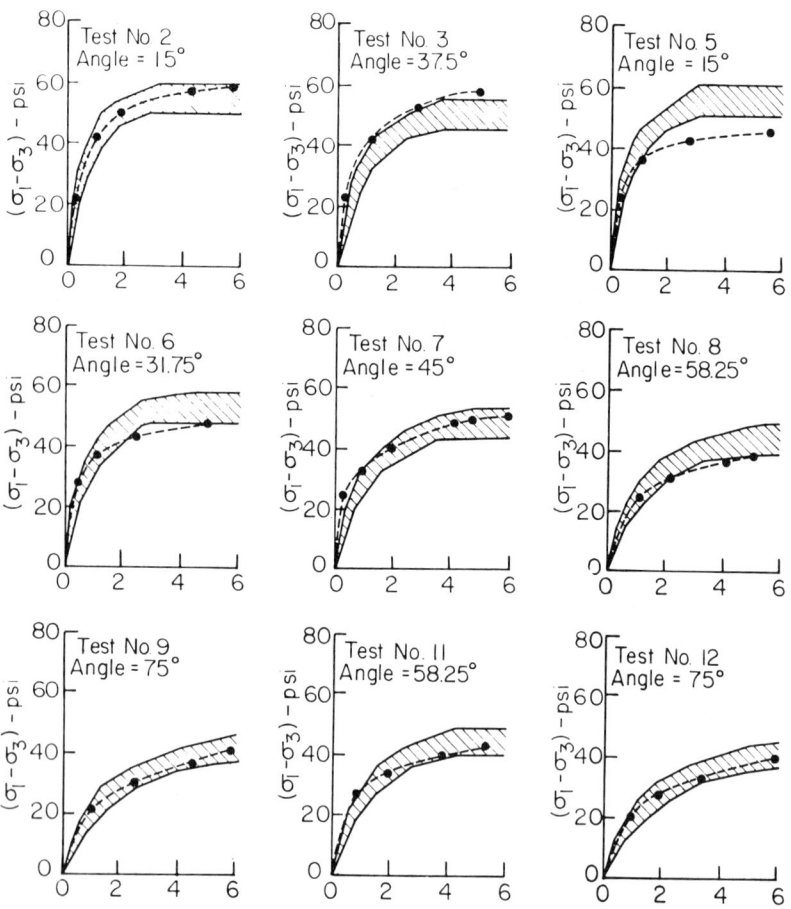

Fig. 3 REMOULDED KAOLINITE - HYPERBOLIC MODEL - COMPARISONS OF PREDICTIONS AND MEASUREMENTS

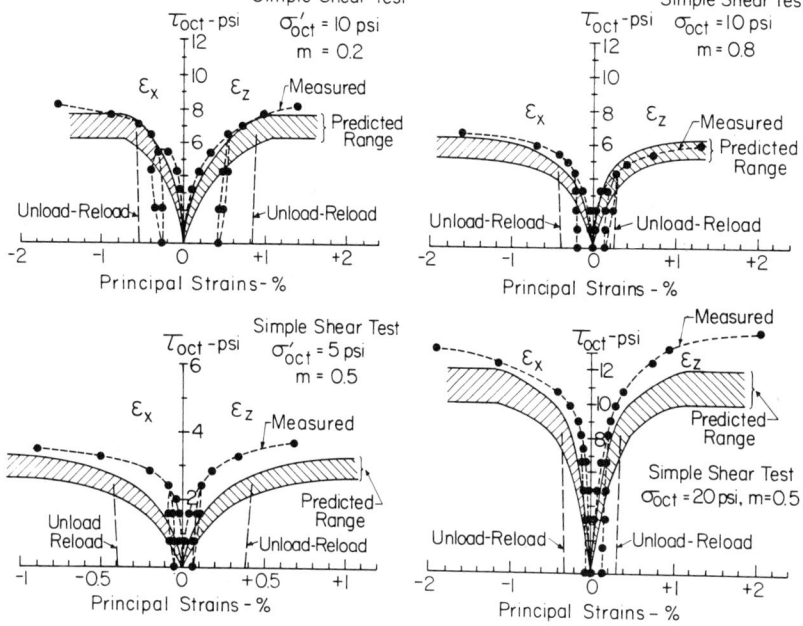

Fig. 4 OTTAWA SAND - HYPERBOLIC MODEL - COMPARISONS OF PREDICTIONS AND MEASUREMENTS

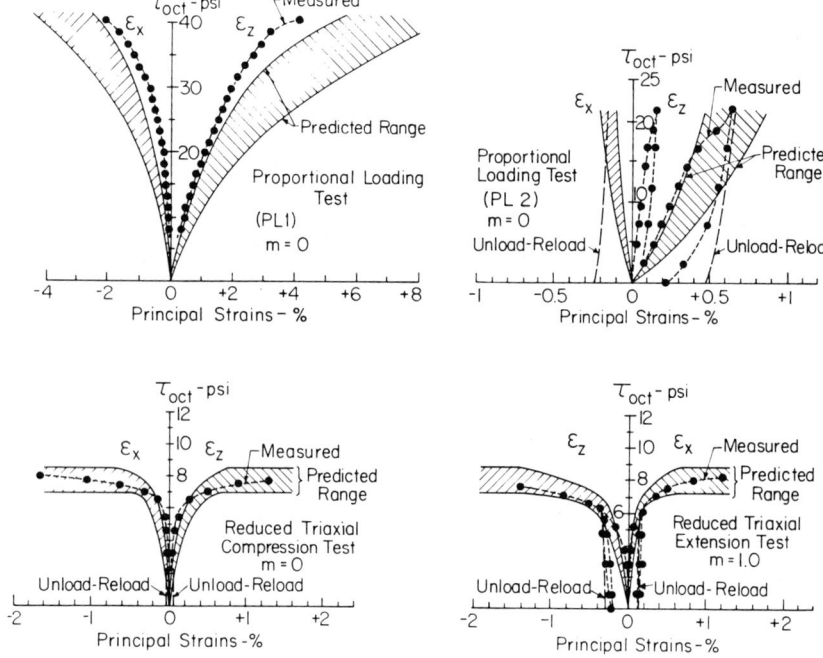

Fig. 5 OTTAWA SAND - HYPERBOLIC MODEL - COMPARISONS OF PREDICTIONS AND MEASUREMENTS

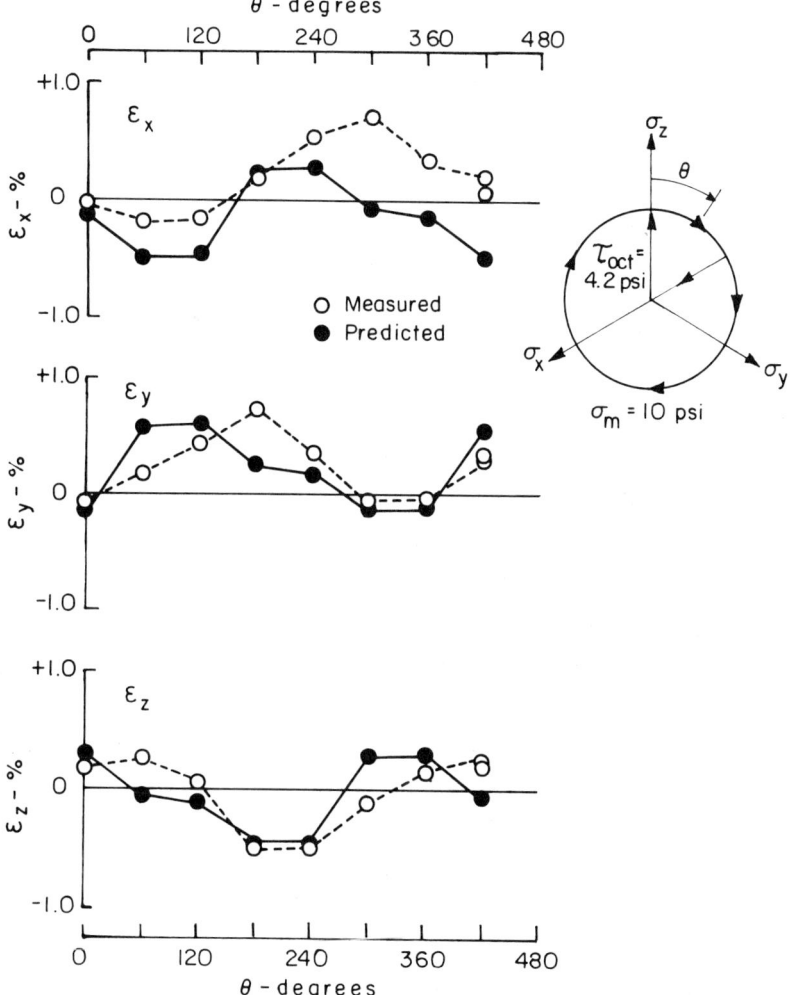

Fig. 6 OTTAWA SAND - HYPERBOLIC MODEL - COMPARISONS FOR CIRCULAR STRESS PATH TEST

EVALUATION OF PREDICTIONS

by

Edward Kavazanjian, Jr.[*] James K. Mitchell,[**] and Rudolph Bonaparte[***]

SUPPLEMENT TO THE
POSITION PAPER FOR NSF/NSERC SYMPOSIUM ON CONSTITUTIVE
RELATIONSHIPS FOR SOILS

McGill University, May 28,29,30, 1980

This supplement to our paper compares the stress-deformation predictions made by the authors for this symposium with test results provided by the organizers of the symposium. In the position paper, the authors develop their model and describe the method used to make these predictions

CLAY X

Figures 1 - 6 compare the predictions made for Clay X with the test results. Figure 1 shows the results of σ'_c = 40 psi, m = 0. The prediction made for this test is an extension of Ladd and Foott's normalized soil properties concept (15)[†] to include the results of drained tests. The excellent agreement between observed and predicted results is evidence that, at least for this soil, the extension is valid.

---

[*]Assistant Professor, Dept. of Civil Engineering, Stanford University
Stanford, California
[**]Professor, Chairman, Dept. of Civil Engineering, University of California, Berkeley, California
[***]Graduate Student, Research Assistant, Dept. of Civil Engineering
University of California, Berkeley, California
[†]Reference numbers refer to the references list in the position paper.

Figures 2, 3, and 4 compare results and predictions for three tests at σ'_c = 30 psi. Although the observed behavior for m = 3/4 deviates sharply from predicted deviator stress-axial strain behavior at higher stress levels (Figure 2), the agreement for all three tests can be described as from good to excellent, and in no case is the observed discrepancy greater than that which might be expected due to natural scatter in test results.

Figures 5 and 6 show the comparisons for the tests at σ'_c = 20 psi. Once again, excellent agreement is observed.

In no case was there greater than 20% discrepancy between the observed and predicted behavior of Clay X, and in most cases the discrepancy was less than 5%. Because there are several interrelated factors involved in making these predictions, the close agreement cannot serve to validate any one of them. However, it can be said that the authors' general theory as a whole works well on drained tests for this type of soil, and in particular that the unique volume strain-axial strain relationship predicted by the model gave good results for Clay X.

EDGAR PLASTIC KAOLIN

Figures 7 - 9 compare observed and predicted p'-q stress paths for Edgar Plastic Kaolin. The $\sigma'_{oct}-\tau_{oct}$ stress paths, although not shown, gave results similar to those shown in Figures 7 - 9. The predicted and observed stress paths compare very well. The prediction of pore pressure for some of the tests at constant mean normal stress (test numbers 5, 7 and 8) do show some systematic differences from measured values at higher stress levels. The good agreement for all tests in the low to intermediate stress level range, however, indicates that the pore pressure model is valid for the Edgar Plastic Kaolin at these stress levels.

Figures 10 - 14 show observed and predicted stress-strain behavior for the Edgar Plastic Kaolin. Here too, there seems to be a systematic error in the predictions, with all tests except tests 2 and 3 showing much more ductile behavior than predicted. One possible source of this systematic error is the strength interpolation function that was used. Comparisons were best where the ultimate strength was accurately predicted. As with the stress paths, all tests showed excellent agreement with predicted results at low to intermediate stress levels. Despite the apparent systematic error, all comparisons showed from very good to excellent agreement.

CONCLUSIONS

The authors are quite satisfied with the agreement between the predictions they made for this symposium and the test results furnished by the organizers. The generally excellent agreement observed is taken as evidence that the immediate deformation portion of the authors extended model can accurately describe the stress-deformation behavior of some soils under both drained and undrained conditions and that this extended model deserves further study. The authors plan to continue refinement of the general phenomenological model and, once refined, to place the model within the framework of some numerical method in order to predict the performance of full scale geotechnical systems.

The authors wish to congratulate the organizers on a stimulating and informative symposium. They wish to thank the organizing committee for their many hours of hard work.

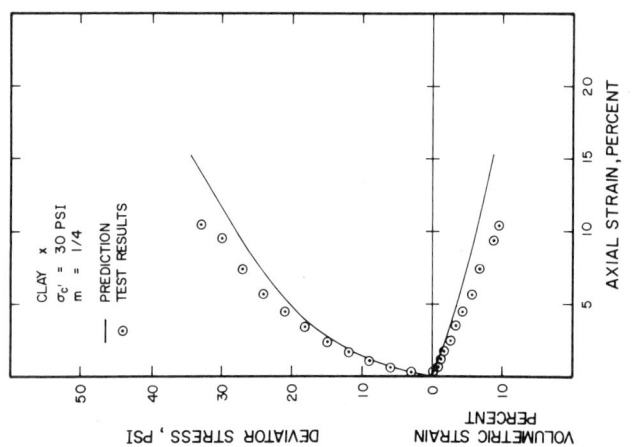

FIG. 2 COMPARISON OF PREDICTED AND OBSERVED BEHAVIOR OF CLAY X.
$\sigma_c' = 30$ PSI, $m = 1/4$

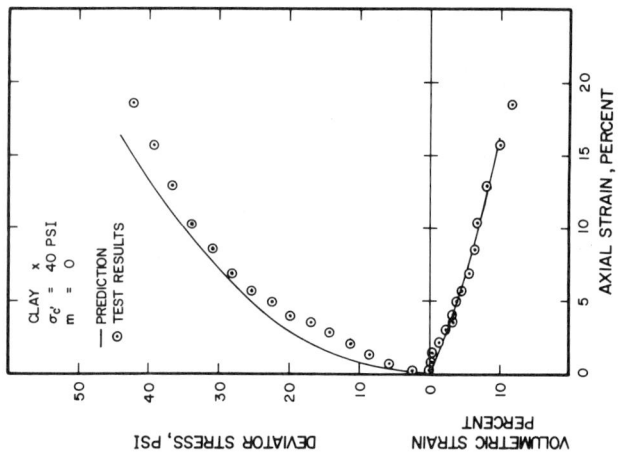

FIG. 1 COMPARISON OF PREDICTED AND OBSERVED BEHAVIOR OF CLAY X.
$\sigma_c' = 40$ PSI, $m = 0$

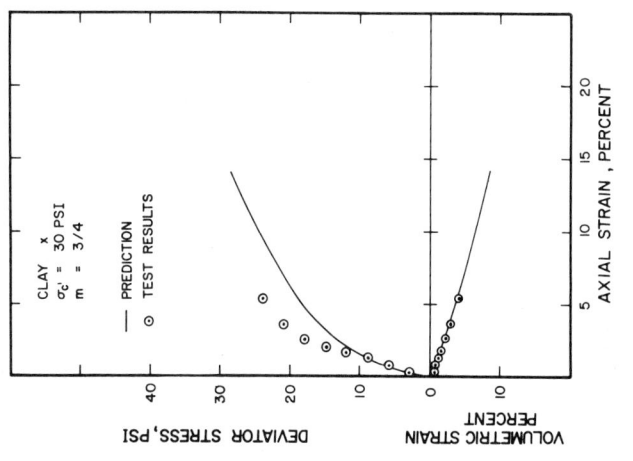

FIG. 4 COMPARISON OF PREDICTED AND OBSERVED BEHAVIOR OF CLAY X. $\sigma_c' = 30$ PSI, $m = 3/4$

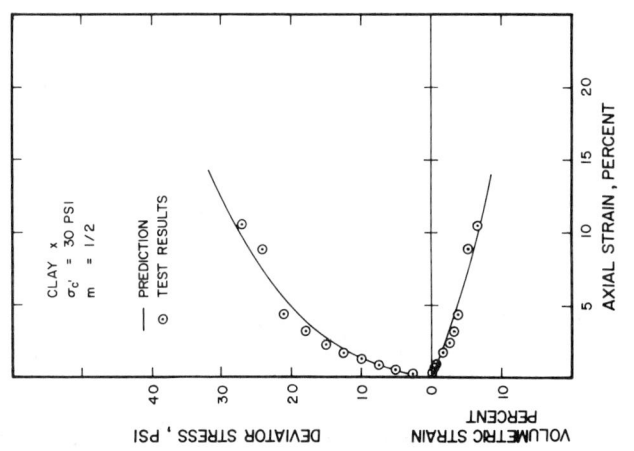

FIG. 3 COMPARISON OF PREDICTED AND OBSERVED BEHAVIOR OF CLAY X. $\sigma_c' = 30$ PSI, $m = 1/2$

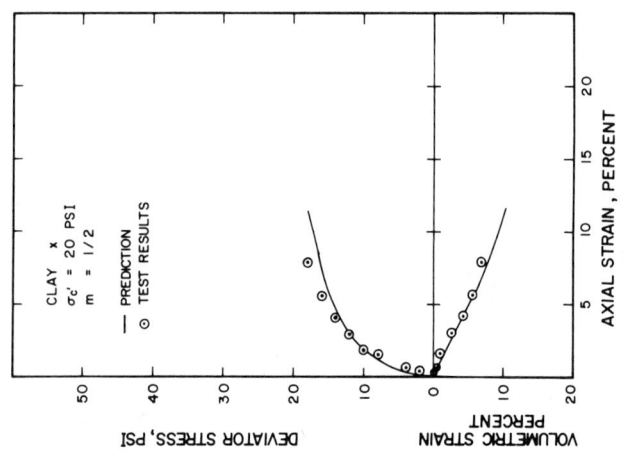

FIG. 6. COMPARISON OF OBSERVED AND PREDICTED BEHAVIOR OF CLAY X. $\sigma_c' = 20$ PSI, $m = 1/2$

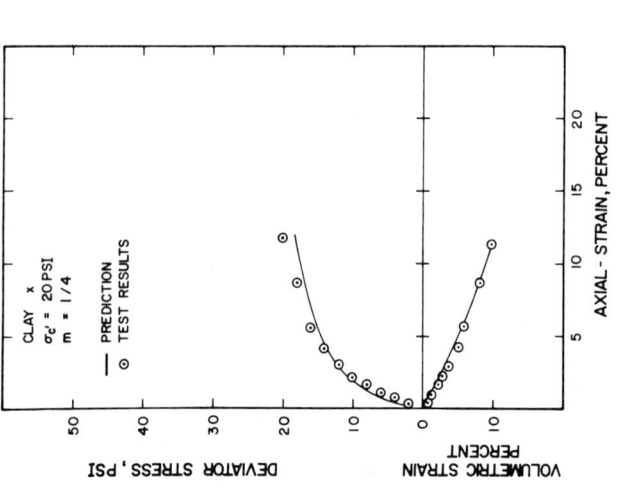

FIG. 5. COMPARISON OF PREDICTED AND OBSERVED BEHAVIOR OF CLAY X. $\sigma_c' = 20$ PSI, $m = 1/4$

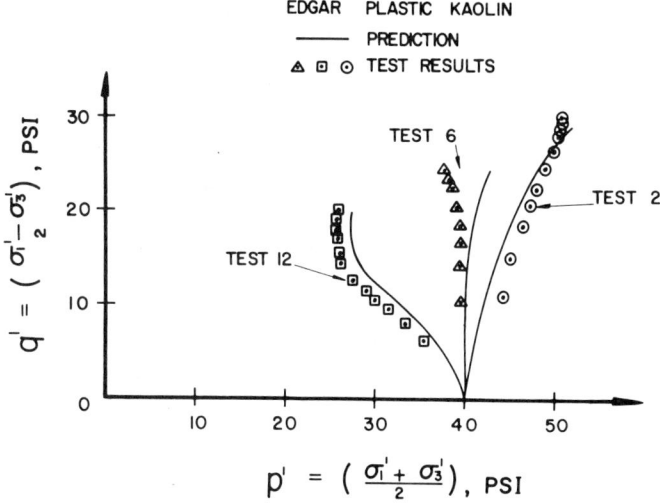

FIG. 7 COMPARISON OF PREDICTED AND OBSERVED STRESS PATHS FOR EDGAR PLASTIC KAOLIN

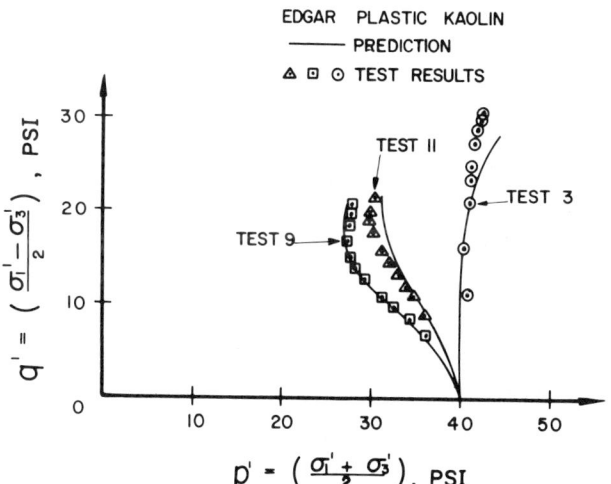

FIG. 8 COMPARISON OF PREDICTED AND OBSERVED STRESS PATHS FOR EDGAR PLASTIC KAOLIN

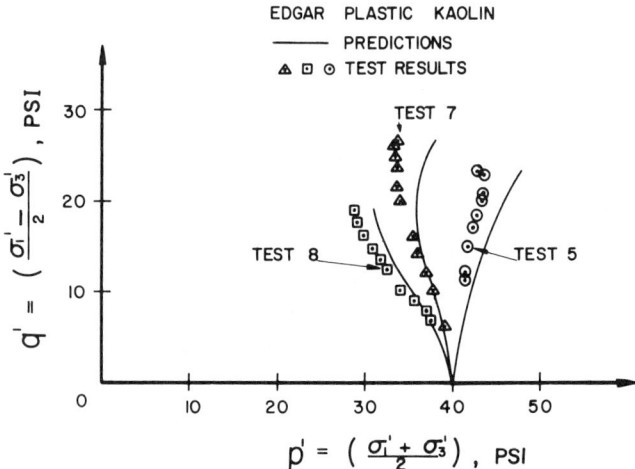

FIG. 9 COMPARISON OF PREDICTED AND OBSERVED STRESS PATHS FOR EDGAR PLASTIC KAOLIN

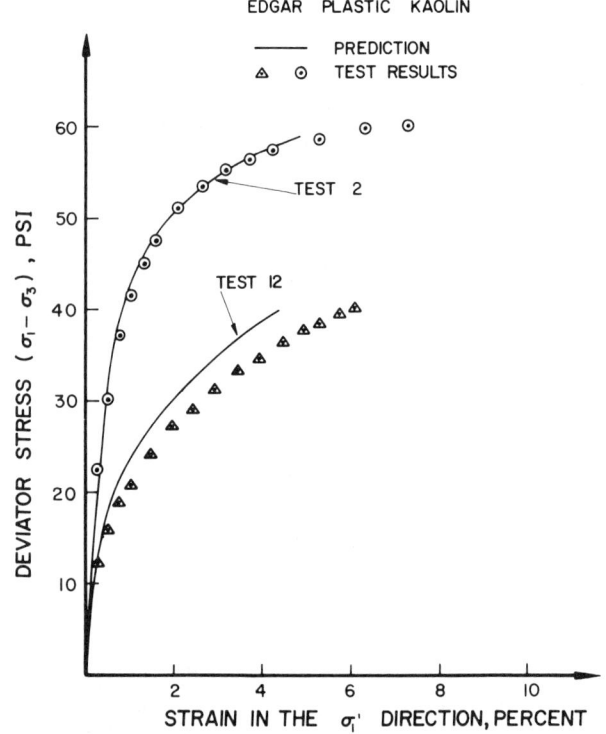

FIG. 10 COMPARISON OF PREDICTED AND OBSERVED STRESS-STRAIN CURVES FOR EDGAR PLASTIC KAOLIN

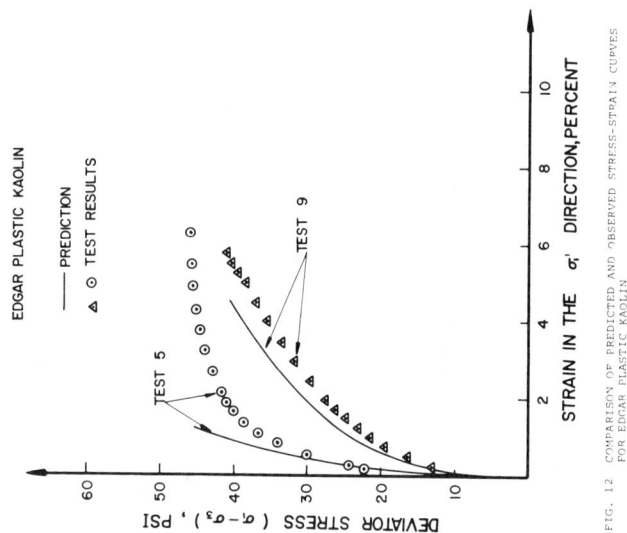

FIG. 12 COMPARISON OF PREDICTED AND OBSERVED STRESS-STRAIN CURVES FOR EDGAR PLASTIC KAOLIN

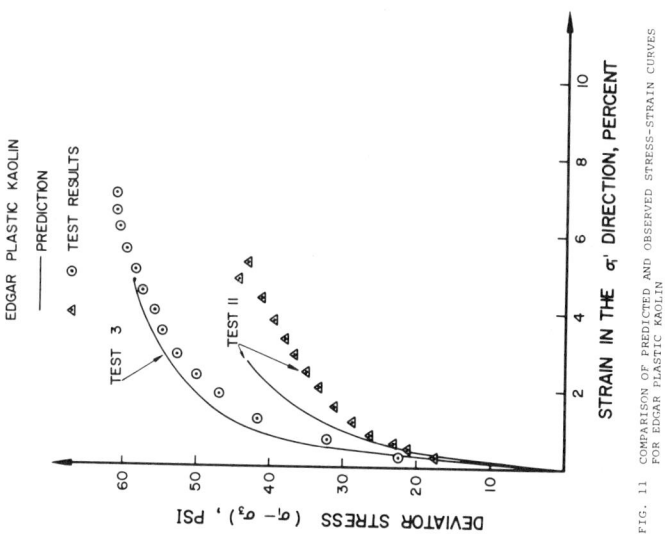

FIG. 11 COMPARISON OF PREDICTED AND OBSERVED STRESS-STRAIN CURVES FOR EDGAR PLASTIC KAOLIN

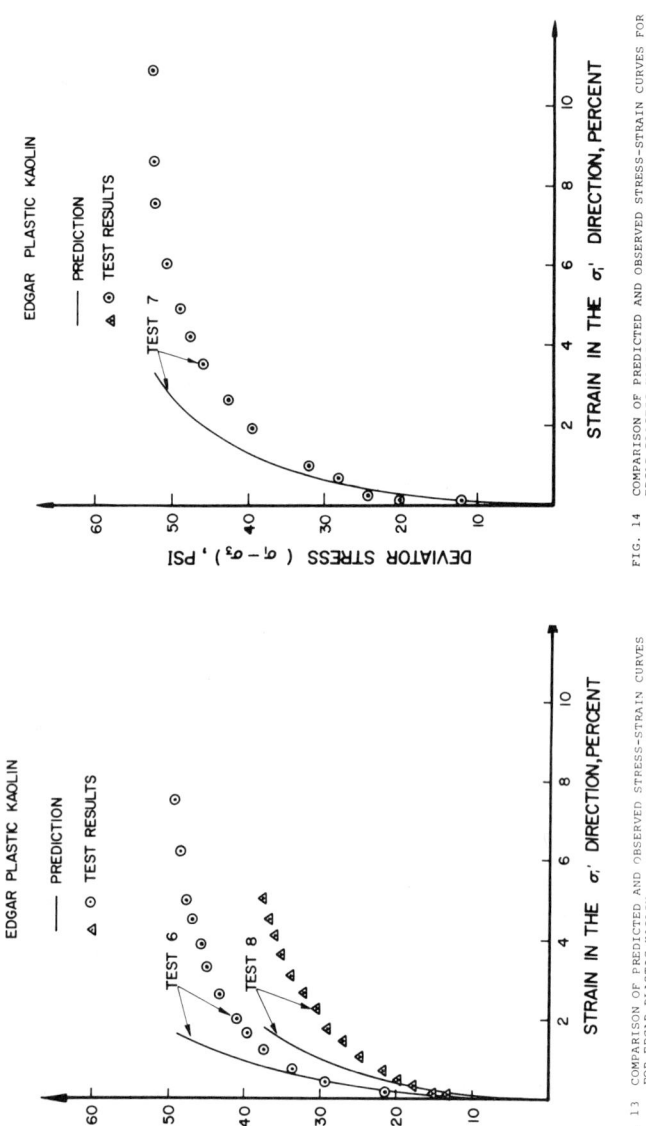

FIG. 13 COMPARISON OF PREDICTED AND OBSERVED STRESS-STRAIN CURVES FOR EDGAR PLASTIC KAOLIN

FIG. 14 COMPARISON OF PREDICTED AND OBSERVED STRESS-STRAIN CURVES FOR EDGAR PLASTIC KAOLIN

<u>Nonlinear Hyperelastic (Green) Constitutive Models</u>

<u>for Soils: Predictions and Comparisons</u>

By: A. F. Saleeb and W. F. Chen

School of Civil Engineering

Purdue University

West Lafayette, IN

1. Introduction

In the first part of the present paper (theory and calibration), the formulation of a third-order nonlinear elastic constitutive model based on Green type of formulation has been given, and the model was used to describe the behavior of three different soils: clays "X", "Y" and Ottawa sand. For the different stress paths used in determining the material constants in the model, the theoretical and experimental results have been compared. Based on these comparisons, it was found that the model qualitatively gives good overall fit to the test results; it produces essentially most of the important features of the behavior observed experimentally such as nonlinearity, stress-path dependency, dilatation, stress-induced anisotropy, effect of the intermediate principal stress and the hydrostatic stress, etc. Quantitatively, the agreement between the calculated and the measured values is reasonably good. However, the theoretical model fails to predict the initial anisotropy observed in tests for clays "X" and "Y". Moreover, the model underestimated the volumetric strains for clay X. For Ottawa sand, it has been demonstrated that the model behavior is very sensitive to changes in the higher order material constants in loading paths near hydrostatic stress path when large stress levels are generally encountered.

Herein, comparisons of the predicted theoretical and experimental results are given for the three soils along test stress paths different from those used in the determination of the material constants in the model. This provides the necessary verification of the applicability of the constitutive model in describing the soil behavior under general three-dimensional states of stress. In the forthcoming, a summary of the comparisons and a number of conclusions will be given for the three types of soils investigated.

2. Comparisons of Theoretical and Experimental Results

(a) Clay "X"

The calculated and measured stress-strain curves along the different stress paths unsed in the prediction for clay "X" are shown in Fig. 1(a) to (p). By examining these curves, the following conclusions may be made:

1. In almost all the cases investigated, good qualitative agreement is observed between the theoretical and the test results, except for the volumetric strains in the case σ_c = 10 psi, m = 0.25, where the model indicates compressive strains while the test gives extensional values, as illustrated in Fig. 1(b).

2. As for the stress paths used in determining the material constants, the model underestimates the volumetric strains in all the cases, and the values of the intermediate principal strain, ε_2, are not correctly predicted. This is mainly due to the inability of the model to take into account the effect of initial anisotropy observed experimentally.

3. In general, the theoretical stress-strain curves for the axial strain ε_1 show stiffer behavior than that of the measured ones, particularly

at low stress levels. The most pronounced discrepancies between the calculated and the test results are observed for σ_c = 10 psi, m = 0.5; and for σ_c = 20 psi, m = 0.25 and 0.5, where the stiffer behavior extends up to the failure stress levels. However, for these cases, the lateral strains, ε_3, agree well with the measured values, as shown in Figs. 1(d) and (j), for example.

4. For the axial strains, ε_1, the best agreement is obtained for the cases of σ_c = 30 psi, and for σ_c = 10 psi, m = 0.25, where the maximum discrepancy is approximately 30%.

5. In almost all the cases, Mohr-Coulomb failure criterion provides reasonably good predictions for failure stresses.

(b) Clay "Y"

Based on the comparisons of the predicted theoretical and experimental stress-strain curves illustrated in Figs. 2(a) to (r), the following conclusions can be made:

1. As for clay "X", good overall qualitative fit is obtained in most cases, However, the predicted values for the volumetric strains, ε_v do not agree, even qualitatively, with the laboratory results for the cases: σ_c = 2.5 psi, m = 0.75; σ_c = 5 psi, m = 0.25; and σ_c = 10 psi, m = 0.5, specially for σ_c = 10 psi, m = 0.5 where the model shows dilatant behavior near failure which is not observed in the test data, as given in Fig. 2(p).

2. The volumetric strains for all other cases are better predicted than those of clay "X", as shown in Figs. (b), (j) and (n), for example.

3. In most cases, the axial strain values are well predicted at low stress levels below failure, except possibly for σ_c = 2.5 psi, m = 0.25, where the theoretical curve is stiffer than the experimental one.

As the stresses increase approaching failure levels, the discrepancies increase. The most notable discrepancy is observed for $\sigma_c = 10$ psi, $m = 0.75$, Fig. 2(q).

4. Because of the initial isotropy of the present model, the calculated values of the intermediate principal strain, ε_2, do no agree with the test results in many cases, as was the case for clay "X", since test data showed strong initial anisotropy for both clays "X" and "Y".

5. The accuracy of Mohr-Coulomb condition in predicting failure stresses for clay "Y" is not as good as for clay "X". For instance, the error in calculated value for σ_1' at failure in the case $\sigma_c = 10$ psi, $m = 0.75$ is approximately 11%. This is mainly because Mohr-Coulomb criterion does not take into account the effect of the intermediate principal stress, σ_2', on failure.

(c) <u>Ottawa Sand</u>

In Figs. 3 to 8, the results of the comparisons for Ottawa sand are illustrated. The curves are plotted in terms of the octahedral shear stress, $\tau_{oct.} = \sqrt{\frac{2}{3} J_2}$, where $J_2 = \frac{1}{2} s_{ij} s_{ij}$ is the second invariant of the stress deviator tensor, s_{ij}. The following observations and conclusions may be made:

1. Reasonably good overall qualitative fit is observed in all cases investigated except for the small compressive lateral strains $\varepsilon_x = \varepsilon_y$ observed experimentally in RTC test at low stress levels. Quantiatively, the predicted values for the SS tests with $\sigma_{oct.} = 5$ and 10 psi agree well with the test results at low stress levels. At high stress levels near failure, the calculated strains are too small compared to the measured values. For the SS test with $\sigma_{oct} = 20$ psi and $m = 0.5$, the theoretical curves are too soft compared to the experimental curves, as

shown in Figs. 5(a) and (b). The axial strain, ε_z, is better predicted in the RTC test than in the RTE test. However, better agreement is obtained for the lateral strains, $\varepsilon_x = \varepsilon_y$, in the RTE test, Figs. 7 and 8.

2. Although only an approximate modeling of unloading-reloading behavior has been used, the calculated slopes describing the unloading-reloading behavior agree well in most cases with those determined experimentally, as shown in Figs. 3, 4, 5, 6 and 8, for example.

3. The calculated values of failure stresses using Mohr-Coulomb condition are less accurate for Ottawa sand for the stress paths described in Figs. 3 to 8 than for those used in determining the material constants. The theoretical values underestimate the measured failure stresses. In general, the accuracy of predicting failure stresses is less for Ottawa sand than for clays "X" and "Y".

4. At high stress levels in increasing proportional loading stress paths (particularly near HC path), the model predicts very large values of strains compared to those obtained in the actual tests. This can be directly seen from the general expressions of strains in Eq. (16a) in the first part of the paper; although the coefficients of the quadratic and cubic terms, C_2 and C_3, respectively, are generally small for such stress paths, the results obtained are considerably large when these coefficients are multiplied by large values of the parameter λ. As has been shown in Fig. 10 in the first part, the model becomes sensitive to changes in the higher order material constants for large values of λ. The behavior of the model with the changed values of B_6 and B_7 agrees better with the test results provided than the behavior when using the original set of the material constants. The latter greatly overestimates the strain values at high stress levels. More work and further refinement

are needed in determining the material constants to overcome this difficulty.

3. Summary

The formulation of an isotropic third-order hyperelastic constitutive model has been given, and the model has been applied to three types of soil. Theoretical and experimental stress-strain curves were compared for different stress paths for each of the three soils. It was found that the model is capable of describing most of the salient features of soil stress-strain behavior such as nonlinearity, stress-path dependency, dilatation, stress-induced anisotropy, effect of the hydrostatic stress, and the effect of intermediate principal stress and third stress invariant. For monotonically increasing loading conditions, the model satisfies all the rigorous mathematical requirements such as uniqueness, stability and continuity. Reasonably good overall fit has been obtained in most of the cases investigated. However, for clays "X" and "Y", the initial anisotropy observed experimentally cannot be predicted since the present model is initially isotropic. Moreover, it has been found that for Ottawa sand the model, as presently formulated, may exhibit questionable behavior in stress paths near hydrostatic compression stress path where large stress levels are generally encountered. This is mainly because of the large effects of the higher order terms in the constitutive law. More refinement and adjustment of the material constants in the model are needed in order to reduce these effects.

Acknowledgments

This material is based upon work supported by the National Science Foundation under Grant No. PFR-7809326 to Purdue University.

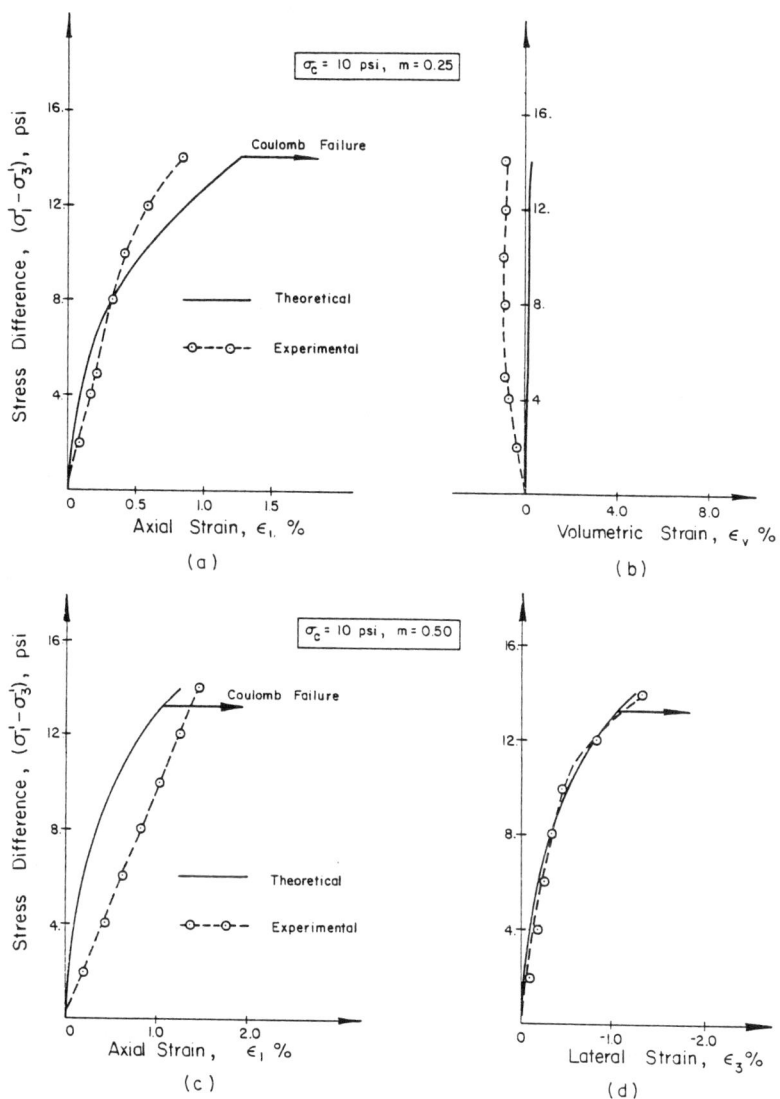

FIG I. Comparison of the Predicted Theoretical and Experimental Stress-Strain Curves for Clay "X" (TC Tests with Constant Mean Normal Stress)

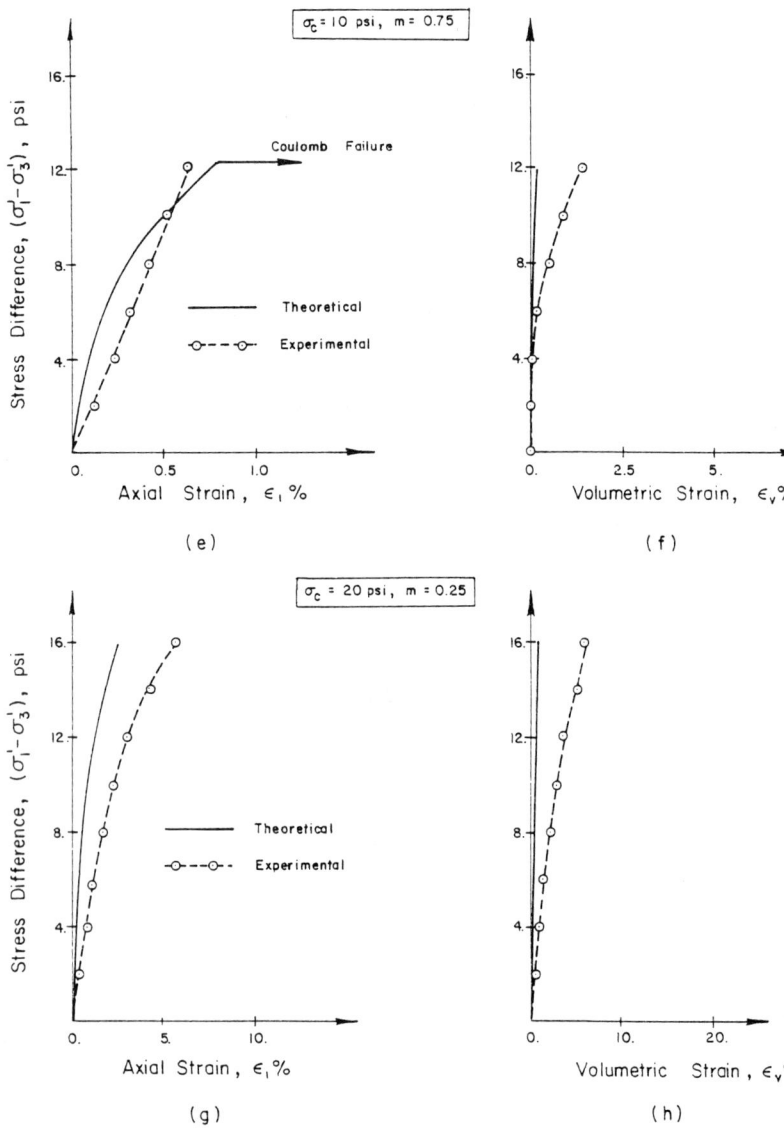

FIG 1. (Cont'd)

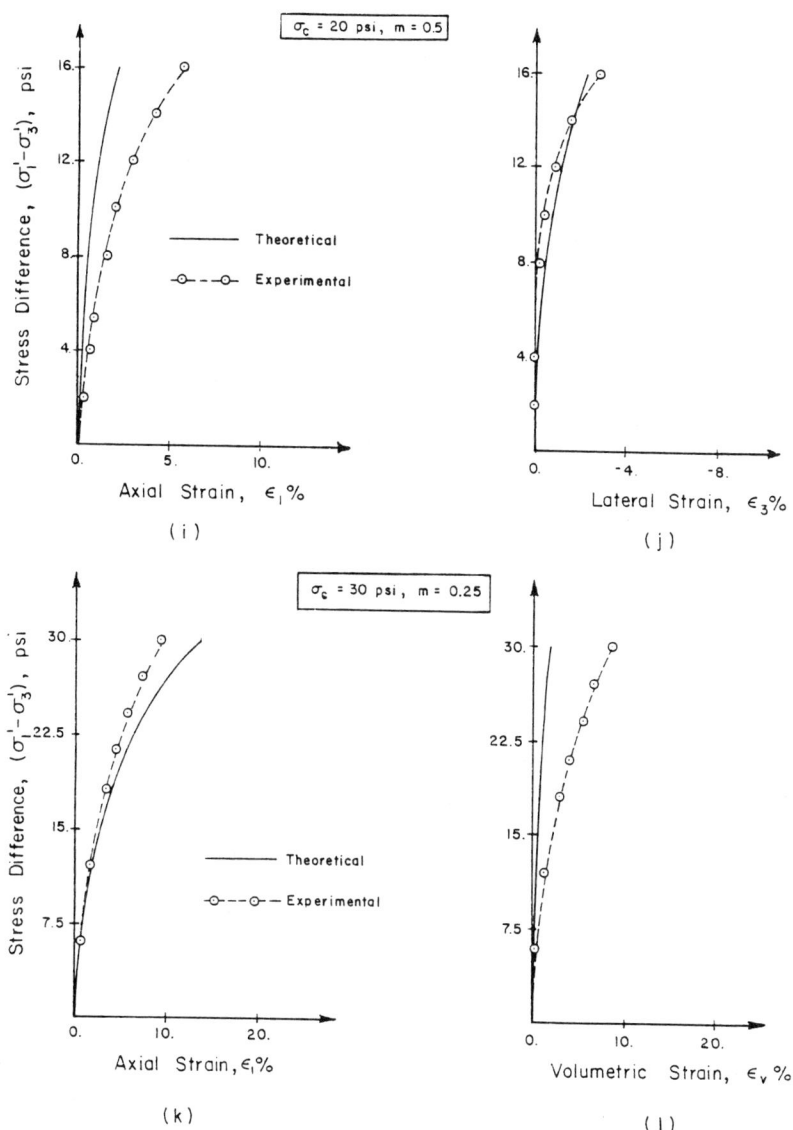

FIG 1. (Cont'd)

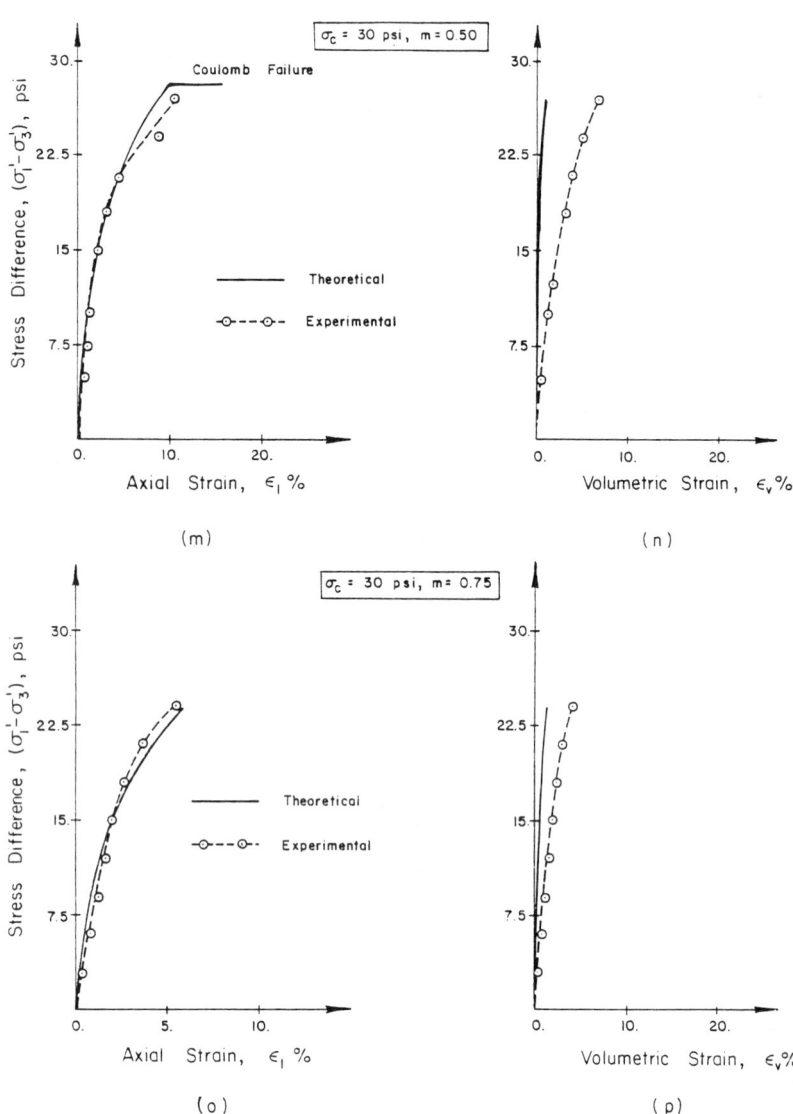

FIG 1. (Cont'd)

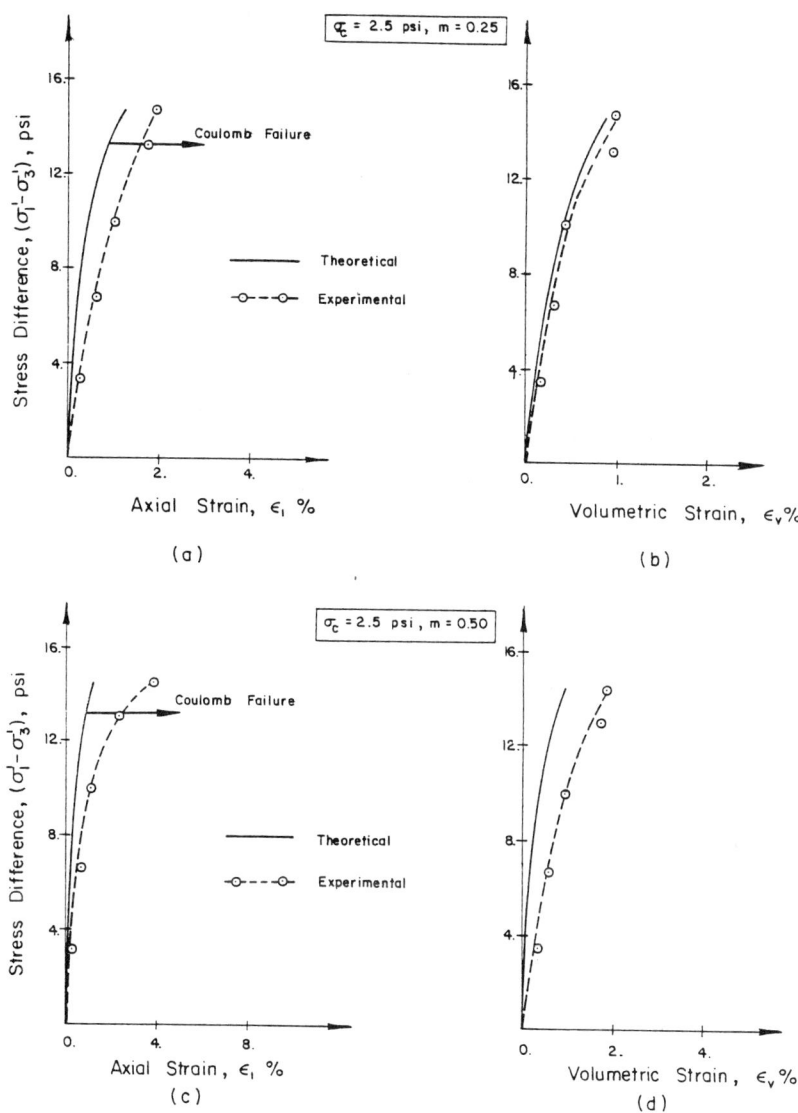

FIG 2. Comparison of the Predicted Theoretical and Experimental Stress-Strain Curves for Clay "Y" (CTC Tests)

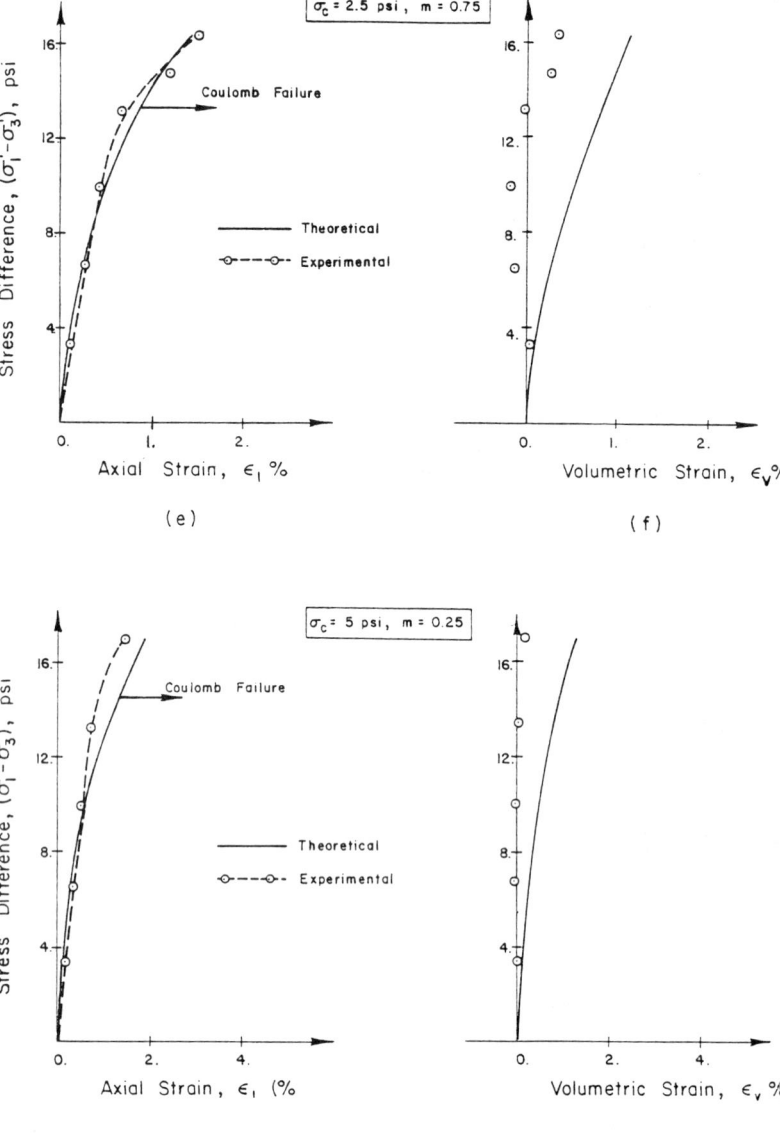

FIG 2. (Cont'd)

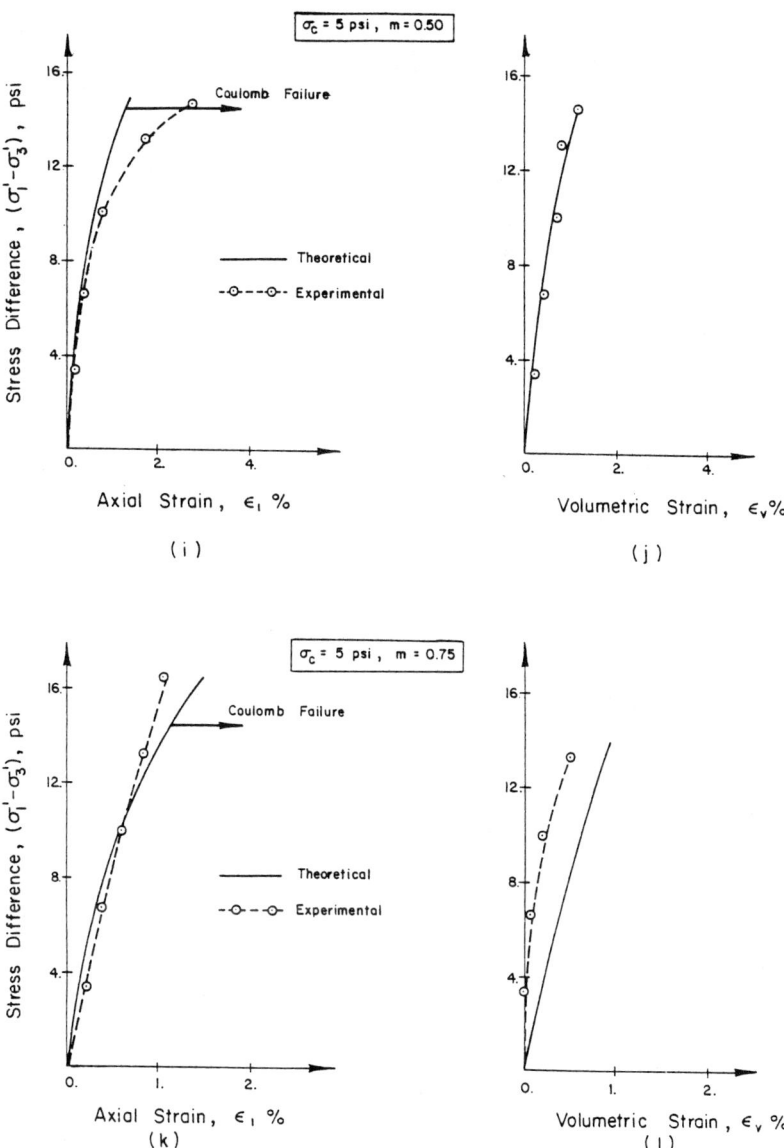

FIG 2. (Cont'd)

278 STRESS STRAIN FOR SOILS

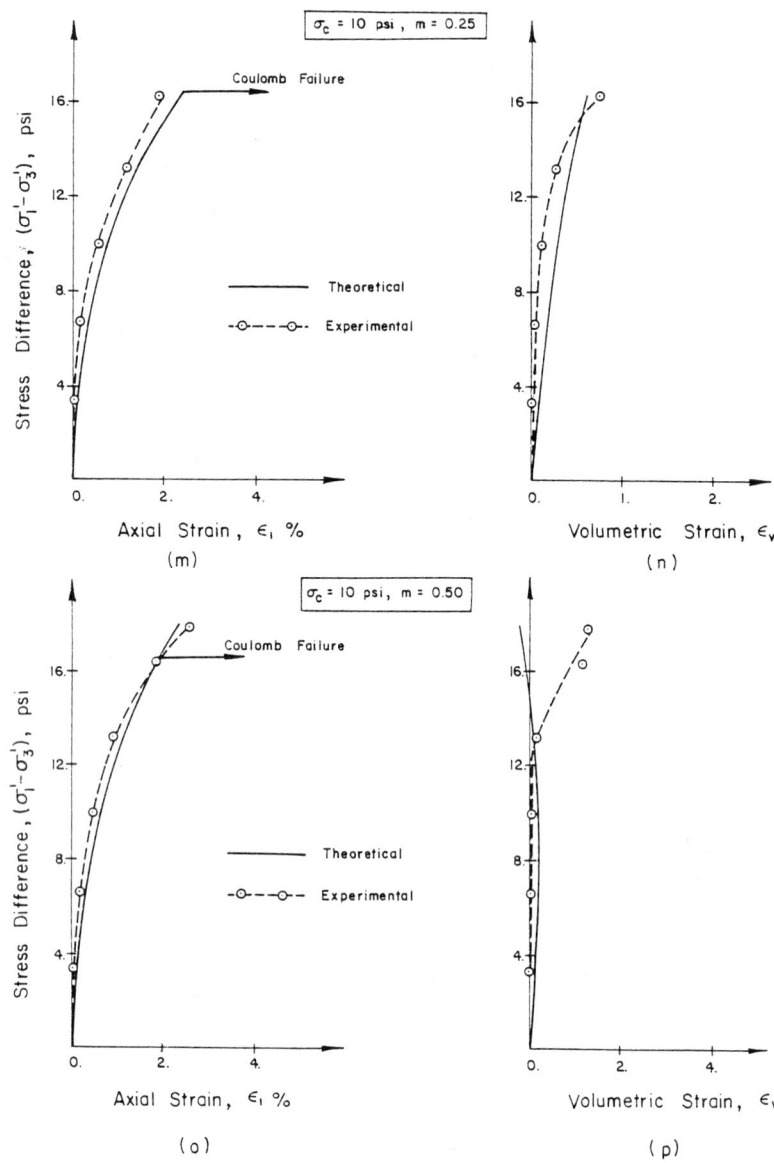

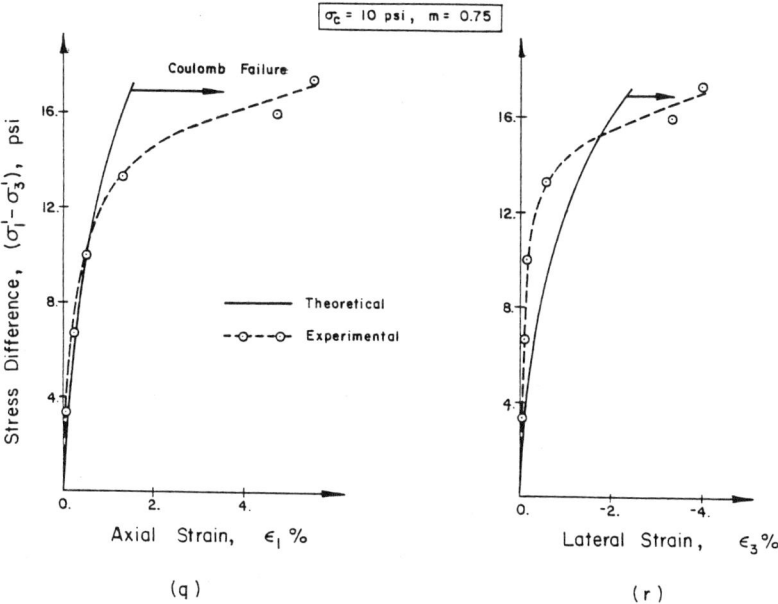

FIG 2. (Cont'd)

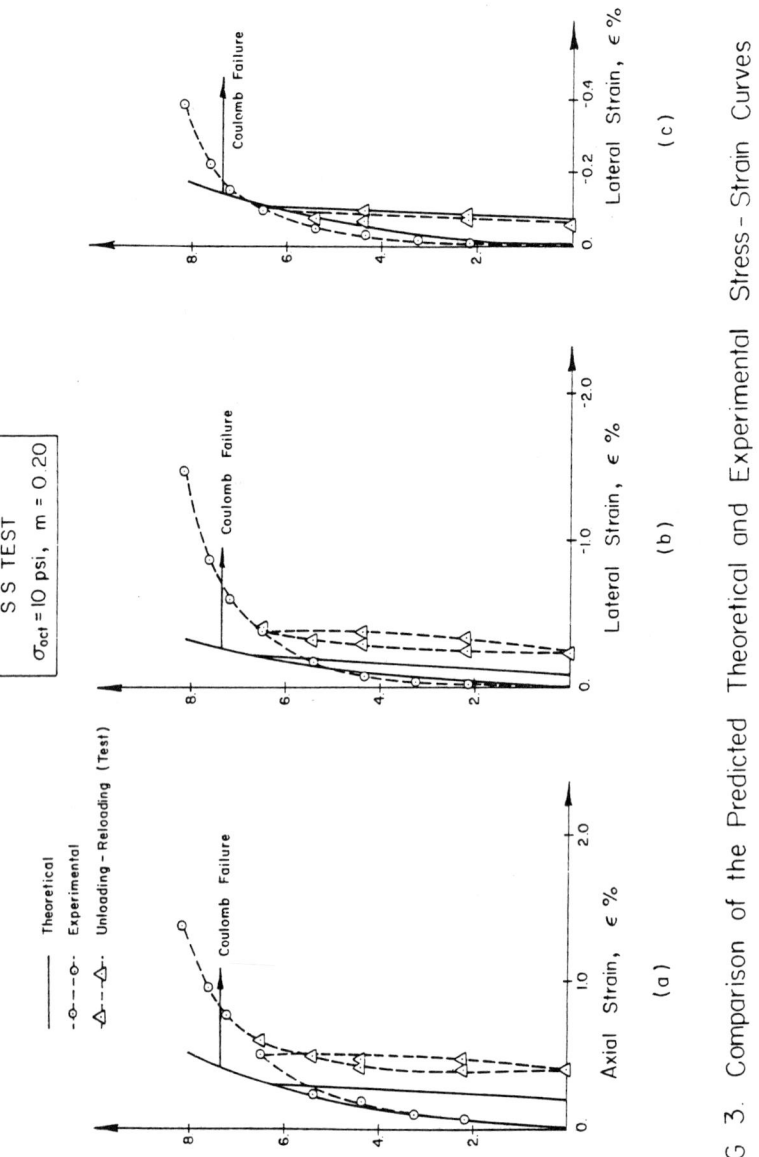

FIG 3. Comparison of the Predicted Theoretical and Experimental Stress-Strain Curves for Ottawa Sand in S S Test (σ_{oct} = 10 psi, m = 0.20)

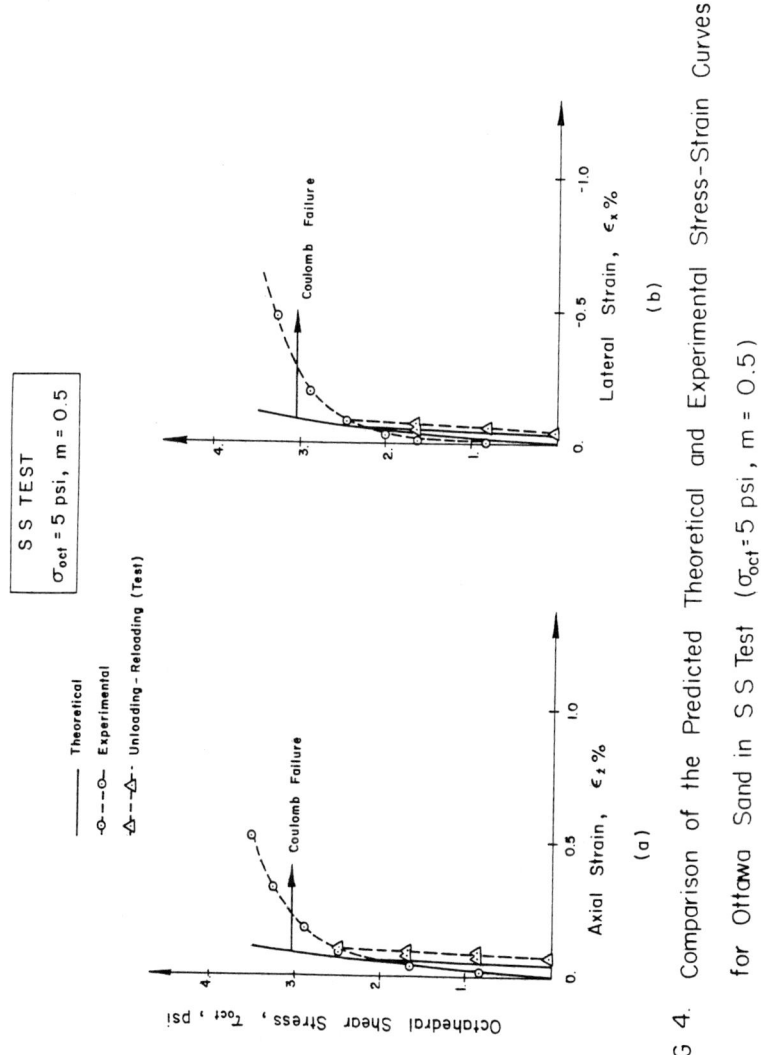

FIG 4. Comparison of the Predicted Theoretical and Experimental Stress-Strain Curves for Ottawa Sand in S S Test (σ_{oct} = 5 psi, m = 0.5)

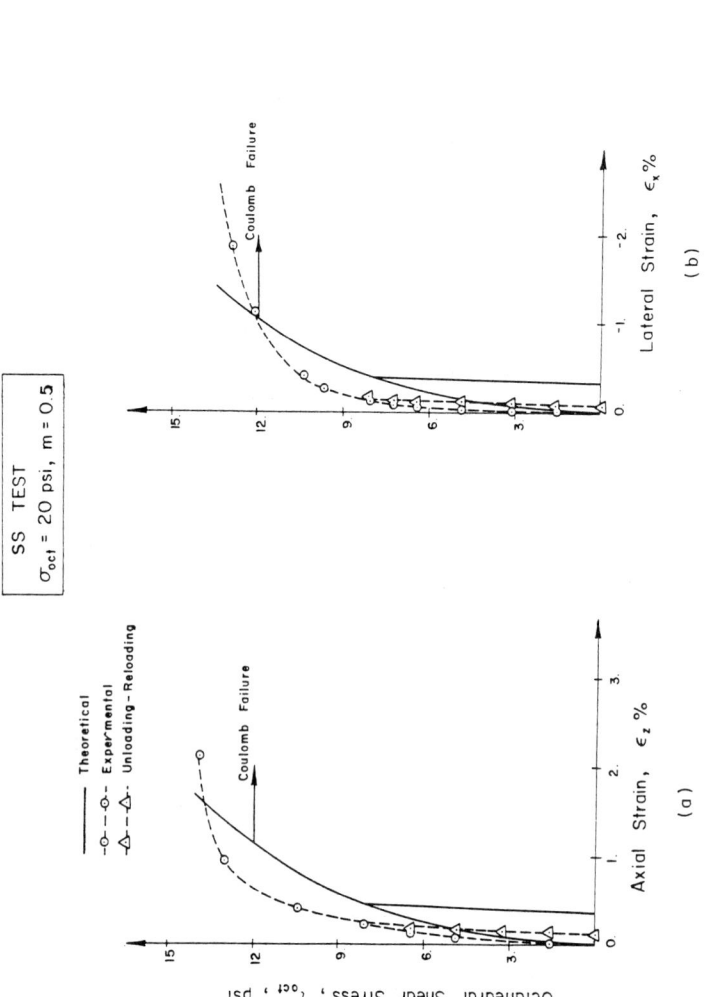

FIG 5. Comparison of the Predicted Theoretical and Experimental Stress-Strain Curves for Ottawa Sand in S S Test ($\sigma_{ocf} = 20$ psi, $m = 0.5$)

MODELS FOR SOILS 283

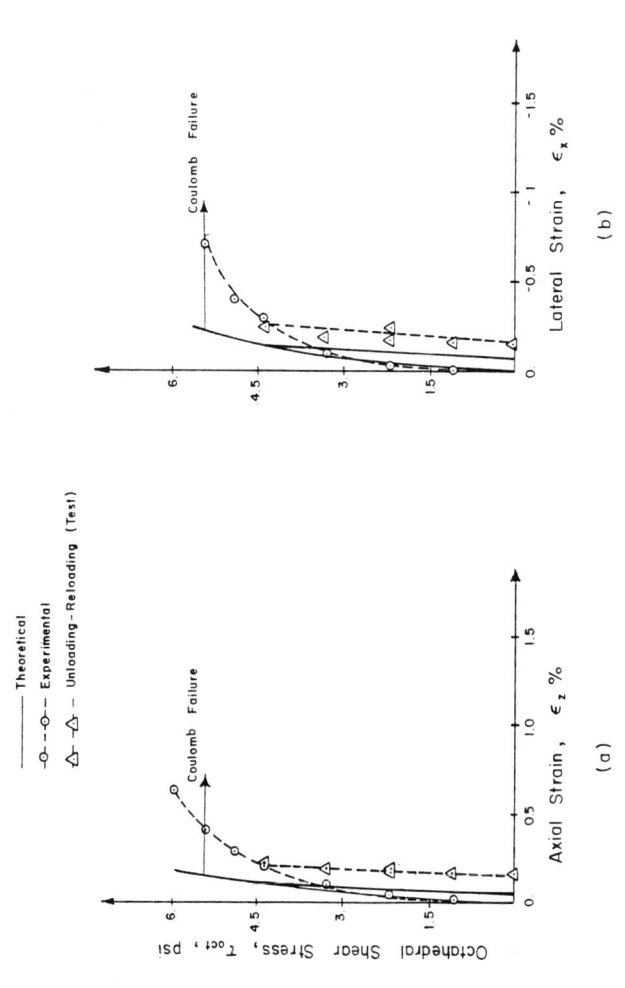

FIG 6. Comparison of the Predicted Theoretical and Experimental Stress-Strain Curve for Ottawa Sand in S S Test ($\sigma_{oct} = 10$ psi, $m = 0.8$)

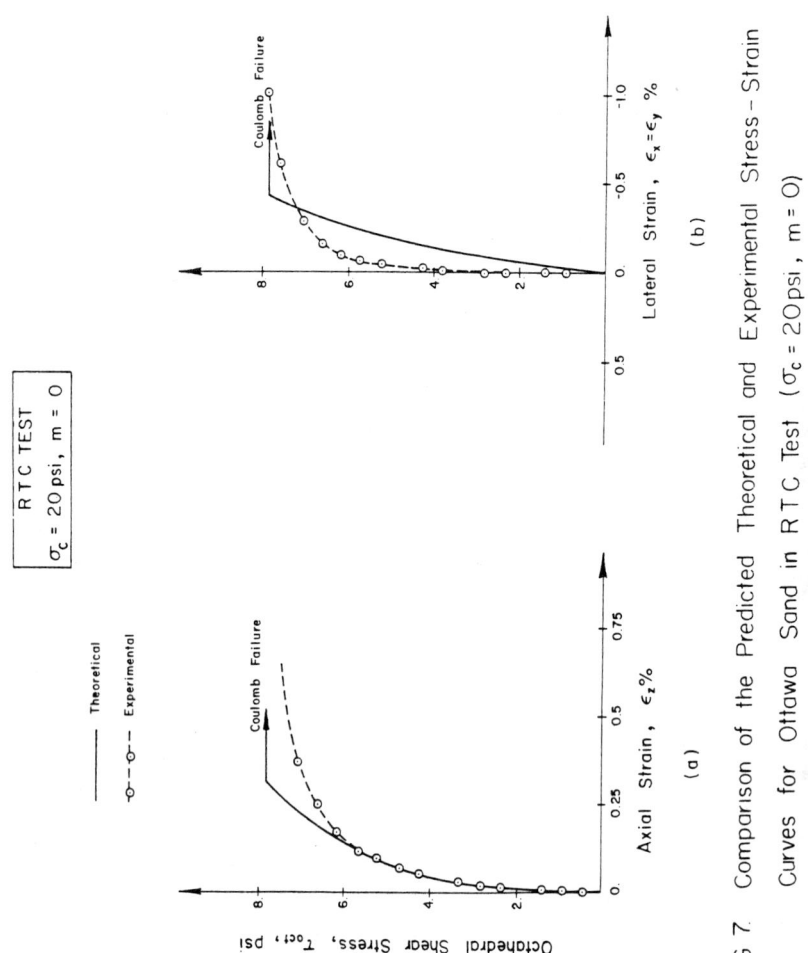

FIG 7. Comparison of the Predicted Theoretical and Experimental Stress-Strain Curves for Ottawa Sand in RTC Test (σ_c = 20 psi, m = 0)

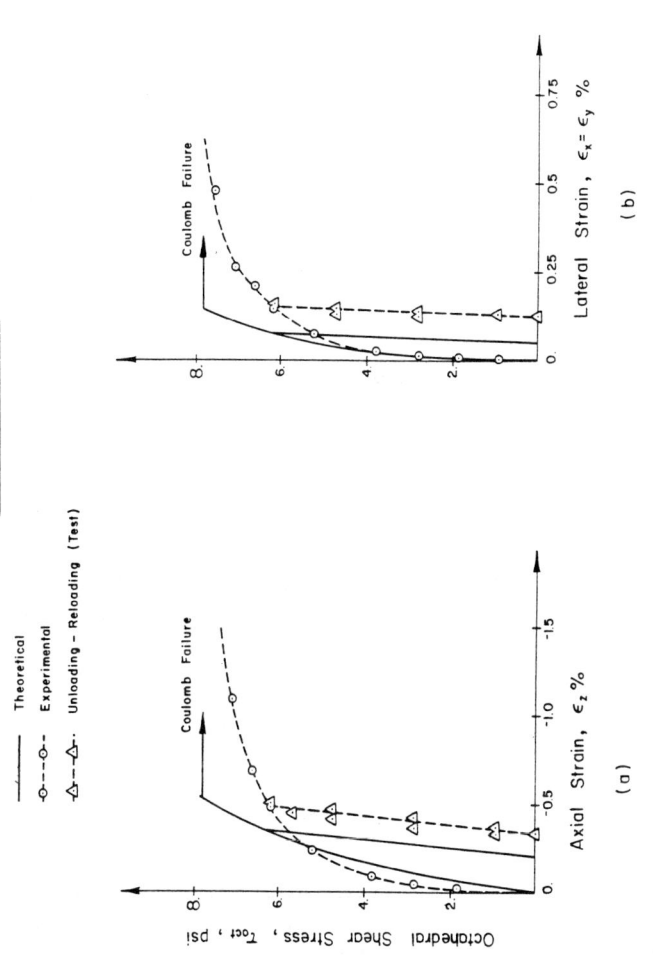

FIG 8. Comparison of the Predicted Theoretical and Experimental Stress-Strain Curves for Ottawa Sand in RTE Test (σ_c = 20 psi, m = 1).

PREDICTION OF SOIL BEHAVIOR BY ENDOCHRONIC THEORY

by

Atilla M. Ansal[1], A. M. ASCE

Raymond J. Krizek[2], M. ASCE

Zdeněk P. Bažant[2], M. ASCE

---

[1] Assistant Professor of Geotechnical Engineering, Macka Civil Engineering Faculty, Istanbul Technical University, Istanbul, Turkey.

[2] Professor of Civil Engineering, The Technological Institute, Northwestern University, Evanston, Illinois, U.S.A.

INTRODUCTION

The ultimate usefulness of any particular constitutive relationship is dictated in large part by a proper balance among (a) the versatility of the theory to characterize experimental data obtained from a variety of different tests, (b) the ability of the resulting relationship to predict behavior for conditions other than those which were used to "calibrate" the model, and (c) the ease with which the formulation can be adopted to the solution of practical boundary value problems. Exercises in fitting data and predicting response patterns therefore provide valuable comparisons among different theories and serve to identify and clarify the advantages and disadvantages of each.

However, when undertaking such comparisons, it must be recognized that most data sets are rather limited and frequently do not include the types of tests that would be required to pass judgment on the fundamental hypotheses and concepts of various constitutive models, and this is indeed the case here. As explained in a companion paper (5), a proper comparison of the basic ideas underlying different constitutive models would require (a) tests for which the stress path in stress space is highly nonproportional, including "loading to the side" (which is difficult to achieve experimentally), and (b) tests which involve unloading and cyclic loading. Since such data are not included in the set provided for this study, it will be impossible to adequately evaluate the validity and utility of various theories. In fact, the predictions requested for this study are primarily interpolations between extreme cases that were given for the same type of test, rather than for a basically different type of test.

In particular, the selected types of tests comprise a data set that happens to be unfavorable for an examination of endochronic theory and favors other theories. This is because endochronic theory was developed primarily to represent the response for unloading and cyclic loading and is most effective in such cases. Other theories which were developed mainly to model monotonic proportional loading response are favored by the data set provided and their performance under conditions of cyclic loading is not tested. This must be taken into account when forming judgments on the relative merits of the various theories based on the present data set. For the types of tests used (which do not include "loading to the side" or cyclic loading), it is, in principle, possible to adequately describe all of the given data and obtain reasonable predictions with nearly all available theories, and the closeness of the fits and predictions is largely a measure of the skill, intuition, experience, and persistence of the predictor rather than the features of the model.

SUMMARY OF BASIC EQUATIONS

The basic concept of endochronic theory is that of an intrinsic time scale, which is formulated to account for the dissipative effects of inelastic strain. Intrinsic time is defined as a monotically increasing scalar function of strain and time, and one suitable definition (9) is

$$(dz)^2 = \left[\frac{d\zeta}{z_1}\right]^2 + \left[\frac{dt}{\tau_1}\right]^2 \tag{1a}$$

$$d\zeta = F(\underline{\varepsilon}, \underline{\sigma}) \, d\xi \tag{1b}$$

$$d\xi = \sqrt{\tfrac{1}{2} \, de_{ij} \, de_{ij}} \tag{c}$$

where ε_{ij} is the strain tensor in cartesian coordinates $x_i (i = 1,2,3)$; z_1 and τ_1 are constant material parameters; $e_{ij} = \varepsilon_{ij} - \delta_{ij}\varepsilon$ is the deviator of ε_{ij};

$\epsilon = \epsilon_{kk}/3$ is the volumetric (mean) strain; δ_{ij} is Kronecker delta; and F is a strain hardening-softening function.

Assuming that the source of inelasticity in soils is the irreversible rearrangement of grain configurations associated with deviatoric strains, it is convenient to characterize the accumulation of grain rearrangements by an appropriate variable, ζ, termed the rearrangement measure, which is expressed in terms of the distortion measure, ξ, as given in Equations 1b and 1c. Since the increments of irreversible (inelastic) strain are caused by interparticle rearrangements, they must be proportional to the increments of the rearrangement measure, and the proportionality coefficient may, in general, depend on both the state of stress and strain. However, it has been observed in both quasi-static and cyclic tests (2,7) that these irreversible rearrangements diminish with the number of cycles. To express this phenomenon, it is suitable to consider the strain hardening-softening relationship as a composite function, such as

$$F(\underline{\epsilon}, \underline{\sigma}, \zeta)d\xi = d\eta/f(\eta) \tag{2a}$$

$$d\eta = F_\eta(\underline{\epsilon}, \sigma)d\xi \tag{2b}$$

The introduction of a new parameter, η, in terms of the function F, adjusts the rate of accumulation of inelastic strains, and the function $f(\eta)$ serves as a hardening function which causes a stiffening of the response for repeated loading (that is, contraction of hysteresis loops and an increase of their slope as the number of load cycles increases).

INTRINSIC MATERIAL FUNCTIONS

The forms of the material functions are assumed to be the same for both isotropic and anisotropic soils, and the anisotropy is accounted for by replacing the stress and strain invariants in these relationships with the appropriate transversely isotropic or orthotropic invariants (3). The material functions were introduced so as to reflect various obvious governing factors. In particular, F_η is assumed to have the form (1, 2).

$$F_\eta(\underset{\sim}{\varepsilon}, \underset{\sim}{\sigma}) = F_{\eta 1}(I_1^e) F_{\eta 2}(I_1^{\bar{\sigma}}) F_{\eta 3}(J_2^e) \qquad (3)$$

Since a change in interparticle distances, reflected as a change in volume, must result in a change in both deviatoric and volumetric stiffnesses, the softening-hardening function should increase as the volume increases. Thus, $F_{\eta 1}$ may be expressed in terms of the first strain invariant as $F_{\eta 1}(I_1^e) = |1 - a_1 I_1^e|$, in which a_1 is a positive material parameter. The frictional aspect of soils suggests that the deviatoric strain softening and hardening should depend on the effective confining stress, and this dependency is achieved by using the first invariant of the effective stress in the from $F_{\eta 2}(I^{\bar{\sigma}}) = [0.01 + a_2(I_1^{\bar{\sigma}}/P_a)]^{-1}$, in which a_2 is a positive material parameter and p_a is atmospheric pressure (expressed in the same units as the stress invariant and introduced to achieve a dimensionless relationship). The effects of shear strains are introduced into the relationship in terms of the second deviatoric strain invariant, J_2^e, in the form $F_{\eta 2}(J_2^e) = 1 + a_3 J_2^e$, in which a_3 is a material parameter.

Aside from the foregoing factors that control strain softening and hardening, there is an upper limit for softening and hardening, and this upper limit depends on the accumulated inelastic strains but seems to be essentially independent of the factors that cause them. This limit corresponds to what is known as the critical state, for which both strain hardening and strain softening diminish. A general limiting relation which incorporates this phenomenon can be formulated in terms of the accumulated grain rearrangements, and this formulation is developed by considering another variable, η, which represents continuous rearrangements, such that

$$d_\eta = \left[a_o + \frac{|1 - a_1 \, I_1^\varepsilon| \, (1 + a_3 J_2^3)}{0.01 + a_2(I_1^{\bar\sigma}/p_a)} \right] d\xi \tag{4}$$

in which a is a positive material parameter that is necessary to determine the irreversible strain increment for the critical case (where there is no hardening or softening). This limiting function is chosen as

$$f(\eta) = 1 + \beta_1 \eta/(1 + \beta_2 \eta) \tag{5}$$

in which β_1 and β_2 are positive material parameters that depend on the stress history.

Most soils also manifest inelastic volumetric strains, termed densification or dilatancy, as a result of shear. It is assumed here that the inelastic volume changes due to shear and those due to changes in the hydrostatic stress can be treated separately. The inelastic volume change (volumetric strain) due to shear is denoted λ and is called the densification and dilatancy measure. It can be expressed as a function of the stress and strain invariants and the accumulated value of λ as follows:

$$d\lambda = L(\underset{\sim}{\varepsilon}, \underset{\sim}{\sigma}, \lambda) d\xi , \tag{6}$$

The major factors affecting dilatancy and densification can be expressed as

$$L(\underset{\sim}{\varepsilon}, \underset{\sim}{\sigma}, \lambda) = L_1(I_1^\varepsilon) L_2(I_1^{\bar\sigma}) L_3(J_2^\varepsilon) L_4(\lambda). \tag{7}$$

Following arguments similar to those outlined for the strain hardening-softening relationship, the increment of inelastic volumetric strain due to shear can be expressed as

$$d\lambda = \left[\frac{C_o\left(1 + C_1 I_1^e\right)}{(1 + C_2 I_1^{\bar{\sigma}}/P_a)(1 + C_3 J_2^e)(1 + C_4 \lambda)}\right] d\xi \tag{8}$$

in which coefficients C_i are material parameters. Inelastic volume changes are also produced by changes in the effective volumetric stress. This phenomenon is taken into account by defining a second intrinsic time, $\bar{z}$, which involves a term called the compaction measure, $\bar{\xi}$, that is expressed in terms of volumetric strains as

$$d\bar{\xi} = \left|d\epsilon_1 + d\epsilon_2 + d\epsilon_3\right| \tag{9a}$$

$$(d\bar{z})^2 = \left[\frac{d\bar{\zeta}}{z_2}\right]^2 + \left[\frac{dt}{\tau_2}\right]^2 \tag{9b}$$

$$d\bar{\zeta} = d\bar{\eta}/h(\bar{\eta}) \tag{9c}$$

$$d\bar{\eta} = H(\underset{\sim}{\sigma}) d\bar{\xi} \tag{9d}$$

in which $h(\bar{\eta})$ is the compaction hardening function and $H(\underset{\sim}{\sigma})$ is the compaction softening function. Following a line of reasoning similar to that outlined before, Equation 9d can be written in an analogous form (8), where

$$d\bar{\eta} = \frac{b_1 \left|I_1^{\bar{\sigma}}/P_a\right|}{1 + b_2 \left|I_1^{\bar{\sigma}}/P_a\right|} d\bar{\xi} \tag{10a}$$

$$h(\bar{\eta}) = 1 + b_3 \bar{\eta} + b_4 \bar{\eta}^2 \tag{10b}$$

in which b_1, b_2, b_3, and b_4 are constant material parameters.

The concept of a two-phase medium is a natural choice for analyzing the undrained behavior of saturated soils. Since the compressibility of the soil grains is about thirty times less than the compressibility of water, the assumption of incompressible soil grains will not introduce any significant error. On the other hand, the fluid phase (pore water) must be treated as compressible, and the solid structure (matrix of soil particles) is even more compressible than the pore water. The tendency of the solid structure to change volume is coupled with the pore pressure, and the solid structure will deform only as much as the pore water permits. If no drainage is allowed, the volume change of the solid structure will equal the volume change of the pore fluid, and the total volumetric strain of the pore water can be expressed in terms of the total volumetric strain of the soil structure; thus, the pore pressure increment can be expressed as (1, 2)

$$du = B_w(d\varepsilon_1 + d\varepsilon_2 + d\varepsilon_3)/n \tag{11}$$

where n is the porosity of the soil and B_w is the bulk modulus of water.

The variations of the elastic moduli along the stress path are taken into consideration and formulated as functions of the initial magnitudes of the moduli and the ratios of the changes that take place in certain important parameters. Two major factors that change along a given stress path are the void ratio and the effective normal stress. These variables, in turn, depend on the accumulated densification-dilatancy measure, λ, and the first effective stress invariant, which is adopted to represent the change in the elastic moduli, and all of the independent elastic moduli are expressed in the form (1)

$$E_r = E_o \left[1 + M_1(I_1^{\bar{\sigma}} - I_1^{\bar{\sigma}_o})/I_1^{\bar{\sigma}_o} + 3M_2\lambda_r/n \right] \tag{12}$$

where E_o is the initial modulus; $I_1^{\bar{\sigma}_o}$ is the initial first effective stress invariant; and M_1 and M_2 are empirically determined material constants which are assumed here to be 0.1.

STRESS-STRAIN RELATIONS

The endochronic constitutive model was originally developed for isotropic soils (2, 4, 6), and the stress-strain relations involved only two independent elasticity coefficients; however, it appears more realistic to consider most soils to be transversely isotropic or orthotropic. An endochronic constitutive model for transversely isotropic clays has been developed (3) and involves five elasticity coefficients. This approach is extended here to model the behavior of orthotropic soils and the following stress-strain relations are proposed

$$d\epsilon_{11} = c_{11}d\bar{\sigma}_{11} + c_{12}d\bar{\sigma}_{22} + c_{13}d\bar{\sigma}_{33} + d\epsilon_{11}'' \tag{13a}$$

$$d\epsilon_{22} = c_{12}d\bar{\sigma}_{11} + c_{22}d\bar{\sigma}_{22} + c_{23}d\bar{\sigma}_{33} + d\epsilon_{22}'' \tag{13b}$$

$$d\epsilon_{33} = c_{13}d\bar{\sigma}_{11} + c_{23}d\bar{\sigma}_{22} + c_{33}d\bar{\sigma}_{33} + d\epsilon_{33}'' \tag{13c}$$

$$d\epsilon_{12} = c_{44}d\sigma_{12} + d\epsilon_{12}'' \tag{13d}$$

$$d\epsilon_{13} = c_{55}d\sigma_{13} + d\epsilon_{13}'' \tag{13e}$$

$$d\epsilon_{23} = c_{66}d\sigma_{23} + d\epsilon_{23}'' \tag{13f}$$

Here, superimposed bars denote effective stresses; $\bar{\sigma}_{ij}$ are components of the effective stress tensor; ϵ_{ij} are the strain components; ϵ_{ij}'' are components of the inelastic strains; and c_{ij} are elasticity coefficients. The inelastic strain increments are assumed to have the same form as previously derived for transversely isotropic soils (3) and are introduced as

$$d\epsilon_{11}'' = D_1(\bar{\sigma}_{11}-\bar{\sigma}_{22})dz + D_2(\bar{\sigma}_{11}-\bar{\sigma}_{33})dz + d\epsilon''/(1 + r_1 + r_2) \tag{14a}$$

$$d\epsilon_{22}'' = D_1(\bar{\sigma}_{22}-\bar{\sigma}_{11})dz + D_3(\bar{\sigma}_{22}-\bar{\sigma}_{33})dz + r_1 d\epsilon''/(1 + r_1 + r_2) \tag{14b}$$

$$d\epsilon_{33}'' = D_2(\bar{\sigma}_{33}-\bar{\sigma}_{11})dz + D_3(\bar{\sigma}_{33}-\bar{\sigma}_{22})dz + r_2 d\epsilon''/(1 + r_1 + r_2) \quad (14c)$$

$$d\epsilon_{12} = c_{44}\sigma_{12}dz \quad (14d)$$

$$d\epsilon_{13} = c_{55}\sigma_{13}dz \quad (14e)$$

$$d\epsilon_{23} = c_{66}\sigma_{23}dz \quad (14f)$$

where it is assumed that $D_1 = c_{44}/3$, $D_2 = c_{55}/3$, and $D_3 = c_{66}/3$; coefficients r_1 and r_2 characterize anisotropy; and $d\epsilon''$ represents the inelastic volume change, expressed as

$$d\epsilon'' = d\lambda + \frac{\bar{\sigma}}{3K} d\bar{z} \quad (15)$$

where $d\lambda$ is the inelastic volume change increment due to deviatoric distortion and ($\bar{\sigma} d\bar{z}/3K$) is the inelastic volume change due to a change in the effective volumetric stress. The stress-strain relations for the simpler case of a transversely isotropic soil are listed in the Appendix.

FITTING OF TEST DATA

When fitting test data with the foregoing constitutive equations, the test specimens are assumed to be in a homogeneous state of stress and strain. Due to the complex nature of the differential equations, a step-by-step process is employed and strains or stresses are increased in small increments. Values from each previous step provide initial estimates, and inner iterations within a step are used to obtain improved response increments within the step. For the initial iteration of the first loading increment, all incremental values must be estimated (can be taken as zero). Experience indicates that this step-by-step integration method is reasonably stable, and, provided loading steps are sufficiently small, convergence is usually achieved after a couple of iterations for strain-controlled tests. For the stress-controlled tests convergence is also achieved after a couple of iterations during the initial stage of loading, but, as the stress-strain curve approaches its peak point, it is necessary to decrease the loading increment continuously in order to prevent instability.

The final form of the equations and the values of the material parameters were obtained by using a mathematical optimization procedure to minimize the differences between observed and calculated responses. A finite difference Levenberg-Marquart subroutine for solving nonlinear least squares problems (developed by T. J. Aird of International Mathematical and Statistical Library package) was utilized for this purpose. Due to the complex nonlinear form of the constitutive relations, it is probably possible to have different sets of material parameter values that give fairly good results; however, the optimization routine only identifies local optimums in the vicinity of the initial estimates.

Due partly to the versatility of endochronic theory and partly to the large number of material parameters used in the constitutive relations, it is always possible to get fairly accurate fits (2, 3, 7). However, the purpose here was to find a set of values which would give reasonable fits for all of the given stress paths and which would yield logical correlations with material properties and the stress and strain state of the soil. This set is not necessarily the global optimum. Many stages of optimization were required to determine the effects of different parameters and to achieve realistic fits and correlations. Although the process of accomplishing this task is based somewhat on previous knowledge about the behavior of soils and the characteristics of the model, it is still largely a trial-and-error procedure.

EXPRESSIONS FOR DATA FITS

Essentially four different soils were investigated in this study.

The first two were sensitive clays which appear to have very similar index properties; however, the tests were performed under different initial confining stresses such that the overconsolidation ratio, p/p_o, was less than unity for one of the clays. For these two clays, which were termed "X" and "Y", the reported response indicated anisotropy with respect to all three axes. To be able to model such a behavior it was necessary to extend the previously developed formulation to handle orthotropic stress-strain relations, which require nine elasticity coefficients. The expressions were developed using arguments previously outlined for transversely isotropic soils (3).

The third soil was a laboratory prepared kaolinite clay which was assumed to be transversely isotropic due to the K_o consolidation that was used in sample preparation. The fourth soil was a dry Ottawa sand, which also showed transversely isotropic response in one test (which was the only test that could reflect this property of the sand). The transversely isotropic stress-strain relation outlined in the Appendix was adopted to predict the stress-strain behavior.

Even though these four soil types are very different from one another, all of the given test data were modeled by using the same material functions, and no attempt was made to modify the general form of the material relationships to improve the accuracy of the fits for each type of soil separately. This demonstrates the generality and versatility of the endochronic constitutive equations. The chosen material functions are given below in incremental form:

$$d\eta = \left[4 + \frac{|1 - 500\ I_1^e|\ (1 + a_d J_2^e)}{0.01 + 0.75\ I_1^{\bar{\sigma}}/P_a} \right] d\xi , \qquad (16)$$

$$dz = \frac{d\eta}{z_1 \left[1 + \beta_1 \eta/(1 + \beta_2 \eta) \right]} \quad , \tag{17}$$

$$d\lambda = \frac{c_1 \left| 1 + 2500\, I_1^e \right|}{(1 + 3000\, J_2^e)(1 + 0.25 I_1^{\bar{\sigma}}/P_a)(1 + c_2 \lambda)} \quad , \tag{18}$$

$$d\bar{\eta} = \frac{b \left| I_1^{\bar{\sigma}}/P_a \right|}{1 + \left| I_1^{\bar{\sigma}}/P_a \right|}\, d\bar{\varsigma} \quad , \tag{19}$$

$$d\bar{z} = \frac{d\bar{\eta}}{z_2(1 + 2500\bar{\eta} + 1000\bar{\eta}^2)} \quad . \tag{20}$$

Only eight soil-specific material parameters were used to model the given test data; all others were held constant for all soils and all stress or strain histories. These eight parameters were initially determined separately for each test; then, trial-and-error procedures were employed to identify a suitable set that would yield realistic fits for all of the tests performed under the same consolidation stress, and the predictions were based on these values.

ANALYSIS OF TEST DATA

Clays X and Y

As mentioned earlier, orthotropic stress-strain relations were used to model the behavior of these clays; these relations required nine elasticity coefficients and two material coefficients to define the appropriate stress-strain invariants. An attempt was made to decrease the number of elasticity coefficients by assuming that (a) values for Poisson's ratio in all three directions are equal ($\mu_{21} = \mu_{32} = \mu_{31}$) and the other values in the corresponding planes ($\mu_{12}, \mu_{23}, \mu_{13}$) can be determined from the symmetry condition of the stiffness matrix, and (b) compression and shear moduli in the minor and intermediate principal planes can be expressed in terms of compression and shear moduli in the major plane, where the proportionality constants are the same for both moduli. With these assumptions it was possible to express elasticity coefficients in terms of five parameters as

$$\mu_{21} = \mu_{32} = \mu_{31} = \mu \tag{21}$$

$$E_2 = r_1 E_1 \; ; \quad E_3 = r_2 E_1 \tag{22}$$

$$G_2 = r_1 G_1 \; ; \quad G_3 = r_2 G_1 \tag{23}$$

where G is the shear modulus; E is the compression modulus; μ is Poisson's ratio; and r_1 and r_2 are proportionality constants which characterize the anisotropy of the material. The variations of the major principal elastic moduli, E_1 and G_1, along a particular stress path were formulated as given by Equation 12, and variations of the other moduli were calculated by Equations 22 and 23.

It was also necessary to express the strain and stress invariants with two proportionality constants in order to include the effect of anisotropy and establish invariance with respect to the orthogonal transformations. As in the case of transversely isotropic clays (3), it was assumed that the first stress invariant retains the same form and the strain invariants are defined in terms of two proportionality constants as

$$I_1^e = \epsilon_1 + r_3 \epsilon_2 + r_4 \epsilon_3 \tag{24}$$

$$J_2^e = \left[(\epsilon_1 - r_3 \epsilon_2)^2 + (\epsilon_1 - r_4 \epsilon_3)^2 + (r_3 \epsilon_2 - r_4 \epsilon_3)^2 \right] \Big/ 6 \tag{25}$$

where $r_3 = 1/r_1$ and $r_4 = 1/r_2$. With the introduction of two proportionality or anisotropy constants (r_1, r_2), three elasticity coefficients (E_1, G_1, μ), and eight material constants (a_d, b, c_1, c_2, z_1, z_2, β_1, β_2), thirteen constants must be determined.

The curves shown in Figure 1 for Clay X were obtained by fitting each test separately; however, every effort was made to hold most of the material parameters constant for the entire data set, and only four (c_1, c_2, E_1, z_1) of the thirteen parameters were allowed to vary within a limited range. The values of the constant and variable parameters used to obtain the fits are summarized in Table 1.

Table 1. Variable and Constant Material Parameters for Clay X

Constant material parameters: a_d = 380, β_1 = 24; β_2 = 5; b = 0.1; z_2 = 50,000; r_1 = 0.8; r_2 = 0.9; G_1/P_a = 48; μ = 0.18					
Variable material parameters:					
		z_1	E_1/P_a	c_1	c_2
σ_c = 10 psi	m = 0	0.0696	88	0.7500	30,000
	m = 1	0.0861	88	0.7500	230,000
σ_c + 20 psi	m = 0	0.0600	88	0.7500	500
	m = 1	0.0660	30	0.0021	230,000
σ_c = 30 psi	m = 0	0.0600	34	0.7500	500
	m = 1	0.1140	29	0.0021	230,000

As can be seen from the given curves, the stress-strain fits are reasonably good, but the model generally yields smaller volumetric strains. It is believed that this is partially a result of keeping certain parameters constant, as given in Equations 16 to 20, because the model was originally developed for normally consolidated clays. In the case of sensitive clays the collapse of the soil structure appears to give large volumetric strains and a more linear stress-strain response in the initial portion followed by a sudden increase in strain. Unlike normally consolidated insensitive clays, there appears to be a distinct yield point. Due to the flexibility of endochronic theory, it is possible to model the stress-strain behavior of sensitive clays very accurately, but it may be necessary to increase the range of the variable parameters or to change the values of the constants and even the functional forms of the material relationships given in Equations 16 to 20.

After the workshop at McGill University, an attempt was made to improve some of the fits for Clay X. As shown typically in Figure 2, it was possible to obtain excellent fits for the first two tests performed at σ_c = 10 psi, but there were large variations in the values of some of the variable parameters, even for the two tests considered. This suggests the need to modify some of the material relationships to handle sensitive clays. The curves given in Figure 2 were calculated by changing only the values of the four variable parameters given in Table 1 to the values presented in Table 2, while keeping all others constant and equal to their values given in Table 1.

Table 2. Variable Parameters Used for Improved Fits for Clay X

		z_1	E_1/P_a	c_1	c_2
σ_c = 10 psi	m = 0	0.038	77	0.750	30,000
	m = 1	1800	26	0.091	30,000

In the case of Clay Y the curves shown in Figure 3 were obtained by optimizing the fits with respect to the four variable parameters given in Table 2 to get the values summarized in Table 3, while keeping all others equal to their values given in Table 1. The model appears to give better results for Clay Y than for Clay X, but there are still large differences between observed and calculated volumetric strains. The initial fits shown in Figure 4 can be compared to the improved fits shown in Figure 3a to evaluate the degree of improvement that has been achieved. As can be seen, a major improvement was realized in modeling the volumetric change.

Table 3. Variable Parameters Used for Improved Fits for Clay Y

		z_1	E_1/P_a	c_1	c_2
σ_c = 2.5 psi	m = 0	0.026	136	0.042	30,000
	m = 1	0.042	129	0.007	30,000
σ_c = 5 psi	m = 0	0.028	133	1.970	30,000
	m = 1	0.046	111	0.660	30,000
σc = 10 psi	m = 0	0.060	681	0.750	500
	m = 1	0.060	681	0.750	500

The stress-strain curves shown in Figures 5 and 6 are typical predicted tests results for Clays X and Y, respectively. These predictions have been obtained by using the same material parameters used in the initial predictions, the only difference between these curves and those presented at the workshop being that the loading was stopped as soon as strain softening started, because the tests were stress-controlled. Due to an error in the computer program, strain hardening previously occurred at this point, and this deficiency was corrected prior to obtaining the new plots.

Kaolinite Clay

A transversely isotropic stress-strain relation was adopted for the kaolinite clay because samples were subjected to K_o consolidation before they were sheared. It is generally accepted that K_o consolidation will result in some particle reorientation, which is a cause of anisotropy. In the case of transversely isotropic soils, five elasticity coefficients are needed. However, expressing these coefficients in terms of ratios makes it possible to decrease this number. As in the case of orthotropic soils, it is assumed here that (a) Poisson's ratio in all planes is the same ($\mu_{21} = \mu_{13} = \mu_{23} = \mu$), and (b) the compression modulus, E_1 in the plane of isotropy can be expressed in the terms of the compression modulus in the vertical direction, E_3, in the form $E_1 = r_1 E_3$. As a result, the number of elasticity coefficients may be reduced to three (μ, E_1, G_{32}), and r_1 is the proportionality parameter representing the degree of anisotropy. The transversely isotropic strain invariants may be expressed (3) as

$$I_1^\epsilon = \epsilon_{11} + \epsilon_{22} + r_2 \epsilon_{33} \tag{26}$$

$$J_2^\epsilon = \left[(\epsilon_{11}-\epsilon_{22})^2 + (\epsilon_{11}-r_2\epsilon_{33})^2 + (\epsilon_{22}-r_2\epsilon_{33})^2\right]\big/6 + \epsilon_{12}^2 + r_3(\epsilon_{13}^2 + \epsilon_{23}^2) \tag{27}$$

where $i = 3$ represents the direction perpendicular to the plane of isotropy; r_3 is a proportionality constant; and $r_2 = (c_1 + c_2 + c_4)/(c_3 + 2c_4)$, where c_i are the elasticity coefficients, as given in the Appendix. The value for r_2 is determined by use of the assumption that hydrostatic stress changes will not produce any distortional inelastic strains (3). In the case of transversely isotropic soils the proposed constitutive model requires three elasticity coefficients (μ, E_1, G_{32}), three proportionality parameters (r_1, r_3, r_4), and eight material constants (a_d, b, c_1, c_2, z_1, z_2, β_1, β_2), where r_4 is incorporated into the constitutive equations as shown in the Appendix.

These parameters were determined by trial and error, and the optimization technique was used to get fits for the compression tests with one set of numbers and fits for the extension tests with another set of numbers. Without further complications it was not possible to incorporate into the formulation the differences in the test data between compression and extension tests and obtain a single set of parameters for all four tests. It also was not possible to obtain the shear modulus, G_{32}, in the plane of anisotropy because no torsional test data were given; hence, G_{32} was estimated from the observed degree of anisotropy and other elasticity moduli. The values of the material constants obtained from the given compression and extension test data are summarized in Table 4.

Table 4. Optimized Parameters for Kaolinite Clay

Tests	a_d	β_1	β_2	z_1	c_1	c_2	E_3/P_a	r_1	r_4
1 & 4	5.0	3.84	6.65	0.01660	1.4800	41,000	709	0.68	0.51
10 & 13	12.0	1.50	5.00	0.00914	0.0012	2,000	386	0.98	1.42

Note: For all tests $b = 0.5$, $\mu = 0.21$, $z_2 = 10,000$, and $r_3 = 0.57$.

The main reason for the differences in the parameters shown in Table 4 is probably due to the difference in the behavior of kaolinite in compression and tension, and it was not possible to handle this phenomenon with the present formulation. The optimized curves for the given test data are shown in Figure 7, and the predictions are shown in Figures 8, 9, 10, and 11. The predictions were based on the average values of the parameters given in Table 4, but, since the behavior is so different in tension and compression, it did not appear realistic to use one value for the elasticity coefficient E_3. Instead, E_3 was calculated separately for each test with respect to the inclination of the stress path compared to the stress path in a standard compression test by using a quarter of an ellipse. The ellipse was drawn with its minor radius as E_3 from extension tests and its major radius as E_3 obtained from compression tests. Then, E_3 for the test was determined graphically by measuring the appropriate radius of the ellipse. The value of the shear modulus, G_{32}, was estimated from the values for E_3.

Ottawa Sand

As mentioned previously, transversely isotropic constitutive relations were used to model the tests on dry Ottawa sand. In particular, the same material relationships were adopted to demonstrate the flexibility of the approach, and, even though sensitive Clay X and dry Ottawa sand manifest different behavioral patterns, it was possible to obtain realistic fits for the data supplied on both soils.

The data set included tests with unloading and reloading cycles. Based on the method explained earlier, jump-kinematic hardening was incorporated into the formulation in terms of the deviatoric component of the stress tensor. This was accomplished by defining the deviatoric stresses as

$$\bar{s}_{ij} = s_{ij} - \alpha_{ij} \tag{28}$$

and
$$d\bar{z} = c_u dz \tag{29}$$

where α_{ij} and c_u must be redefined at each load reversal point. For the given test data $c_u = 1$ and $\alpha_{ij} = 0$ for virgin loading, $c_u = 0.5$ and $\alpha_{ij} = s_{ij_{max}}$ for unloading, and $c_u = 0.7$ and $\alpha_{ij} = 0$ for reloading, because loading and reloading started from the hydrostatic stress state. Even though jump-kinematic hardening is in essence just an empirical technique to improve the model, it also reflects some of the salient features found in the stress-strain behavior of soil under repeated loading. Most soils manifest considerable inelastic response in the loading branch, but their response is much more elastic upon unloading; hence, it is logical to decrease the value of the intrinsic time increment, dz, since it is the controlling factor for accumulating plastic strains. On the other hand, it has been observed that the response in the reloading branch up to the previous maximum stress level is more or less elastic, after which the stress-strain behavior continues almost on the path obtained in the virgin loading branch. Therefore, it appears logical to reset c_u and α_{ij} at these points.

The fits were obtained by using trial-and-error and the optimization scheme, and values of the parameters and constants are summarized in Table 6.

Table 6. Material Parameters Used to Obtain Fits for Ottawa Sand

	c_1	a_d	β_1	β_2	z_1	E_3/P_a	G_{32}/P_a	r_3	r_4
σ_c = 5 psi									
CTC m = 0	0.000720	0.800	3.8	5	0.00280	1034	748	0.80	1.3
TE m = 1	0.000720	0.800	3.8	5	0.00330	517	340	0.80	0.8
σ_c = 10 psi									
CTC m = 0	0.000014	1.230	3.8	28	0.00510	1177	953	0.50	1.3
TC m = 0	0.000014	1.230	3.8	28	0.00510	1177	953	0.35	1.3
TE m = 1	0.000014	2.500	3.8	28	0.00510	1177	953	0.20	1.3
σ_c = 20 psi									
TC m = 0	0.000014	0.034	20.0	5	0.00556	734	510	0.80	0.8
TE m = 1	0.057700	0.048	20.0	5	0.00313	374	210	0.80	0.8

Constants: $\mu = 0.3$; $c_2 = 2{,}000$; $b = 0.01$; $z_2 = 1.7 \times 10^6$

As can be seen, it was possible to obtain the fits shown in Figures 12, 13, and 14 by changing the values of only a few parameters. However, in the case of Ottawa sand it appears necessary with the present intrinsic relationships to have different sets of values for the material parameters for each confining stress. The predictions shown in Figures 15 to 21 were calculated by using a different average set of material parameters for each confining stress, with estimates being based on the values in Table 6.

APPENDIX

STRESS-STRAIN RELATIONS FOR TRANSVERSELY ISOTROPIC SOILS

When the soil is transversely isotropic, the constitutive relations simplify to the form:

$$d\epsilon_{11} = c_1 d\bar{\sigma}_{11} + c_2 d\bar{\sigma}_{22} + c_4 d\bar{\sigma}_{33} + d\epsilon_{11}'' \tag{A-1}$$

$$d\epsilon_{22} = c_2 d\bar{\sigma}_{11} + c_1 d\bar{\sigma}_{22} + c_4 d\bar{\sigma}_{33} + d\epsilon_{22}'' \tag{A-2}$$

$$d\epsilon_{33} + c_4 d\bar{\sigma}_{11} + c_4 d\bar{\sigma}_{22} + c_3 d\bar{\sigma}_{33} + d\epsilon_{33}'' \tag{A-3}$$

$$d\epsilon_{12} = (c_1 - c_2) d\sigma_{12} + d\epsilon_{12}'' \tag{A-4}$$

$$d\epsilon_{23} = c_5 d\sigma_{23} + d\epsilon_{23}'' \tag{A-5}$$

$$d\epsilon_{13} = c_5 d\sigma_{13} + d\epsilon_{13}'' \tag{A-6}$$

Here superimposed bars denote effective stresses; ϵ_{ij} are components of total strains; ϵ_{ij}'' are components of inelastic strains; and c_i are elasticity coefficients. Equations A-1 to A-6 are referred to cartesian axes x_1, x_2 and x_3, of which x_3 is normal to the plane of isotropy. The inelastic strain increments can be given as

$$d\epsilon_{11}'' = D_1(\bar{\sigma}_{11} - \bar{\sigma}_{22})dz + D_2(\bar{\sigma}_{11} - \bar{\sigma}_{33})dz + (1 - r_4)d\epsilon'' \tag{A-7}$$

$$d\epsilon_{22}'' = D_1(\bar{\sigma}_{22} - \bar{\sigma}_{11})dz + D_2(\bar{\sigma}_{22} - \bar{\sigma}_{33})dz + (1 - r_4)d\epsilon'' \tag{A-8}$$

$$d\epsilon_{33}'' = D_2(\bar{\sigma}_{33} - \bar{\sigma}_{11})dz + D_2(\bar{\sigma}_{33} - \bar{\sigma}_{22})dz + (1 + 2r_4)d\epsilon'' \tag{A-9}$$

$$d\epsilon_{12}'' = (c_1 - c_2)\sigma_{12} dz \tag{A-10}$$

$$d\epsilon_{13}'' = c_5 \sigma_{13} dz \tag{A-11}$$

$$d\epsilon_{23}'' = c_5 \sigma_{23} dz \tag{A-12}$$

where $D_1 = (c_1-c_2)/3$, $D_2 = (c_3-c_4)/3$, r_4 is an anisotropy coefficient, and $d\varepsilon''$ represents the inelastic volume change as given by Equation 15.

REFERENCES

1. Ansal, A. M. (1977), "An Endochronic Constitutive Law for Normally Consolidated Cohesive Soils," Ph.D. Dissertation, Department of Civil Engineering, Northwestern University, Evanston, Illinois, 166 pp.

2. Ansal, A. M., Bažant, Z. P. and Krizek, R. J. (1979), "Viscoplasticity of Normally Consolidated Clays," Journal of the Geotechnical Engineering Division, American Society of Civil Engineers, Volume 105, Number GT4, pp. 519-537.

3. Bažant, Z. P., Ansal, A. M. and Krizek, R. J. (1979), "Viscoplasticity of Transversely Isotropic Clays," Journal of the Engineering Mechanics Division, American Society of Civil Engineers, Volume 105, Number EM4, pp. 549-569.

4. Bažant, Z. P. and Krizek, R. J. (1976), "Endochronic Constitutive Law for Liquefaction of Sand," Journal of the Engineering Mechanics Division, American Society of Civil Engineers, Volume 102, Number EM2, pp. 225-238.

5. Bažant, Z. P., Ansal, A. M. and Krizek, R. J. (1980), "Critical Appraisal of Endochronic Theory for Soils," Proceedings of Workshop on Constitutive Modeling of Soils," McGill University, Montreal, Canada.

6. Cuellar, V., Bažant, Z. P., Krizek, R. J. and Silver, M. L. (1977), "Densification and Hysteresis of Sand under Cyclic Shear," Journal of the Geotechnical Engineering Division, American Society of Civil Engineers, Volume 103, Number GT5, pp. 399-416.

7. Krizek, R. J., Ansal, A. M. and Bažant, Z. P. (1978), "Constitutive Equation for Cyclic Behavior of Cohesive Soils," Geotechnical Engineering Division, Proceedings of the Specialty Conference on Earthquake Engineering and Soil Dynamics, American Society of Civil Engineers, Volume 2, pp. 557-568.

8. Sener, C. (1979), "An Endochronic Nonlinear Inelastic Constitutive Law for Cohesionless Soils Subjected to Dynamic Loading," Ph.D. Dissertation, Department of Civil Engineering, Northwestern University, Evanston, Illinois.

9. Valanis, K. C. (1971), "A Theory of Viscoplasticity Without a Yield Surface Part I. General Theory; Part II. Application to Mechanical Behavior of Metals," Archives of Mechanics (Archiwum Mechaniki Stosowanej), Volume 23, pp. 517-555.

SOIL BEHAVIOR PREDICTION

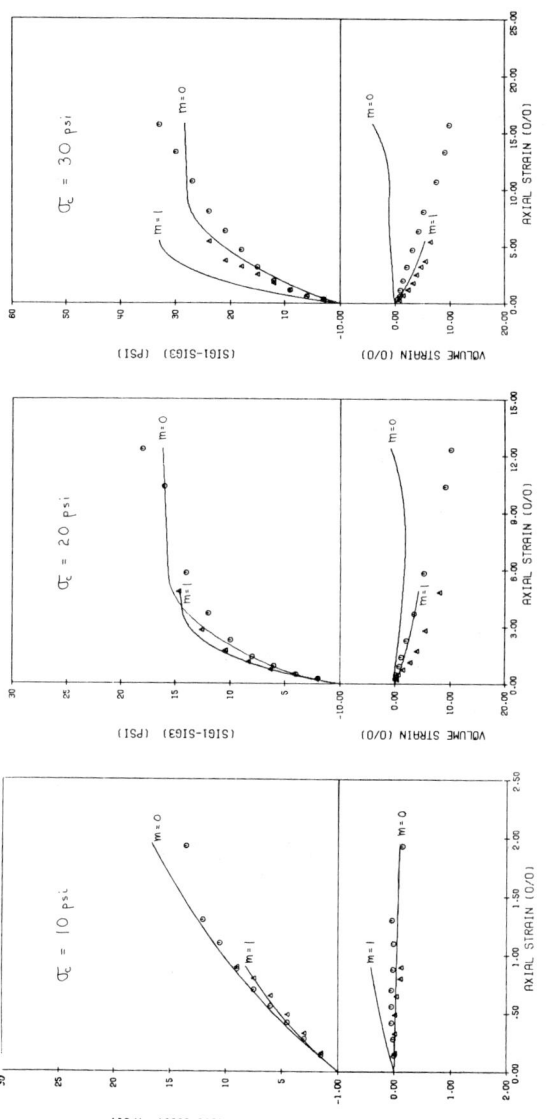

Figure 1. Initial Fits for Clay X

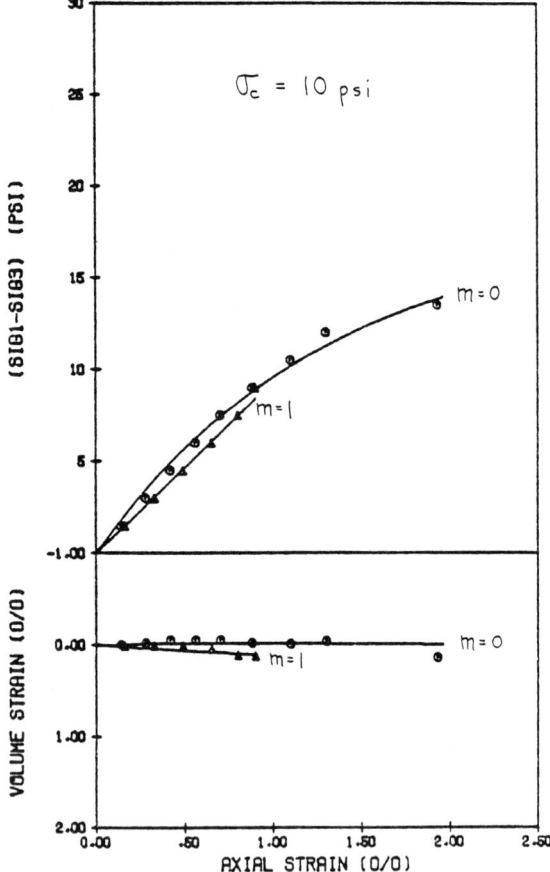

Figure 2. Improved Fits for Clay X

SOIL BEHAVIOR PREDICTION   311

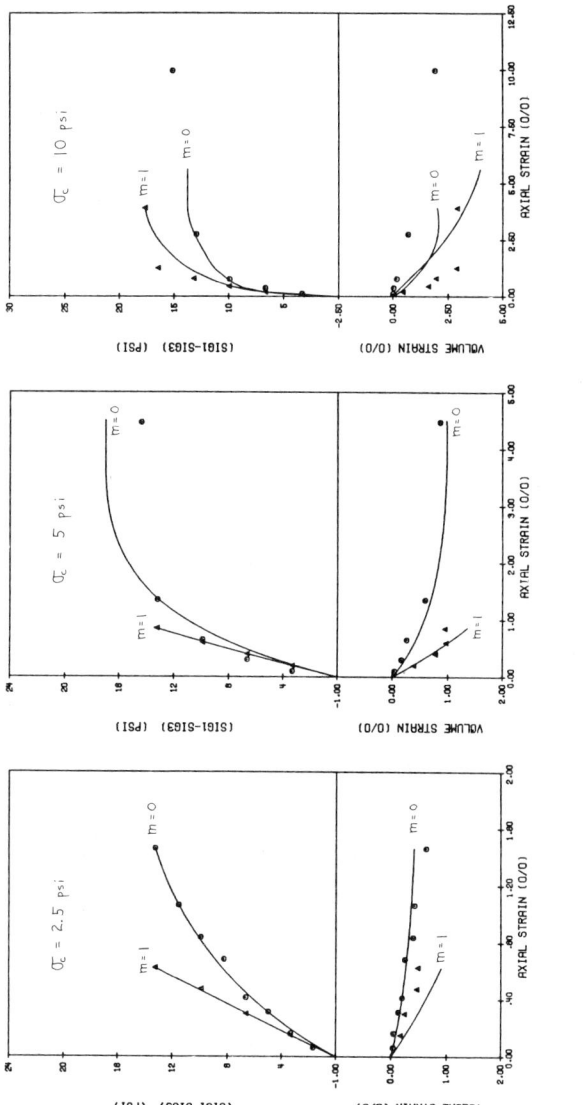

Figure 3. Improved Fits for Clay Y

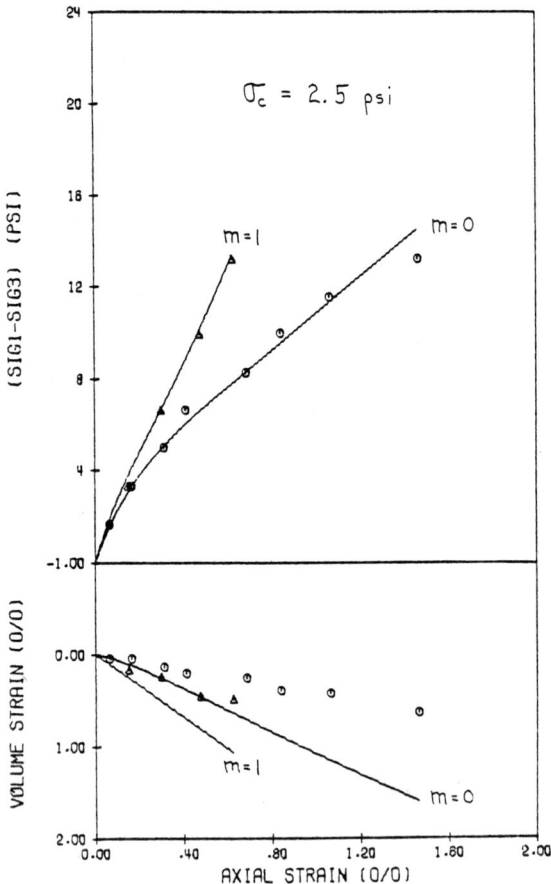

Figure 4. Initial Fits for Clay Y

SOIL BEHAVIOR PREDICTION

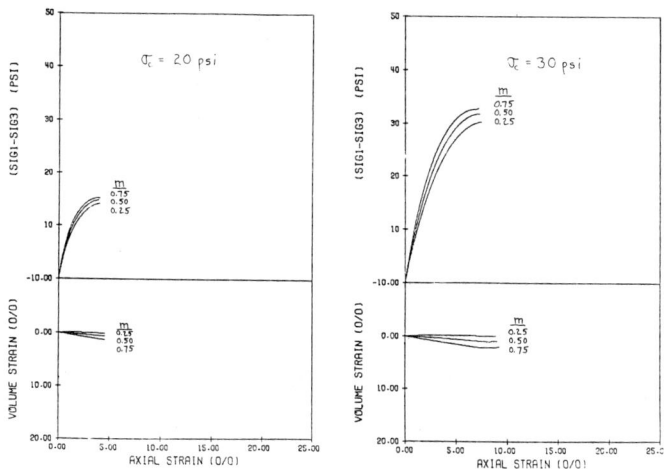

Figure 5. Typical Predictions for Clay X

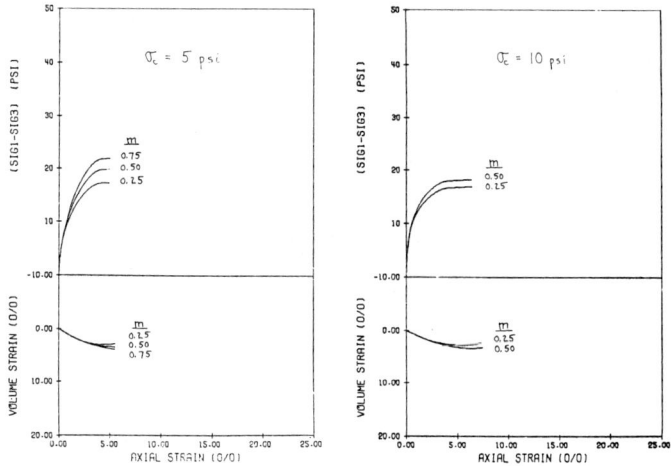

Figure 6. Typical Predictions for Clay Y

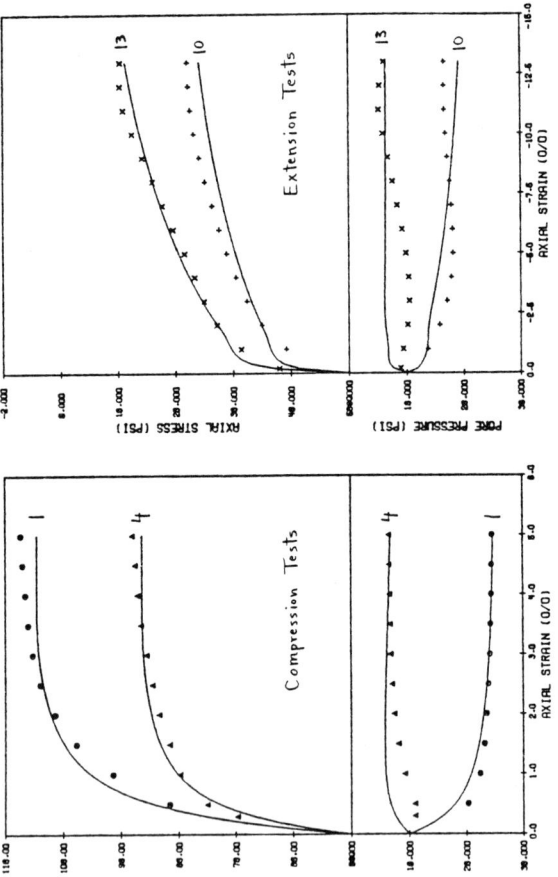

Figure 7. Optimized Fits for Kaolinite

SOIL BEHAVIOR PREDICTION

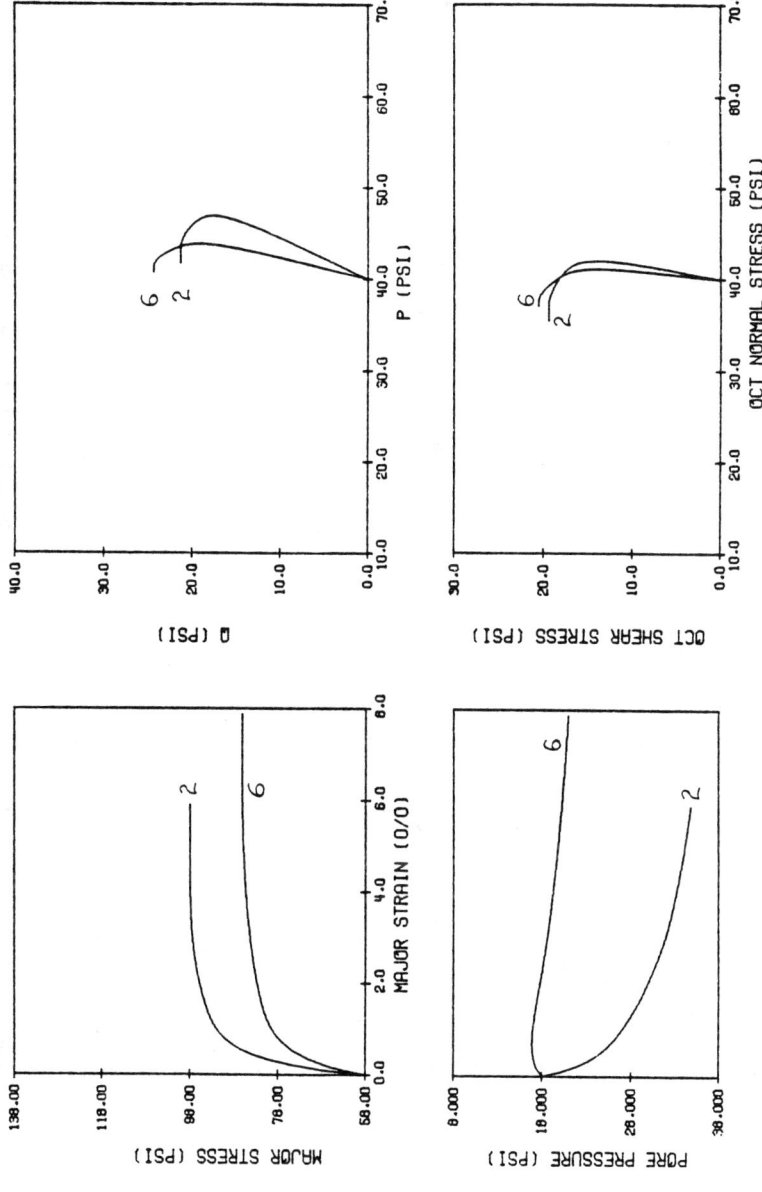

Figure 8. Predicted Response for Kaolinite (Tests 2 and 6)

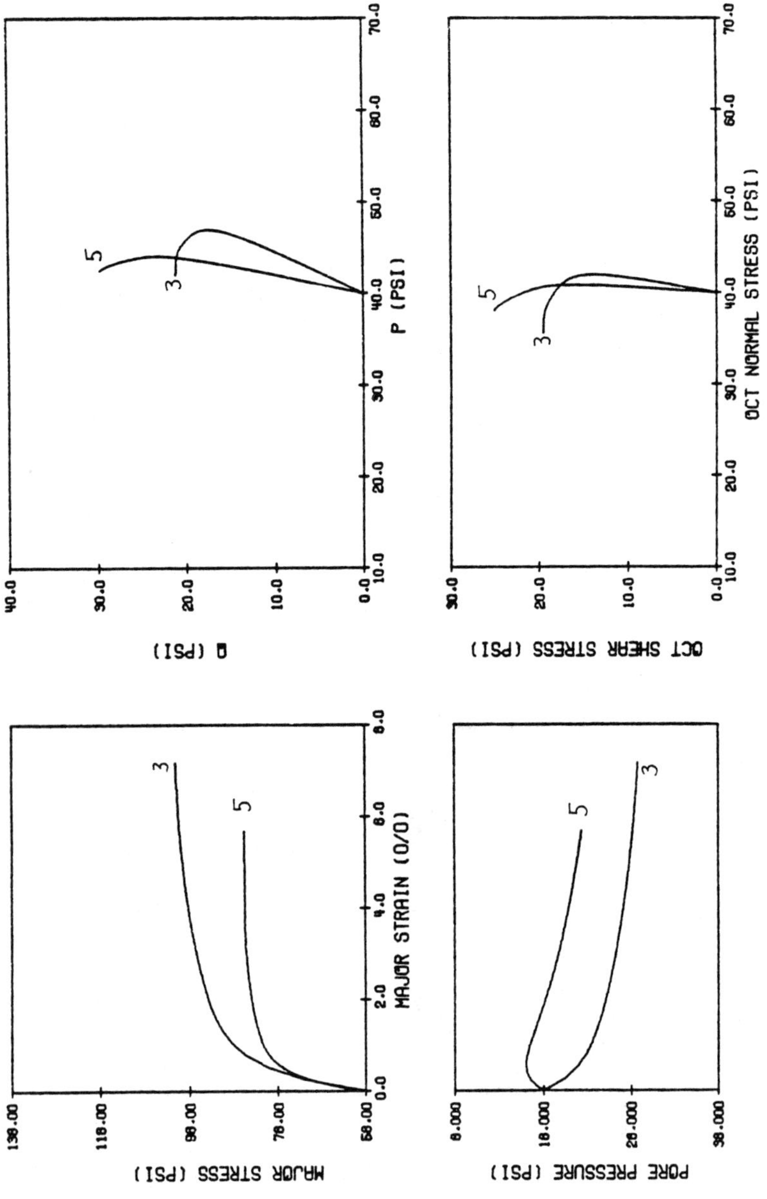

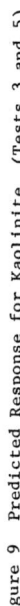

Figure 9 Predicted Response for Kaolinite (Tests 3 and 5)

SOIL BEHAVIOR PREDICTION 317

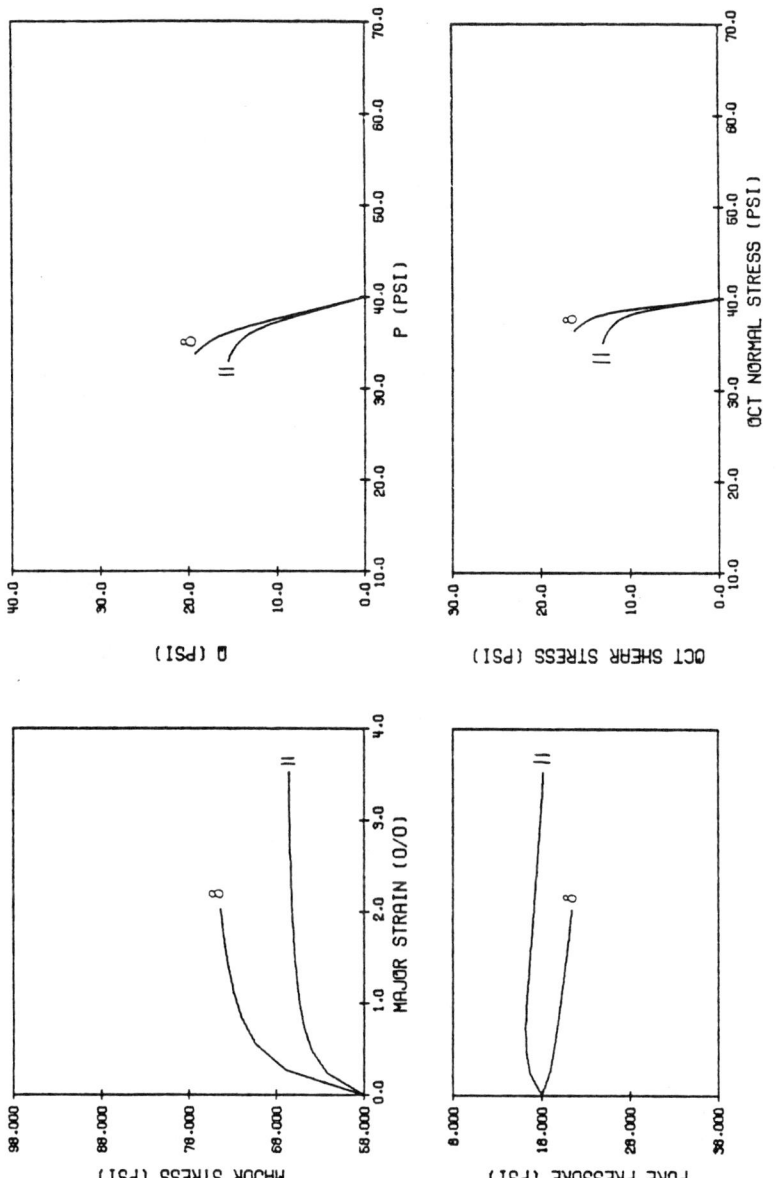

Figure 10. Predicted Response for Kaolinite (Tests 8 and 11)

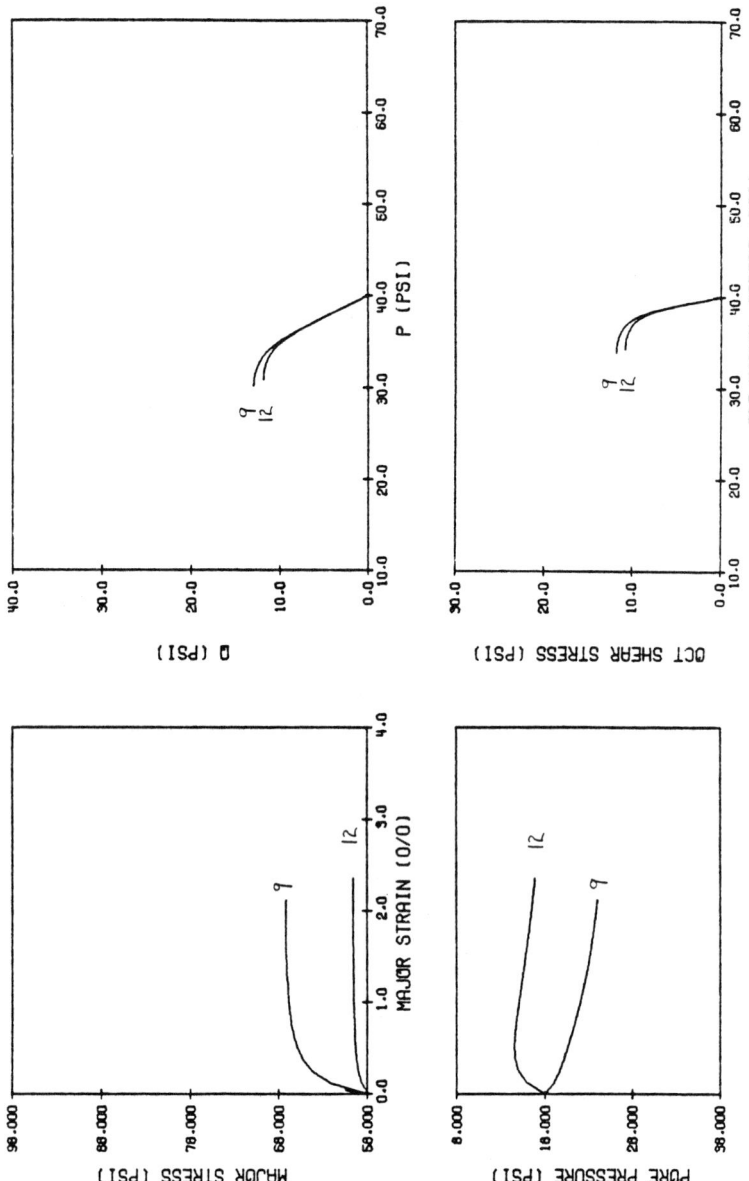

Figure 11. Predicted Response for Kaolinite (Tests 9 and 12)

SOIL BEHAVIOR PREDICTION

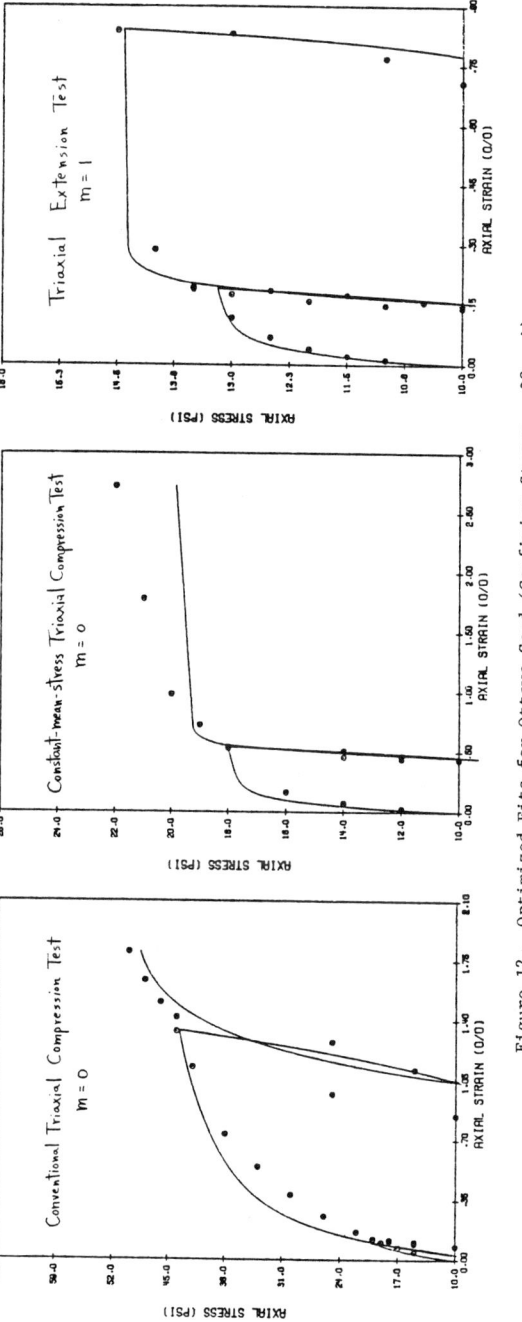

Figure 12. Optimized Fits for Ottawa Sand (Confining Stress = 10 psi)

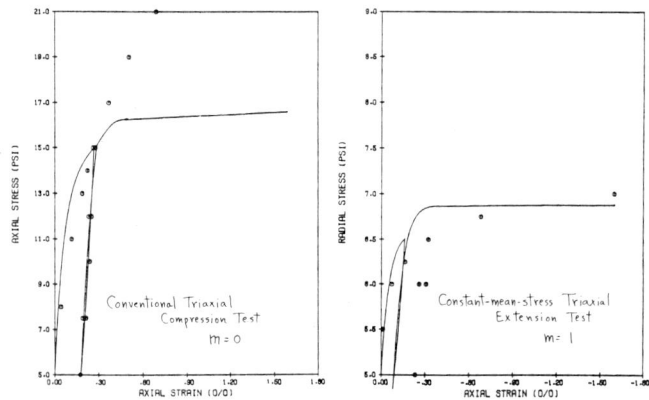

Figure 13. Optimized Fits for Ottawa Sand (Confining Stress = 5 psi)

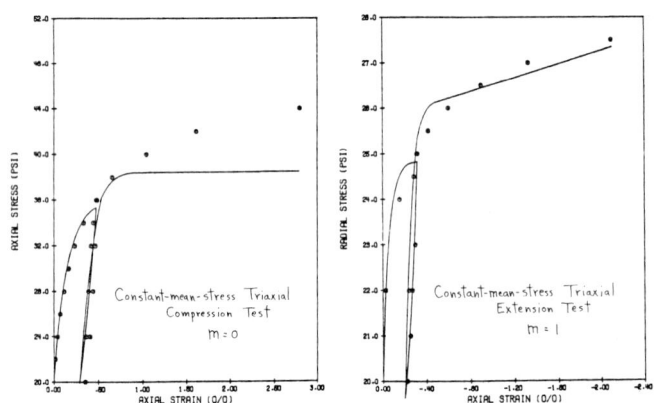

Figure 14. Optimized Fits for Ottawa Sand (Confining Stress = 20 psi)

SOIL BEHAVIOR PREDICTION

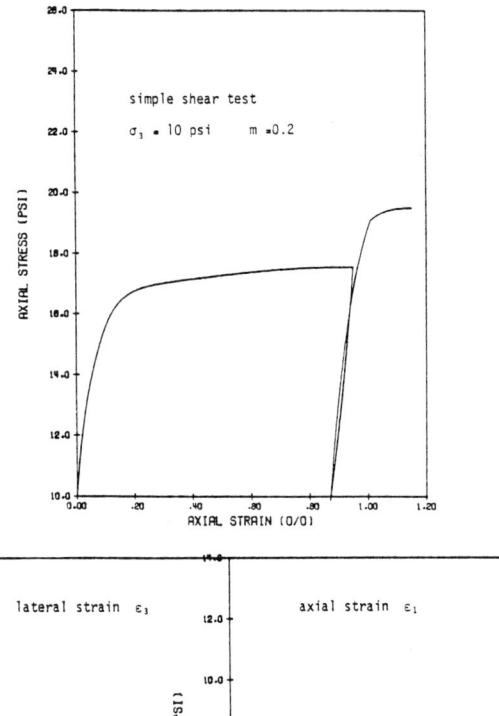

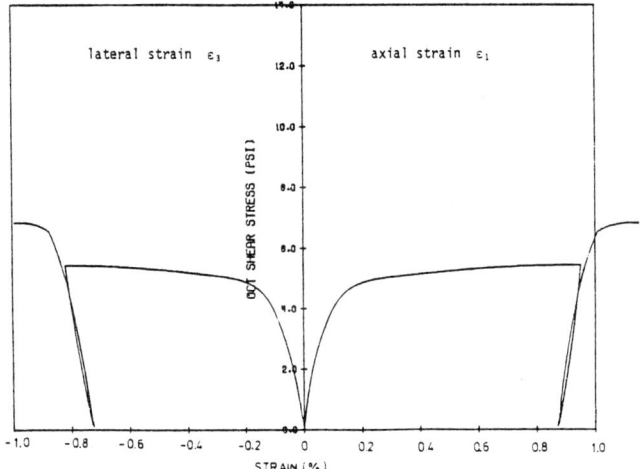

Figure 15. Predicted Response for Ottawa Sand in Simple Shear (Mean Stress = 10 psi; m = 0.2)

322 STRESS STRAIN FOR SOILS

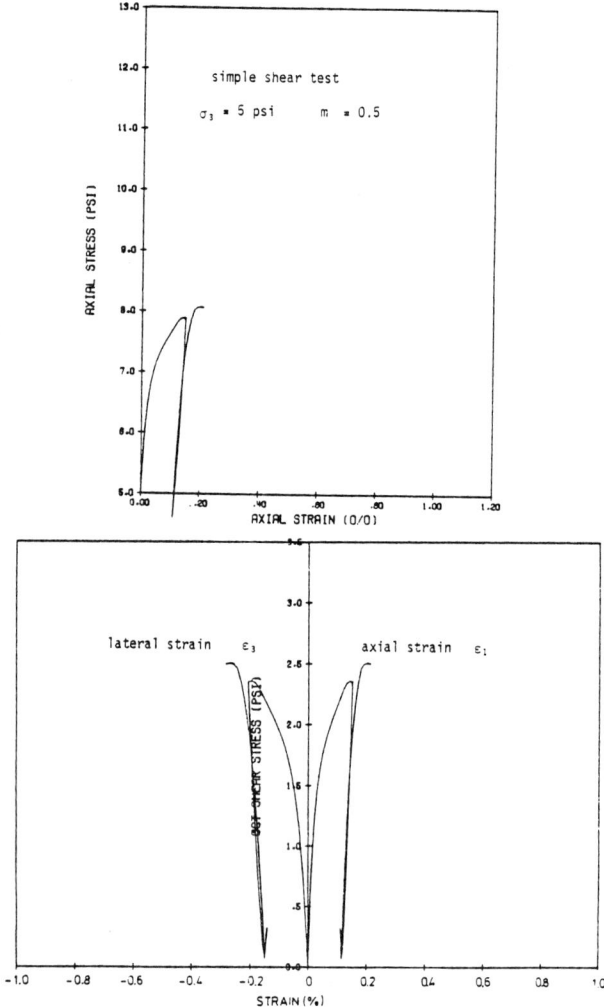

Figure 16. Predicted Response for Ottawa Sand in Simple Shear (Mean Stress = 5 psi; m = 0.5)

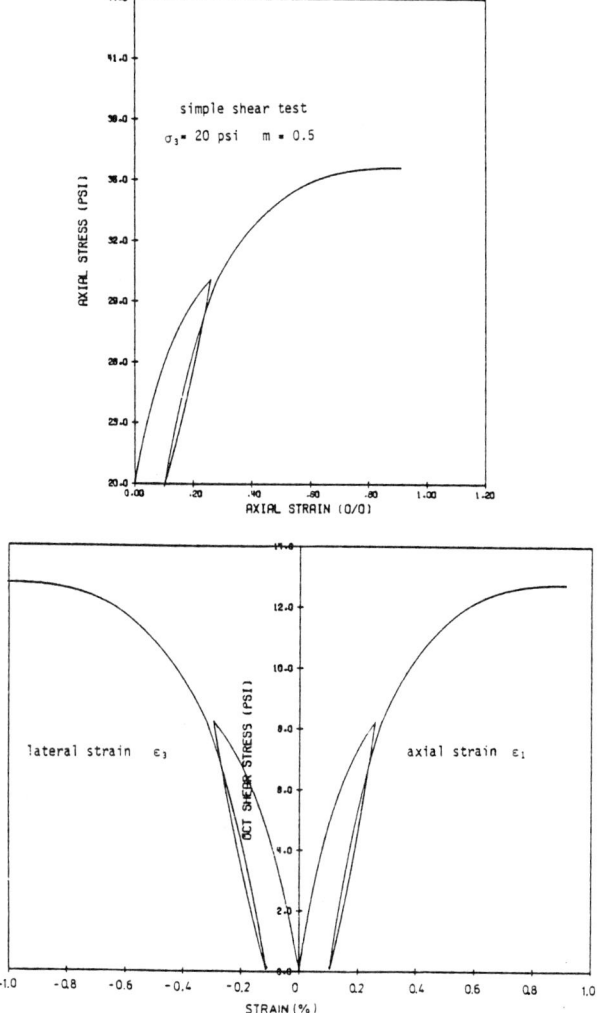

Figure 17. Predicted Response for Ottawa Sand in Simple Shear (Mean Stress = 20 psi; m = 0.5)

324 STRESS STRAIN FOR SOILS

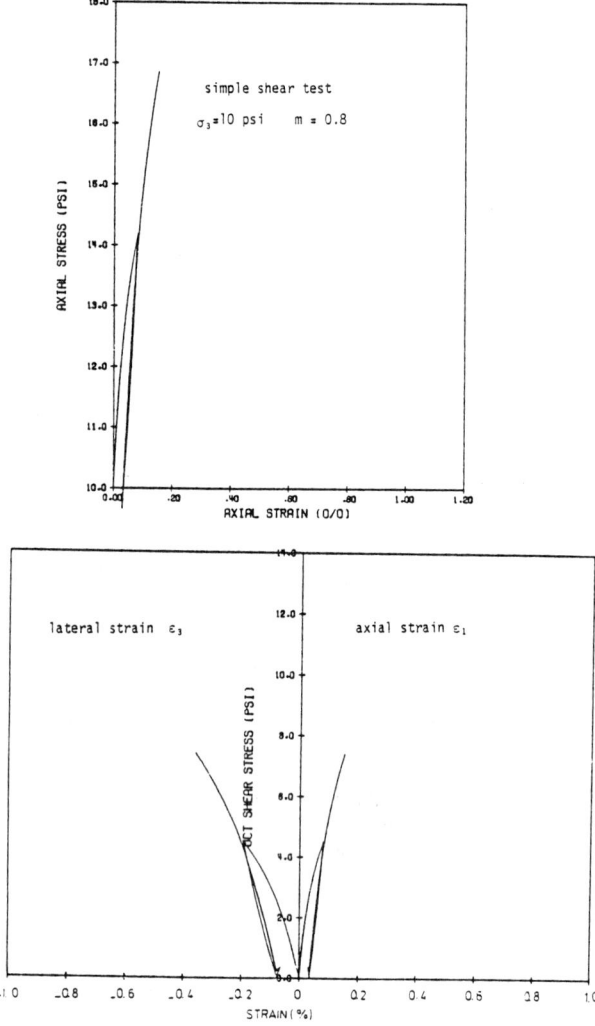

Figure 18. Predicted Response for Ottawa Sand in Simple Shear
(Mean Stress = 10 psi; m = 0.8)

SOIL BEHAVIOR PREDICTION

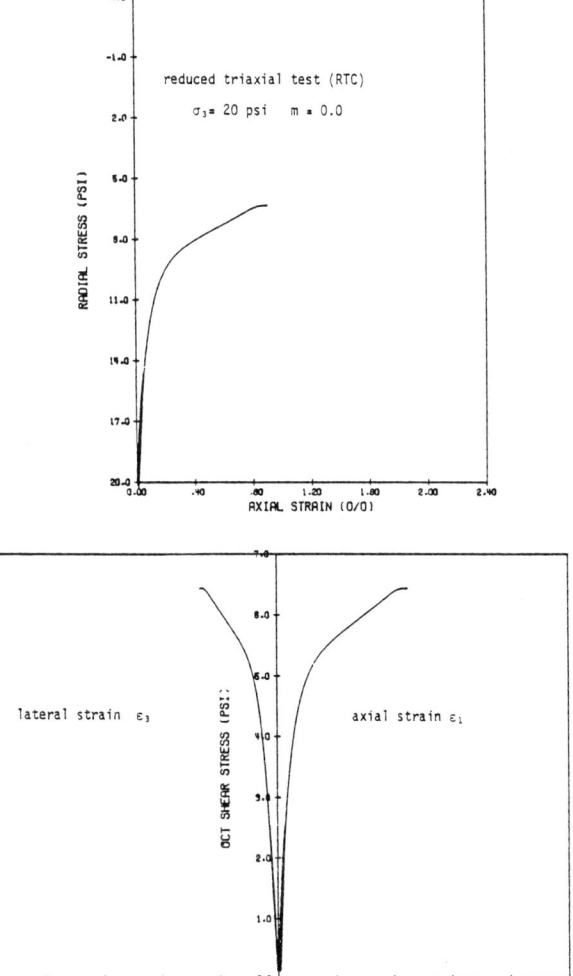

Figure 19. Predicted Response for Ottawa Sand in Reduced Triaxial Compression (Mean Stress=20 psi; m = 0)

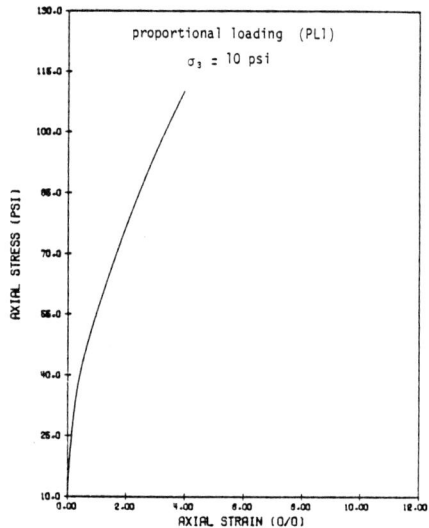

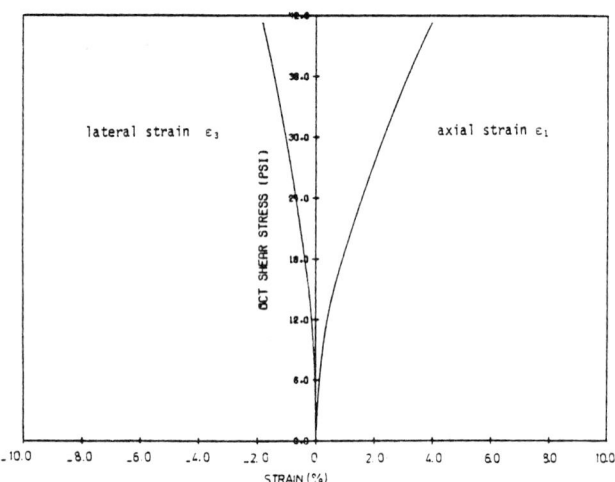

Figure 20. Predicted Response for Ottawa Sand Subjected to Proportional Loading PL 1 (Mean Stress = 10 psi; m = 0)

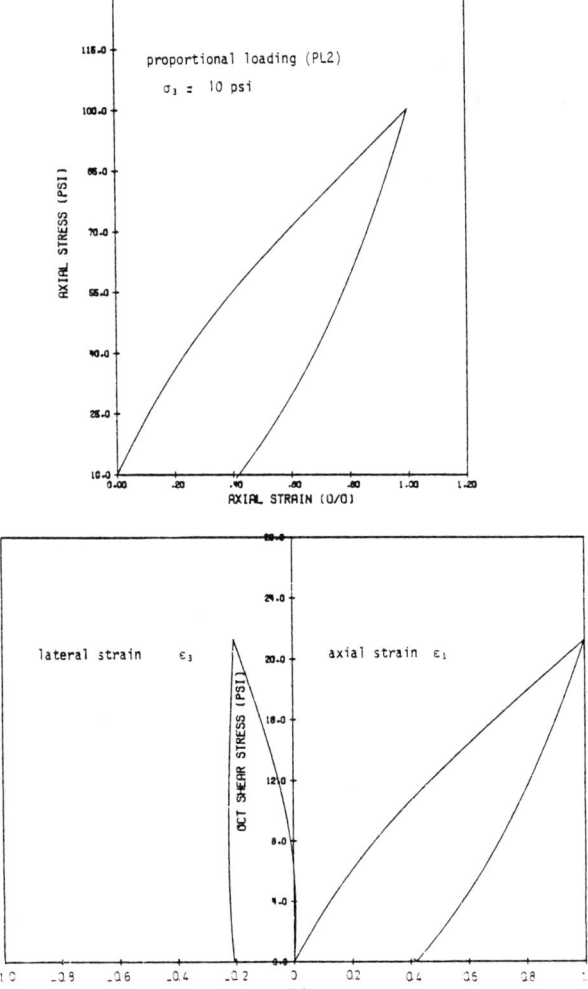

Figure 21. Predicted Response for Ottawa Sand Subjected to Proportional Loading PL 2 (Mean Stress = 10 psi; m = 0)

PLASTICITY MODELS FOR SOILS

Comparison and Discussion

by

E. Mizuno[1] and W. F. Chen[2], M. ASCE

Introduction

Herein, the predictions are made using the following three material models:

(i) Nonlinear elasticity model with the Mohr-Coulomb or the Drucker-Prager criterion as failure surface. This model is applied to all predictions.

(ii) The Mohr-Coulomb type of elastic-plastic material model with two different elliptical hardening caps. One on tensile meridian plane ($\theta = 0°$) and the other on compressive meridian plane ($\theta = 60°$). (Cap Model I). This model is applied to "Clay X" and "Clay Y".

(iii) The Mohr-Coulomb type of elastic-plastic material model with a three-dimensional elliptical cap whose shape depends on the Lode angle θ. (Cap Model II). This model is applied to "Clay X" and "Clay Y".

The comparison and discussion are given in the forthcoming.

Clay X

The comparison of the predictions with the experimental data on the simple shear tests with constant mean stresses at 10 psi, 20 psi and

---

[1] Research Assistant, School of Civil Engineering, Purdue University, West Lafayette, IN 47907

[2] Professor of Structural Engineering, School of Civil Engineering, Purdue University, West Lafayette, IN 47907

30 psi are presented in Figs. 1 through 4 for each of the stress ratio m = 0.25, 0.5, 0.75. In these figures, the stress difference-axial strain relation, and the axial strain-volumetric strain relation are presented. The experimental data are shown by the open circles and the predictions are denoted by lines.

As can be seen from the figures, the predictions by the nonlinear elasticity model are off somewhat from the experimental data with σ_c = 20 psi and 30 psi. The model, however, gives a good prediction for the tests with σ_c = 10 psi and for σ_c = 30 psi, m = 0.75. Since the model is assumed to be isotropic, it is difficult to predict the behavior of the anisotropic clay X. Also, it should be noted that the volumetric strain ϵ_v on the simple shear tests can not be predicted by this model.

Cap models are seen to predict the axial strain-stress difference relation and the axial strain-volumetric strain relation well for all experiments shown in figures. The tests at σ_c = 10 psi show elastic. In the prediction (σ_c = 10 psi), the cap models show some plastic strains in the region near failure surface. For the experiment with σ_c = 10 psi, m = 0.5, the cap model I predicts only elastic strains because the decomposed stresses do not reach the hardening caps. Therefore, there is no plastic strains.

Clay Y

The comparison of prediction with experimental data on the triaxial tests under initial confining pressures σ_c = 2.5, 5.0 and 10 psi is presented in Figs. 5 through 7 for each of the stress ratio m = 0.25, 0.5, 0.75. As the initial confining pressures lie on the swelling line, clay Y behaves first like an elastic material in the tests. Nonlinear elasticity model predicts well the experiments with σ_c = 2.5 and 5.0 psi. However, the predictions for the tests with σ_c = 10 psi, m = 0.5 and σ_c = 10 psi, m = 0.75 are off somewhat from the actual data. This implies that plastic strains must occur in the tests. Cap models reflect this as shown in Fig. 7. The predictions by cap models for the triaxial tests with σ_c = 2.5 psi and 5 psi are almost the same as that of nonlinear elasticity model. For these cases, either the stress paths have not reached the hardening caps, or the plastic strains are small at this stage of loading.

Kaolinite Clay

The comparison of the predictions with the data on the triaxial or simple shear tests under the undrained condition is presented in Figs. 8 through 12. The total axial stress vs. axial strain curves are shown in the figures. As can be seen, the nonlinear elasticity model predicts the experiments well. In particular, the predictions for Test NO. 7, 8, 9, 11 and 12 are very good. In this model, the pore water pressure increment is taken to be equivalent to the increment of the first invariant of the stress tensor I_1.

Ottawa Sand

The comparison of predictions with experimental data on various tests is presented in Figs. 13 through 21. In each figure, the strains in x, y and z directions are plotted against the octahedral shear stress τ_{oct}. Except the circular stress path, all predictions are good. In particular, the initial slope of the actual behavior of Ottawa sand and the failure load are accurately predicted by the nonlinear elasticity model combined with the Mohr-Coulomb criterion. As for the circular stress path, the model prediction is different significantly from the actual data in the region of Lode angle θ between 60° and 420° (Fig. 21).

Acknowledgments

This material is based upon work supported by the National Science Foundation under Grant No. PFR-7809326 to Purdue University.

PLASTICITY MODELS FOR SOILS 331

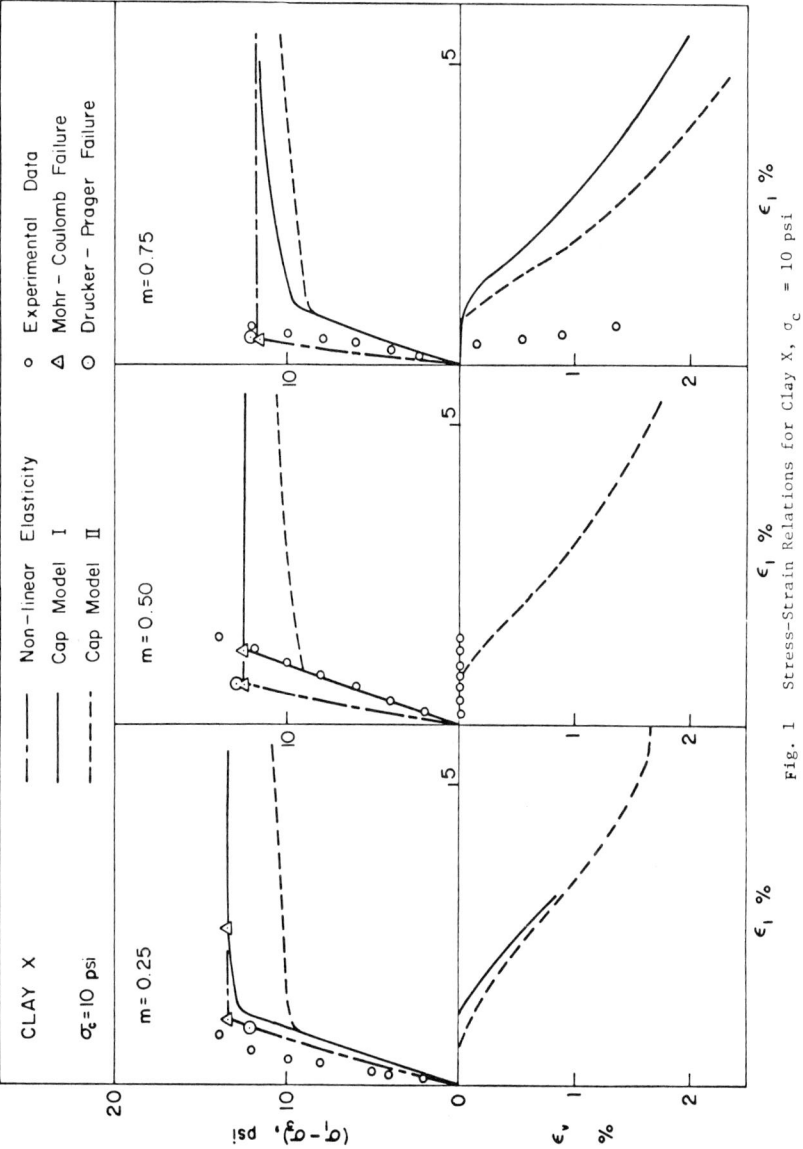

Fig. 1 Stress-Strain Relations for Clay X, σ_c = 10 psi

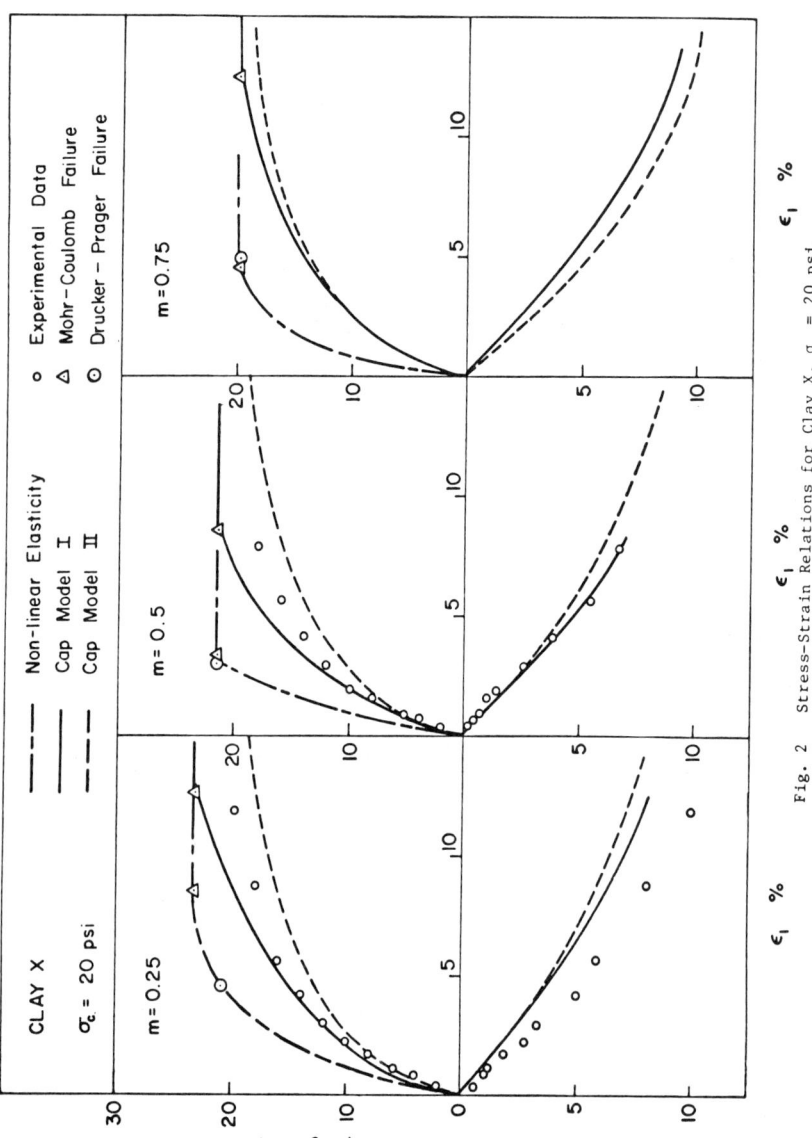

Fig. 2 Stress-Strain Relations for Clay X, $\sigma_c = 20$ psi

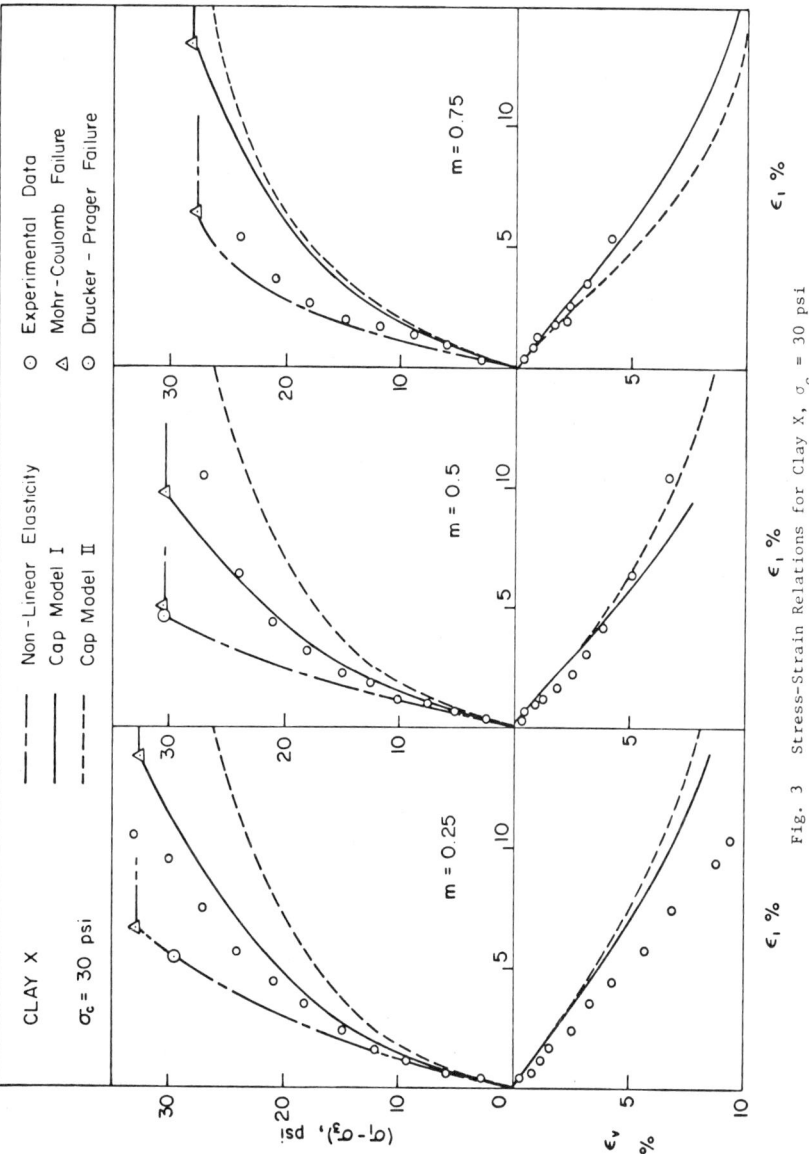

Fig. 3 Stress-Strain Relations for Clay X, $\sigma_c = 30$ psi

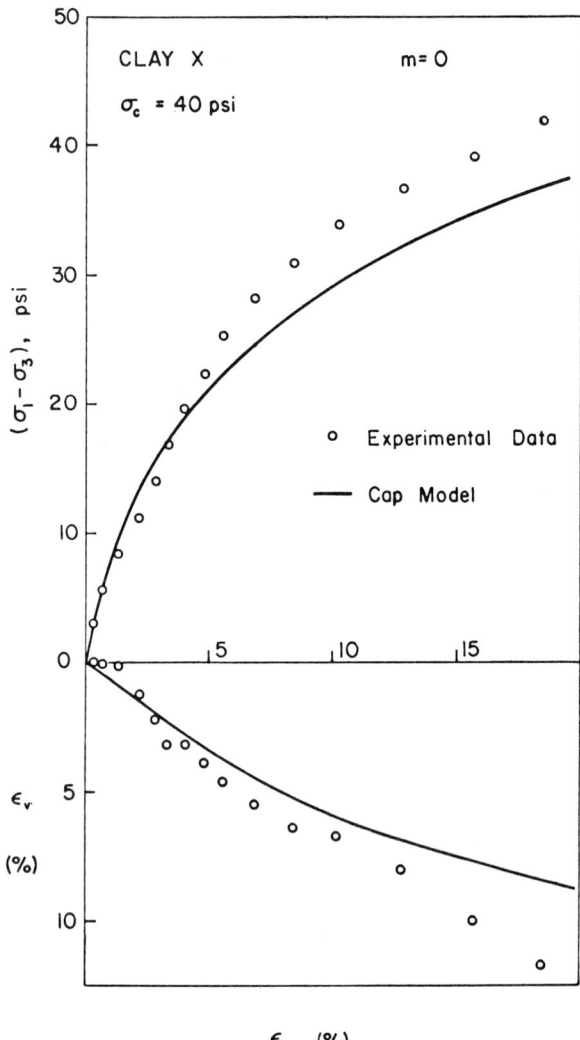

Fig. 4 Stress-Strain Relations for Clay X, σ_c = 40 psi

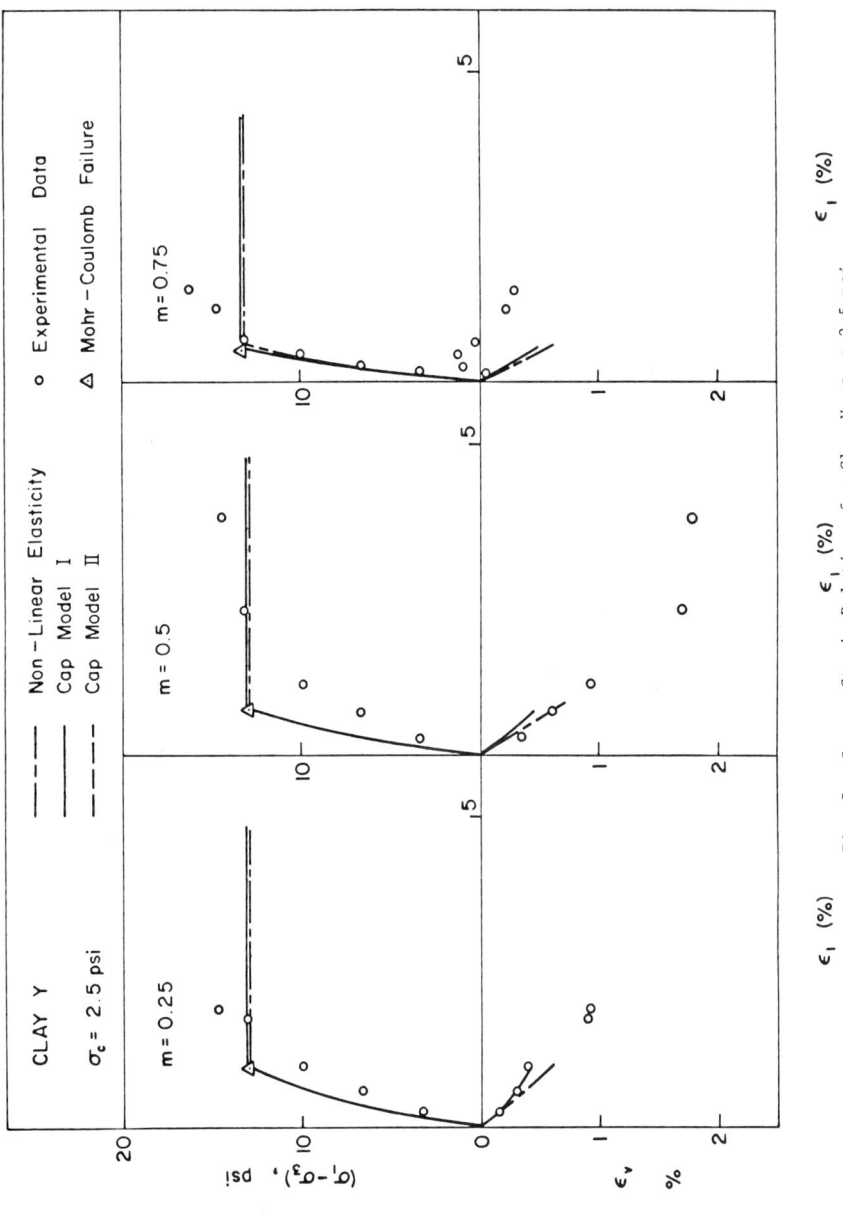

Fig. 5 Stress-Strain Relations for Clay Y, $\sigma_c = 2.5$ psi

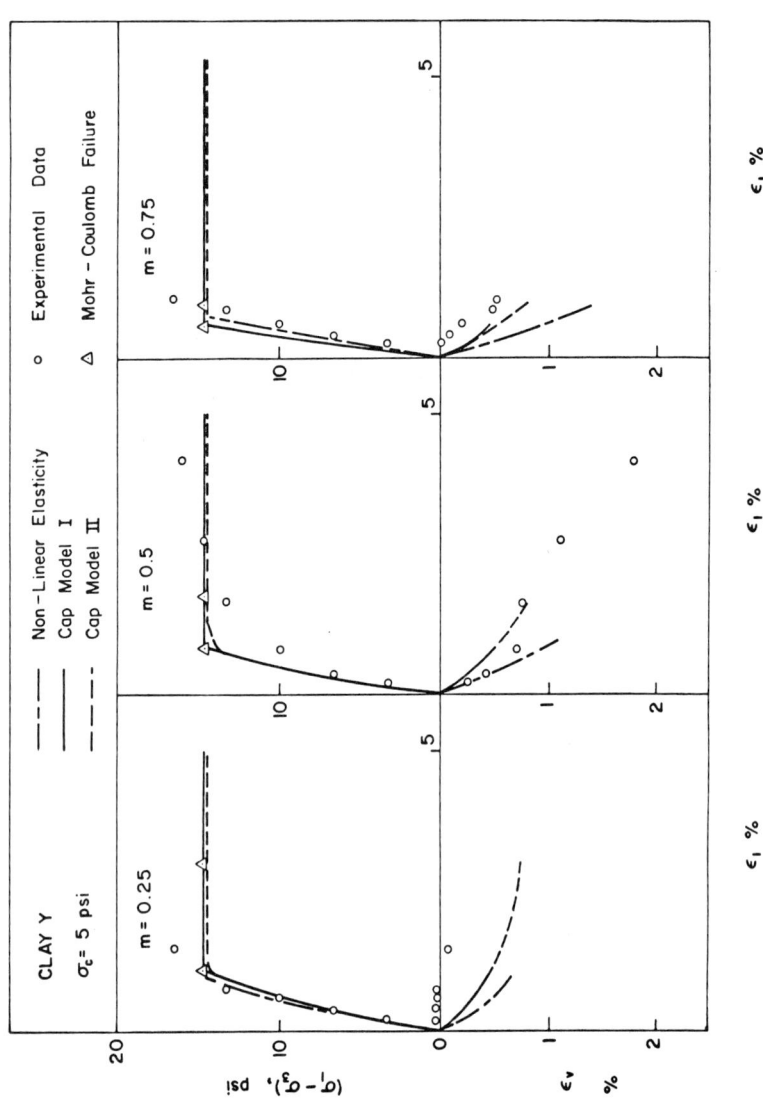

Fig. 6 Stress-Strain Relations for Clay Y, $\sigma_c = 5.0$ psi

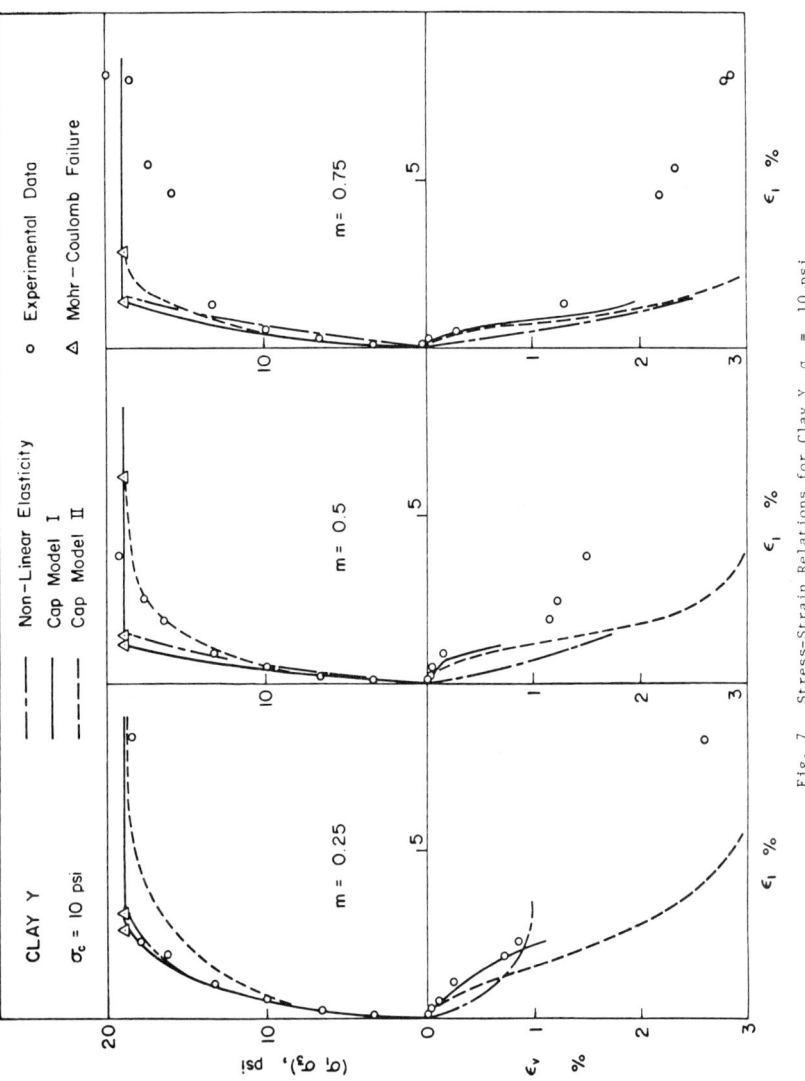

Fig. 7 Stress-Strain Relations for Clay Y, $\sigma_c = 10$ psi

338 STRESS STRAIN FOR SOILS

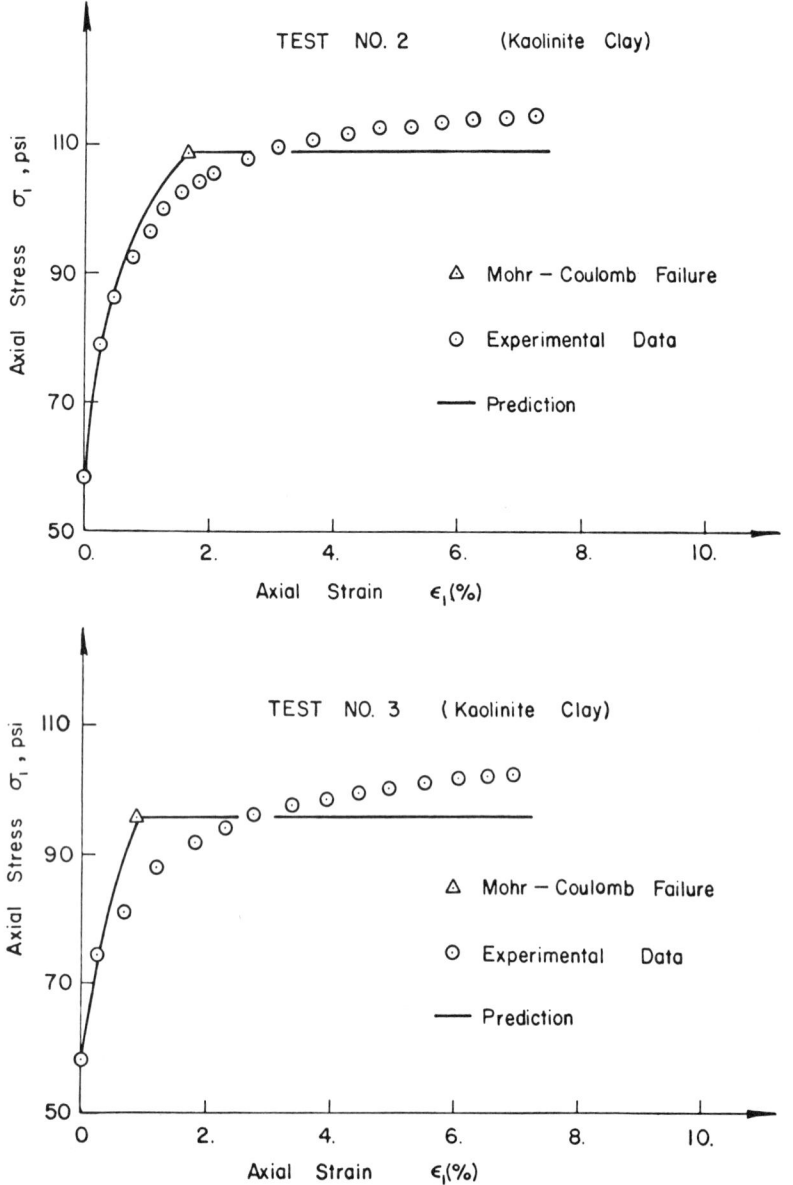

Fig. 8 Axial Stress-Strain Relations for Kaolinite Clay, Test No. 2 and

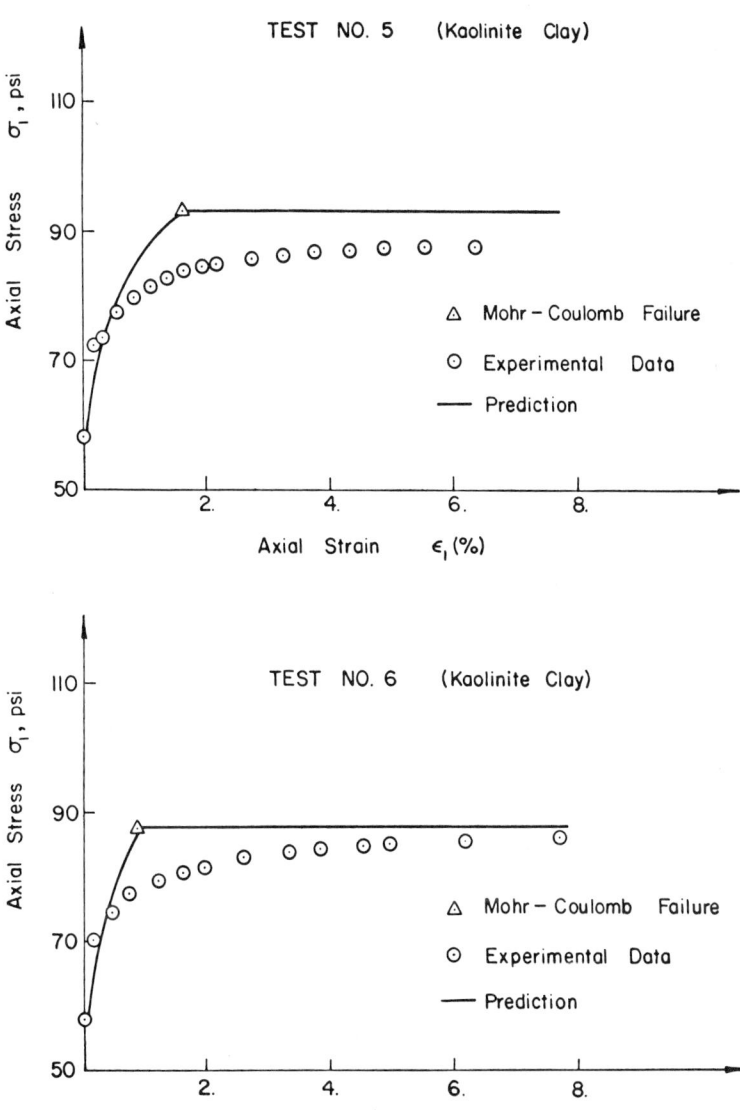

Fig. 9 Axial Stress-Strain Relations for Kaolinite Clay, Test No. 5 and 6

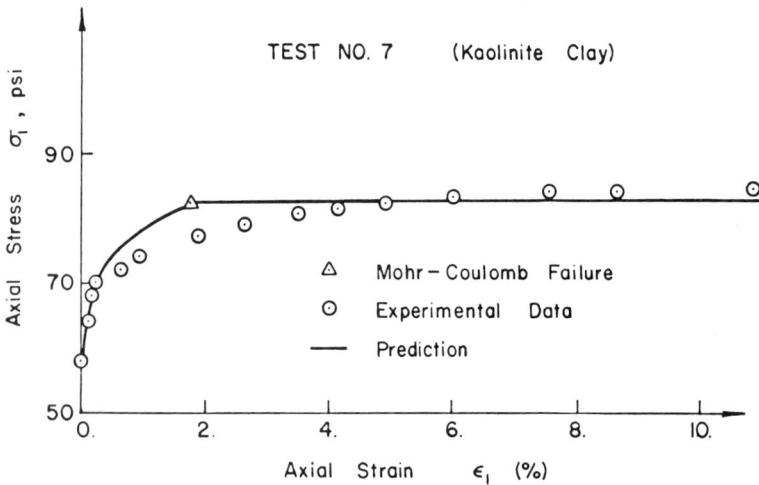

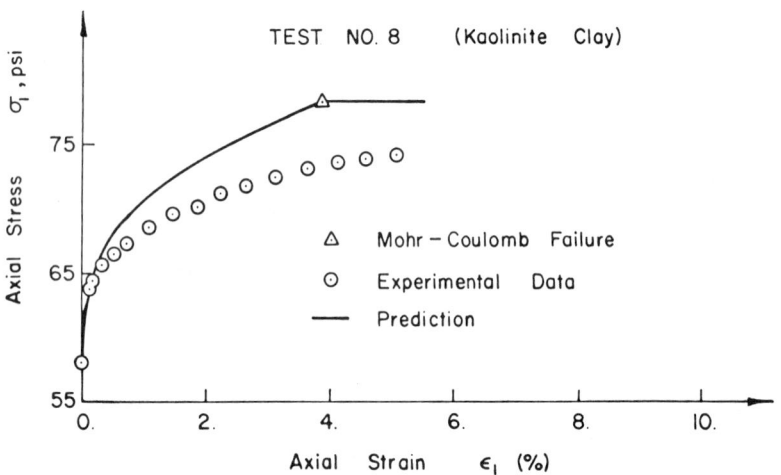

Fig. 10 Axial Stress-Strain Relations for Karolinite Clay, Test No. 7 and 8

PLASTICITY MODELS FOR SOILS 341

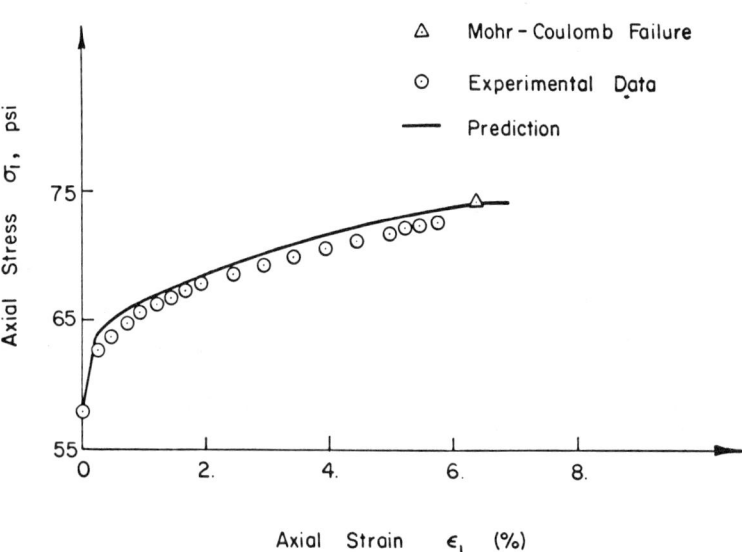

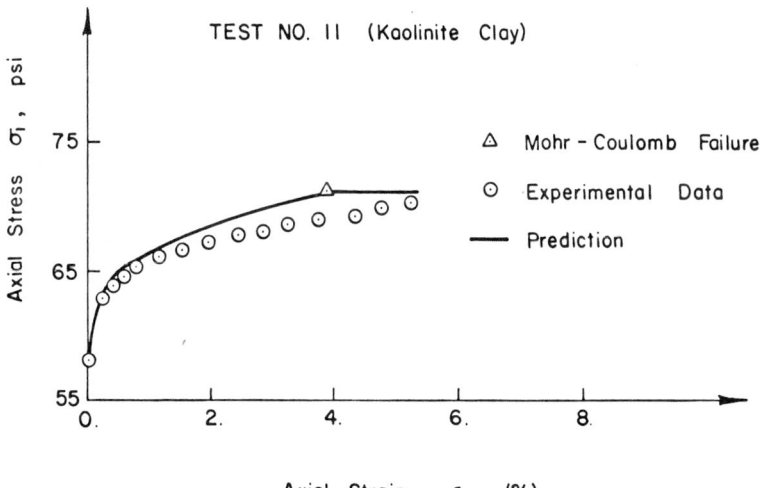

Fig. 11 Axial Stress-Strain Relations for Karolinite Clay, Test No. 9 and 11

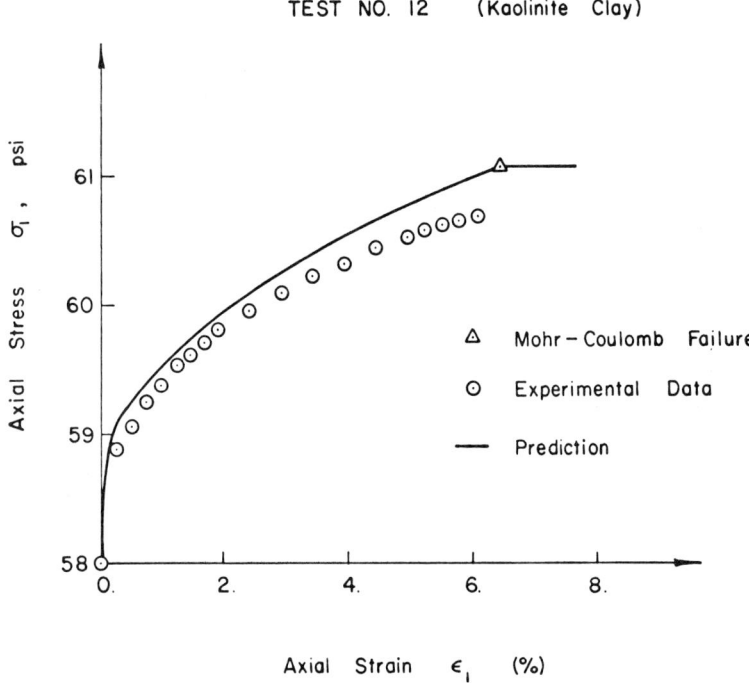

Fig. 12 Axial Stress-Strain Relation for Karolinite Clay, Test No. 12

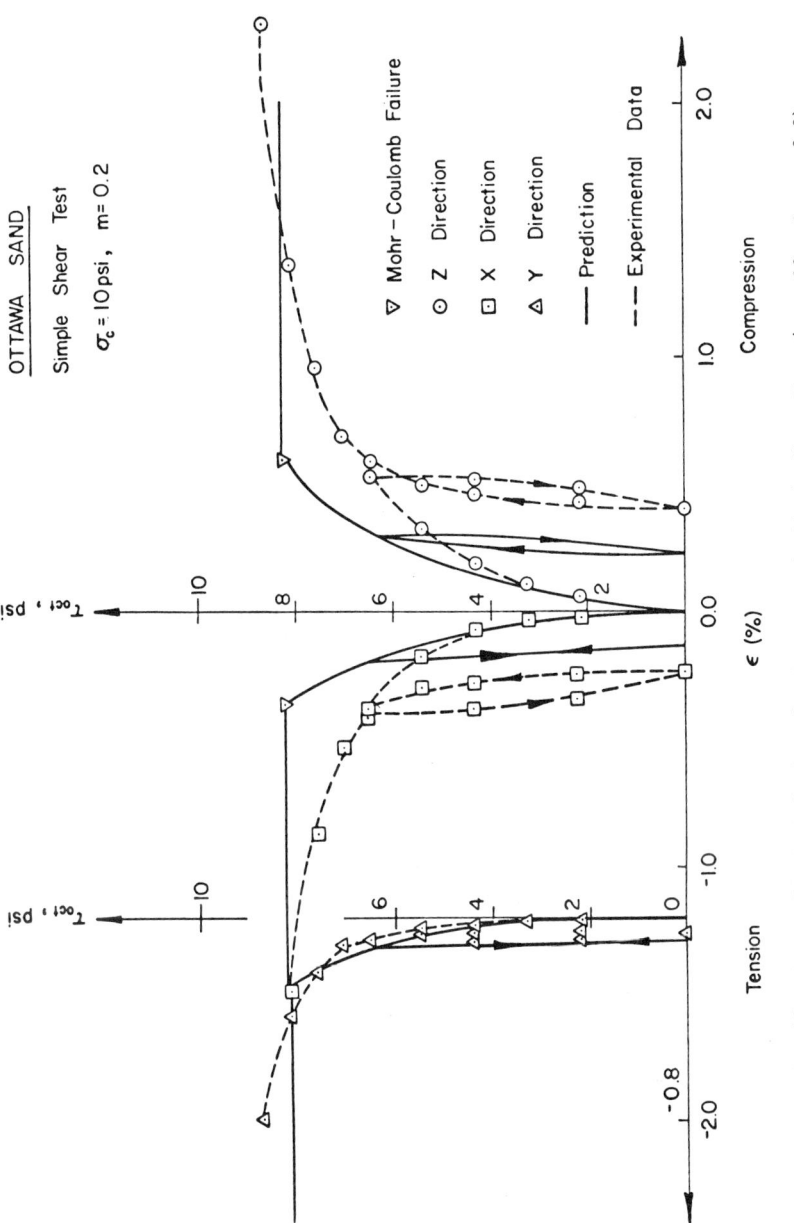

Fig. 13 Stress-Strain Relations for Ottawa Sand, Simple Shear Test ($\sigma_c = 10$ psi, $m = 0.2$)

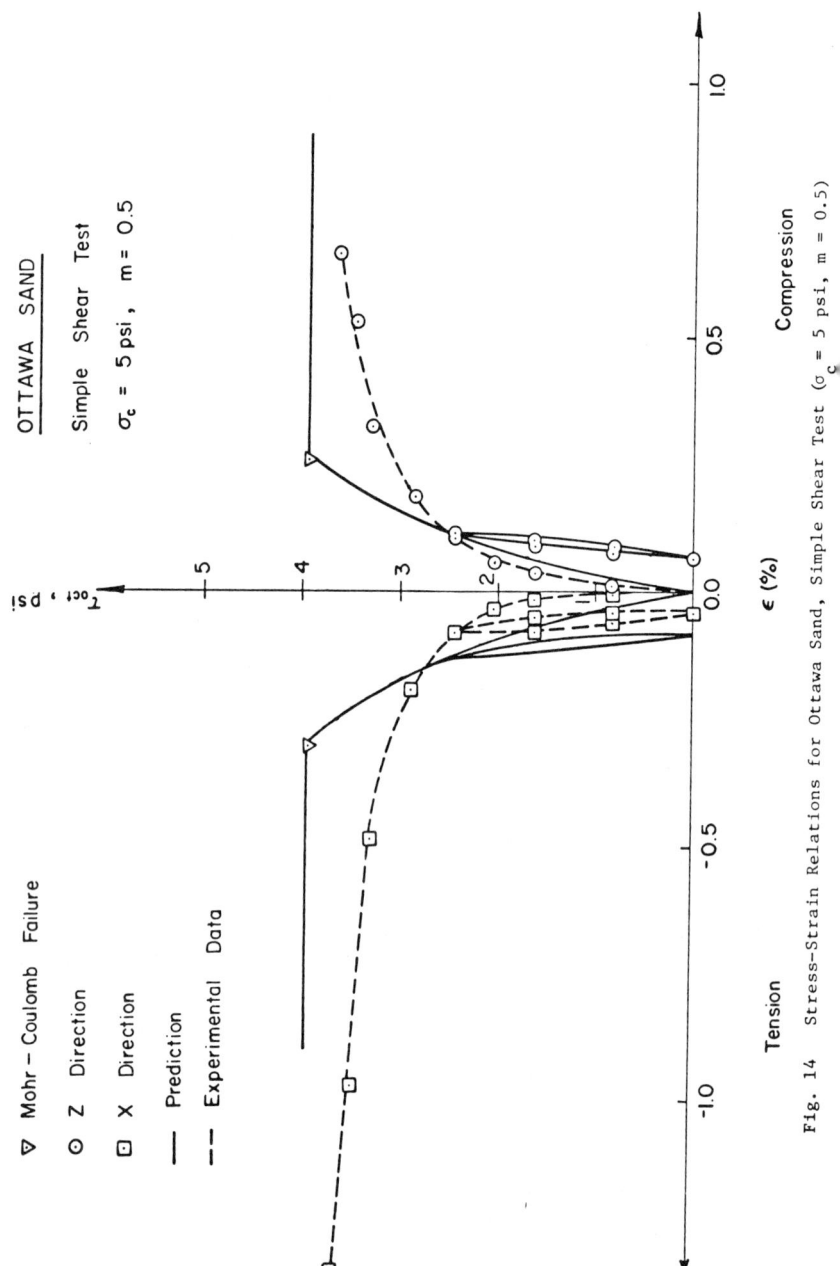

Fig. 14 Stress-Strain Relations for Ottawa Sand, Simple Shear Test (σ_c = 5 psi, m = 0.5)

PLASTICITY MODELS FOR SOILS 345

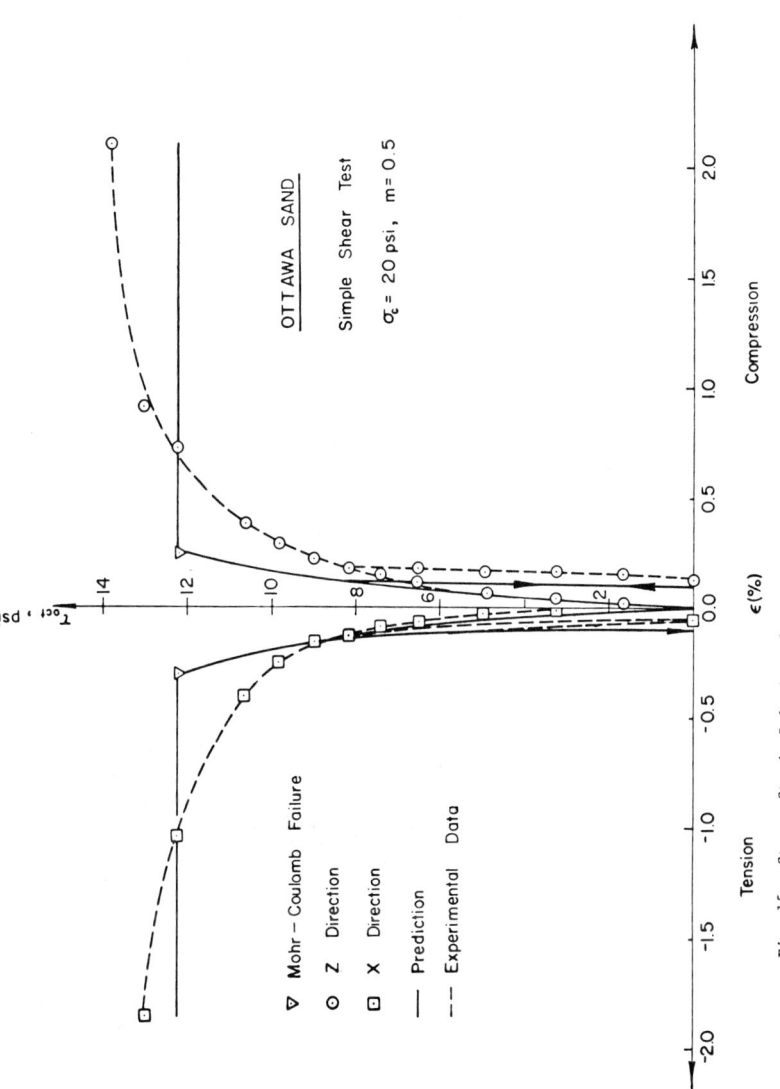

Fig. 15 Stress-Strain Relations for Ottawa Sand, Simple Shear Test (σ_c = 20 psi, m = 0.5)

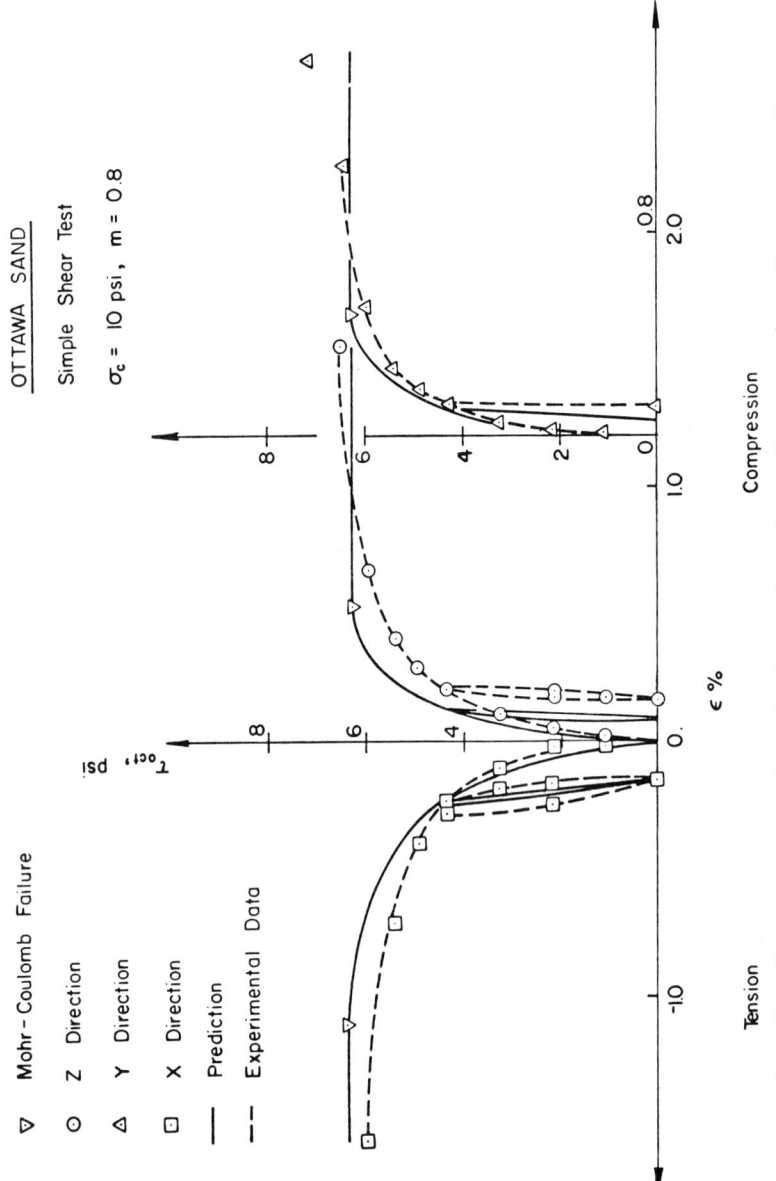

Fig. 16 Stress-strain Relations for Ottawa Sand, Simple Shear Test (σ_c = 10 psi, m = 0.8)

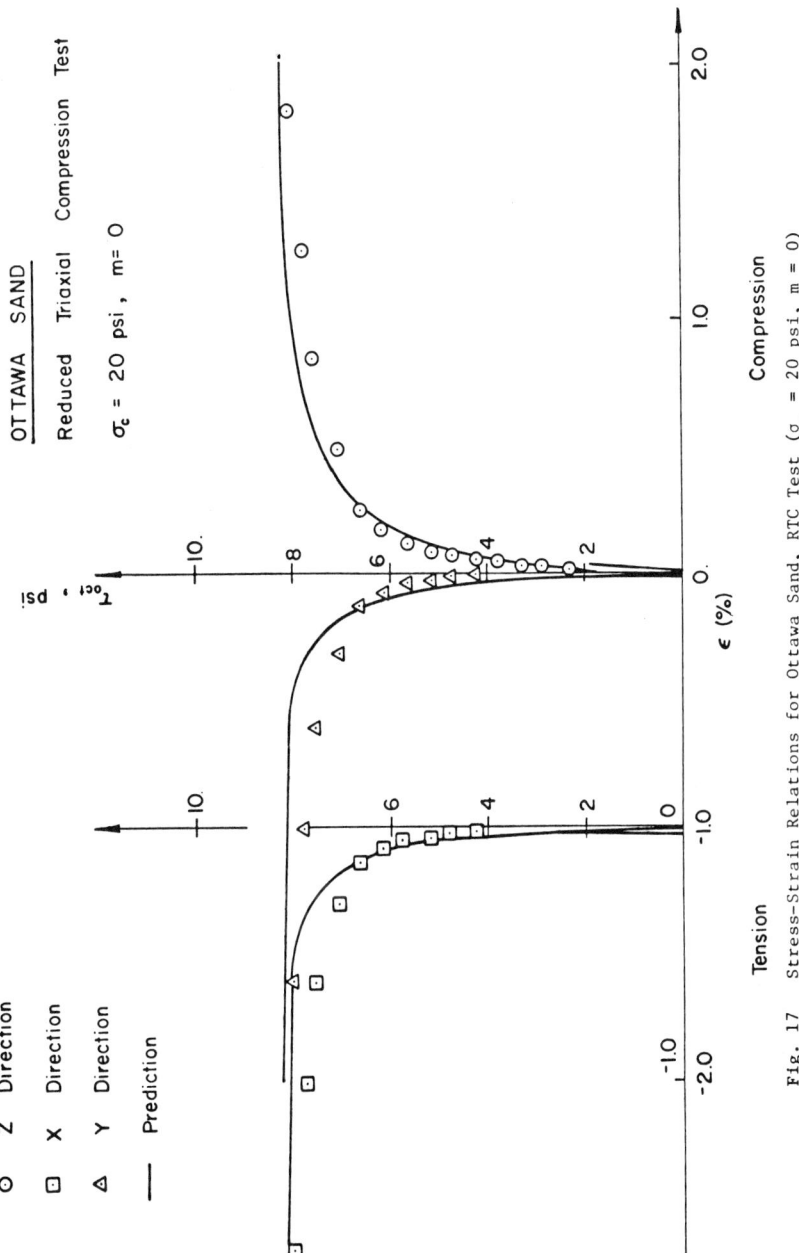

Fig. 17 Stress-Strain Relations for Ottawa Sand, RTC Test (σ_c = 20 psi, m = 0)

348 STRESS STRAIN FOR SOILS

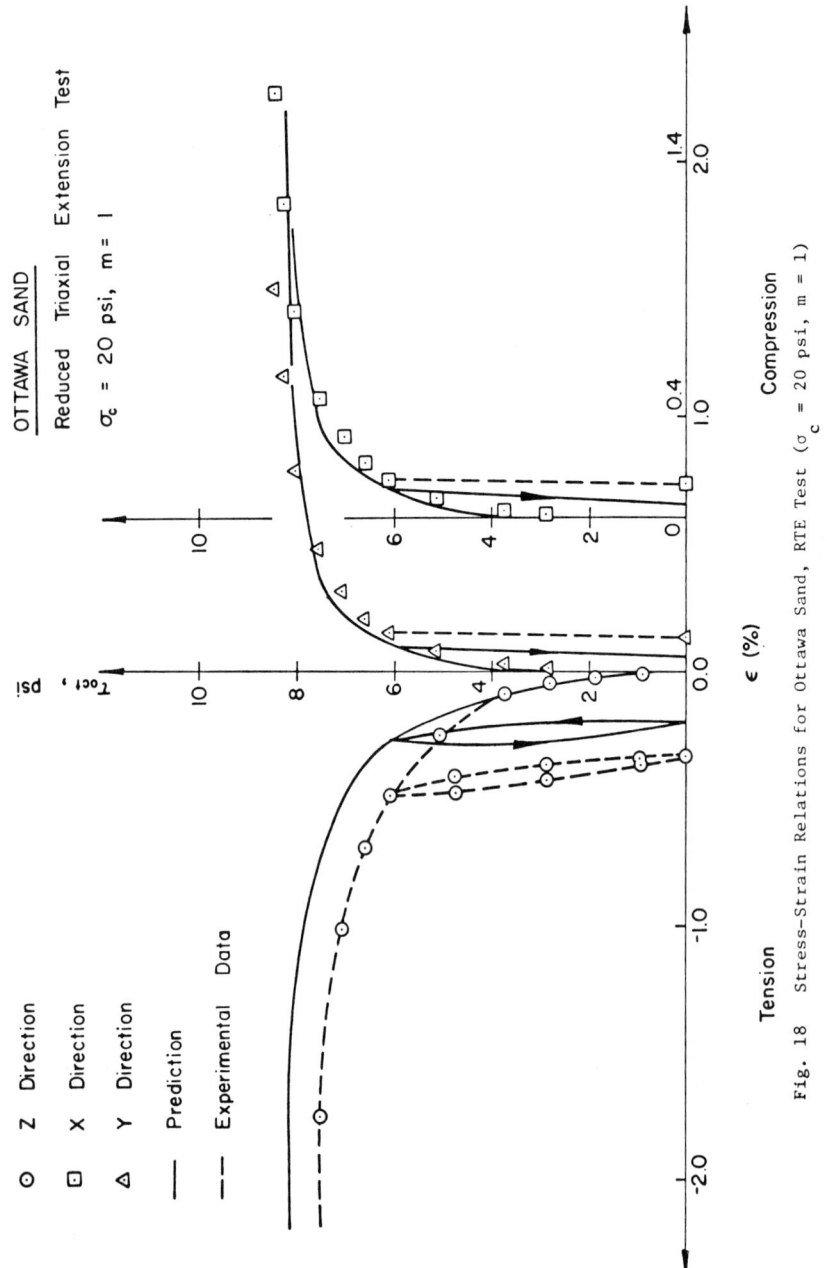

Fig. 18 Stress-Strain Relations for Ottawa Sand, RTE Test (σ_c = 20 psi, m = 1)

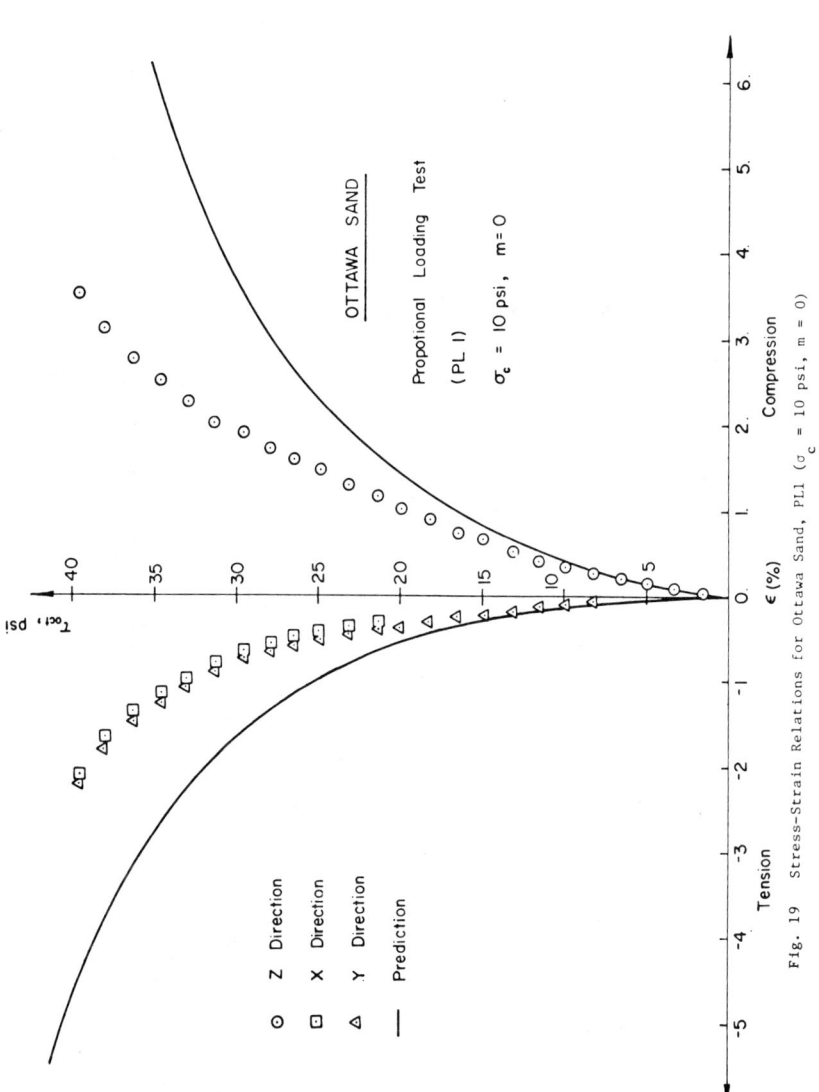

Fig. 19 Stress-Strain Relations for Ottawa Sand, PL1 ($\sigma_c = 10$ psi, m = 0)

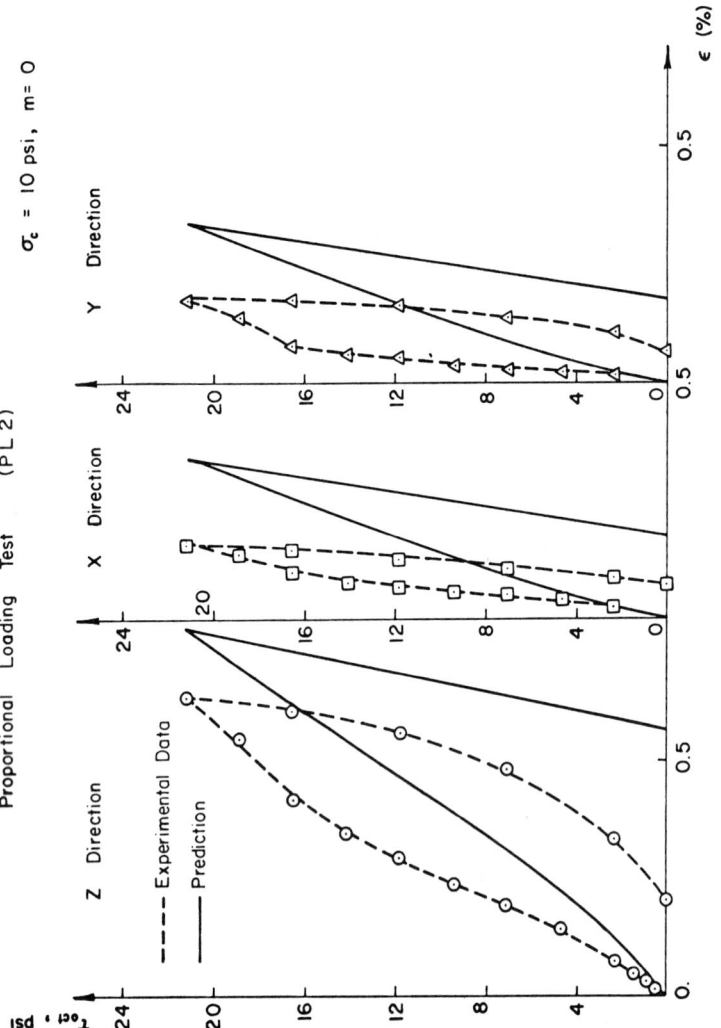

Fig. 20 Stress-Strain Relations for Ottawa Sand, PL2 (σ_c = 10 psi, m = 0)

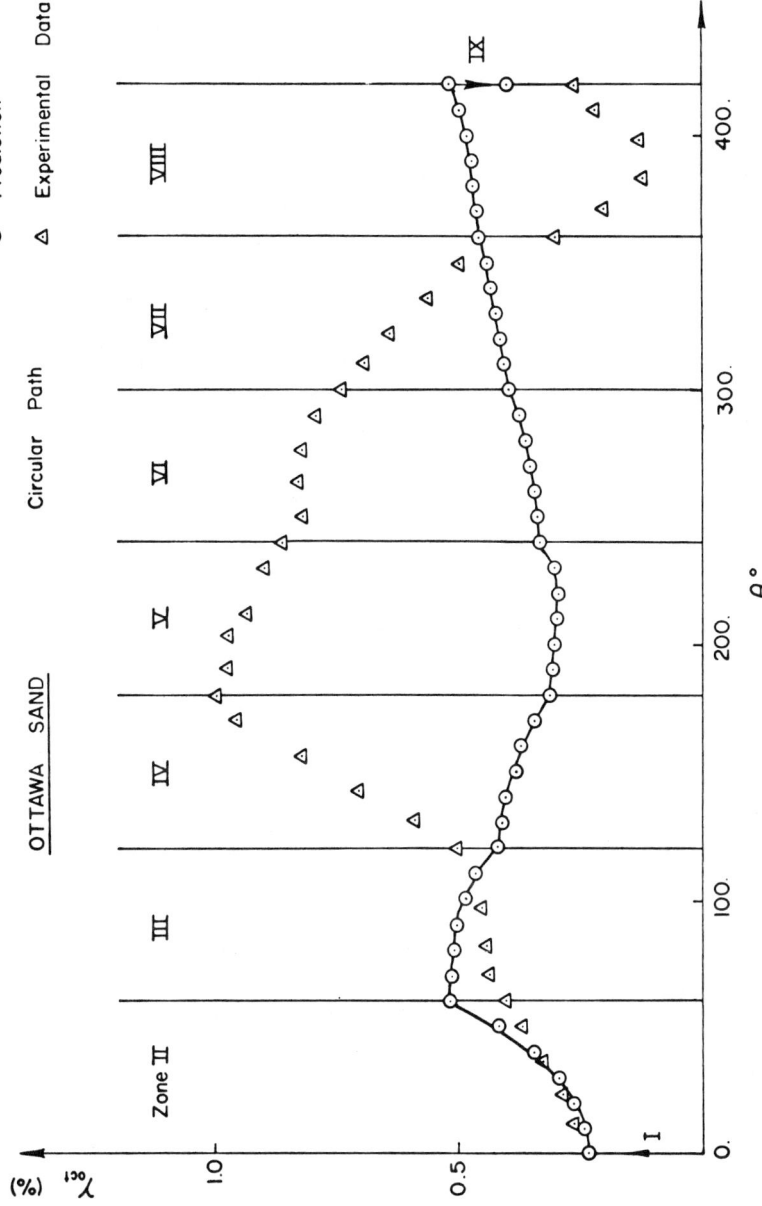

Fig. 21 Strain-Degree Relation for Ottawa Sand, Circular Path

PREDICTIONS OF STRESS-STRAIN BEHAVIOR FOR OTTAWA SAND

by Poul V. Lade[1]

The development of the elasto-plastic model used for predictions of stress-strain behavior of Ottawa Sand as well as the determination of appropriate values of the soil parameters are presented in a companion paper contained elsewhere in these Proceedings. The soil parameters were essentially determined from the triaxial compression tests with constant confining pressure and the isotropic compression tests. The predicted behavior is tabulated in Tables 1-9 at the end of this paper. Figures 1-5 show the comparisons between measured and predicted stress-strain behavior for the nine different stress-paths specified for prediction. The respective stress-paths are indicated on inserts in these figures.

Figure 1 shows the simple shear tests performed with constant mean normal stress σ_m = 10 psi and m - values of 0.2 and 0.8. The predicted response for these tests compare very well with the measured behavior, both in terms of the stress-strain behavior and in terms of the strength measured in these tests.

The two simple shear tests shown on Figure 2 were both performed with m = 0.5, but with σ_m = 5 psi and 20 psi, respectively. It is interesting to note that the relative magnitude of the measured strains in the y-direction are quite different in the two cases. Whereas this difference is not accurately predicted by the model, the predictions are considered to be within the scatter experienced in tests of this type.

Figure 3 shows comparisons between measured and predicted stress-strain behavior for reduced triaxial compression (m = 0) and reduced triaxial extension (m = 1) tests. The predicted behavior is in both cases very favorable, both in terms of stress-strain behavior and in terms of strength.

Figure 4 shows the two tests incorporating "proportional" loading. The results indicated in Figure 4(a) shows that the predicted behavior at low values of octahedral shear stress matches the observed behavior very well, whereas the predicted strains are larger than those measured at high values of

---

[1] Associate Professor, Mechanics and Structures Department, School of Engineering and Applied Science, University of California, Los Angeles, California.

octahedral shear stress. It should be noted that failure occurs in this test according to the model at a lower value of octahedral shear stress than indicated by the experimental results. Thus, the predicted strains tend to become larger at lower values of octahedral shear stress.

The predictions in Figure 4(b) indicate strains that are somewhat lower than those measured. This is opposite the predictions in Figure 4(a). It should be noted that the soil parameters for the model were derived from isotropic compression and triaxial compression tests with constant confining pressure. Thus, the stress-paths for these tests are slightly outside the stress-paths for the "proportional" loading tests, and it would therefore be expected that the model would predict a regular pattern of behavior between these two "outside" stress-paths. However, comparison with the experimental results indicates that the predicted pattern of behavior is different from that observed in the tests. Previous attempts at prediction of proportional loading behavior (see ref. 2 in companion paper) have been more successful than those shown in Figure 4. It is possible that the experimental results for Ottawa Sand may indicate a degree of scatter which cannot be predicted by the model.

Figure 5 shows the comparison between measured and predicted results for the circular stress-path contained in one octahedral plane. Whereas the trends for the predicted behavior correctly reflect the observed behavior, the magnitude of the predicted strains is much smaller than the strains measured in the experiments. According to the model in its present form, plastic strains are predicted for only the first $60°$ of the circular stress-path for which increasing stress levels are encountered. The order of magnitude of the predicted strains within the first $60°$ corresponds reasonably well with the measured strains. However, only elastic strains are predicted during the remainder of the stress-path, and this is obviously not correct.

In conclusion, it appears that the model is able to predict the behavior of Ottawa Sand with reasonable accuracy for the stress-paths of practical importance. Thus, the model is applicable to general three-dimensional conditions with the additional advantage that only conventional triaxial compression tests and an isotropic compression test are required for determination of the material parameters.

354 STRESS STRAIN FOR SOILS

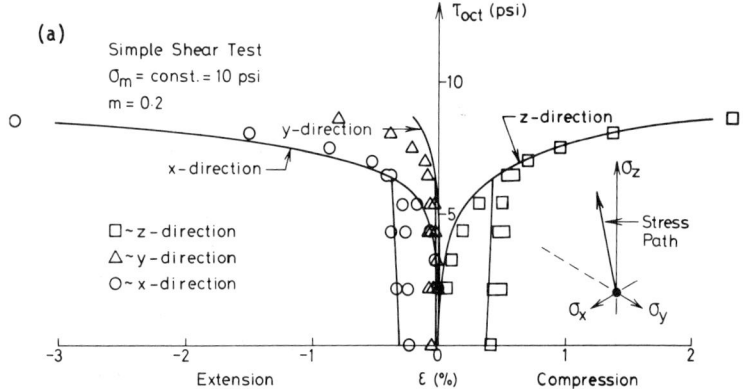

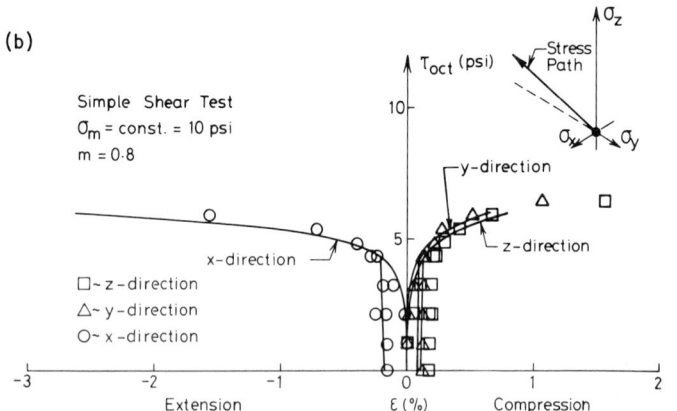

Figure 1. Predictions of Simple Shear Stress with σ_m = 10 psi and (a) m = 0.2 snd (b) m = 0.8.

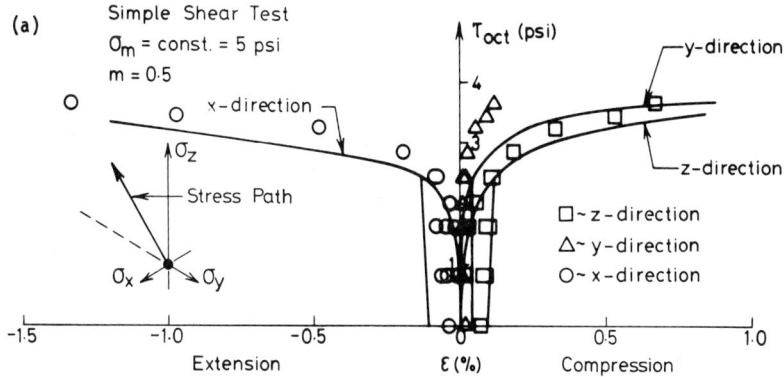

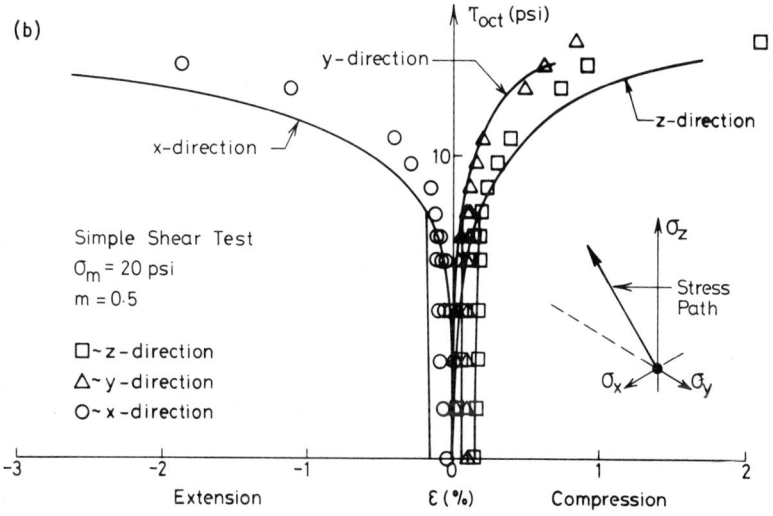

Figure 2. Predictions of Simple Shear Tests with m = 0.5 and (a) σ_m = 5 psi and (b) σ_m = 20 psi.

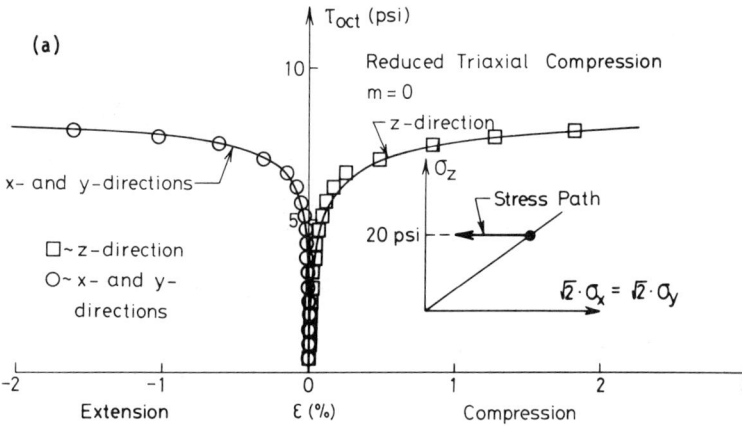

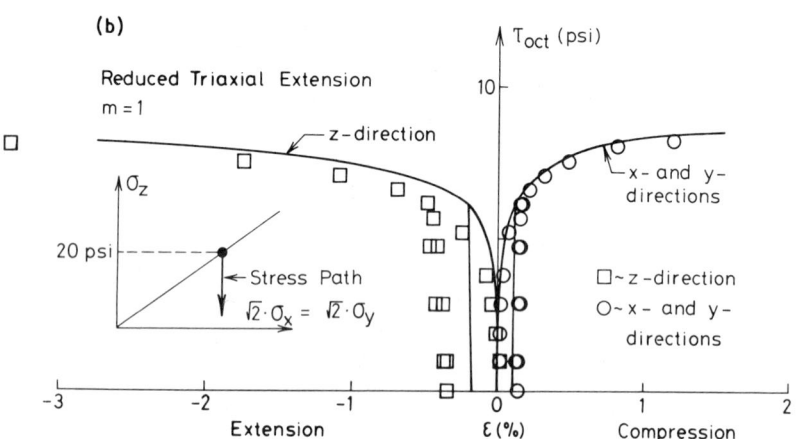

Figure 3. Predictions of (a) Reduced Triaxial Compression Tests and (b) Reduced Triaxial Extension Tests.

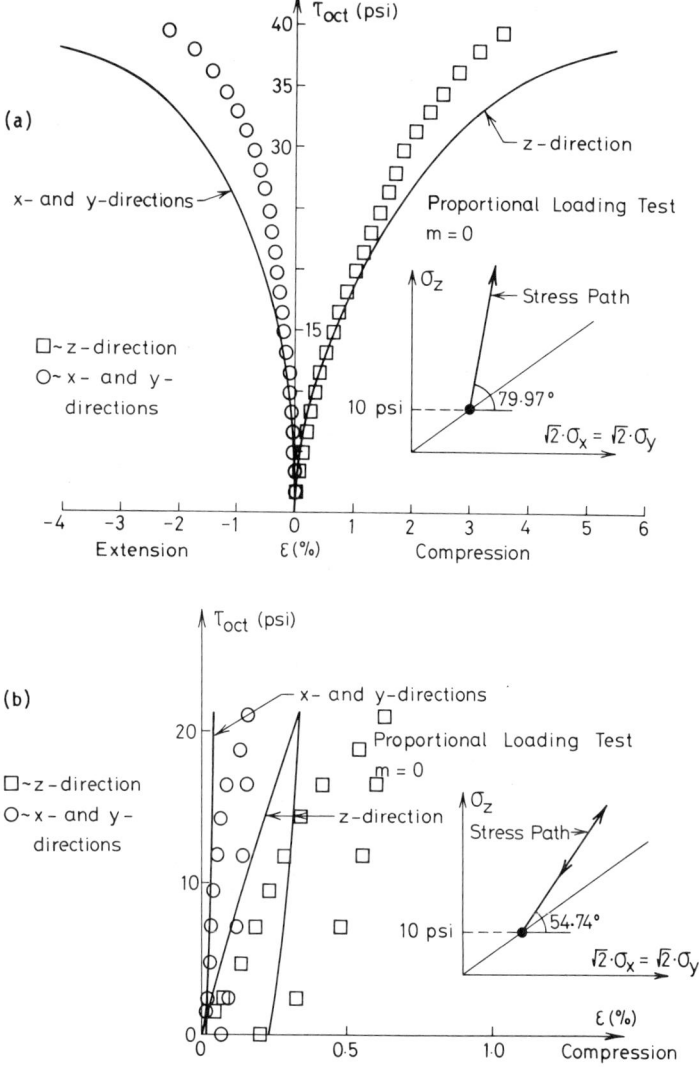

Figure 4. Predictions of Proportional Loading Tests.

STRESS STRAIN FOR SOILS

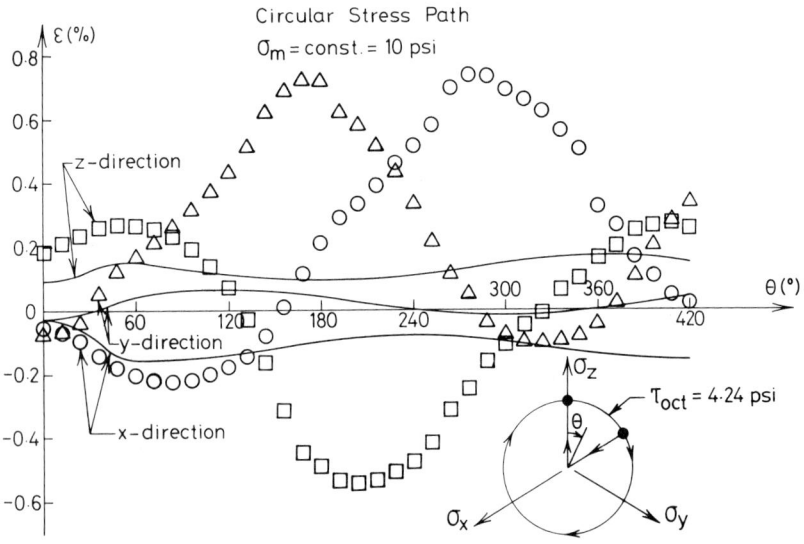

Figure 5. Predictions of Test with Circular Stress Path.

Table 1. SIMPLE SHEAR TEST (SS) WITH CONSTANT MEAN STRESS AT 10 PSI

Stress Ratio, m = 0.2

Principal Stresses			Principal Strains		
σ_x	σ_y	σ_z	ε_x, %	ε_y, %	ε_z, %
10	10	10	0	0	0
8	9	13	-0.006	0.000	0.023
7	8.5	14.5	-0.020	-0.002	0.052
6	8	16	-0.060	-0.007	0.111
5	7.5	17.5	-0.155	-0.017	0.222
4	7	19	-0.389	-0.036	0.424
6	8	16	-0.377	-0.030	0.405
8	9	13	-0.367	-0.025	0.391
10	10	10	-0.359	-0.021	0.379
8	9	13	-0.367	-0.025	0.391
6	8	16	-0.377	-0.030	0.405
5	7.5	17.5	-0.382	-0.033	0.414
4	7	19	-0.389	-0.036	0.424
3.5	6.75	19.75	-0.806	-0.060	0.711
3	6.5	20.5	-1.250	-0.082	0.968
2.5	6.25	21.25	-2.157	-0.121	1.396
2	6	22	-4.651	-0.209	2.319

STRESS STRAIN BEHAVIOR 359

Table 2. SIMPLE SHEAR TEST (SS) WITH CONSTANT MEAN STRESS AT 10 PSI

Stress Ratio, m = 0.8

Principal Stresses			Principal Strains		
σ_x	σ_y	σ_z	ε_x, %	ε_y, %	ε_z, %
10	10	10	0	0	0
8.5	10.5	11.0	-0.002	0.006	0.008
7.0	11.0	12.0	-0.013	0.013	0.019
5.5	11.5	13.0	-0.055	0.035	0.050
4.0	12.0	14.0	-0.207	0.096	0.132
7.0	11.0	12.0	-0.190	0.091	0.121
10.0	10.0	10.0	-0.177	0.087	0.113
8.5	10.5	11.0	-0.183	0.089	0.117
7.0	11.0	12.0	-0.190	0.091	0.121
5.5	11.5	13.0	-0.198	0.093	0.126
4.0	12.0	14.0	-0.207	0.097	0.133
3.25	12.25	14.5	-0.564	0.212	0.280
2.5	12.50	15.0	-1.044	0.338	0.437
1.75	12.75	15.5	-2.481	0.619	0.780

Table 3. SIMPLE SHEAR TEST (SS) WITH CONSTANT MEAN STRESS AT 5 PSI

Stress Ratio, m = 0.5

Principal Stresses			Principal Strains		
σ_x	σ_y	σ_z	ε_x, %	ε_y, %	ε_z, %
5	5	5	0	0	0
4	5	6	-0.005	0.002	0.009
3	5	7	-0.023	0.009	0.033
2.5	5	7.5	-0.055	0.020	0.064
2	5	8	-0.133	0.043	0.120
3	5	7	-0.122	0.043	0.110
4	5	6	-0.114	0.043	0.102
5	5	5	-0.108	0.043	0.095
4	5	6	-0.114	0.043	0.102
3	5	7	-0.122	0.043	0.110
2	5	8	-0.133	0.043	0.120
1.5	5	8.5	-0.481	0.125	0.299
1.0	5	9	-1.181	0.241	0.539
.75	5	9.25	-2.399	0.396	0.831
.50	5	9.5	-7.678	0.868	1.693

Table 4. SIMPLE SHEAR TEST (SS) WITH CONSTANT MEAN STRESS AT 20 PSI

Stress Ratio, $m = 0.5$

Principal Stresses			Principal Strains		
σ_x	σ_y	σ_z	ϵ_x, %	ϵ_y, %	ϵ_z, %
20	20	20	0	0	0
18	20	22	0.005	0.010	0.014
16	20	24	-0.002	0.011	0.024
14	20	26	-0.021	0.017	0.046
12	20	28	-0.069	0.030	0.092
11	20	29	-0.116	0.042	0.130
10	20	30	-0.187	0.060	0.182
12	20	28	-0.180	0.060	0.175
14	20	26	-0.174	0.060	0.169
16	20	24	-0.169	0.060	0.163
18	20	22	-0.164	0.060	0.158
20	20	20	-0.159	0.060	0.154
18	20	22	-0.164	0.060	0.158
14	20	26	-0.174	0.060	0.169
12	20	28	-0.180	0.060	0.175
11	20	29	-0.183	0.060	0.178
10	20	30	-0.187	0.060	0.182
9	20	31	-0.368	0.103	0.297
8	20	32	-0.531	0.139	0.390
7	20	33	-0.789	0.192	0.519
5	20	35	-2.018	0.399	0.988
4	20	36	-3.998	0.683	1.565

Table 5. REDUCED TRIAXIAL COMPRESSION TEST (RTC)

Stress Ratio, $m = 0$

Principal Stresses			Principal Strains		
σ_x	σ_y	σ_z	ϵ_x, %	ϵ_y, %	ϵ_z, %
20	20	20	0	0	0
19	19	20	0.008	0.008	0.010
18	18	20	0.006	0.006	0.011
17	17	20	0.005	0.005	0.012
16	16	20	0.003	0.003	0.013
15	15	20	0.001	0.001	0.016
17	17	20	0.005	0.005	0.014
19	19	20	0.008	0.008	0.012
20	20	20	0.009	0.009	0.012
19	19	20	0.008	0.008	0.013
17	17	20	0.005	0.005	0.014
15	15	20	0.001	0.001	0.016
14	14	20	-0.002	-0.002	0.024
13	13	20	-0.005	-0.005	0.030
12	12	20	-0.009	-0.009	0.040
11	11	20	-0.015	-0.015	0.055
10	10	20	-0.024	-0.024	0.078
9	9	20	-0.039	-0.039	0.112
8	8	20	-0.063	-0.063	0.163
7	7	20	-0.102	-0.102	0.241
6	6	20	-0.172	-0.172	0.366
5	5	20	-0.312	-0.312	0.581
4	4	20	-0.651	-0.651	1.007
3.5	3.5	20	-1.049	-1.049	1.435
3	3	20	-1.939	-1.939	2.254
2.75	2.75	20	-3.043	-3.043	3.157

Table 6. REDUCED TRIAXIAL EXTENSION TEST (RTE)

Stress Ratio, m = 1.0

Principal Stresses			Principal Strains		
σ_x	σ_y	σ_z	ε_x, %	ε_y, %	ε_z, %
20	20	20	0	0	0
20	20	18	0.010	0.010	0.005
20	20	16	0.011	0.011	0.001
20	20	14	0.015	0.015	-0.006
20	20	12	0.023	0.023	-0.020
20	20	9	0.057	0.057	-0.082
20	20	7	0.113	0.113	-0.205
20	20	10	0.111	0.111	-0.195
20	20	14	0.108	0.108	-0.184
20	20	18	0.107	0.107	-0.175
20	20	20	0.106	0.106	-0.172
20	20	18	0.107	0.107	-0.175
20	20	14	0.109	0.109	-0.184
20	20	10	0.111	0.111	-0.195
20	20	8	0.112	0.112	-0.201
20	20	7	0.113	0.113	-0.205
20	20	6	0.202	0.202	-0.427
20	20	5	0.279	0.279	-0.648
20	20	4	0.407	0.407	-1.076
20	20	3	0.652	0.652	-2.085
20	20	2.5	0.901	0.901	-3.323
20	20	2	1.445	1.445	-6.547

Table 7. PROPORTIONAL LOADING TEST (PL1)

Stress Ratio, m = 0

Principal Stresses			Principal Strains		
σ_x	σ_y	σ_z	ϵ_x, %	ϵ_y, %	ϵ_z, %
10	10	10	0	0	0
10.5	10.5	14	0.004	0.004	0.018
11	11	18	0.002	0.002	0.043
11.5	11.5	22	-0.007	-0.007	0.086
12	12	26	-0.025	-0.025	0.147
12.5	12.5	30	-0.053	-0.053	0.225
13	13	34	-0.091	-0.091	0.317
13.5	13.5	38	-0.140	-0.140	0.423
14	14	42	-0.199	-0.199	0.541
14.5	14.5	46	-0.268	-0.268	0.671
15	15	50	-0.349	-0.349	0.813
15.5	15.5	54	-0.441	-0.441	0.967
16	16	58	-0.545	-0.545	1.134
16.5	16.5	62	-0.662	-0.662	1.314
17	17	66	-0.795	-0.795	1.509
17.5	17.5	70	-0.943	-0.943	1.721
18	18	74	-1.111	-1.111	1.953
18.5	18.5	78	-1.303	-1.303	2.208
19	19	82	-1.523	-1.523	2.494
19.5	19.5	86	-1.780	-1.780	2.820
20	20	90	-2.090	-2.090	3.202
20.5	20.5	94	-2.481	-2.481	3.671
21	21	98	-3.023	-3.023	4.307
21.5	21.5	102	-4.027	-4.027	5.455

Table 8. PROPORTIONAL LOADING TEST (PL2)

Stress Ratio, m = 0

Principal Stresses			Principal Strains		
σ_x	σ_y	σ_z	ϵ_x, %	ϵ_y, %	ϵ_z, %
10	10	10	0	0	0
11	11	12	0.006	0.006	0.009
12	12	14	0.008	0.008	0.015
13	13	16	0.010	0.010	0.021
15	15	20	0.014	0.014	0.033
20	20	30	0.022	0.022	0.066
25	25	40	0.027	0.027	0.102
30	30	50	0.030	0.030	0.141
35	35	60	0.033	0.033	0.181
40	40	70	0.035	0.035	0.220
45	45	80	0.037	0.037	0.260
50	50	90	0.038	0.038	0.300
55	55	100	0.040	0.040	0.339
45	45	80	0.036	0.036	0.324
35	35	60	0.032	0.032	0.307
25	25	40	0.027	0.027	0.286
15	15	20	0.019	0.019	0.257
10	10	10	0.014	0.014	0.236

Table 9. CIRCULAR STRESS PATH (CSP) WITH CONSTANT MEAN STRESS AT 10 PSI

Principal Stresses			Principal Strains		
σ_x	σ_y	σ_z	ε_x, %	ε_y, %	ε_z, %
10	10	10	0	0	0
9	9	12	0.000	0.000	0.012
8	8	14	-0.006	-0.006	0.033
7	7	16	-0.026	-0.026	0.091
6.07	8.04	15.89	-0.034	-0.022	0.097
5.18	9.31	15.51	-0.060	-0.013	0.115
4.49	10.69	14.81	-0.102	0.007	0.137
4.11	11.96	13.93	-0.141	0.028	0.150
4	13.0	13.0	-0.156	0.041	0.150
4.11	13.93	11.96	-0.155	0.048	0.142
4.49	14.81	10.69	-0.152	0.054	0.133
5.18	15.51	9.31	-0.148	0.058	0.125
6.07	15.89	8.04	-0.143	0.060	0.118
7.00	16.0	7.0	-0.138	0.061	0.112
8.04	15.89	6.07	-0.133	0.060	0.108
9.31	15.51	5.18	-0.126	0.058	0.103
10.69	14.81	4.49	-0.117	0.054	0.098
11.96	13.93	4.11	-0.108	0.048	0.096
13.00	13.0	4.0	-0.101	0.041	0.095
13.93	11.96	4.11	-0.094	0.033	0.096
14.81	10.69	4.49	-0.088	0.025	0.098
15.51	9.31	5.18	-0.084	0.016	0.103
15.89	8.04	6.07	-0.081	0.009	0.108
16	7.0	7.0	-0.081	0.004	0.112
15.89	6.07	8.04	-0.081	-0.001	0.118
15.51	5.18	9.31	-0.084	-0.006	0.125
14.81	4.49	10.69	-0.088	-0.011	0.133
13.93	4.11	11.96	-0.094	-0.013	0.142
13	4.0	13.0	-0.101	-0.014	0.150
11.96	4.11	13.93	-0.108	-0.013	0.156
10.69	4.49	14.81	-0.117	-0.011	0.162
9.31	5.18	15.51	-0.126	-0.006	0.167
8.04	6.07	15.89	-0.133	-0.001	0.169
7.00	7.0	16.0	-0.138	0.004	0.170
6.07	8.04	15.89	-0.143	0.009	0.169
5.18	9.31	15.51	-0.148	0.016	0.167
4.49	10.69	14.81	-0.152	0.025	0.162
4.11	11.96	13.93	-0.155	0.033	0.156
4	13.0	13.0	-0.156	0.041	0.150
6	12	12	-0.143	0.035	0.144
8	11	11	-0.134	0.030	0.139
10	10	10	-0.126	0.026	0.135

COMPARISON OF CAP MODEL PREDICTIONS WITH LABORATORY DATA FOR SOILS

By George Y. Baladi[1] and Ivan S. Sandler,[2] Members, ASCE

Introduction

The materials considered in the NSF/NSERC Workshop on Plasticity Theories and Generalized Stress-Strain Modeling of Soil were (1) two natural clays (designated as Clay X and Clay Y, but collectively called Materials A), (2) a laboratory-prepared kaolinite clay (Material B), and (3) an Ottawa sand (Material C). For each of these materials two generic sets of laboratory stress-strain data were generated. The first generic set for each material was mailed to each predictor several months before the workshop for model calibration and/or development. In addition, each predictor was furnished the stress path associated with the second set of stress-strain data and was asked to predict the stress-strain responses. The actual stress-strain data for the second set were revealed to the predictor at the time of the workshop for use in evaluating the predictive capabilities of their models.

The first sets of data indicated that each of the above materials required different features of the cap model. Therefore, a transversely isotropic version of the cap model was used to simulate the behavior of Materials A, a kinematically hardening version was utilized for Material B, and an isotropic hardening version was used for Material C. These versions are described in detail in Volume II of the NSF/NSERC workshop proceedings together with their calibrations against the first sets of data.

The following three parts present comparisons of the model <u>predictions</u> with the second sets of data <u>which were not known when the above three different versions of the cap model were constructed</u>. Part I contains the comparison of the model predictions with the laboratory data for Clay X and Clay Y (Materials A). Comparison of the model predictions with the laboratory data for kaolinite clay (Material B) is presented in Part II. Part III presents comparison of the model predictions with the laboratory data for Ottawa sand (Material C).

---

[1] Research Civ. Engr., Geomechanics Div., Structures Lab, Waterways Experiment Station, Vicksburg, Miss.

[2] Partner of Weidlinger Associates, Consulting Engineers, New York, NY.

PART I: COMPARISON OF CAP MODEL PREDICTIONS WITH LABORATORY DATA FOR CLAY (MATERIALS A)

By George Y. Baladi

The model predictions for Clay X are shown in Figures 1 through 5 for various constant mean normal stresses and stress ratios. The scales and format of these figures were requested by the workshop's organizers.

The stress-strain behavior of the model is compared to the Clay X behavior in Figures 6 through 8. The format of these figures is chosen so that the ability of the model to predict the shear behavior and the shear-induced volume change of Clay X can be easily examined. It is clear from these figures that the model does a very good job both qualitatively and quantitatively to represent Clay X.

The model predictions for Clay Y are shown in Figures 9 through 14 for various confining pressure levels and stress ratios. Like Figures 1 through 5, the scales and format of these figures were requested by the workshop's organizers.

The Clay Y behavior is compared with the stress-strain behavior of the model in Figures 15 and 16. Figure 15 compares octahedral shear stress and volumetric strain versus octahedral shear strain for a confining pressure of 5 psi and stress ratios of 0.5 and 0.75. Figure 16 shows a similar figure for a confining pressure of 10 psi. Here again the comparisons are fairly good, both qualitatively and quantitatively. These comparisons could be improved however, if some additional test results were available to fit the model such as the results of isotropic consolidation tests with major and minor principal strain measurements, and triaxial tests conducted on cubical specimens in which the axial stress is applied parallel to the plane of isotropy.

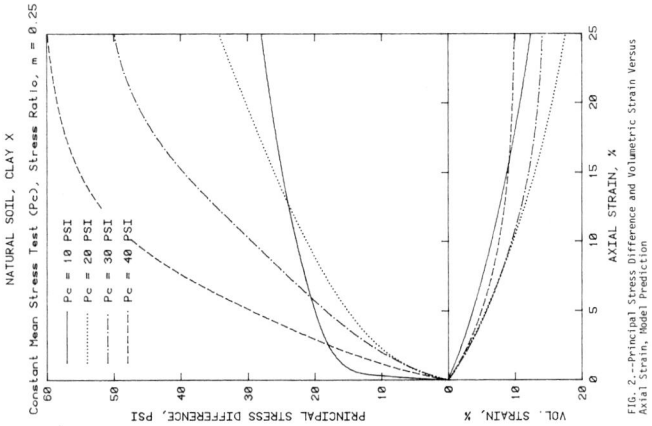

FIG. 1.—Principal Stress Difference and Volumetric Strain Versus Axial Strain, Model Prediction

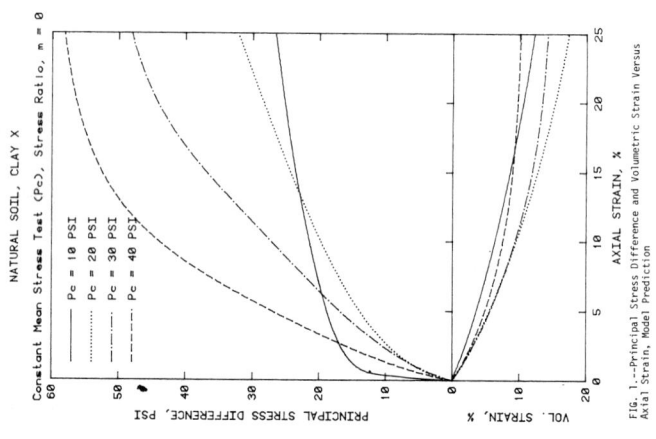

FIG. 2.—Principal Stress Difference and Volumetric Strain Versus Axial Strain, Model Prediction

MODEL/LABORATORY COMPARISONS

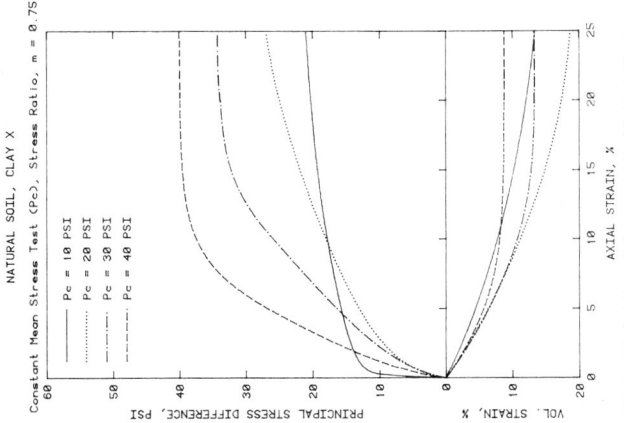

FIG. 4.—Principal Stress Difference and Volumetric Strain Versus Axial Strain, Model Prediction

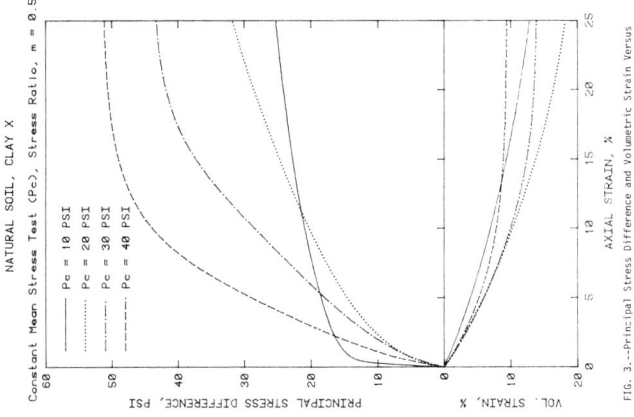

FIG. 3.—Principal Stress Difference and Volumetric Strain Versus Axial Strain, Model Prediction

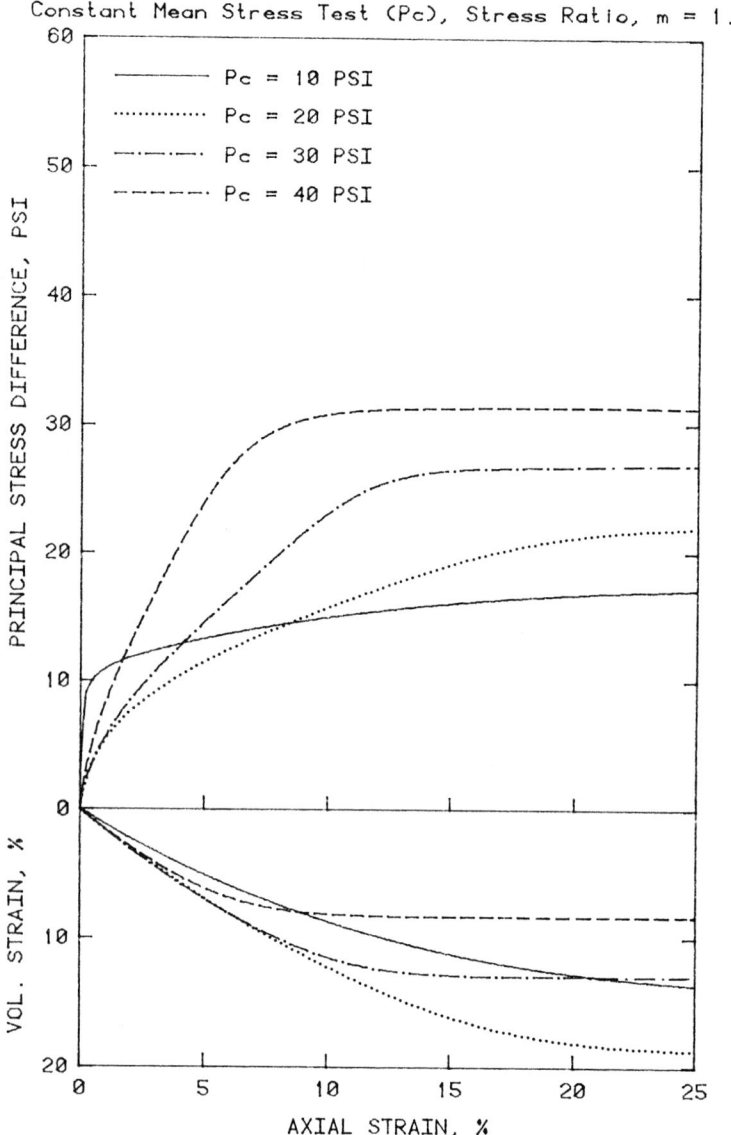

FIG. 5.--Principal Stress Difference and Volumetric Strain Versus Axial Strain, Model Prediction

NATURAL SOIL, CLAY X

Constant Mean Stress Test (Pc) With Pc = 20.0 PSI

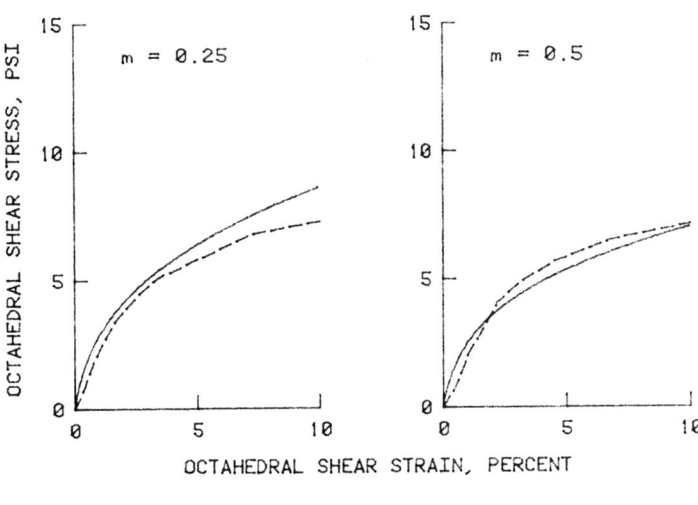

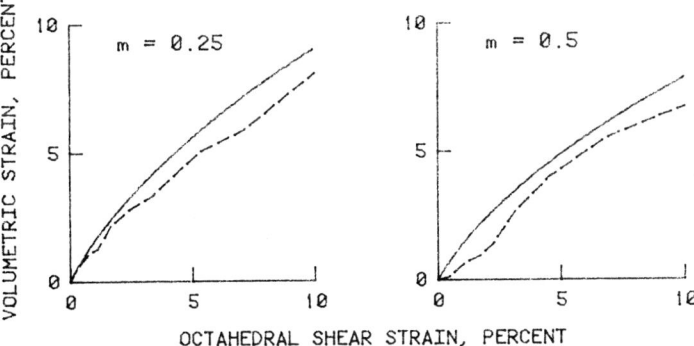

FIG. 6.--Octahedral Shear Stress and Volumetric Strain Versus Octahedral Shear Strain, Model Prediction Versus Laboratory Measurements

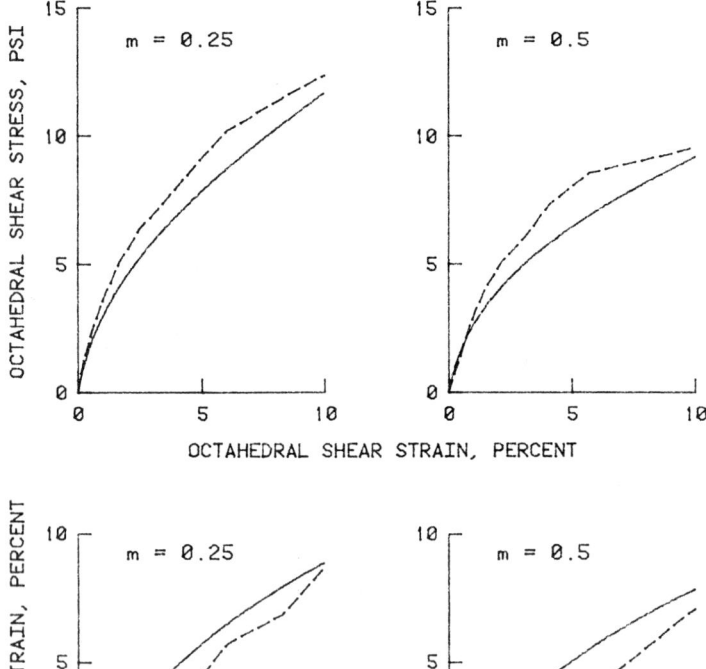

FIG. 7.--Octahedral Shear Stress and Volumetric Strain Versus Octahedral Shear Strain, Model Prediction Versus Laboratory Measurements

NATURAL SOIL, CLAY X

Constant Mean Stress Test (Pc) With Pc = 40.0 PSI

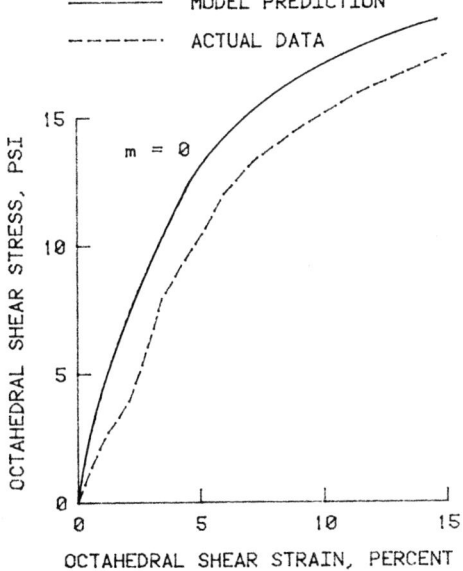

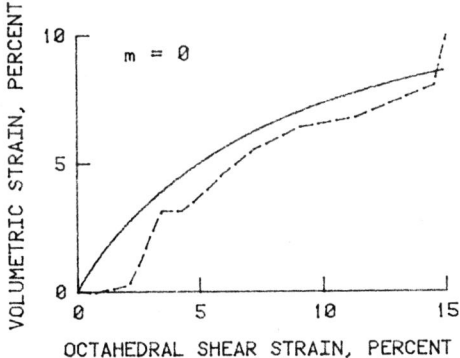

FIG. 8.--Octahedral Shear Stress and Volumetric Strain Versus Octahedral Shear Strain, Model Prediction Versus Laboratory Measurements

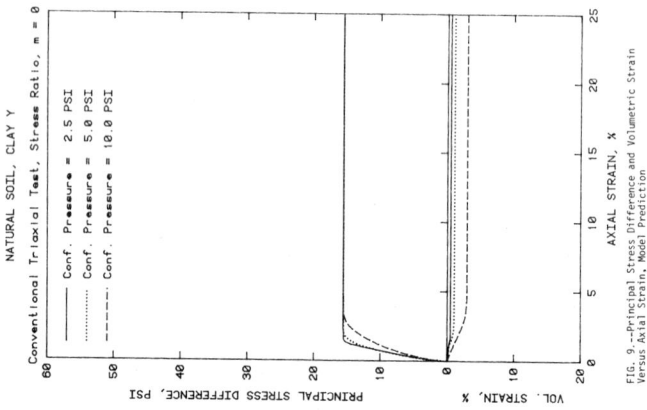

FIG. 9.—Principal Stress Difference and Volumetric Strain Versus Axial Strain, Model Prediction

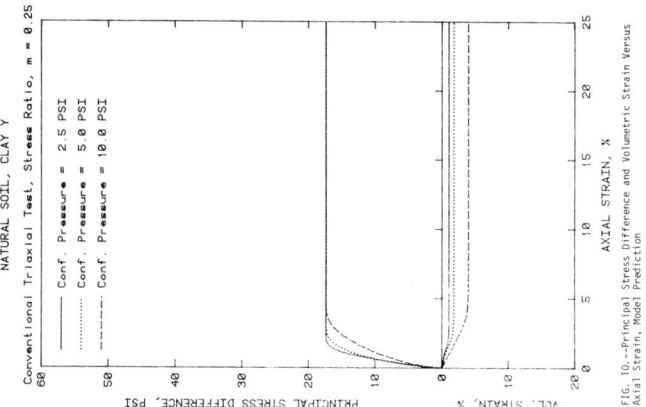

FIG. 10.—Principal Stress Difference and Volumetric Strain Versus Axial Strain, Model Prediction

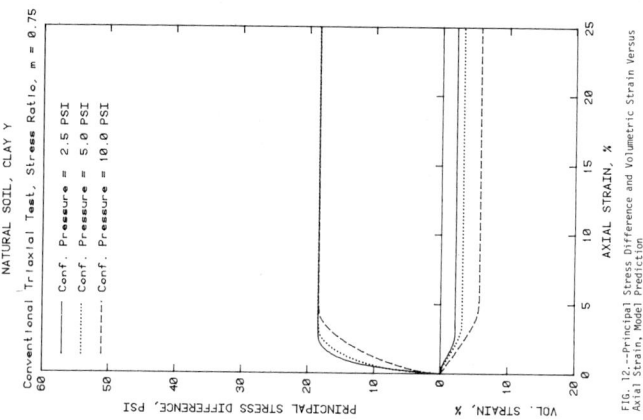

FIG. 12.—Principal Stress Difference and Volumetric Strain Versus Axial Strain, Model Prediction

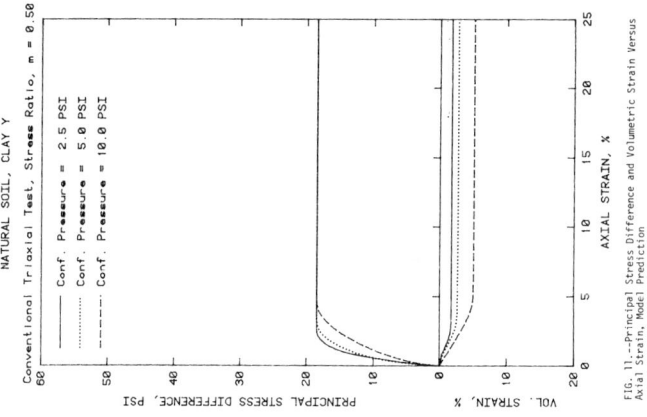

FIG. 11.—Principal Stress Difference and Volumetric Strain Versus Axial Strain, Model Prediction

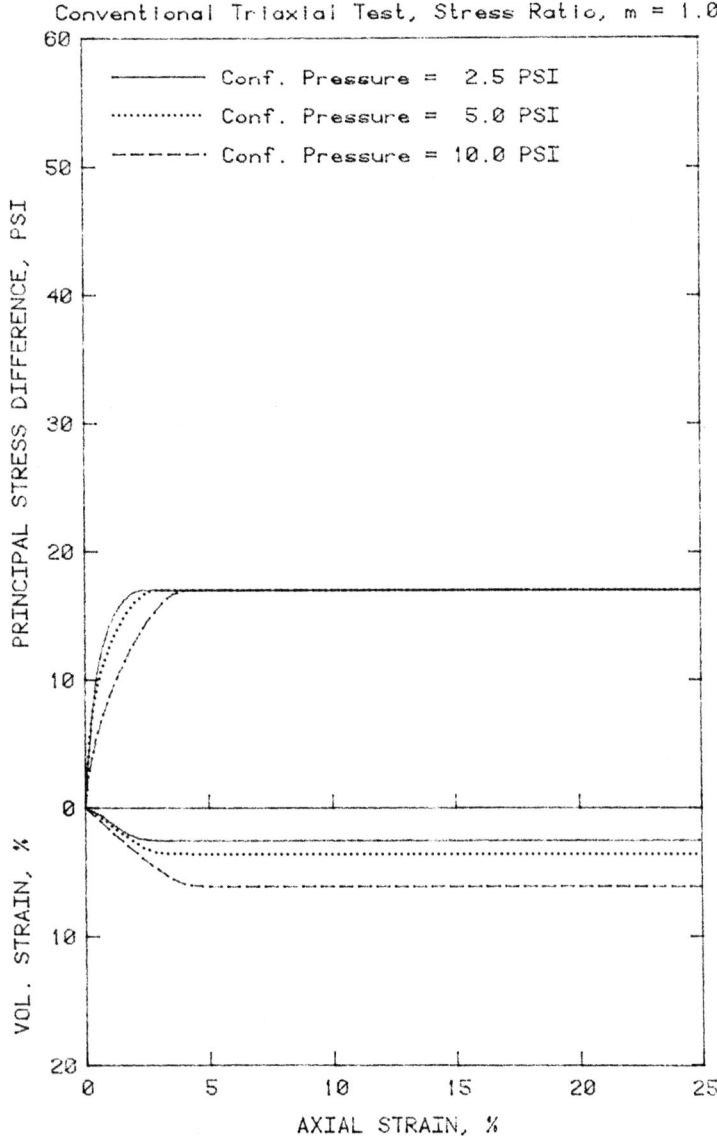

FIG. 13.--Principal Stress Difference and Volumetric Strain Versus Axial Strain, Model Prediction

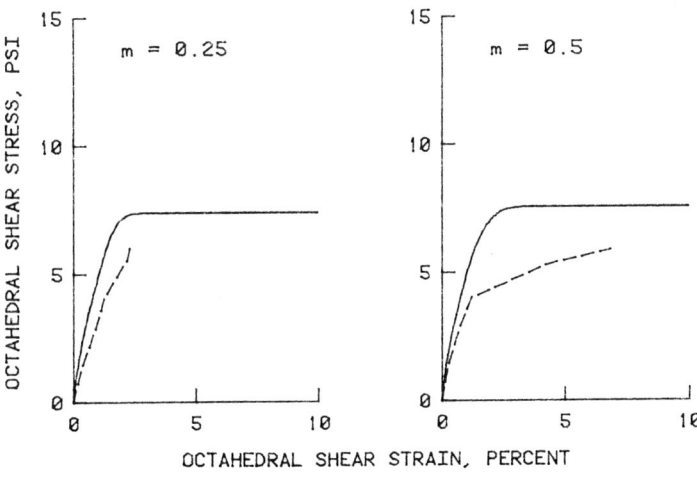
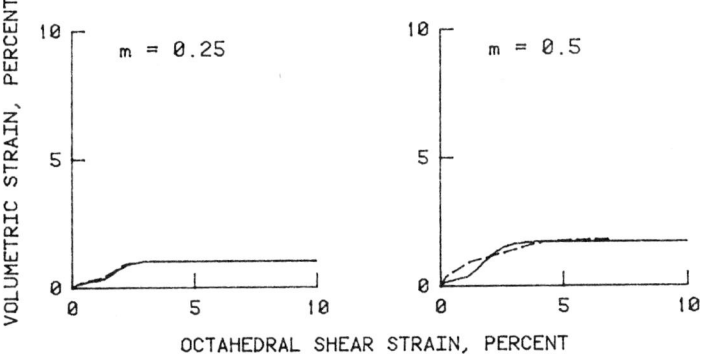

FIG. 14.--Octahedral Shear Stress and Volumetric Strain Versus Octahedral Shear Strain, Model Prediction Versus Laboratory Measurements

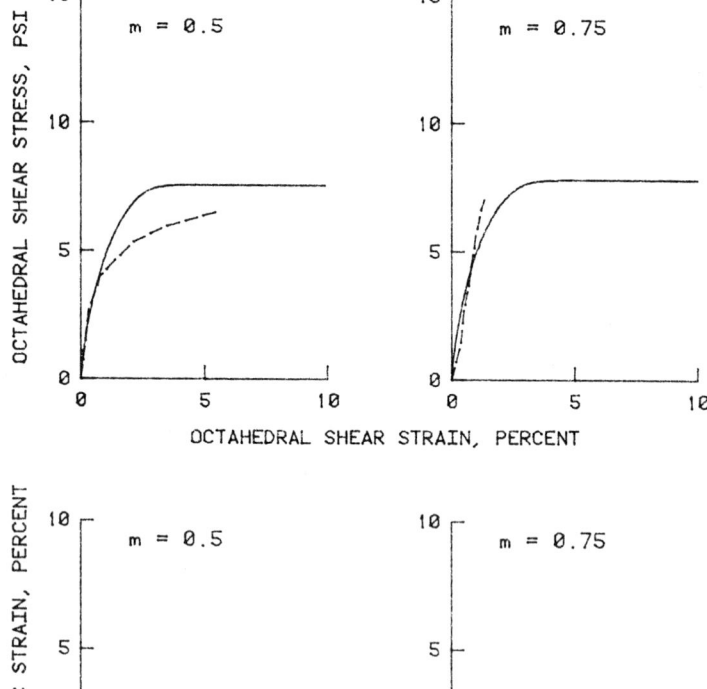

FIG. 15.--Octahedral Shear Stress and Volumetric Strain Versus Octahedral Shear Strain, Model Prediction Versus Laboratory Measurements

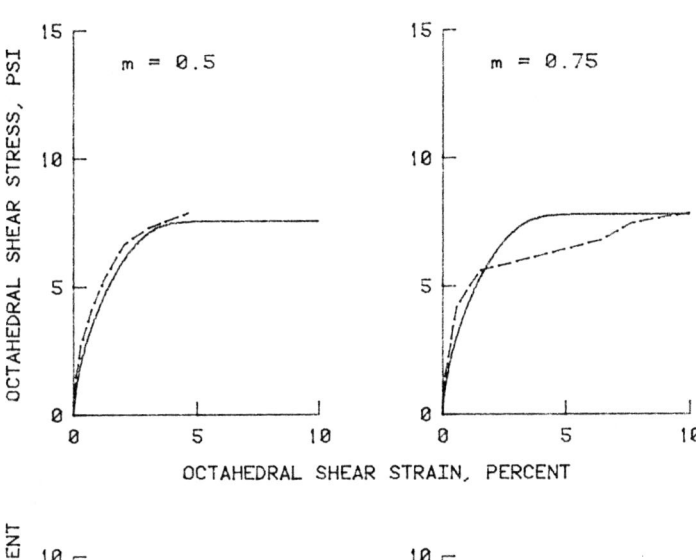
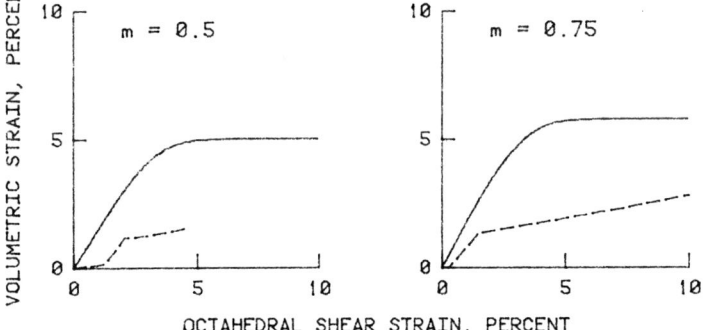

FIG. 16.—Octahedral Shear Stress and Volumetric Strain Versus Octahedral Shear Strain, Model Prediction Versus Laboratory Measurements

PART II: COMPARISON OF CAP MODEL PREDICTIONS WITH LABORATORY DATA FOR REMOLDED KAOLINITE CLAY (MATERIAL B)

By Ivan S. Sandler

The model predictions for undrained triaxial loading tests are shown in Figures 1 through 11. The effective stress paths for the various tests to be predicted (Nos. 2, 3, 5, 6, 7, 8, 9, 11, and 12) are shown in Figures 1 and 2, and the total stress versus strain curves for those tests are shown in Figures 3 through 11. The scales and format of these figures were requested by the workshop's organizers.

The stress-strain behavior of the model is compared to the kaolinite behavior in Figures 12 through 14. As can be seen from the figures, the model does a very good job both qualitatively and quantitatively to represent kaolinite behavior, except for tests Nos. 5 and 6. Even for these tests, however, the stress-strain curves would bend over (to indicate failure) at only slightly higher stress levels so that the overall representations of these tests are better than would ordinarily be judged from the figures.

The pore pressure response in the various tests is shown in Figures 15 through 17. Here again the agreement between prediction and measurement is quite good both qualitatively and quantitatively.

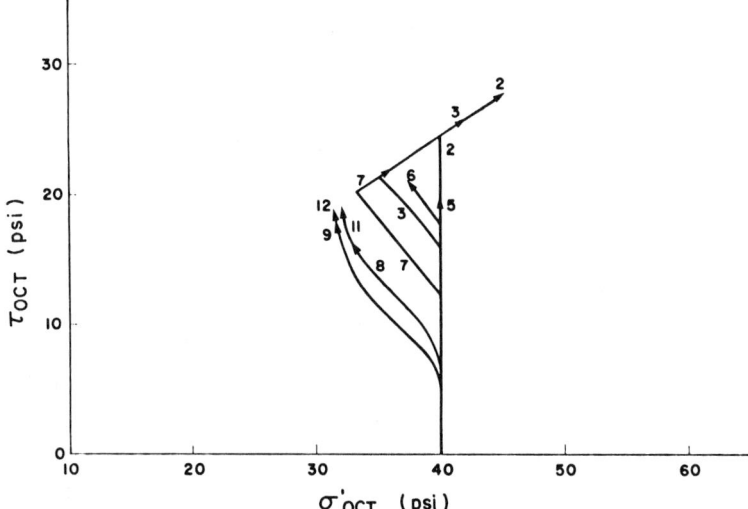

FIG. 1.—Effective Stress Paths in Tests Nos. 2, 3, 5, 6, 7, 8, 9, 11, and 12 for Material B

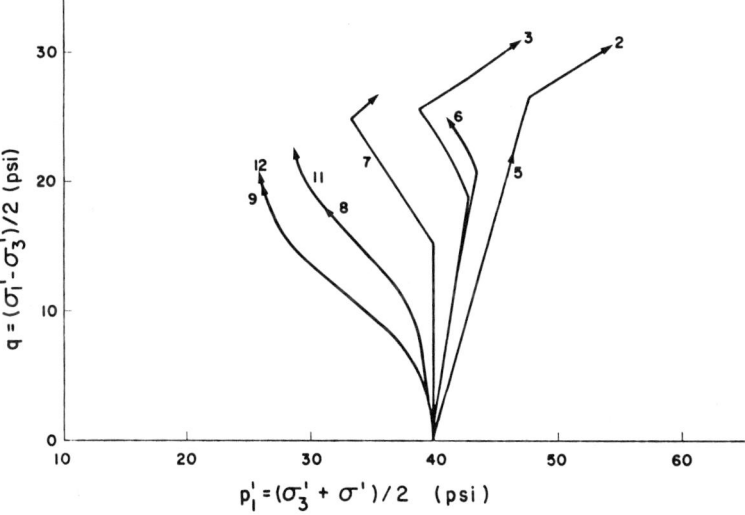

FIG. 2.—Effective Stress Paths in Tests Nos. 2, 3, 5, 6, 7, 8, 9, 11, and 12 for Material B

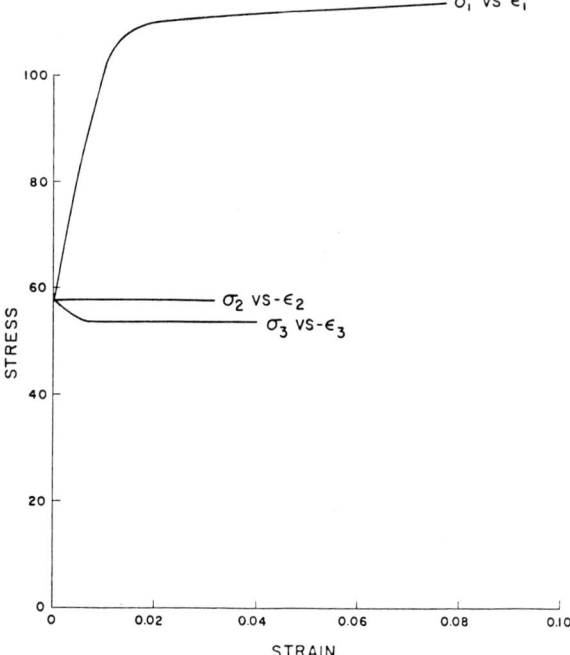

FIG. 3.--Stress-Strain Curves for Test No. 2 on Material B

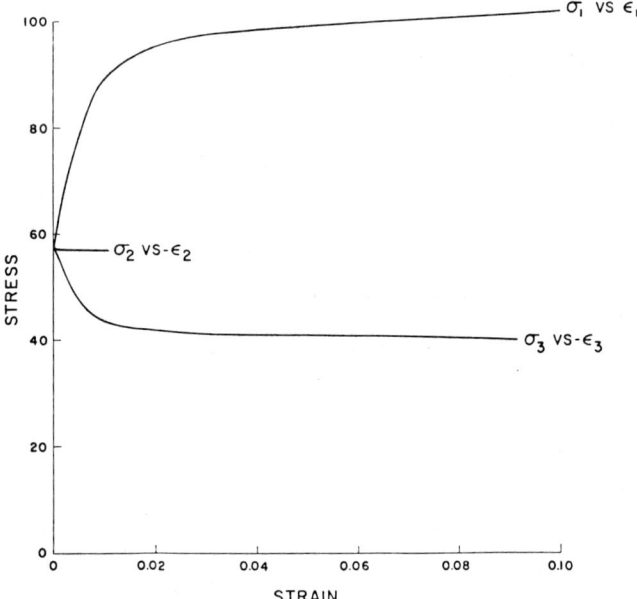

FIG. 4.--Stress-Strain Curves for Test No. 3 on Material B

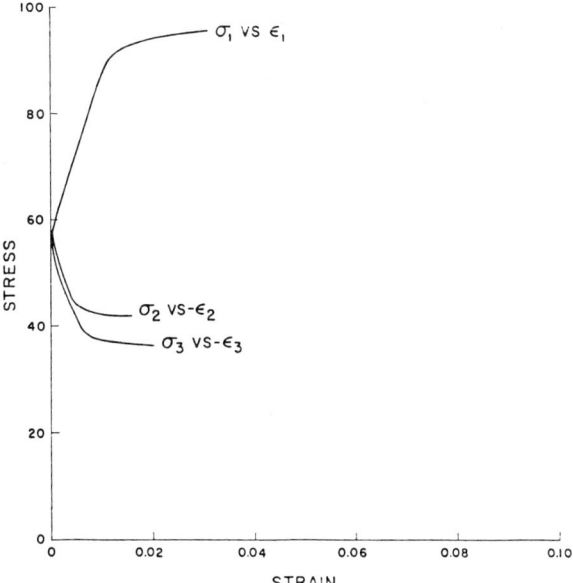

FIG. 5.--Stress-Strain Curves for Test No. 5 on Material B

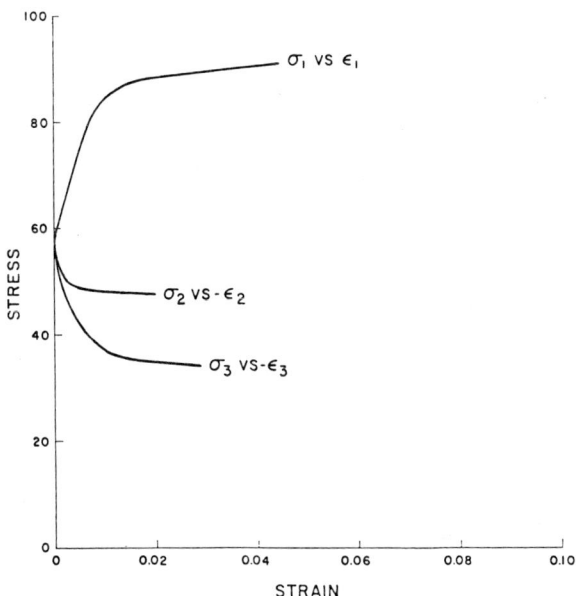

FIG. 6.--Stress-Strain Curves for Test No. 6 on Material B

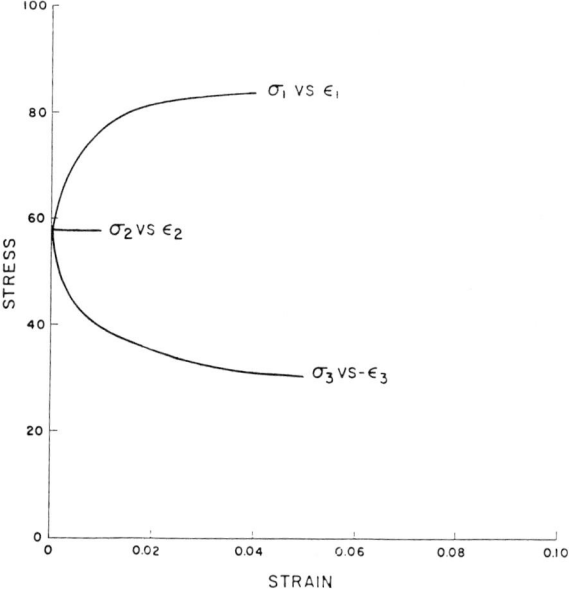

FIG. 7.--Stress-Strain Curves for Test No. 7 on Material B

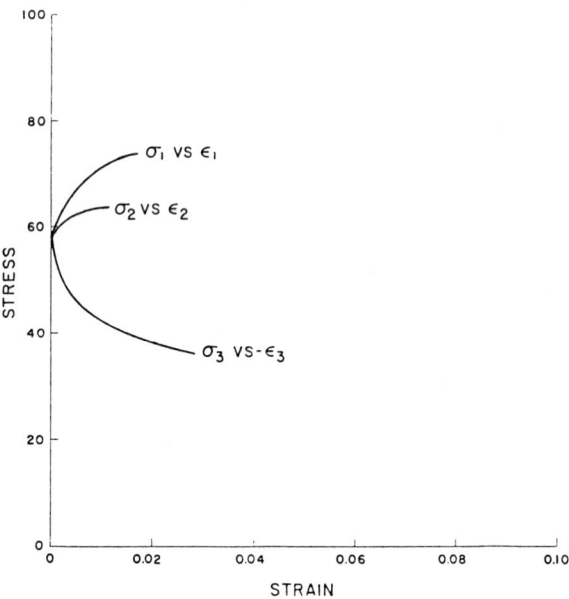

FIG. 8.--Stress Strain Curves for Test No. 8 on Material B

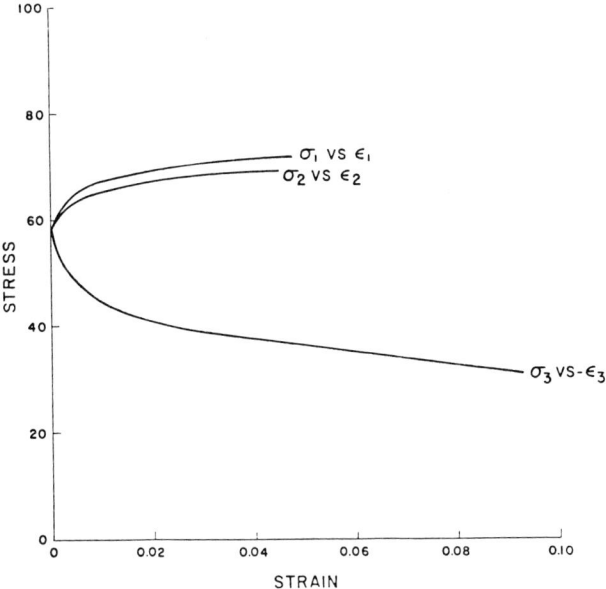

FIG. 9.--Stress-Strain Curves for Test No. 9 on Material B

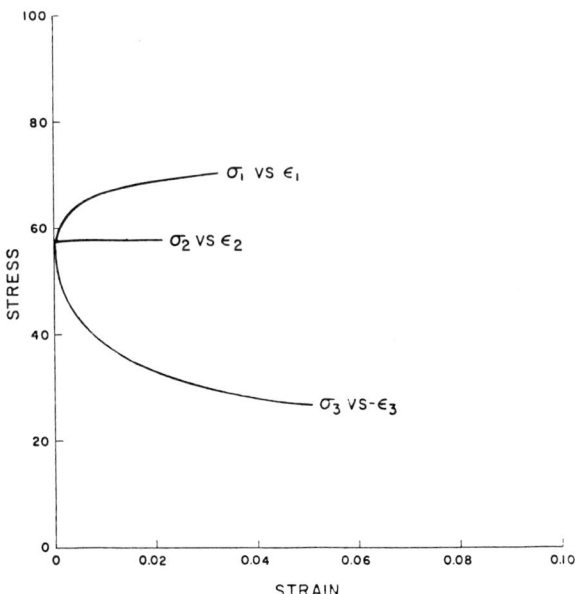

FIG. 10.--Stress-Strain Curves for Test No. 11 on Material B

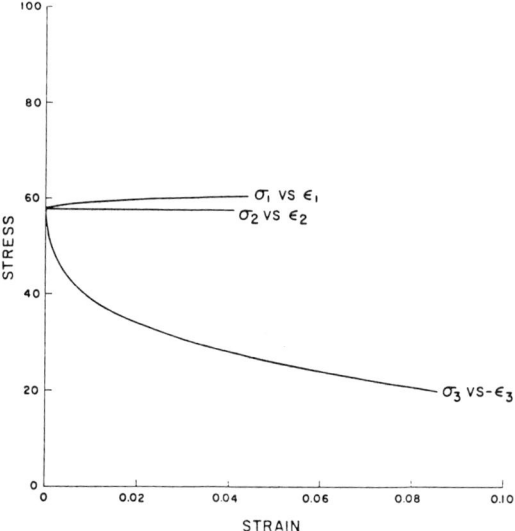

FIG. 11.--Stress Strain Curves for Test No. 12 on Material B

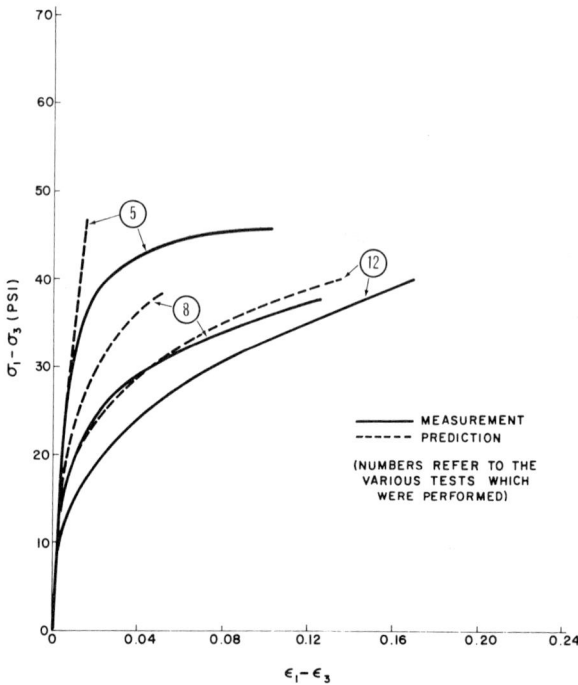

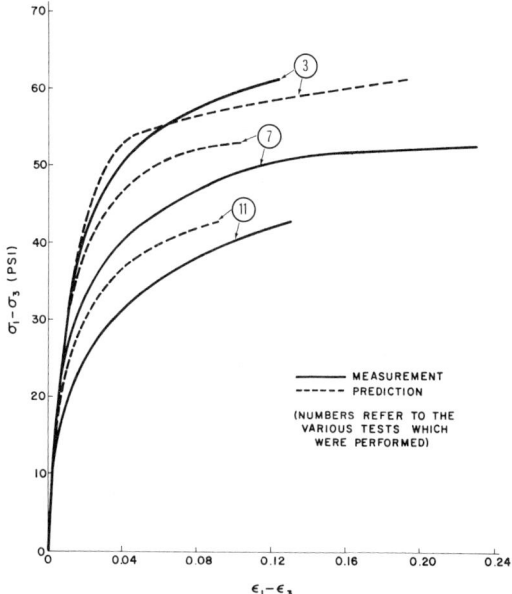

FIG. 13.—Comparison of Predicted and Observed Stress-Strain Differences for Kaolinite

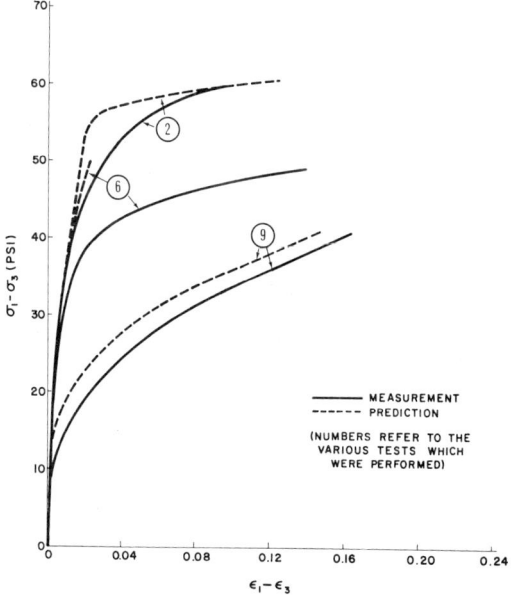

FIG. 14.—Comparison of Predicted and Observed Stress-Strain Differences for Kaolinite

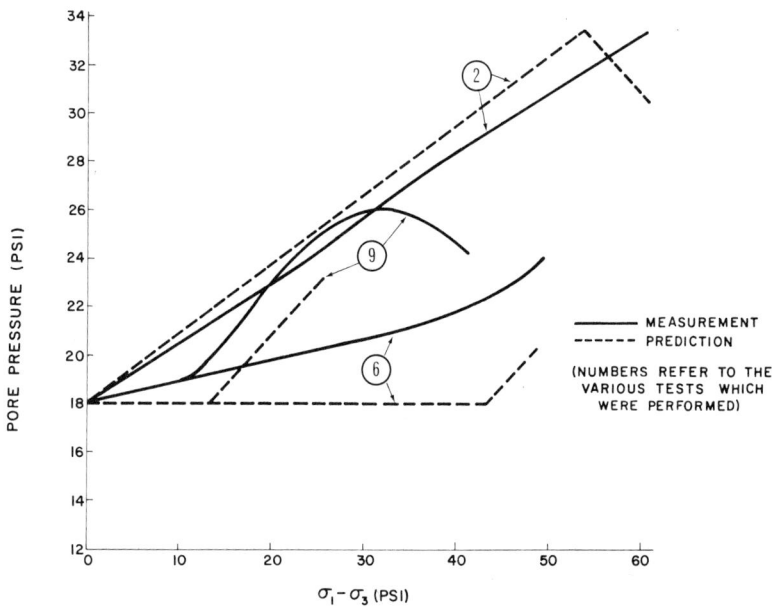

FIG. 15.--Comparison of Predicted and Observed Pore Pressure for Kaolinite

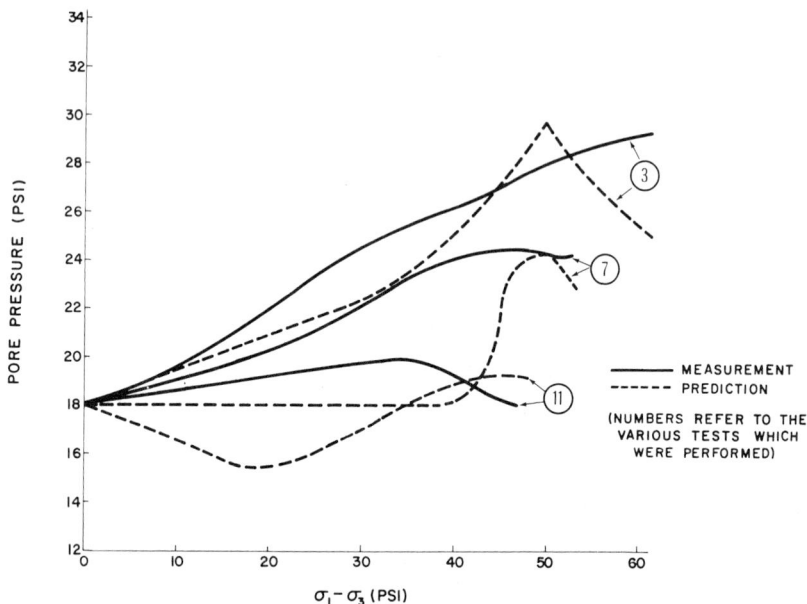

FIG. 16. Comparison of Predicted and Observed Pore Pressure for Kaolinite

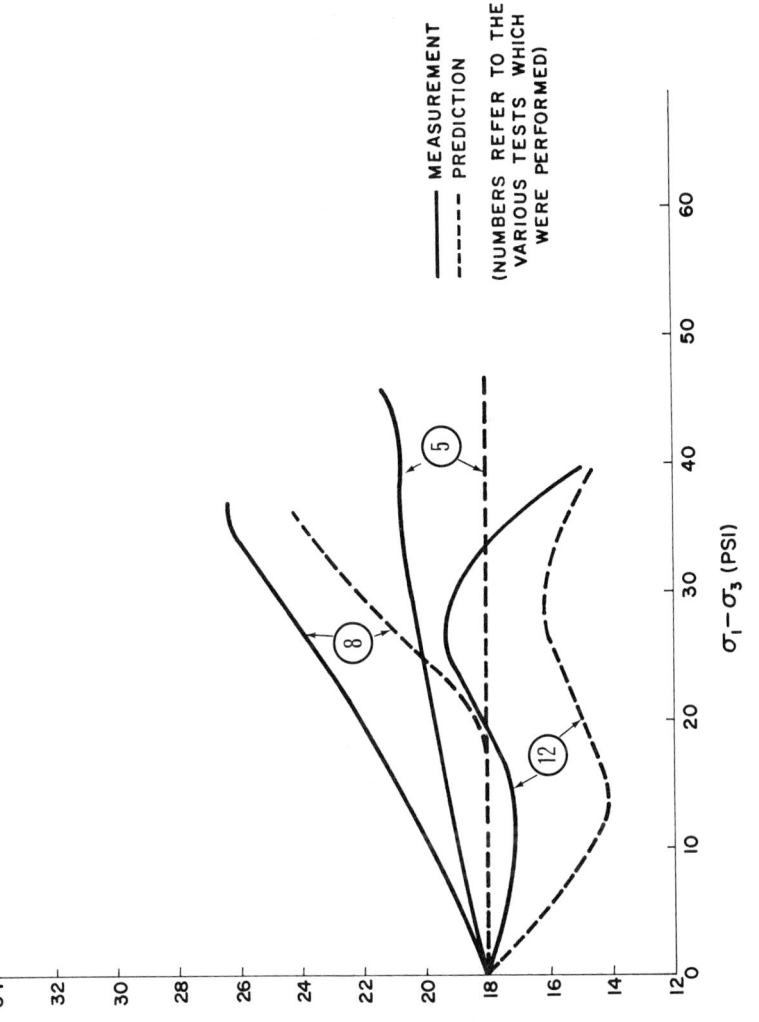

FIG. 17.—Comparison of Predicted and Observed Pore Pressure for Kaolinite

PART III: COMPARISON OF CAP MODEL PREDICTIONS WITH LABORATORY DATA
FOR OTTAWA SAND (MATERIAL C)

By George Y. Baladi

The model predictions according to scales and format requested by the workshop's organizers are shown in Figures 1 through 9. These predictions include several simple shear tests with constant mean normal stresses, a reduced triaxial compression test with a stress ratio of 0, a reduced triaxial extension test with a stress ratio of 1, two proportional loading tests with a stress ratio of 0, and a circular stress path with a constant mean normal stress of 10 psi.

The stress-strain behavior of the model is compared to the Ottawa sand behavior in Figures 10 through 16. The format of these figures is chosen so that the ability of the model to predict the shear behavior and the shear-induced volume change of Ottawa sand can be easily examined. It is clear from these figures that the model does a very good job both qualitatively and quantitatively to represent the behavior of Ottawa sand, especially in the octahedral shear stress-octahedral shear strain space. The prediction in volumetric strain-octahedral shear strain space could be improved, however, if unloading isotropic consolidation test results were available.

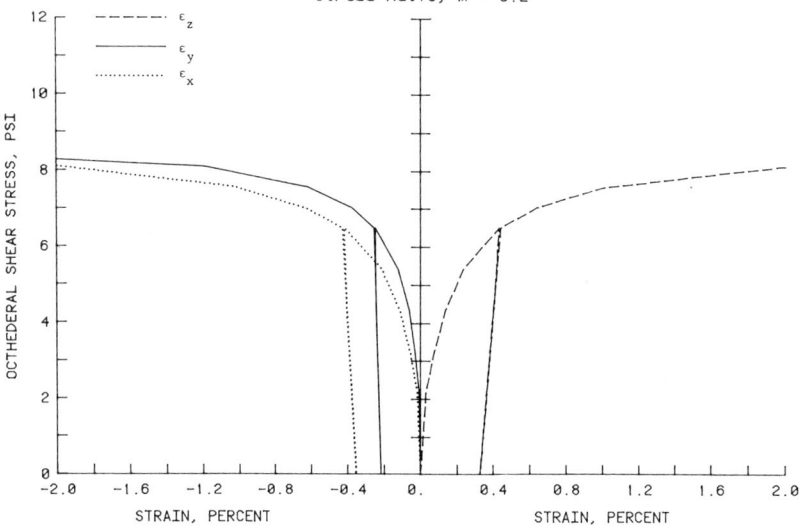

FIG. 1.--Octahedral Shear Stress Versus Principal Strains, Model Prediction

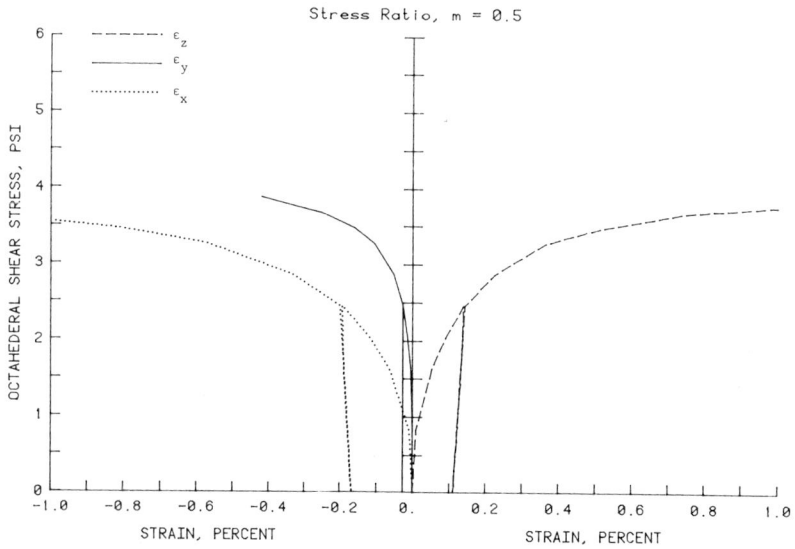

FIG. 2.--Octahedral Shear Stress Versus Principal Strains, Model Prediction

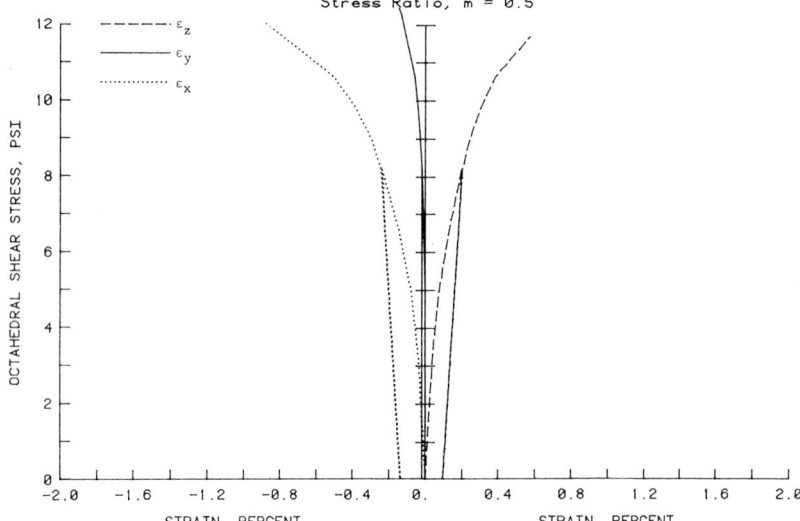

FIG. 3.--Octahedral Shear Stress Versus Principal Strains, Model Prediction

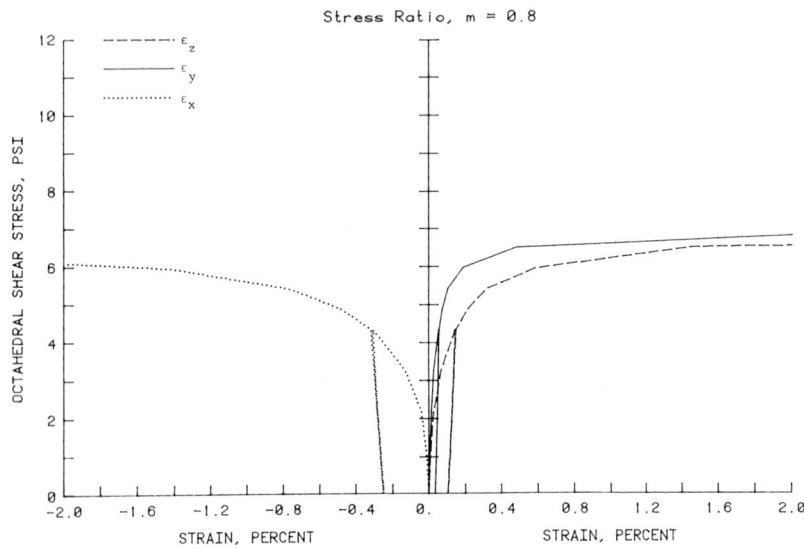

FIG. 4.--Octahedral Shear Stress Versus Principal Strains, Model Prediction

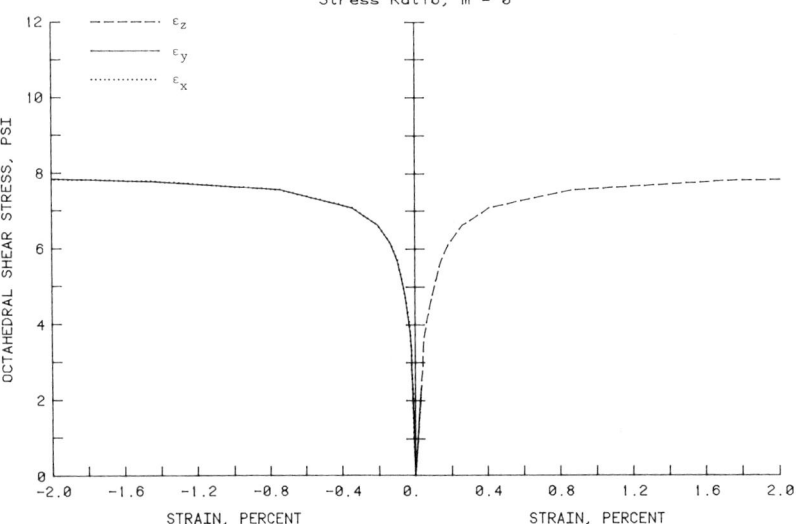

FIG. 5.--Octahedral Shear Stress Versus Principal Strains, Model Prediction

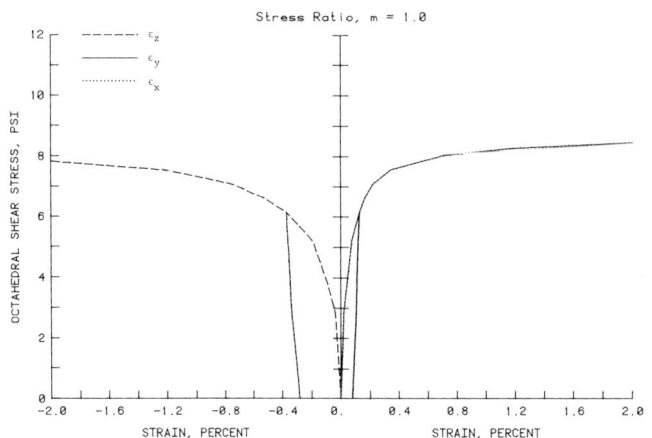

FIG. 6.--Octahedral Shear Stress Versus Principal Strains, Model Prediction

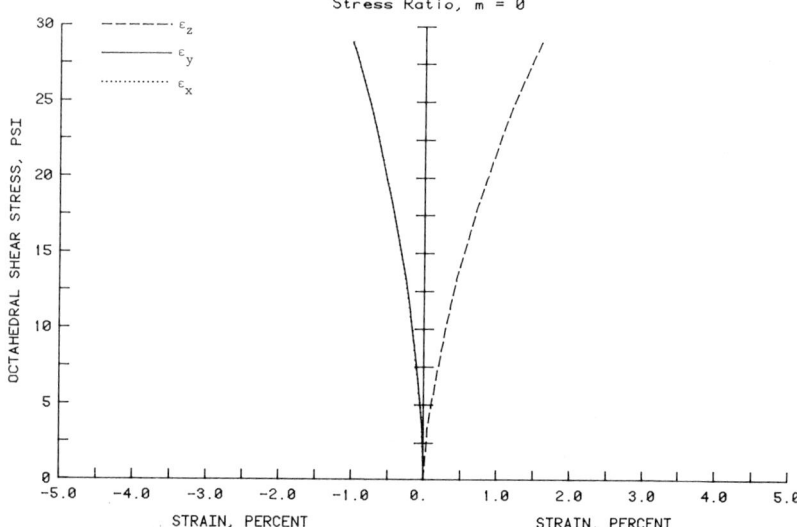

FIG. 7.—Octahedral Shear Stress Versus Principal Strains, Model Prediction

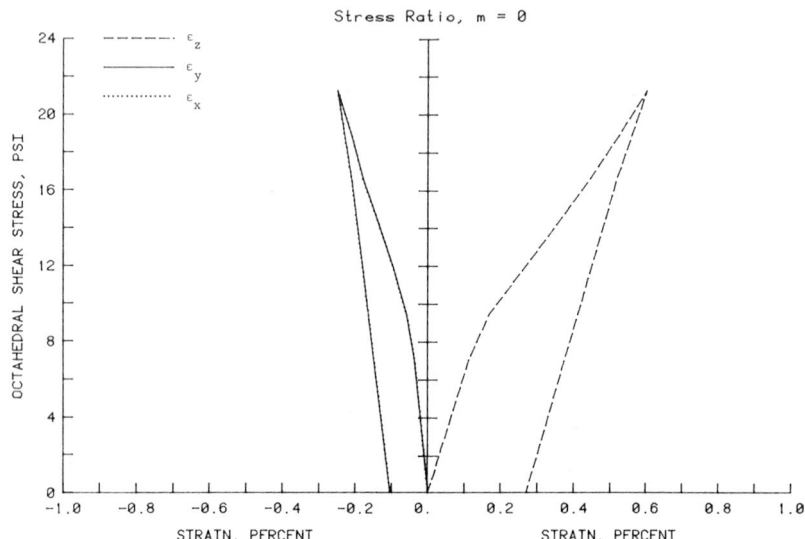

FIG. 8.—Octahedral Shear Stress Versus Principal Strains, Model Prediction

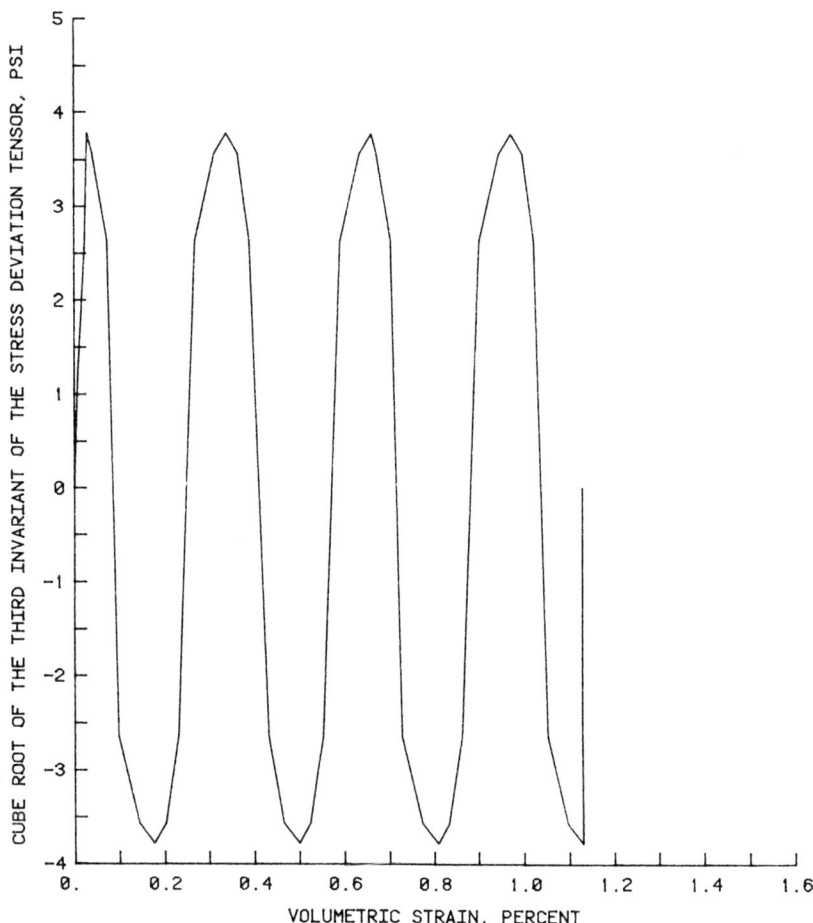

FIG. 9.---Cube Root of the Third Invariant of the Stress Deviation Tensor Versus Volumetric Strain, Model Prediction

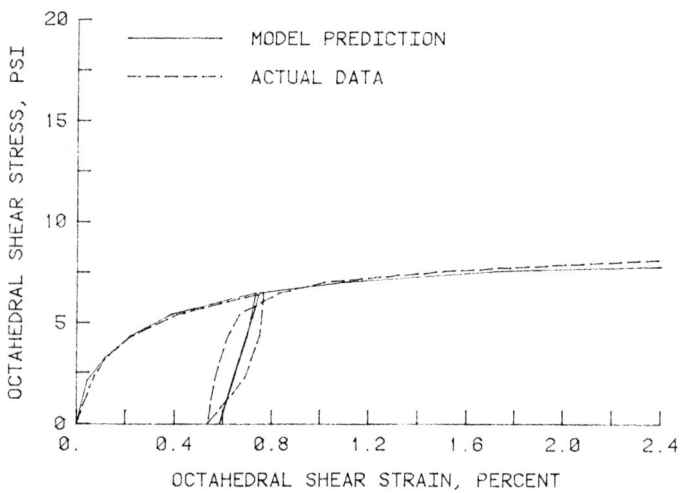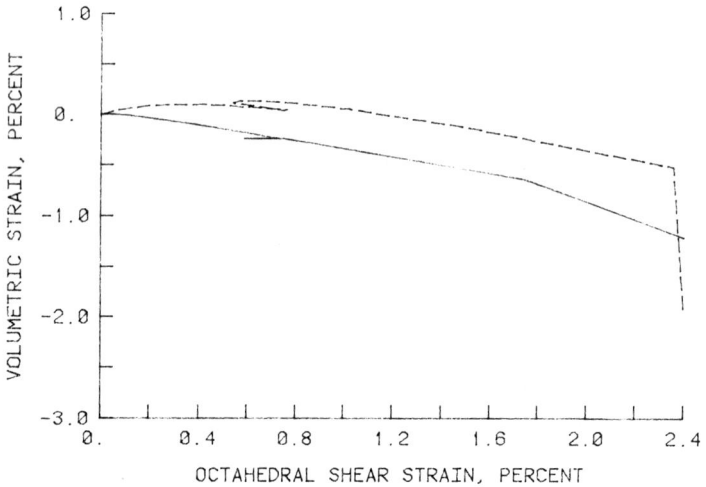

FIG. 10.—Octahedral Shear Stress and Volumetric Strain Versus Octahedral Shear Strain, Model Prediction Versus Laboratory Measurements

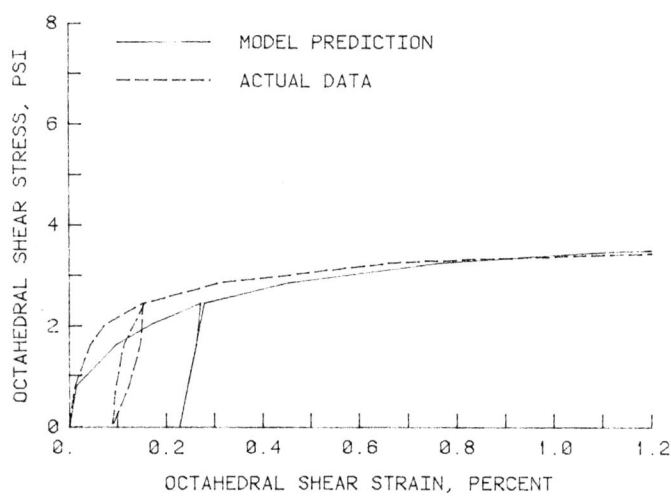
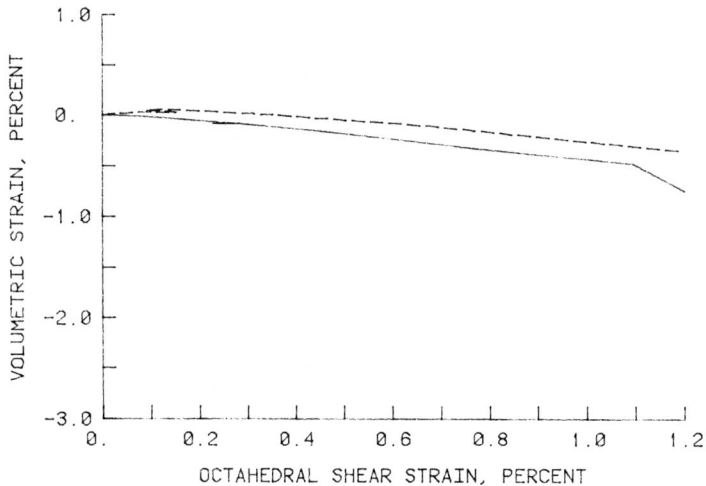

FIG. 11.—Octahedral Shear Stress and Volumetric Strain Versus Octahedral Shear Strain, Model Prediction Versus Laboratory Measurements

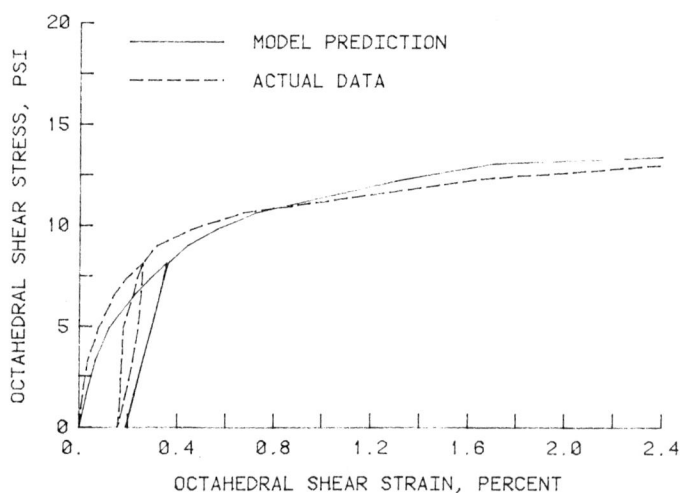
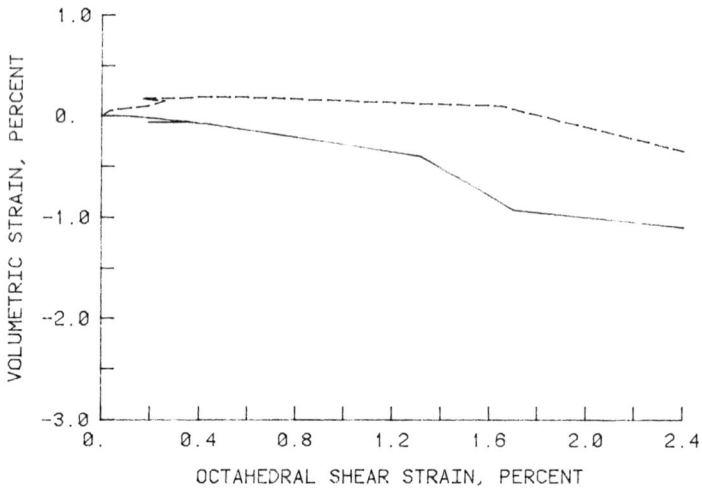

FIG. 12.—Octahedral Shear Stress and Volumetric Strain Versus Octahedral Shear Strain, Model Prediction Versus Laboratory Measurements

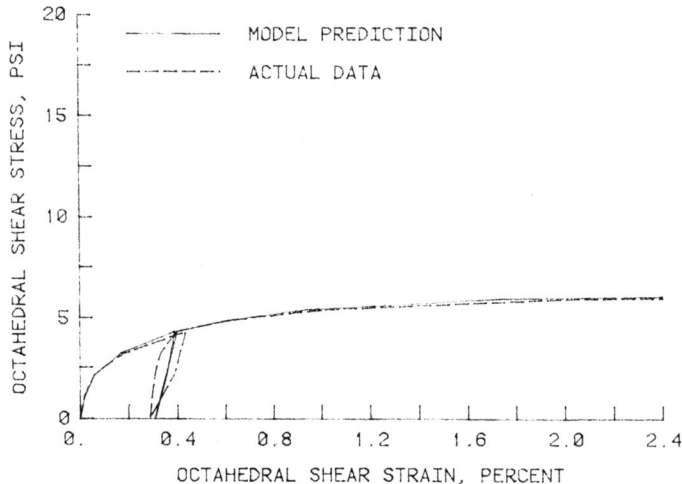
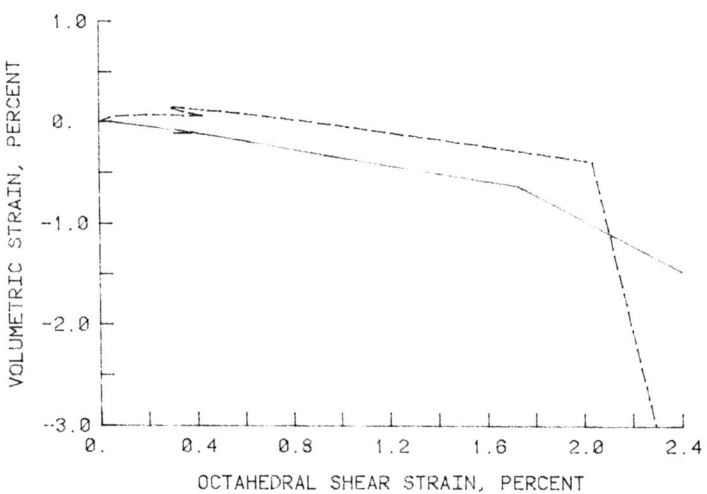

FIG. 13.--Octahedral Shear Stress and Volumetric Strain Versus Octahedral Shear Strain, Model Prediction Versus Laboratory Measurements

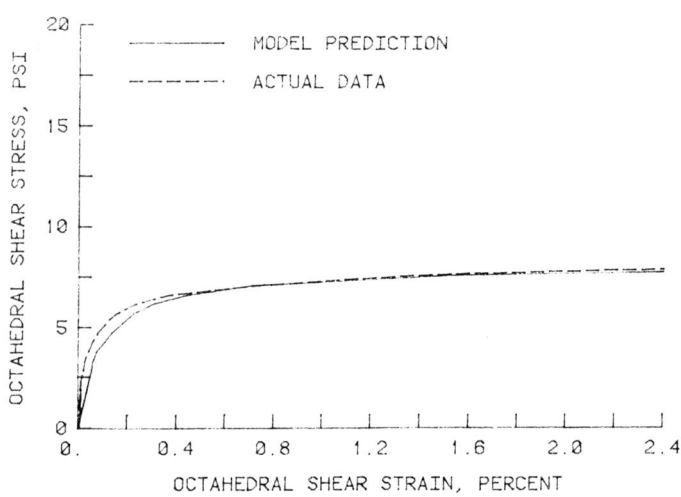
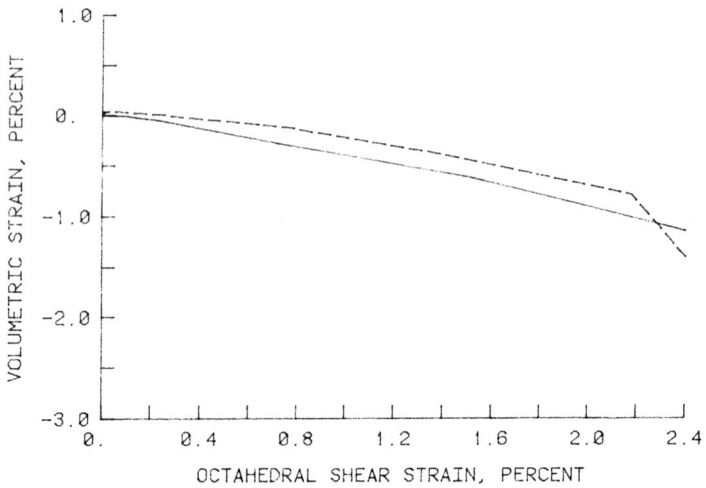

FIG. 14.—Octahedral Shear Stress and Volumetric Strain Versus Octahedral Shear Strain, Model Prediction Versus Laboratory Measurements

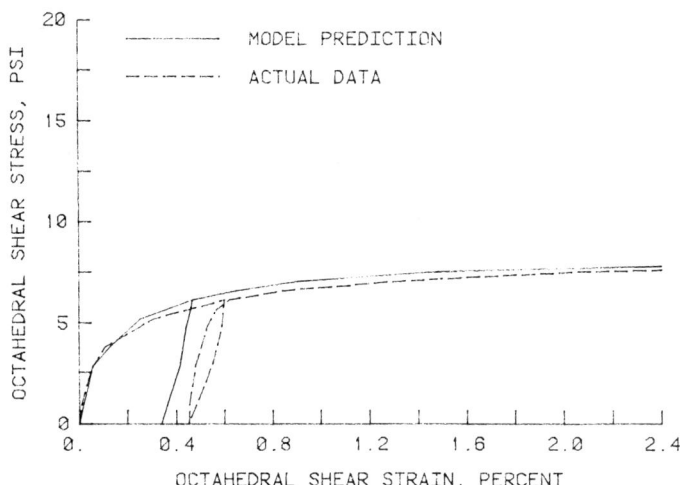

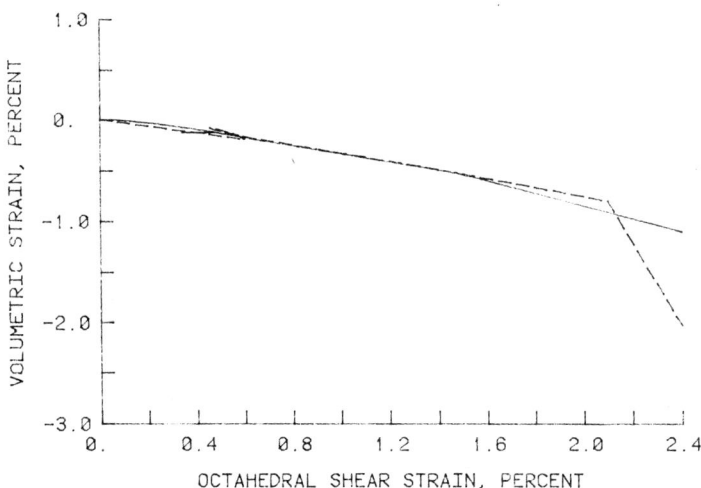

FIG. 15.--Octahedral Shear Stress and Volumetric Strain Versus Octahedral Shear Strain, Model Prediction Versus Laboratory Measurements

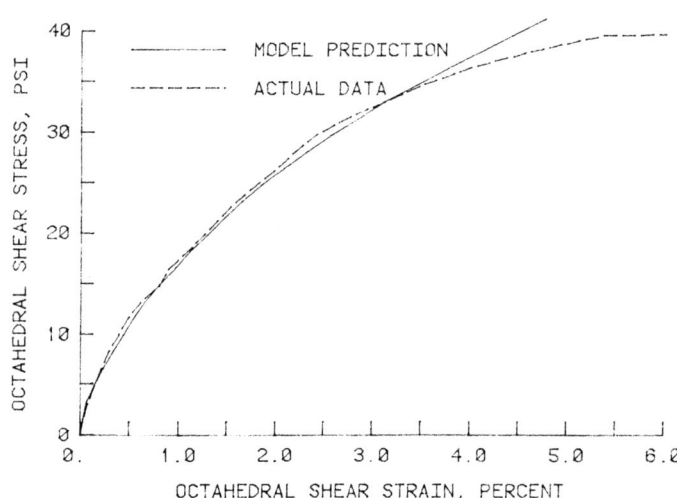
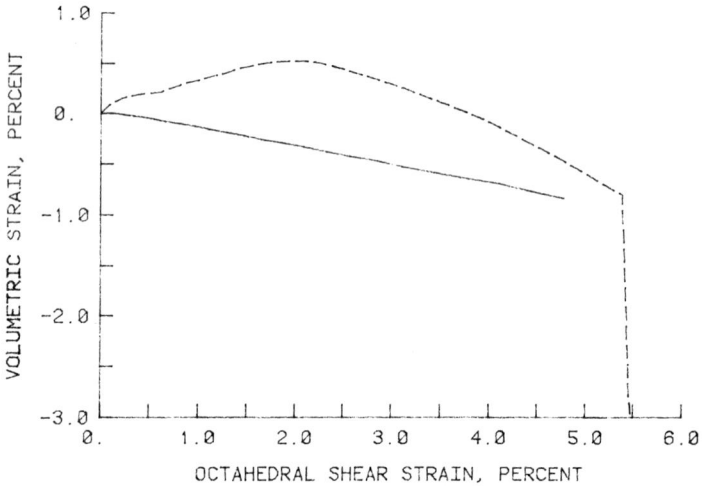

FIG. 16.--Octahedral Shear Stress and Volumetric Strain Versus Octahedral Shear Strain, Model Prediction Versus Laboratory Measurements

Conclusion

The cap model is capable of representing the behavior of a wide class of geological materials. In this paper several predictions and comparisons with actual test results are presented. It is unfortunate that repetitions of the laboratory experiments were not performed. The authors believe that if this were done the model predictions would fall well within the data scatter, indicating that the cap model is truly adequate to represent these diverse situations.

PREDICTION OF THE RESPONSE OF THE NATURAL CLAYS X AND Y USING THE BOUNDING SURFACE MODEL

By Yannis F. Dafalias[1], A.M. ASCE, Leonard R. Herrmann[2], M. ASCE, and Jay Scott DeNatale[3]

INTRODUCTION

It is assumed that the reader is familiar with the bounding surface model and the corresponding calibration procedures presented in the position paper in the second volume of these proceedings. For completeness, however, a few important comments are briefly presented here as well. The formulation of the bounding surface model used in this workshop is appropriate for clay behavior within the general framework of critical state soil mechanics. Hence, the sand behavior cannot be predicted by this formulation. The predictions of the laboratory prepared clay were mainly an exercise in anisotropy and were not attempted because it is shown in the introduction of the position paper that insufficient experimental data were provided for a rigorous characterization of the transverse isotropy which develops with the laboratory K_o consolidation. Similarly, it is shown in the position paper that although the natural clay samples X and Y are in general anisotropic and exhibit different properties in compression-extension, these two features cannot be uniquely characterized for the following two reasons: first, the given experimental data for calibration mix the effect of anisotropy with that of the different response in compression-extension, rendering impossible to isolate and characterize either one. Secondly, apparently the natural clay samples were tested at random and unspecified orientations with respect to the vertical direction in the field, therefore,

---

[1] Assoc. Prof., Department of Civil Engineering, Univ. of California, Davis, CA.
[2] Professor, Department of Civil Engineering, Univ. of California, Davis, CA.
[3] Research Asst., Department of Civil Engineering, Univ. of California, Davis, CA.

precluding the anisotropic characterization even if the effect of the compression-extension differences was not present simultaneously. Since, however, the important point for the response of clays X and Y is the effect of overconsolidation, it was decided to treat the clays as isotropic with no difference in the compression-extension response, and use a corresponding simple isotropic version of the bounding surface model in the space of the two stress invariants.

Finally, a very important point which must be mentioned at the outset is that for the predictions presented at the workshop, the overconsolidation ratios (OCR) of the initial state of the samples were overestimated due to a misinterpretation of the definition of the preconsolidation pressure in the provided e-ℓnp curves. Details can be found in the introduction and the calibration section of the position paper and here it suffices to say that while the "correct" overconsolidation ratios (the " indicates that correctness depended on an assumed value of K_o = 0.50) for clay X and for confining pressures σ_c(psi) = 10, 20, 30 and 40 are OCR = 1.55, 1, 1 and 1, they were overestimated as OCR = 2.35, 1.18, 1 and 1 respectively. Similarly, for clay Y and for σ_c(psi) = 2.5, 5 and 10, the "correct" OCR = 7.92, 3.96 and 1.48 and the overestimated OCR = 8.90, 4.45 and 2.23, respectively. This effected to a certain degree (if not drastically) the calibration and predictions, and in what follows the predictions for both the correct OCR and the overestimated OCR (shown at the workshop) are presented.

PREDICTIONS AND DISCUSSION

The predictions are shown in the following figures with continuous curves properly identified by numbers. On the same figures, the corresponding experimental data which were to be predicted are also shown by appropriate symbols. The stress strain curves are the deviatoric stress q = $\sigma_1-\sigma_3$ versus the axial strain ε_1 and the volumetric strain $\varepsilon_v = \varepsilon_1 + \varepsilon_2 + \varepsilon_3$ versus ε_1 according to the provided instructions. Alternatively, one could plot all the $\sigma_i-\varepsilon_i$ (i = 1,2,3) curves. The predictions were obtained by the general bounding surface three dimensional constitutive relations which have been numerically implemented. The stress history was the input in terms of principal stress variation, and the output was all the strain components (no shear components due to assumed isotropy).

In each figure, the predictions and the experimental results are grouped according to the stess ratio m = $(\sigma_2-\sigma_3)/(\sigma_1-\sigma_3)$ for all confining pressures. When two prediction curves are almost indistinguishable, especially for the volumetric strains, their numbers are placed close to each other. The predictions-experimental results for clay X are shown in Fig. 1-8 and these for clay Y in Figs. 10-15. The predictions for the correct OCR and the overestimated OCR are placed next to each other for easy comparison, with the correct OCR preceeding the overestimated. For example for m = 0.25, Fig. 3 shows the response of clay X for the correct OCR and Fig. 4 the same response on the basis of the overestimated OCR. Finally, Figures 9 (clay X) and 16 (clay Y) show the predicted range of variation of the corresponding stress-strain curves as m changes from 0 to 1 for all confining pressures σ_c, except σ_c=40 psi for clay X. The curves for m=0, 1 are taken from the corresponding figures of the position paper where they were used for the calibration.

An important point deserving attention is the comparison between the predictions for the same m and the two sets of OCR. In general, one may say that the predictions improve for clay X but are slightly less accurate for clay Y as the OCR change from the overestimated to their correct values.

Consider first clay X. The improvement is evident for both q-ε_1 and ε_v-ε_1 curves, especially for m=0, 0.25 and high confining pressures σ_c = 20, 30, 40 psi (the q-ε_1 curves at σ_c = 20 psi are almost identical). For m=0.50, 0.75 the predictions are satisfactory for both sets of OCR. A plausible question now is why the initial predictions for the overestimated OCR were unsatisfactory for σ_c=40, 30 psi and m=0, 0.25, Figs. 2 and 4, for which the OCR were estimated equal to 1 as for the correct set of values. In other words, since the overestimation did not affect the OCR=1 for these cases, why the predictions are not accurate? The answer can be better described if one refers to Fig. 4 of the position paper which was used for calibration with m=0 but, for convenience, the arguments can be presented also with reference to Fig. 4 of the present part where m=0.25 is close to m=0. The model could not accomodate the experimental results showing almost identical volumetric strains for σ_c=20 psi and 30 psi or even greater for 20 psi than 30 psi, because the σ_c=20 psi was considered a stiffer overconsolidated state with OCR = 1.18 and, logically, the model was predicting a smaller ε_v than for σ_c=30 psi (compare curves 2 and 3 in ε_v-ε_1 plot). Any effort to increase the ε_v by changing the parameters resulted in a decrease of q and a corresponding greater deviation of the q-ε_1 curve from the experimental data. Thus, the final choice of the parameters was done with the objective to minimize the deviation for both curves q-ε_1 and ε_v-ε_1. As a result, the predictions were unsatisfactory not only for σ_c=20, 30 psi but also for σ_c=40 psi due to this "compromised" choice of the parameters. When later it was recognized that σ_c=20 psi corresponds to OCR=1, the parameters were properly chosen and the considerable improvement for the smaller m's is shown in the figures with the correct OCR. One can say that in general for m=0, 0.25 the prediction curves slightly undershoot the experimental data while for m=0.50, 0.75 and 1 are quite accurate. One also can say that the initial part of the curves for up to about ε_1=3% are better described for the correct OCR than the overestimated, and this is due to the better description of the plastic response only, since the used shear modulus G was the same.

Coming now to clay Y, it appears that the "correct" OCR predictions are less accurate than the ones based on overestimated OCR. In general, the model indicates smaller q at failure for σ_c=2.5, 5 psi and greater for σ_c=10 psi compared with the experimental data for both sets of OCR, and also a greater ε_v especially for the correct OCR. Recall, however, that the "correct" OCR were based on an assumed K_o=0.50 which may not be entirely accurate if K_o changes, especially for such small σ_c. Definitely, the large OCR exhibited by the initial state of clay Y renders the predictions much more difficult than for normally consolidated states and the deviation tolerence greater, especially if one recalls that some models did not even attempt to predict the behavior of clay Y although they did so for clay X. Finally, the following general observations can be made for the experiments and predictions of clays X and Y:

a. With respect to the inital slope of the q-ε_1 curves, the experimental scatter in the clay X tests is more pronounced than that in the clay Y. This can be attributed to the larger range of consolidation stresses applied to X than to Y. Nevertheless, it shows the importance of further investigation of the elastic soil response with a variable G.

b. Both clays X and Y show the expected trend that q_{max} increases as σ_c is increased.

c. For both clays X and Y, the stress-strain response at different stress ratios m is not very much different. This can be clearly observed in Figures 9 and 16 where the corresponding "bands" for each σ_c shows the variation of the response as m changes from 0 to 1. In addition, the limiting stress-strain responses occur at the limiting stress ratios. That is, a stress ratio of m=0 results in the smallest q_{max} and ε_{v-max}, whereas a stress ratio of m=1 produces the greatest q_{max} and ε_{v-max}. This is expected on physical ground because m=1 represents extension where the two maximum and equal principal stresses increase while the third decreases (clay X) or remains constant (clay Y), and m=0 represents compression where the one maximum stress increases while the two minimum and equal stresses decrease or remain constant, thus stressing the material less than for m=1.

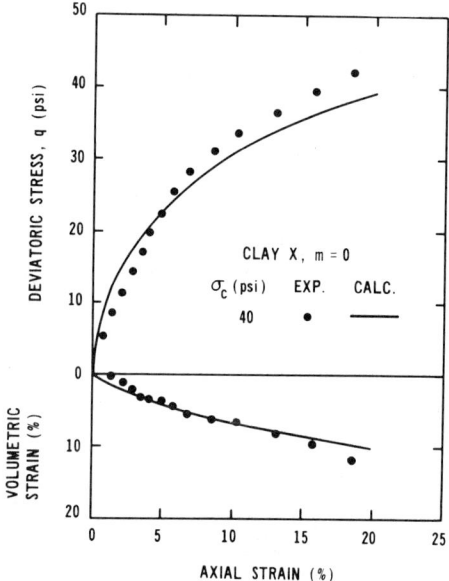

FIG. 1 - Experimental data and prediction curves for correct OCR.

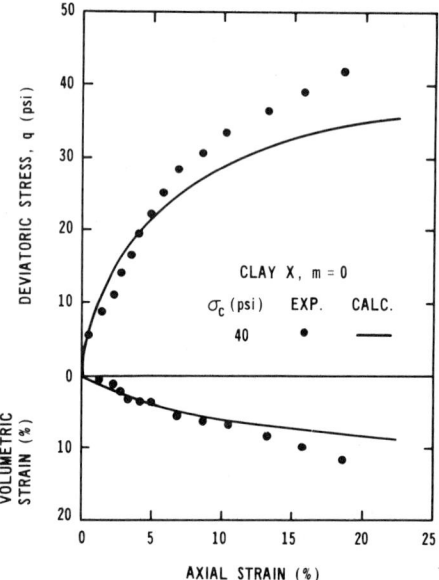

FIG. 2 - Experimental data and prediction curves for overestimated OCR.

NATURAL CLAYS RESPONSE

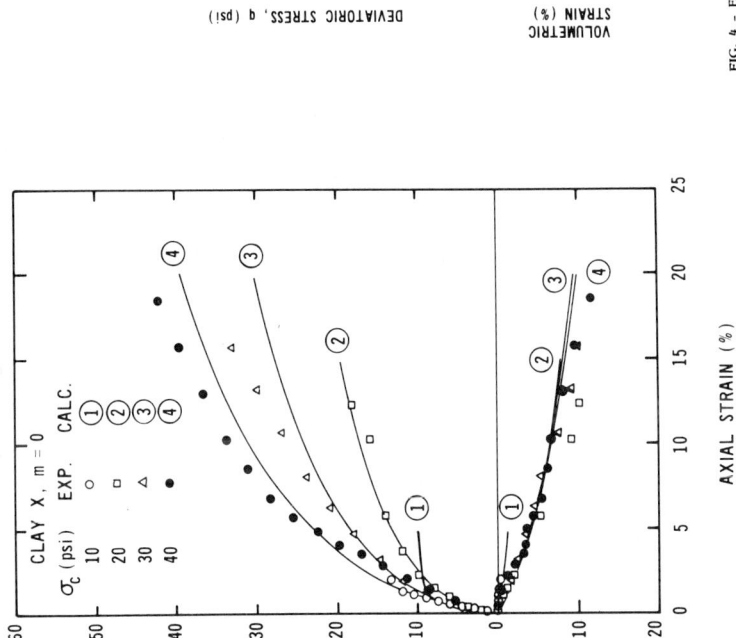

FIG. 3 – Experimental data and calibration curves for correct OCR.

FIG. 4 – Experimental data and prediction curves for overestimated OCR.

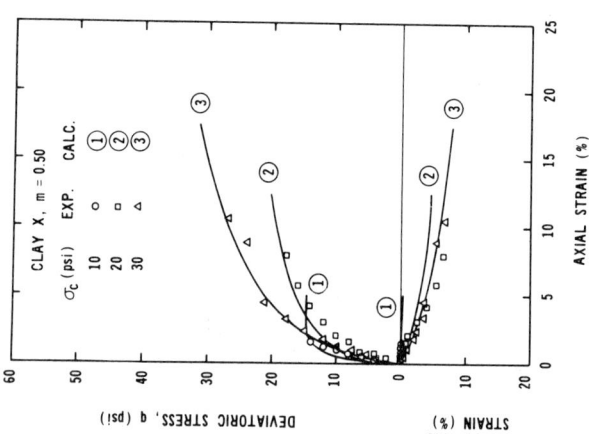

FIG. 6 – Experimental data and prediction curves for overestimated OCR.

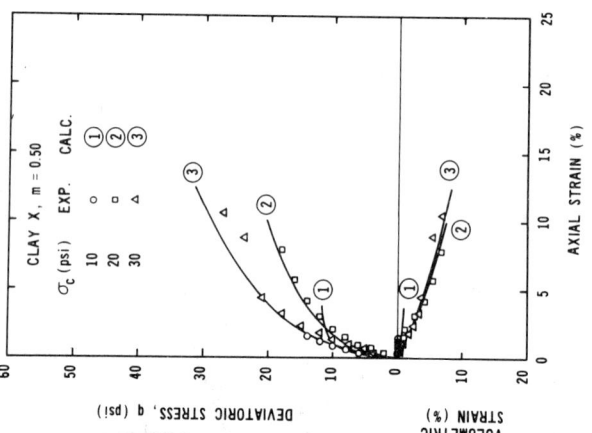

FIG. 5 – Experimental data and prediction curves for correct OCR.

NATURAL CLAYS RESPONSE

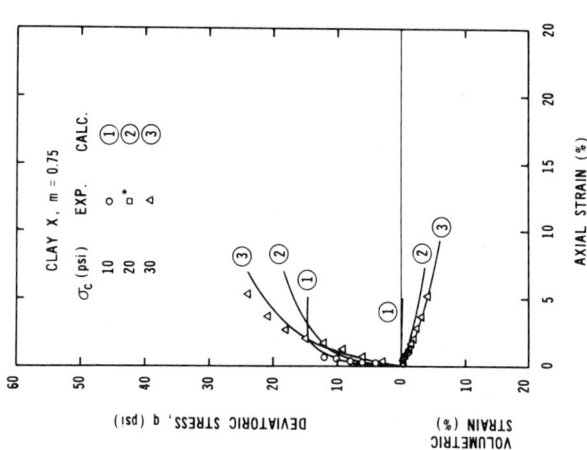

FIG. 7 – Experimental data and prediction curves for correct OCR.

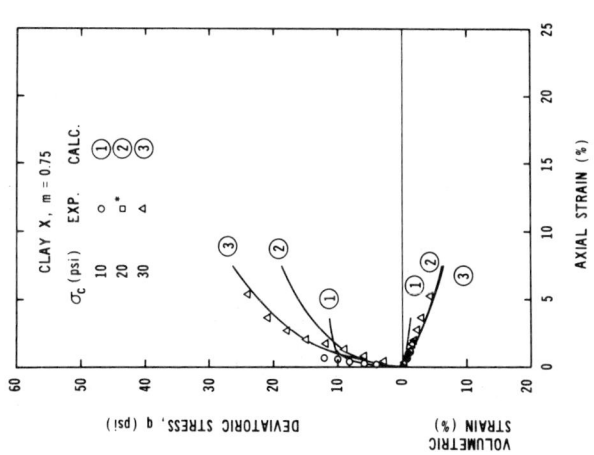

FIG. 8 – Experimental data and prediction curves for overestimated OCR.

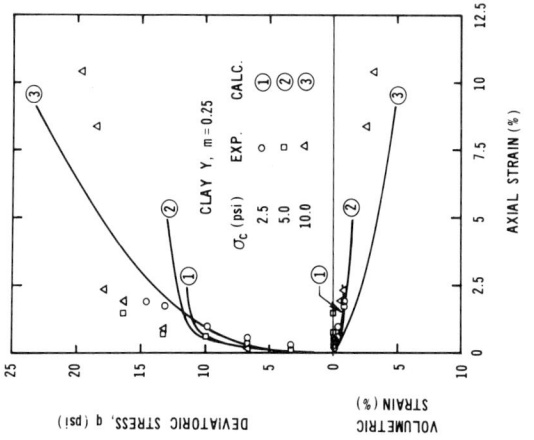

FIG. 10 - Experimental data and prediction curves for correct OCR.

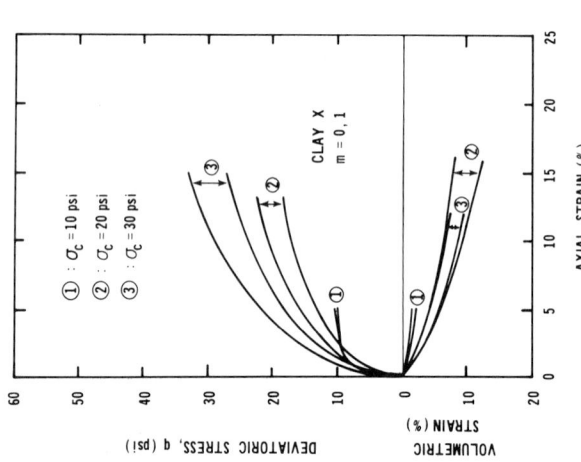

FIG. 9 - The range of variation of the stress-strain curves for clay X as m changes from 0 to 1 for the different confining pressures.

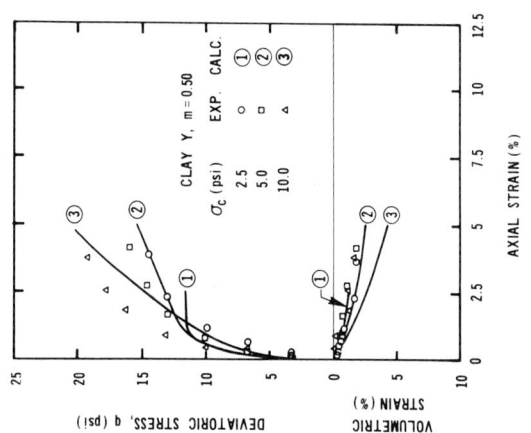

FIG. 12 - Experimental data and prediction curves for correct OCR.

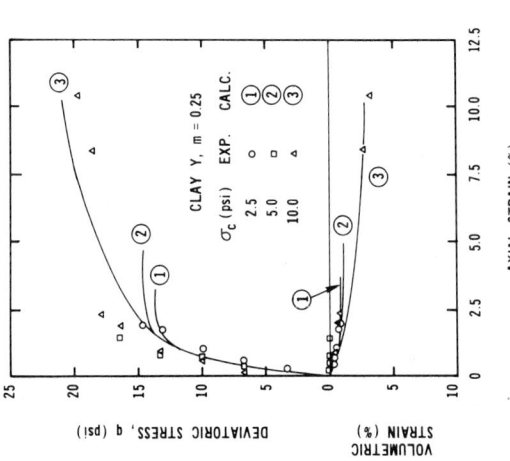

FIG. 11 - Experimental data and prediction curves for overestimated OCR.

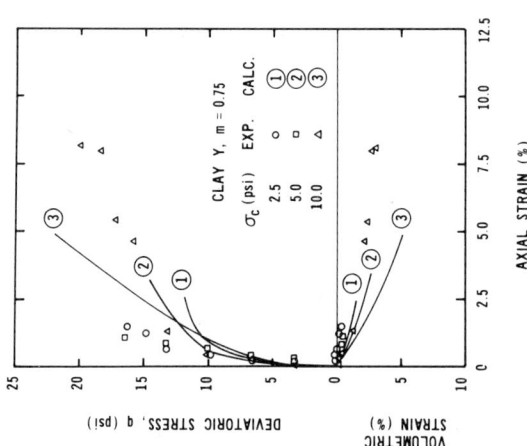

FIG. 14 – Experimental data and prediction curves for correct OCR.

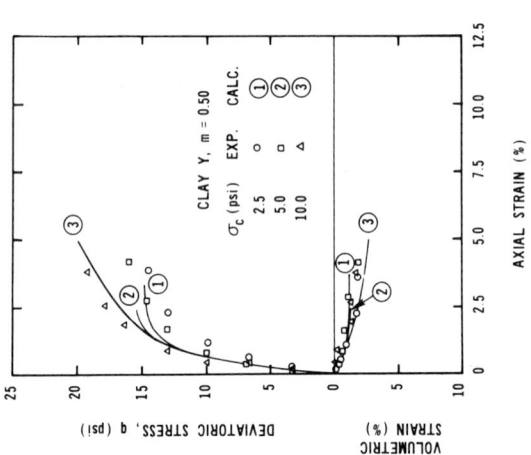

FIG. 13 – Experimental data and prediction curves for overestimated OCR.

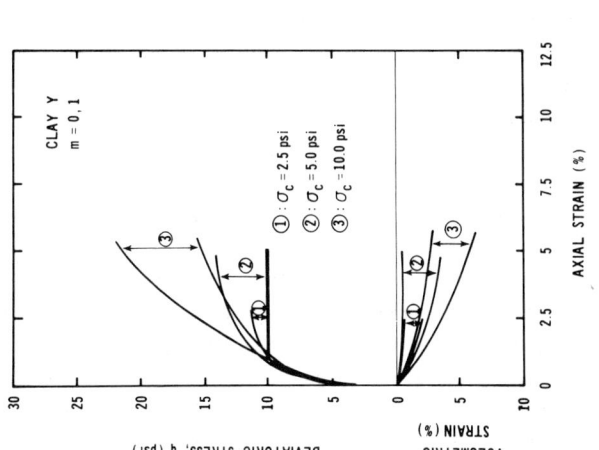

FIG. 16 - The range of variation of the stress-strain curves for clay Y as m changes from 0 to 1 for the different confining pressures.

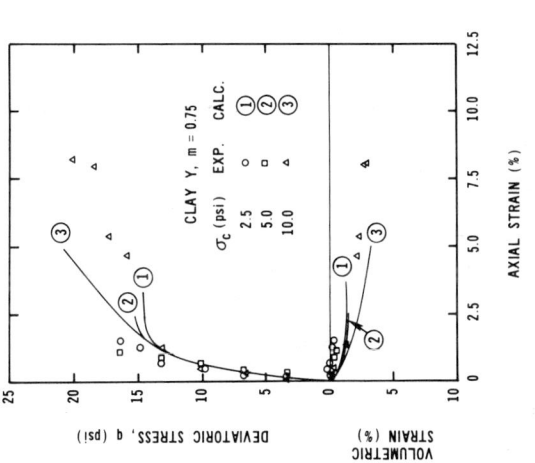

FIG. 15 - Experimental data and prediction curves for overestimated OCR.

CONCLUSION

The reader is referred to the introduction and last section of the position paper for general conclusions and observations regarding the soil constitutive models and in particular that of the bounding surface, with regards to different features including the easiness of calibration and numerical implementation. With respect to the present predictions, the following points can summarize the results:

1. In general the predictions of the model for both clays X and Y can be considered satisfactory, especially because of the very simple formulation which although is particularly suited for laboratory prepared isotropic clays with equal response in compression-extension, managed to predict qualitatively and quantitatively the response of anisotropic natural clays at different OCR and consolidation pressures <u>with one single set</u> of material constants. It is expected that much better predictions can be obtained if a more general formulation of the model is used which includes the third stress invariant. This would allow to account for the different response in compression-extension (i.e., the effect of intermediate principal stress), even if it is not possible to separate rigorously this effect from existing anisotropy for calibration purposes. Work towards this direction is under way.

2. The overestimation of OCR which is associated with the model calibration and predictions presented during the workshop resulted in an unsatisfactory prediction for clay X at m=0 and 0.25 and for the larger σ_c, while for m=0.50, 0.75 and 1 the predictions were satisfactory as well as for all m for clay Y. The recalibration of the model on the basis of the "correct" OCR considerably improved the predictions for the lower m's to a satisfactory level, while kept the good predictions at the larger m's for clay X and slightly decreased the quality of the predictions for clay Y.

3. It is considered very important and in fact is the very essence of the bounding surface concept that the model can predict the clay response for all OCR which essentially represent different initial states of the same material.

We would like to close this presentation commenting on the following point. What was at stake in this workshop was not the credibility of the simple models but that of the more sophisticated ones. The simple and rather empirical models were built to describe the material response under simple loading conditions and especially for clays X and Y any curve fitting and interpolation between m=0 and m=1 would easily provide good answers for intermediate values of m. The sophisticated models were built with greater aspirations, were necessarily more complex and therefore what seems an easy task for a simple model here, may not be so for the sophisticated one. One important conclusion reached in this workshop is that many of the complex models performed satisfactorily in this respect. But this is just the beginning. Because next remains the challenge to stand up to the greater expectations of predictive capabilities that these models possess under cyclic, drained, undrained, etc., loading conditions, a realm where the simple models cannot exist by their own definition. Another workshop towards this goal should be the next step.

PREDICTION AND COMPARISON

By Koichi Akai[1], M. JSCE, Toshihisa Adachi[2], M. JSCE and Fusao Oka[3], M. JSCE

As discussed in our position paper, we used two different types of constitutive models to carry out the assigment of this workshop. The material parameters required to be obtained are listed in Table A-1(Clay-X), Table A-2(Clay-Y), Table A-4(Kaolinite Clay) and Table A-5(Ottawa Sand) in the position paper.

Our predictions of soil behaviors are given as stress-strain curves. In the case of Kaolinite clay, the effective stress paths are also predicted. The predicted results are compared with the actual test results in Fig.1 to Fig.4 for Clay-X, in Fig.5 to Fig.7 for Clay-Y, in Fig. 8 to Fig.12 for Kaolinite Clay and in Fig.13 to Fig.30 for Ottawa Sand. For convenience, the list of figures is given in Table 1.

By comparing the predicted values with the actual test results, the following conclusions are summarized.

(1) For Clay-X, the stress-strain relations are underestimated by the model especially under lower confining pressure. However, under higher confining pressure, they show good agreement. On the other hand, the volumetric strain is always underestimated.

(2) For Clay-Y, the stress-strain relations are underestimated by the models, but the volumetric strain is well predicted.

(3) For Kaolinite Clay, the predictions of stress-strain relations are acceptable, but the effective stress paths(which are shown in those figures (b) of Fig.8 to Fig.12) are not well predicted.

(4) For Ottawa Sand, the predictions in such cases of larger m values(i.e., m= 0.8 and 1.0) are reasonable, however when smaller m values, the difference between the predictions and the actual test results become larger. It is seen in Fig.21 that our model qualitatively predict the behaviors of sand

---
[1]Prof., Dept. of Transportation Engrg., Kyoto Univ., Kyoto, Japan.
[2]Associate Prof., Disaster Prevention Research Inst., Kyoto Univ., Kyoto, Japan.
[3]Assistant Prof., Dept. of Civil Engrg., Gifu Univ., Gifu, Japan.

under circular stress path loading, however improvement of our model is required in order to quantitatively predict the behaviors. The original predictions had been done by the way discussed in our position paper. Afterwards, we re-predicted the stress-strain relations of Ottawa Sand by using the average values of material parameters given in Table A-5 in the position paper. The repredicted results are shown in Fig.22 to Fig.30. Some improvements are found in those figures

Table 1. List of figures for the predictions

Soils	Test number and testing conditions	Figure number
Clay-X	σ_c= 10 psi, m = 0.25, 0.50, 0.75	Fig.1 (a), (b), (c)
	σ_c= 20 psi, m = 0.25. 0.50	Fig.2 (a), (b)
	σ_c= 30 psi, m = 0.25, 0.50, 0.75	Fig.3 (a), (b), (c)
	σ_c= 40 psi, m = 0	Fig.4
Clay-Y	σ_c= 2.5 psi, m = 0.25, 0.50, 0.75	Fig.5 (a), (b), (c)
	σ_c= 5.0 psi, m = 0.25, 0.50, 0.75	Fig.6 (a), (b), (c)
	σ_c= 10.0psi, m = 0.25, 0.50, 0.75	Fig.7 (a), (b), (c)
Kaolinite Clay	tests No.2, No.5, 15°	Fig.8 (a), (b)
	tests No.3, No.6, 37.5°	Fig.9 (a), (b)
	test No.7, 45°	Fig.10 (a), (b)
	tests No.8, No.11 58.25°	Fig.11 (a), (b)
	tests No.9, No.12 75°	Fig.12 (a), (b)
Ottawa Sand	P-1, SS σ_m= 10 psi, m=0.2	Fig.13 & Fig.22
	P-2, SS σ_m= 5 psi, m=0.5	Fig.14 & Fig.23
	P-3, SS σ_m= 20 psi, m=0.5	Fig.15 & Fig.24
	P-4, SS σ_m= 10 psi, m=0.8	Fig.16 & Fig.25
	P-5, RTC σ_x= 20 psi, m=0	Fig.17 & Fig.26
	P-6, RTE σ_x= σ_y= 20 psi, m=1	Fig.18 & Fig.27
	P-7 PL1 m=1	Fig.19 & Fig.28
	P-8 PL2 m=0	Fig.20 & Fig.29
	P-9 CSP	Fig.21 & Fig.30

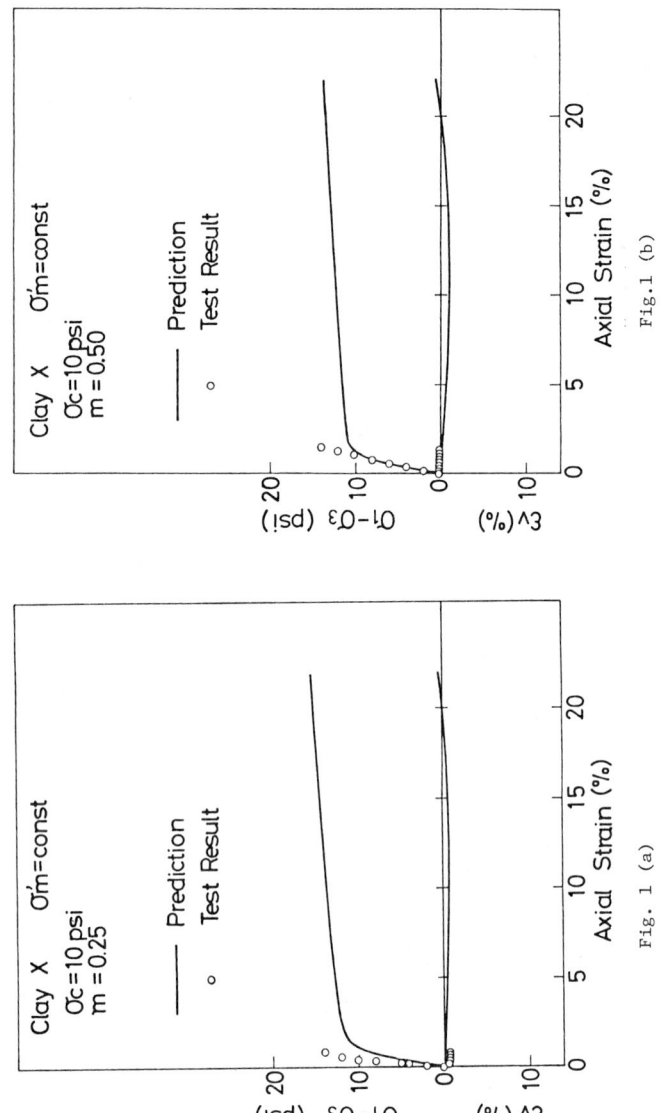

Fig. 1 (a)

Fig. 1 (b)

PREDICTION AND COMPARISON 419

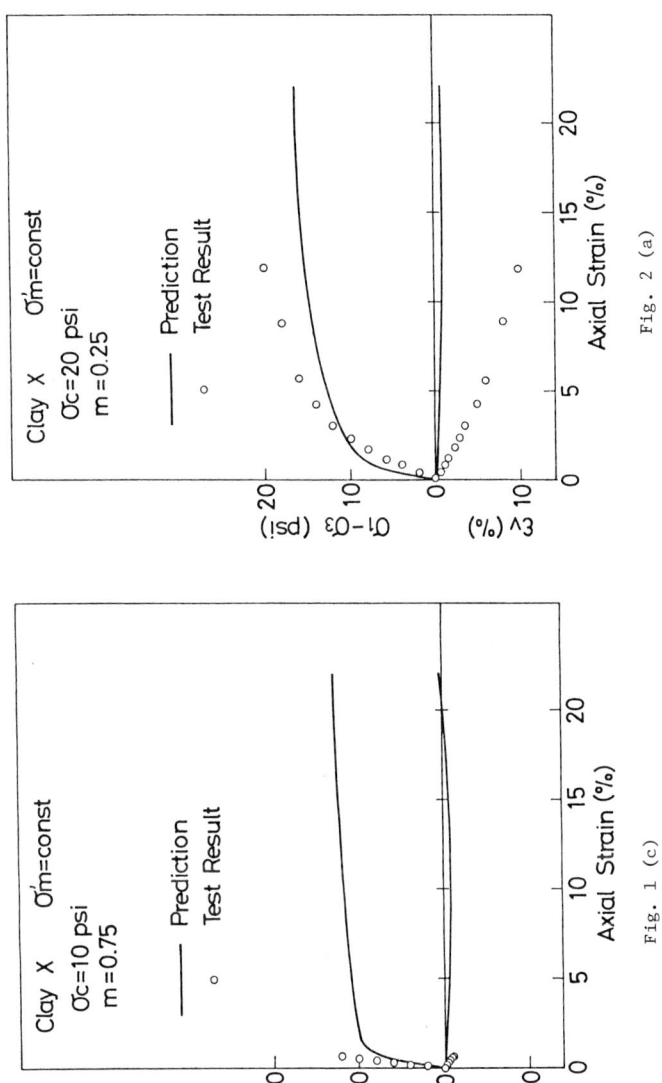

Fig. 1 (c)

Fig. 2 (a)

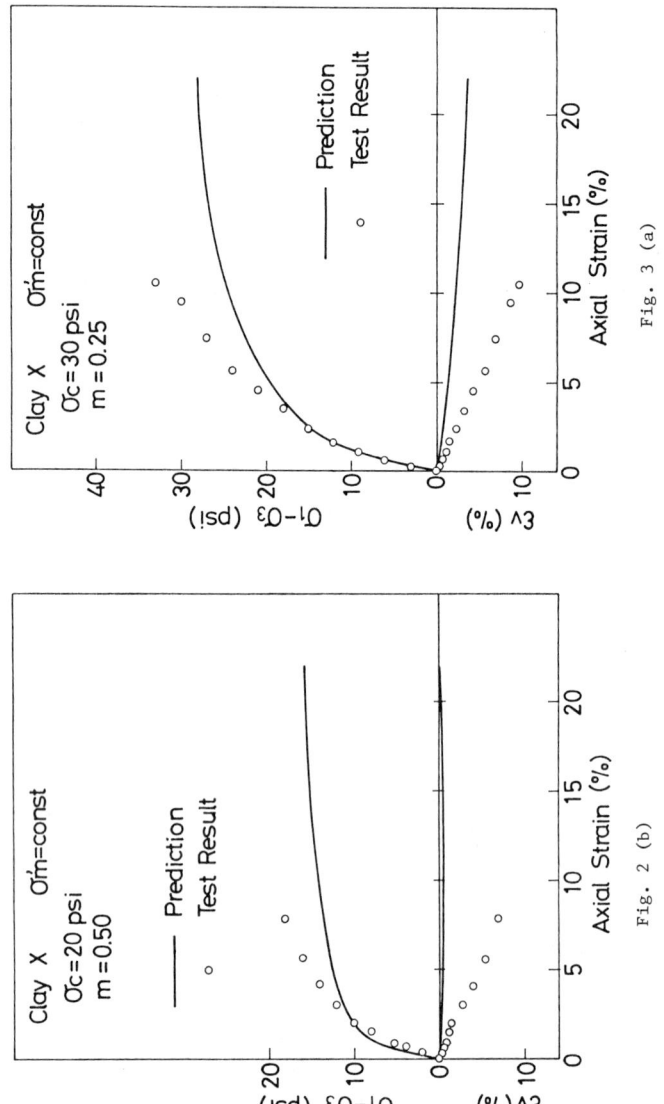

Fig. 2 (b)

Fig. 3 (a)

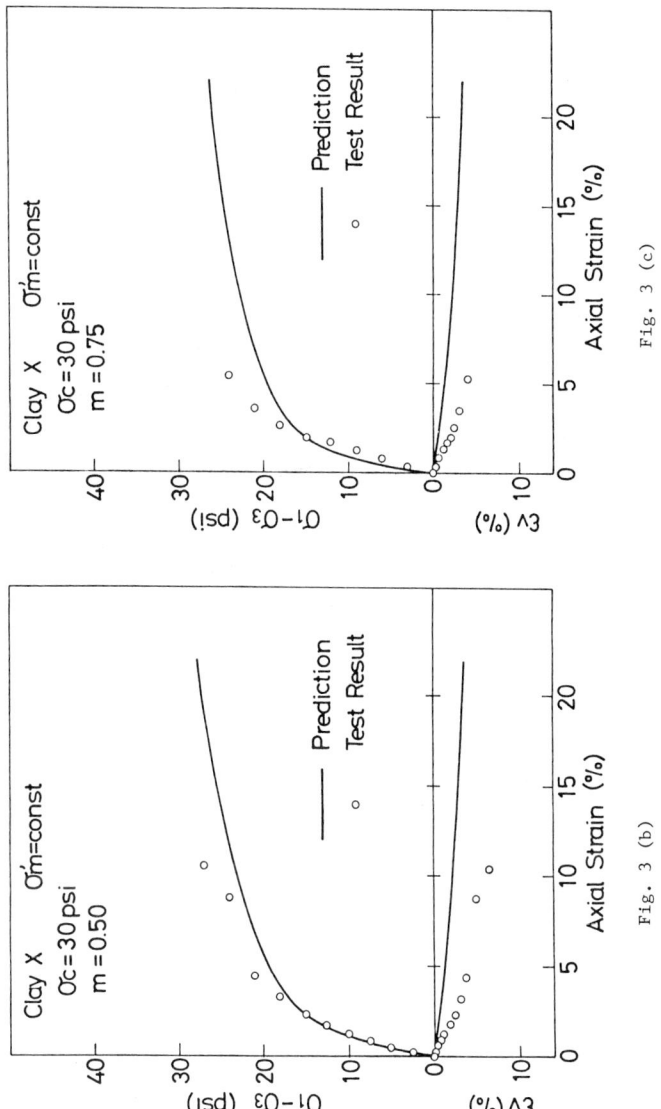

Fig. 3 (b)

Fig. 3 (c)

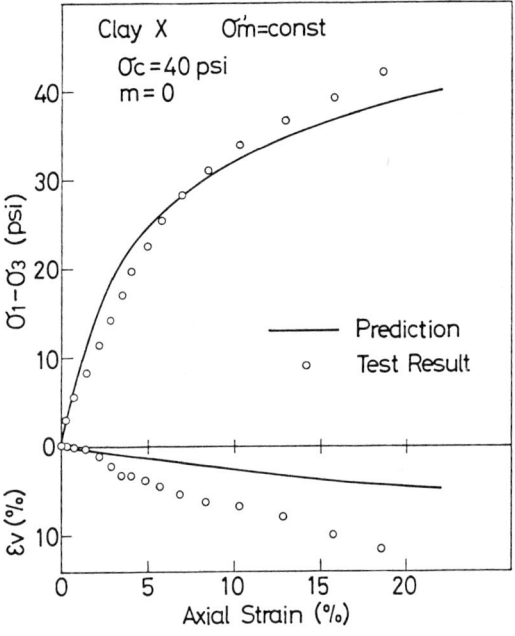

Fig. 4

PREDICTION AND COMPARISON

423

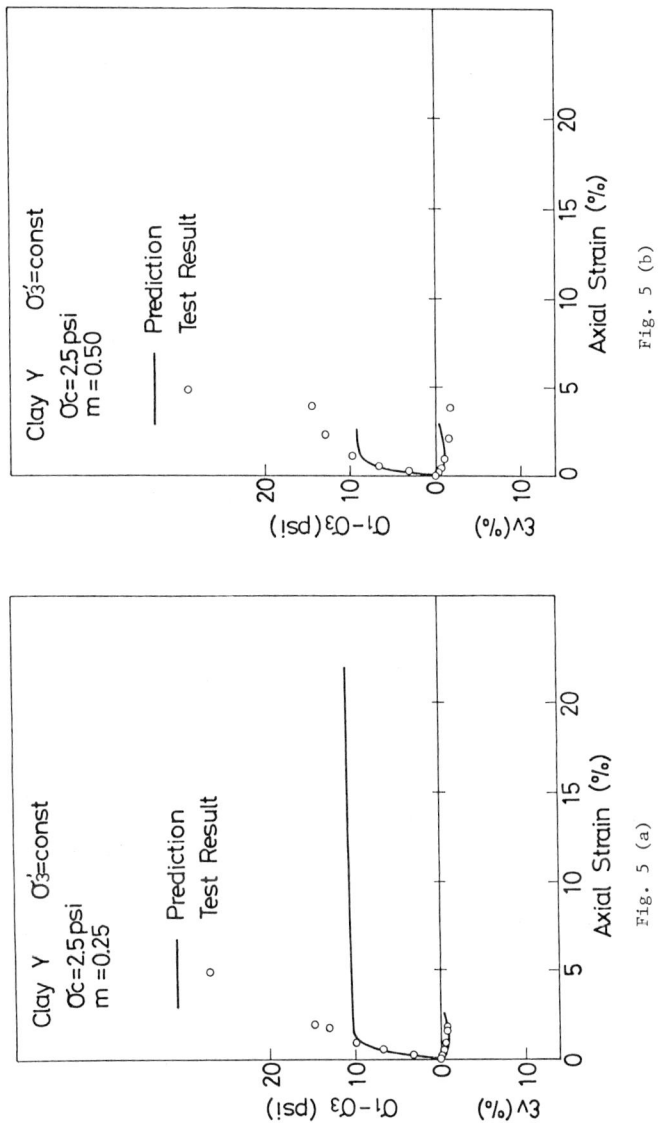

Fig. 5 (a)

Fig. 5 (b)

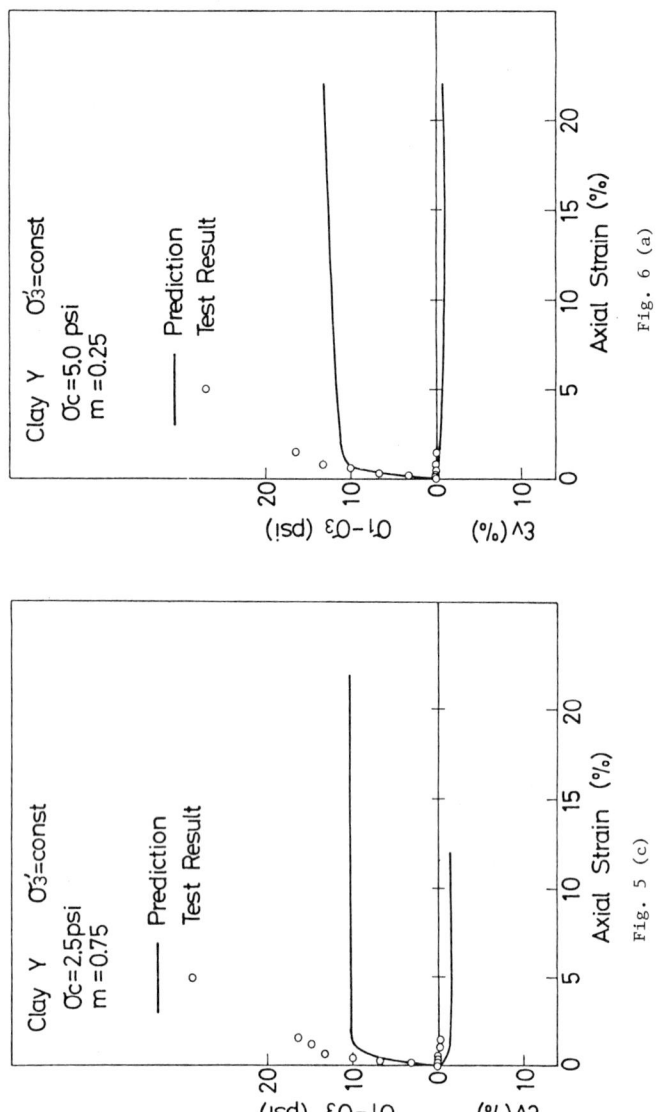

Fig. 6 (a)

Fig. 5 (c)

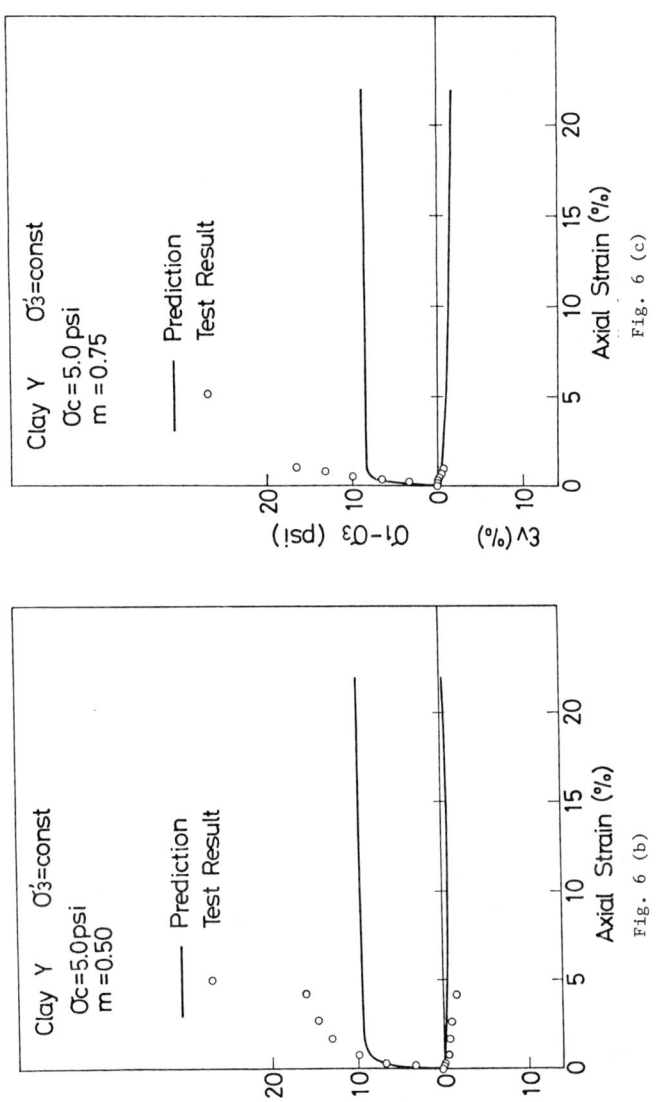

Fig. 6 (c)

Fig. 6 (b)

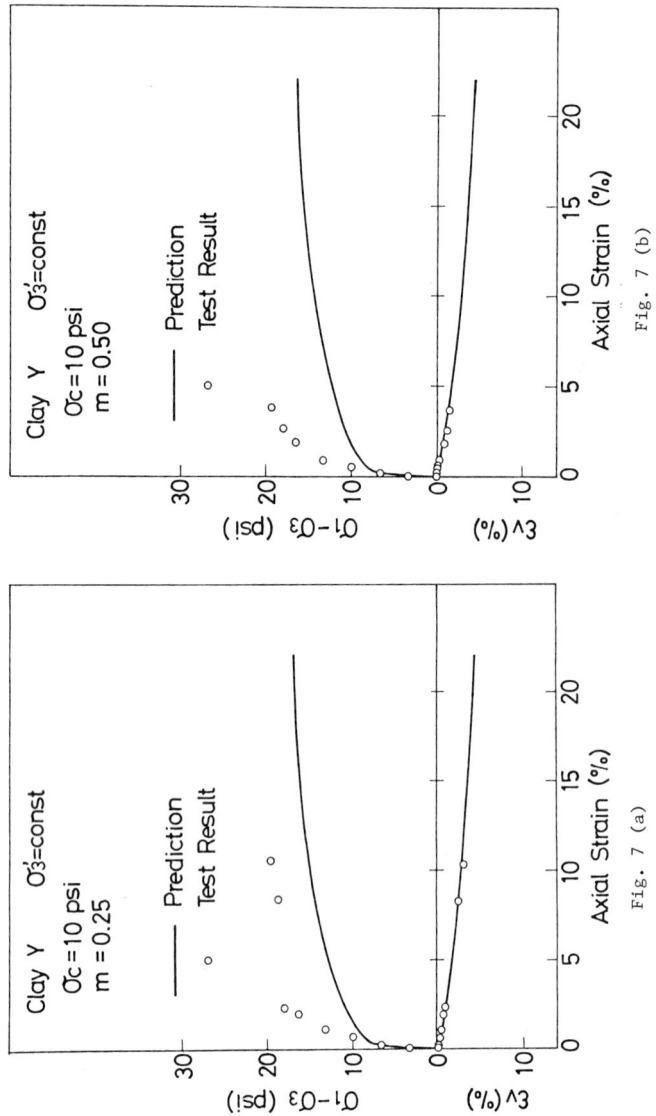

Fig. 7 (a)

Fig. 7 (b)

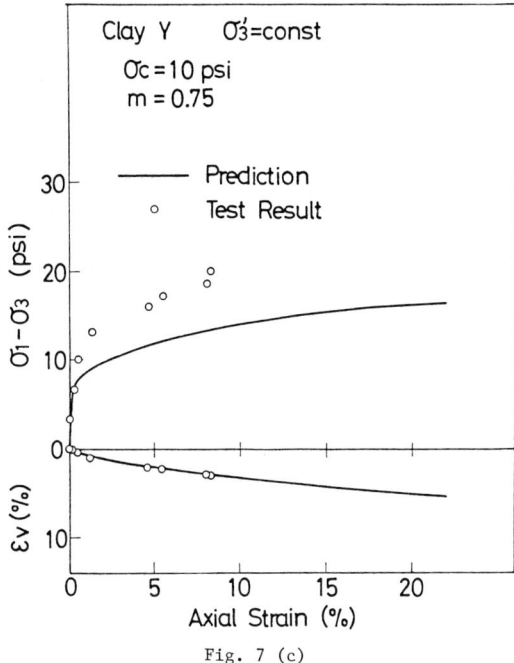

Fig. 7 (c)

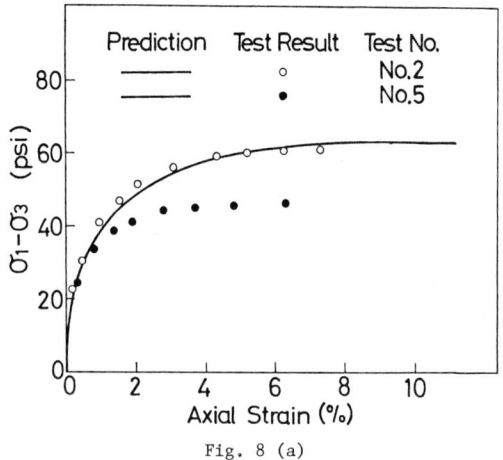

Fig. 8 (a)

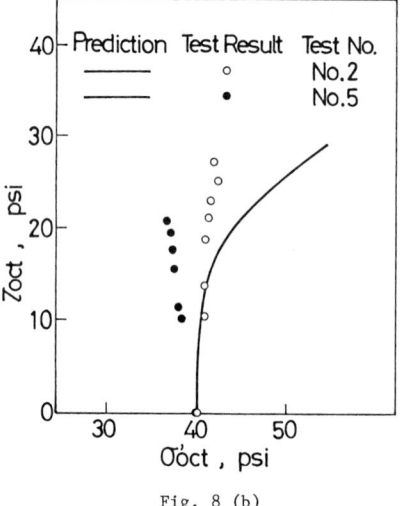

Fig. 8 (b)

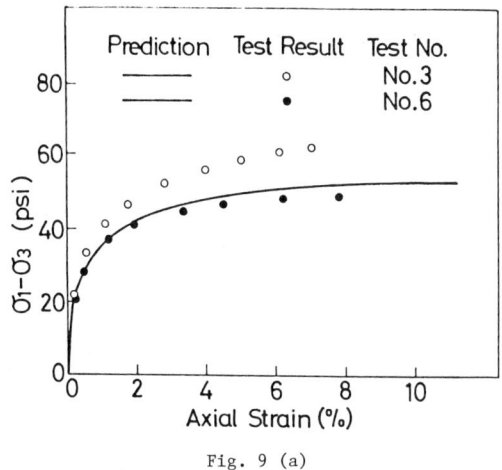

Fig. 9 (a)

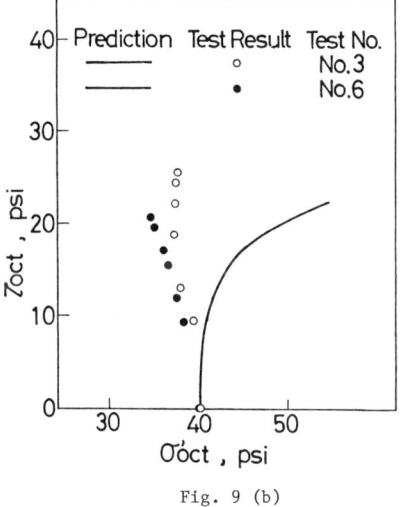

Fig. 9 (b)

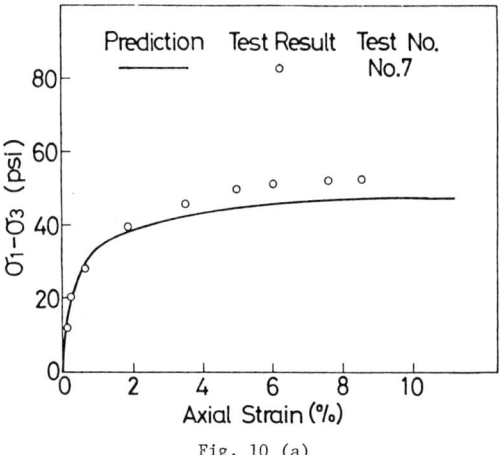

Fig. 10 (a)

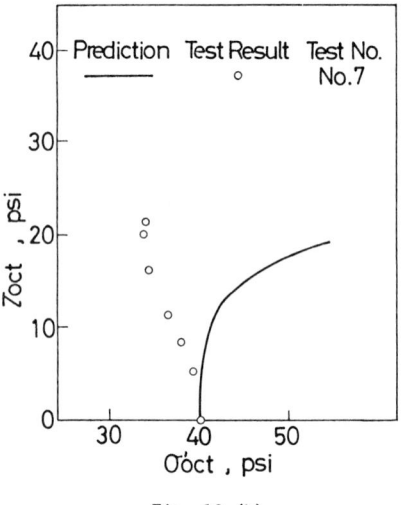

Fig. 10 (b)

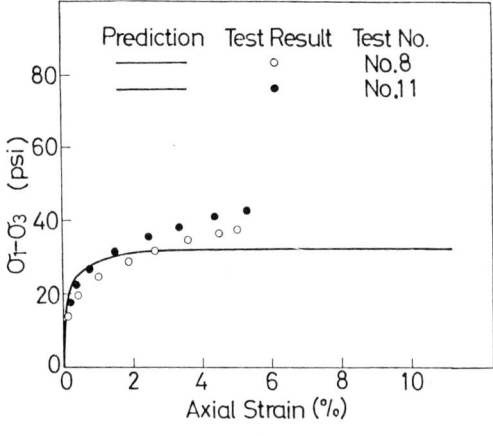

Fig. 11 (a)

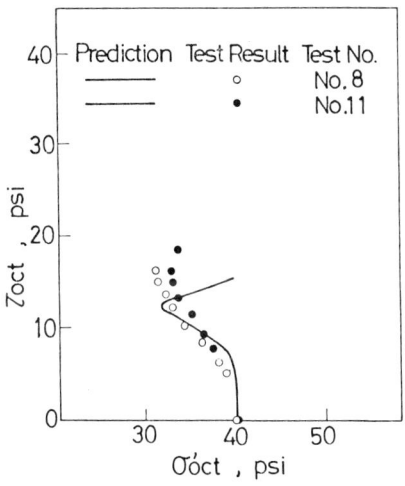

Fig. 11 (b)

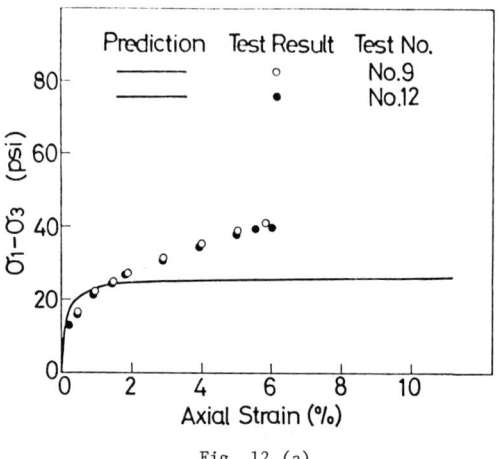

Fig. 12 (a)

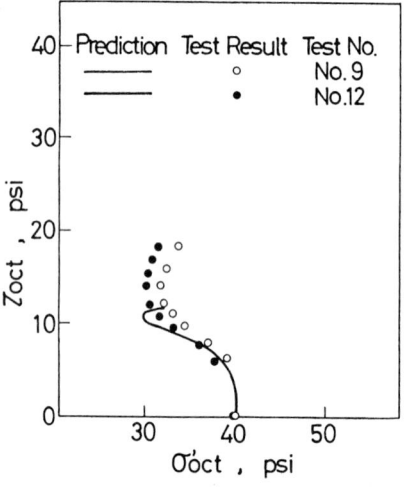

Fig. 12 (b)

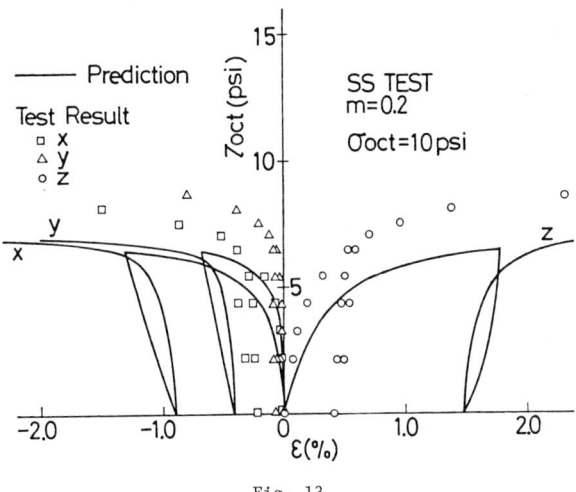

Fig. 13

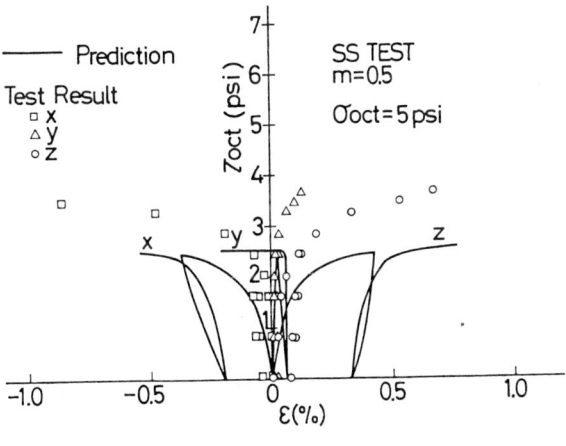

Fig. 14

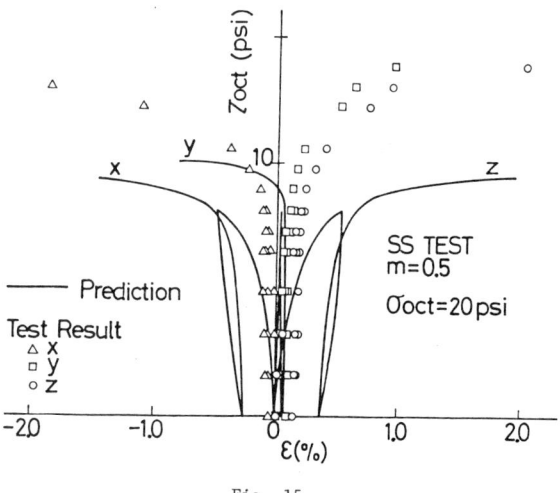

Fig. 15

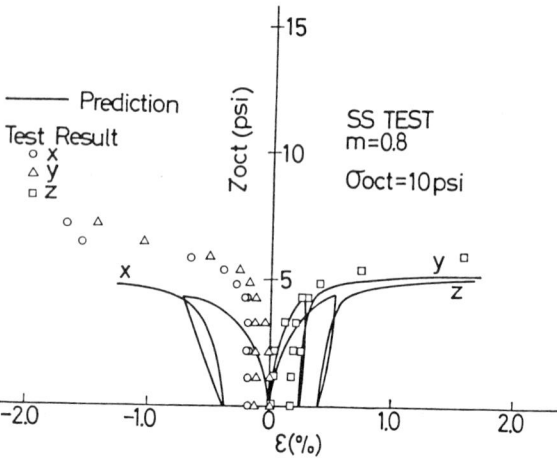

Fig. 16

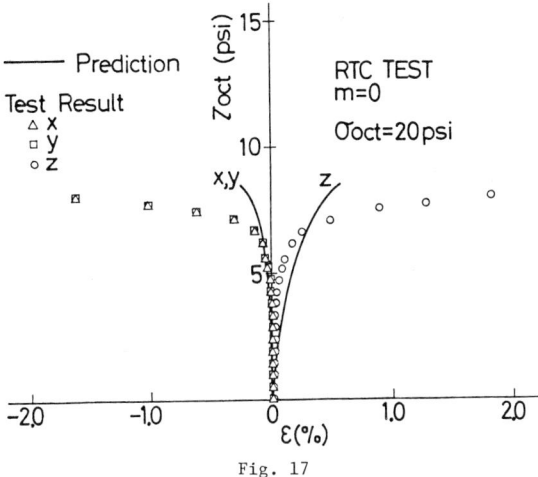

Fig. 17

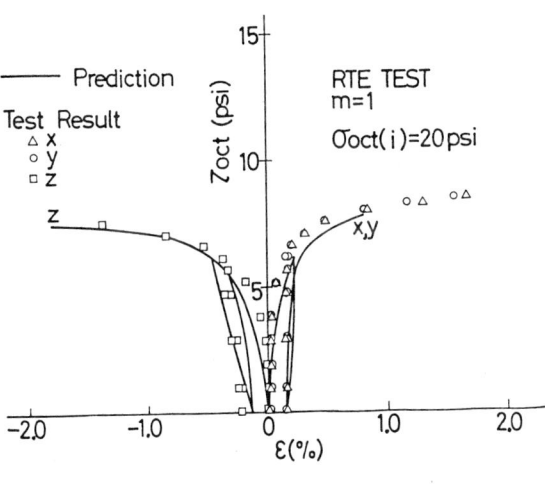

Fig. 18

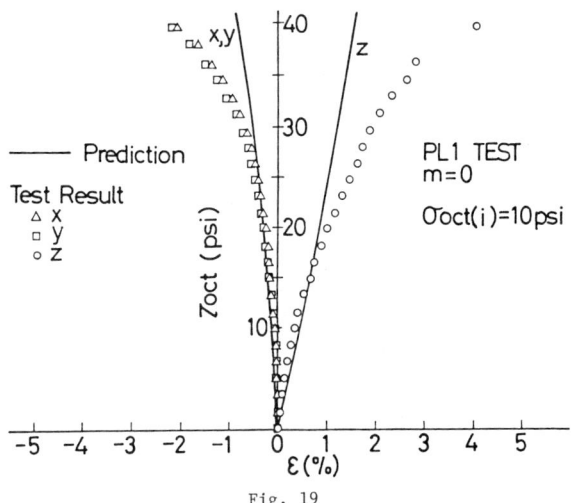

Fig. 19

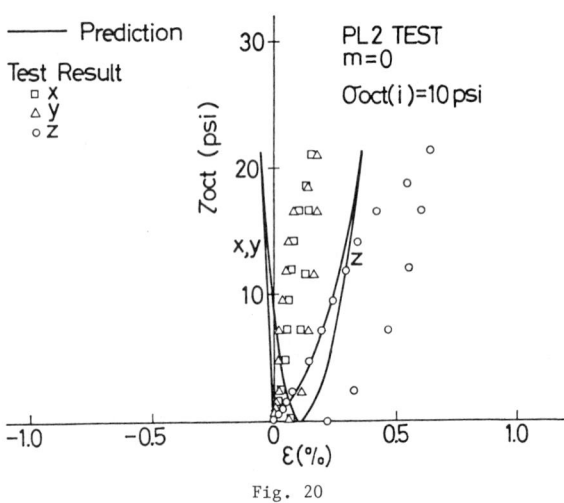

Fig. 20

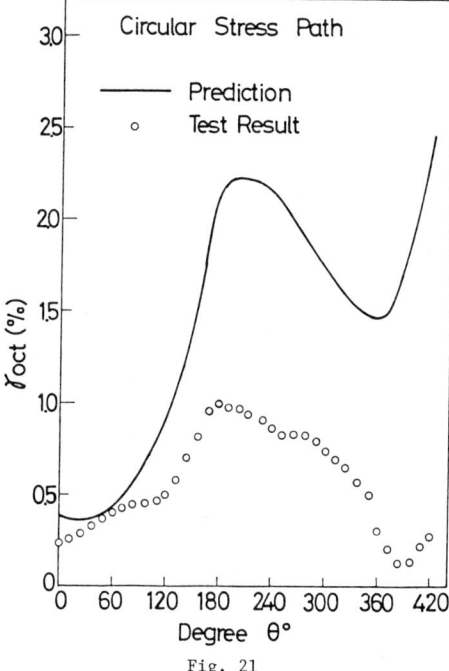
Fig. 21

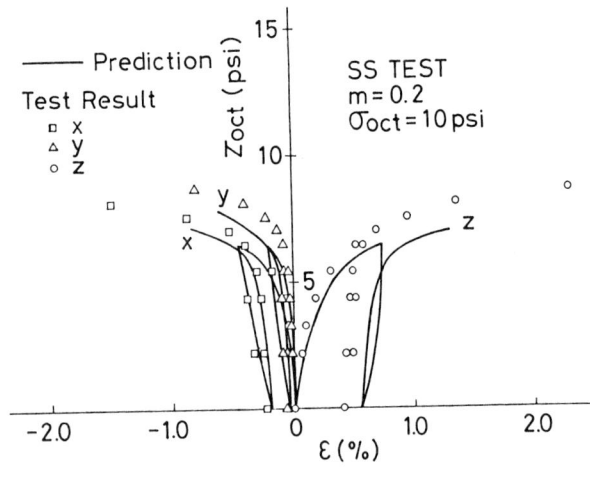

Fig. 22

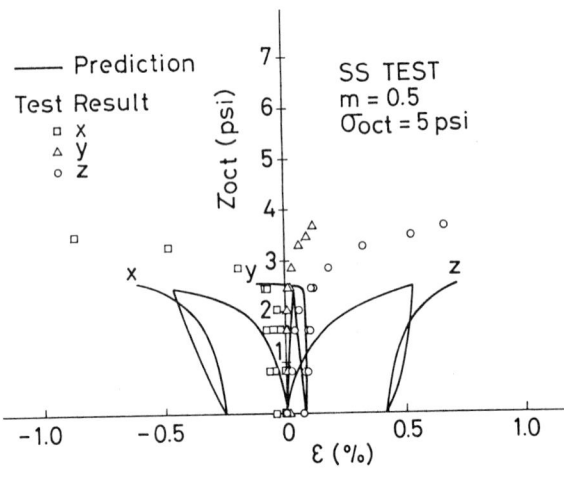

Fig. 23

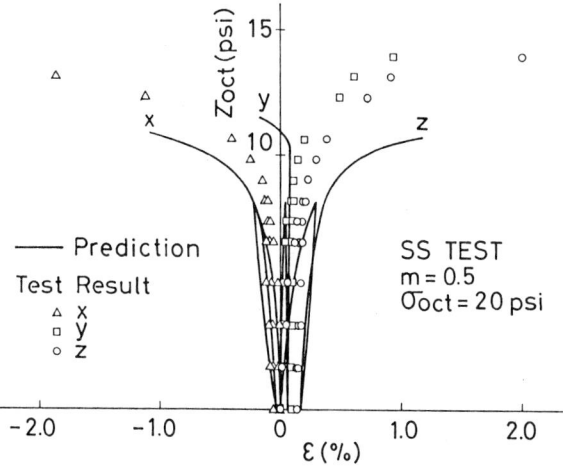

Fig. 24

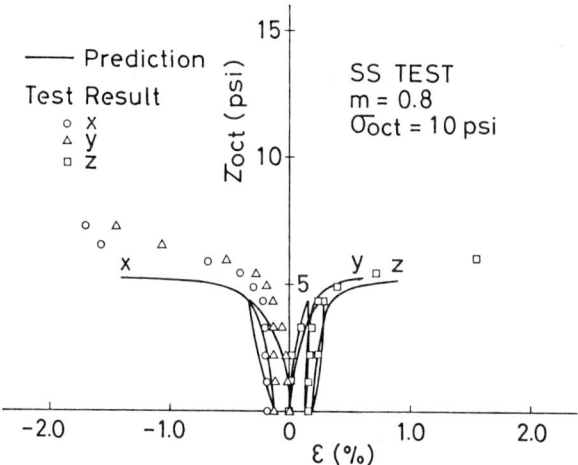

Fig. 25

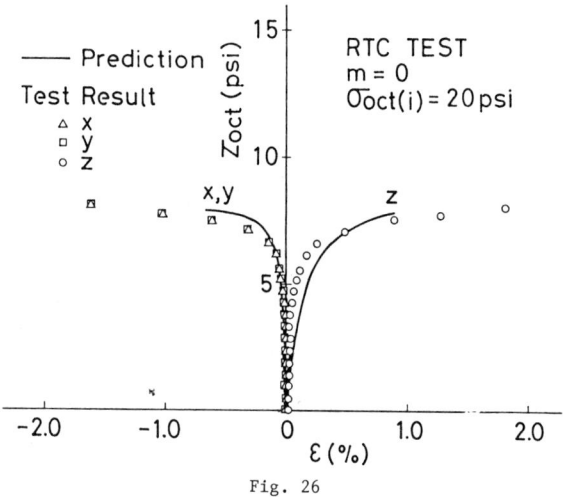

Fig. 26

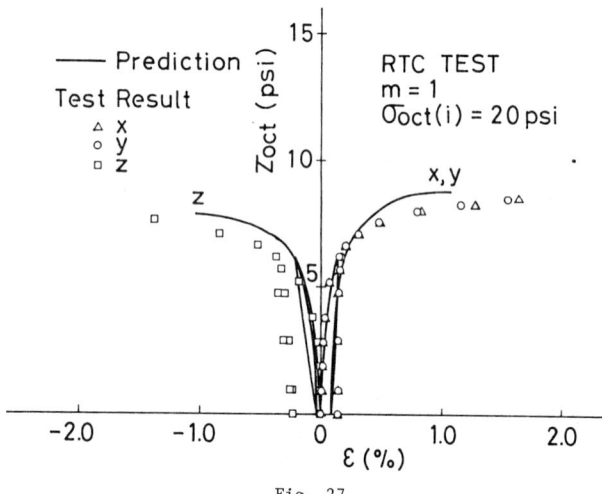

Fig. 27

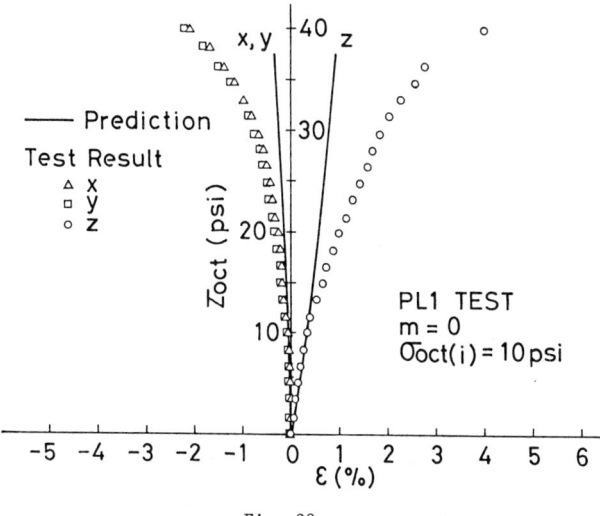

Fig. 28

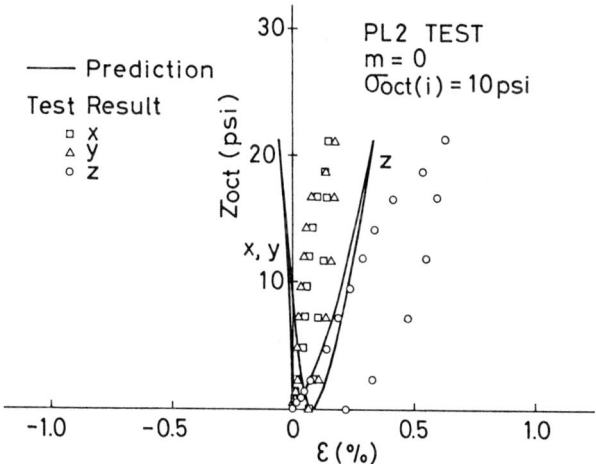

Fig. 29

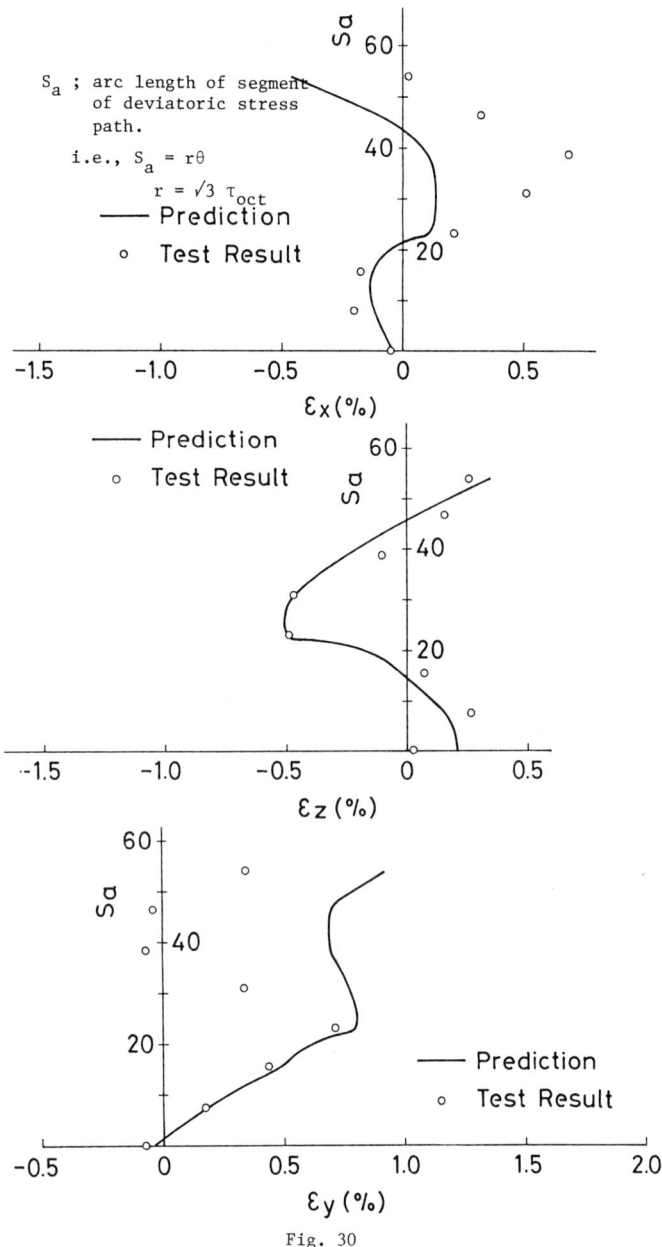

Fig. 30

HYPERBOLIC STRESS-STRAIN RELATIONSHIPS

by James M. Duncan,[1] M. ASCE

INTRODUCTION

The hyperbolic stress-strain relationship was developed for use in finite element analyses of stresses and movements in earth masses. In the ten years since its development, the model has been used in analyses of a large number of dams, of braced and open excavations, and a variety of types of soil-structure interaction problems.

In its original form, as described by Duncan and Chang (1970), the hyperbolic model employed tangent values of Young's modulus (E_t), which varied with the magnitudes of the stresses, and constant values of Poisson's ratio. More recently the accuracy of the model has been improved by using values of bulk modulus which vary with confining stress rather than constant values of Poisson's ratio. The Young's modulus relationships remain the same as described by Duncan and Chang (1970). The formulations in current use are described in the following section of this report.

The principal advantage of the hyperbolic model is its generality. It can be used to represent the stress-strain behavior of soils ranging from clays and silts through sands, gravels and rockfills. It can be used for partly saturated or fully saturated soils, and for either drained or undrained loading conditions in compacted earth materials or naturally-occurring soils. Experience with treating these various types of problems, and the accumulated background of stress-strain

---

[1] Prof. of Civ. Engrg., Univ. of California, Berkeley, CA

parameter values for a wide variety of soils, provide a useful base for further applications.

STRESS-STRAIN RELATIONSHIPS

The hyperbolic stress-strain relationships were developed for incremental analyses of soil deformations where nonlinear behavior is modeled by a series of linear increments. The relationship between stress and strain is assumed to be governed by the generalized Hooke's Law of elastic deformations. For plane strain conditions this relationship may be expressed in terms of Young's modulus and bulk modulus as follows:

$$\begin{Bmatrix} \Delta\sigma_x \\ \Delta\sigma_y \\ \Delta\tau_{xy} \end{Bmatrix} = \frac{3B}{9B - E} \begin{bmatrix} (3B + E) & (3B - E) & 0 \\ (3B - E) & (3B + E) & 0 \\ 0 & 0 & E \end{bmatrix} \begin{Bmatrix} \Delta\varepsilon_x \\ \Delta\varepsilon_y \\ \Delta\gamma_{xy} \end{Bmatrix} \qquad (1)$$

in which $\Delta\sigma_x$, $\Delta\sigma_y$ and $\Delta\tau_{xy}$ are stress increments; $\Delta\varepsilon_x$, $\Delta\varepsilon_y$ and $\Delta\gamma_{xy}$ are strain increments; B is bulk modulus; and E is Young's modulus.

By varying the values of Young's modulus and bulk modulus appropriately as the stresses vary within the soil, it is possible using the simple equation (1) to model three important characteristics of the stress-strain behavior of soils, namely, nonlinearity, stress-dependency, and inelasticity. The procedures used to account for these characteristics are described in the following paragraphs.

Nonlinear Stress-Strain Curves Represented by Hyperbolas. Kondner and his co-workers (Kondner, 1963; Kondner and Zelasko, 1963) have shown that the stress-strain curves for a number of soils could be approximated reasonably accurate by hyperbolas like the one shown in Fig. 1. This hyperbola can be represented by an equation of the form:

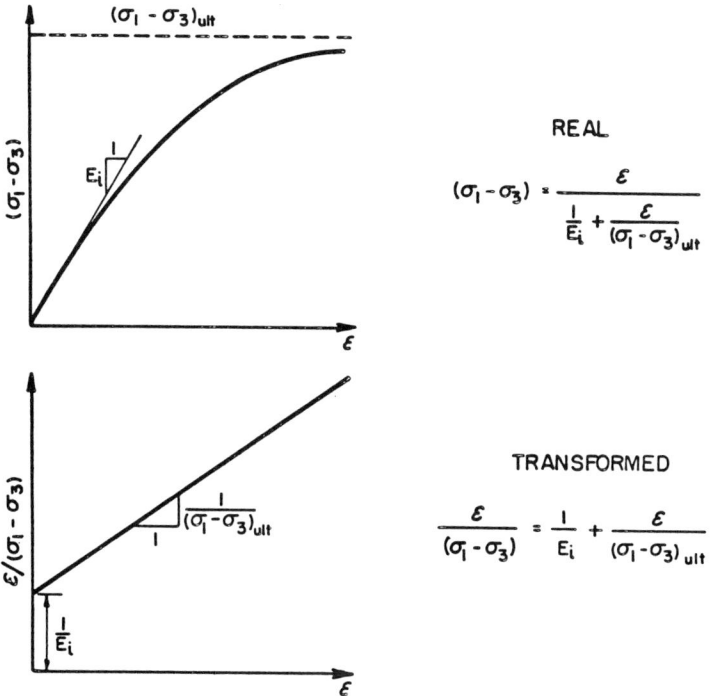

Fig. 1 HYPERBOLIC REPRESENTATION OF A STRESS-STRAIN CURVE

$$(\sigma_1 - \sigma_3) = \frac{\varepsilon}{\dfrac{1}{E_i} + \dfrac{\varepsilon}{(\sigma_1 - \sigma_3)_{ult}}} \qquad (2)$$

While other types of curves could also be used, these hyperbolas have two characteristics which make their use convenient:

(1) The parameters which appear in the hyperbolic equation have physical significance. E_i is the initial tangent modulus or initial slope of the stress-strain curve, and $(\sigma_1 - \sigma_3)_{ult}$ is the asymptotic value of stress difference which is related closely to the strength of the soil. The value of $(\sigma_1 - \sigma_3)_{ult}$ is always greater than the compressive strength of the soils, as discussed subsequently.

(2) The values of E_i and $(\sigma_1 - \sigma_3)_{ult}$ for a given stress-strain curve can be determined easily. If the hyperbolic equation is transformed as shown in the lower part of Fig. 1, it represents a linear relationship between $\varepsilon/(\sigma_1 - \sigma_3)$ and ε. Thus, to determine the best-fit hyperbola for the stress-strain curve, values of $\varepsilon/(\sigma_1 - \sigma_3)$ are calculated from the test data and are plotted against ε. The best-fit straight line on this transformed plot corresponds to the best-fit hyperbola on the stress-strain plot.

When data from actual tests are plotted on the transformed plot, the points frequently are found to deviate from the ideal linear relationship. The data for stiff soils, such as dense sands, usually plot on a mild curve which is concave upward, whereas the data for soft soils, such as loose sands, usually plot on a mild curve which is concave downward. Experience with several hundred stress-strain curves for well over a hundred different soils indicates that a good match is usually achieved by selecting the straight line so that it passes through the points where 70% and 95% of the strength are mobilized (Duncan and

Chang, 1970; Kulhawy, et al., 1969). Thus, in practice, only two points for each stress-strain curve (the 70% point and the 95% point) are plotted on the transformed diagram.

<u>Stress Dependent Stress-Strain Behavior Represented by Varying E_i and $(\sigma_1 - \sigma_3)_{ult}$ with Confining Pressure</u>. For all except fully saturated soils tested under unconsolidated-undrained conditions, an increase in confining pressure will result in a steeper stress-strain curve and a higher strength, and the values of E_i and $(\sigma_1 - \sigma_3)_{ult}$ therefore increase with increasing confining pressure. This stress-dependency is taken into account by using empirical equations to represent the variations of E_i and $(\sigma_1 - \sigma_3)_{ult}$ with confining pressure.

The variation of E_i with σ_3 is represented by an equation of the following form, which was suggested by Janbu (1963):

$$E_i = K p_a \left(\frac{\sigma_3}{p_a}\right)^n \tag{3}$$

The variation of E_i with σ_3 corresponding to this equation is shown in Fig. 2. The parameter K in equation (3) is the modulus number, and n is the modulus exponent. Both are dimensionless numbers. p_a is atmospheric pressure, introduced into the equation to make conversion from one system of units to another more convenient. The values of K and n are the same for any system of units, and the units of E_i are the same as the units of p_a. To change from one system of units to another it is only necessary to introduce the appropriate value of p_a in equation (3).

The variation of $(\sigma_1 - \sigma_3)_{ult}$ with σ_3 is accounted for as shown in Fig. 3 by relating $(\sigma_1 - \sigma_3)_{ult}$ to the compressive strength or stress difference at failure, $(\sigma_1 - \sigma_3)_f$, and then using the Mohr-Coulomb strength equation to relate $(\sigma_1 - \sigma_3)_f$ to σ_3. The values of $(\sigma_1 - \sigma_3)_{ult}$ and $(\sigma_1 - \sigma_3)_f$ are related by:

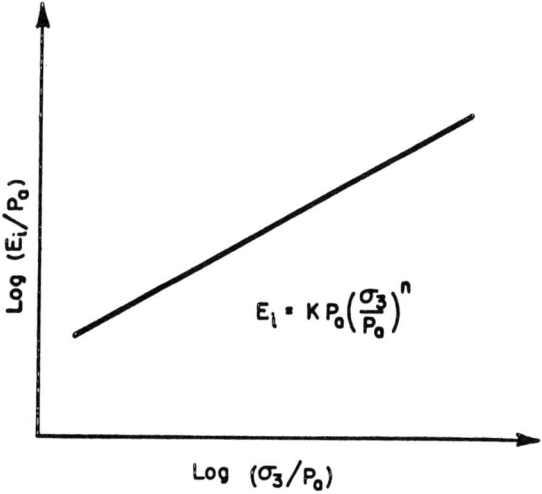

Fig. 2 VARIATION OF INITIAL TANGENT MODULUS WITH CONFINING PRESSURE

$$(\sigma_1 - \sigma_3)_f = R_f(\sigma_1 - \sigma_3)_{ult} \qquad (4)$$

in which R_f is the failure ratio. Because $(\sigma_1 - \sigma_3)_f$ is always smaller than $(\sigma_1 - \sigma_3)_{ult}$, the value of R_f is always smaller than unity, and varies from 0.5 to 0.9 for most soils.

The variation of $(\sigma_1 - \sigma_3)_f$ with σ_3 is represented by the familiar Mohr-Coulomb strength relationship, which can be expressed as follows:

$$(\sigma_1 - \sigma_3)_f = \frac{2c \cos\phi + 2\sigma_3 \sin\phi}{1 - \sin\phi} \qquad (5)$$

in which c and ϕ are the cohesion intercept and the friction angle, as shown in Fig. 3.

Relationship Between E_t and the Stresses. The instantaneous slope of the stress-strain curve is the tangent modulus, E_t. By differentiating equation (2) with respect to ε and substituting the expressions of equations (3), (4), and (5) into the resulting expression for E_t, the following equation can be derived:

$$E_t = \left[1 - \frac{R_f(1 - \sin\phi)(\sigma_1 - \sigma_3)}{2c \cos\phi + 2\sigma_3 \sin\phi}\right]^2 K p_a \left(\frac{\sigma_3}{p_a}\right)^n \qquad (6)$$

This equation can be used to calculate the appropriate value of tangent modulus for any stress conditions [σ_3 and $(\sigma_1 - \sigma_3)$] if the values of the parameters K, n, c, ϕ, and R_f are known.

Inelastic Behavior Represented By Use of Different Modulus Values for Loading and Unloading. If a triaxial specimen is unloaded at some stage during a test, the stress-strain curve followed during unloading is steeper than the curve followed during primary loading, as shown in Fig. 4. If the specimen is subsequently reloaded, the stress-strain

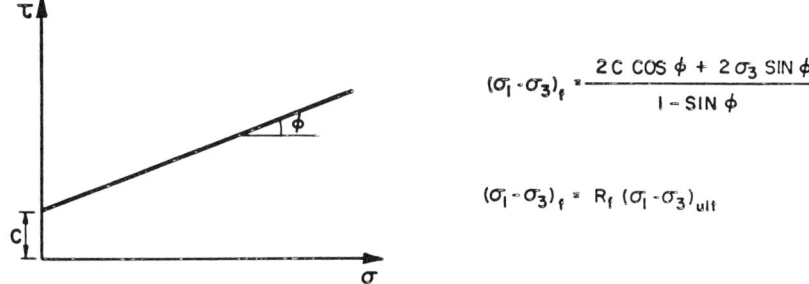

$$(\sigma_1 - \sigma_3)_f = \frac{2C \cos \phi + 2\sigma_3 \sin \phi}{1 - \sin \phi}$$

$$(\sigma_1 - \sigma_3)_f = R_f (\sigma_1 - \sigma_3)_{ult}$$

Fig. 3 VARIATION OF STRENGTH WITH CONFINING PRESSURE

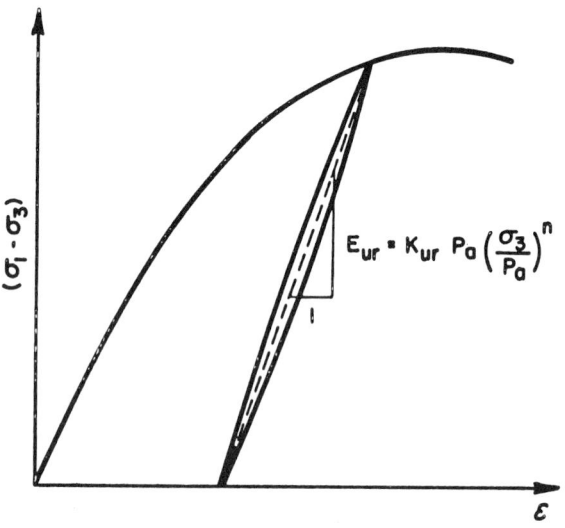

$$E_{ur} = K_{ur} P_a \left(\frac{\sigma_3}{P_a}\right)^n$$

Fig. 4 UNLOADING-RELOADING MODULUS

curve followed is also steeper than the curve for primary loading and is quite similar in slope to the unloading curve. Thus the soil behavior is inelastic, because the strains occurring during primary loading are only partially recoverable on unloading. On subsequent reloading there is always some hysteresis, but it is usually reasonably accurate to approximate the behavior during unloading-reloading stress changes as linear and elastic, in effect ignoring any hysteresis effects.

In the hyperbolic stress-strain relationships, the same value of modulus (E_{ur}) is used for both unloading and reloading. The value of E_{ur} is related to the confining pressure by an equation of the same form as equation (3):

$$E_{ur} = K_{ur} p_a \left(\frac{\sigma_3}{p_a}\right)^n \qquad (7)$$

In this equation K_{ur} is the unloading-reloading modulus number. The value of K_{ur} is always larger than the value of K (the modulus number for primary loading). K_{ur} may be 20% greater than K for stiff soils such as dense sands. For soft soils, like loose sands, K_{ur} may be three times as large as K. The value of the exponent n is always very similar for primary loading and unloading, and in the hyperbolic relationships it is assumed to be the same.

Nonlinear Volume Change Accounted for By Using Constant Bulk Modulus. Many soils exhibit nonlinear and stress-dependent volume change characteristics, as illustrated by the volume change curves shown in Fig. 5. The assumption that the bulk modulus of the soil is independent of stress level ($\sigma_1 - \sigma_3$) and that it varies with confining pressure provides reasonable approximations to the shapes of these volume change curves. Furthermore, the assumption that the bulk modulus is independent of stress level provides perhaps the best representation of soil behavior which is possible within the framework of incremental elasticity, because

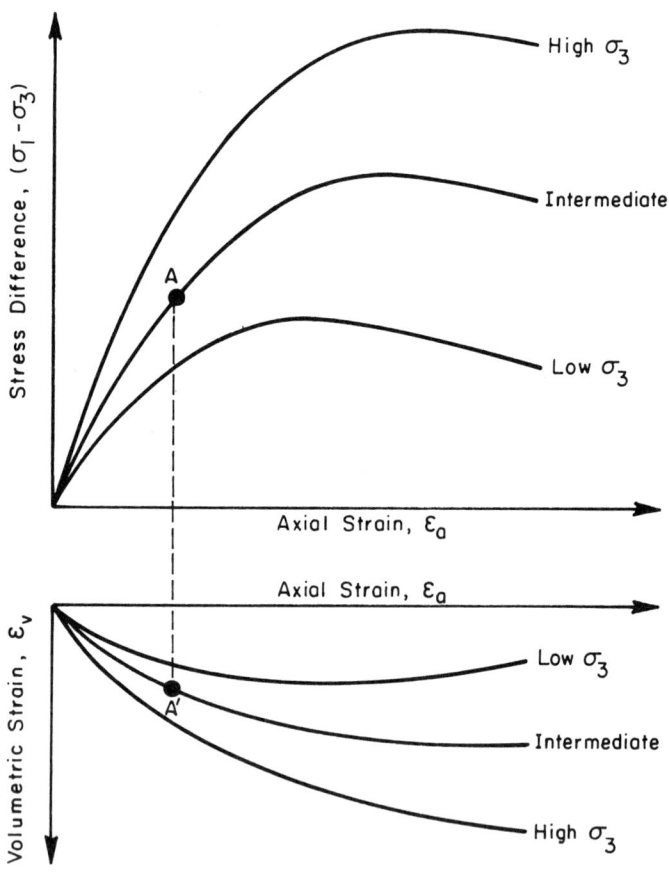

Fig. 5 NONLINEAR AND STRESS-DEPENDENT STRESS-STRAIN AND VOLUME CHANGE CURVES

it correctly reflects the fact that the response of the soil to changes in mean stress is virtually unaffected by the value of $(\sigma_1 - \sigma_3)$.

According to the theory of elasticity, the value of bulk modulus is defined by:

$$B = \frac{\Delta\sigma_1 + \Delta\sigma_2 + \Delta\sigma_3}{3\varepsilon_v} \quad (8)$$

in which B is the bulk modulus; $\Delta\sigma_1$, $\Delta\sigma_2$, and $\Delta\sigma_3$ are the changes in the values of the principal stresses, and $\Delta\varepsilon_v$ is the corresponding change in volumetric strain. For a conventional triaxial test, in which the deviator stress $(\sigma_1 - \sigma_3)$ increases from zero while the confining pressure is held constant, equation (8) may be expressed:

$$B = \frac{(\sigma_1 - \sigma_3)}{3\varepsilon_v} \quad (9)$$

The value of bulk modulus for a conventional triaxial compression test may be calculated using the value of $(\sigma_1 - \sigma_3)$ corresponding to any point on the stress-strain curve, such as point A in Fig. 5, and the corresponding point on the volume change curve (A').

Because real soils undergo some volume change as a result of changes in shear stress in addition to those caused by changes in normal stresses, the values of B calculated using equation (9) vary somewhat depending on which points on the stress-strain and volume change curves are employed in the calculation. Study of the volume change behavior of a wide variety of soils has led to the following criteria for selecting which points to use in calculating the value of B:

(1) If the volume change curve does not reach a horizontal tangent prior to the stage at which 70% of the strength is mobilized,

use the points on the stress-strain and volume change curves corresponding to a stress level of 70%.

(2) If the volume change curve does reach a horizontal tangent prior to the stage at which 70% of the strength is mobilized, use the point on the volume change curve where the curve becomes horizontal, and the corresponding point on the stress-strain curve.

Variation of B with Confining Pressure. When values of B are calculated for tests on the same soil at various confining pressures, the bulk modulus will usually be found to increase with increasing confining pressure. As shown in Fig. 6, the variation of B with confining pressure can be approximated by an equation of the form:

$$B = K_b \, p_a \left(\frac{\sigma_3}{p_a}\right)^m \qquad (10)$$

in which K_b is the bulk modulus number and m is the bulk modulus exponent, both of which are dimensionless. p_a is atmospheric pressure, expressed in the same units as σ_3 and B. For most soils the values of m vary between 0.0 and 1.0. In the case of undrained tests on clays compacted dry of optimum, values of m less than zero have been determined. Values of m less than zero correspond to a decrease in the value of B as the confining pressure increases. This apparently anomalous behavior is believed to result from breakdown in the structural arrangement of the soil particles due to the application of larger confining pressures.

Restrictions on the Range of Values of B. As the value of B approaches $E_t/3$, the corresponding value of Poisson's ratio approaches zero. Therefore in finite element computer programs, the values of Poisson's ratio may be restricted to positive values by using $B = E_t/3$ in cases where equation (10) indicates lower values. Similarly, by

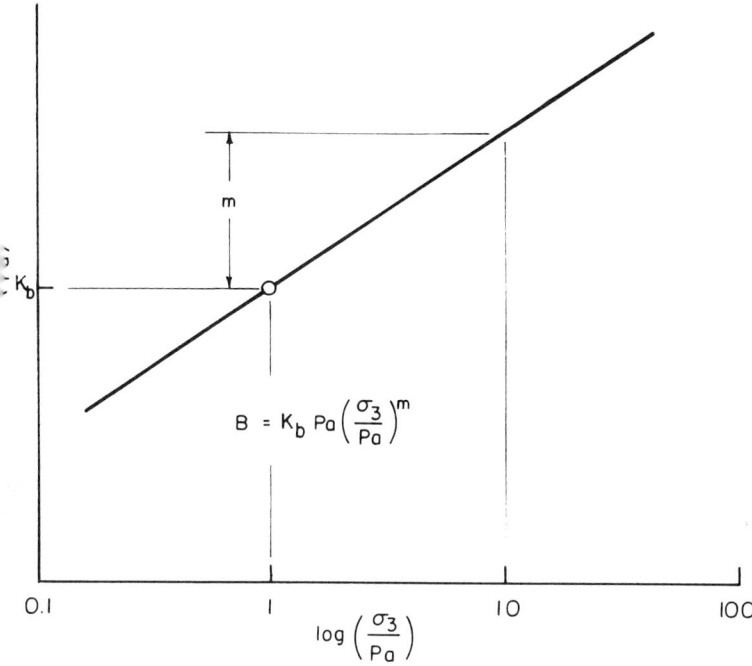

FIG. 6 VARIATION OF BULK MODULUS WITH CONFINING PRESSURE

TABLE 1. SUMMARY OF THE HYPERBOLIC PARAMETERS

Parameter	Name	Function
K, K_{ur}	Modulus number	Relate E_i and E_{ur} to σ_3
n	Modulus exponent	
c	Cohesion intercept	Relate $(\sigma_1 - \sigma_3)_f$ to σ_3
ϕ, $\Delta\phi$	Friction angle parameters	
R_f	Failure ratio	Relates $(\sigma_1 - \sigma_3)_{ult}$ to $(\sigma_1 - \sigma_3)_f$
K_b	Bulk modulus number	Value of B/P_a at $\sigma_3 = P_a$
m	Bulk modulus exponent	Change in B/P_a for ten-fold increase in σ_3

using $B = 17 E_t$ where equation (10) indicates higher values, the value of ν_t may be restricted to values less than or equal to 0.49.

Summary of Hyperbolic Parameters. In all, nine parameters are employed in the hyperbolic stress-strain relationships described in this report. These parameters and their functions within the relationships, are listed in Table 1.

Advantages and Limitations. The advantages of the hyperbolic stress-strain relationships stem from their simplicity and their previous wide use for a variety of different types of practical problems. The principal advantages of these relationships are:

(1) The values of the required parameters can be readily determined from the results of a series of conventional triaxial compression tests. In fact, if triaxial tests results are not available, the required parameter values can be determined from direct shear and consolidation tests, or they can be estimated from a knowledge of the soil type and in-situ density. Use of the model thus does not require more extensive or more exotic testing programs than those which are routinely performed for other soil engineering purposes. This is an important consideration, because the cost of a few unusual or unconventional laboratory tests may easily exceed the cost of the finite element analysis for which they are performed.

(2) The parameters of the model have readily understood physical significance. This is helpful because engineering judgment is enhanced when each element of the model is comprehensible in physical terms, and the effects of changes in the parameter values can be readily anticipated.

(3) Parameter values have been determined for many soils, under both drained and undrained conditions. These data are useful for estimating parameter values when only soil type and density are known, and for judging the correctness of laboratory test results by means of comparisons with results for similar soils.

The simple hyperbolic relationships have some significant limitations which should be understood by anyone who uses them:

(1) Being based on the generalized Hooke's Law, the relationships are most suitable for analysis of stresses and movements prior to failure. The relationships are capable of predicting accurately nonlinear relationships between loads and movements, and it is possible to continue the analyses up to the stage where there is local failure in some elements. However, when a stage is reached where the behavior of the soil mass is controlled to a large extent by the properties assigned to elements which have already failed, the results will no longer be reliable, and they may be unrealistic in terms of the behavior of real soils at and after failure. These relationships are not useful, therefore, for analyses extending up to the stage of instability of a soil mass. They are useful for predicting movements in stable earth masses.

(2) The hyperbolic relationships do not include volume changes due to changes in shear stress, or "shear dilatancy." They may therefore be limited in the accuracy with which they can be used to predict deformations in dilatant soils, such as dense sands under low confining pressures.

(3) The parameters are not fundamental soil properties, but only values of empirical coefficients which represent the behavior of the soil under a limited range of conditions. The values

of the parameters depend on the density of the soil, its
water content, the range of pressures used in testing, and
the drainage conditions. In order that the parameters will
be representative of the behavior of the soil in the field
condition, the laboratory test conditions must correspond
to the field conditions with regard to these factors.

DETERMINATION OF PARAMETER VALUES FOR WORKSHOP PREDICTIONS

Hyperbolic parameter values for the four soils (clay X, clay Y, remoulded kaolinite, and Ottawa sand) were determined using a process of repeated trial in order to make best use of all of the available data. Parameter values were estimated using data from the conventional tests, and these were then used to calculate strains for the non-conventional tests for which data were provided. Comparisons of the calculated and measured strains provided a means of evaluating the scatter in the test results and the degree of accuracy of the calculated strains. These comparisons were used as a basis for establishing the range of uncertainty in the predictions. In the case of clay X, for which no conventional triaxial test data were provided, the parameters were evaluated using the data for the constant mean normal stress compression tests, and were checked against the remaining test results.

The ranges of parameter values used in the predictions are listed in Table 2. The ranges of stiffness and strength parameters were selected so that the calculated stress-strain curves for the available tests covered the range of measured values.

SUMMARY

The hyperbolic model is a simple stress-strain relationship based on the concept of incrementally nonlinear elastic behavior. It is applicable to virtually any type of soil and to drained or undrained conditions. Experience in applying the hyperbolic model to analyses of dams, excavations and various types of soil-structure interaction

TABLE 2. HYPERBOLIC PARAMETER VALUES FOR WORKSHOP PREDICTIONS

Parameter	SOIL							
	Clay X		Clay Y		Kaolinite		Ottawa Sand	
	Average Value	Range	Average Value	Range	Average Value	Range	Average Value	Range
K	55	±25	165	±65	800 (θ = 0) 90 (θ = 90°)	±250 ± 30	400	±120
K_{ur}	not required	--	not required	--	not required	--	2000	--
n	0	--	0	--	0	--	0.85	--
c	0	--	8.0 psi	±1.5 psi	28.5 psi (θ = 0) 20.0 psi (θ = 90°)	±3.0 psi ±2.0 psi	0.5 psi	--
ϕ	30°	±6°	0	--	0	--	43°	±3°
R_f	0.95*	--	0.6	±0.2	0.72	±0.05	0.86	--
K_b	45	±20	80	±30	10,000	--	700	--
m	0	--	0	--	0	--	0.5	--

*For σ_c = 10 psi, R_f = 0.3

problems has shown that it is useful for calculating movements in stable earth masses, and is not suitable for predicting instability or collapse loads. Like any theory or hypothesis of soil behavior, its successful application requires the exercise of engineering judgment.

APPENDIX I.--REFERENCES

1. Duncan, J. M. and Chang, Y-Y. (1970) "Nonlinear Analysis of Stress and Strain in Soils," Journal of the Soil Mechanics and Foundations Division, ASCE, Vol. 96, No. SM5, September.

2. Janbu, Nilmar (1963) "Soil Compressibility as Determined by Oedometer and Triaxial Tests," European Conference on Soil Mechanics and Foundation Engineering, Wissbaden, Germany, Vol. 1, pp. 19-25.

3. Kondner, R. L. (1963) "Hyperbolic Stress-Strain Response: Cohesive Soils," Journal of the Soil Mechanics and Foundations Division, ASCE, Vol. 89, No. SM1, February, p. 115.

4. Kondner, R. L. and Zelasko, J. S. (1963) "A Hyperbolic Stress-Strain Formulation of Sands," Proceedings of the 2nd PanAmerican Conference on Soil Mechanics and Foundation Engineering, Vol. 1, Brazil, p. 289.

5. Kulhawy, F. H., Duncan, J. M., and Seed, H. B. (1969) "Finite Element Analysis of Stresses and Movements in Embankments During Construction," Report No. TE 69-4, Office of Research Services, University of California, Berkeley, California.

6. Wong, Kai S. and Duncan, J. M. (1974) "Hyperbolic Stress-Strain Parameters for Nonlinear Finite Element Analyses of Stresses and Movements in Soil Masses," Report No. TE 74-3, University of California, Berkeley, July.

STRESS-DEFORMATION PREDICTIONS

USING A GENERAL PHENOMENOLOGICAL MODEL

by

Edward Kavazanjian, Jr.,[*] *James K. Mitchell*,[**] *and Rudolph Bonaparte*[***]

POSITION PAPER FOR NSF/NSERC SYMPOSIUM ON CONSTITUTIVE
RELATIONSHIPS FOR SOILS

McGill University, May 28,29,30, 1980

INTRODUCTION

The constitutive model as originally developed by the authors (12) is a phenomenological model for the time-dependent stress-strain-time behavior of normally consolidated to lightly overconsolidated saturated cohesive soils of low to intermediate sensitivity. Tentative extensions of the model to consider other than triaxial stress states, the effect of anisotropy, and rotation of principal planes were developed for this symposium. The predictions made herein are used to evaluate for the first time the applicability of this extended model to anisotropic and stiff, sensitive soils.

The original model combined existing phenomenological relationships for various aspects of soil behavior within a consistent framework to form a comprehensive model for the time-dependent behavior of soft clays. Bjerrum's

---

[*]Assistant Professor, Dept. of Civil Engineering, Stanford University
Stanford, California
[**]Professor, Chairman, Dept. of Civil Engineering, University of California,
Berkeley, California
[***]Graduate Student, Research Assistant, Dept. of Civil Engineering,
University of California, Berkeley, California

one-dimensional compression model (4), hypothesizing separate time-independent (immediate) and time-dependent (delayed) components of deformation, provided this framework. Volumetric and deviatoric deformations were considered separately, though not independently.

The concepts of normalized soil properties (14,15) were used to describe deviatoric behavior and to extend Bjerrum's one-dimensional volumetric model to encompass all principal stress ratios. The deviatoric behavior of slightly overconsolidated soil was accounted for by normalizing the stress-strain curve with respect to the deviator stress at failure, and by hypothesizing a unique relationship between void ratio and deviator stress at failure (10). Time-dependent deformations were described by the Singh-Mitchell creep function (22) and the hypothesis of a constant coefficient of secondary compression, C_α (16). Changes in void ratio were related to changes in effective stress by assuming a unique relationship among void ratio, deviatoric stress level, octahedral effective stress, and time.

In the following pages are presented the essential elements of the original model, the extensions developed for this symposium, and predictions for the behavior of Clay X and Edgar Plastic Kaolin. A more complete description of the original model can be found in the June, 1980 issue of the Journal of Geotechnical Division of the American Society of Civil Engineers (13).

THE ORIGINAL MODEL

In his 1967 Rankine Lecture on "Engineering Geology of Normally Consolidated Marine Clays as Related to Settlements of Buildings," Bjerrum (3) postulated separate time-independent (immediate) and time-dependent (delayed) components of deformation. Because of the hydrodynamic lag, these components are not directly obtainable from the results of consolidation tests, and verification of Bjerrum's hypothesis is difficult. However, Garlanger (7) has shown that by using backfigured parameters, this concept successfully predicts aspects of the time-dependent behavior of soft clays observed in one-dimensional compression tests by Berre and Iverson (1) that were not otherwise explainable.

This hypothesis of distinct immediate and delayed contributions to deformation and the one-dimensional compression model with which it was presented provided the basic structure around which the authors' model is constructed. Deviatoric and volumetric components to the deformation were considered separately, though not independently. Each component is assumed to consist of an immediate, time-independent contribution and a delayed, time-dependent contribution. This concept can be described mathematically as follows:

$$\underline{\varepsilon} = \underline{\varepsilon}_V + \underline{\varepsilon}_D \qquad (1)$$

$$\underline{\varepsilon}_V = \underline{\varepsilon}_{V_i} + \underline{\varepsilon}_{V_d} \qquad (2)$$

$$\underline{\varepsilon}_D = \underline{\varepsilon}_{D_i} + \underline{\varepsilon}_{D_d} \qquad (3)$$

where $\underline{\varepsilon}$ is the general strain tensor, the subscripts V and D denote the volumetric and deviatoric strain tensor components, and the subscripts i and d denote the immediate and delayed contributions to the deformation.

Bjerrum's model described the immediate and delayed components of deformation for the special case of one-dimensional compression. In the authors' original model, existing phenomenological relationships were used to extend Bjerrum's one-dimensional model to provide the components of equations 1-3 for all triaxial stress states.

The one-dimensional compression behavior of a soil represents deformation at a unique principal stress ratio, $K' = \sigma'_3/\sigma'_1 = K_o$ or constant deviatoric stress level, $\overline{D} = (\sigma_1-\sigma_3)/(\sigma_1-\sigma_3)_f$ (assuming an undrained shear strength line parallel to the primary consolidation line on a void ratio versus log stress diagram). Ladd and Foott's concept of normalized soil properties (15) provides a means of extending Bjerrum's one-dimensional compression model to other stress levels.

If, as Rendulic suggested (20), an effective stress path for time-independent deformations represents a unique contour of water content, or void ratio, in a principal stress space or on a p'-q diagram, and if all such effective stress paths are similar (have the same shape) (Fig. 1), as suggested by normalized soil properties, then all contours of constant principal stress ratio (or constant deviatoric stress level) would be parallel on a void ratio-log effective stress plot (Fig. 2). This pattern of behavior is in substantial agreement with the void ratio-log stress relationship developed by Taylor (23) in 1948 (Fig. 3) and with the undrained shear strength line ($\overline{D} = 1.0$) parallel to the compression isochrones on Bjerrum's one-dimensional compression plot (Fig. 4).

The type of plot shown in Fig. 2 along with a normalized stress-strain curve completely describe the immediate, time-independent behavior of a soil based on stress-history, the applied stress, and either the void ratio or pore pressure. If the void ratio is known, the deviator stress at failure can be directly determined from Fig. 2. If total stresses and pore pressure are known, void ratio can be calculated based on octahedral effective stress and principal stress ratio, and then deviator stress at failure can be calculated. Knowing the deviator stress at failure, $\overline{D}$, the deviatoric stress level, can be calculated. Given the stress-history of the soil, deviatoric strains can be calculated from the normalized stress-strain curve (Fig. 5) once $\overline{D}$ is known. For the cases in which void ratio is given, e.g., undrained tests, pore pressure is calculated based on the assumed unique relationship among void ratio, deviatoric stress level, and octahedral effective stress. If the void ratio is unknown, e.g., drained tests, changes in void ratio are determined based on stress history and the change in effective stress.

For the predictions required for the symposium, only the time-independent immediate compression portion of the authors' model was required, although the primary contribution of our approach is that it includes the effect of time-dependent delayed deformations. To account for delayed deformations, the model assigns both volumetric and deviatoric "age" or "intrinsic time" variables to each element of soil. These ages describe the stress history of the soil, and thus are used to compute both delayed deformations and immediate deformations due to superposition of stress increments.

Delayed volumetric deformations are computed using C_α, the coefficient of secondary compression, as shown by equation 4.

$$\varepsilon_{V_d} = C_\alpha \log t_V \qquad (4)$$

t_V is the volumetric age of the soil element.

The value of C_α describes the spacing of the constant volumetric age isochrones on a void ratio-log stress plot for a constant deviatoric stress level such as Bjerrum's one-dimensional plot (Fig. 4).

Delayed deviatoric deformation components are computed with equation 5, the Singh-Mitchell creep function (22).

$$\varepsilon_{1D_d} = Ae^{\overline{\alpha}\overline{D}} t_D^{-m} \qquad (5)$$

ε_{1D_d} is the delayed component of the major principal strain. A, $\overline{\alpha}$, and m are constants determined from laboratory tests. t_D is the deviatoric age of the soil element. For an isotropic material, $\varepsilon_{2D_d} = \varepsilon_{3D_d} = \frac{1}{2} \varepsilon_{1D_d}$.

Volumetric soil behavior due to superposition of stress increments was determined in the same manner as in Bjerrum's one-dimensional model, using the concept of a quasi-preconsolidation pressure (19). Superposition rules for deviatoric behavior were fashioned along the same lines.

Using the components described above, the state of an element of soil is described by five variables. These state variables are void ratio, deviatoric stress level, octahedral effective stress, and volumetric and deviatoric age. If any four of these variables are known, the fifth can be computed.

To facilitate computations and to minimize the total number of model parameters, a series of simplifying assumptions were made in the original model. These simplifying assumptions included hyperbolic stress-strain curves, constant values of the virgin compression ratio C_c, the recompression ratio C_r, and the coefficient of secondary consolidation C_α, and a deviatoric unload-reload modulus equal to the initial tangent modulus for the soil. In total, fourteen parameters are required to describe the stress-strain-time behavior using this simplified general model. Table 1 indicates the number and type of tests required to determine each of the 14 model parameters.

References 11, 12, and 13 give more detailed information about the original model, along with discussion on justifications for the choice of components, simplifying assumptions, limitations, and experimental evidence of its validity.

THE EXTENDED MODEL

All but one of the predictions requested for this symposium required consideration of either the effect of the intermediate principal stress, σ_2, the effect of anisotropy, or both. To make these predictions, the original model was extended to consider these effects. The extensions to the model were developed in the same manner as the original model components in that existing soil behavior relationships were chosen on the basis of compatibility, simplicity, and a good fit with observed phenomena whenever possible. The predictions made for the symposium represent the first time the validity of the extended model has been investigated.

The original model established a unique relationship among void ratio, effective stress, and time using five state variables; void ratio, deviatoric stress level, octahedral effective stress, and deviatoric and volumetric age. In developing the extended model for this symposium, the authors assumed that only deviatoric stress level, $\bar{D}$, need be redefined to account for the effect of σ_2, but that both $\bar{D}$ and the corresponding deviatoric stress-strain and pore pressure-strain relationships need be redefined to account for anisotropic behavior.

In the original model, $\bar{D}$ was defined as $(\sigma_1-\sigma_3)/(\sigma_1-\sigma_3)_f$. This definition ignores the effect of σ_2 on stress level. To extend the model to general three-dimensional stress states, $\bar{D}$ was redefined to include the effect of σ_2.

Based on the established selection criteria for model components, the ratio of the octahedral shear stress τ_{oct} to the octahedral shear stress at failure $(\tau_{oct})_f$ was chosen as the means of including the effect of σ_2 on $\overline{D}$. This ratio is not only a logical definition of stress level, but also reduces to the original definition of $\overline{D}$ in the special cases of triaxial stress states for which the original model was developed. However, in order to use $\tau_{oct}/(\tau_{oct})_f$ as the definition of $\overline{D}$, a method for defining $(\tau_{oct})_f$ as a function of void ratio and principal effective stresses had to be developed. The Mohr-Coulomb failure criterion cannot be used, as it was in the original model, to determine $(\tau_{oct})_f$ because it ignores the effect of σ_2.

The three-dimensional failure criterion developed by Lade and Duncan (17), $I'^3_1/I'_3 = K$, where I'_1 is the first principal effective stress invariant, I'_3 is the third principal effective stress invariant, and K is a constant, was chosen to relate octahedral shear stress at failure to octahedral normal effective stress. Figure 6 shows the projection of this criterion on the octahedral plane along with several other common failure criteria. Although Lade and Duncan developed this criterion for isotropic cohesionless soils, Lade and Musante (18) have shown it to work reasonably well for normally consolidated, isotropic, remolded cohesive soils under undrained test conditions. Comparisons performed by the authors on the results of K_o consolidated undrained triaxial compression and plane strain active tests for a number of natural "undisturbed" soils suggest this criterion may be valid for some natural soils as well.

Although there are other three-dimensional failure criteria which might have been used, time did not permit an evaluation of alternative choices. The Lade and Duncan criterion was chosen because it seemed reasonable and was compatible with the definition of stress level. Lade and Duncan defined stress level as a constant value of I'^3_1/I'_3. For an isotropically consolidated soil specimen, I'^3_1/I'_3 increases from a value of 27 at the start of a test up to the value of K at failure. If $m = \frac{\sigma_2 - \sigma_3}{\sigma_1 - \sigma_3}$ is constant, then $\tau_{oct}/(\tau_{oct})_f$ is uniquely related to I'^3_1/I'_3. Thus, for each constant value of m, a plot similar to Fig. 2 can be constructed with parallel contours of I'^3_1/I'_3 and/or $\tau_{oct}/(\tau_{oct})_f$.

Therefore, using the Lade and Duncan failure criterion and given effective principal stresses, the octahedral effective stress σ'_{oct} can be calculated, $\bar{D} = \tau_{oct}/(\tau_{oct})_f$ can be determined based on I'^3_1/I'_3, and the void ratio can be determined by the intersection of $\bar{D}$ and σ'_{oct} on the plot for the appropriate value of m. Given the void ratio and total stresses, $\bar{D}$ can be calculated based on τ_{oct} and an assumed unique relationship between $(\tau_{oct})_f$ and void ratio, and σ'_{oct} can be found from the intersection of $\bar{D}$ and void ratio.

Thus by generalizing $\bar{D}$ to $\tau_{oct}/(\tau_{oct})_f$, by hypothesizing a unique relationship between void ratio and $(\tau_{oct})_f$, and using Lade and Duncan's failure criterion $I'^3_1/I'_3 = K$, the original model was extended to consider the effect of σ_2 for isotropic soils.

Prediction of the behavior of anisotropic soils involved consideration of not only the variation of $(\tau_{oct})_f$ due to anisotropy in order to predict $\bar{D}$, but also the anisotropic variation in the parameters correlated to $\bar{D}$, i.e., the deviatoric pore pressures Δu_d and the deviatoric strain in the direction of σ_1, ε_{A_D}.

The conventional method of accounting for the effect of anisotropic strength variations is to fit a strength variation function to experimentally derived data points. Several such functions have been proposed (9). The most general of these require tests at three different principal stress orientations. Since the data provided for the symposium had tests at only two different orientations, the orientation of the principal anisotropic strength directions must be assumed in order to use these general formulations. A less general formulation proposed by Casagrande and Carillo (5) which assumes vertical and horizontal principal anisotropic strength directions was used in the extended model. Casagrande and Carillo assumed the undrained strength of the soil on a plane at an angle θ to the major principal strength plane was equal to

$$(S_u)_\theta = (S_u)_{min} + ((S_u)_{max} - (S_u)_{min}) \cos^2\theta \qquad (6)$$

where $(S_u)_{max}$ and $(S_u)_{min}$ are the maximum and minimum undrained strengths. The value of $(\tau_{oct})_f$ was substituted for S_u in this equation for purposes of the extended model.

Dunlop et al. (6) have developed an anisotropy function to describe the variation of the initial tangent modulus, E_i, with θ. This anisotropic modulus function takes the form of

$$\frac{E_\theta - E_0}{E_{90} - E_0} = \sin^2\theta + A_E \sin 2\theta \tag{7}$$

In this equation, E_0 and E_{90} are the initial tangent moduli of the vertically and horizontally oriented samples, respectively. A_e is a skewness parameter which equals zero for a soil in which E is symmetrical about $\theta = 45°$.

Hansen and Clough (9) have found that for values of $\overline{D}$ less than 0.6 to 0.8, the ratio of E_0/E_{90} is approximately constant. If E_0/E_{90} is assumed constant, and $A_E = 0$, then according to equation 7, ε_{AD}, the deviatoric strain in the σ_1 direction at any stress level $\overline{D}$ can be evaluated as

$$\varepsilon_{AD} = \varepsilon_0 / \left(1 + \left(\frac{\varepsilon_0}{\varepsilon_{90}} - 1\right) \sin^2\theta\right) \tag{8}$$

where ε_0 and ε_{90} are the deviatoric strains in the vertical and horizontally oriented samples respectively at the desired stress level.

The prediction of pore pressure changes during undrained tests are made by assuming that a change in pore pressure, Δu, is made up of two components: Δu_V, the change in pore pressure associated with a change in volumetric stress, and Δu_D, the change in pore pressure associated with a change in deviatoric stress. Mathematically, the relationship between Δu, Δu_V, and Δu_D is expressed as

$$\Delta u = \Delta u_V + \Delta u_D \qquad (9)$$

In the authors' model Δu_V is equal to the change in octahedral total stress. Δu_D is assumed to be a function of deviatoric stress level, effective confining pressure, and the overconsolidation ratio (OCR). For the tests on the Edgar Plastic Kaolin the OCR on any plane with a normal making an angle of θ with the vertical was defined as σ'_{max}/σ'_o, where σ'_{max} is the greatest normal effective stress the soil has been subjected to on the given plane and σ'_o is the effective isotropic confining pressure at the start of the undrained portion of the test. In the tests on the Edgar Plastic Kaolin, the OCR in the vertical direction was 2.08 while the OCR in the horizontal direction was 1.0. The OCR on planes between the horizontal and vertical can be determined from statics.

For the special case of K_o consolidation followed by rebound of the axial stress to a pressure equal to the cell pressure, the OCR on any plane having a normal at an angle of θ to the vertical can be calculated using the equation

$$\text{OCR}(\theta) = \frac{1}{K_o}\cos^2\theta + \sin^2\theta. \qquad (10)$$

If the direction of σ_1 during the undrained portion of the test makes an angle of θ with the vertical, the deviatoric pore pressure, Δu_D

at any value of $\bar{D}$ can be determined by

$$\Delta u_D = \Delta u_1 - \left[\frac{\Delta u_1 - \Delta u_{k_o}}{(1 - 1/k_o)}\right](1 - OCR^*) \quad (11)$$

where Δu_1 is the deviatoric pore pressure at an OCR of 1, Δu_{k_o} is the deviatoric pore pressure at an OCR of $1/k_o$, and OCR^* is the OCR on the plane normal to the direction of σ_1.

Therefore, the extended model describes the behavior of anisotropic soil by first applying an anisotropy strength factor to $(\tau_{oct})_f$ for an isotropic soil, and then applying anisotropy functions to the $\bar{D}$ versus Δu_D and $\bar{D}$ versus ε_A relationships determined from the results of undrained laboratory tests.

SYMPOSIUM PREDICTIONS

The original model, being restricted to triaxial stress states, was applicable to only one of the predictions requested for the symposium; the prediction of the drained behavior of Clay X at m = 0 and σ'_o = 40 psi from the results of a drained test at σ'_o = 30 psi with m = 0. To make this prediction, the immediate compression diagram for Clay X shown in Fig. 7, was developed. This diagram was generated by first drawing lines parallel to the virgin compression curve through the initial ($\bar{D}$ = 0, $I'_1{}^3/I'_3$ = 27) and final ($\bar{D}$ = 1, $I'_1{}^3/I'_3$ = 38) points from the test at σ'_o = 30 psi, m = 0.

For each point in the test at 30 psi, the value of e was computed from the volumetric strain, the value of $(\sigma_1-\sigma_3)_f$ was computed from e and the $\bar{D} = 1$ line, the value of $\bar{D}$ was calculated from $(\sigma_1-\sigma_3)_f$ and $(\sigma_1 - \sigma_3)$. Parallel contours of constant $\bar{D}$, or $I'_1{}^3/I'_3$, were then drawn through the points from the test at 30 psi and extended into the region of $\sigma'_{oct} = 40$ psi.

For each given set of applied stresses in the 40 psi test, the values of σ'_{oct} and $I'_1{}^3/I'_3$ were calculated, the state of the soil was located on the immediate compression plot, and the volumetric strain was calculated based on the new equilibrium void ratio. The major principal deviatoric strain was calculated from a normalized plot of $\bar{D}$ versus major principal deviatoric strain developed by subtracting one-third of the volumetric strain from the axial strain in the test at 30 psi.

The results of the prediction for the test on Clay X at $\sigma'_o = 40$ psi with $m = 0$ are presented in Fig. 8.

The remainder of the required predictions for Clay X afford the opportunity to evaluate the use of Lade and Duncan's failure criterion to redefine $\bar{D}$, taking into account the effect of σ_2. The results of the tests at $\sigma'_o = 20$ psi and at $\sigma'_0 = 30$ psi were predicted using the same procedure described above. The immediate compression plot shown in Fig. 7 was also used for these predictions. Volume change and deviatoric strain were calculated in exactly the same manner as for the test at $\sigma'_o = 40$ psi.

The predictions of the results of the tests on Clay X at σ'_o = 30 psi and at σ'_o = 20 psi are shown in Figs. 9 and 10, respectively. At both of these initial effective stresses, the value of $(I'^3_1/I'_3)_f$ from triaxial compression (m = 0) agreed quite well with the value of $(I'^3_1/I'_3)_f$ from triaxial extension (m = 1) and the soil was assumed to behave isotropically, even though the axial strain varied anisotropically

Because both ε_{1d} and e are uniquely related to $\overline{D}$ for a given value of σ'_o, the authors' model predicts a unique relationship between ε_{AD} and e_V, as seen in Figs. 9 and 10.

The value of $(I'^3_1/I'_3)_f$ varied significantly between m = 0 and m = 1 for Clay X at 10 psi. Because of this variation, Clay X behavior at this value of σ'_o must be considered using the anisotropic features of the extended model. Sufficient time was not available to develop the procedures required to predict drained, anisotropic behavior using the extended model. Therefore, no predictions were made for the tests on Clay X at σ'_o = 10 psi.

Predictions made for Clay Y required consideration of not only anisotropy and the effect of σ_2, but also the effects of sensitivity and overconsolidation. For this reason, the authors felt that predictions made using their model were not justifiable.

The anisotropic features of the extended model were used to predict the undrained behavior of the Edgar Plastic Kaolin. Since tests at only two different major principal stress orientations were available, the

Casagrande and Carillo two-parameter strength variation equation (equation 6) was used to predict the anisotropic variation of $(\tau_{oct})_f$. The strength variation obtained using this equation is plotted in Fig. 11, along with experimental results for Edgar Plastic Kaolin reported by Saada and Bianchini (21) and the best fit strength curve to Saada's data points using Bishop's (2) general three-parameter strength variation function. While it is likely that better agreement between observed and predicted results may have been obtained by using the Bishop curve, the authors did not feel justified in using Saada and Bianchini's data.

For each undrained test on Edgar Plastic Kaolin $(\tau_{oct})_f$ was calculated by applying the Casagrande and Carillo strength variation function to assess the effect of the orientation of the principal planes. Since tests were undrained and the void ratio remained constant, $(\tau_{oct})_f$ depended only on θ.

Once $(\tau_{oct})_f$ was calculated, $\overline{D}$ could be calculated knowing the applied stresses. Knowing $\overline{D}$, the deviatoric contributions to axial strain (strain in the σ_1 direction) and pore pressure could be calculated using the results of triaxial extension and compression tests and the appropriate directional property variation functions (equations 8 and 11).

The predicted deviatoric stress-strain behavior for the Edgar Plastic Kaolin is shown in Figs. 12a and 12b. The numbers on the figures refer to the test numbers from the data provided. Note that at present, the authors can suggest no phenomenological method for determining the ratio of the strain in the σ_2 direction to the strain in the σ_3 direction.

Furthermore, the major principal strain, ε_1, will generally not be in the same direction as the major principal stress in an anisotropic material, and thus ε_1 is not predicted by the authors' theory.

The predicted p-q and $\tau_{oct} - \sigma'_{oct}$ effective stress paths for the Edgar Plastic Kaolin are shown in Figs. 13 and 14.

SUMMARY AND CONCLUSIONS

In their original constitutive model, the authors developed a theory for the time-dependent stress-deformation behavior of cohesive soils under triaxial stress states. This original model was developed by extending the conventional hypothesis of a unique relationship between void ratio and effective stress to become a unique relationship among void ratio, effective stress, and time, or age. The model used two time variables (volumetric age and deviatoric age) and used octahedral effective stress and deviatoric stress level as the stress variables. Volumetric and deviatoric deformation components were computed separately, though not independently. Each deformation component was assumed to consist of time-independent and time dependent contributions. The immediate component of the original model was used to predict the behavior of Clay X at σ'_o = 40 psi and m = 0 from the results of a test at σ'_o = 30 psi and m = 0 for the symposium.

For the purposes of this symposium, the original model was extended to consider the effect of σ_2 and of anisotropy. Lade and Duncan's failure criterion was used to redefine $\overline{D}$ to account for the effect of

of σ_2. Directional variation functions were fit to the data to
account for the effect of anisotropy.

The extended model developed for this symposium represents a first
attempt by the authors at incorporating the effect of σ_2 and anisotropy
in their constitutive equations. The predictions made for the symposium
are the first set of predictions made using the extended model.

The authors believe that the predictions requested for this
symposium test only limited portions of any general constitutive model.
The ability to predict drained behavior from the results of drained tests
and undrained behavior from the results of undrained tests while accounting
for the effects of σ_2 and anisotropy is doubtless a good indication of
the reliability of a constitutive model. On the other hand, inability to
predict the effect of σ_2 or anisotropy does not completely invalidate
any model. Furthermore, the ability of various constitutive models to
predict isotropic soil behavior under triaxial or plane strain stress
states has not yet been clearly demonstrated in the opinion of the authors.
The ability to predict drained behavior from undrained tests (and vice-
versa), behavior under superimposed volumetric stress increments, and
time-dependent deformations must also be established before a model can
be acknowledged as generally acceptable.

The constitutive model developed by the authors is a phenomenological
model for cohesive soil behavior. The purpose of such a phenomenological
model is to elucidate what variables control the behavior of the soil and
in what manner. To predict the behavior of full scale engineering systems,

a mathematical constitutive model must be used within a numerical method. The phenomenological model provides a basis for choosing which mathematical model to use. The mathematical model which can describe the behavioral concepts of the phenomenological model most accurately and most efficiently should be selected for use in the numerical method. At this stage, the authors express no opinion of what type of mathematical model best embodies the concepts of their phenomenological model, and hope to gain valuable insight into this problem from the symposium.

Continued refinement of the authors' extended model, and selection of a mathematical model and development of a numerical method are planned for the near future.

The authors wish to thank the sponsors and organizers of the symposium for the opportunity to present their model and to participate as predictors. The problems posed for the predictors have stimulated the thinking of the authors about the effects of σ_2 and anisotropy on soil behavior. The authors expect that many valuable lessons will be learned as a result of the exercises performed for the symposium.

REFERENCES

- Berre, J. and Iverson, K. (1972) "Oedometer Tests with Different Specimen Heights on a Clay Exhibiting Large Secondary Compression," Geotechnique, Vol. 22, No. 1, pp. 53-70, March.

- Bishop, A. W. (1966) "The Strength of Soils as Engineering Materials," Sixth Rankine Lecture, Geotechnique, Vol. 16, No. 2, pp. 91-130, June.

- Bjerrum, L. (1967) "Engineering Geology of Normally-Consolidated Marine Clays as Related to Settlements of Buildings," Seventh Rankine Lecture, Geotechnique, Vol. 17, No. 2, pp. 82-118, June.

- Bjerrum, L. (1973) "Problems of Soil Mechanics and Construction on Soft Clays and Structurally Unstable Soils," General Report, Session 4, Proceedings, 8th ICSMFE, Vol. 3, pp. 111-159.

- Casagrande, A. and Carillo, N. (1944) "Shear Failure of Anisotropic Soils," Contributions to Soil Mechanics, 1941-1953, Boston Society of Civil Engineers.

- Dunlop, P., Duncan, J. M. and Seed, H. B. (1968) "Finite Element Analyses of Slopes in Soil," Contract Report S-68-6, U.S. Army Engineers Waterways Experiment Station, Corps of Engineers, Vicksburg, M.S., May.

- Garlanger, J. E. (1972) "The Consolidation of Soils Exhibiting Creep Under Constant Effective Stress," Geotechnique, Vol. 22, No. 1, pp. 71-78, March.

- Hansen, L. A. (1980) "Prediction of the Behavior of Braced Excavations in Anisotropic Clay," Ph.D. Dissertation, Stanford University, Stanford, CA.

- Hansen, L. A. and Clough, G. W. (1980) "Characterization of the Undrained Anisotropy of Clays," paper submitted for publication to the ASCE Journal of the Geotechnical Division, March.

- Henkel, D. J. (1960) "The Relationship Between Effective Stresses and Water Content in Saturated Soils," Geotechnique, Vol. 10, No. 1, pp. 41-54, February.

- Kavazanjian, Jr., E. (1978) Ph.D. Dissertation, University of California, Berkeley, CA.

- Kavazanjian, Jr., E. and Mitchell, J. K. (1977) "A General Stress-Strain-Time Formulation for Soils," Specialty Session 9, Ninth International Conference on Soil Mechanics and Foundation Engineering, Tokyo.

- Kavazanjian, Jr., E. and Mitchell, J. K. (1980) "Time-Dependent Deformation of Clays," Journal of the Geotechnical Division, ASCE, Vol. 106, No. GT6, June.

14. Ladd, C. C. (1971) "Strength Parameters and the Stress-Strain Behavior of Saturated Clays," Dept. of Civil Engineering, M.I.T., Report R71-23, Soils Publication 278.

15. Ladd, C. C. and Foott, R. (1974) "New Design Procedure for Stability of Soft Clays," Journal of the Geotechnical Division, ASCE, Vol. 100, No. GT7, pp. 763-786, July.

16. Ladd, C. C. and Preston, W. B. (1965) "On the Secondary Compression of Saturated Clays," Research in Earth Physics, Phase Report No. 6, Dept. of Civil Engineering, M.I.T., Research Report R65-59.

17. Lade, P. V. and Duncan, J. M. (1975) "Elastoplastic Stress-Strain Theory for Cohesionless Soil," Journal of the Geotechnical Division, ASCE, Vol. 101, No. GT10, pp. 1037-1053, October.

18. Lade, P. V. and Musante, H. M. (1976) "Three-Dimensional Behavior of Normally Consolidated Cohesive Soil," School of Engineering and Applied Science, Research Report UCLA-ENG-7626, University of California, Los Angeles, April.

19. Leonards, G. A. and Ramiah, B. K. (1960) "Time Effects in the Consolidation of Clays," ASTM Special Technical Publication 254, pp. 116-130

20. Rendulic, L. (1936) "Pore-Index and Pore-Water Pressure," Bauingenier, Vol. 17, p. 559.

21. Saada, A. S. and Bianchini, G. F. (1975) "Strength of One-Dimensionally Consolidated Clay," Journal of the Geotechnical Division, ASCE, Vol. 101, No. GT11, November.

22. Singh, A. and Mitchell, J. K. (1968) "General Stress-Strain-Time Function for Soils," Journal of the Soil Mechanics and Foundation Division, ASCE, Vol. 94, No. SM1, pp. 21-46, January.

23. Taylor, D. W. (1948) <u>Theoretical Soil Mechanics</u>, John Wiley and Sons, New York.

TABLE 1 THE FOURTEEN MODEL PARAMETERS

PARAMETER	SYMBOL	TYPE OF TEST
VIRGIN COMPRESSION INDEX	C_c	ISOTROPIC OR ANISOTROPIC CONSOLIDATION (IC or AC)
SECONDARY COMPRESSION INDEX	C_α	
RECOMPRESSION INDEX	C_R	
HYPERBOLIC STRESS-STRAIN PARAMETERS	a, b	ISOTROPICALLY CONSOLIDATED UNDRAINED TRIAXIAL COMPRESSION (CIU)
HYPERBOLIC PORE PRESSURE PARAMETERS	a_p, b_p	
SINGH-MITCHELL CREEP PARAMETERS	A, $\bar{\alpha}$, m	ISOTROPICALLY CONSOLIDATED UNDRAINED CREEP (ICC)
PRIMARY COMPRESSION VOID RATIO AT 1.0 KSC	Γ_p	ISOTROPIC OR ANISOTROPIC CONSOLIDATION (IC or AC)
FAILURE VOID RATIO AT $(\sigma_1 - \sigma_{3f}) = 1.0$ KSC	λ	CONSOLIDATED UNDRAINED TRIAXIAL TEST (CIU)
IMMEDIATE VOID RATIO AT 1.0 KSC	Γ_o	BACKFIGURED FROM CONSOLIDATION TEST (IC or AC) WITH TIMDEP
HYDRAULIC CONDUCTIVITY (PERMEABILITY)	k	

*For Remolded San Francisco Bay Mud

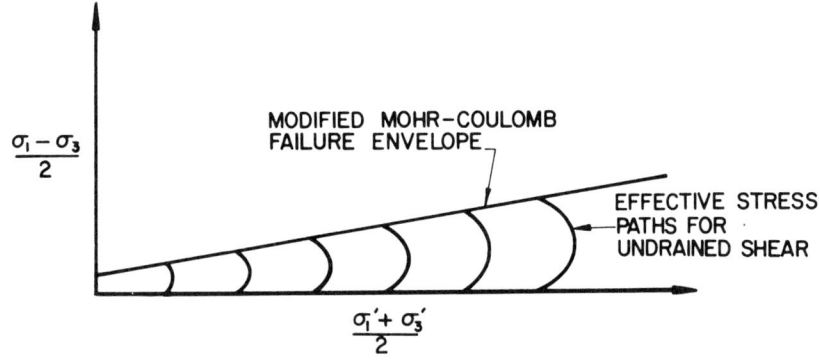

FIG. 1 SIMILAR EFFECTIVE STRESS PATHS

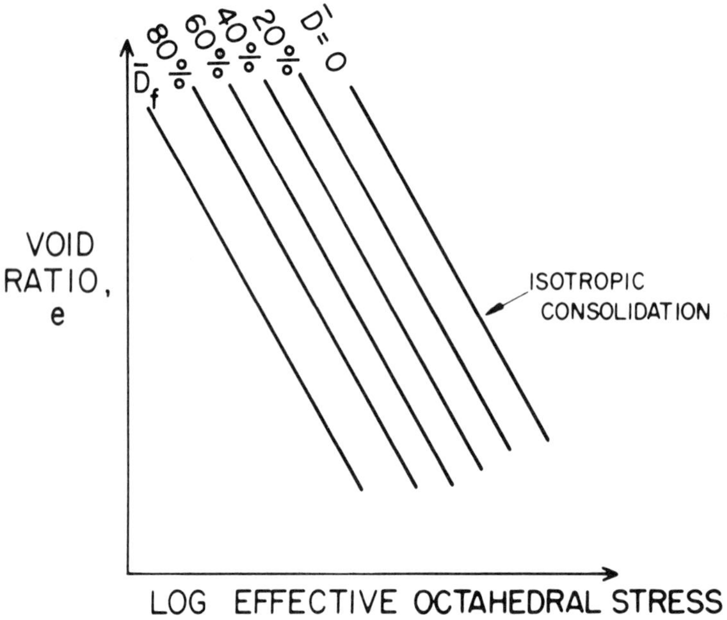

FIG. 2 IMMEDIATE COMPRESSION VOID RATIO-LOG EFFECTIVE STRESS CONTOURS FOR DIFFERENT PRINCIPAL STRESS PATHS

STRESS DEFORMATIONS

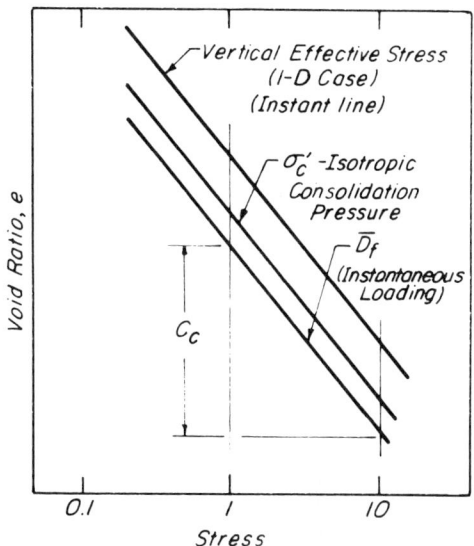

FIG. 3 TAYLOR'S VOID RATIO-LOG STRESS PLOT

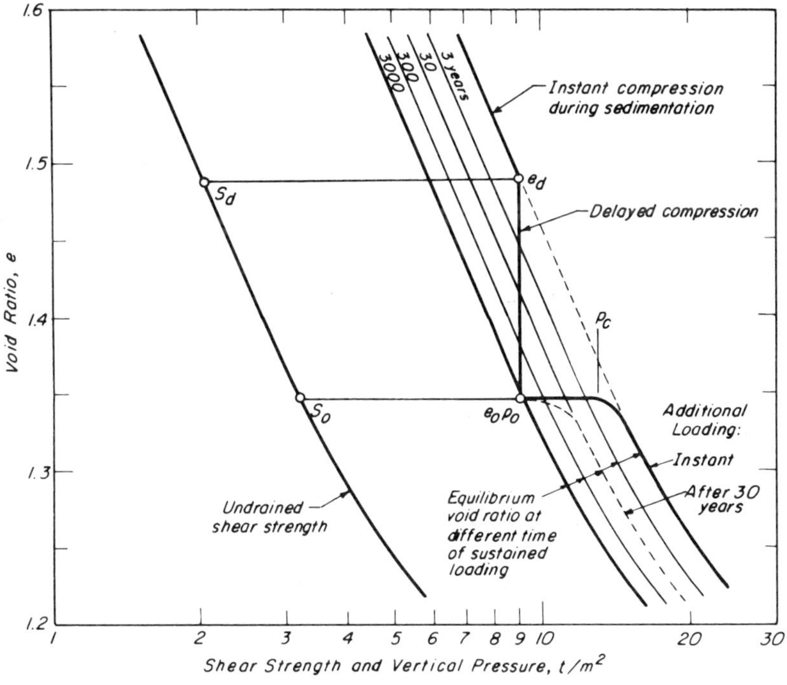

FIG. 4 BJERRUM'S MODEL FOR ONE-DIMENSIONAL COMPRESSION

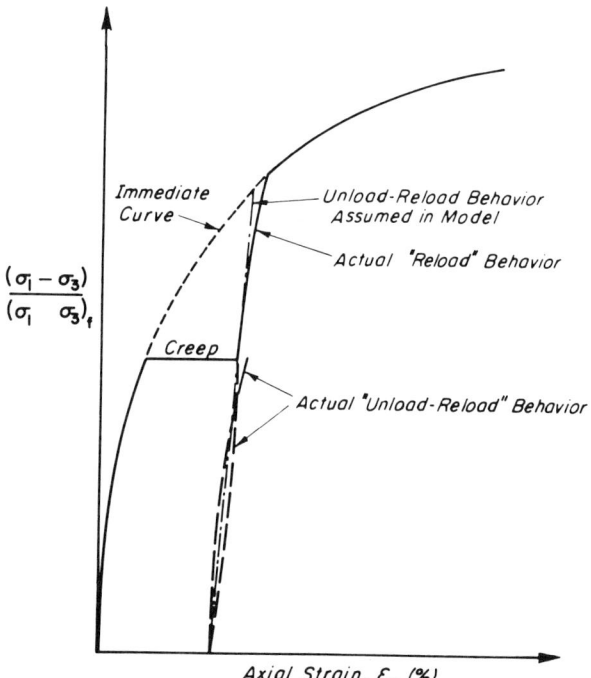

FIG. 5 SUPERPOSITION OF DEVIATORIC LOADS AFTER UNDRAINED CREEP

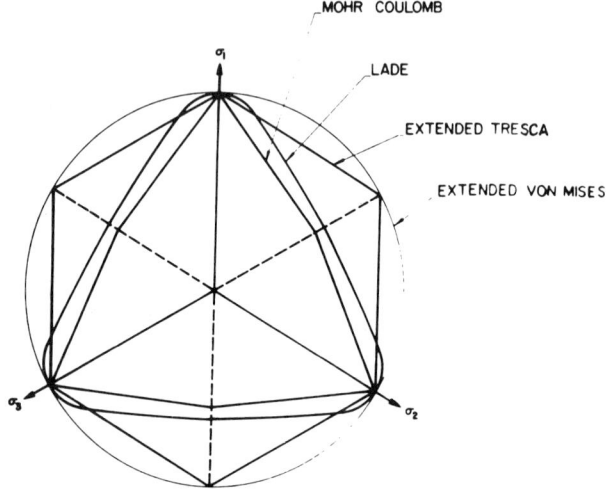

FIG. 6 PROJECTION OF LADE'S FAILURE CRITERION ON THE OCTAHEDRAL PLANE

STRESS DEFORMATIONS

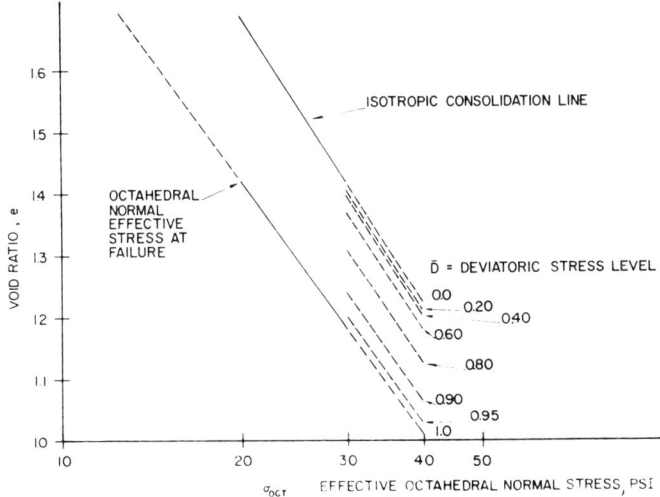

FIG. 7 IMMEDIATE COMPRESSION PLOT OF CLAY X FOR TRIAXIAL STRESS STATES

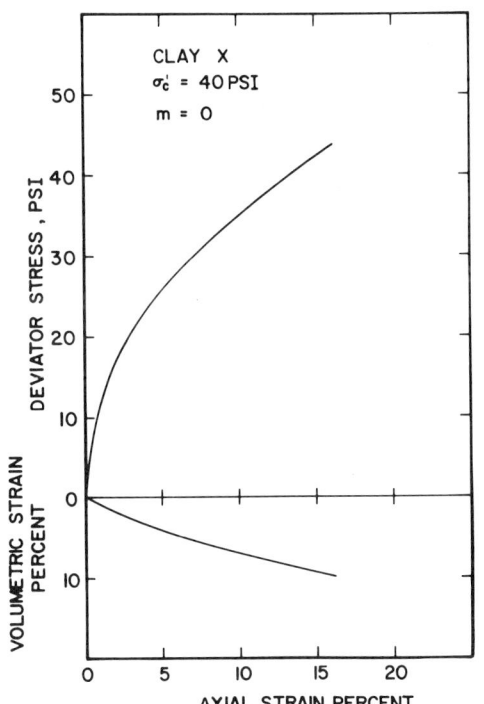

FIG. 8 PREDICTED BEHAVIOR OF CLAY X AT AN ISOTROPIC CONSOLIDATION PRESSURE OF 40 PSI (M=0)

488 STRESS STRAIN FOR SOILS

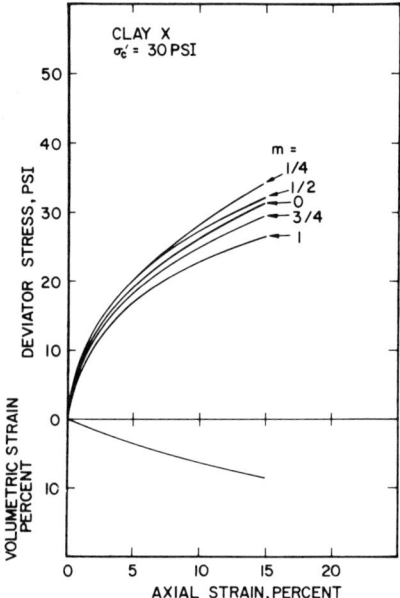

FIG. 9 PREDICTED BEHAVIOR OF CLAY X AT AN ISOTROPIC CONSOLIDATION PRESSURE OF 30 PSI

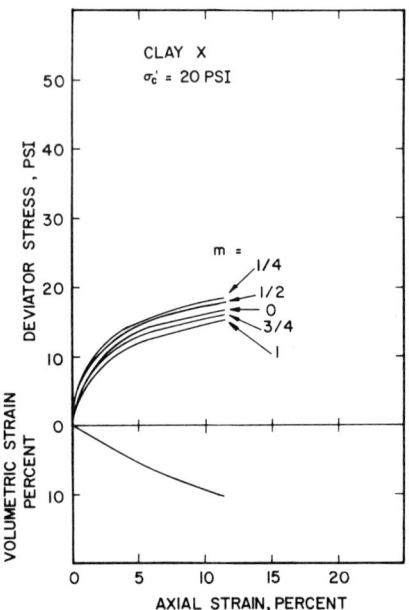

FIG. 10 PREDICTED BEHAVIOR OF CLAY X AT AN ISOTROPIC CONSOLIDATION PRESSURE OF 20 PSI

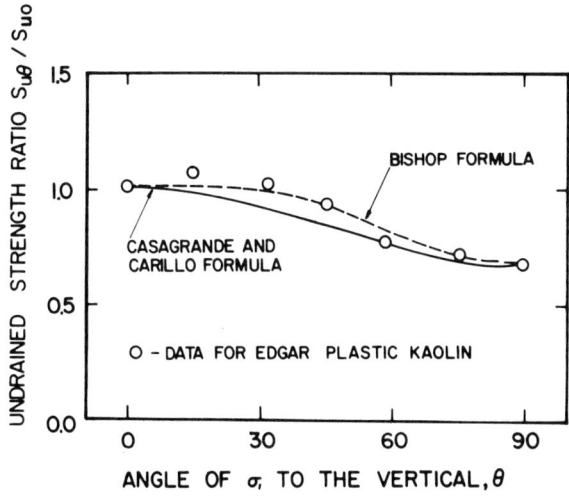

FIG. 11 ANISOTROPIC UNDRAINED STRENGTH OF EDGAR PLASTIC KAOLIN (DATA FROM SAADA AND BIANCHINI (1975), AS SUMMARIZED BY HANSEN (1980))

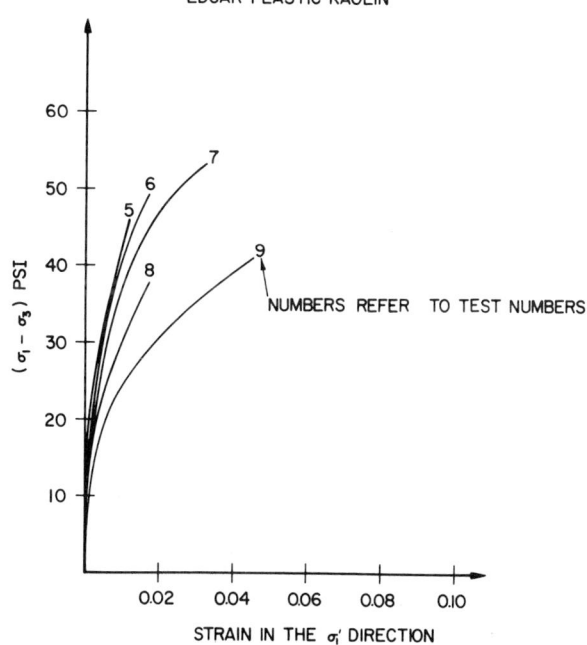

FIG. 12a DEVIATORIC STRESS STRAIN BEHAVIOR OF EDGAR PLASTIC KAOLIN AT AN EFFECTIVE ISOTROPIC CONFINING PRESSURE OF 40 PSI (CONSTANT MEAN TOTAL STRESS TESTS)

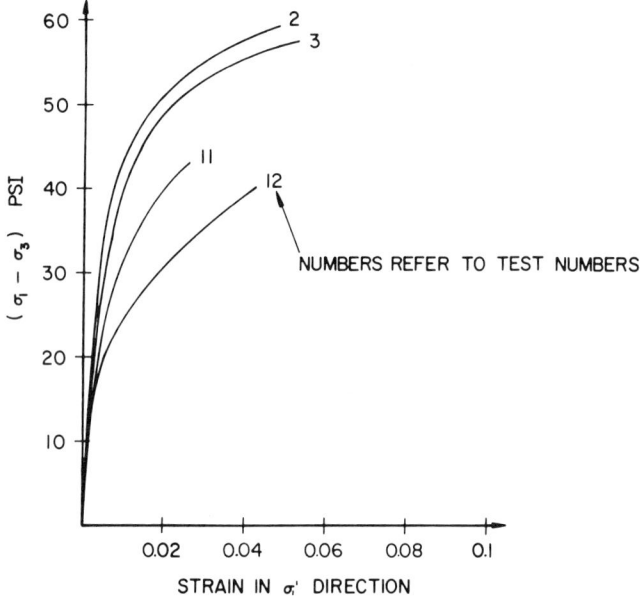

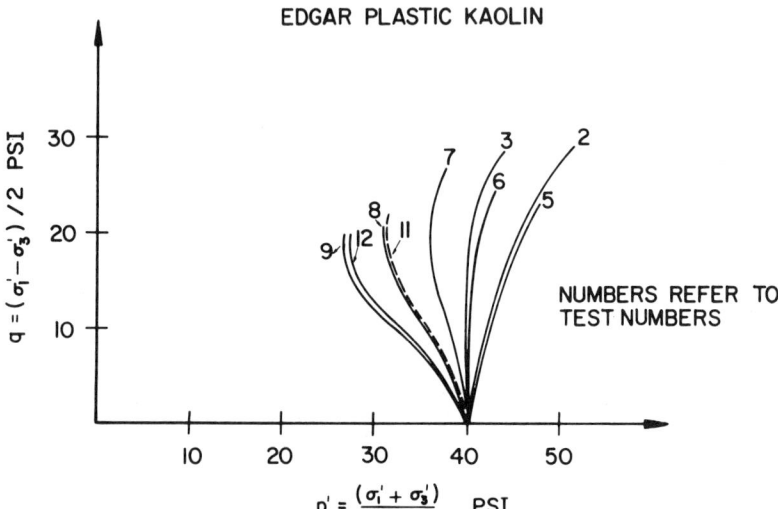

FIG. 12b DEVIATORIC STRESS STRAIN BEHAVIOR OF EDGAR PLASTIC KAOLIN AT AN EFFECTIVE ISOTROPIC CONFINING PRESSURE OF 40 PSI

STRESS DEFORMATIONS 491

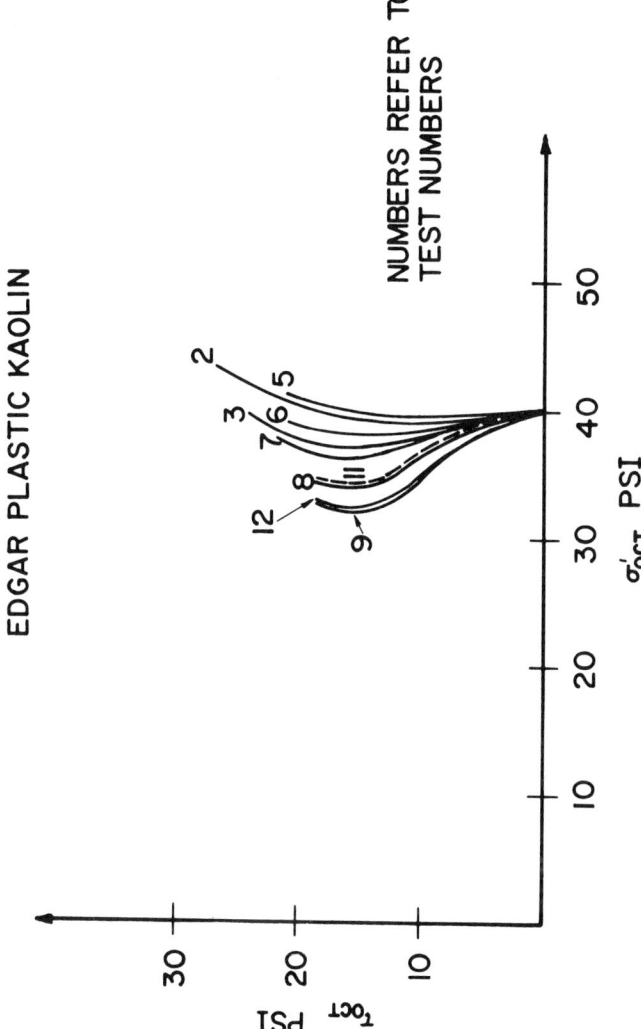

FIG. 14 $\tau_{OCT} - \sigma'_{OCT}$ EFFECTIVE STRESS PATHS FOR EDGAR PLASTIC KAOLIN

Nonlinear Hyperelastic (Green) Constitutive Models

for Soils: Theory and Calibration

By: A. F. Saleeb and W. F. Chen

School of Civil Engineering

Purdue University

West Lafayette, IN

1. Introduction

1.1 General

The stress-strain behavior of any type of soil subjected to externally applied loads is quite complicated and depends on many factors. This has been a subject of research for many years, and the advent of the numerical technique of finite element has given added impetus to these efforts [29]. The problems stem mainly from the fact that, unlike the properties of most engineering materials, soil stress-strain responses are greatly affected by such factors as soil structure (grain size, grain shape, surface texture, mineralogy, cementation or bonding, etc.), density, water content, drainage conditions, degree of voids saturation, loading rate, confining pressure, loading history, and current stress state [8, 11, 13, 30]. Clearly, the number of variables is far too extensive to offer any encouraging hope for developing a simple, yet realistic, constitutive relation that is capable of modelling the behavior of all soils under general loading conditions. Drastic idealizations and simplifications are essential in order to model mathematically and approximately the real behavior for the solution of the problem at hand. For example, in most of the presently available constitutive relations, soil behavior has been drastically idealized as time-independent, such as elastic and elastic-plastic models where time effects are neglected. In addition, interaction between mechanical and thermal processes is usually neglected.

Recent work [6, 14, 18, 19, 21, 22] has indicated that the stress-strain behavior of most granular materials may be separated into recoverable and irrecoverable components, and attempts have been made to treat each component individually. The recoverable behavior is treated within the framework of elasticity theory; while the irrecoverable part is based on plasticity theory. Such separation is necessary if cyclic loading and unloading are encountered. However, for problems in which a monotonically increasing load prevails, elasticity based models provide a much simpler approach. Most of the commonly used plasticity models for soils are summarized in the recent paper by Chen [2]. In the forthcoming, three general methods of formulation for elasticity based stress-strain relations are reviewed. Later in this paper, a complete development of the proposed isotropic third-order hyperelastic model will be presented and its application to three different types of soils: clay "X", clay "Y" and Ottawa Sand will be discussed.

1.2 Review of Elasticity-Based Stress-Strain Relations

Three different types of elasticity-based constitutive models are presently available for general formulation. These are summarized in the following [4, 15]:

(1) _Cauchy type_ in which the current state of stress depends only on the current state of deformation, i.e., stress is a function of strain. Mathematically, the constitutive equations for this material are given by:

$$\sigma_{ij} = F_{ij}(\varepsilon_{kl}) \tag{1}$$

where F_{ij} is the elastic response function of the material, σ_{ij} and ε_{kl} are the components of stress and strain tensors, respectively. The elastic behavior described by Eq. (1) is both reversible and path independent in the sense that stresses are uniquely determined from the current state of strain or vice versa. There is no dependence of the behavior on the stress or strain histories followed to reach the current state of stress or strain. It can be shown [4] that Cauchy elastic material may generate energy under certain loading-unloading cycles. Such behavior is inadmissible since it violates the laws of thermodyanmics. This leads naturally to the consideration of the second type of formulation, Green hyperelastic type. Many commonly used constitutive models for soils are based on Cauchy type of formulation. For example, different incremental isotropic nonlinear elastic stress-strain relations have been formulated based on the modification of the isotropic linear elastic relations with two of the elastic moduli (Young's modulus E, Poisson's ratio ν, shear modulus G, constrained modulus M, and bulk modulus K) being taken to be sealer functions of stress and/or strain states [5, 12, 17, 18, 19, 20].

(2) <u>Hyperelastic (Green) type</u>. This is based on the assumption of the existence of a strain energy density function W (or a complementary energy density function Ω) such that [4, 15]:

$$\sigma_{ij} = \frac{\partial W}{\partial \varepsilon_{ij}} \tag{2a}$$

or,

$$\varepsilon_{ij} = \frac{\partial \Omega}{\partial \sigma_{ij}} \tag{2b}$$

in which $W = \int_0^{\varepsilon_{ij}} \sigma_{ij} d\varepsilon_{ij}$ and $\Omega = \int_0^{\sigma_{ij}} \varepsilon_{ij} d\sigma_{ij}$ are functions of the current components of the strain and stress tensors, respectively. This ensures that no energy can be generated through load cycles and laws of thermodynamics are always satisfied.

For an initially isotropic elastic material, W or Ω are expressed in terms of <u>any three independent</u> invariants of strain or stress tensors ε_{ij} or σ_{ij}, respectively. In general, if Ω is expressed in terms of the three stress invariants:

$$\bar{I}_1 = \sigma_{kk}$$
$$\bar{I}_2 = \frac{1}{2}\sigma_{km}\sigma_{km} \tag{3}$$
$$\bar{I}_3 = \frac{1}{3}\sigma_{km}\sigma_{kn}\sigma_{mn}$$

then, Eq. (2b) yields the following constitutive law

$$\varepsilon_{ij} = \frac{\partial\Omega}{\partial\bar{I}_1}\frac{\partial\bar{I}_1}{\partial\sigma_{ij}} + \frac{\partial\Omega}{\partial\bar{I}_2}\frac{\partial\bar{I}_2}{\partial\sigma_{ij}} + \frac{\partial\Omega}{\partial\bar{I}_3}\frac{\partial\bar{I}_3}{\partial\sigma_{ij}} = \phi_1\delta_{ij} + \phi_2\sigma_{ij} + \phi_3\sigma_{im}\sigma_{jm} \tag{4}$$

where the material response functions, ϕ_i, are defined as

$$\phi_i = \phi_i(\bar{I}_j) = \frac{\partial\Omega}{\partial\bar{I}_i} \tag{5}$$

and these functions are related through the three equations: [4]

$$\frac{\partial\phi_i}{\partial\bar{I}_j} = \frac{\partial\phi_j}{\partial\bar{I}_i} \tag{6}$$

In Eq. (4), δ_{ij} is the Kronecker Delta ($\delta_{11} = 1$, $\delta_{12} = 0$, etc.).

The choice of the three independent stress invariants appearing in Eqs. (3 and 4) is arbitrary. Instead, one may use the invariants $J_1 = s_{kk}$, $J_2 = \frac{1}{2}s_{ij}s_{ij}$ and $J_3 = \frac{1}{3}s_{ij}s_{ik}s_{jk}$ of the stress deviator tensor $s_{ij} = \sigma_{ij} - \frac{1}{3}\sigma_{kk}\delta_{ij}$, or even mixed invariants such as $\bar{I}_1$, J_2 and J_3. The particular advantage of the choice here is the separation of the functions

ϕ_i in a simple convenient manner. Based on assumed polynomial expansions of the function Ω in terms of the three invariants, different constitutive models can be developed. In particular, Evans and Pister [16] developed a general third-order stress-strain law using Eq. (4) and retaining terms in Ω from second to fourth order in stress. Ko and Masson [23] used this law and described the fitting procedure of the model and applied it to describe the behavior of Ottawa sand. This same model and fitting procedure [23] will be used in the present paper to describe the behavior of three different types of soils.

(3) <u>Incremental (Hypoelastic) type</u>. This type of formulation is often used to describe the mechanical behavior of a class of materials in which the state of stress depends on the current state of strain as well as on the stress path followed to reach that state [4, 6, 15, 31]. In general, the incremental constitutive relations for time-independent materials are written as [6, 12]:

$$F(\sigma_{ij}, \dot{\sigma}_{kl}, \varepsilon_{mn}, \dot{\varepsilon}_{pq}) = 0 \tag{7}$$

provided that this equation is homogeneous in time (i.e., time occurs to the same order in all terms of the equation and therefore, may be eliminated). In Eq. (7), $\dot{\sigma}_{kl}$ and $\dot{\varepsilon}_{pq}$ are the stress increment and strain increment tensors, respectively, and F is a tensor function. Equation (7) is very general, but because of its complexity it is not possible to indicate in which manner the total and incremental stresses and strains are related and therefore, for simplicity, special cases of the general law are usually used. In particular, four special cases of the general law, in which strain increments are linearly related to stress increments through the material response moduli which depend on a single state variable, will be given. These cases are described by the relations [6]:

$$\dot{\sigma}_{ij} = A_{ijkl}(\sigma_{mn}) \dot{\varepsilon}_{kl} \quad (8a)$$

$$\dot{\sigma}_{ij} = B_{ijkl}(\varepsilon_{mn}) \dot{\varepsilon}_{kl} \quad (8b)$$

$$\dot{\varepsilon}_{ij} = C_{ijkl}(\varepsilon_{mn}) \dot{\sigma}_{kl} \quad (8c)$$

$$\dot{\varepsilon}_{ij} = D_{ijkl}(\sigma_{mn}) \dot{\sigma}_{kl} \quad (8d)$$

where A_{ijkl} through D_{ijkl} are general functions of their indicated argument. The behavior described by any of Eqs. (8) is <u>infinitesimally</u> (or <u>incrementally</u>) <u>reversible</u>. This justifies the use of the suffix "elastic" in the term hypoelastic used by Truesdell [31] to describe the constitutive relations in Eq. (8a). Based on the degree of dependence of the tensorial functions A_{ijkl} through D_{ijkl} upon the components of the corrresponding tensor argument, different types of constitutive laws are obtained. For example, in a grade one (or first order) constitutive law, the tensorial functions in Eqs. (8) are linear functions of their arguments. A hypoelastic material of grade zero (zero order) is equivalent to anisotropic elastic Cauchy material. Hook's Law is representative of this type of behavior in case of isotropic materials. For <u>isotropic</u> materials, the tensorial response functions in Eqs. (8) are further restricted to be <u>form-invariant</u> under the full group of transformation of the coordinate axes [4].

Several incremental constitutive relations have been used in modelling the behavior of soils and rocks [1, 6, 7, 19, 28, 32]. More recently, incremental stress-strain relations have been formulated separately for a special class of the hypoelastic materials, in which the response tensors in Eqs. (8) are assumed to depend on the invariants, but not on the stress (or strain) tensor itself. However, in these later models, different forms for the material response functions apply in initial loading, and in subsequent unloading and reloading, i.e., the models are generally irreversible, even for incremental loading. These models are now known as variable-moduli models, and they have been extensively used to describe the behavior of soils in ground shock studies [26, 27]. In Ref. [4], a complete critical theoretical study of the three types of elasticity formulation has been made.

In the first part of the following discussions, a third-order hyperelastic (Green) constitutive model is formulated (sec. 2), and specific fitting procedure to determine the material constants together with several numerical examples are subsequently described (sec. 3). Based on this formulation, explicit expression is developed for the incremental stress-strain relation (sec. 4). This model is subsequently refined by introducing a loading criterion and a failure criterion, thus, extending the range of application to reversed loading case (sec. 5). In the second part, theoretical considerations of the model for uniqueness and stability of numerical solutions are examined (sec. 6). Comparisons of the model predictions with experimental results are made in sec. 7. Advantages and limitations of the deformational types of plasticity models, which are identical with nonlinear elastic types of stress-strain relations as long as unloading does not occur, are emphasized. The model described in this paper can be readily applied to nonlinear stress analysis of geotechnical problems involving three-dimensional stress and strain components.

2. Formulation of the Proposed Third-Order Hyperelastic (Green) Constitutive Model

2.1 General

For an initially isotropic material, if the complementary energy function $\Omega(\bar{I}_1, \bar{I}_2, \bar{I}_3)$ is expressed as a fourth order polynominal in the stress components, one can write [16, 23]:

$$\Omega(\bar{I}_1, \bar{I}_2, \bar{I}_3) = A_0 + A_1\bar{I}_1 + \frac{1}{2}B_1\bar{I}_1^2 + \frac{1}{3}B_2\bar{I}_1^3 + B_3\bar{I}_1\bar{I}_2 + B_4\bar{I}_2 + B_5\bar{I}_3$$
$$+ \frac{1}{4}B_6\bar{I}_1^4 + B_7\bar{I}_1^2\bar{I}_2 + \frac{1}{2}B_8\bar{I}_2^2 + B_9\bar{I}_1\bar{I}_3 \qquad (9)$$

where the stress invariants $\bar{I}_1$, $\bar{I}_2$ and $\bar{I}_3$ are defined in Eqs. (3) and A_0, A_1 and B_1 to B_9 are material constants. The numerical values of the coefficients in Eq. (9) above are inserted for convenience in subsequent derivation. Then, using the normality condition of Eq. (2b), the constitutive relations may be written as:

$$\varepsilon_{ij} = A_1\delta_{ij} + [B_1\bar{I}_1 + B_2\bar{I}_1^2 + B_3\bar{I}_2 + B_6\bar{I}_1^3 + 2B_7\bar{I}_1\bar{I}_2 + B_9\bar{I}_3]\delta_{ij}$$
$$+ [B_3\bar{I}_1 + B_4 + B_7\bar{I}_1^2 + B_8\bar{I}_2]\sigma_{ij} + [B_5 + B_9\bar{I}_1]\sigma_{im}\sigma_{jm} \qquad (10)$$

Assuming that the initial stress-free state corresponds to initial strain-free state, the constant A_1 is equal to zero, and we have from Eq. (10),

$$\varepsilon_{ij} = \phi_1\sigma_{ij} + \phi_2\sigma_{ij} + \phi_3\sigma_{im}\sigma_{jm} \qquad (11a)$$

where the material response functions, ϕ_i, of Eq. (4), are given by

$$\phi_1 = B_1\bar{I}_1 + B_2\bar{I}_1^2 + B_3\bar{I}_2 + B_6\bar{I}_1^3 + 2B_7\bar{I}_1\bar{I}_2 + B_9\bar{I}_3 \qquad (11b)$$

$$\phi_2 = B_3\bar{I}_1 + B_4 + B_7\bar{I}_1^2 + B_8\bar{I}_2 \qquad (11c)$$

$$\phi_3 = B_5 + B_9\bar{I}_1 \qquad (11d)$$

and obviously the relations between these functions, Eqs. (6), are satisfied. Equations (11a to d) are the general third-order hyperelastic (Green) stress-strain relations. It only remains to determine the nine material constants B_1 to B_9 from experimental results in order to complete the model formulation. This will be explained in subsequent sections. The general form of the proposed third-order constitutive law, Eqs. (11), was formulated by Evans and Pister [16]. The present formulation of the model and the procedure of determining the nine material constants B_1 to B_9 follow the method originally developed by Ko and Masson [23].

As mentioned in [23], based on the experimental test results, stress-strain curves describing the behavior of many soils especially loose sands and soft clays in shearing along conventional soil tests (e.g., triaxial compression, triaxial extension, and simple shear tests) can be best represented by odd functions. For instance, typical stress-strain curves for soil along different shearing stress path (Fig. 1a, b) are illustrated in Fig. 1(c) for Ottawa sand [23]. This is the reason for the particular choice of the third-order stress-strain law.

For most soils (and particularly for cohesionless soils) initial natural state includes a nonzero reference state of stress, σ_{ij}^0, corresponding to a zero reference state of strain $\varepsilon_{ij}(\sigma_{kl}^0)$. Therefore, the components of the incremental strains $\Delta\varepsilon_{ij}$ corresponding to the incremental stresses $\Delta\sigma_{ij}$ measured from the initial reference state of stress, σ_{ij}^0, are calculated as:

$$\Delta\varepsilon_{ij}(\sigma_{kl}^0, \Delta\sigma_{kl}) = \varepsilon_{ij}(\sigma_{kl}^0 + \Delta\sigma_{kl}) - \varepsilon_{ij}(\sigma_{kl}^0) \quad (12)$$

Substituting σ_{kl}^0 and $\Delta\sigma_{kl}^0$ into Eq. (11a) to calculate the corresponding strains $\varepsilon_{ij}(\sigma_{kl}^0)$ and $\varepsilon_{ij}(\sigma_{kl}^0 + \Delta\sigma_{kl})$, Eq. (12) can be written in the form:

$$\Delta\varepsilon_{ij}(\sigma_{kl}^0 + \Delta\sigma_{kl}) = (\tilde{\phi}_1 - \phi_1)\delta_{ij} + (\tilde{\phi}_2 - \phi_2)\sigma_{ij}^0 + \tilde{\phi}_2\Delta\sigma_{ij}$$
$$+ (\tilde{\phi}_3 - \phi_3)\sigma_{im}^0\sigma_{jm}^0 + \tilde{\phi}_3(\sigma_{im}^0\Delta\sigma_{jm} + \sigma_{jm}^0\Delta\sigma_{im} + \Delta\sigma_{im}\Delta\sigma_{jm}) \quad (13a)$$

where,

$$\tilde{\phi}_1 = \phi_1(\sigma_{kl}^0 + \Delta\sigma_{kl}) \;;\; \tilde{\phi}_2 = \phi_2(\sigma_{kl}^0 + \Delta\sigma) \;;\; \tilde{\phi}_3 = \phi_3(\sigma_{kl}^0 + \Delta\sigma_{kl})$$
$$\phi_1 = \phi_1(\sigma_{kl}^0) \qquad\qquad\;;\; \phi_2 = \phi_2(\sigma_{kl}^0) \qquad\;;\; \phi_3 = \phi_3(\sigma_{kl}^0) \quad (13b)$$

i.e., the functions $\tilde{\phi}_i$ and ϕ_i are computed by substituting the corresponding states of stress, $(\sigma_{kl}^0 + \Delta\sigma)$ and σ_{kl}^0, respectively, into the expressions of Eqs. (11b to d). It is to be emphasized that $\Delta\varepsilon_{ij}(\sigma_{kl}^0 + \Delta\sigma)$ are not incremental strains in a complete sense; they represent total strains measured from the initial reference state, σ_{kl}^0, with the components $\varepsilon_{ij}(\sigma_{kl}^0)$ taken as zero. The stress components $\Delta\sigma_{ij}$ are not necessarily small.

In order to study the behavior of the stress-strain model of Eq. (13) under the conditions of conventional soil tests, it is necessary to reduce the relations for the particular loading paths followed in these tests. Since the stress paths in most soil tests are <u>straight line paths</u> in the principal stress space (e.g., CTC, CTE, SS, HC, etc.), it is more convenient to develop the stress-strain relations of the model under a general straight line loading path in the principal stress space. Then, for a particular loading path in any test, the relations can be formulated very easily as a special case from the general straight line loading path. In the forthcoming, formulation of the constitutive relations of the model under a general straight line (proportional) loading path is given. In addition, examples of specific relations in a number of conventional tests are presented.

2.2 Stress-Strain Relation for a General Straight Line Loading Path in the Principal Stress Space

In the formulation presented here, the initial reference state of stress, σ_{kl}^0, is assumed to be a hydrostatic stress state (since all the tests performed satisfy this condition); i.e.,

$$\sigma_{ij}^0 = \sigma_c \quad (\text{or } \sigma_1^0 = \sigma_2^0 = \sigma_3^0 = \sigma_c); \text{ for } i = j$$
$$\sigma_{ij}^0 = 0; \text{ for } i \neq j \tag{13}$$

where σ_1, σ_2 and σ_3 are the principal major, intermediate and minor stresses, respectively, and σ_c denotes the initial consolidation pressure.

Denote the increment stress $\Delta\sigma_1$ in the major principal stress measured from the initial hydrostatic state of stress by λ. Then, for a general straight line stress path, the increments $\Delta\sigma_1$ and $\Delta\sigma_2$ in the other two principal stress are $\alpha_1\lambda$ and $\alpha_2\lambda$, respectively; i.e.,

$$\Delta\sigma_1 : \Delta\sigma_2 : \Delta\sigma_3 = 1 : \alpha_1 : \alpha_2 \tag{14}$$

where α_1 and α_2 are parameters determining the direction of the straight line path (Fig. 2). Thus, the invariants $\bar{I}_1$, $\bar{I}_2$ and $\bar{I}_3$ of Eqs. (3) for the intial and current states of stress σ_{kl}^0 and $\sigma_{kl} + \Delta\sigma_{kl}^0$, respectively, as shown in Fig. 2, are given by

$$\bar{I}_1(\sigma_{kl}^0) = 3\sigma_c, \quad \bar{I}_2(\sigma_{kl}^0) = \frac{3}{2}\sigma_c^2, \quad \bar{I}_3(\sigma_{kl}^0) = \sigma_c^3$$

$$\bar{I}_1(\sigma_{kl}^0 + \Delta\sigma_{kl}) = 3\sigma_c + 3\lambda k_1$$

$$\bar{I}_2(\sigma_{kl}^0 + \Delta\sigma_{kl}) = \frac{3}{2}\sigma_c^2 + \sigma_c \lambda k_1 + \frac{1}{2}\lambda^2 k_2 \tag{15a}$$

$$\bar{I}_3(\sigma_{kl}^0 + \Delta\sigma_{kl}) = \sigma_c^3 + \sigma_c^2 \lambda k_1 + \sigma_c \lambda^2 k_2 + \frac{1}{3}\lambda^3 k_3$$

in which the parameters k_1, k_2 and k_3 are defined as

$$k_1 = 1 + \alpha_1 + \alpha_2$$

$$k_2 = 1 + \alpha_1^2 + \alpha_2^2 \tag{15b}$$

$$k_3 = 1 + \alpha_1^3 + \alpha_2^3$$

Hence, substituting the expressions (15a, b) into Eqs. (13b) and using Eq. (13a) to calculate the principal strains $\Delta\varepsilon_1$, $\Delta\varepsilon_2$ and $\Delta\varepsilon_3$, we finally get

$$\Delta\varepsilon_1 = C_1^{(1)}\lambda + C_2^{(1)}\lambda^2 + C_3^{(1)}\lambda^3$$

$$\Delta\varepsilon_2 = C_1^{(2)}\lambda + C_2^{(2)}\lambda^2 + C_3^{(2)}\lambda^3 \tag{16a}$$

$$\Delta\varepsilon_3 = C_1^{(3)}\lambda + C_2^{(3)}\lambda^2 + C_3^{(3)}\lambda^3$$

in which the $C_j^{(i)}$ coefficients are given by

$$C_1^{(1)} = (k_1 B_1 + B_4) + [6k_1 B_2 + (2k_1 + 3)B_3 + 2B_5]\sigma_c$$
$$+ [27k_1 B_6 + (15k_1 + 9)B_7 + (k_1 + \frac{3}{2})B_8 + (2k_1 + 6)B_9]\sigma_c^2$$

$$C_2^{(1)} = [k_1^2 B_2 + (k_1 + \frac{1}{2}k_2)B_3 + B_5] + [9k_1^2 B_6 + (6k_1 + 3k_1^2 + 3k_2)B_7$$
$$+ (k_1 + \frac{1}{2}k_2)B_8 + (2k_1 + k_2 + 3)B_9]\sigma_c$$

$$C_3^{(1)} = k_1^3 B_6 + (k_1^2 + k_1 k_2) B_7 + \tfrac{1}{2} k_2 B_8 + (k_1 + \tfrac{1}{3} k_3) B_9$$

$$C_1^{(2)} = (k_1 B_1 + \alpha_1 B_4) + [6 k_1 B_2 + (2 k_1 + 3 \alpha_1) B_3 + 2 \alpha_1 B_5] \sigma_c$$
$$+ [27 k_1 B_6 + (15 k_1 + 9 \alpha_1) B_7 + (k_1 + \tfrac{3}{2} \alpha_1) B_8 + (2 k_1 + 6 \alpha_1) B_9] \sigma_c^2$$

$$C_2^{(2)} = [k_1^2 B_2 + (\alpha_1 k_1 + \tfrac{1}{2} k_2) B_3 + \alpha_1^2 B_5] + [9 k_1^2 B_6 + (6 \alpha_1 k_1 + 3 k_1^2 + 3 k_2) B_7$$
$$+ (\alpha_1 k_1 + \tfrac{1}{2} k_2) B_8 + (3 \alpha_1^2 + 2 \alpha_1 k_1 + k_2) B_9] \sigma_c$$

$$C_3^{(2)} = k_1^3 B_6 + (k_1 k_2 + \alpha_1 k_1^2) B_7 + \tfrac{1}{2} \alpha_1 k_2 B_8 + (\alpha_1^2 k_1 + \tfrac{1}{3} k_3) B_9 \qquad (16b)$$

$$C_1^{(3)} = (k_1 B_1 + \alpha_2 B_4) + [6 k_1 B_2 + (2 k_1 + 3 \alpha_2) B_3 + 2 \alpha_2 B_5] \sigma_c$$
$$+ [27 k_1 B_6 + (15 k_1 + 9 \alpha_2) B_7 + (k_1 + \tfrac{3}{2} \alpha_2) B_8 + (2 k_1 + 6 \alpha_2) B_9] \sigma_c^2$$

$$C_2^{(3)} = [k_1^2 B_2 + (\alpha_2 k_1 + \tfrac{1}{2} k_2) B_3 + \alpha_2^2 B_5] + [9 k_1^2 B_6 + (6 \alpha_2 k_1 + 3 k_1^2 + 3 k_2) B_7$$
$$+ (\alpha_2 k_1 + \tfrac{1}{2} k_2) B_8 + (3 \alpha_2^2 + 2 \alpha_2 k_1 + k_2) B_9] \sigma_c$$

$$C_3^{(3)} = k_1^3 B_6 + (k_1 k_2 + \alpha_2 k_1^2) B_7 + \tfrac{1}{2} \alpha_2 k_2 B_8 + (\alpha_2^2 k_1 + \tfrac{1}{3} k_3) B_9$$

Clearly, the advantage of performing the tests along straight line loading paths is that the changes in the principal stress and strain components during the tests are conveniently expressed in terms of a single parameter, λ, as given in Eqs. (14) and (16a). Furthermore, the components of the principal strain increments are given as cubic functions in this parameter (see Eq. 16a). Indeed, these cubic relations are the basis for the determination of the nine material constants, B_i, from the experimental results, as will be explained later.

In the following, special cases of the relations of Eq. (16) are given for some of the stress paths in the conventional soil tests.

(a) <u>Hydrostatic Compression Path (HC)</u>: Fig 1(a)

In this case, the stress components $\sigma_1 = \sigma_2 = \sigma_3$ are increasing and $\alpha_1 = \alpha_2 = 1$, $k_1 = k_2 = k_3 = 3$. Thus, Eqs. (16) simplify to

$$\begin{aligned}
\Delta\varepsilon_1 = \Delta\varepsilon_2 = \Delta\varepsilon_3 = &\ [(3B_1 + B_4) + (18B_2 + 9B_3 + 2B_5)\sigma_c \\
&+ (81B_6 + 54B_7 + \tfrac{9}{2}B_8 + 12B_9)\sigma_c^2]\lambda + [(9B_2 + \tfrac{9}{2}B_3 + B_5) \\
&+ (81B_6 + 54B_7 + \tfrac{9}{2}B_8 + 12B_9)\sigma_c]\lambda^2 \\
&+ [27B_6 + 18B_7 + \tfrac{3}{2}B_8 + 4B_9]\lambda^3
\end{aligned} \qquad (17)$$

(b) <u>Conventional Triaxial Compression Path (CTC)</u>: Fig. 1(a)

The components $\sigma_2 = \sigma_3$ are constant, and σ_1 is increasing; i.e., $\alpha_1 = \alpha_2 = 0$, $k_1 = k_2 = k_3 = 1$. Hence, for $\sigma_1^0 = \sigma_2^0 = \sigma_3^0 = \sigma_c$ and $\Delta\sigma_1 = \lambda$, we have from Eqs. (16)

$$\begin{aligned}
\Delta\varepsilon_1 = &\ [(B_1 + B_4) + (6B_2 + 5B_3 + 2B_5)\sigma_c + (27B_6 + 24B_7 + \tfrac{5}{2}B_8 + 8B_9)\sigma_c^2]\lambda \\
&+ [(B_2 + \tfrac{3}{2}B_3 + B_5) + (9B_6 + 12B_7 + \tfrac{3}{2}B_8 + 6B_9)\sigma_c]\lambda^2 \\
&+ [B_6 + 2B_7 + \tfrac{1}{2}B_8 + \tfrac{4}{3}B_9]\lambda^3 \\
\Delta\varepsilon_2 = \Delta\varepsilon_3 = &\ [B_1 + (6B_2 + 2B_3)\sigma_c + (27B_6 + 15B_7 + B_8 + 2B_9)\sigma_c^2]\lambda \\
&+ [(B_2 + \tfrac{1}{2}B_3) + (9B_6 + 6B_7 + \tfrac{1}{2}B_8 + B_9)\sigma_c]\lambda^2 \\
&+ [B_6 + B_7 + \tfrac{1}{3}B_9]\lambda^3
\end{aligned} \qquad (18)$$

(c) <u>Simple Shear Path (SS)</u>: Fig. 1(b)

The stress component σ_1 is increasing, and σ_3 is decreasing while σ_2 is held constant; i.e., $\alpha_1 = 0$, $\alpha_2 = -1$. Therefore, the equations in (16) are reduced to: ($k_1 = k_3 = 0$, $k_2 = 2$)

$$\Delta\varepsilon_1 = [B_4 + (3B_3 + 2B_5)\sigma_c + (9B_7 + \tfrac{3}{2}B_8 + 6B_9)\sigma_c^2]\lambda$$
$$+ [(B_3 + B_5) + (6B_7 + B_8 + 5B_9)\sigma_c]\lambda^2 + [B_8]\lambda^3$$

$$\Delta\varepsilon_2 = [B_3 + (6B_7 + B_8 + 2B_9)\sigma_c]\lambda^2 \qquad (19)$$

$$\Delta\varepsilon_3 = -[B_4 + (3B_3 + 2B_5)\sigma_c + (9B_7 + \tfrac{3}{2}B_8 + 6B_9)\sigma_c^2]\lambda$$
$$+ [(B_3 + B_5) + (6B_7 + B_8 + 5B_9)\sigma_c]\lambda^2 - [B_8]\lambda^3$$

(d) <u>Triaxial Compression Path With Constant Mean Normal Stress (TC)</u>:

Fig. 1(a, b)

In this case, the component σ_1 is increasing while $\sigma_2 = \sigma_3$ are decreasing such that $\sigma_1 + \sigma_2 + \sigma_3 = $ const. $= 3\sigma_c$; i.e., $\Delta\sigma_1 + \Delta\sigma_2 + \Delta\sigma_3 = 0$; or $\alpha_1 = \alpha_2 = -\tfrac{1}{2}$ and $k_1 = 0$, $k_2 = \tfrac{3}{2}$, $k_3 = \tfrac{3}{4}$. Then, we have

$$\Delta\varepsilon_1 = [B_4 + (3B_3 + 2B_5)\sigma_c + (9B_7 + \tfrac{3}{2}B_8 + 6B_9)\sigma_c^2]\lambda$$
$$+ [(\tfrac{3}{4}B_3 + B_5) + (\tfrac{9}{2}B_7 + \tfrac{3}{4}B_8 + \tfrac{9}{2}B_9)\sigma_c]\lambda^2 + [\tfrac{3}{4}B_8 + \tfrac{1}{4}B_9]\lambda^3$$

$$\Delta\varepsilon_2 = \Delta\varepsilon_3 = -[\tfrac{1}{2}B_4 + (\tfrac{3}{2}B_3 + B_5)\sigma_c + (\tfrac{9}{2}B_7 + \tfrac{3}{4}B_8 + 3B_9)\sigma_c^2]\lambda \qquad (20)$$
$$+ [(\tfrac{3}{4}B_3 + \tfrac{1}{4}B_5) + (\tfrac{9}{2}B_7 + \tfrac{3}{4}B_8 + \tfrac{9}{4}B_9)\sigma_c]\lambda^2$$
$$+ [-\tfrac{3}{8}B_8 + \tfrac{1}{4}B_9]\lambda^3$$

(e) <u>Triaxial Extension Path With Constant Mean Normal Stress (TE)</u>:

Fig. 1(a, b)

The component σ_1 is decreasing while $\sigma_2 = \sigma_3$ are increasing such that $\sigma_1 + \sigma_2 + \sigma_3 = $ const. $= 3\sigma_c$. Denoting the change (decrease) in σ_1 by λ (i.e., $\Delta\sigma_1 = -\lambda$), then $\alpha_1 = \alpha_2 = -\tfrac{1}{2}$ and $k_1 = 0$, $k_2 = \tfrac{3}{2}$, $k_3 = \tfrac{3}{4}$ and the strain increments are:

$$\Delta\varepsilon_1 = -[B_4 + (3B_3 + 2B_5)\sigma_c + (9B_7 + \tfrac{3}{2}B_8 + 6B_9)\sigma_c^2]\lambda$$
$$+ [(\tfrac{3}{4}B_3 + B_5) + (\tfrac{9}{2}B_7 + \tfrac{3}{4}B_8 + \tfrac{9}{2}B_9)\sigma_c]\lambda^2 - [\tfrac{3}{4}B_8 + \tfrac{1}{4}B_9]\lambda^3$$

$$\Delta\varepsilon_2 = \Delta\varepsilon_3 = [\tfrac{1}{2}B_4 + (\tfrac{3}{2}B_3 + B_5)\sigma_c + (\tfrac{9}{2}B_7 + \tfrac{3}{4}B_8 + 3B_9)\sigma_c^2]\lambda \qquad (21)$$
$$+ [(\tfrac{3}{4}B_3 + \tfrac{1}{4}B_5) + (\tfrac{9}{2}B_7 + \tfrac{3}{4}B_8 + \tfrac{9}{4}B_9)\sigma_c]\lambda^2 + [\tfrac{3}{8}B_8 - \tfrac{1}{4}B_9]\lambda^3$$

3. Fitting Procedure to Determine Material Constants

3.1 Fitting Procedure

The procedure of the determination of the nine material constants, B_i, is outlined in Ref. [23]. In this section, this procedure is summarized and the results obtained for the three different soils: clay "X", clay "Y" and Ottawa sand are given.

(a) Along a straight line stress path in any test, the principal strain increments $\Delta\varepsilon_1$, $\Delta\varepsilon_2$ and $\Delta\varepsilon_3$ are expressed as <u>cubic</u> functions of the parameter λ, as given in Eq. (16a), with the appropriate constants C_1, C_2 and C_3 according to the particular type of the test. Moreover, the constants C_i, which depend on the reference initial state of stress in the test, σ_{ij}^0, and the orientation of the loading path (defined by α_1 and α_2), are then linearly related to the nine material constants, B_i, using Eqs. (16b).

(b) By fitting <u>cubic</u> curves to the experimentally obtained stress-strain curves for each component of the principal strain increments, the numerical values of the three distinct constants C_1, C_2 and C_3 for this component are determined. Standard procedures of regression analysis may be used in fitting the cubic curves. However, it was found that the initial slopes of the curves determined from such analyses usually deviate too much from the measured slopes. Therefore, the procedure employed here is based on matching the initial slope and two points for each curve to determine the three constants C_1, C_2 and C_3, as shown in Fig. 3. In some cases, adjustment of the determined constants may be made to give a better overall fit to the experimental curves.

(c) From the results obtained in (a) and (b) above, we have a number of linear simultaneous equations in the nine unknowns, B_i (three equations for each distinct stress-strain curve). Theoretically, since nine unknown independent material constants are present, only three independent stress-strain curves are required to give nine equations in nine independent unknowns. These three curves are completely arbitrary and they could be from any type of test. However, it is not expected that a constitutive model with its constants determined from one test is capable of predicting the behavior of the material under arbitrary loading paths different from that of the test. Thus, it is better to make use of most of the available test results, in which case the number of linear equations will exceed the number of unknown independent material constants.

(d) The system of linear equations obtained in (c) (which generally exceeds the number of unknown material constants, B_i) is solved using a least-square solution technique to determine the nine material constants B_i.

3.2 Numerical Results

The procedure described above has been applied to the three types of soils mentioned earlier. The test conditions and the numerical results obtained for each are summarized in the forthcoming.

(a) <u>Clay "X"</u>

The experimental data for this soil were obtained from triaxial tests with constant mean normal stress under consolidated drained conditions conducted on prismatic samples trimmed from block samples. The samples were 100% saturated. During the tests, the measured quantities were the principal stress components σ_1, σ_2 and σ_3 (the major principal stress σ_1 is in the vertical direction while σ_2 and σ_3 are in the lateral directions), the principal strains ε_1 and ε_3, and the volumetric strains $\varepsilon_v = \varepsilon_{kk}$ (the signs follow the soil mechanics sign convention, i.e., compressive stresses and strains are positive).

The tests were performed under constant stress ratio, m, where m is defined as

$$m = \frac{\sigma_2 - \sigma_3}{\sigma_1 - \sigma_3} \quad (22)$$

six different sets of data are provided for clay "X": two sets with m = 0, 1 for each one of the initial hydrostatic (consolidation) stress σ_c = 10, 20, 30 psi. Note that for drained conditions total stresses, σ_{ij}, and effective stresses, σ'_{ij}, are equal. The effective stresses, σ'_{ij}, are given by

$$\sigma'_{ij} = \sigma_{ij} - u\delta_{ij} \quad (23)$$

where u is the pore water pressure.

As can be noted from Eqs. (20), for the TC tests, only six material constants, B_3, B_4, B_5, B_7, B_8 and B_9 are used. Thus, the least-square solution procedure is used for 36 equations in 6 unknowns (six equations for each set of data after allowing for symmetry in the tests). The results obtained are:

$B_3 = 4.4073 \times 10^{-5} \text{ psi}^{-2}$, $B_4 = 8.5 \times 10^{-5} \text{ psi}^{-1}$,

$B_5 = -5.861 \times 10^{-5} \text{ psi}^{-2}$, $B_7 = -4.3667 \times 10^{-6} \text{ psi}^{-3}$,

$B_8 = 2.8092 \times 10^{-5} \text{ psi}^{-3}$, $B_9 = 3.478 \times 10^{-7} \text{ psi}^{-3}$

These values of material constants are used in the prediction of stress-strain curves for m = 0.25, 0.50, 0.75 for each value of σ_c = 10, 20, 30 psi, as shown in Fig. 4(a).

(b) Clay "Y"

As for clay "X", 100% saturated prismatic samples were tested under consolidated drained conditions. The tests were of the conventional triaxial type in which the minor principal stress, σ_3, was kept constant.

Six different sets of data were available for clay "Y": two sets with m = 0, 1 for σ_c = 2.5, 5, 10 psi. The results obtained for the solution of 36 equations in 9 unknown material constants are:

$B_1 = -3.7425 \times 10^{-4} \text{ psi}^{-1}$, $B_2 = -6.69 \times 10^{-7} \text{ psi}^{-2}$,

$B_3 = 5.5416 \times 10^{-5} \text{ psi}^{-2}$, $B_4 = 6.913 \times 10^{-4} \text{ psi}^{-1}$,

$B_5 = -1.3109 \times 10^{-4} \text{ psi}^{-2}$, $B_6 = 1.164 \times 10^{-6} \text{ psi}^{-3}$,

$B_7 = -3.954 \times 10^{-6} \text{ psi}^{-3}$, $B_8 = 1.254 \times 10^{-5} \text{ psi}^{-3}$,

$B_9 = 3.9257 \times 10^{-6} \text{ psi}^{-3}$

These values are used to predict the material behavior for m = 0.25, 0.5, 0.75 under the initial consolidation stresses σ_c = 2.5, 5, 10 psi, as shown in Fig. 4(b).

(c) Ottawa Sand

Samples of Ottawa sand were compacted by aerial pluviation to a relative density of 87% and they were tested in the dry state three-dimensionally under different stress paths. The data sets provided were for different stress paths: CTC, CTE, HC, TC and TE, for varieties of initial hydrostatic stresses, σ_c. For example, the CTC tests were carried out for σ_c = 5 and 10 psi, and σ_c = 10 psi was used for the CTE test. Both the TC and TE tests were performed for 3 values of the initial equal hydrostatic pressure, σ_c = 5, 10, 20 psi. The principal axes are denoted by z (vertical) and x, y (horizontal).

For Ottawa sand, the material constants reported in Ref. [20] were tried and reasonably good overall agreement of the theoretical curves and the experimental results provided was obtained. Therefore, it was decided to use these same values instead of the values initially obtained from separate least square solutions of the equations. This provides means for further investigation of the range of applicability of the model. For instance, although these values of the constants yielded reasonably good results for most of the cases discussed in [23] and for those of the test cases for model calibration investigated here, it was found that the behavior of the model is sensitive to the small changes of the constants of the higher order terms (e.g., B_6, B_7, B_8 and B_9). Slight changes in the values of these constants will change greatly the behavior of the model under certain stress paths. This will be illustrated in the second part of the present paper. The values of the constants reported in [23] are:

B_1 = - 4.431 x 10^{-5} psi^{-1}, B_2 = 1.685 x 10^{-6} psi^{-2}, B_3 = -3.107 x 10^{-6} psi^{-2}

B_4 = 1.885 x 10^{-4} psi^{-1} , B_5 = 1.725 x 10^{-6} psi^{-2}, B_6 = 0.1237 x 10^{-6} psi^{-3}

B_7 = -0.4018 x 10^{-6} psi^{-3}, B_8 = 2.578 x 10^{-7} psi^{-3}, B_9 = 5.597 x 10^{-9} psi^{-3}

For the study of the effect of the change in the numerical values of the material constants on the results, the values of the constants B_6 and B_7

were changed to $B_6 = 0.1687 \times 10^{-6}$ and $B_7 = -0.4690 \times 10^{-6}$ without changing the other constants. The stress paths used in the prediction are shown in Fig. 4(c).

4. Incremental Form of the Stress-Strain Relations

The ultimate goal of developing constitutive models is its use in the solution of a boundary value problem to predict the behavior of the structure. Most often, numerical techniques such as the finite element method are used in the solution, and incremental stress-strain relations are needed to solve problems involving material and/or geometrical non-linearities. In the following, an incremental form of the nonlinear general constitutive law of Eqs. (11) is formulated. In addition, the material compliance matrix relating the increments of the principal stresses and strains is developed.

Differentiating Eq. (11a), the strain increment tensor $\dot{\varepsilon}_{ij}$ can be written as:

$$\dot{\varepsilon}_{ij} = [\frac{\partial \phi_1}{\partial \sigma_{kl}} \delta_{ij} + \phi_2 \frac{\partial \sigma_{ij}}{\partial \sigma_{kl}} + \sigma_{ij} \frac{\partial \phi_2}{\partial \sigma_{kl}} + \phi_3 \frac{\partial (\sigma_{im}\sigma_{jm})}{\partial \sigma_{kl}} + \sigma_{im}\sigma_{jm} \frac{\partial \phi_3}{\partial \sigma_{kl}}] \dot{\sigma}_{kl} \quad (24)$$

where $\dot{\sigma}_{kl}$ is the stress increment tensor, and the functions ϕ_i are given in Eqs. (11b to d). The partial derivatives in the equation above are calculated using the expressions for ϕ_i, and the results are given by

$$\frac{\partial \phi_1}{\partial \sigma_{kl}} = (B_1 + 2B_2 \bar{I}_1 + 3B_6 \bar{I}_1^2 + 2B_7 \bar{I}_2)\delta_{kl} + (B_3 + 2B_7 \bar{I}_1)\sigma_{kl} + B_9 \sigma_{kn}\sigma_{ln}$$

$$\frac{\partial \phi_2}{\partial \sigma_{kl}} = (B_3 + 2B_7 \bar{I}_1)\delta_{kl} + B_8 \sigma_{kl} \quad (25)$$

$$\frac{\partial \phi_3}{\partial \sigma_{kl}} = B_9 \delta_{kl} \; ; \; \frac{\partial \sigma_{ij}}{\partial \sigma_{kl}} = \delta_{ik}\delta_{jl} \; ; \; \frac{\partial \sigma_{im}\sigma_{jm}}{\partial \sigma_{kl}} = \sigma_{il}\delta_{jk} + \sigma_{jl}\delta_{ik}$$

Combining the results of Eq. (24) into Eq. (25), we finally obtain

$$\dot{\varepsilon}_{ij} = [\frac{\partial \phi_1}{\partial \sigma_{kl}} \delta_{ij} + \phi_2 \delta_{ik} \delta_{jl} + \sigma_{ij} \frac{\partial \phi_2}{\partial \sigma_{kl}} + \phi_3(\sigma_{il}\delta_{jk} + \sigma_{jl}\delta_{ik})$$
$$+ \sigma_{im}\sigma_{jm} \frac{\partial \phi_3}{\partial \sigma_{kl}}]\dot{\sigma}_{kl} \qquad (26)$$

This equation represents the general incremental form of the proposed nonlinear third-order hyperelastic constitutive model. This equation can always be written in a matrix form

$$\{\dot{\varepsilon}\} = [C]\{\dot{\sigma}\} \qquad (27)$$

where $\{\dot{\varepsilon}\}$ and $\{\dot{\sigma}\}$ are the strain and stress increment vectors, respectively, and [C] is the material tangential compliance symmetric matrix which depends on the current state of stress σ_{ij} and the material constants B_i. Special cases such as plane stress, plane strain and axisymmetric, can be readily developed from the general form of Eq. (27). As an example, the matrix equation relating the principal stress and strain increments is written as

$$\begin{Bmatrix} \dot{\varepsilon}_1 \\ \dot{\varepsilon}_2 \\ \dot{\varepsilon}_2 \end{Bmatrix} = \begin{bmatrix} C_{11} & C_{12} & C_{13} \\ C_{21} & C_{22} & C_{23} \\ C_{31} & C_{32} & C_{33} \end{bmatrix} \begin{Bmatrix} \dot{\sigma}_1 \\ \dot{\sigma}_2 \\ \dot{\sigma}_3 \end{Bmatrix} \qquad (28)$$

where the elements of symmetrix matrix [C] are given by

$$C_{11} = [(B_1 + B_4) + (2B_2 + B_3)\bar{I}_1 + (3B_6 + B_7)\bar{I}_1^2 + (2B_7 + B_8)\bar{I}_2]$$
$$+ [2(B_3 + B_5) + 2(2B_7 + B_9)\bar{I}_1]\sigma_1 + (B_8 + 2B_9)\sigma_1^2$$

$$C_{22} = [B_1 + B_4) + (2B_2 + B_3)\bar{I}_1 + (3B_6 + B_7)\bar{I}_1^2 + (2B_7 + B_8)\bar{I}_2]$$
$$+ [2(B_3 + B_5) + 2(2B_7 + B_9)\bar{I}_1]\sigma_2 + (B_8 + 2B_9)\sigma_2^2$$

$$C_{33} = [(B_1 + B_4) + (2B_2 + B_3)\bar{I}_1 + (3B_6 + B_7)\bar{I}_1^2 + (2B_7 + B_8)\bar{I}_2]$$
$$+ [2(B_3 + B_5) + 2(2B_7 + B_9)\bar{I}_1]\sigma_3 + (B_8 + 2B_9)\sigma_3^2$$

$$C_{12} = C_{21} = [B_1 + 2B_2\bar{I}_1 + 3B_6\bar{I}_1^2 + 2B_7\bar{I}_2] + (B_3 + 2B_7\bar{I}_1)(\sigma_1 + \sigma_2)$$
$$+ B_8\sigma_1\sigma_2 + B_9(\sigma_1^2 + \sigma_2^2) \quad (28a)$$

$$C_{13} = C_{31} = [B_1 + 2B_2\bar{I}_1 + 3B_6\bar{I}_1^2 + 2B_7\bar{I}_2] + (B_3 + 2B_7\bar{I}_1)(\sigma_1 + \sigma_3)$$
$$+ B_8\sigma_1\sigma_3 + B_9(\sigma_1^2 + \sigma_3^2)$$

$$C_{23} = C_{32} = [B_1 + 2B_2\bar{I}_1 + 3B_6\bar{I}_1^2 + 2B_7\bar{I}_2] + (B_3 + 2B_7\bar{I}_1)(\sigma_2 + \sigma_3)$$
$$+ B_8\sigma_2\sigma_3 + B_9(\sigma_2^2 + \sigma_3^2).$$

It is to be emphasized that the previous equations are valid only for the case when the principal axes of stresses and strains coincide and do not rotate as the material element deforms. In such cases, the principal axes of strain and stress increments also coincide. In general, the principal axes of stress and strain increments do not coincide; shearing stress increments will produce volumetric strains in addition to the shearing strains, and deviatoric and hydrostatic components of the response are always coupled. Such interaction and cross effects between the deviatoric and hydrostatic responses are extremely important in modeling such phenomena like dilatation or compaction and stress-(or strain-)induced anisotropy for granular materials. Moreover, as can easily be seen from Eqs. (11) and (26), the effect of the intermediate principal stress, σ_2, which is related to the direction of the stress path in the deviatoric plane [12], is accounted for by the inclusion of the third stress invariant $\bar{I}_3$. The importance of these phenomena has been supported by experimental results, and it is desirable to include them in the mathematical model.

It is a limitation of most elasticity-based models that behavior in unloading is not correctly described; these models are basically intended for use in cases where monotonically increasing loads prevail. Since there is no explicit yield (or loading) surface in the elasticity models, the definition of loading and unloading has no clear cut meanings. This has naturally led to the introduction of loading functions in the theory of plasticity. However, in the present formulation, an approximate method of modeling unloading and reloading behavior will be described and used in the prediction for some tests on sand. Furthermore, a failure condition will be postulated in order to determine the limiting state of stress.

5. Unloading - Reloading Behavior and Failure Condition

5.1 Approximate Modeling of Unloading - Reloading

The observations and studies made in Refs. [8, 10, 15, 28] have shown that unloading and reloading behavior of many soils is very nearly linear and elastic in nature. Further, this behavior is independent of the stress and strain levels at which unloading starts. For example, in conventional triaxial compression tests (Fig. 5), unloading and reloading at different stress levels, A and B, will have essentially the same slope which is nearly the same as the slope of the initial tangent, as shown in Fig. 5. Actual behavior of soils will show a small hysteresis loop as that shown at point A in Fig. 5. Based one these observations, the unloading and reloading (to the maximum previous stress level) is approximated herein as being linear elastic. This behavior is completely described by any two of the familiar elastic moduli (E, ν, K, G, and M).

The values of these two parameters are approximately chosen to be those of the initial tangent moduli. In the following two expressions for the initial tangent Young's modulus, E_i, and Poisson's ratio, ν_i, are developed from the general expressions in the conventional tests.

Considering the expressions given in Eqs. (18) for the CTC test, it can easily be shown that the initial moduli E_i and ν_i are given by:

$$\frac{1}{E_i} = [\frac{\partial \Delta \varepsilon_1}{\partial \lambda}]_{\lambda=0} = (B_1 + B_4) + (6B_2 + 5B_3 + 2B_5)\sigma_c + (27B_6 + 24B_7 + \frac{5}{2}B_8 + 8B_9)\sigma_c^2 \quad (29a)$$

and

$$\nu_i = -[\frac{\partial \Delta \varepsilon_2}{\partial \Delta \varepsilon_1}]_{\lambda=0} = -[\frac{\partial \Delta \varepsilon_2}{\partial \lambda} \frac{\partial \lambda}{\partial \Delta \varepsilon_1}]_{\lambda=0} \quad ; \text{ i.e.}$$

$$\nu_i = -[\frac{B_1 + (6B_2 + 2B_3)\sigma_c + (27B_6 + 15B_7 + B_8 + 2B_9)\sigma_c^2}{(B_1+B_4) + (6B_2+5B_3+2B_5)\sigma_c + (27B_6+24B_7+\frac{5}{2}B_8+8B_9)\sigma_c^2}] \quad (29b)$$

These expressions will be used in modeling the unloading-reloading behavior. As can be seen from Eqs. (29a, b), both elastic moduli depend on the initial value of the hydrostatic (confining) pressure as has been experimentally demonstrated.

Finally, in order to complete the formulation of the approximate method of modeling the loading-unloading behavior, it is necessary to postulate a loading criterion. In simple test cases, unloading and reloading can be easily visualized from the examination of the stress-strain curves. However, for a completely general case, a clear well-defined criterion for unloading that is the same in any coordinate system (i.e., invariant) is needed. Herein, a simple unloading and reloading condition is used. This condition is expressed in terms of the complementary energy function Ω defined earlier which is invariant with respect to coordinate transformation. Unloading is indicated by the condition $\dot{\Omega} < 0$, where $\dot{\Omega} = \varepsilon_{ij} d\sigma_{ij}$

is the incremental change in Ω. The condition $\dot{\Omega} > 0$ indicates loading. Reloading is defined by the condition $\dot{\Omega} > 0$ and $\Omega < \Omega_{max}$, where Ω_{max} is the maximum previous value of Ω at the material point. Mathematically, these general conditions may be written as:

$$\begin{aligned}&\text{Loading:} \quad \text{when } \Omega = \Omega_{max} \text{ and } \dot{\Omega} > 0 \\ &\text{Unloading:} \quad \text{when } \Omega \leq \Omega_{max} \text{ and } \dot{\Omega} < 0 \\ &\text{Reloading:} \quad \text{when } \Omega < \Omega_{max} \text{ and } \dot{\Omega} > 0\end{aligned} \quad (30)$$

For cases of unloading or reloading the moduli of Eqs. (29a, b) apply, while for loading the expressions of Eqs. (26) or (27) are used. In terms of stresses, $\dot{\Omega}$ can be calculated using Eq. (11a) as

$$\dot{\Omega} = \frac{\partial \Omega}{\partial \sigma_{ij}} \dot{\sigma}_{ij} = \phi_1 \dot{\bar{I}}_1 + \phi_2 \dot{\bar{I}}_2 + \phi_3 \dot{\bar{I}}_3 \quad (31)$$

in which the following relations for the increment invariants $\dot{\bar{I}}_1$, $\dot{\bar{I}}_2$ and $\dot{\bar{I}}_3$

$$\dot{\bar{I}}_1 = \dot{\sigma}_{kk} \; ; \quad \dot{\bar{I}}_2 = \sigma_{ij} \dot{\sigma}_{ij} \quad \text{and} \quad \dot{\bar{I}}_3 = \sigma_{im} \sigma_{jm} \dot{\sigma}_{ij} \quad (31a)$$

have been used. The only objection of the present definition of loading and unloading, as for most variable-moduli models, is the ambiguity encountered at the neutral loading condition $\dot{\Omega} = 0$, where one may arbitrarily assign either value of the loading or unloading moduli. The result is that infinitesimal stress changes near neutral loading may produce finite strain changes, and continuity condition may be violated which is not physically acceptable. However, apart from severe multi-dimensional loading conditions, many practical solutions involve moderate loading conditions and loading paths near neutral loading are not likely to occur most often.

5.2 Mohr-Coulomb Failure Condition

For a general practical finite element, a failure condition must be postulated in order to determine the limiting state of stress. For granular materials, different failure conditions have been proposed and used, such as Mohr-Coulomb and Drucker-Prager conditions [3, 9, 24]. Experimental results have shown that Mohr-Coulomb criterion is among the best failure conditions and it yields reasonably accurate results for most soils. Herein, this failure criterion is employed for the three soils described.

In the π- (deviatoric) plane, the Mohr-Coulomb condition is represented by an irregular hexagon, as shown in Fig. 6. The general expression for Mohr-Coulomb condition may be written as [3]:

$$f = \sigma_1(1 - \sin\phi) - \sigma_3(1 + \sin\phi) - 2c \cos\phi = 0 \quad (32)$$

where σ_1 = the major (max.) principal stress, σ_3 = the minor (min.) principal stress, c = cohesion and ϕ = angle of internal friction of the soils, and soil mechanics sign convention is used (compression is positive). For undrained conditions, effective stresses and effective values for c and ϕ are used.

6. Theoretical Conditions: (Uniqueness and Stability)

It is a desirable feature for any mathematical theory describing the mechanical behavior of materials that the resulting solutions for practical problems are unique and exhibit stable equilibrium configurations. These characteristics are generally to be expected for most actual physical situations. The stability and uniqueness requirements and their implications for elasticity based constitutive models are discussed in Ref. [4], based on Drucker's material stability postulate [10]. The implications of Drucker's stability postulate on the constitutive relations presented may be summarized in the following:

(1) Although the present formulation is based on an assumed function for Ω, stability postulate assures the existence of the strain energy density function, W such that $W + \Omega = \sigma_{ij}\varepsilon_{ij}$, and the laws of thermodynamics are always satisfied. Moreover, the Hessian matrices [H] and [H'] for both functions W and Ω, respectively, are positive definite, where the components of these matrices are defined as [12]:

$$H_{ijkl} = \frac{\partial^2 W}{\partial \varepsilon_{ij} \partial \varepsilon_{kl}} \quad ; \quad H'_{ijkl} = \frac{\partial^2 \Omega}{\partial \sigma_{ij} \partial \sigma_{kl}} \tag{33}$$

(2) The surfaces W = const. and Ω = const. are always convex in the strain or stress space, respectively.

(3) Based on the positive definite character of the Hessian matrices, the inverse constitutive relations always exist. For example, the inverse relation of Eq. (27) exists and stress increments, $\dot{\sigma}_{ij}$, can always be uniquely determined in terms of the strain increments, $\dot{\varepsilon}_{ij}$. This is an extremely important requirement in the finite element formulation which always requires the material stiffness matrix $[D] = [C]^{-1}$.

Satisfaction of the above requirements can be achieved numerically during the solution process. This will guarantee the uniqueness of the results obtained for each step in incremental finite element solutions.

7. Comparison of Experimental With Theoretical Results

Herein, the numerical results obtained using the proposed model and the determined material constants will be compared to the experimental results used as data base for model formulation. Based on these comparisons and the discussion made earlier, a number of conclusions will be summarized in the following section.

For clay "X", typical stress-strain curves for some of the loading paths in the tests provided as data base are shown in Fig. 7. Good agreement is obtained between experimental and theoretical results in most of the cases. However, there are some discrepancies between calculated and measured strains for the case of σ_c = 20 psi and m = 1; the model underestimates the axial and lateral strains at high stress levels, as shown in Figs. 7(c) and (d). Considering the volumetric strains observed under constant mean normal stresses, the model correctly predicts volume changes. However, this only qualitatively, since the calculated volumetric strains are too small compared to the measured values, as can be seen from Fig. 7(h). This is expected in the present formulation since the model in its present form cannot account for material initial (inherent) anisotropy which has been observed for clays "X" and "Y". For example, for the case σ_c = 30 psi and m = 0, the model correctly predicts the axial strain, ε_1, and lateral strain, ε_3, but it gives extensional strains for $\varepsilon_2 (= \varepsilon_3)$ while the test measurements indicate compressive values at high stress levels. Obviously, this causes reduction in the calculated values for ε_v.

Comparison of the experimental with theoretical stress-strain curves for clay "Y" are illustrated in Fig. 8 for some of the tests provided. Again, a reasonably good overall agreement is observed in most cases. Unlike clay "X", the measured values of ε_v for clay "Y" are small, and the model reproduces them with a better agreement with the test results, as shown in Figs. 8(b), (d) and (f). As for clay "X", the calculated values for the lateral strains, ε_2, do not agree with the experimental results in many cases because of the initial anisotropy exemplified previously.

Different comparisons of the measured and calculated stress-strain curves for Ottawa sand are made in Fig. 9. The results in the CTE and TE tests, Figs. 9(a) to (d), are in good agreement with the measured values, both in terms of the axial and lateral strains. In the CTC tests, the discrepancies are more pronounced. Although, the initial soil behavior, at low stress levels, appears to be adequately represented by the model, the test data depart significantly from the calculated curves for large stress levels, particularly for the case σ_c = 5 psi, as shown in Figs. 9(g) and (h). However, the overall agreement for the cases shown and those which are not included here is reasonably good. Also, the approximate modeling of loading-unloading behavior is seen to be adequate, as can be observed in Figs. 9(c) and (d). The predicted failure stresses using Mohr-Coulomb condition agree very well with the test values, as can be seen in Figs. 9(e) and (g), for example. In Ref. [20], other cases were investigated using the same values of the material constants employed here, and good agreement with experiments was obtained (in the cases reported in [23], stress-strain curves were shown only for low stress levels below the failure values). Hence, based on the cases investigated both in Ref. [23] and in the present paper, it is believed that the values of the constants used will give the best results at low stress levels compared to the failure values.

In order to study the effect of the change in the values of the material constants on the behavior of the model, comparison is made in Fig. 10 for the two sets of constants given in sec. 3. The results shown are for a proportional loading stress path with $\Delta\sigma_x : \Delta\sigma_y : \Delta\sigma_z$ = 0.5 : 0.5 : 1, and σ_c = 10 psi. The results shown indicate that the behavior of the model becomes sensitive to changes in the constants as the stress level increases, and completely different results are obtained. This will be more pronounced for increasing proportional loading paths where large stress levels are generally expected before failure (e.g., near HC path). However, the strains occurring in stress paths near the HC path are generally small compared to those observed under most other loading conditions in practical engineering problems so that discrepancies of the results for these loading conditions may be less significant to overall soil behavior than discrepancies in other stress paths. The important point to make is that it is necessary to appreciate the conditions under which the model will be used, and to determine the material constants from tests performed under conditions selected to duplicate as many as possible of the expected field conditions.

8. Summary and Conclusions

The formulation of nonlinear hyperelastic constitutive model, originally developed by Ko and Masson [23], has been presented and applied to three different types of soils: clays "X" and "Y", and Ottawa sand. Detailed description of the procedure of determining the nine material constants in the model has been made and the stress-strain relations for general straight line stress paths and for examples of stress paths in conventional soil tests are given. Incremental forms of the constitutive relations, approximate method of modeling unloading-reloading behavior, and a failure condition are also included for general nonlinear finite element analyses. Finally, comparisons of the results obtained with the experimental measurements are made.

Against the background of the discussion and comparisons made in the present study, the following conclusions, concerning the advantages and limitations of the proposed procedures, can be made:

(1) The proposed constitutive relations can model many of the characteristics of soil behavior such as: nonlinearity, stress-path dependency, dilatation, stress-induced anisotropy, effect of the confining (or hydrostatic) stress, the effect of the third stress invariant, and the noncoincidence of the principal axes of stress and strain increment tensors, especially near failure. However, the present formulation is limited to initially isotropic materials.

(2) When used for monotonically increasing loading conditions, the present formulation satisfies all the rigorous mathematical requirements such as uniqueness, stability and continuity. For cases where general unloading-reloading conditions are expected, the present approximate criterion proposed for loading and unloading fails to satisfy the continuity condition at or near neutral loadings. Further refinement is needed concerning this aspect.

(3) Once the nine material constants are determined, the incremental form of the model can be easily implemented in finite element codes for general analyses. The method described for material constants determination allows a great flexibility for inclusion as many test data as possible in the fitting procedure. However, the procedure is not easy to apply; many trials for fitting the cubic curves may be needed in order to obtain reasonable results.

(4) For the cases investigated, the model gives a reasonably good overall agreement with the experimental results for the tests used as data base in the model formulation. But the model fails to predict the large volumetric strains for clay X, and the behavior near failure for Ottawa sand in CTC and TC tests. For almost all of the cases studied, the model does not correctly predict the values of the principal strain ε_2 for

clays "X" and "Y". This is expected because of the initial anisotropy observed for both soils which cannot be accounted for in the present formulation.

(5) The model behavior for Ottawa sand is found to be sensitive to changes in values of the material constants for increasing proportional loading stress paths where large stress levels are generally expected before failure. Duplication of as many expected field conditions as possible in the tests used for model calibration is generally recommended to reduce such effects.

(6) Best results from the model are generally expected at low stress levels below failure. This is usually the range where most of the elasticity-based models are frequently used.

(7) The present formulation cannot account for post-failure behavior in strain-softening materials (e.g., dense sand) since it indicates increasing strains for increasing stresses (work-hardening type).

Acknowledgments

This material is based upon work supported by the National Science Foundation under Grant No. PFR-7809326 to Purdue University.

References

1. Chang, T. Y., et al., "An Integrated Approach to Stress Analysis of Granular Materials," Lab. Report, California Institute of Technology, Soil Mechanics Laboratory, 1967.

2. Chen, W. F., "Plasticity in Soil Mechanics and Landslides," Journal of the Engineering Mechanics Division, ASCE, vol. 106, no. EM3, June, 1980, pp. 443-464.

3. Chen, W. F., Limit Analysis and Soil Plasticity, Scientific Publishing Publishing Co., Elsevier, Amsterdam, The Netherlands, 1975.

4. Chen, W. F., and Saleeb, A. F., "Constitutive Equations for Engineering Materials," vol. 1. "Elasticity and generalized stress-strain models." Lecture Notes in Structural Engineering, School of Civil Engineering, Purdue University, West Lafayette, Indiana, 1980.

5. Clough, R. W., and Woodward, R. J. III, "Analysis of Embankment Stress and Deformations," Journal of Soil Mechanics and Foundations Division, ASCE, vol. 93, no. SM4, July, 1967, pp. 529-549.

6. Coon, M. D., and Evans, R. J., "Recoverable Deformation of Cohesionless Soils," Journal of the Soil Mechanics and Foundations Division, ASCE, vol. 97, no. SM2, Feb., 1971, pp. 375-391.

7. Corotis, R. B., Frazin, M. H., and Krizek, R. J., "Nonlinear Stress-Strain Formulation for Soils," Journal of the Geotechnical Engineering Division, ASCE, vol. 100, No. GT9, Sept., 1974, pp. 993-1008.

8. Domaschuk, L., and Wade, N. H., "A Study of Bulk and Shear Moduli of Sand," Journal of the Soil Mechanics and Foundations Divison, ASCE, vol. 95, no. SM2, March, 1969, pp. 561-582.

9. Drucker, D. C., and Prager, W., "Soil Mechanics and Plastic Analysis or Limit Design," Quarterly of Applied Mathematics, vol. 10, no. 2, 1963, pp. 157-165.

10. Drucker, D. C., "A More Fundamental Approach to Plastic Stress-Strain Relations," Proceedings, 1st U.S. National Congress on Applied Mechanics, 1951, pp. 487-491.

11. Duncan, J. M., "Finite Element Analyses of Stresses and Movements in Dams, Excavation and Slopes," Proc. Symposium on Applications of the Finite Element Method in Geotechnical Engineering, Vicksburg, Mississippi, USA, May, 1972, pp. 267-326.

12. Duncan, J. M., and Chang, C-Y, "Nonlinear Analysis of Stress and Strain in Soils," Journal of the Soil Mechanics and Foundations Division, ASCE, vol. 96, no. SM5, September, 1970, pp. 1629-1653.

13. El-Sohby, M. A., "Deformation of Sands Under Constant Stress Ratios," Proc. of the 7th Int. Conference on Soil Mechanics and Foundation Engineering, Int. Soc. of Soil Mechanics and Foundation Engineering, vol. 1, 1969, pp. 111-119.

14. El-Sohby, M. A., "Elastic Behavior of Sand," *Journal of the Soil Mechanics and Foundations Division*, ASCE, vol. 95, no. SM6, Nov., 1969, pp. 1393-1409.

15. Eringen, A. C., *Nonlinear Theory of Continuous Media*, McGraw-Hill Book Co. Inc., New York, N.Y., 1962.

16. Evans, R. J., and Pister, K. S., "Constitutive Equations for a Class of Nonlinear Elastic Solids," *International Journal of Solids and Structures*, vol. 2, no. 3, 1966, pp. 427-445.

17. Girijavallabhen, C. V., and Reese, L. C., "Finite Element Method for Problems in Soil Mechanics," *Journal of the Soil Mechanics and Foundations Divison*, ASCE, vol. 94, no. SM2, March, 1968, pp. 473-496.

18. Hardin, B. O., "The Nature of Stress-Strain Behavior for Soils," State-of-the-Art *Report for the ASCE Specialty Conference on Earthquake Engineering and Soil Dynamics*, Pasadena, California, Proceedings, vol. 1, June, 1978, pp. 3-90.

19. Holubec, I., "Elastic Behavior of Cohesionless Soil," *Journal of the Soil Mechanics and Foundations Divison*, ASCE, vol. 94, no. SM6, Nov., 1968, pp. 1215-1231.

20. Janbu, N., "Soil Compressibility as Determined by Oedometer and Triaxial Tests," *Proceedings, European Conference on Soil Mechanics and Foundation Engineering*, Wiesbaden, Germany, vol. 1, 1963, pp. 19-25.

21. Ko, H. Y., and Scott, R. F., "Deformations of Sand in Hydrostatic Compression," *Journal of the Soil Mechanics and Foundations Division*, ASCE, vol. 93, no. SM3, May, 1967, pp. 137-156.

22. Ko, H. Y., and Scott, R. F., "Deformation of Sand in Shear," *Journal of the Soil Mechanics and Foundations Divison*, ASCE, vol. 93, no. SM5, Sept., 1967, pp. 283-310.

23. Ko, H. Y., and Masson, R. M., "Nonlinear Characterization and Analysis of Sand," *Numerical Methods in Geomechanics*, ASCE, 1976, pp. 294-305.

24. Lade, P. V., "The Stress-Strain and Strength Characteristics of Cohesionless Soils," Thesis presented to the University of California at Berkeley, in 1972, in partial fulfillment of the requirements for the degree of Doctor of Philosophy.

25. Makhlouf, H. M., and Stewart, J. J., "Factors Influencing the Modulus of Elasticity of Dry Sand," *Proceedings, 6th International Conference on Soil Mechanics and Foundations Engineering*, Montreal, vol. 1, 1965, pp. 298-302.

26. Nelson, I., and Baladi, G. Y., "Outrunning Ground Shock Computed with Different Models," *Journal of the Engineering Mechanics Division*, ASCE, vol. 103, no. EM3, June, 1977, pp. 377-393.

27. Nelson, I., and Baron, M. L., "Application of Variable Moduli Models to Soil Behavior," *Int. Journal of Solids and Structures*, vol. 7, 1971, pp. 399-417.

28. Romano, M., "A continuum Theory for Granular Media with a Critical State," *Archives of Mechanics*, vol. 26, no. 6, 1973, pp. 1011-1028.

29. Scott, R. F., and Ko, H. Y., "Stress Deformation and Strength Characteristics," *Proceedings of the 7th Int. Conference on Soil Mechanics and Foundation Engineering*, International Society for Soil Mechanics and Foundation Engineering, vol. 1, 1969, pp. 1-49.

30. Smith, I. M., and Kay, S., "Stress Analysis of Contractive or Dilative Soil," *Journal of the Soil Mechanics and Foundations Division*, ASCE, vol. 97, no. SM7, July, 1971, pp. 981-997.

31. Truesdell, C., "Hypo-elasticity," *Journal of Rational Mechanics and Analysis*, vol. 4, no. 1, 1955, pp. 83-133.

32. Vagneron, J., Lade, P. V., and Lee, K. L., "Evaluation of Three Stress-Strain Models for Soils," *Numerical Methods in Geomechanics*, ASCE, 1976, pp. 1329-1351.

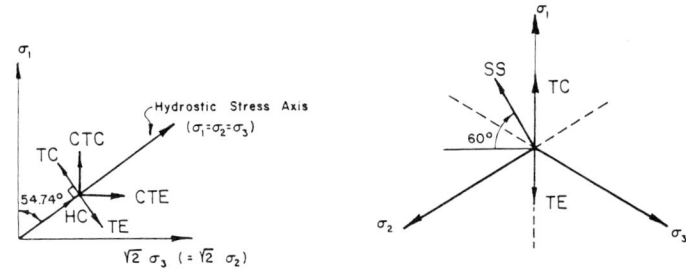

(a) Stress Paths in the Triaxial Stress Plane

(b) Stress Paths in the Deviatoric (π-) Plane

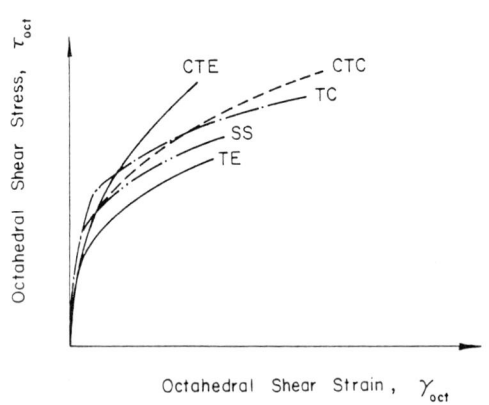

(c) Octahedral Shear Stress-Strain Curves

FIG 1. Typical Octahedral Shear Stress-Strain Curves for Ottawa Sand for Different Stress Paths.

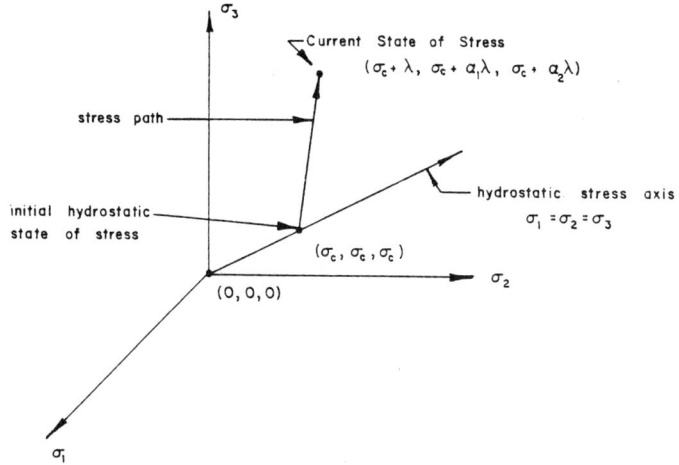

FIG 2. General Straight Line Stress Path in Principle Stress Space.

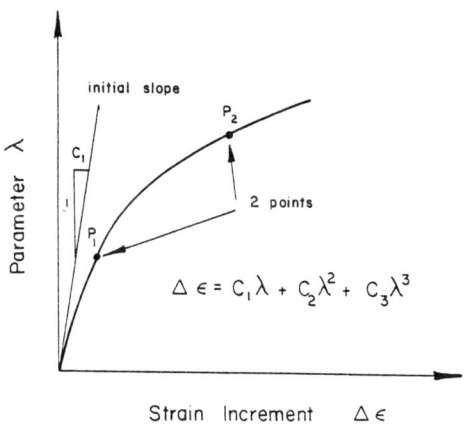

FIG 3. Fitting Procedure for Cubic Curves

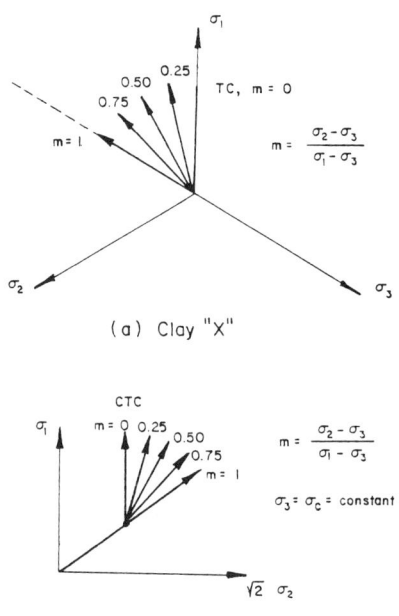

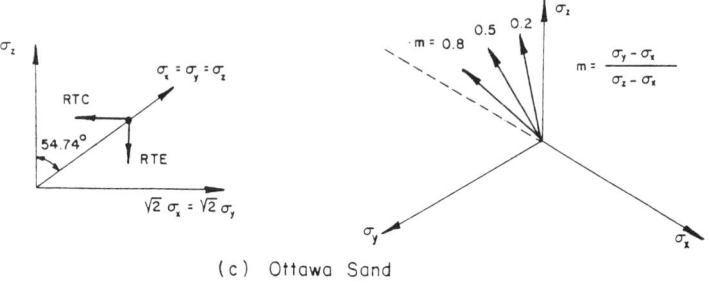

FIG 4. Stress Paths for Prediction for Clay "X", Clay "Y" and Ottawa Sand.

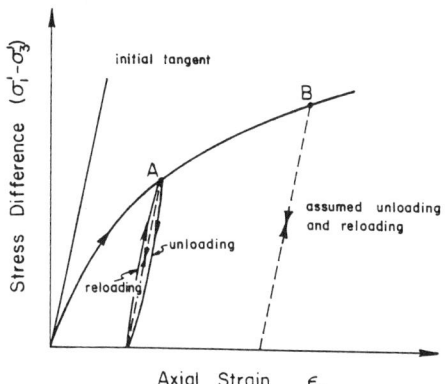

FIG 5. Approximate Representation of Loading-Unloading Behavior in Conventional Triaxial Compression Test

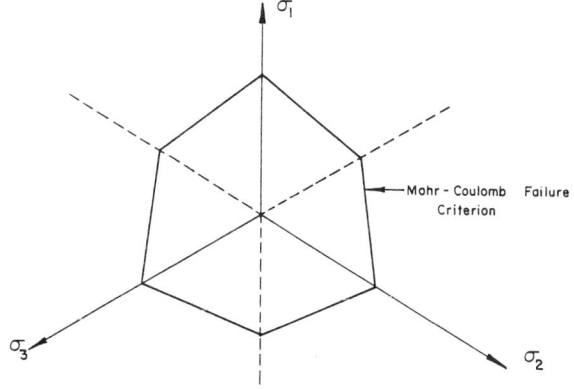

FIG 6. Mohr-Coulomb Failure Criterion in the Deviatoric Plane

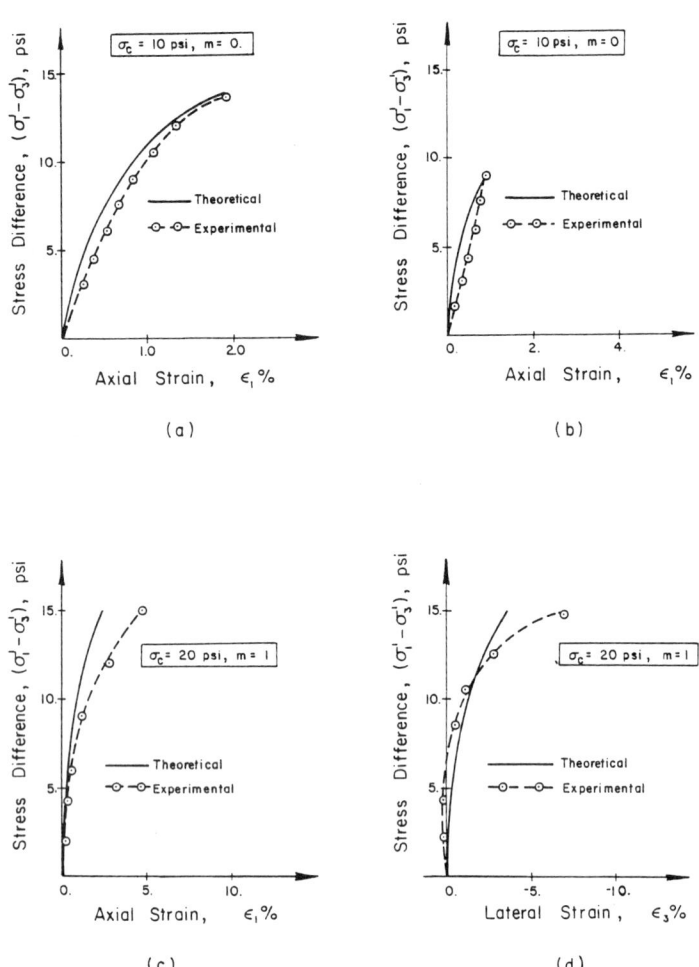

FIG 7. Comparison of Theoretical and Experimental Stress-Strain Curves For Clay "X", (TC Tests with Const. Mean Normal Stress)

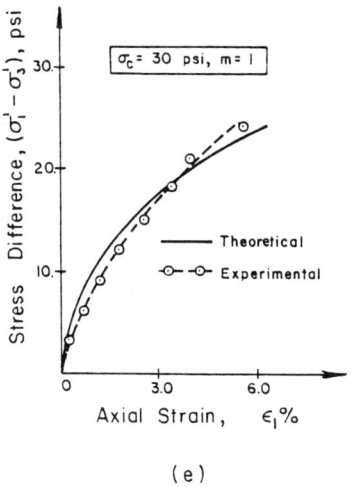

(e)

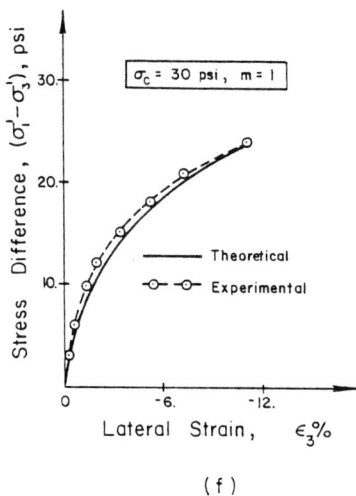

(f)

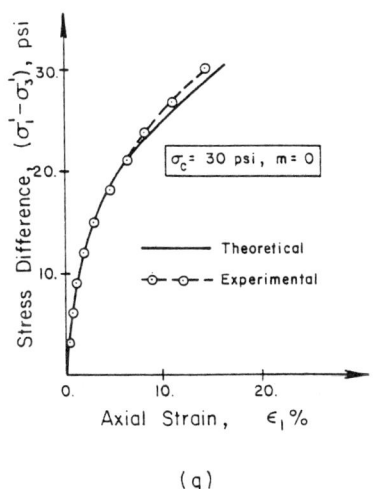

(g)

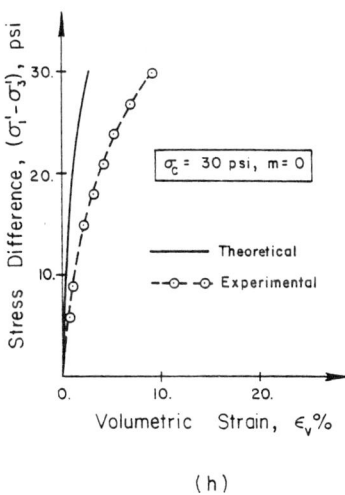

(h)

FIG 7. (Cont'd)

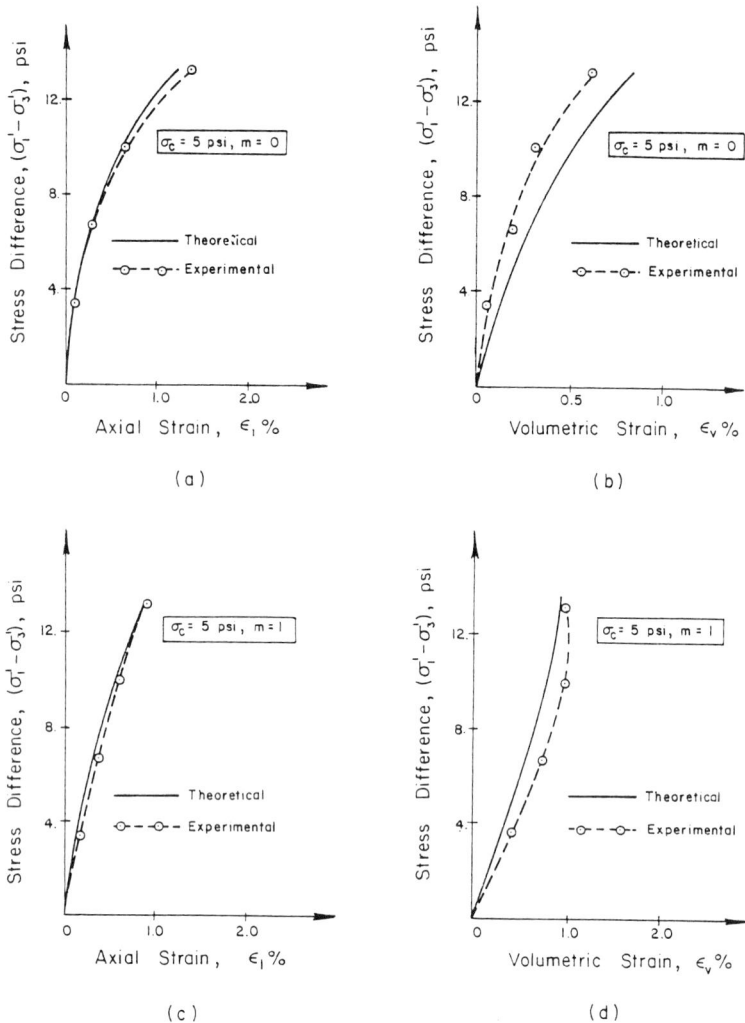

FIG 8. Comparison of Theoretical and Experimental Stress-Strain Curves for Clay "Y", (CTC Tests)

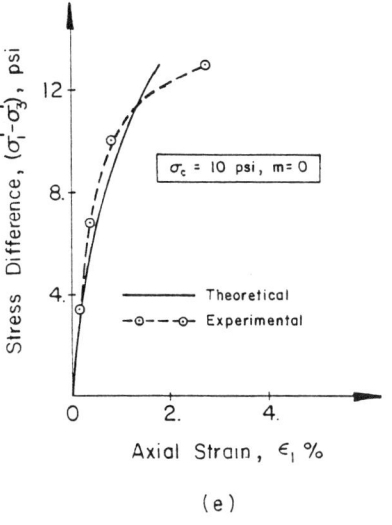
(e)

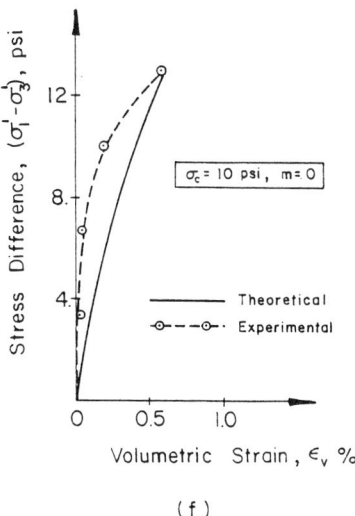
(f)

FIG 8. (Cont'd)

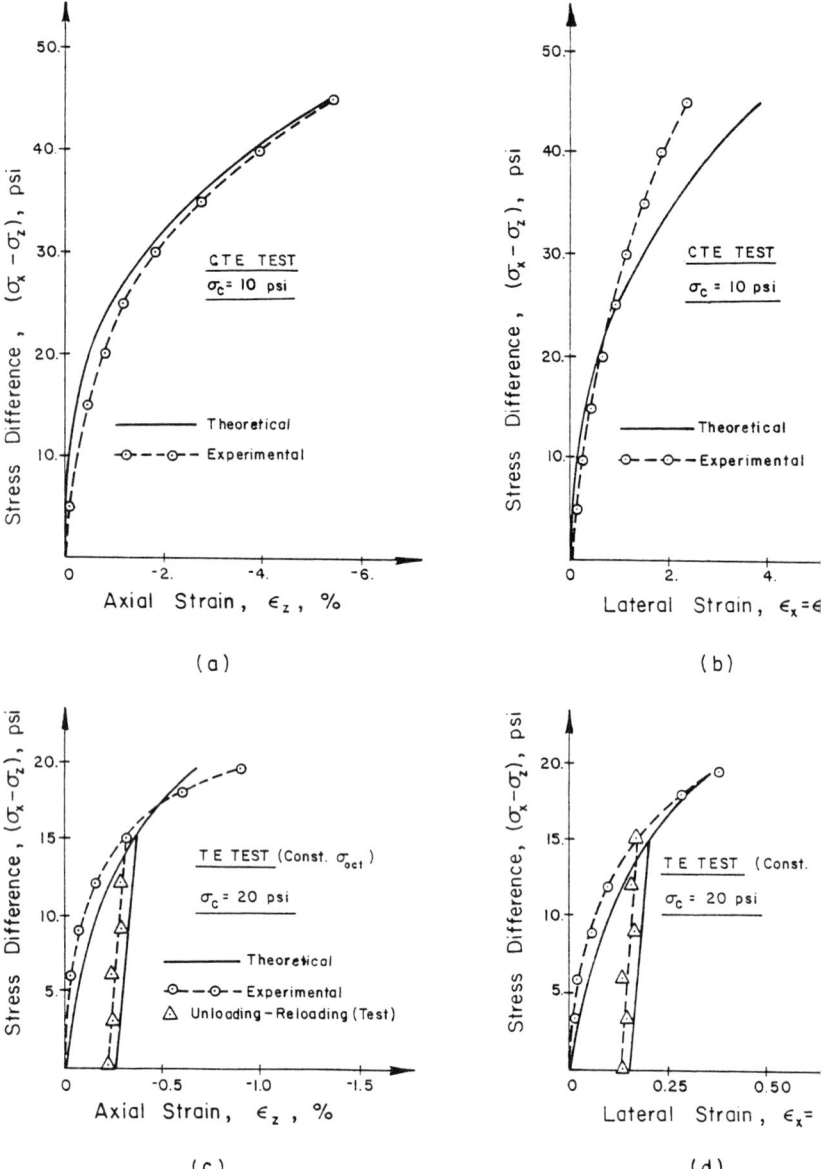

FIG 9. Comparison of Theoretical and Experimental Stress-Strain Curves for Ottawa Sand.

CONSTITUTIVE MODELS FOR SOILS 537

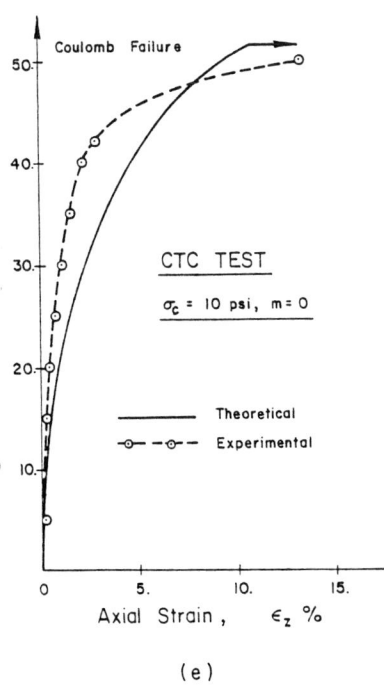

(e)

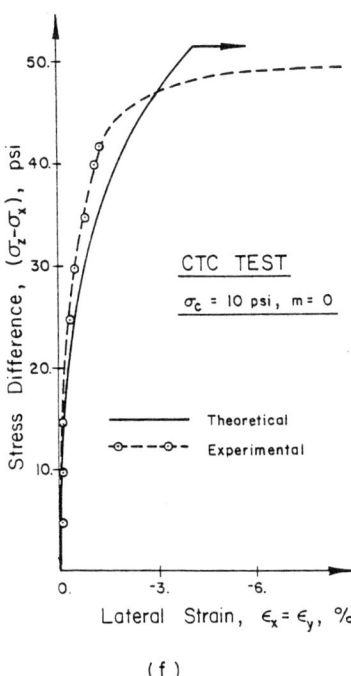

(f)

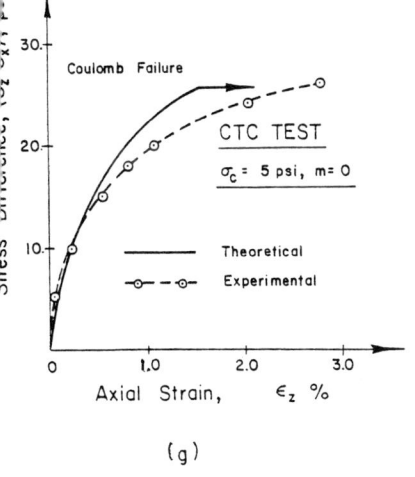

(g)

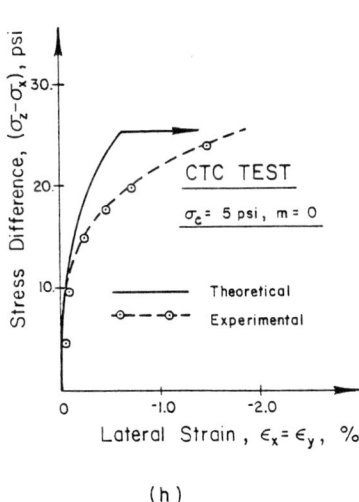

(h)

FIG 9. (Cont'd)

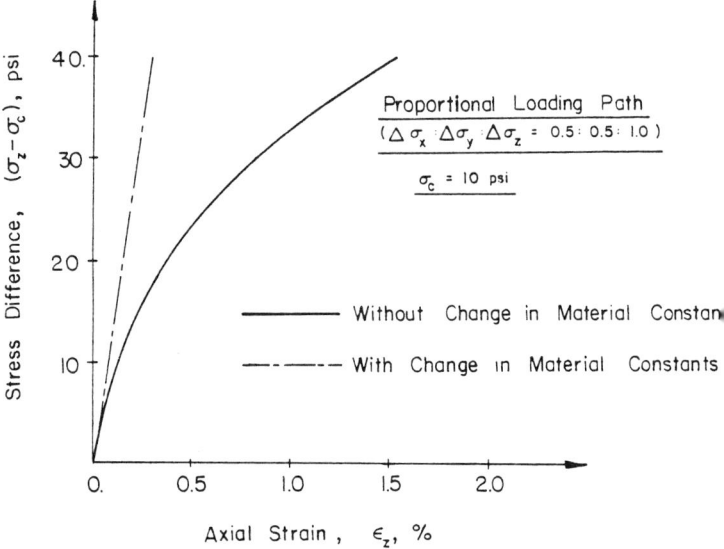

FIG. 10. Effect of Changes in Material Constants on the Model Behavior in Proportional Loading Path for Ottawa Sand

CRITICAL APPRAISAL OF ENDOCHRONIC THEORY FOR SOILS

by Zdeněk P. Bažant[1] M.ASCE, Atilla M. Ansal[2], and
Raymond J. Krizek[1] M.ASCE

INTRODUCTION

Although endochronic (endo = internal + chronos = time) theory may not follow strictly from the basic principles of continuum mechanics and thermodynamics, this novel approach to modeling the nonlinear constitutive behavior of soils and other similar materials is quite effective and versatile. Therefore, it is worthwhile to examine closely the capabilities and limitations of this theory and to determine its relationship to other established constitutive theories with emphasis on the advantages it offers. Accordingly, the basic features of the theory will be summarized and explained, and, after calibrating the model by use of a limited set of test data, predictions of the response to certain specified types of loading will be presented and critically interpreted in a companion paper included in Section 4 of Volume 1.

GENERAL FORM OF THE THEORY

Endochronic theory is best regarded as a special form of viscoplasticity in which the viscosity coefficients depend on the strain rate, as proposed by Schapery (1968). Thus, the general form of the resulting constitutive equation is

$$d\underline{\varepsilon} = \underline{C}\, d\underline{\sigma} + \frac{\partial F}{\partial \underline{\sigma}}\, d\zeta \tag{1}$$

in which $\underline{\varepsilon}$ represents a (6 x 1) column matrix of the components of the strain tensor, $\underline{\sigma}$ represents a similar column matrix of the stress components, $\underline{C} = \underline{C}(\underline{\sigma},\underline{\varepsilon})$ is a (6 x 6) incremental elastic stiffness matrix, $F = F(\underline{\sigma},\underline{\varepsilon})$ is the loading function (Figure 1a), and ζ is a non-decreasing independent scalar variable, termed the intrinsic time (Valanis, 1971), that depends on the

---

[1] Professor of Civil Engineering, Northwestern University, Evanston, Illinois, USA

[2] Assistant Professor of Geotechnical Engineering, Macka Civil Engineering Faculty, Istanbul Technical University, Istanbul, Turkey

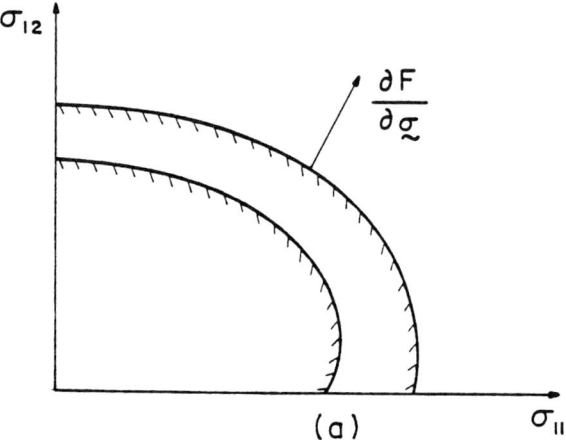

(a)

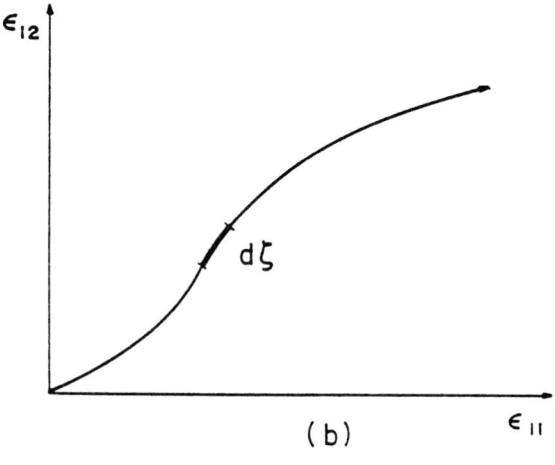

(b)

Figure 1

deformation. It is the use of this latter variable which distinguishes endochronic theory from other constitutive theories, such as incremental plasticity or hypoelasticity. The intrinsic time may be defined as the length of the path (Figure 1b) traced by successive states of the material in strain space and it may be expressed as

$$d\zeta = \sqrt{d\underline{\varepsilon}^T \underline{p} \, d\underline{\varepsilon}} \qquad (2)$$

in which $\underline{p}$ represents a (6×6) matrix of the coefficients characterizing the proper metric of the strain space and the superscript T indicates a transpose of the matrix In applications to soils containing pore water, the stress σ must be interpreted in accordance with a Biot-type two-phase medium model and may be associated with the effective stress commonly used in soil mechanics.

The forms of the metric matrix, $\underline{p}$, stiffness matrix, $\underline{C}$, and loading function, F, may be simplified considerably by evoking the conditions of initial isotropy of the material. In this case, the function F must be of the form:

$$F = F(I_1^\sigma, \; J_2^\sigma, \; J_3^\sigma, \; I_1^\varepsilon, \; J_2^\varepsilon, \; J_3^\varepsilon) \qquad (3)$$

in which I_1, J_2, and J_3 are the first invariant, second deviator invariant, and third deviator invariant, respectively, of either the stress tensor or the strain tensor, as indicated by superscripts σ and ε. In all practical forms of endochronic theory applied thus far to soils and concrete, the dependence of F on the third invariants, J_3^σ and J_3^ε, has not been considered; hence, equation (3) simplifies to

$$F = F(I_1^\sigma, \; J_2^\sigma, \; I_1^\varepsilon, \; J_2^\varepsilon) \qquad (4)$$

Despite omission of the third invariants, the failure envelope obtained from the peak points of the response curves for simulated triaxial tests with proportional loading at various stress ratios manifests the form illustrated in Figure 2. Thus, a rounded triangular shape of the failure surface in the octahedral plane does not necessarily indicate an influence of the third invariants. Such a shape of the loading surface may be interpreted as an indication of the simultaneous influence of the first and second invariants of both stress and strain. Indeed, when F depends only on the first and second stress

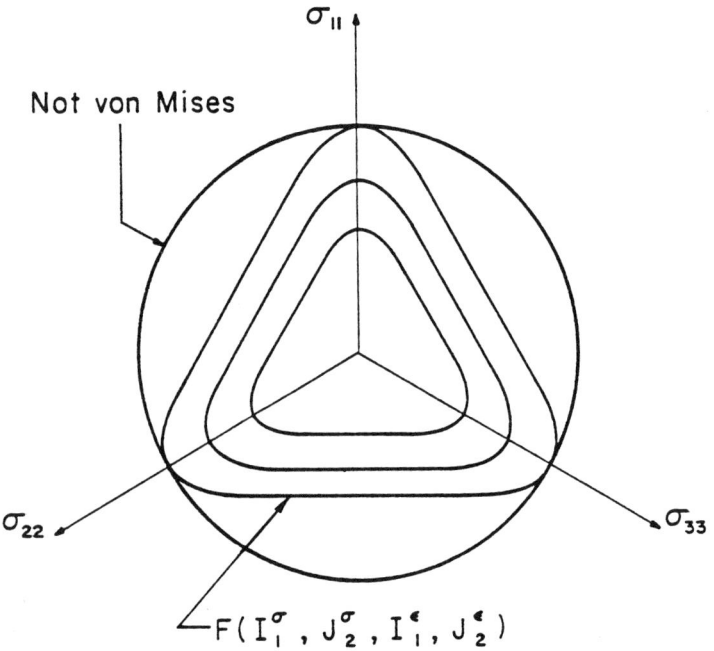

Figure 2

invariants, the projection of the failure surface in the octahedral plane is a circle, but it cannot remain a circle when F depends also on the first and second strain invariants because the strain components are generally not proportional to the stress components.

One important aspect worth noting is the fact that endochronic theory can not be brought to the incrementally linear form:

$$d\underline{\sigma} = \underline{D}(\underline{\sigma},\underline{\varepsilon})\ d\underline{\varepsilon} \tag{5}$$

in which $\underline{D}$ is the incremental stiffness matrix for the total (elastic plus inelastic) strains. This simple fact distinguishes endochronic theory from most other constitutive theories, such as hypoelasticity, incremental hardening plasticity, plastic-fracturing theory, and total strain theory. However, if loading directions in the vicinity of a certain fixed direction are restricted, it is possible to linearize endochronic theory and obtain equation (5). In such a case a different incremental stiffness matrix, $\underline{D}$, is obtained for each choice of the straining direction in whose vicinity the theory is to be linearized.

TREATMENT OF UNLOADING AND RELOADING

Much of the effectiveness and flexibility of endochronic theory is due to the fact that it can model the irreversibility associated with unloading without the use of any inequalities. The feature which renders this possible is the use of the square root in a quadratic form in the definition of intrinsic time (equation 2). In the case of shear straining, this expression reduces to $|d\gamma|$ and the increment of shear stress, $d\tau$, may be expressed as

$$d\tau = G d\gamma - \tau F_1 |d\gamma| \tag{6}$$

in which γ is the shear angle, G is the shear modulus, and F_1 is a certain function of stress and strain. If loading ($d\gamma > 0$) changes to unloading ($d\gamma < 0$), the first (elastic) term of equation (6) changes sign while the second (inelastic) part of the stress increment retains the same sign, as indicated by the vertical downward arrows in Figure 3a. This illustrates visually that the unloading slope given by endochronic theory must be smaller than the previous loading slope. In the case of alternating loads, this property enables endochronic theory to describe hysteresis loops without any use of an unloading criterion (Figure 3b).

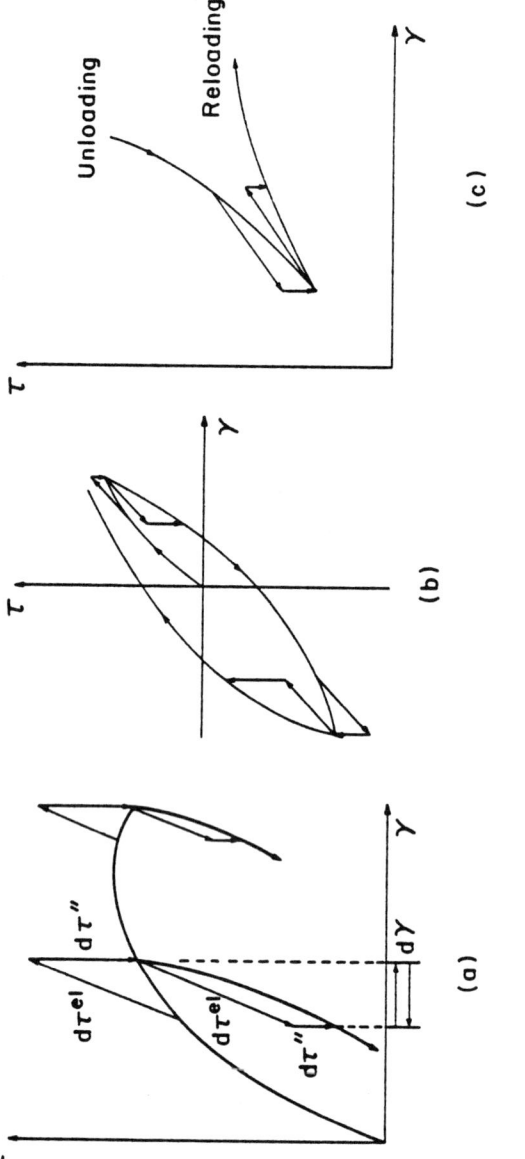

Figure 3

The foregoing attractive property, however, works to the disadvantage of endochronic theory when a small unloading is followed by reloading. In this case, one obtains a reloading slope that is smaller than the previous unloading slope and the response does not form a closed loop (Figure 3c). Although this is not strictly unacceptable in all situations, a reloading slope that is steeper than the unloading slope should be obtained in most cases (for example, pure deviatoric straining). The only way to achieve such behavior with endochronic theory is to avoid relying on the intrinsic time alone to model the irreversibility phenomenon and to introduce an unloading criterion.

One method to obtain closed hysteresis loops for small unload-reload cycles and assure fulfillment of Drucker's postulate was described by Bažant (1978) and applied by Bažant, Krizek, and Shieh (1980). This method consists of two relatively simple corrections. First, the intrinsic time increment, $d\zeta$, must be replaced by $c\,d\zeta$, in which c is a correction coefficient which is taken as unity for virgin loading and less than unity for unloading and reloading. The fulfillment of a certain condition on the values of c for unloading, reloading, and virgin loading is also necessary to satisfy Drucker's stability postulate (Bažant, 1978). Furthermore, a three-way loading-unloading-reloading criterion is needed (Figure 4a). Unloading is characterized by $dW < 0$, in which W is some loading function (for example, the work stored in the material), and virgin loading, as well as reloading, are characterized by $dW > 0$. The distinction between the latter two conditions may be made on the basis of the maximum energy, W_o, stored in the material up to the current time. If W_o is larger than the current W, we have reloading, and, if W_o is equal to W, we have virgin loading. The second correction which must be introduced in endochronic theory is a certain particular form of kinematic hardening, called jump-kinematic hardening (Bažant, 1978) which involves moving the center of the loading surface to the last extreme stress point whenever loading changes to unloading or unloading to reloading. The points to which the center of the loading surface is "jumped" are indicated in the stress-strain diagram in Figure 4b, and in the stress-space plots of the subsequent loading surfaces in Figures 4c, 4d, and 4e. With the use of kinematic hardening it is impossible for the current state point to move inward from the loading surface; only outward movement from the current loading surface can be obtained.

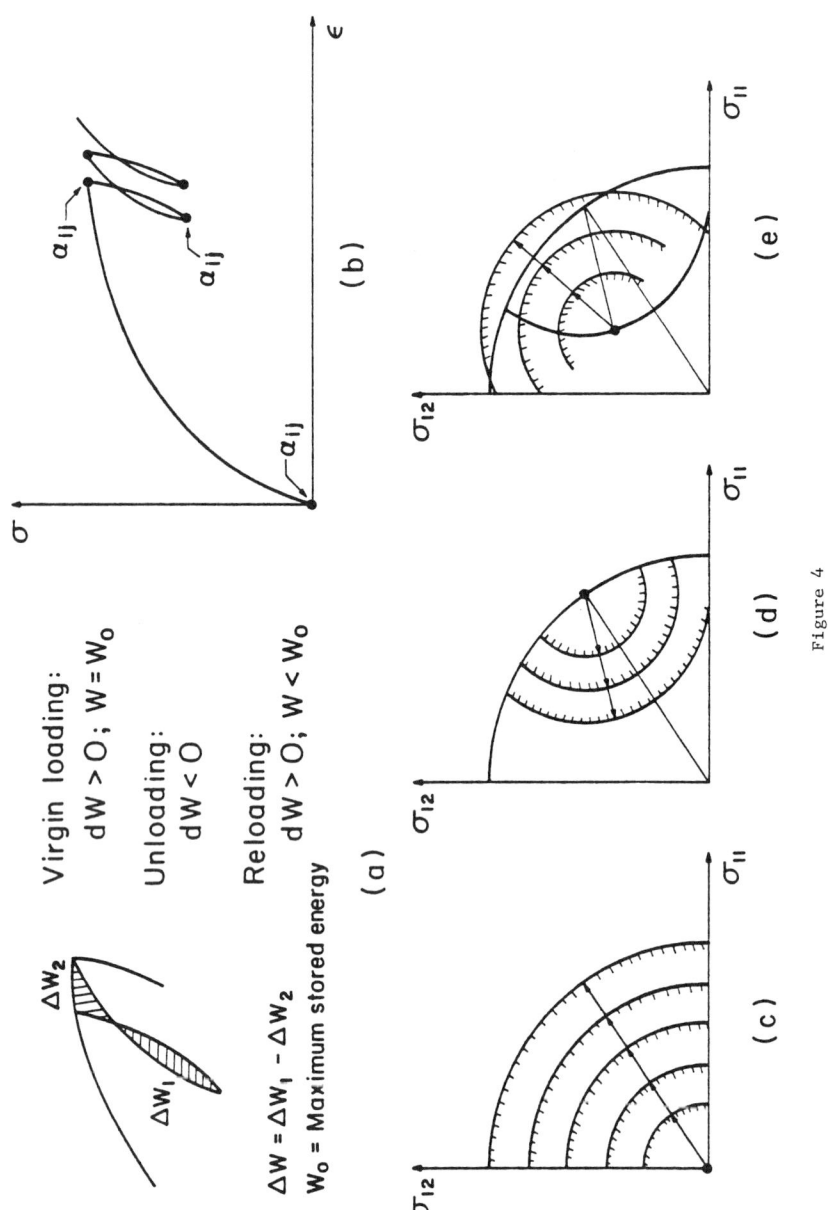

Figure 4

ENDOCHRONIC SOILS THEORY

ESSENTIAL DIFFERENCES FROM OTHER THEORIES

The most significant difference between various constitutive models, such as plastic, endochronic, and total strain (deformation) models, is the behavior for so-called loading to the side (that is, loading which is parallel to the loading surface and approximately normal to the proportional loading path in stress space). The stiffness of the inelastic response for loading to the side (and generally the stiffness of the response for loading of any direction) may be characterized and instructively visualized by means of the inelastic stiffness locus (Bažant, 1978). This locus is defined as the locus of all strain increment vectors, $d\varepsilon_{ij}$ (that is, a nine-dimensional vector formed of the components of $d\varepsilon_{ij}$), that give the same magnitude $\|d\varepsilon''_{ij}\|$ of the inelastic strain increments, $d\varepsilon''_{ij}$. The distance of a point on the locus for any loading direction is proportional to the stiffness modulus for inelastic strain in that direction.

It can be shown (Bažant, 1978) for incremental plasticity that the inelastic stiffness locus is a straight line parallel to the tangent to the loading surface (Figure 5a). The fact that this locus must be a straight line is also evident from the linearity property of incremental plasticity. Since the loading direction parallel to the loading surface intersects the elastic stiffness locus at infinity, the response to such loading is obtained as perfectly elastic. Recently, however, it has become widely accepted that the actual response resulting from loading to the side is not or should not be perfectly elastic, but softer than elastic. For this reason, various theories which introduce inelastic strain due to loading to the side are being developed. This trend is reflected in the vertex-hardening models for the plasticity of metals and other materials. In general, the vertex concept does not involve a fixed vertex (corner) at a predetermined place on the loading surface, but a vertex which is always superimposed on the current stress point of the loading surface and moves jointly with this point. Conceptually, the most simple and effective vertex model appears to be that of Rudnicki and Rice (1975) for which the inelastic stiffness locus is obtained as the smooth curve shown in Figure 5b.

For endochronic theory it can be shown that the inelastic stiffness locus is either a circle centered around the current stress point on the loading

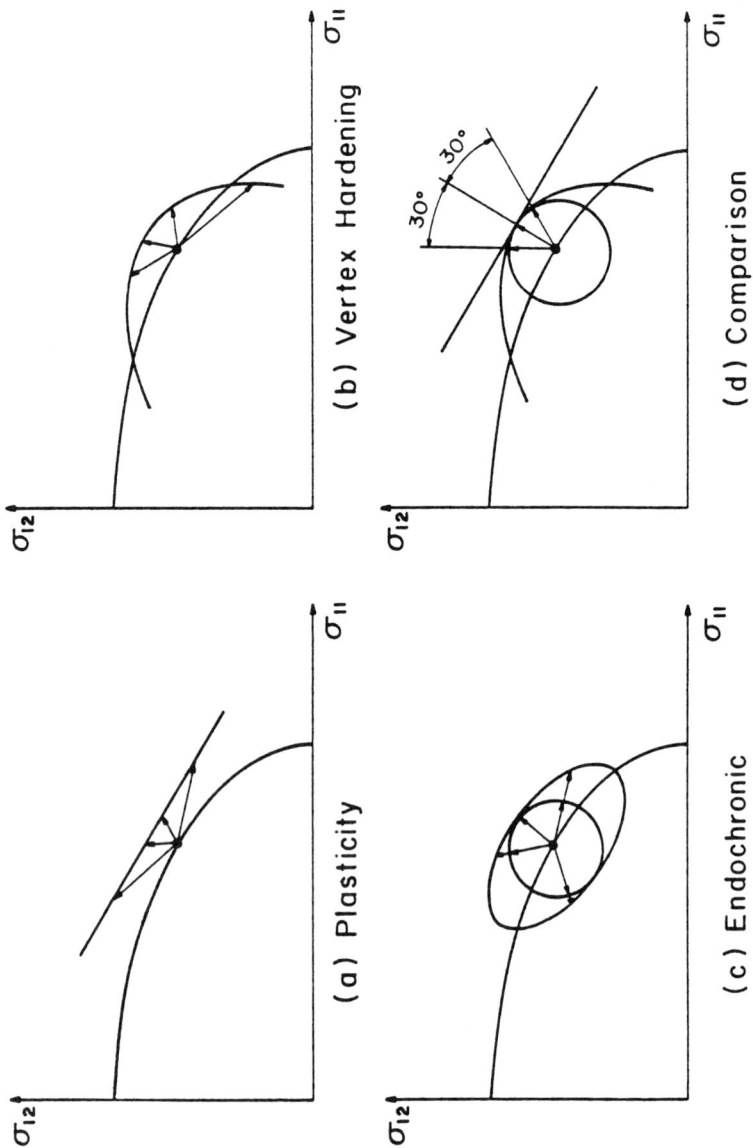

Figure 5

surface or an ellipse (Figure 5c). Since for this shape of the inelastic stiffness locus the tip of the vector $d\varepsilon_{ij}$ for loading to the side or for unloading is at a finite distance from the loading surface, the response for these loading directions is inelastic. Accordingly, given that one accepts the widespread opinion that the response for loading tangent to the loading surface is not perfectly elastic, endochronic theory appears to be more realistic. It should also be noted that the curved form of the inelastic stiffness locus for endochronic theory is similar to that for vertex-hardening plasticity, as well as for the total strain (deformation) theory. In view of the curved form of the inelastic stiffness locus, endochronic theory represents a development in the same direction as the introduction of vertex-hardening models in plasticity.

It has been mentioned that endochronic theory can not be expressed in an incrementally linear form (equation 5), even if unloading is excluded. In this respect, endochronic theory differs from most existing nonlinear constitutive theories, including incremental plasticity, hypoelasticity, and the total strain (deformation) theory. Nevertheless, it is possible to linearize endochronic theory in the vicinity of a chosen fixed loading direction for the given stress state under consideration, and this brings the theory to the form of equation (5). This is achieved by replacing the increment of intrinsic time, $d\zeta$, given by equation (2), by the linearized expression

$$d\zeta = \frac{\underline{\varepsilon}^T \underline{p} d\underline{\varepsilon}}{|e|} \qquad (7)$$

This linearization transforms endochronic theory to a stress-strain relation of essentially the same form as in incremental plasticity. Geometrically, this linearization corresponds to the replacement of the circular or curved inelastic stiffness locus by a tangent straight line. The response of this linearized theory is then very close to the response indicated by endochronic theory for all straining directions that deviate by less than $30°$ from the chosen stress direction about which the linearization is made (Figure 5d). Within this fan of directions it is obviously impossible to find much difference between the various theories which lead to a different form of the inelastic stiffness locus.

The foregoing consideration shows that, in order to obtain experimental information to prove or refute the validity of endochronic theory or any other theory, it is necessary to conduct experiments in which the loading path forms a sharp corner with a sudden and large change in the loading direction (that is, so-called loading to the side). For loading paths which do not involve such sharp changes in the loading direction and remain close to a proportional loading path, the test data (with experimental scatter) can propably be represented equally well by various theories and it will be virtually impossible to evaluate the advantages of any particular theory. However, highly nonproportional loading paths with sudden changes in the loading direction are often typical of final failure modes; for example, the stress condition in the soil mass directly under a footing prior to failure consists essentially of a vertical stress, whereas at the moment of failure the strain consists chiefly of shear on the plane of the previous vertical stress. Significant differences between various theories are also obtained in unloading and cyclic loading.

Another property for which various theories manifest significant differences are the so-called cross effects, such as the effects of the shear strain increment, $d\gamma_{12}$, on the normal stress increment, $d\sigma_{11}$, the shear stress increment, $d\tau_{12}$, on the normal strain increment, $d\epsilon_{11}$, or one shear strain increment, $d\gamma_{12}$, on the shear stress increment, $d\tau_{23}$, on another plane. In the linear or linearized form of the incremental constitutive equation (equation 5), this is manifested by non-zero values of the stiffness matrix coefficients corresponding to these components (the upper right corner and lower left corner of the stiffness matrix). In endochronic theory, as well as certain other formulations, these cross effects are rather significant and their presence is required to model various salient constitutive properties, such as densification or dilatancy due to monotonic and cyclic shear strain. This phenomenon leads to noncoaxiality of stress and strain and precludes the use of an orthotropic form of the incremental stiffness matrix (in particular, an isotropic form of this matrix). Thus, the differences between endochronic theory and the so-called incremental orthotropic or isotropic models are rather significant, although these differences can be clearly discerned only for certain loading paths.

ADVANTAGES OF ENDOCHRONIC THEORY

From a practical point of view, the following useful features of endochronic theory may be summarized:

1. The theory is rather convenient and effective for representing unloading irreversibility, and this makes it particularly useful for cyclic loading.

Since the intrinsic time increases with the number of loading cycles, this variable is suitable as a measure of the softening or hardening produced in the soil by cyclic loading. This fact, which was first exploited by Valanis (1971), enables endochronic theory to represent rather simply the stiffening and contraction exhibited by hysteresis loops from one cycle to the next.

2. Compared to classical plasticity, the fact that the intrinsic time is independent of the loading surface and its evolution lends the theory considerable flexibility and enables it to represent diverse phenomena more easily. Furthermore, certain of the characteristic material functions, such as the hardening function, softening function, and dilatancy function, have a relatively simple, albeit intuitive, physical interpretation, and this provides endochronic theory with an advantage in helping to understand the behavior of the model.

3. From the fundamental theoretical point of view, the main difference between endochronic theory and other theories is obtained for loading to the side. In this case endochronic theory always exhibits a softer response for loading than incrementally linear theories, such as classical incremental plasticity. Since a softer response is obviously safer with regard to failure predictions (particularly because failure often occurs with a stress path of this type), endochronic theory will yield conservative predictions. At the same time, a softer or inelastic response for loading to the side is probably more correct, since all microstructural models for inelastic behavior of materials point to the lack of purely elastic response for this type of loading.

4. Endochronic theory is particularly effective for modeling cross effects, such as dilatancy due to shear, and cross hardening (for example, hardening of the volumetric response due to accumulated shear).

5. Finally, endochronic theory provides a relatively simple way to introduce strain-rate effects.

There is, however, one major disadvantage to endochronic theory. The constitutive equation can not be integrated explicitly to give the response curves for various basic types of tests; this complicates data fitting and requires the use of trial-and-error iterative approaches, possibly combined with optimization computer techniques. One noteworthy exception is the case of cyclic pure shear straining, for which endochronic theory yields relatively simple explicit expressions that can be fitted to test data quite easily (Cuellar et al, 1977). As a consequence of these difficulties in the data fitting procedure, it has thus far been necessary to use a relatively large number of material parameters to represent the hardening, softening, and dilatancy functions, and this has made it impossible to attach a clear physical significance to each numerical parameter. Hopefully, further work will lead to improvements in this regard.

CONCLUSION

Endochronic theory provides a very flexible and effective approach for modeling the nonlinear behavior of soils. It is able to handle a wide range of phenomena (such as strain rate effects, dilatancy due to shear, hardening and softening, etc) and seems to predict the proper behavior for cases involving highly non-proportional loading with a sharp corner on the stress path.

ACKNOWLEDGMENT

Financial support by the U.S. National Science Foundation under Grant No ENG-7807777 is gratefully acknowledged.

REFERENCES

1. Bažant, Z. P. (1978), "Endochronic Inelasticity and Incremental Plasticity," *International Journal of Solids and Structures*, Volume 14, Number 9, pp. 691-714.

2. Bažant, Z. P., Krizek, R. J. and Shieh, C. L. (1979), Hysteretic Endochronic Theory for Sand, *Technical Report*, Department of Civil Engineering, Northwestern University, Evanston, Illinois.

3. Cuellar, V., Bažant, Z. P., Krizek, R. J. and Silver, M. L. (1977), "Densification and Hysteresis of Sand under Cyclic Shear," *Journal of the Geotechnical and Engineering Division*, American Society of Civil Engineers, Volume 103, GT5, pp. 399-416.

4. Rudnicki, J. W. and Rice, J. R. (1975), "Conditions of the Localization of Deformation in Pressure-Sensitive Dilatant Materials," *Journal of the Mechanics and Physics of Solids*, Volume 23, pp. 371-394.

5. Schapery, R. A. (1968), "On a Thermodynamic Constitutive Theory and Its Application to Various Nonlinear Materials," *Proceedings of the International Union of Theoretical and Applied Mathematics Symposium*, East Kilbride, B. A. Boley, Editor, Springer-Verlag, New York, New York, pp. 259-285.

6. Valanis, K. C. (1971), "A Theory of Viscoplasticity Without a Yield Surface; Part I. General Theory; Part II. Application to Mechanical Behavior of Metals," *Archives of Mechanics*, (Archiwum Mechaniki Stosowanej), Volume 23, pp. 517-555.

PLASTICITY MODELS FOR SOILS

Theory and Calibration

by

E. Mizuno[1] and W. F. Chen[2], M. ASCE

1. Introduction

The mechanical behavior of soil and rock is complicated and they can not be modelled accurately as a continuum. At present, however, the concept of continuum mechanics has been used extensively in the mathematical modelling of these materials. These include the applications of linear elastic models, nonlinear elastic models, and elastic-plastic models to geotechnical engineering problems. Although the models such as hyperelastic or hypoelastic can represent the phenomena such as dilitancy and hardening or softening of soil behavior, the effect of plastic strain induced during loading can not be predicted within the framework of an incremental Hooke's law with variable moduli which are functions of the stress and/or strain levels.

Current research in soil constitutive modelling is moving toward the development of three-dimensional stress-strain relations based on the principles of plasticity as well as elasticity.

Herein, three types of soil models are described. The first type was used for prediction before the workshop was held, thus without the benefit of the test results. The second and third types are subsequently developed and used after the workshop.

(i) Nonlinear elasticity material model with the Mohr-Coulomb or the Drucker-Prager surface as failure criterion.

(ii) Mohr-Coulomb type of elastic-plastic material model with two different sizes of elliptical hardening cap which are defined respectively on the tensile meridian plane ($\theta = 0°$) and the compressive meridian plane ($\theta = 60°$). (Cap Model I)

---

[1] Research Assistant, School of Civil Engineering, Purdue University, West Lafayette, IN 47907

[2] Professor of Structural Engineering, School of Civil Engineering, Purdue University, West Lafayette, IN 47907

(iii) Mohr-Coulomb type of elastic-plastic material model with an elliptical hardening cap whose size depends on the Lode angle θ. (Cap Model II).

2. A Brief Historical Review

The Mohr-Coulomb criterion of failure is certainly the best known in soil mechanics. This criterion states that failure occurs when the shear stress τ and the normal stress σ acting on any element in the material satisfy the linear equation.

$$\tau + \sigma \tan \phi - c = 0 \qquad (1)$$

where c and φ denote the cohesion and the angle of internal friction, respectively.

Although this criterion has in the past been used by necessity and simplicity to obtain reasonable solutions to important, practical problems in geotechnical engineering, the following limitations should be noticed: (1) This criterion neglects the influence of intermediate principal stress on shear strength; and (2) the failure surface of the Mohr-Coulomb criterion exhibits corners or singularities in the three dimensional principal stress space. From the second limitation, these singularities are difficult to handle in a numerical analysis.

The Drucker-Prager surface [7] can then be considered as a three dimensional approximation to the Mohr-Coulomb failure criterion with a simple smooth surface. This criterion is expressed as a linear combination of the first invariant of stress tensor I_1 and the square root of the second invariant of the deviatoric stress tensor $\sqrt{J_2}$ together with two material constants α and k. The material constants α, k can be related to the Coulomb's c and φ constants in several ways. The Drucker-Prager yield surface with an associated flow rule, however, can not predict the plastic volumetric strain observed in experiments. To improve this, extended von Mises model with convex end cap was proposed by Drucker, Gibson and Kenkel [6].

Following the concept of Drucker et al., subsequent strain hardening plasticity models using the critical state concept were developed by researchers at Cambridge [12], and a specific Cam-Clay model based on normally consolidated or lightly overconsolidated clay was suggested by Roscoe,

Schofield, and Thurairajah [9]. However, failure surface used in this model is still the Drucker-Prager type which results in a much greater dilatancy prediction than that observed in experiments. As a result, modified failure or yield criterion with an elliptical hardening cap which controls the dilatancy was subsequently proposed by DiMaggio and Sandler [5]. In recent years, the cap model has been further modified and refined by Sandler [10,11] and Baladi [1]. Various advanced version of this model including such features as kinematic hardening, double hardening etc. have recently been proposed. The historical review of the plasticity modelling of geotechnical materials has been given in a recent paper by Chen [3].

3. Notations

The state of stress at a point inside a soil medium can be completely determined by the stress tensor σ_{ij} in a three dimensional space. In general, the stress tensor can be decomposed into two parts: (1) the hydrostatic pressure part, where the off-diagonal terms are identically zeros and the diagonal terms are equal to mean normal stress; and (2) the deviatoric part S_{ij}. Thus,

$$\sigma_{ij} = \frac{1}{3} I_1 \delta_{ij} + S_{ij} \tag{2}$$

where I_1, the first invariant of the stress tensor, is the sum of the diagonal stress componsnts and δ_{ij} is the Kronecker delta. The hydrostatic pressure and the deviatoric stress, respectively, cause volumetric change and shape change of the material element.

Fig. 1 shows the view of the state of stress in the principal stress coordinate system (σ_1, σ_2, σ_3). Stress vector $\overrightarrow{OA}$ can be decomposed into $\overrightarrow{OB}$ in the ξ-axis which is called hydrostatic axis ($\sigma_1 = \sigma_2 = \sigma_3$) and $\overrightarrow{BA}$ in the deviatoric plane (π-plane) which is perpendicular to the ξ-axis. The component vector $\overrightarrow{OB}$ represents the mean normal stress p ($I_1/3$) and the component $\overrightarrow{BA}$ represents the deviatoric stress S_{ij}. The length of $\overrightarrow{OB}$ and $\overrightarrow{BA}$ are $\sqrt{3}$ p and $\rho = \sqrt{S_{11}^2 + S_{22}^2 + S_{33}^2}$, respectively. If the stress vector $\overrightarrow{OA}$ is viewed from the hydrostatic ξ-axis, the actual length and direction of it can be represented respectively by ρ and the Lode angle θ. The Lode angle θ is given by

$$\theta = \frac{1}{3} \cos^{-1} \left(\frac{3\sqrt{3}}{2} \frac{J_3}{J_2^{3/2}}\right) \tag{3}$$

where J_3 is the third invariant of the deviatoric stress tensor.

In this paper, the typical continuum mechanics sign convention (tensile stress positive) is utilized in the theoretical development.

4. Failure and Yield Functions

The following failure and yield functions are used in the three proposed models to be described in the subsequent sections.

Mohr-Coulomb Criterion

The Mohr-Coulomb criterion given by Eq. 1 can now be written more generally in terms of stress invariants [8,14].

$$F = I_1 \sin \phi + \frac{3(1-\sin \phi) \sin \theta + \sqrt{3}(3+\sin \phi) \cos \theta}{2} \sqrt{J_2} - 3c \cos \phi = 0 \tag{4}$$

where θ is the Lode angle (Eq. 3).

The cross sectional shape of the Mohr-Coulomb surface on the deviatoric plane is an irregular hexagon as shown in Fig. 2.

Drucker-Prager Criterion

This criterion has the simple form:

$$F = \alpha I_1 + \sqrt{J_2} - k = 0 \tag{5}$$

where α and k are material constants which can be related to cohesion c and the angle of internal friction ϕ of the Mohr-Coulomb criterion in several ways. For example, if the Drucker-Prager criterion is matched with the Mohr-Coulomb criterion in three dimensional principal stress space (Fig. 2) along the compressive meridian (point A) or tensile meridian (point B), the two sets of familiar material constants α and k can be obtained. For the compressive meridian matching, substituting $\theta = \pi/3$ into Eq. 4 and rearranging it, we obtain

$$\left.\begin{array}{c}\alpha = \dfrac{2 \sin \phi}{\sqrt{3}\,(3-\sin \phi)} \\ \\ k = \dfrac{6c \cos \phi}{\sqrt{3}\,(3-\sin \phi)}\end{array}\right\} \qquad (6)$$

These material constants are identical to those given by Zienkiewicz [13]. For the tensile meridian matching, substituting $\theta = 0°$ in Eq. 4 and we obtain

$$\left.\begin{array}{c}\alpha = \dfrac{2 \sin \phi}{\sqrt{3}\,(3+\sin \phi)} \\ \\ k = \dfrac{6c \cos \phi}{\sqrt{3}\,(3+\sin \phi)}\end{array}\right\} \qquad (7)$$

Various matchings between the Drucker-Prager surface and the Mohr-Coulomb surface for material constants are given by Chen and Mizuno [4]. In general, if α is zero, Eq. 5 reduces to the well known von Mises yield condition for metals. The Drucker-Prager surface is used here as the failure surface for cap models described in what follows. Eq. 5 represents an axisymmetric cone-shaped surface with respect to $\sigma_1 = \sigma_2 = \sigma_3$ axis in principal stress space (Fig. 2).

5. Conventional Cap Models

The loading functions are usually assumed to be isotropic and to consist of the following three parts:

(i) An ultimate failure envelope can be either of the simple linear Drucker-Prager form or the nonlinear form assumed by Sandler [11].

$$F_L = \sqrt{J_2} - (A - C\,e^{BI_1}) \qquad (8)$$

in which A, B and C are material constants. Here, the failure equation becomes parallel to I_1 axis under large value of I_1, and this results in a limited dilatancy under high pressure I_1.

(ii) Strain-hardening cap function has the form of a quarter of an ellipse (Fig. 3)

$$F_c = F_c(I_1, \sqrt{J_2}, \varepsilon_{kk}^P) = (I_1-L)^2 + R^2 J_2 - (x-L)^2 = 0 \qquad (9)$$

in which x is the intersection of cap with I_1 axis and x is also a hardening function which depends on plastic volumetric change $d\varepsilon_{kk}^P$. The location of the cap x is related to the plastic volumetric strain function ε_{kk}^P with the material constants W and D according to

$$\varepsilon_{kk}^P = W(e^{Dx} - 1) \qquad (10)$$

and R is the ratio of the major to the minor axis of the cap ellipse which may be a function of L and the Lode angle θ, and L is the value of I_1 at the center of the elliptic cap; and

(iii) Tension cutoff limit plane is introduced

$$F_t = I_1 - T = 0 \qquad (11)$$

where T is tension cutoff limit.

The cap can control the dilatancy of soils under hydrostatic pressure I_1. Although the cap can predict not only strain-hardening of soils, but also strain-softening, this type of model can not predict exactly the hysteresis loop under shear loading. This is because the hardening function in this model is assumed to be controlled by plastic volumetric strain.

Each of these models mentioned above contains several material constants which can be determined from data of standard simple shear test, isotropic consolidation test, uniaxial strain test, and triaxial compression, tension tests.

The determination of these material constants will be given in the part on model calibration.

6. Basic Concepts of Models Developments

The following assumptions are made for the three types of models considered here [2]

(i) Linear elastic, hypoelastic or hyperelastic function is used in the elastic range for the isotropic or anisotropic material element,

(ii) Incremental plasticity theory is applied to calculate plastic strain increment during loading range,

(iii) The Mohr-Coulomb or the Drucker-Prager criterion is used as failure criterion. Effect of strain hardening on this portion of surface is not considered,

(iv) Associated flow rule is assumed for the cap hardening portion of the surface.

$$d\varepsilon^P_{ij} = d\lambda \frac{\partial F_c}{\partial \sigma_{ij}} \qquad (12)$$

where $d\lambda$ is positive scalar function.

In the following, the concept of "decomposition" of stress state onto tensile meridian plane and compressive meridian plane is described.

Suppose that σ_{ij} is the principal stress state acting on an element in soil mass and $d\sigma_{ij}$ is the principal stress increment after the application of an external load increment. The representation of the state of stress as viewed in $I_1 - \sqrt{J_2}$ space is shown in Fig. 4. The CTE line in Fig. 4 is on the $\theta = 0°$ plane and represents the conventional triaxial extension test. The CTC line is on the $\theta = 60°$ plane and represents the conventional triaxial compression test. If stress path is along I_1 axis, it represents the isotropic consolidation test. These three tests are commonly performed tests in geotechnical field. The strengths obtained by the compression test and tension test for soils are different, and the bulk moduli K and shear moduli G determined from these tests are also different. Thus, in the proposed modelling, different material constants (bulk modulus K and shear modulus G) for CTE and CTC tests are introduced. The following items are taken into consideration in the present developments.

(i) The state of stress σ_{ij} as represented in $I_1 - \sqrt{J_2}$ space lies in the range of Lode angle from $\theta = 0°$ and $\theta = 60°$.

(ii) The behaviors of material corresponding to the paths lying on the tensile and compressive meridian planes ($\theta = 0°$ and $\theta = 60°$) are determined first from CTE and CTC tests, respectively.

(iii) The behavior of material corresponding to a stress path lying on a plane making an angle $0° \leq \theta \leq 60°$ is determined from the combined CTE and CTC tests.

Herein, the combined concept of CTE and CTC tests for item (iii) is explained.

The points A and A' in Fig. 4 denote the present state of principal stress σ_{ij} and the subsequent state of principal stress $(\sigma_{ij} + d\sigma_{ij})$, respectively. The vector $\overrightarrow{AA'}$ or $d\sigma_{ij}$ is now projected onto the deviatoric stress plane (π-plane) as the vector $\overrightarrow{BB'}$, and onto the CTC-CTE plane as the vector $\overrightarrow{CC'}$. The vector $\overrightarrow{CC'}$ is on the intersecting line, which is the intersection between the CTC-CTE plane and the plane passing through the hydrostatic axis I_1 and the stress vector $\overrightarrow{AA'}$.

The deviatoric stress vector $\overrightarrow{BB'}$ on π-plane can now be further decomposed into two parts: $\overrightarrow{EE'}$ and $\overrightarrow{DD'}$ along the OE' and OD' axes respectively, as

$$dS_{ij} = dS_{ij}^0 + dS_{ij}^1 \tag{13}$$

where dS_{ij}^0 and dS_{ij}^1 are the components of the stress in the $\theta = 60°$ and $0°$ planes, respectively. From the geometry, the magnitudes of these components of dJ_2^1 on $0°$-plane and dJ_2^0 on $60°$-plane can be calculated from the total dJ_2 as

$$\left.\begin{aligned} dJ_2^0 &= dJ_2 [\cos(\pi/3-\theta) - \frac{1}{\sqrt{3}} \sin(\pi/3-\theta)]^2 \\ dJ_2^1 &= dJ_2 (\cos\theta - \frac{1}{\sqrt{3}} \sin\theta)^2 \end{aligned}\right\} \tag{14}$$

Since the deviatoric stresses dS_{ij}^0 and dS_{ij}^1 have the following characteristics:

$$\left. \begin{aligned} dS_{ij}^0 &= (dS_{11}^0, dS_{22}^0, dS_{33}^0) = (dS_{11}^0, -\tfrac{1}{2} dS_{11}^0, -\tfrac{1}{2} dS_{11}^0) \\ dS_{ij}^1 &= (dS_{11}^1, dS_{22}^1, dS_{33}^1) = (dS_{11}^1, dS_{11}^1, -2dS_{11}^1) \end{aligned} \right\} \quad (15)$$

Therefore, for the case of loading, the components of dS_{ij}^0 and dS_{ij}^1 can be calculated from Eqs. 14 and 15 as

$$(dS_{11}^0, dS_{22}^0, dS_{33}^0) = \left(-2\sqrt{\frac{dJ_2^0}{3}}, \sqrt{\frac{dJ_2^0}{3}}, \sqrt{\frac{dJ_2^0}{3}} \right)$$

and

$$(dS_{11}^1, dS_{22}^1, dS_{33}^1) = \left(-\sqrt{\frac{dJ_2^1}{3}}, -\sqrt{\frac{dJ_2^1}{3}}, 2\sqrt{\frac{dJ_2^1}{3}} \right) \quad (16)$$

Similarly, the vector $\overrightarrow{CC'}$ with length dI_1 on the CTC-CTE plane can be decomposed as dI_1^0 and dI_1^1 onto the CTC line and CTE line respectively. From the geometry of Fig. 4, we have

$$\left. \begin{aligned} dI_1^0 &= dI_1 \left[\frac{\cos(\pi/3 - \theta) - \frac{1}{\sqrt{3}} \sin(\pi/3 - \theta)}{2 \cos \theta} \right] \\ dI_1^1 &= dI_1 \left(1 - \frac{1}{\sqrt{3}} \tan \theta \right) \end{aligned} \right\} \quad (17)$$

Therefore, the decomposed principal stresses $d\sigma_{ij}^0$ and $d\sigma_{ij}^1$ can be rewritten as

$$d\sigma_{ij}^0 = \left(\frac{dI_1^0}{3} + dS_{11}^0, \frac{dI_1^0}{3} + dS_{22}^0, \frac{dI_1^0}{3} + dS_{33}^0\right)$$

$$d\sigma_{ij}^1 = \left(\frac{dI_1^1}{3} + dS_{11}^1, \frac{dI_1^1}{3} + dS_{22}^1, \frac{dI_1^1}{3} + dS_{33}^1\right) \quad (18)$$

In the following, explicit expressions for calculating the strains are developed for the three proposed models: (1) nonlinear elasticity model with the Mohr-Coulomb or the Drucker-Prager failure surface; (2) cap model I; and (3) cap model II.

7. Incremental Constitutive Equations

7.1 Nonlinear Elasticity Model

In this modelling, behavior of materials is assumed to be elastic until the state of stress reaches the failure surface. For an isotropic, the bulk moduli K_0, K_1 and the shear moduli G_0, G_1 may be determined from the triaxial compression and triaxial tension tests. Two types of bulk moduli and shear moduli are considered.

(i) Bulk moduli and shear moduli are constant. This reduces to the linear elastic model.

(ii) Elastic bulk moduli K_0 and K_1 are assumed to be functions of the first invariant of the decomposed stress tensor I_1^0 and I_1^1, respectively,

$$K_0 = K_0(I_1^0) \quad \text{and} \quad K_1 = K_1(I_1^1) \quad (19)$$

and elastic shear moduli G_0 and G_1 are assumed to be functions of the second invariant of the decomposed deviatoric stress tensor J_2^0 and J_2^1, respectively.

$$G_0 = G_0\left(\sqrt{J_2^0}\right) \quad \text{and} \quad G_1 = G_1\left(\sqrt{J_2^1}\right) \quad (20)$$

According to the incremental Hooke's law, the strain increments $d\varepsilon_{ij}^0$ and $d\varepsilon_{ij}^1$ are written in terms of the decomposed stresses $d\sigma_{ij}^0$ and $d\sigma_{ij}^1$ as

$$d\varepsilon_{ij}^0 = \frac{1}{9K_0} dI_1^0 \delta_{ij} + \frac{1}{2G_0} dS_{ij}^0$$

$$d\varepsilon_{ij}^1 = \frac{1}{9K_1} dI_1^1 \delta_{ij} + \frac{1}{2G_1} dS_{ij}^1$$

(21)

and the total strain increment $d\varepsilon_{ij}$ is the sum of the two components.

$$d\varepsilon_{ij} = d\varepsilon_{ij}^0 + d\varepsilon_{ij}^1 \qquad (22)$$

It should be noted that the direction of the total strain increment $d\varepsilon_{ij}$ is not necessary in the same direction as that of the total stress increment $d\sigma_{ij}$. The following shortcomings are noted in this modelling:

(i) The decomposition of dI_1 into two parts is not unique. Here, for convenience, we use the CTC-CTE planes.

(ii) The decomposition of dI_1 for the case of isotropic compression test can not be made because many combinations of decomposition can be considered.

(iii) For two stress paths which are extremely close to each other along the hydrostatic pressure axis but lie on two different planes $\theta = 0°$ and $60°$ respectively, the volumetric strains predicted with the corresponding bulk moduli K_0 and K_1 will have different values at the boundary of the hydrostatic axis.

The nonlinear elasticity model described above was used for predictions at the workshop.

7.2 Cap Model I

The conventional cap model used by Baladi [1], among others, has the Drucker-Prager type of yield surface with an elliptic cap. The size of the cap is assumed to have either a constant value or to be a function of the plastic volumetric strain. Since the Mohr-Coulomb failure surface is probably the best among all the failure criteria for soils, it follows that the size of the cap should depend not only on the plastic volumetric strain, but also on the Lode angle θ. In the present modelling, therefore, two different caps are used. One is defined in the tensile meridian plane ($\theta = 0°$) and the other is defined in the compressive meridian plane ($\theta = 60°$). This is illustrated in Fig. 5-a, where the Mohr-Coulomb failure lines on these two planes are also shown.

In the elastic range, two types of bulk moduli K_0, K_1 and shear moduli G_0, G_1 may be considered.

(i) K_0, K_1, G_0 and G_1 are constants.

(ii) K_0, K_1 are functions of the first invariant of the decomposed stress tensor I_1^0 or I_1^1 and the plastic volumetric strains ε_{kk}^{P0} or ε_{kk}^{P1}, respectively,

$$\left. \begin{array}{l} K_0 = K_0 \, (I_1^0, \; \varepsilon_{kk}^{P0}) \\ K_1 = K_1 \, (I_1^1, \; \varepsilon_{kk}^{P1}) \end{array} \right\} \tag{23}$$

and G_0, G_1 are functions of the second invariant of the decomposed stress tensor J_2^0 or J_2^1 and the plastic volumetric strains ε_{kk}^{P0} or ε_{kk}^{P1}.

$$\left. \begin{array}{l} G_0 = G_0 \left(\sqrt{J_2^0}, \; \varepsilon_{kk}^{P0} \right) \\ G_1 = G_1 \left(\sqrt{J_2^1}, \; \varepsilon_{kk}^{P1} \right) \end{array} \right\} \tag{24}$$

where $d\varepsilon_{kk}^{P0}$ and $d\varepsilon_{kk}^{P1}$ are the plastic volumetric strains induced by the caps on $\theta = 60°$ and $0°$ planes, respectively.

The elastic strains $d\varepsilon_{ij}^e$ are calculated in the same manner as that used previously in the nonlinear elasticity model.

In the plastic range, the two elliptic caps located on the $\theta = 0°$ and $60°$ planes will contract or expand as the plastic volumetric strain ε_{kk}^P decreases or increases. The Mohr-Coulomb criterion is used here as the failure envelope which will not harden. The plastic strain $d\varepsilon_{ij}^P$ during the loading is derived from the flow rule (Eq. 12). The total plastic strain increment $d\varepsilon_{ij}^P$ is the sum of the plastic strain increments of $d\varepsilon_{ij}^{P0}$ and $d\varepsilon_{ij}^{P1}$ which are induced by the two caps:

$$d\varepsilon_{ij}^{P} = d\varepsilon_{ij}^{P0} + d\varepsilon_{ij}^{P1} \tag{25}$$

Each of the plastic strain increments can be written as

$$\left. \begin{array}{l} d\varepsilon_{ij}^{P0} = d\lambda^0 \dfrac{\partial F_c^0}{\partial \sigma_{ij}^0} \\[2ex] d\varepsilon_{ij}^{P1} = d\lambda^1 \dfrac{\partial F_c^1}{\partial \sigma_{ij}^1} \end{array} \right\} \tag{26}$$

where $d\lambda^0$, $d\lambda^1$ are different positive scalar functions, F_c^0, F_c^1 are the elliptic cap functions and σ_{ij}^0, σ_{ij}^1 are the decomposed stresses, respectively. From Eqs. 25 and 26, we have

$$\left. \begin{array}{l} d\varepsilon_{ij}^{P0} = d\lambda^0 \left[\dfrac{\partial F_c^0}{\partial I_1^0} \dfrac{\partial I_1^0}{\partial \sigma_{ij}^0} + \dfrac{\partial F_c^0}{\partial \sqrt{J_2^0}} \dfrac{\partial \sqrt{J_2^0}}{\partial J_2^0} \dfrac{\partial J_2^0}{\partial \sigma_{ij}^0} \right] \\[3ex] d\varepsilon_{ij}^{P1} = d\lambda^1 \left[\dfrac{\partial F_c^1}{\partial I_1^1} \dfrac{\partial I_1^1}{\partial \sigma_{ij}^1} + \dfrac{\partial F_c^1}{\partial \sqrt{J_2^1}} \dfrac{\partial \sqrt{J_2^1}}{\partial J_2^1} \dfrac{\partial J_2^1}{\partial \sigma_{ij}^1} \right] \end{array} \right\} \tag{27}$$

The total strain increment $d\varepsilon_{ij}$ is the sum of the elastic strain and plastic strain increments.

$$d\varepsilon_{ij} = \left[\frac{dI_1^0}{9K_0(I_1^0, \varepsilon_{kk}^{P0})} + \frac{dI_1^1}{9K_1(I_1^1, \varepsilon_{kk}^{P1})} \right] \delta_{ij} +$$

$$\left[\frac{dS_{ij}^0}{2G_0\left(\sqrt{J_2^0}, \varepsilon_{kk}^{P0}\right)} + \frac{dS_{ij}^1}{2G_1\left(\sqrt{J_2^1}, \varepsilon_{kk}^{P1}\right)} \right] +$$

$$d\lambda^0 \left[\frac{\partial F_c^0}{\partial I_1^0} \delta_{ij} + \frac{1}{2\sqrt{J_2^0}} \frac{\partial F_c^0}{\partial \sqrt{J_2^0}} S_{ij}^0 \right] +$$

$$d\lambda^1 \left[\frac{\partial F_c^0}{\partial I_1^1} \delta_{ij} + \frac{1}{2\sqrt{J_2^1}} \frac{\partial F_c^1}{\partial \sqrt{J_2^1}} S_{ij}^1 \right] \tag{28}$$

Eq. 28 is the incremental stress-strain relations corresponding to the proposed cap model I. In order to use this relation (Eq. 28) for a stress analysis, the scalar functions $d\lambda^0$ and $d\lambda^1$ must be determined. Using the consistency condition $dF_c = 0$ during loading and Eq. 12, $d\lambda^0$, $d\lambda^1$ can be derived in a straightforward manner as

$$d\lambda^0 = \frac{3K_0 \dfrac{\partial F_c^0}{\partial I_1^0} de_{kk}^0 + \dfrac{G_0}{\sqrt{J_2^0}} \dfrac{\partial F_c^0}{\partial \sqrt{J_2^0}} S_{ij}^0 de_{ij}^0}{9K_0 \left[\dfrac{\partial F_c^0}{\partial I_1^0}\right]^2 + G_0 \left[\dfrac{\partial F_c^0}{\partial \sqrt{J_2^0}}\right]^2 - 3 \dfrac{\partial F_c^0}{\partial I_1^0} \dfrac{\partial F_c^0}{\partial \varepsilon_{kk}^{P0}}} \tag{29}$$

where de_{ij}^0 is the strain corresponding to the decomposed deviatoric stress dS_{ij}^0. Also, $d\lambda^1$ has the same form as that of Eq. 29 except changing the subscript from 0 to 1. For the states of stresses on the Mohr-Coulomb surface, the corresponding plastic strain increment $d\varepsilon_{ij}^P$ is derived from the flow rule:

$$d\varepsilon_{ij}^P = d\lambda \frac{\partial F_L}{\partial \sigma_{ij}} \tag{30}$$

where F_L is the Mohr-Coulomb failure function, σ_{ij} is the total stress and $d\lambda$ is a positive scalar function.

The main characteristics of this model are:

(i) The direction of the strain increment $d\varepsilon_{ij}$ is not necessary in the same direction as that of the stress increment $d\sigma_{ij}$ even with elastic region.

(ii) The hardening caps exist only in the $\theta = 0°$ and $\theta = 60°$ planes. Two caps control the hardening of the isotropic materials within the range of $0°$ to $60°$.

The limitations of this model are:

(i) The same limitations as that of nonlinear elasticity model described previously.

(ii) Because this model assumes two independent caps, the model may predict a total plastic volumetric strain ε_{kk}^P that may exceed the maximum plastic strain W.

(iii) The intersections of the caps with I_1-axis are not the same. The model does not satisfy the continuity condition along the hydrostatic axis.

7.3 Cap Model II

This model can be considered as a generalization of the cap model I described in the preceding section. The two caps in the $\theta = 0°$ and $60°$ planes are now connected by a three-dimensional cap as shown in Fig. 5-b.

In the elastic range, the same procedure as that of Cap Model I is used for the elastic strain calculation. However, the bulk modulus K_0 or K_1 is now a function of I_1^0 or I_1^1 and the total plastic volumetric strain ε_{kk}^P, while the shear modulus G_0 or G_1 is a function of $\sqrt{J_2^0}$ or $\sqrt{J_2^1}$ and the total plastic volumetric strain ε_{kk}^P, respectively. Similarly, the plastic strain increment is determined from Eq. 12. Thus, the total strain increment $d\varepsilon_{ij}$ is written as

$$d\varepsilon_{ij} = \left[\frac{dI_1^0}{9K_0(I_1^0, \varepsilon_{kk}^P)} + \frac{dI_1^1}{9K_1(I_1^1, \varepsilon_{kk}^P)}\right]\delta_{ij} + \left[\frac{dS_{ij}^0}{2G_0(\sqrt{J_2^0}, \varepsilon_{kk}^P)}\right]$$

$$+ \frac{dS_{ij}^1}{2G_1\left(\sqrt{J_2^1}, \epsilon_{kk}^P\right)}\right] + d\lambda\left[\frac{\partial F_c}{\partial I_1}\delta_{ij} + \frac{1}{2\sqrt{J_2}}\frac{\partial F_c}{\partial\sqrt{J_2}}S_{ij}\right] \qquad (31)$$

where $d\lambda$ is derived using the same procedure as in Cap Model I.

For the stress state on the Mohr-Coulomb failure surface, the plastic strain increment is given by Eq. 30.

8. Model Calibration

General

In the workshop, three sets of soil materials are available for prediction under different stress paths. These are "Clay X", "Clay Y", "Kaolinite Clay" and "Ottawa Sand". Herein, nonlinear elasticity model is applied to predict the behavior of all three materials. Further, plasticity models (Cap Model I and Cap Model II) are applied only to "Clay X" and "Clay Y".

In the following, the stress paths used in the experiments and in the predictions are first defined. Then, the general procedure for determining the material constants for the three models is explained, and the stress-strain relations corresponding to particular stress paths are derived. Herein, typical soil mechanics sign convention (compressive stress positive) is used in the model calibration.

8.1 Stress Paths in Experiments

The stress path used in the experiments can be described by the stress ratio m.

$$m = \frac{\sigma_2 - \sigma_3}{\sigma_1 - \sigma_3} \qquad (32)$$

where σ_1, σ_2 and σ_3 are the principal stresses applied to the cylindrical or the cubic soil spedimen and σ_1 is the stress in the vertical or axial direction.

The stress path with the ratio $m=0$ or $m=1$ is on the plane $\theta = 60°$ or $\theta = 0°$ respectively. The stress path corresponding to a simple shear test as viewed in the π-plane is shown in Fig. 6-a. If the Lode angle θ is defined from the plane, $m=0$, then, the relation between θ and m is given by

$$\cos \theta = \frac{(2-m)(1-2m)(1+m)}{2(m^2-m+1)^{3/2}} \tag{33}$$

The stress path corresponding to conventional triaxial test lies between the stress path CTC test and the path CTE test as shown in Fig. 6-b in $I_1 - \sqrt{J_2}$ space.

8.2 Determination of Material Constants and Analysis

General

For the nonlinear elasticity model, we need to determine the bulk moduli K_0 and K_1, shear moduli G_0 and G_1. The bulk modulus K is a function of I_1 and is determined by an isotropic compression test. The shear moduli G_0 and G_1 are functions of $\sqrt{J_2^0}$ and $\sqrt{J_2^1}$ and are determined by the stress difference-strain difference curves from drained triaxial compression and tension tests conducted at different levels of confining pressure. If the model is applied to problems involved cyclic and reversed loading, bulk modulus K, and the shear moduli G_0 and G_1 can be determined respectively by the unloading curves of the tests mentioned above. Hence, the variable moduli model is used.

For the plasticity models, the bulk modulus K and the shear moduli G_0 and G_1 can be determined from the slopes of the unloading curves of an isotropic compression test; and from the slopes of the unloading stress difference-strain difference curves of a drained triaxial compression tests and tension tests at different levels of confining pressure, respectively. In the plastic range, the values of α and k are obtained from c and ϕ values associated with the Mohr-Coulomb failure envelope, which is constructed through simple shear tests. The material constants W and D in Eq. 10 associated with the hardening function are determined by isotropic compression and unloading tests. The constant R associated with cap shape is determined from simple shear and uniaxial strain tests. The choice of material constants R from experimental data requires a considerable experience.

Clay X

The experimental data on the simple shear tests with the stress ratio $m = 0$ and 1 under confining pressure σ_c = 10, 20 and 30 psi are available. The nonlinear elasticity model, cap model I and cap model II are used for prediction. In the case of nonlinear elasticity model, the stress difference-strain difference curves are drawn for each confining pressure as shown in Figs. 7 through 9 where we have used the average values of ε_1 and ε_2, or ε_2 and ε_3. These curves are fitted by a function using nonlinear regression analysis. In general, the relation between stress difference and strain difference can be expressed as

$$\sigma_1 - \sigma_3 = f(\varepsilon_1^* - \varepsilon_3 \text{ or } \varepsilon_1 - \varepsilon_3^*) \tag{34}$$

Taking derivative with respect to the strain difference, we have

or

$$\left.\begin{array}{c} G_0 \text{ or } G_1 = g(\sigma_1^0 - \sigma_3^0 \text{ or } \sigma_1^1 - \sigma_3^1) \\ \\ G_0 \text{ or } G_1 = g(\sqrt{3J_2^0} \text{ or } \sqrt{3J_2^1}) \end{array}\right\} \tag{35}$$

Each pair of G_0 and G_1 functions are utilized for the present prediction under different confining pressure. The matchings of the Mohr-Coulomb constants (c,ϕ) and the Drucker-Prager constants (α,k) are listed in Table 1 for all three materials. The principal stress increments $d\sigma_{11}$, $d\sigma_{22}$ and $d\sigma_{33}$ can be written in the form as

$$\left.\begin{array}{c} d\sigma_{11} = d\sigma_{11} \\ \\ d\sigma_{22} = \dfrac{2m-1}{2-m} d\sigma_{11} \\ \\ d\sigma_{33} = \dfrac{m+1}{m-2} d\sigma_{11} \end{array}\right\} \tag{36}$$

from which dJ_2 can be expressed in terms of $d\sigma_{11}$. Thus, for a given $d\sigma_{11}$, the corresponding strain increment $d\varepsilon_{ij}$ can be computed. The process continues until the stress path reaches the Mohr-Coulomb or Drucker-Prager failure surface.

In the case of plasticity models, clay X is assumed to be an anisotropic material. Fig. 10 shows the relation between the stress increment $d\sigma_{11}^0$ or $d\sigma_{11}^1$ and each strain increment.

The stress-strain relation can be regarded as linear up to $d\sigma_{11}^0$ of 5 psi or $d\sigma_{11}^1$ of 2 psi, respectively. Thus, the following relation is assumed in the leastic range, for $m = 0$

$$d\varepsilon_{11}^0 = 0.0014 \, d\sigma_{11}^0, \quad d\varepsilon_{22}^0 = -0.00156 \, d\sigma_{11}^0 \text{ and } d\varepsilon_{33}^0 = 0.0002 \, d\sigma_{11}^0 \quad (37)$$

and for $m = 1$,

$$d\varepsilon_{11}^1 = 0.00325 \, d\sigma_{11}^1, \quad d\varepsilon_{22}^1 = -0.00267 \, d\sigma_{11}^1 \text{ and } d\varepsilon_{33}^1 = -0.00055 \, d\sigma_{11}^1 \quad (38)$$

The constants W and D are estimated to be 0.3174 and 0.0087, respectively. Therefore, Eq. 10 is

$$\varepsilon_{kk}^P = 0.3174 \, [1 - e^{-0.0087(x-46.5)}] \quad (39)$$

where the value 46.5 is three times of the preconsolidation pressure σ_p (compressive stress taken as positive). The material constants R (cap shape) are determined to be 4.7 from the simple shear test data with the stress ratio $m = 0$ and to be 5.7 from those with the stress ratio $m = 1$. Fig. 11 shows the location of hardening caps for the planes of $m = 0$ and $m = 1$ in $I_1 - \sqrt{J_2}$ space. The plastic strain increments $d\varepsilon_{ij}^P$ are calculated for the cases of Cap Model I and Cap Model II. As the experiments have been conducted under stress control condition, the plastic volumetric strain increment $d\varepsilon_{kk}^P$ can be calculated from Eq. 39 after the value of x is obtained from the subsequent state of stress and elliptic cap equation. Therefore, $d\lambda$ is obtained from Eq. 12.

For cap model I,

$$d\lambda^0 = \frac{d\varepsilon_{kk}^{P0}}{\frac{\partial F_c^0}{\partial \sigma_{ij}^0}} \quad \text{and} \quad d\lambda^1 = \frac{d\varepsilon_{kk}^{P1}}{\frac{\partial F_c^1}{\partial \sigma_{ij}^1}} \tag{40}$$

and for cap model II,

$$d\lambda = \frac{d\varepsilon_{kk}^{P}}{\frac{\partial F_c}{\partial \sigma_{ij}}} \tag{41}$$

Therefore, the plastic strain increment $d\varepsilon_{ij}^{P}$ is obtained from Eq. 12 respectively.

Clay Y

The experimental data are obtained from the triaxial tests with the stress ratio $m = 0$ and $m = 1$ under the initial confining pressure $\sigma_c = 2.5$, 5.0 and 10 psi. The property of clay Y appears to be similar to that of clay X. As the initial location of confining pressure lies on "Dry of Critical" from viewpoint of the critical state soil mechanics, the Coulomb constants c and ϕ (in Table 1) are different from those constants for clay X. The experimental data indicate that the behavior of clay Y appears to be isotropic. Therefore, the stress difference-strain difference relation, and the stress invariant I_1 - the volumetric strain relation are checked for three sets of data as shown in Fig. 12 through 14. For nonlinear elasticity model, the functions of bulk moduli K_0, K_1 and shear moduli G_0, G_1 are obtained using the curve fitting procedure similar to that of clay X. The principal stress increments $d\sigma_{11}$, $d\sigma_{22}$ and $d\sigma_{33}$ with the stress ratio m is expressed in terms of $d\sigma_{11}$ by

$$d\sigma_{11} = d\sigma_{11}, \quad d\sigma_{22} = m d\sigma_{11} \quad \text{and} \quad d\sigma_{33} = 0 \tag{42}$$

The corresponding strain increment can be calculated as that of clay X.

For the plasticity models, the stress-strain relations are again checked for the data with m = 0 and 1 as shown in Fig. 15. The shear modulus G_0 for m = 0 or G_1 for m = 1 is assumed to be a function of $\sqrt{J_2^0}$ or constant, respectively. The bulk moduli K_0 and K_1 are assumed to be two different constants respectively. The material constants W and D are estimated to be 0.135 and 0.009. Therefore, Eq. 10 is

$$\varepsilon_{kk}^P = 0.135 \, [1 - e^{-0.009(x-46.5)}] \tag{43}$$

where the preconsolidation pressure σ_p is assumed to be the same as that in clay X.

The material constants R (cap shape) are determined to be 2.39 from the triaxial test data with m = 0 and to be 0.87 from those with m = 1. The location of hardening caps are shown in Fig. 16. The calculation of the strain increment is the same as that of clay X.

Kaolinite Clay

The available experimental data are the triaxial tests (No. 1 and 10) and the simple shear tests (No. 4 and 13) with m = 0 or m = 1 under an undrained condition. The nonlinear elasticity model is utilized for the prediction.

The experimental data are plotted in the stress difference-the strain difference space as shown in Fig. 17. Here, as in the previous case, the stress difference-the strain difference curves are fitted by functions in order to determine the shear moduli G_0 and G_1 which are functions of J_2. For these tests to be predicted by the model, the major principal stress is inclined at an angle to the vertical axis of the specimen, while the intermediate principal stress remains horizontal. Since the material is assumed to be isotropic for present case, the direction of the applied principal stresses will not affect the results of prediction. The increments of principal stresses with stress ratio m are expressed in terms of $d\sigma_{11}$ by

$$d\sigma_{11} = d\sigma_{11}, \quad d\sigma_{22} = 0 \quad \text{and} \quad d\sigma_{33} = \frac{m}{m-1} d\sigma_{11} \tag{44}$$

for Test No. 2, 3, 7, 11 and 12 and

$$d\sigma_{11} = d\sigma_{11}, \ d\sigma_{22} = \frac{1-2m}{m-2} d\sigma_{11} \text{ and } d\sigma_{33} = \frac{m+1}{m-2} d\sigma_{11} \tag{45}$$

for Test No. 5, 6, 8 and 9.

It should be noted that those principal stresses are the total principal stresses because the tests are conducted under an undrained condition. Therefore, the effective principal stress increments must be known in order to calculate the principal strain increments. In this case, the effective stress increments are the deviatoric stress increments dS_{ij} because the inclusion of hydrostatic pressure increment dI_1 causes a volumetric change. Thus, the effective stress increments are calculated from Eqs. 44 and 45, and the corresponding strain increments can be calculated in a same manner as that described above.

Ottawa Sand

The experimental data available for predictions are the conventional triaxial compression tests under the initial confining pressure of 5, 10 psi, the conventional triaxial tension test under the initial confining pressure of 5 psi, and the simple shear tests with the stress ratio of $m = 0$, 1 under the mean normal stresses of 5, 10 and 20 psi.

Here, as the experimental data for the loading, unloading and reloading cases are given, the variable moduli model is used. In order to determine the shear moduli G_0 and G_1 for the loading case, the experimental data for the loading parts in the simple shear tests are plotted in the space of the strain difference and the stress difference divided by the stress difference at each failure (Fig. 18). The curves used in the prediction are shown by the dotted curves which are obtained by a nonlinear regression analysis. Thus, the shear moduli G_0^L and G_1^L for the loading cases are given by

and
$$\left. \begin{array}{l} G_0^L = G_0^L\left(\sigma_f^0, \sqrt{J_2^0}\right) \\[6pt] G_1^L = G_1^L\left(\sigma_f^1, \sqrt{J_2^1}\right) \end{array} \right\} \tag{46}$$

where σ_f^0 and σ_f^1 are the stress differences at failure in CTC and CTE tests. The stress difference at a failure is given by

$$\sigma_f^0 = \frac{6(c \cos \phi + \sigma_m \sin \phi)}{3 - \sin \phi} \tag{47-a}$$

for the modulus G_0^L and

$$\sigma_f^1 = \frac{6(c \cos \phi + \sigma_m \sin \phi)}{3 + \sin \phi} \tag{47-b}$$

for the modulus G_1^L, where σ_m is the mean normal stress. In this case, the shear moduli are functions of I_1 and J_2^0 or J_2^1.

To determine the shear moduli G_0^{ur}, G_1^{ur} for the unloading and reloading cases, the experimental data for the unloading and reloading parts in the simple shear tests are plotted in Fig. 19. In this case, the stress difference and the strain difference are measured from the unloading and reloading points. The curves used in the prediction are shown by the dotted curves. Thus, the shear moduli G_0^{ur} and G_1^{ur} have the same form as that in Eq. 46.

The bulk modulus K_0 and K_1 are determined from the conventional triaxial compression and tension tests. Although the data on the isotropic consolidation test are given, they are not used here to determine the bulk moduli. The experimental data are plotted in I_1 and ε_{kk} space as shown in Fig. 20. The bulk moduli K_0^L and K_1^L for the loading up to the unloading point, K_0^{ur} and K_1^{ur} for the unloading, reloading up to the point of previous unloading, and K_0^{NL} and K_1^{NL} for the loading starting from the point of the previous unloading are determined from the figure. These bulk moduli appear to depend on the initial confining pressure σ_c. Therefore, the general form for K may be written as

$$K = K(\sigma_c) \tag{48}$$

In general, the bulk modulus of Ottawa sand may be expressed by $K = K(\sigma_{max})$ where σ_{max} is a maximum confining pressure similar to the preconsolidation pressure.

The calculation of the strain can be carried out in the same way as mentioned previously.

9. Acknolwedgments

This material is based upon work supported by the National Science Foundation under Grant No. PFR-7809326 to Purdue University.

References

[1] Baldi, G. Y. and Rohani, B., "An Elastic-Plastic Constitutive Model for Saturated Sand Subjected to Monotonic and/or Cyclic Loadings," Third International Conference on Numerical Method in Geomechanics, Aachen, 2-6 April 1979, pp. 389-404.

[2] Chen, W. F., "Limit Analysis and Soil Plasticity," Elsevier, Amsterdam, The Netherlands, 1975.

[3] Chen, W. F., "Plasticity in Soil Mechanics and Landslides," Engineering Mechanics Division, ASCE, Vol. 106, No. EM3, June, 1980, pp. 443-464.

[4] Chen, W. F. and Mizuno, E., "On Material Constants for Soil and Concrete Models," Third ASCE/EMD Specialty Conference, 1979, pp. 539-542.

[5] DiMaggio, F. L. and Sandler, I. S., "Material Models for Granular Soils," Journal of the Engineering Mechanics Division, ASCE, Vol. 97, No. EM3, 1971, pp. 936-950.

[6] Drucker, D. C., Gibson, R. E., and Henkel, D. J., "Soil Mechanics and Work-Hardening Theories of Plasticity," Transactions, ASCE, Vol. 122, 1957, pp. 338-346.

[7] Drucker, D. C. and Prager, W., "Soil Mechanics and Plastic Analysis or Limit Design," Quarterly of Applied Mathematics, Vol. 10, No. 2, July 1952, pp. 157-175.

[8] Mizuno, E. and Chen, W. F., "Analysis of Soil Response with Different Plasticity Models," ASCE Symposium in Florida, 1980.

[9] Roscoe, K. H., Schofield, A. N., and Thurairajah, A., "Yielding of Clays in State Wetter than Critical," Géotechnique, Vol. 13, No. 3, 1963, pp. 211-240.

Table 1 Material Constants c, ϕ, α and k

Material Constants	Clay X	Clay Y	Kaolinite Clay	Ottawa Sand
C (psi)	2.0	4.0	13.75	2.0
ϕ (°)	26.57	16.88	17.04	42.34
$\dfrac{2 \sin \phi}{\sqrt{3}(3 - \sin \phi)}$	0.2023	0.1221	0.1250	0.2117
$\dfrac{6c \cos \phi}{\sqrt{3}(3 - \sin \phi)}$	2.4275	4.892	16.823	1.394
$\dfrac{2 \sin \phi}{\sqrt{3}(3 + \sin \phi)}$	0.1498	0.10	0.1027	0.334
$\dfrac{6c \cos \phi}{\sqrt{3}(3 + \sin \phi)}$	1.7976	4.304	13.830	2.201

[10] Sandler, I. S., DiMaggio, F. L., and Baladi, G. Y., "Generalized Cap Model for Geological Materials," Geotechnical Division, ASCE, Vol. 102, No. GT. 7, 1976, pp. 683-699.

[11] Sandler, I. S. and Melvin, L. B., "Material Models of Geological Materials in Ground Shock," Numerical Method in Geomechanics. Edited by C. S. Desai, Vol. 1, 1976, pp. 219-231.

[12] Schofield, A. N. and Wroth, P., "Critical State Soil Mechanics," McGraw-Hill, New York, 1968.

[13] Zienkiewicz, O. C., "The Finite Element Method," McGraw-Hill, 1978 (Chap. 18).

[14] Zienkiewicz, O. C., Humpheson, C., and Lewis, R. W., "Associated and Non-Associated Visco-Plasticity and Plasticity in Soil Mechanics," Géotechnique, 25, No. 4, 1975, pp. 671-689.

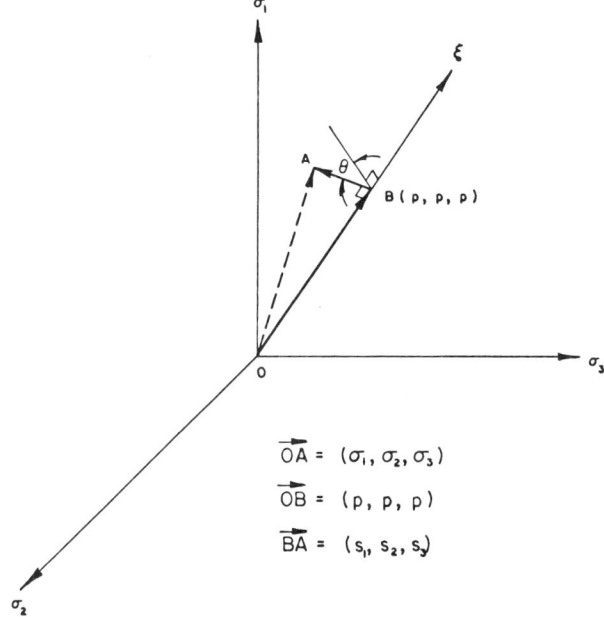

Fig. 1   Stress State in Principal Stress Space

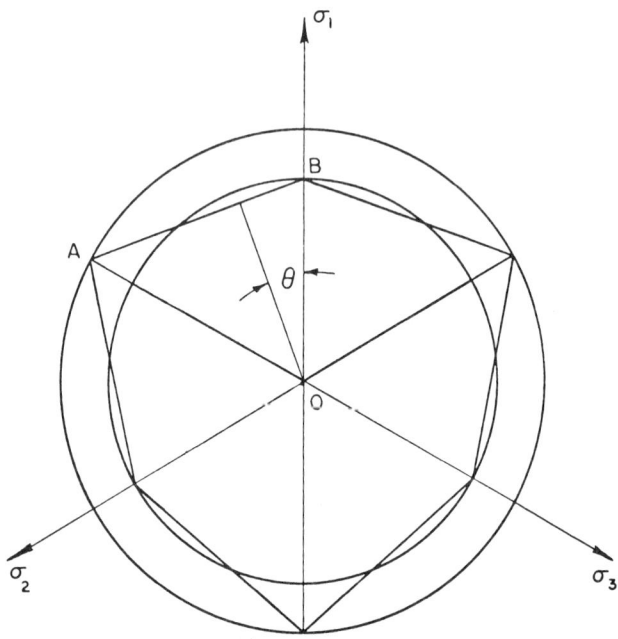

Fig. 2   Shape of Yield Criterion on π-Plane

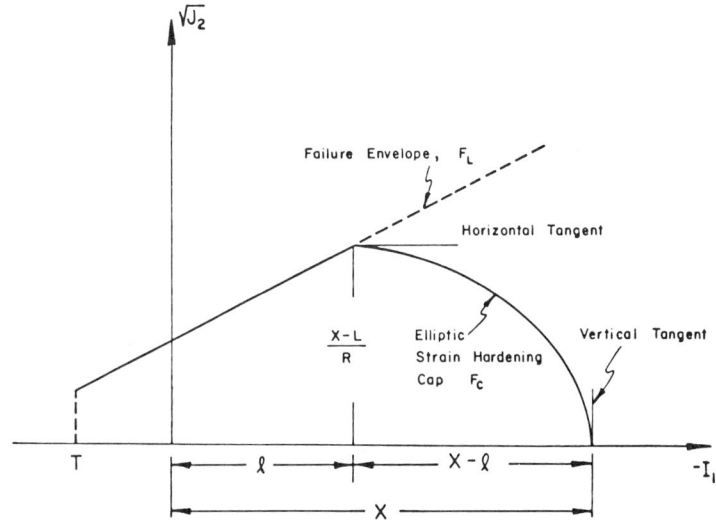

Fig. 3   Elliptic Cap Model

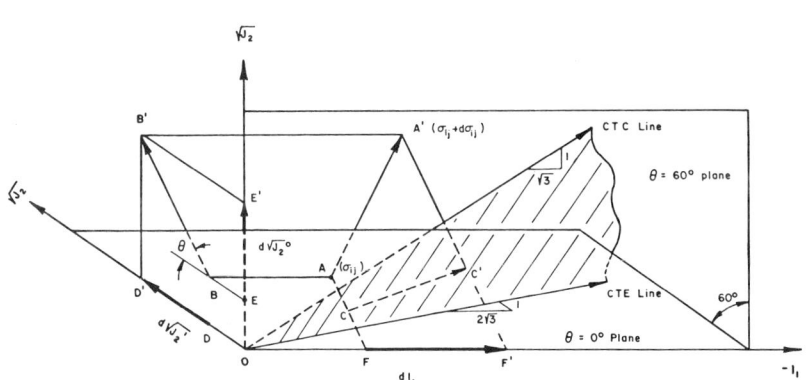

Fig. 4   General Stress Path in $I_1 - \sqrt{J_2}$ Space

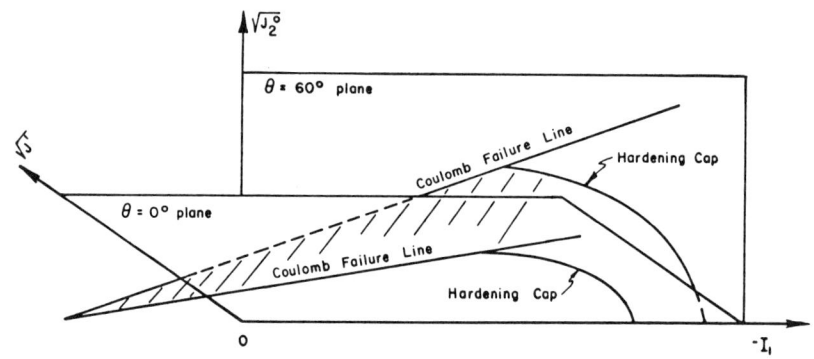

a) Cap Model I

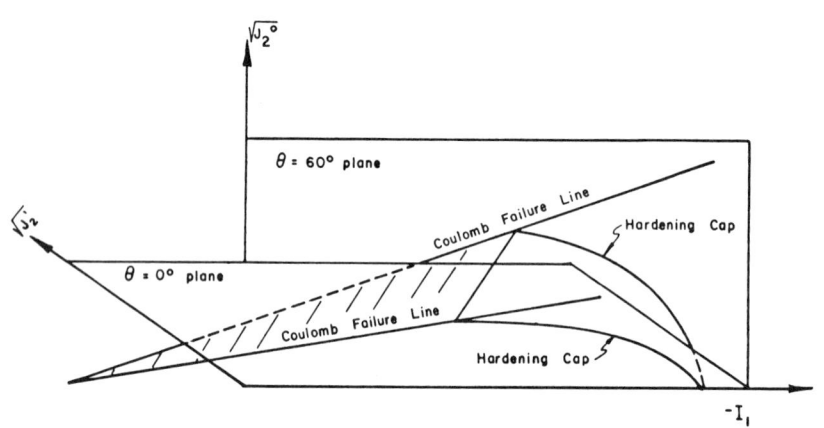

b) Cap Model II

Fig. 5 Scheme of Cap Model I and Cap Model II

PLASTICITY MODELS FOR SOILS   581

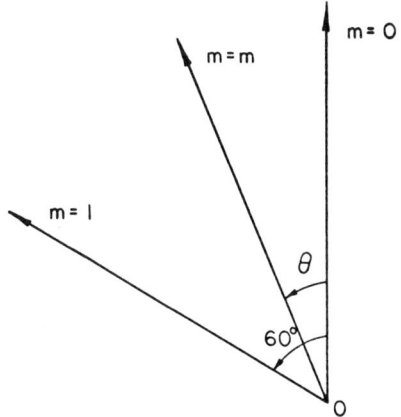

a) Stress Path on π-Plane

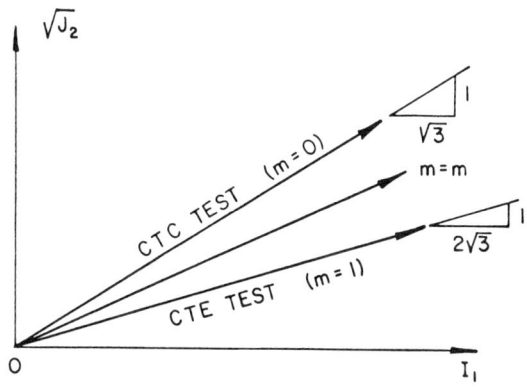

b) Stress Path in $I_1 - \sqrt{J_2}$ Space

Fig. 6 Stress Path in Experiments

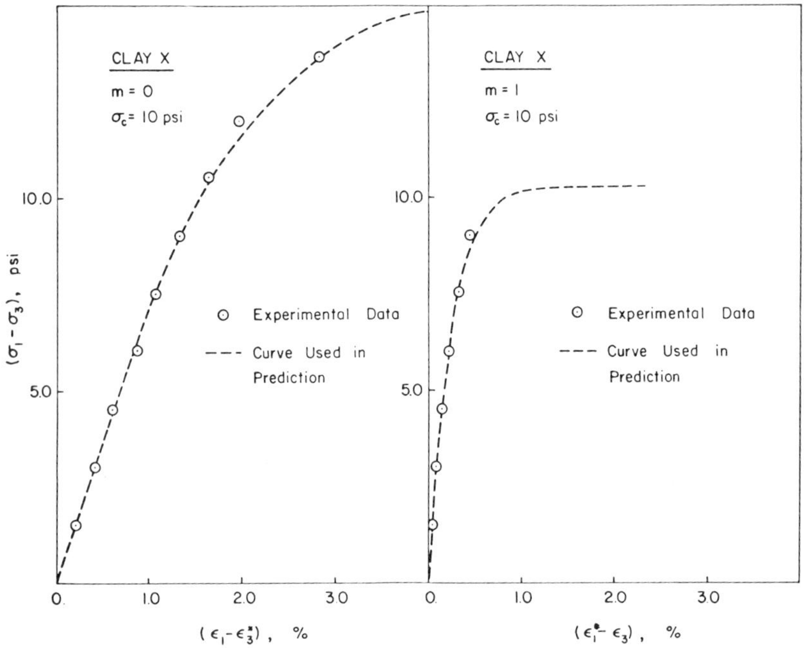

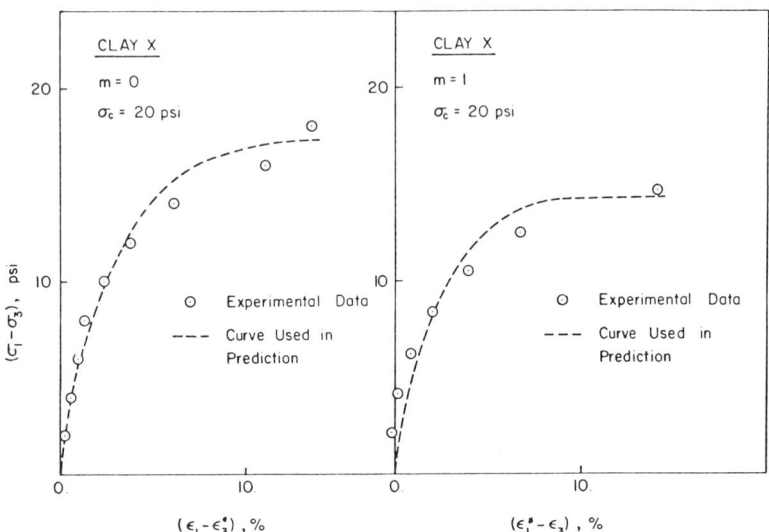

Fig. 7 Calibration of Shear Moduli for Clay X (σ_c = 10 psi, Nonlinear Elasticity Model)

Fig. 8 Calibration of Shear Moduli for Clay X (σ_c = 20 psi, Nonlinear Elasticity Model)

PLASTICITY MODELS FOR SOILS

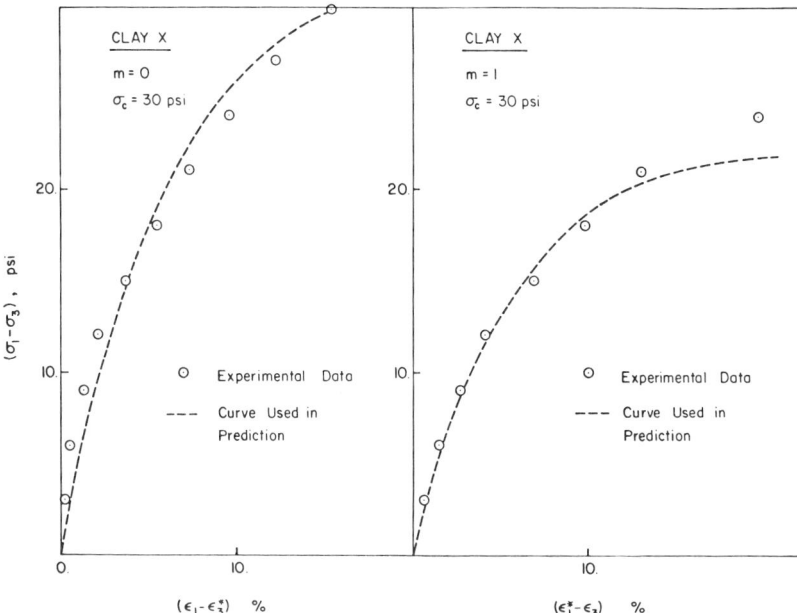

Fig. 9 Calibration of Shear Moduli for Clay X (σ_c = 30 psi, Nonlinear Elasticity Model)

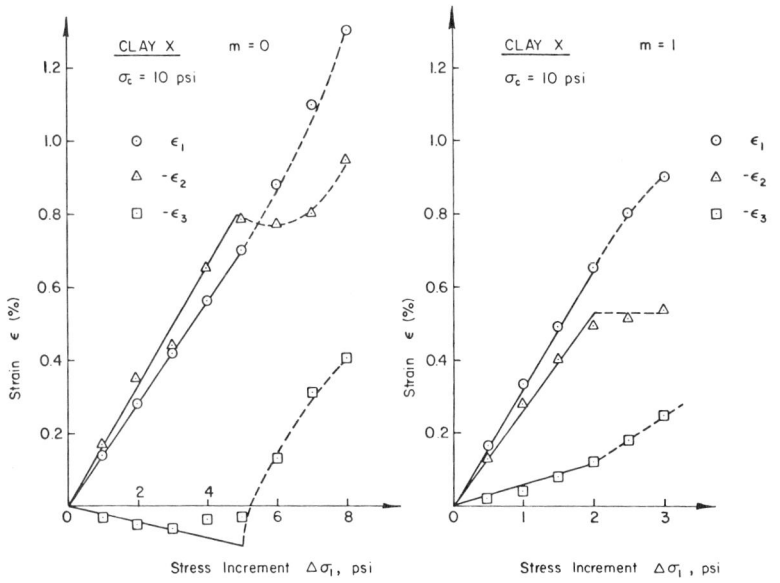

Fig. 10 Calibration of Orthotropic Material

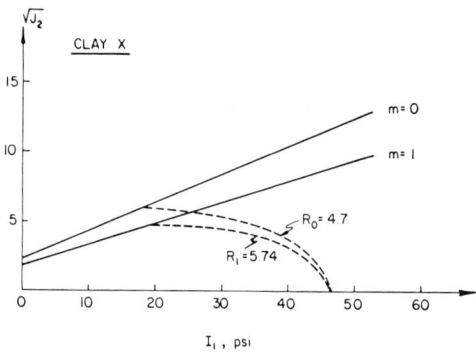

Fig. 11 Location of Hardening Caps for Clay X

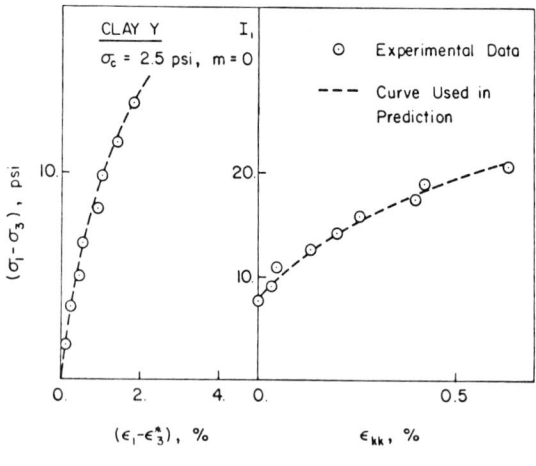

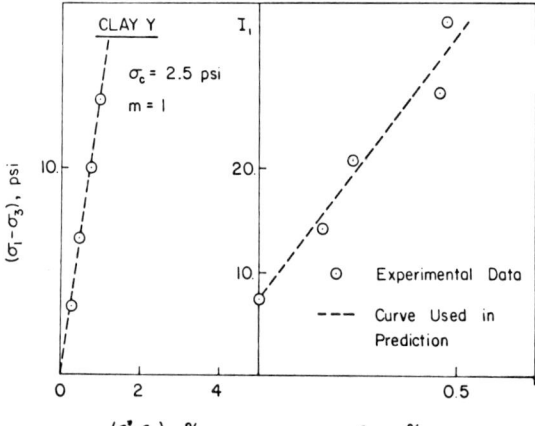

Fig. 12 Calibration of Shear and Bulk Moduli for Clay Y
(σ_c = 2.5 psi, Nonlinear Elasticity Model)

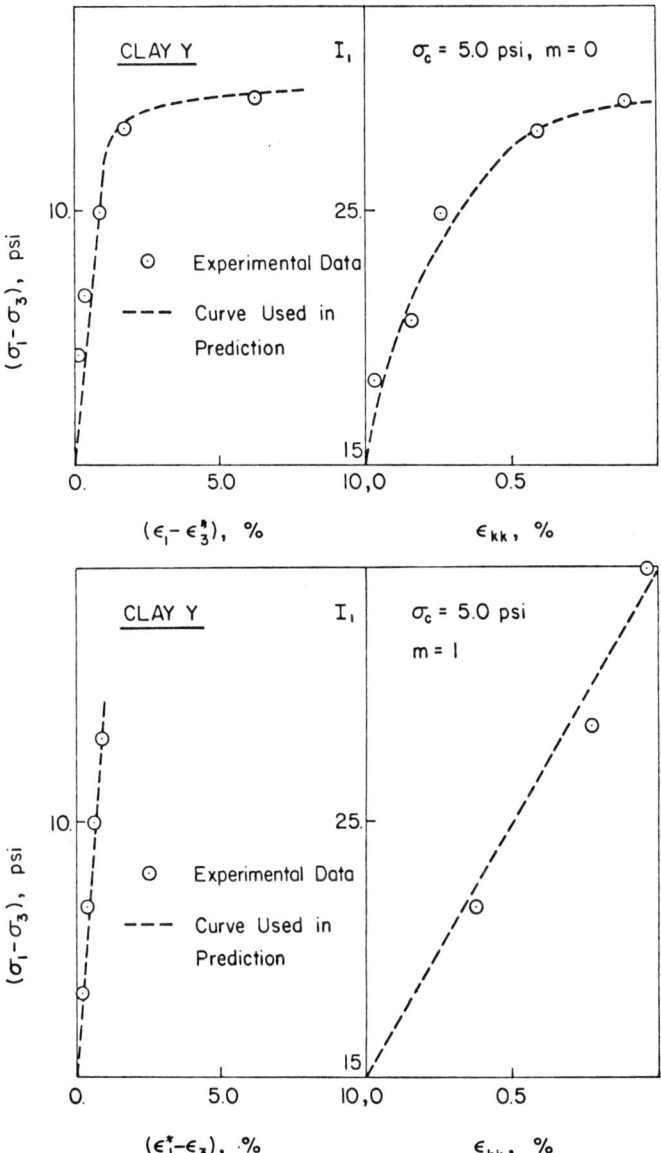

Fig. 13 Calibration of Shear and Bulk Moduli for Clay Y (σ_c = 5.0 psi, Nonlinear Elasticity Model)

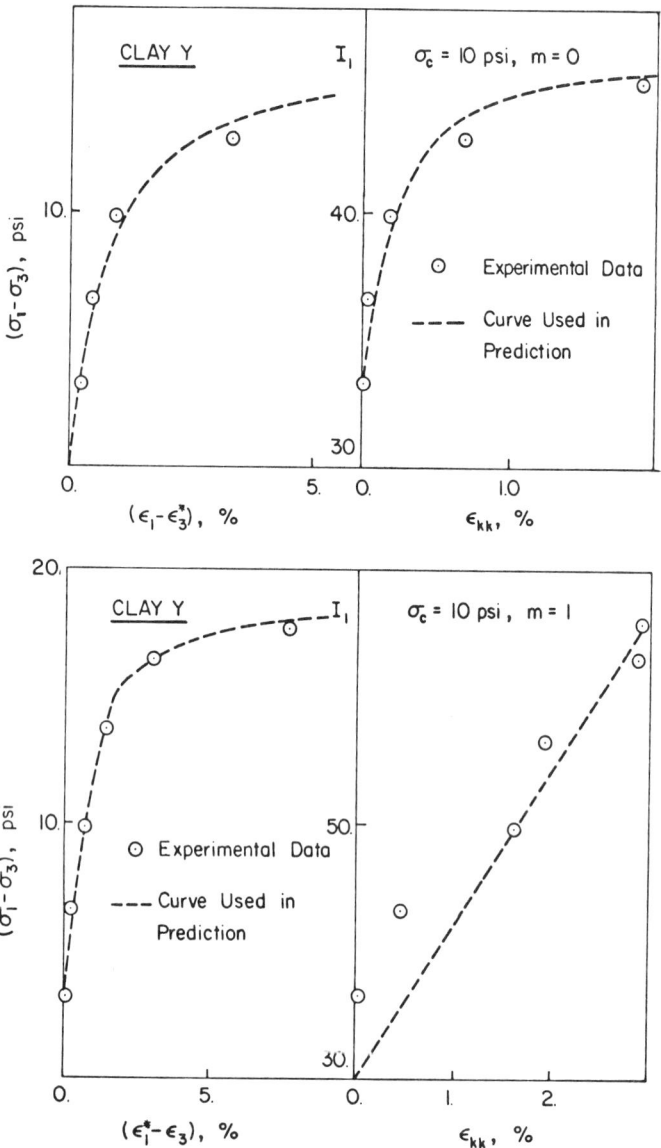

Fig. 14 Calibration of Shear and Bulk Moduli for Clay Y (σ_c = 10 psi, Nonlinear Elasticity Model)

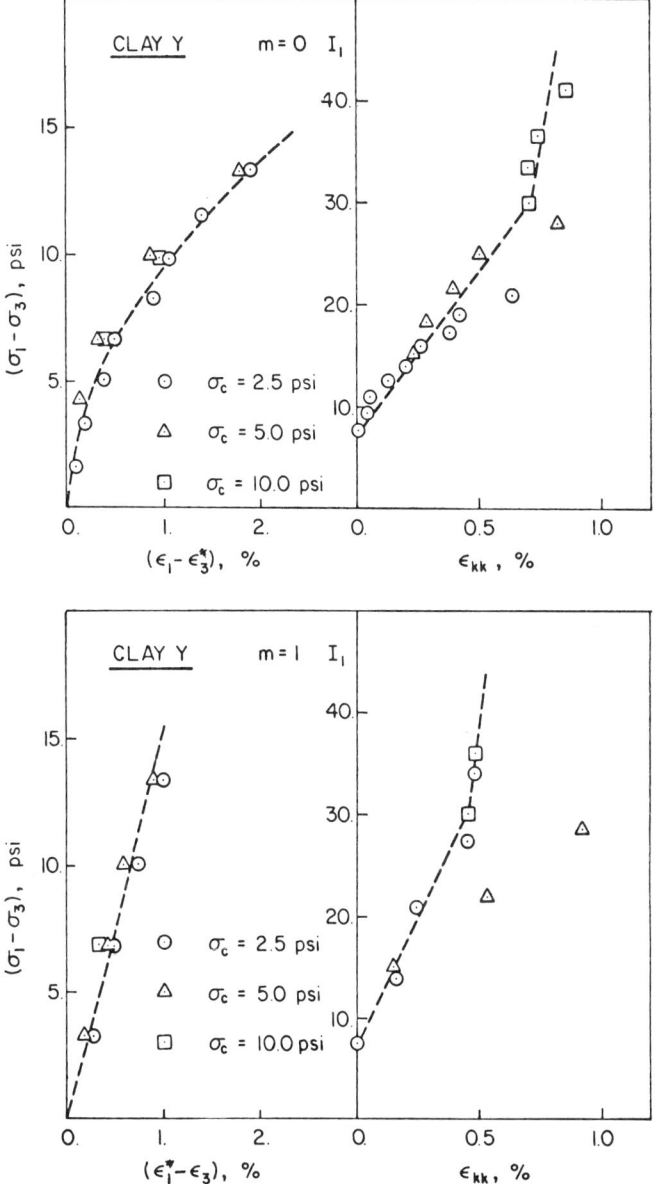

Fig. 15 Calibration of Shear and Bulk Moduli for Clay Y
(Cap Model I and Cap Model II)

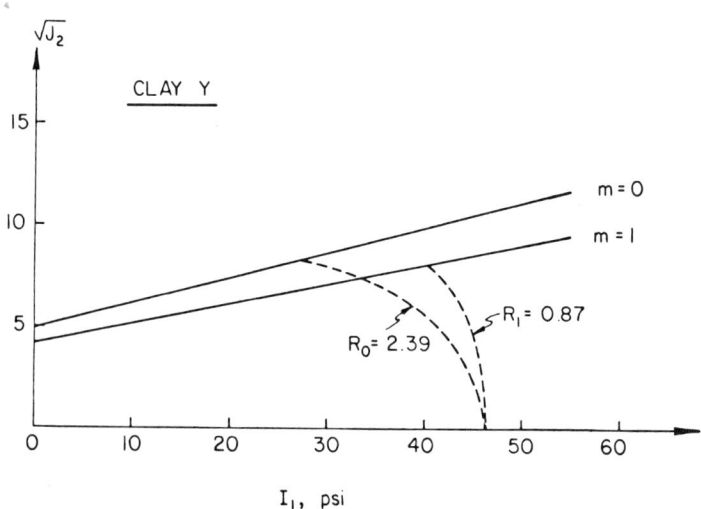

Fig. 16 Location of Hardening Caps for Clay Y

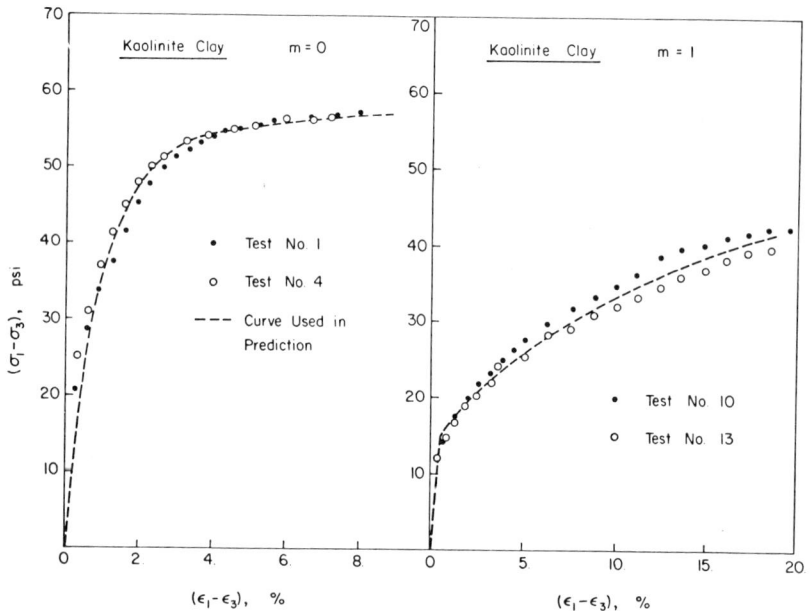

Fig. 17 Calibration of Shear Moduli for Kaolinite Clay

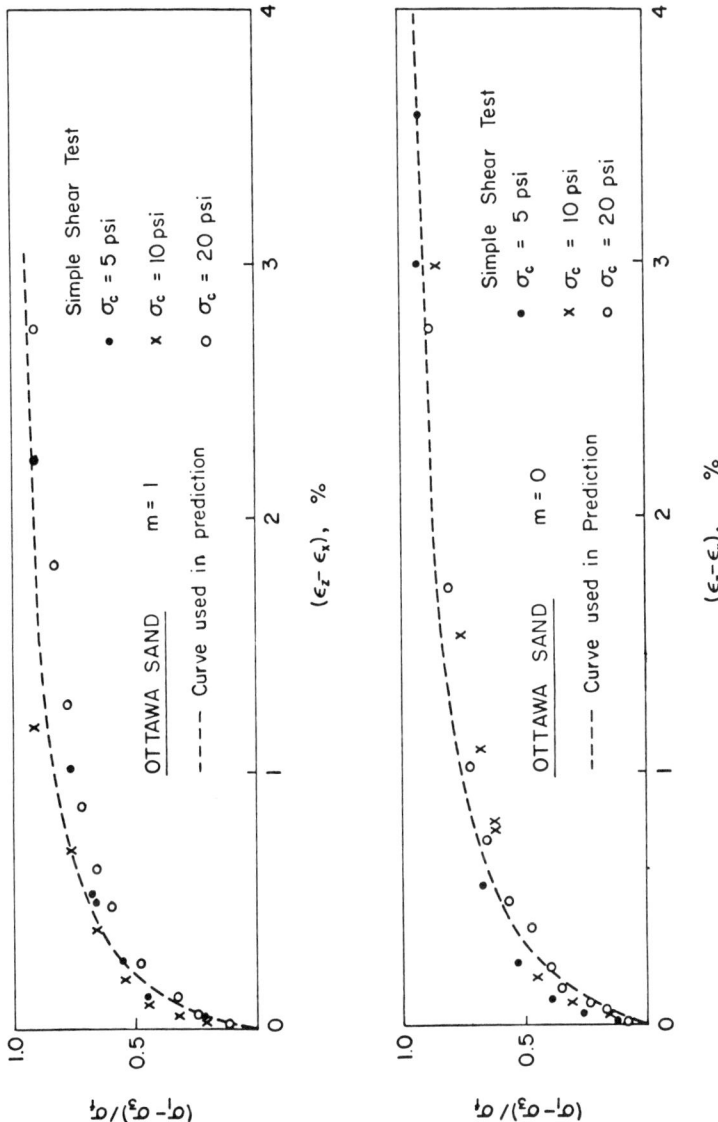

Fig. 18 Calibration of Shear Moduli for Ottawa Sand (Loading Case)

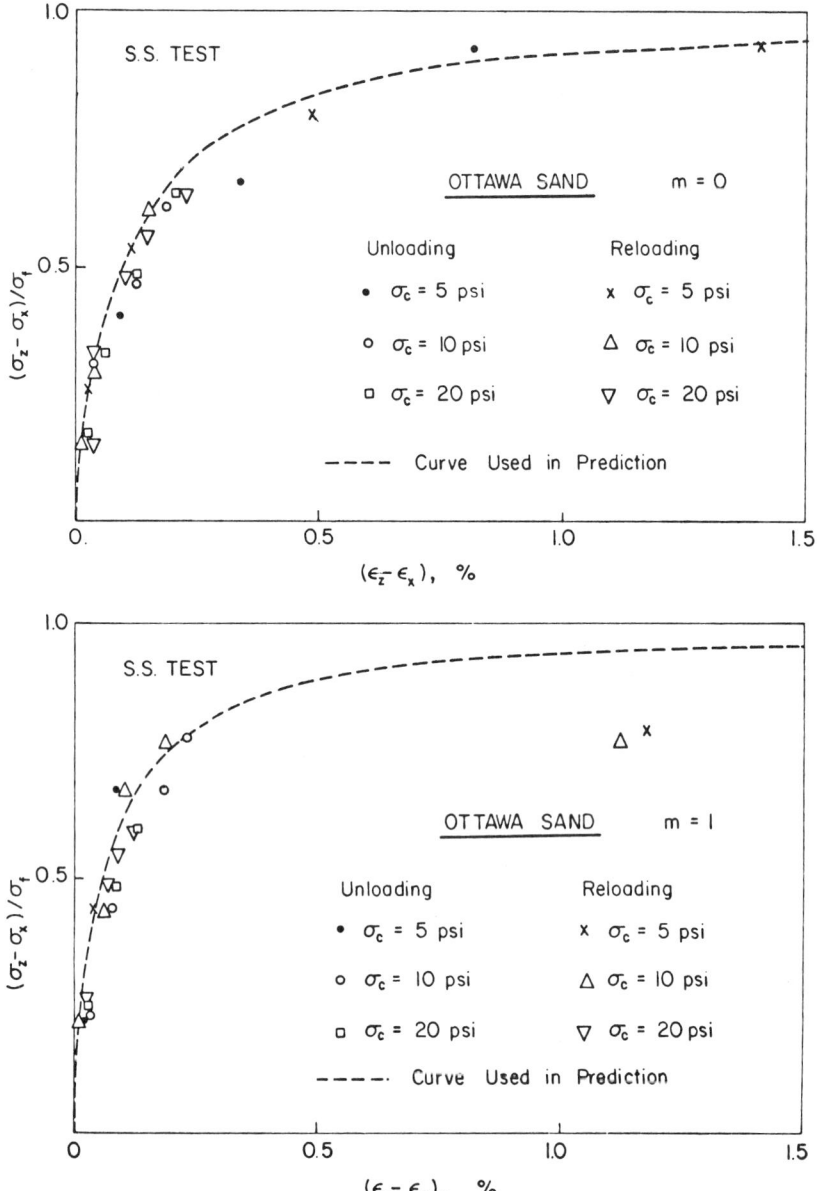

Fig. 19 Calibration of Shear Moduli for Ottawa Sand (Unloading and Reloading Cases)

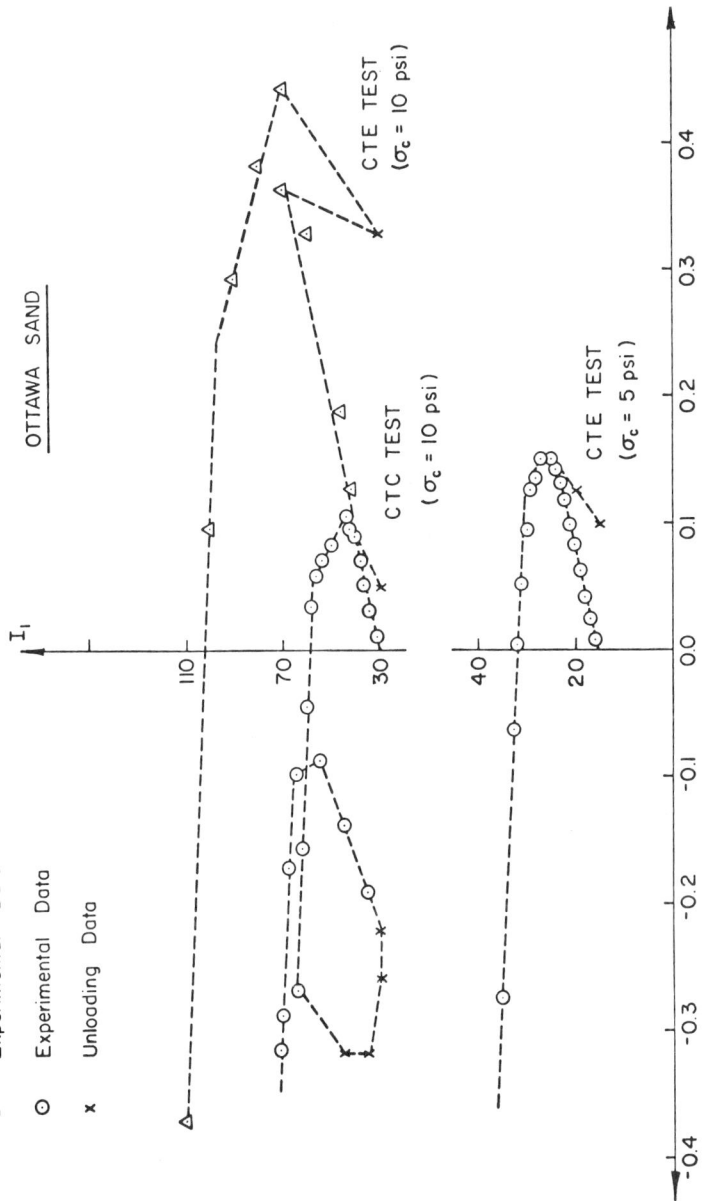

Fig. 20 Calibration of Bulk Moduli for Ottawa Sand

A CRITICAL STATE MODEL FOR PREDICTING THE BEHAVIOUR OF CLAYS

By C. Peter Wroth[1] and Guy T. Houlsby[2]

INTRODUCTION

The role of a mathematical model for the description of soil behaviour is twofold: firstly, as a simple qualitative framework against which soil behaviour may be assessed and, secondly, as a complete quantitative model for analysis and design. In its first role the model is of use both in judging the quality and consistency of data, and in predicting the character and trends of behaviour of a soil under a variety of circumstances. For this purpose the model must represent the essential features of soil behaviour in as simple a manner as possible.

A simple model is also desirable for the second role; this is because, in a detailed numerical calculation (such as one carried out by means of the finite element method), an excessively complex model may be difficult to implement and expensive to use. Very often the quality of the available data about the actual soil from the field will not warrant the additional precision of calculation achieved using a complex model. If, however, the model is to be used for mathematical calculations, it is important that it should be properly founded in the theory of continuum mechanics and should be internally consistent.

---

[1] Professor of Engineering Science, University of Oxford, Parks Road, Oxford OX1 3PJ, UK.
[2] Research Student, Engineering Department, University of Cambridge, Trumpington Street, Cambridge CB2 1PZ, UK.

A simple model will require few parameters for the description of a soil, and each of these should have a definite physical significance so that an assessment can easily be made of its importance and the likely results of any changes in its value. The parameters should be measurable directly in a small number of simple tests.

Both the purpose and the limitations of the model should be well understood. The extensive use of one model should lead to an understanding of its range of applicability and of the types of problem for which it may be useful. The trends of behaviour predicted by the model under extreme conditions should be explored and their importance assessed.

Some essential features of soil behaviour under conditions of monotonic loading which should be included in a model are listed below. The importance lies not in the fact that these features should be included but that the model should reproduce well-established experimentally determined behaviour.

1. A two (or more) phase material (material properties must be expressed in terms of effective stresses)
2. Both non-linear response and irrecoverable strains
3. Plastic dilation or compression
4. Failure conditions (e.g. Mohr-Coulomb)
5. The influence of consolidation history
6. Time effects
7. Anisotropy
8. Certain well-established empirical relationships which should be reproduced by the model, e.g. the ratio of undrained shear strength c_u to effective overburden pressure σ_v' may be expressed as a function of overconsolidation ratio.

All the above features are included in the model described below, except that:

1. Only the time effects due to primary consolidation are considered, whereas creep, secondary consolidation, aging and cementation are excluded;
2. The model is isotropic, so that certain effects due to the reversal or rotation of the principal stress axes are not included.

The model described in this paper, a minor variant of Modified Cam-Clay, is based on the theory of elasto-plasticity. An outline of the model is given in the next section, and a full mathematical derivation is given in Appendix I.

The model was chosen for use in the prediction workshop since it is a well-established and understood model which satisfies the general requirements of simplicity and completeness, which are essential for the economic solution of real problems. The model has not been altered to include any special features for the particular workshop problems. The model is suitable for the analysis of normally consolidated and highly overconsolidated clays, but is not thought to be suitable for sands.

THE MODIFIED CAM-CLAY MODEL

The theory of perfect plasticity is founded on the hypothesis that, during plastic deformation of a material, the strain increments are functions of the absolute stresses. The theory evolved from careful observations of the behaviour of ductile metals whose response to loading is typified by an initial phase of elastic, recoverable deformation until a yield condition is reached; after yield, a second phase of behaviour occurs which is a combination of elastic and plastic irrecoverable deformation. In general, the plastic deformations are greater by an order of magnitude than the accompanying elastic ones.

The study of a typical stress-strain curve for a soil in which the stress is cycled, for example, a one-dimensional consolidation test on a clay, reveals qualitatively similar behaviour. However, there is one essential difference and that is that soils experience irrecoverable *volumetric* as well as irrecoverable shear strains. The classical theory of plasticity, developed for describing the large strain response of ductile metals, has had to be extended to account for plastic volumetric strains in order to become applicable to soils.

Making use of this analogy with metal behaviour, and coupled with the experimental evidence for a critical state for deforming soil, a family of elasto-plastic stress-strain models has been developed at Cambridge. One such model, known as Modified Cam-Clay, initially proposed by Burland (2) has been used extensively in finite element computations. A description of this model follows.

In detail, the typical behaviour of a soil specimen in a triaxial compression test can be characterised by the curve of Fig. 1(a). If the specimen is loaded along the path OA and then unloaded along ABC, the path ABC can be approximated as elastic behaviour, with the strain represented by OC being a permanent, irrecoverable plastic strain denoted by ε_p. On reloading, the

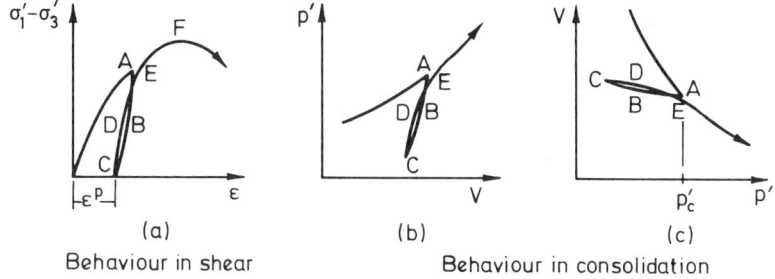

(a) Behaviour in shear
(b), (c) Behaviour in consolidation

FIG. 1. - TYPICAL BEHAVIOUR OF SOIL SPECIMEN IN CONSOLIDATION AND SHEAR TESTS

specimen behaves essentially elastically along CDE (displaying a small amount of hysteresis) until at point E it experiences the previous maximum deviator stress; it yields and undergoes further plastic strain. Point F denotes failure (a unique condition for this particular test); failure must be distinguished from yield which is a progressive phenomenon and which may occur at any point along the primary loading curve OAEF depending upon the exact loading sequence followed in the test.

Analogous behaviour will be displayed by a soil specimen if tested in consolidation, as shown in Fig. 1(b), which is an unconventional plot of effective pressure against specific volume (V) (V = 1 + e, where e is the voids ratio). If this diagram is rotated clockwise through 90° to give Fig. 1(c), the more usual plot of consolidation is obtained.

It should be realised that for a specimen that is in the overconsolidated state, represented by point C in Fig. 1(c), its preconsolidation pressure p_c', given by point A, is the current yield stress for further consolidation; it is the pressure at which further plastic volumetric strains begin to occur.

Now consider a soil specimen that has been isotropically normally consolidated (with $\sigma_1' = \sigma_2' = \sigma_3'$) to point A in Fig. 2(a), and then allowed to swell to some state along the unloading curve ABC. If the specimen were subjected to a variety of effective stress paths, it is assumed that there would be a well-defined region of stress states* for which the specimen would

---

* The stress space used for representing the data of triaxial tests is the mean normal principal effective stress $p' = \frac{1}{3}(\sigma_1' + \sigma_2' + \sigma_3')$ and deviator stress $q = \sigma_1' - \sigma_3'$.

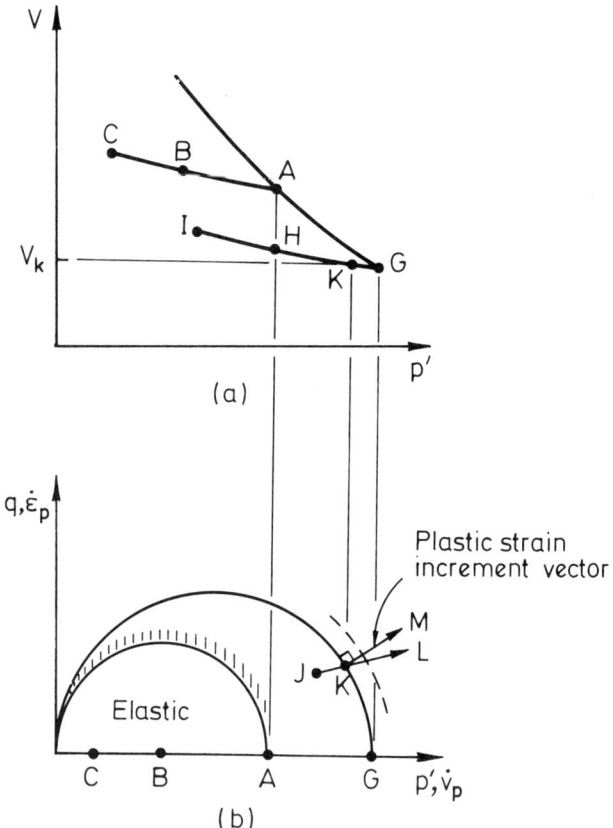

FIG. 2. - YIELD CURVES FOR SPECIMENS WITH DIFFERENT CONSOLIDATION HISTORIES

remain elastic. This region is bounded by a yield curve. If the stresses applied to the specimen take its state outside the current yield curve, it will yield and experience both plastic volumetric and plastic shear strains.

Consider a second specimen that has been normally consolidated to G and then allowed to swell to some point on the unloading curve GHI. Associated with this specimen is a larger yield curve, Fig. 2(b), but one that has the same shape as that for the first specimen.

The sizes of the yield curves are dictated by the points A and G, which lie on the normal isotropic consolidation curve. The choice of yield curve appropriate for any specimen depends on the maximum consolidation pressure (i.e. the pre-consolidation pressure, p_c').

The shape of the yield curve, Fig. 3(a), is assumed to be elliptical; this choice is based on considerations of energy dissipated plastically within the specimen. One axis of the ellipse BA is fixed by the consolidation history relevant to the specimen. The other axis BX_1 is given by the assumption that the point X_1 is on the line of critical states, or failure states of normally consolidated specimens. This line is parallel to the normal consolidation line in the logarithmic plot of Fig. 3(b).

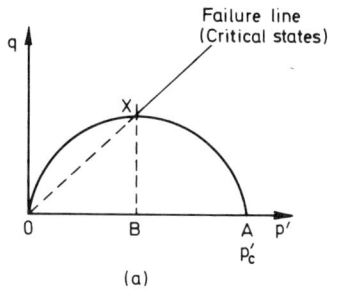
(a)

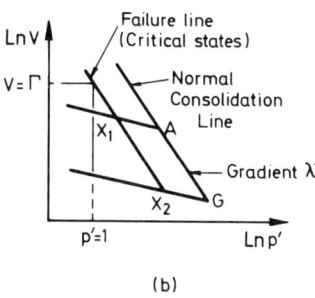
(b)

FIG. 3. - DETAILS OF YIELD CURVE FOR MODIFIED CAM-CLAY MODEL

As shown in Fig. 3(b), both consolidation and swelling lines are considered as straight in $\ln V - \ln p'$ space. This represents a very minor alteration of the model from previously published descriptions of Modified Cam-Clay, for which the lines were taken as straight in $V - \ln p'$ space. Consequences of the change are:

1. The numerical values of the parameters λ^* and κ^* (see below) are reduced by a factor approximately equal to V from the original parameters λ and κ;

2. The bulk modulus becomes truly proportional to pressure rather than depending also upon specific volume;

3. Evidence in support of this change is given by Butterfield (3), and a further advantage is a slight simplification of the mathematics of the model.

For a specimen undergoing *small* strains the new version of the model is identical to the old one.

The complete description of the model requires five parameters to specify the shape and size of the yield locus for a soil specimen at a given pressure and specific volume, as well as the elastic properties of the material. The parameters are:

1. λ^*, the gradient of the consolidation line in $\ln V - \ln p'$ space. This is related directly to the conventional compression index C_c' by $\lambda^* = C_c'/(2.303\,V)$;

2. κ^*, the gradient of the swelling line, similarly related to the swelling index C_s' by $\kappa^* = C_s'/2.303\,V)$;

3. M, the gradient of the critical state line in $q - p'$ space. This is related to the angle of shearing resistance of a normally consolidated clay by $M = 6 \sin \phi'/(3 - \sin \phi')$;

4. G, the shear modulus, taken to be constant;

5. Γ.

The last parameter Γ is required solely to locate either the critical state line or the isotropic normal consolidation line in Fig. 3(b); it is the voids ratio on the critical state line for which p' has unit value. It is in some ways analogous to the liquid limit (see Schofield and Wroth(10)) and increases markedly with the plasticity index of the clay being considered.

The elastic behaviour of the specimen for stress states within the current yield locus is assumed to be isotropic, but with the bulk modulus directly dependent upon the current mean stress p'. This latter feature is a consequence of the volumetric strains associated with the swelling and recompression lines AX_1 and GX_2 in Fig. 3(b).

Simple calculation shows that for a specimen experiencing a pressure p' the incremental bulk modulus, derived from the local gradient of the swelling line, can be written

$$K = p'/\kappa^* \quad \quad \quad \quad \quad \ldots \ldots (1)$$

The elastic shear response is specified by assuming a constant shear modulus, G.

Now, suppose the state of stress experienced by the specimen is at point J in Fig. 2(b) close to the relevant yield curve. If the stress increment JKL is applied to the specimen, the increment JK will cause elastic strain increments only, whereas the increment KL will cause both elastic and plastic strain increments. As the specimen yields at K, the yield curve is expanded as the specimen undergoes consolidation. The plastic volumetric strain increment* that occurs from K to L is given by a hardening law,

$$\dot{v}_p = -\frac{\dot{V}_p}{V} = (\lambda^* - \kappa^*)\frac{\dot{p}_c'}{p_c'} \qquad \ldots \ldots (2)$$

which is derived from the normal consolidation line

$$\ln(V/V_c) = \lambda^* \ln(p_c'/p') \qquad \ldots \ldots (3)$$

The details of this derivation are not given here but can be found in Schofield and Wroth (10). The important point is that no additional soil parameter is required.

The flow rule governing the ratio of the plastic strain increments is given by the condition of normality. This condition stipulates that if the associated plastic strain increments $\dot{v}_p$ and $\dot{\varepsilon}_p$ are plotted on the same axes as the stresses in Fig. 2(b), the vector KM of plastic strain increment is normal to the yield curve at K. The gradient of the curve at K is known from its elliptical shape, so that the ratio $\dot{v}_p/\dot{\varepsilon}_p$ can be calculated and since $\dot{v}_p$ is already established (from the hardening law), $\dot{\varepsilon}_p$ can be evaluated.

Consider now a specimen undergoing a conventional undrained triaxial compression test, with initial state as shown by point A in Fig. 4. The condition of the test is such that the total volume must stay constant, i.e. that the path in Fig. 4(a) coincides with a line of constant voids ratio, that is, constant specific volume. The initial response of the specimen must be elastic as the Point A lies within the current yield curve, Fig. 4(b). There can be no change of p' during elastic deformation as the point A in Fig. 4(a) is constrained to lie on both the elastic swelling line and the constant voids ratio line. When the effective stress path reaches the point B, the specimen yields. Beyond this point, both plastic and elastic volumetric strains occur, of equal magnitude but opposite sign so as to maintain the total volume constant. The ratio of the plastic volume change to the plastic shear strain is calculated as before, and the test proceeds to failure at point D. The effective stress path ABCD is as shown in Fig. 4(b). The total stress path for a triaxial compression test (with constant cell pressure) is given by AE.

---

* The dot notation of plasticity theory has been adopted to indicate a small (time-independent) increment.

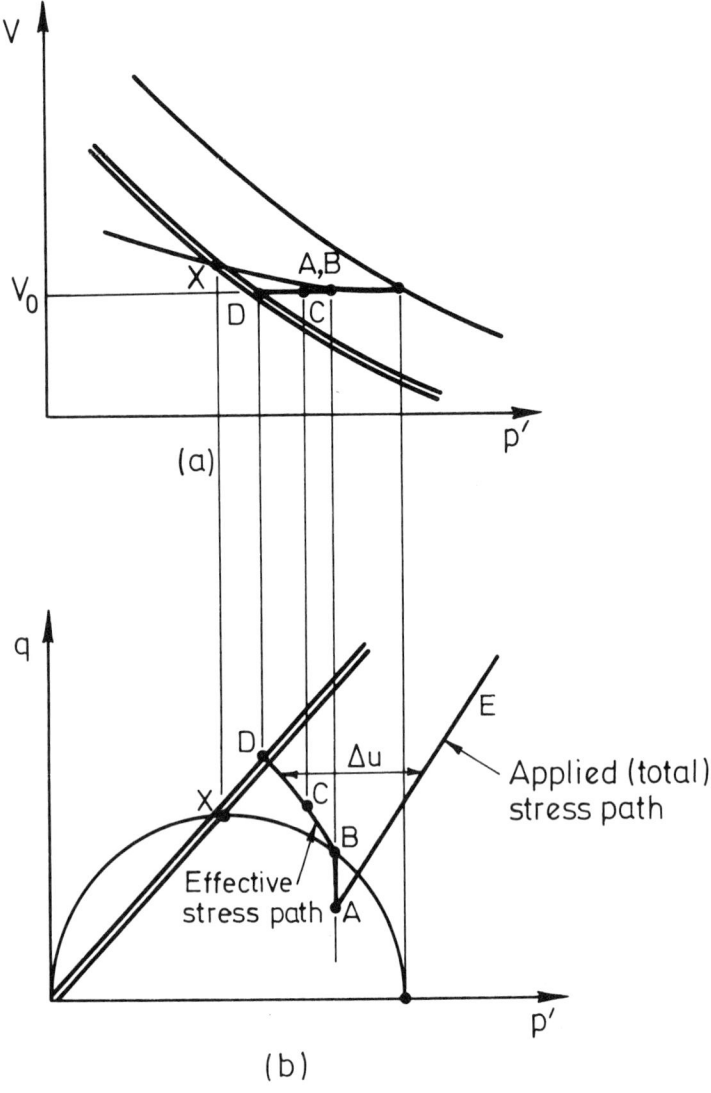

FIG. 4. - STRESS PATHS FOR CONVENTIONAL UNDRAINED TRIAXIAL COMPRESSION TEST

The pore pressure at any stage of the test is simply the difference in mean stress between the two curves, as shown in Fig. 4(b).

This model successfully reproduces the major deformation characteristics of soft clay, and is expressed in terms of effective stress allowing predictions to be made of pore pressures in undrained tests.

THE ANALYSIS

For the problems posed to the predictors, the analysis of all tests was carried out using a computer program which solves the constitutive equations incrementally in the same way as is used in a finite element analysis. The number of increments used in any given test is usually of the order of two hundred and this has been verified to be sufficient by checking cases for which an exact closed-form solution exists. Each test is analysed as a single homogeneous element, and any combination of stress and strain control can be applied. Drainage of the pore fluid is properly accounted for.

The tests on clays 'X' and 'Y' involve load control in which an increment of load is applied to the sample, and the recorded strains then represent the values at which the sample finally comes into equilibrium. Although this is a 'drained' test, this particular loading sequence involves a slightly different stress path from a slow strain controlled drained test, in that each application of load takes place under essentially undrained conditions, followed by a period of consolidation as the sample comes into equilibrium. The process is illustrated in Fig. 5.

Consider a specimen of clay that has been isotropically normally consolidated to point A. In a conventional slow drained triaxial compression test the path AE (of gradient 3) could be followed but, in the incremental test, the first load application results in plastic undrained deformation AB, followed by consolidation BC. Note that not only will the strains developed on this path be different from the slow drained case, but the final strength at D is considerably lower.

In contrast, a specimen which starts in an overconsolidated state, such as G, within the current yield envelope will behave elastically. The effective stress path during the undrained phase will be at constant p' (i.e. GH) with subsequent consolidation HI. The strains experienced by this specimen will be independent of the path between states G and I, provided that the path remains within the yield locus.

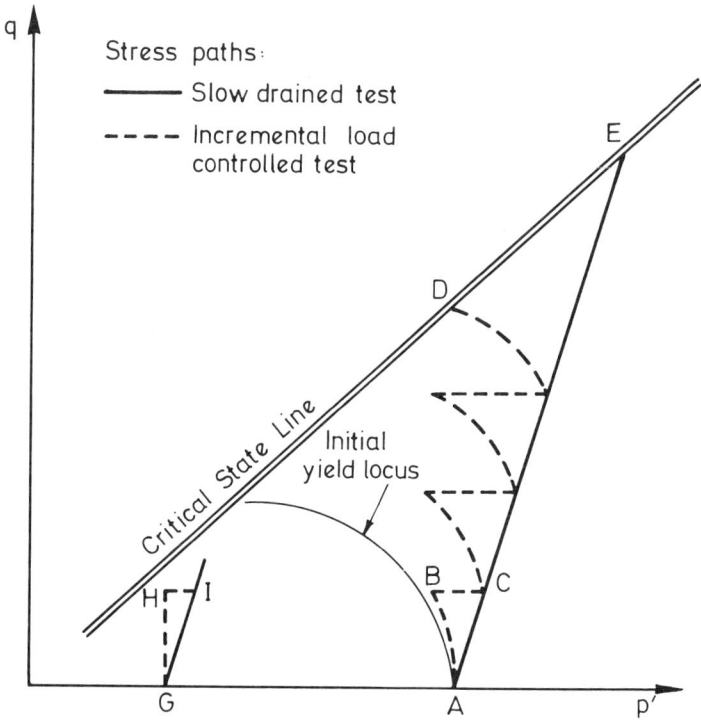

FIG. 5. – COMPARISON OF SLOW DRAINED TEST AND
INCREMENTAL LOAD CONTROLLED TEST

MODELLING OF THE NATURAL CLAYS 'X' AND 'Y'

The first stage in the modelling of the natural clays is an assessment of their index properties. The two clays appear to be very similar, with the main features being their very high values of liquidity index and sensitivity. The positions of clays 'X' and 'Y' may be plotted on a liquidity index - pressure diagram, and it is seen in Fig. 6 (for the range of pressures used in the tests) that they plot well outside the region which usually applies for remoulded clays (Wroth (12)). Since the Modified Cam-Clay model was originally developed for remoulded clays, some caution must be exercised in applying the model to these natural materials.

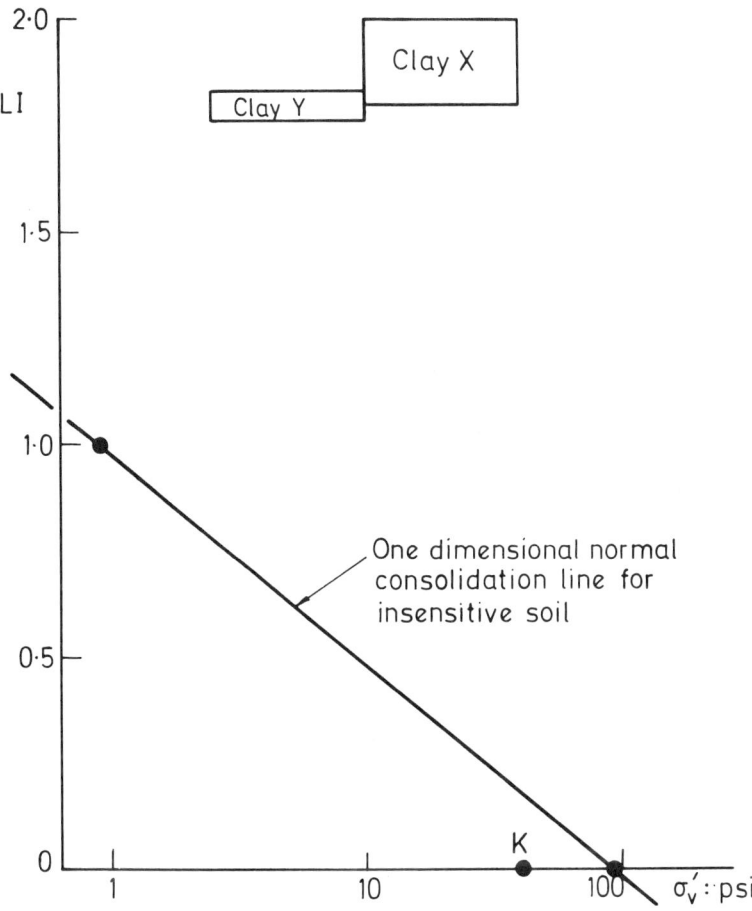

FIG. 6. - INITIAL STATES OF CLAY IN RELATION TO CONSOLIDATION OF INSENSITIVE SOIL

In the following sections, the procedure adopted for modelling clay 'X' will be described. An exactly similar method was used in the modelling of clay 'Y'.

Consolidation. - Fig. 7 shows data for one-dimensional and isotropic consolidation of clay 'X', together with the theoretical fit which has been used to match the data. Clearly a lower value of p_c' would have provided a better fit to the experimental data, but the pattern of the predictions is basically correct. The p_c' value was chosen to give the best fit to the triaxial tests, and it may be that the p_c' value varies for different samples of the clay; the two tests in Fig. 7 clearly show different p_c' values (with the difference being larger than can be accounted for by the difference between isotropic and one-dimensional consolidation).

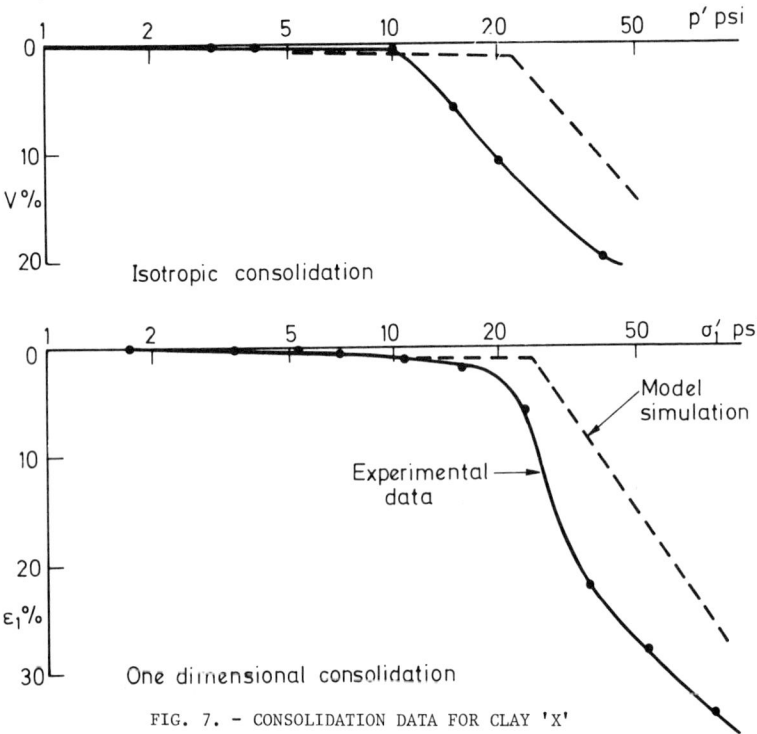

FIG. 7. - CONSOLIDATION DATA FOR CLAY 'X'

The ratio of λ^*/κ^* is unusually high for clays 'X' and 'Y', this also being a feature of a sensitive clay.

Shear Tests. - For each of the compression and extension tests the three measured strains, ε_1, ε_2 and ε_3, were plotted against a measure of the deviator stress, $q = \sigma_1 - \sigma_3$. Fig. 8(a) shows a typical plot for a test at low stress level, with the response being substantially elastic, followed by a brittle failure with no preceding plastic deformation. This behaviour corresponds to that of a material at a pressure on the 'dry' side of the critical state line, i.e. at less than $p_c'/2$. Fig. 8(b) shows the typical response of the clay at a higher pressure, with a small initial elastic portion followed by substantial plastic strains before failure.

In order to establish the initial yield locus for the material, the stress paths for all the shear tests have been plotted in Fig. 9. The parts of the path for which the response is essentially elastic, giving very small strains, is shown as a solid line; the plastic part, with larger and non-linear strains, is shown as a broken line and the range of stress within which failure occurred is shown as a heavy band. (Because the tests were load controlled the precise failure stress cannot be determined.) For convenience the extension tests are plotted in the negative q direction, but these tests were in fact carried out in a mode with a stress path which is only 60° from that for the triaxial compression tests in the octahedral plane, rather than the more conventional 180°.

The initial yield locus used for calculation, and the position of the critical state line are also shown on Fig. 9. Note that the locus approximately encloses the region for which an elastic response was observed. The critical state line appears to indicate an overestimation of strength, but when the load controlled nature of the tests is accounted for (see page 10) approximately the correct strength is modelled.

TABLE 1. - PARAMETERS USED IN MODIFIED CAM-CLAY MODEL

Parameter	Natural Clays		Kaolin
	X	Y	
λ^*	0.2	0.16	0.066
κ^*	0.005	0.005	0.018
M	1.2	1.2	1.28
G/psi	300	450	2000
Γ		Not determined	
p_c'/psi	22	25	74.8

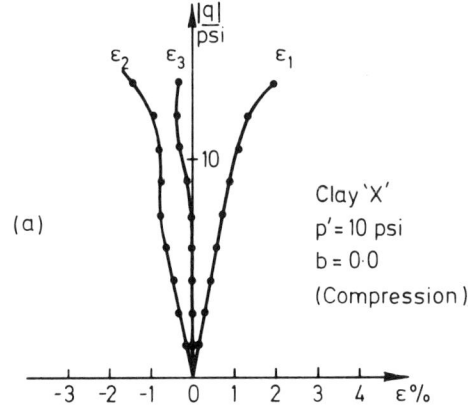

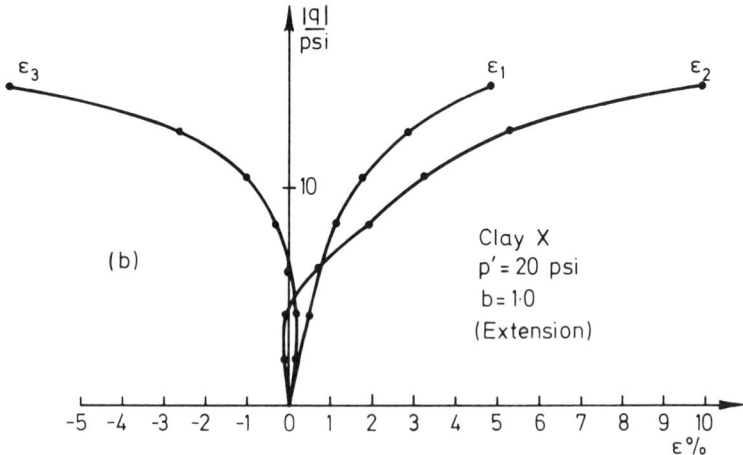

FIG. 8. — TYPICAL STRESS-STRAIN CURVES FOR CLAY 'X'

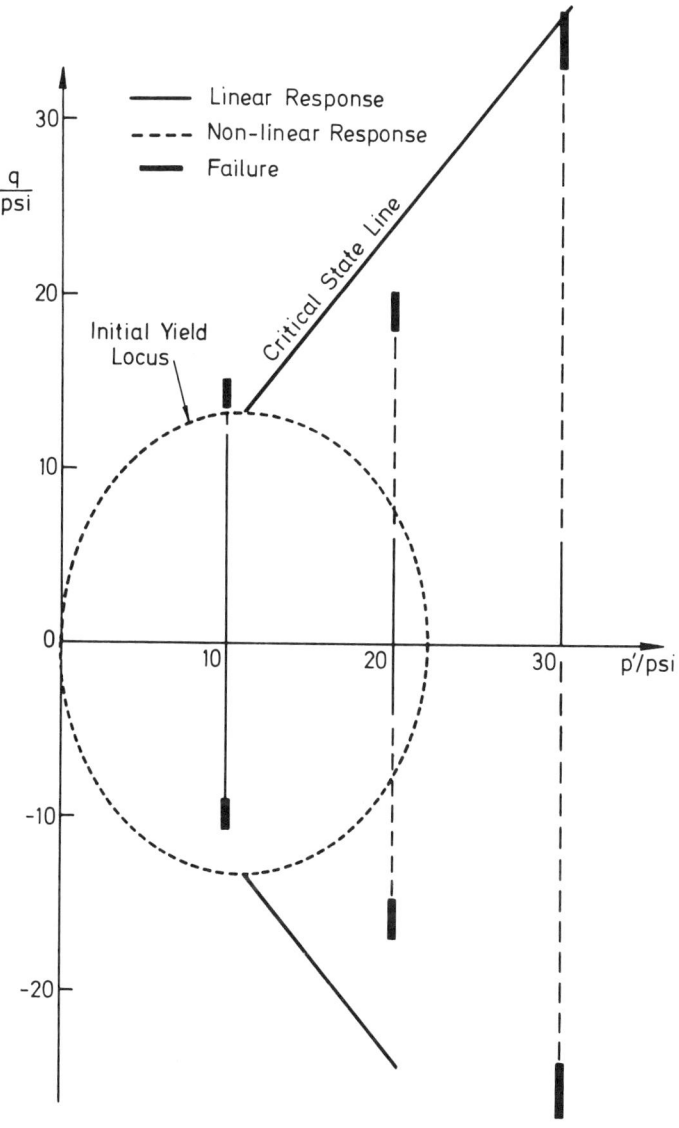

FIG. 9. - YIELD AND FAILURE OF CLAY 'X'

The parameters used to model clays 'X' and 'Y' are given in Table 1. The parameter Γ did not need to be determined since it does not enter the incremental stress-strain relationship and no prediction of specific volume is required. Of the remaining parameters, λ^*, κ^*, M and G are material constants and p_c' describes the initial state of the material and establishes the size of the initial yield locus.

Using these parameters the tests on clays 'X' and 'Y' were modelled both as slow drained tests and as incremental load control tests. Typical results are shown in Figs 10-12. Fig. 10 shows the results for a test at a low pressure (p' = 10 psi) at which the clay is overconsolidated. Both forms of modelling give identical results and indicate a brittle failure after an elastic response. Fig. 11 shows the behaviour at a higher pressure, with an initially elastic response followed by plastic deformation. For this case the load controlled test gives larger strains and a lower strength.

In addition to the load controlled tests a series of slow strain controlled drained triaxial compression tests was carried out on clay 'X'. The modelling of these tests is shown in Fig. 12, and the experimental results in Fig. 13. Although the precise numerical values are not fitted exactly, the general character of the responses and the trends of behaviour at different confining pressures are modelled very satisfactorily.

The family of Cam-Clay models was originally developed for representing the results of conventional triaxial tests. For modelling the results of other types of test, some generalisation into three-dimensional stress space has to be made. In the case of Modified Cam-Clay the simplest isotropic generalisation has been made consisting of the rotation of the elliptical yield locus around the p'-axis (or space diagonal) in principal stress space. This circular generalisation allows no dependence of plastic behaviour on the third stress invariant.

The parameters derived from the compression and extension tests were used to model the tests at intermediate b values.* The character of the response predicted is very similar to that of the compression and extension tests: brittle behaviour at low pressures and plastic deformation at high pressures. The effect of the parameter b is relatively small. Predictions were made

---

* The parameter b is defined as the ratio $(\sigma_2 - \sigma_3)/(\sigma_1 - \sigma_3)$ following Bishop (1).

BEHAVIOUR OF CLAYS 609

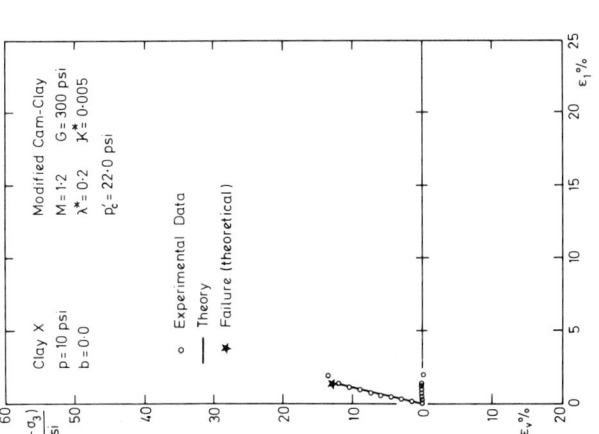

FIG. 10. — COMPRESSION TEST ON CLAY 'X' AT p' = 10 psi

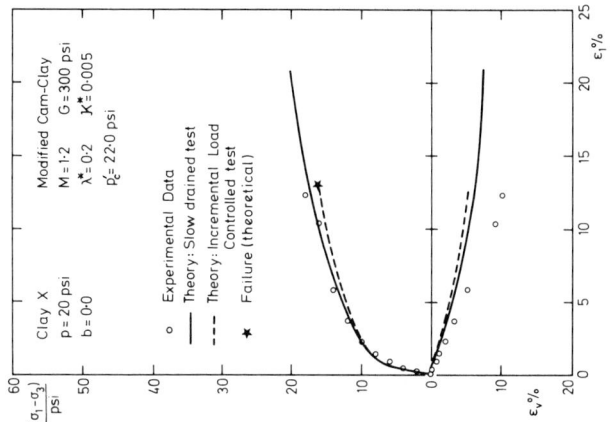

FIG. 11. — COMPRESSION TEST ON CLAY 'X' AT p' = 20 psi

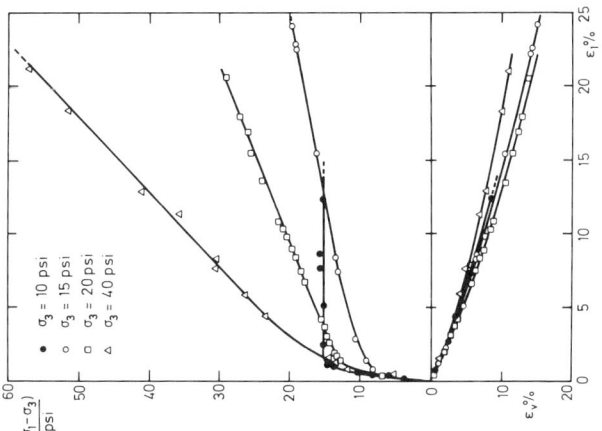

FIG. 13. — EXPERIMENTAL DATA FOR SERIES OF SLOW DRAINED TESTS ON CLAY 'X'

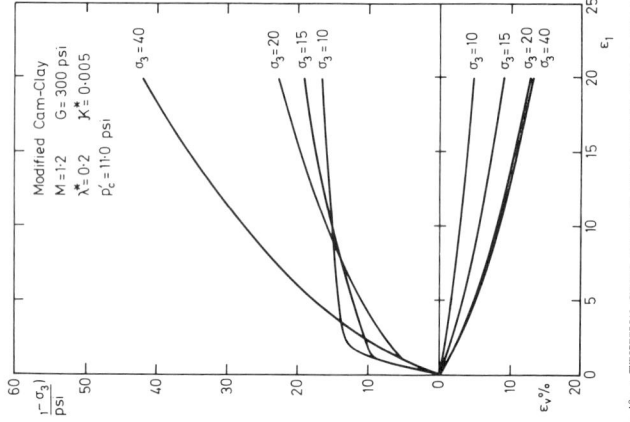

FIG. 12. — THEORETICAL CURVES FOR SERIES OF SLOW DRAINED TESTS ON CLAY 'X'

both for slow drained tests and for incremental load controlled tests. In the latter case the size of the load increment was chosen as approximately the same as that for the compression and extension tests. A different increment size would result in different predictions and, in particular, the failure strength would be affected for the tests at higher pressures. No difference in behaviour between slow strain controlled tests and load controlled tests is predicted for those tests which fail on the 'dry' side of the critical state line; the particular tests which are affected are shown in Table 2.

TABLE 2. — TESTS ON NATURAL CLAYS FOR WHICH LOAD CONTROL GIVES DIFFERENT RESULTS FROM STRAIN CONTROL

		$b = 0.0$	$b = 0.25$	$b = 0.5$	$b = 0.75$	$b = 1.0$
Clay 'X'	$p' = 10$ psi					
	$p' = 20$ psi	✓	✓	✓	✓	✓
	$p' = 30$ psi	✓	✓	✓	✓	✓
	$p' = 40$ psi	✓				
Clay 'Y'	$\sigma'_c = 2.5$ psi					
	$\sigma'_c = 5$ psi			✓	✓	✓
	$\sigma'_c = 10$ psi	✓	✓	✓	✓	✓

Anisotropy. — The early parts of the tests on clays 'X' and 'Y' at low stress levels show a markedly linear response, which may be assumed to be elastic. From the gradients of the graphs of ε_1, ε_2 and ε_3 against deviator stress, together with the conditions of the tests, it should be possible to establish the elastic compliance matrix accurately for these materials. In general, if the material is anisotropic the matrix will involve six variables:

$$\begin{bmatrix} \dot\varepsilon_1 \\ \dot\varepsilon_2 \\ \dot\varepsilon_3 \end{bmatrix} = \begin{bmatrix} c_1 & c_2 & c_3 \\ c_2 & c_4 & c_5 \\ c_3 & c_5 & c_6 \end{bmatrix} \begin{bmatrix} \dot\sigma_1 \\ \dot\sigma_2 \\ \dot\sigma_3 \end{bmatrix}$$

For both materials there are two types of test, for which three slopes ($\varepsilon_1 - q$, $\varepsilon_2 - q$ and $\varepsilon_3 - q$) may be obtained, giving six pieces of information. In fact, in each case the information is such that only five of the above variables may be determined, the sixth gradient providing a check on the consistency of the data.

For clay 'X' the only tests with a substantially linear section are those at 10 psi. For these tests the consistency check shows that the behaviour cannot be modelled by linear anisotropic elasticity. Possibly the strains given do not represent the true deformation of the sample, or the samples exhibit a wide variation in elastic properties, or the axes of the two samples do not coincide.

The data from clay 'Y', obtained from tests at 2.5 and 5.0 psi, show an accurate consistency check, and the compliance matrix (in units of % and psi) may be determined approximately as:

$$\begin{bmatrix} \varepsilon_1 \\ \varepsilon_2 \\ \varepsilon_3 \end{bmatrix} = \begin{bmatrix} 0.07 & -0.02 & -0.02 \\ -0.02 & 0.05 & 0.0 \\ -0.02 & 0.0 & \text{unknown} \end{bmatrix} \begin{bmatrix} \dot{\sigma}_1 \\ \dot{\sigma}_2 \\ \dot{\sigma}_3 \end{bmatrix}$$

It is clear that this behaviour does not represent a high degree of elastic anisotropy, and that the use of isotropic elastic properties is fully justified. The matrices at 2.5 and 5.0 psi used in the modelling may be derived as:

$$\begin{bmatrix} 0.052 & -0.008 & -0.008 \\ -0.008 & 0.052 & -0.008 \\ -0.008 & -0.008 & 0.052 \end{bmatrix} \text{ and } \begin{bmatrix} 0.063 & -0.019 & -0.019 \\ -0.019 & 0.063 & -0.019 \\ -0.019 & -0.019 & 0.063 \end{bmatrix}$$

MODELLING OF KAOLIN

When plotted on a liquidity index against pressure diagram (see point K on Fig. 6) the Kaolin is seen to fall within the usual range for a non-sensitive clay. The isotropic consolidation test (Fig. 14) shows typical consolidation and swelling lines which correspond closely to those used in Modified Cam-Clay. Taken together with the fact that the Kaolin is a laboratory prepared clay at a low overconsolidation ratio, the Modified Cam-Clay model may be expected to give a reasonable modelling of this material.

The four undrained shear tests supplied fall into two pairs - compression and extension. Any model based on effective stress principles would predict

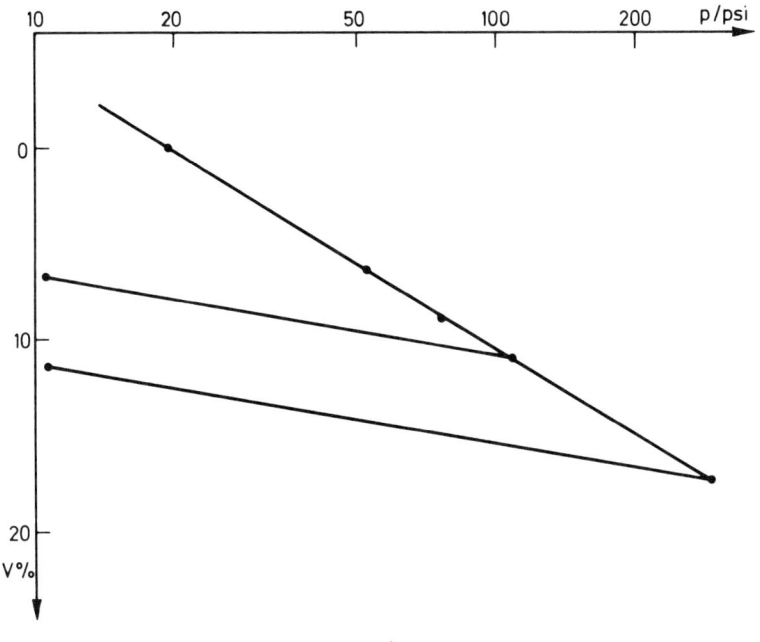

FIG. 14. - ISOTROPIC CONSOLIDATION OF KAOLIN

identical effective stress-strain behaviour for both tests within each pair, since all tests are undrained. The tests are plotted in Figs 15 and 16, showing the effective stress paths in $q - p'$ space and the shear stress-shear strain curve. Within each pair of tests there is virtually identical behaviour, although there are large differences between compression and extension. All four tests terminate at a stress ratio $\eta = q/p' \simeq 1.28$, and this value has been taken for M. In order to establish the preconsolidation pressure p_c' a value of K_o is required. A theoretical value for K_o may be derived as 0.596 at $\sigma_3' = 40$ psi, as compared with the measured value of 0.48.

The slight overestimation of K_o for a normally consolidated sample is not uncommon using the Modified Cam-Clay model. A value of M of 1.28 is equivalent to $\phi' = 32°$, for which Jaky's simplified formula of $K_o = (1 - \sin \phi')$ gives a K_o value of 0.47. For all calculations the maximum consolidation point, which establishes the initial yield locus, was calculated using the experimentally determined K_o value rather than the theoretical one. Using this value a p_c' of 74.8 psi is determined. The values of all the parameters are given in Table 1.

Since all the tests start at a pressure of 40 psi, which is slightly to the 'wet' side of the critical state line, the initial response will be predicted to be elastic. During undrained elastic behaviour Modified Cam-Clay predicts a stress path at constant mean effective pressure, which corresponds closely to the stress path observed in the compression tests (Fig. 15).

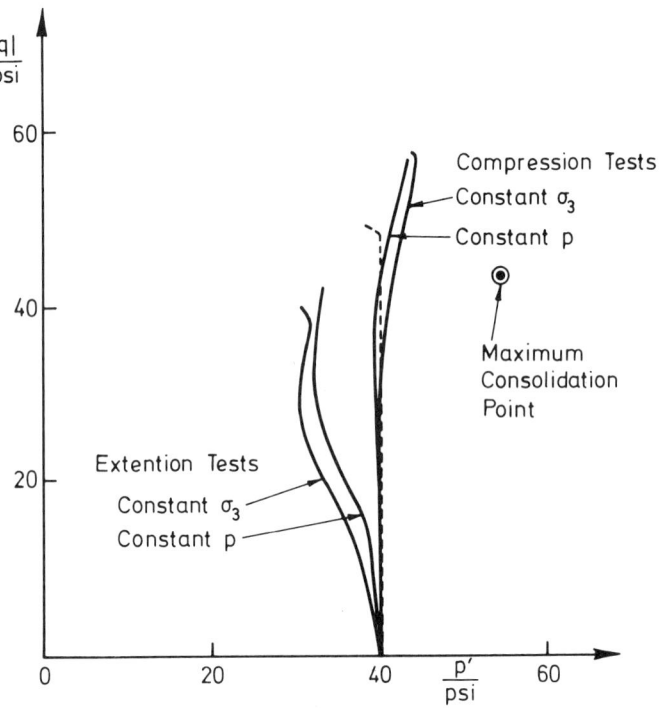

FIG. 15. - UNDRAINED COMPRESSION AND EXTENSION TESTS ON KAOLIN

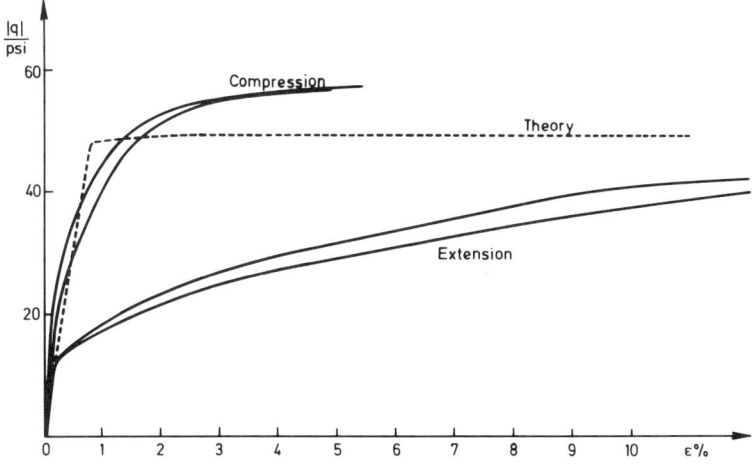

FIG. 16. - UNDRAINED COMPRESSION AND EXTENSION TESTS ON KAOLIN

This stress path intersects the yield locus near the critical state line, so the sample proceeds almost directly to the critical state, with only a very small amount of plastic hardening. The overall behaviour at this pressure is virtually an elastic - perfectly plastic response. Since Modified Cam-Clay is an isotropic model which takes no account of the third stress variant, all undrained stress and strain paths are identical when plotted in terms of

$$p' = (\sigma_1' + \sigma_2' + \sigma_3')/3$$
$$q = [((\sigma_1 - \sigma_2)^2 + (\sigma_2 - \sigma_3)^2 + (\sigma_3 - \sigma_1)^2)/2]^{\frac{1}{2}}$$
and $$\varepsilon = [((\varepsilon_1 - \varepsilon_2)^2 + (\varepsilon_2 - \varepsilon_3)^2 + (\varepsilon_3 - \varepsilon_1)^2)/3]^{\frac{1}{2}}.$$

The predicted strength lies between that measured in compression and that in extension. The actual stress-strain curves show a much smoother response than the predicted elastic-plastic behaviour, and the extension tests clearly show plastic strain early in the test. The stress path, although well matched for compression is not so accurately modelled in extension.

Since the model is isotropic, no account is taken of the fact that the principal axes of stress during the K_o consolidation differ from the axes during subsequent shearing in the tests to be predicted. A further consequence is that the axes of strain and of stress are predicted as coincident for these tests. When plotted in q – p' and q – ε space all the tests to be predicted will follow the paths as shown in Figs 15 and 16. However, in some alternative plots, such as $(\sigma_1 - \sigma_3)$ against ε_1, the predicted paths for different tests would not coincide.

ASSESSMENT OF THE MODIFIED CAM-CLAY MODEL

The model has been developed to describe the behaviour of soft clays, and it is not thought to be suitable for the modelling of sands. Predictions for the behaviour of overconsolidated clays may be made with the model, but these are thought to be less reliable.

Five main regions may be identified in which the model has shortcomings, and these may briefly be defined:

1. The model predicts no plastic strains at low stress ratios in overconsolidated clays. The behaviour of the lightly overconsolidated Kaolin illustrates the presence of plastic strains under these conditions.

2. The circular generalisation of the yield locus in the octahedral plane leads to over-prediction of strength in triaxial extension by comparison with that in compression. This is apparent in the case of some of the tests on both the natural clays and the Kaolin, although there are insufficient data to identify whether in these cases the difference is due to the mode of shearing or to anisotropy.

Although the circular generalisation is the simplest, other relatively simple, but more realistic, shapes have been proposed, e.g. by Lade and Duncan (6) and Matsuoka and Nakai (7). Both these shapes are curvilinear triangles which approximate more closely to the shape of the Mohr-Coulomb irregular hexagon in the octahedral plane. However, neither the Lade nor the Matsuoka shapes may be used to generalise a yield locus which allows any stress to be tensile. The Modified Cam-Clay locus (Fig. 17(a)) involves tensile stresses in the region between the q-axis and a line of gradient at 3, and so this locus must first be changed (for instance, as in Fig. 17(b)) before the non-circular generalisation may be applied.

Although these empirical adjustments may readily be made within plasticity theory, they have not been achieved within the thermodynamic framework for the model (see Appendix I). Some of the advantages of the very simple model would be lost, and the gain in accuracy is uncertain and may not be warranted.

3. Although the model properly predicts the trends of ductile behaviour

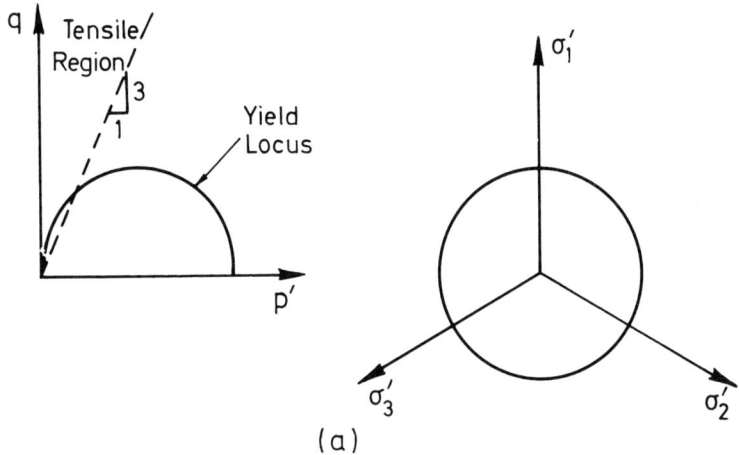

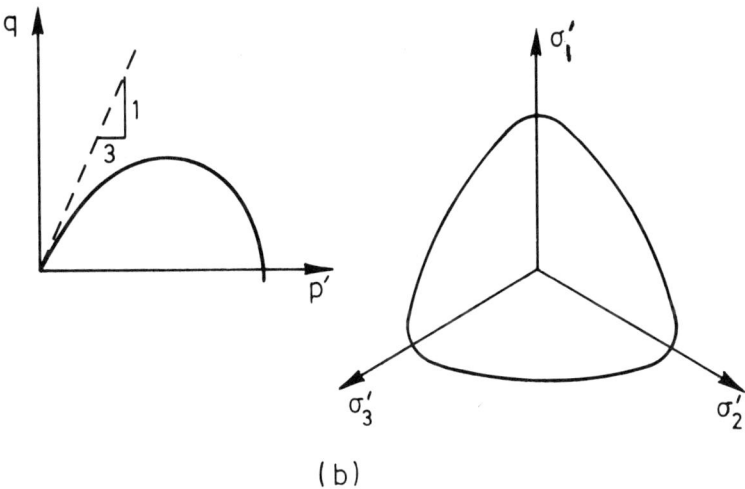

FIG. 17. - CIRCULAR AND NON-CIRCULAR GENERALISATION OF YIELD LOCI

at low overconsolidation ratios and a more brittle strain softening response at high overconsolidation ratios, some of the details are not matched correctly. In particular, the position of the yield locus on the dry side of critical is uncertain and may take the form of a Mohr-Coulomb condition at which the material ruptures. At very high overconsolidation ratios a further regime of behaviour may occur in which splitting of the sample occurs in the direction of the major principal stress.

4. The model is isotropic, whereas both natural clays and one-dimensionally consolidated artificial clays exhibit some anisotropy. Both elastic and plastic anisotropy may be introduced into the model; but a satisfactory method has not been achieved for dealing with anisotropy and its dependence upon stress history.

However, with careful attention to the nature of the stress paths applied to a sample, the isotropic model may still be of value. For instance, the natural clays 'X' and 'Y' probably have a yield locus similar to that shown as the solid curve in Fig. 18, and this is approximated by the dotted Modified Cam-Clay locus. All the tests on this material involved loading in directions which do not vary by more than $60°$ in the octahedral plane, and for these cases the theoretical yield locus probably corresponds quite closely to the actual one. If an extension test in which the path was $180°$ from the original consolidation direction was carried out, then the poor agreement of the two loci in Fig. 18 in the extension region indicates that strains would be underpredicted. The extension tests on the Kaolin were of this type and the strains were indeed underpredicted. This indicates that the different behaviour in compression and extension is probably due to anisotropy rather than the mode of shearing in this case.

5. Hysteresis effects are not modelled, since a large elastic region with fully recoverable behaviour is present. The effects of hysteresis may be important in cases of repeated loading.

CONCLUSIONS

In spite of the known shortcomings in some details of the model, it may be used to predict successfully the overall behaviour of soft clays. The limitations of the model are well understood, and it was chosen for use in the prediction workshop because, not only is it known to fit the behaviour of soft clays reasonably well, but it also has a proven capability for use in the solution of boundary value problems (e.g. Wroth (11)). It is emphasised that

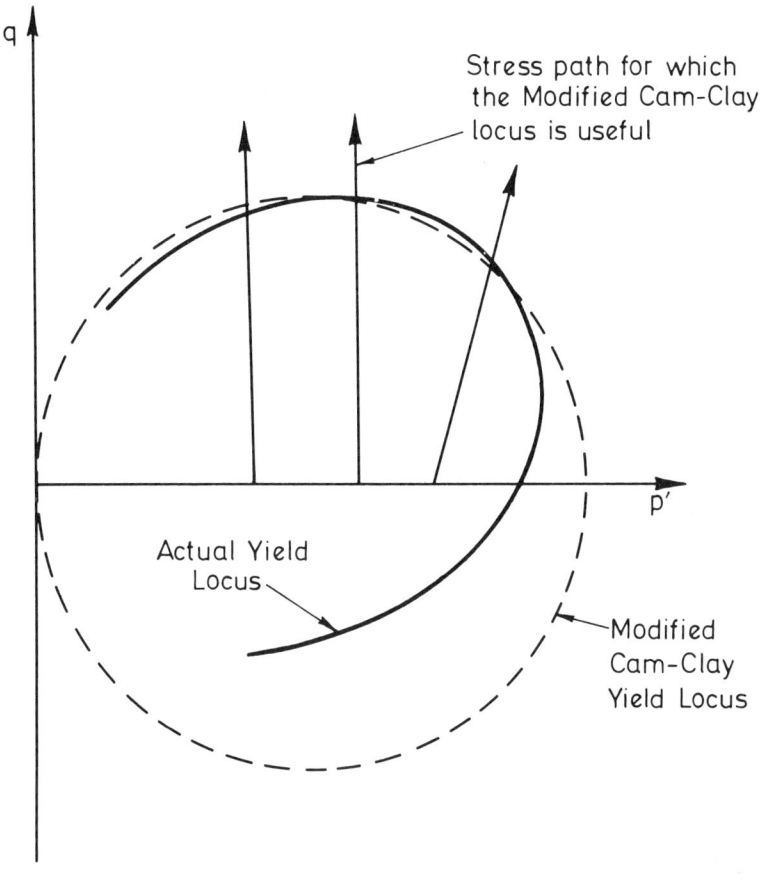

FIG. 18. - THEORETICAL AND ACTUAL YIELD LOCI FOR AN ANISOTROPIC CLAY

the use of a simple model such as Modified Cam-Clay allows the qualitative behaviour under different stress histories to be predicted without mathematical calculation. For example, the elastic-brittle failure behaviour of clay 'X' at low pressures and the more ductile response at higher pressures, are fundamental features of the model which are readily understood.

The success of the model in fitting the behaviour of the natural clays 'X' and 'Y' is in part due to the existence of a well-defined yield locus for these materials. Mitchell (8) reports experimental evidence of yield loci for the Champlain Sea clays, and the data applied to the predictors suggested that the two natural clays were probably of this type.

Although the model may also be expected to predict the behaviour of the Kaolin reasonably well, it was seen that in the results presented to the predictors the material did not exhibit the distinct yield locus which is a feature of the model. In particular, substantial plastic strains were observed early in the extension tests. In spite of the lack of success in fitting closely the supplied data, the model has been used to predict the results of the Kaolin tests. It is realised that the predictions are unlikely to be good; but it is considered to be in the interests of improving the understanding of the stress-strain behaviour of clays that the attempt should be made. The expected shortcomings should then be examined critically and used for improving the model.

The Modified Cam-Clay model is known not to be suitable for modelling sand and, since the authors do not have a similarly well-established model for dealing with this material, no attempt has been made to predict the sand tests.

ACKNOWLEDGEMENTS

The authors are most grateful for the invitation to act as predictors and take part in the Workshop on Plasticity Theories and Generalised Stress-Strain Modelling of Soils. They would like to acknowledge all the contributions made by many research workers to the continuing programme of research into the stress-strain behaviour of soils being conducted by the Soil Mechanics Group at the University of Cambridge, under the direction of Professor A.N. Schofield. This work has led to the family of elastic-plastic work hardening models of the behaviour of clay, one of which (Modified Cam-Clay) has been used for the predictions presented in the Workshop.

G. T. Houlsby is grateful for the financial support provided by the Science Research Council of the United Kingdom and St. John's College, Cambridge.

APPENDIX I. - DERIVATION OF THE 'MODIFIED CAM-CLAY'
SOIL MODEL FROM THERMODYNAMIC FUNCTIONS

The 'Modified Cam-Clay' model (Roscoe and Burland (9)) is usually developed in terms of conventional plasticity theory. The same model may also be derived from a different set of hypotheses, using an approach based on thermodynamic functions (Ziegler (13); Houlsby (5)). A full development of the thermodynamic approach is not given here, but the essential results are presented.

The approach is based on the hypothesis that the state of a material may be completely defined by certain kinematic parameters, for instance, the strains and some additional 'internal' parameters, which might take the form of plastic strains. The constitutive law is then determined by the specification of only two functions:

1. The free energy, ψ, which is a function of the kinematic parameters,

2. The dissipation function, ϕ, which is a function of the kinematic parameters and their rates of change.

For the special case of rate independent materials, the latter function must be of first order in the rates of change. Adopting a tensorial notation the stresses are then given by:

$$\sigma_{ij} = \rho \frac{\partial \psi}{\partial \varepsilon_{ij}} + \rho \frac{\partial \phi}{\partial \dot{\varepsilon}_{ij}} \qquad \ldots \ldots (A1)$$

where ρ is the density and σ_{ij} and ε_{ij} the stresses and strains, respectively. Since additional internal parameters have been introduced, some additional equations are necessary to eliminate these from the constitutive law, and these are of the form:

$$0 = \rho \frac{\partial \psi}{\partial \alpha_{ij}} + \rho \frac{\partial \phi}{\partial \dot{\alpha}_{ij}} \qquad \ldots \ldots (A2)$$

where α_{ij} represents the internal parameters, taken in this case in the form of a second order tensor.

Certain forms of ψ and ϕ give rise to conventional plasticity theories, and this will now be illustrated by the derivation of the Modified Cam-Clay model. The model will be derived for the limited case of the triaxial test, using the strain parameters v and ε, with corresponding stresses p' and q (Schofield and Wroth (10)). In addition, some internal parameters which play the role of plastic strains, v_p and ε_p, will be introduced.

Specification of the model requires five parameters:

M   the slope of the critical state line in p' - q space,
G   the elastic shear modulus,
λ^*, κ^* the slope of consolidation and swelling lines in $\ln p'$ - $\ln V$ space,
Γ the specific volume (V) at p' = 1 on the critical state line.

These parameters are illustrated in Fig. A1, which shows the conventional stress space and consolidation plots for the model.

It should be noted that because of the choice of straight consolidation lines in $\ln p'$ - $\ln V$ rather than $\ln p'$ - V space the values of λ^* and κ^* are slightly altered from those of λ and κ in the original Cam-Clay models (Schofield and Wroth (10); Roscoe and Burland (9)). The parameter Γ does not enter the stress-strain law, but serves only to locate consolidation plots in $\ln p'$ - $\ln V$ space.

The functions which generate the Modified Cam-Clay constitutive law are given by:

$$\rho\psi = \kappa^* \exp((v-v_p)/\kappa^*)) + 3G(\varepsilon-\varepsilon_p)^2/2 + (\lambda^*-\kappa^*) \exp((\ln(\Gamma/V_o)+v_p)/(\lambda^*-\kappa^*)) \quad \ldots \ldots (A3)$$

$$\rho\phi = \exp((\ln(\Gamma/V_o)+v_p)/(\lambda^*-\kappa^*))(\dot{v}_p^2+M^2\dot{\varepsilon}_p^2)^{\frac{1}{2}} \quad \ldots \ldots (A4)$$

Applying equation (A1) the stresses are obtained by differentiating (A3) and (A4) to give:

$$p' = \rho\frac{\partial\psi}{\partial v} + \rho\frac{\partial\phi}{\partial\dot{v}} = \exp((v-v_p)/\kappa^*) \quad \ldots \ldots (A5)$$

$$q = \rho\frac{\partial\psi}{\partial\varepsilon} + \rho\frac{\partial\phi}{\partial\dot{\varepsilon}} = 3G(\varepsilon-\varepsilon_p) \quad \ldots \ldots (A6)$$

The internal parameters v_p and ε_p will be eliminated using expressions in the form of equation (A2):

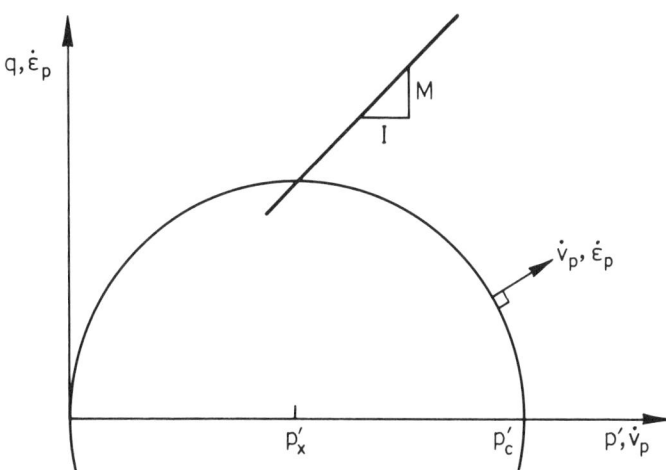

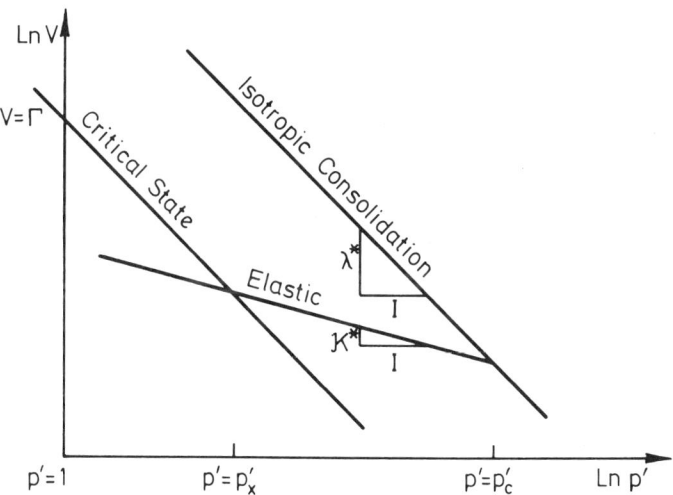

FIG. A1. - YIELD LOCUS AND CONSOLIDATION LINES FOR MODIFIED CAM-CLAY

$$0 = \rho\frac{\partial\psi}{\partial v_p} + \rho\frac{\partial\phi}{\partial\dot{v}_p} = -\exp((v-v_p)/\kappa^*) + \exp((\ln(\Gamma/V_o)+v_p)/(\lambda^*-\kappa^*)) +$$
$$+ \exp((\ln(\Gamma/V_o)+v_p)/(\lambda^*-\kappa^*))\dot{v}_p(\dot{v}_p^2+M^2\dot{\varepsilon}_p^2)^{-\frac{1}{2}} \quad \ldots \ldots (A7)$$

$$0 = \rho\frac{\partial\psi}{\partial\varepsilon_p} + \rho\frac{\partial\phi}{\partial\dot{\varepsilon}_p} = -3G(\varepsilon-\varepsilon_p) +$$
$$+ \exp((\ln(\Gamma/V_o)+v_p)/(\lambda^*-\kappa^*))M^2\dot{\varepsilon}_p(\dot{v}_p^2+M^2\dot{\varepsilon}_p^2)^{-\frac{1}{2}} \quad \ldots \ldots (A8)$$

It is useful to introduce the definition

$$p_x' = \exp((\ln(\Gamma/V_o)+v_p)/(\lambda^*-\kappa^*)) \quad \ldots \ldots (A9)$$

which allows simplification of (A7) and (A8) to:

$$p' - p_x' = \dot{v}_p p_x'(\dot{v}_p^2 + M^2\dot{\varepsilon}_p^2)^{\frac{1}{2}} \quad \ldots \ldots (A10)$$

$$q = \dot{\varepsilon}_p M^2 p_x'(\dot{v}_p^2+M^2\dot{\varepsilon}_p^2)^{\frac{1}{2}} \quad \ldots \ldots (A11)$$

The time differentials of (A5) and (A6) give the incremental relations

$$\dot{p}' = \exp((v-v_p)/\kappa^*)(\dot{v}-\dot{v}_p)/\kappa^* \quad \ldots \ldots (A12)$$

$$\dot{q} = 3G(\dot{\varepsilon}-\dot{\varepsilon}_p) \quad \ldots \ldots (A13)$$

or:

$$\dot{v} = \kappa^*\dot{p}'/p' + \dot{v}_p \quad \ldots \ldots (A14)$$

$$\dot{\varepsilon} = \dot{q}/3G + \dot{\varepsilon}_p \quad \ldots \ldots (A15)$$

The incremental strain is therefore made up from two components, the "elastic" strain which is directly related to the stress increment, with a constant shear modulus, G, and a bulk modulus proportional to pressure, p'/κ^*. The internal variables v_p and ε_p represent additional "plastic" strains which only occur during dissipative processes, i.e. when ϕ is non-zero. If both $\dot{\varepsilon}_p$ and $\dot{v}_p$ are zero the process is conservative and equations (A10) and (A11) are indeterminate. The incremental stress-strain relation is, however, determined.

If either $\dot{v}_p$ or $\dot{\varepsilon}_p$ is non-zero dissipation occurs, and equations (A10) and (A11) impose additional conditions. They may be re-arranged to give:

$$(p'-p_x')^2 + q^2/M^2 = p_x'^2 \quad \ldots \ldots (A16)$$

$$\dot{v}_p/\dot{\varepsilon}_p = M^2(p'-p_x')/q \qquad \ldots \ldots (A17)$$

Equation (A16) requires that the stress point should lie on a given surface: the yield locus. Since p_x' increases with v_p the yield locus may expand (or contract). Equation (A17) establishes a fixed ratio between the plastic strain increments, and clearly represents the flow rule. It may readily be verified that in this case the normality principle (Drucker (4)) is satisfied.

The parameter p_x' is the p' coordinate of the centre of the elliptical yield locus, i.e. the pressure at the intersection of the yield locus and the critical state line. It is related to the pressure on the isotropic, normal consolidation line p_c' by $p_c' = 2p_x'$.

Equations (A14), (A15) and (A17) provide three incremental relations; in order to eliminate $\dot{\varepsilon}_p$ and $\dot{v}_p$ and still provide the stress strain equation a fourth incremental equation is required. This is given by the differential of the yield locus (A16) which gives:

$$(p'-p_x')(\dot{p}-\dot{p}_x') + q\dot{q}/M^2 = p_x'\dot{p}_x' \qquad \ldots \ldots (A18)$$

where

$$\dot{p}_x = p_x \dot{v}_p/(\lambda^*-\kappa^*) \qquad \ldots \ldots (A19)$$

It has been demonstrated that the functions (A3) and (A4) give rise to a plasticity model with the yield locus and flow rule used in Modified Cam-Clay, but it has to be shown that the hardening behaviour of the material is also as required. The hardening behaviour of a Cam-Clay model is most simply described by the condition that the material state should lie in a unique 'State Boundary Surface' in p'-q-V space during plastic deformation.

Equations (A5), (A9) and (A15) may be combined to eliminate v_p and p_x', and substitution of the definition:

$$v = \ln(Vo/V) \qquad \ldots \ldots (A20)$$

then leads to the equation of the state boundary surface:

$$\ln V = \ln\Gamma - \lambda^*\ln p' - (\lambda^*-\kappa^*)\ln((p^2+q^2/M^2)/2p^2) \qquad \ldots \ldots (A21)$$

This equation results in straight consolidation lines, in a $\ln V - \ln p$ plot.

Similarly equations (A5), (A9) and (A20) may be combined to give the equation for the swelling lines, which are also straight in $\ln p'-\ln V$ space:

$$\ln V = \ln \Gamma - \lambda^* \ln p' - \kappa^* \ln(p'/p_x') \qquad \ldots \ldots (A22)$$

For simplicity the above derivation has been limited to the case of the triaxial test, but the model may easily be extended to general stress states. Use will be made of the following definitions of the invariants of a tensor t_{ij}:

$$t_{(1)} = t_{ii} \qquad \ldots \ldots (A23)$$

$$t_{(2)} = \tfrac{1}{2}(t_{ij}t_{ji} - t_{ii}t_{jj}) \qquad \ldots \ldots (A24)$$

and also the definition of the deviator:

$$t_{ij}' = t_{ij} - \tfrac{1}{3} t_{(1)} \delta_{ij} \qquad \ldots \ldots (A25)$$

Let the kinematic parameters defining the state be the strains ε_{ij}, and the parameters which will play the role of plastic strains $\varepsilon_{ij}^{(p)}$. Introducing the definition:

$$\varepsilon_{ij}^{(e)} = \varepsilon_{ij} - \varepsilon_{ij}^{(p)} \qquad \ldots \ldots (A26)$$

then the expressions for the free energy and dissipation for the general model are:

$$\rho\psi = \kappa^* \exp(\varepsilon_{(1)}^{(e)}/\kappa^*) + 2G\varepsilon'^{(e)}_{(2)} +$$
$$+ (\lambda^* - \kappa^*)\exp((\ln(\Gamma/Vo) + \varepsilon_{(1)}^{(p)})/(\lambda^* - \kappa^*)) \qquad \ldots \ldots (A27)$$

$$\rho\phi = \exp((\ln(\Gamma/Vo) + \varepsilon_{(1)}^{(p)})/(\lambda^* - \kappa^*))(\dot\varepsilon_{(1)}^{(p)2} + 4M^2 \dot\varepsilon'^{(p)2}_{(2)}/3)^{\tfrac{1}{2}} \quad \ldots (A28)$$

It is apparent how these expressions relate to equations (A3) and (A4), and the derivation of the generalised behaviour follows an exactly similar pattern to the triaxial analysis. The generalisation of the yield locus results in a circular cross section in the octahedral plane.

APPENDIX II. - REFERENCES

1. Bishop, A. W., "The Strength of Soils as Engineering Materials," Géotechnique, Vol. 16, 1966, pp. 91-128.

2. Burland, J. B., "Deformation of Soft Clay," Ph.D. Thesis, University of Cambridge, 1967.

3. Butterfield, R., "A Natural Compression Law for Soils (An Advance on e-log p')," Géotechnique, Vol. 29, 1979, pp. 469-480.

4. Drucker, D. C., "A More Fundamental Approach to Plastic Stress Strain Relations," Proceedings, 1st US National Congress on Applied Mechanics, ASME, June 1951, pp. 487-491.

5. Houlsby, G. T., "Some Implications of the Derivation of the Small Strain Incremental Theory of Plasticity from Thermomechanics," CUED-10 Soils TR 74, University of Cambridge, 1979.

6. Lade, P. V. and Duncan, J. M., "Elastoplastic Stress-Strain Theory for Cohesionless Soil," Proceedings, ASCE, Journal of Geotechnical Engineering Division, No. GT10, October 1975, pp. 1037-1053.

7. Matsuoka, H. and Nakai, T., "Stress-Strain Relationship of Soil Based on the 'SMP'," Proceedings, Speciality Session No. 9, 9th International Conference on Soil Mechanics and Foundation Engineering, Tokyo, 1977.

8. Mitchell, R. J., "On the Yielding and Mechanical Strength of Leda Clays," Canadian Geotechnical Journal, Vol. 7, 1970, pp. 297-312.

9. Roscoe, K. H. and Burland, J. B., "On the Generalised Behaviour of 'Wet' Clays," Engineering Plasticity, Cambridge University Press, 1968, pp. 535-609.

10. Schofield, A. N. and Wroth, C. P., Critical State Soil Mechanics, McGraw Hill, London, 1968.

11. Wroth, C. P. "The Predicted Performance of Soft Clay Under a Trial Embankment Loading Based on the Cam-Clay Model," Finite Elements in Geomechanics, John Wiley & Son, London, 1975, pp. 191-208.

12. Wroth, C. P. "Correlations of Some Engineering Properties of Soils," Proceedings, 2nd International Conference on Behaviour of Off-Shore Structures, London, 1979, Vol. 1, pp. 121-132.

13. Ziegler, H., An Introduction to Thermomechanics, North-Holland, Amsterdam, 1977.

ELASTO-PLASTIC STRESS-STRAIN MODEL FOR SAND

by Poul V. Lade[1]

Introduction

The basic principles involved in the elasto-plastic stress-strain model for sand under general three-dimensional stress conditions are outlined, and the ability of this model to account for various aspects of observed soil behavior is briefly described. The specific procedures employed for determination of soil parameters are then presented in connection with derivation of parameters for Ottawa Sand. Finally, the predictions of behavior of Ottawa Sand during various specified stress-paths are presented and discussed.

Stress-Strain Model

The elasto-plastic stress-strain model originally developed for cohesionless soil (1,6) reflects many of the characteristics of sand behavior observed in laboratory tests. Results of conventional triaxial tests and cubical triaxial tests on cohesionless soil (1,5) and concepts from elasticity and plasticity were employed in formulating the original model. Further developments involve employment of curved yield and failure surfaces, addition of a cap-type yield surface, and use of work-hardening as well as -softening relationships (2). The working principles of the model in its present form are briefly reviewed below.

For the purpose of modeling the stress-strain behavior of soils by an elasto-plastic theory, the total strain increments, $\{d\varepsilon_{ij}\}$, are divided into an elastic component, $\{d\varepsilon_{ij}^e\}$, a plastic collapse component, $\{d\varepsilon_{ij}^c\}$, and a plastic expansive component, $\{d\varepsilon_{ij}^p\}$, such that

$$\{d\varepsilon_{ij}\} = \{d\varepsilon_{ij}^e\} + \{d\varepsilon_{ij}^c\} + \{d\varepsilon_{ij}^p\} \tag{1}$$

These strain components are calculated separately, the elastic strains by Hooke's law, the plastic collapse strains and plastic expansive strains by a plastic stress-strain theory that involves, respectively, a cap-type yield

---

[1]Associate Professor, Mechanics and Structures Department, School of Engineering and Applied Science, University of California, Los Angeles, California.

surface and a conical yield surface with apex at the origin of the stress space.

Fig. 1(a) shows the parts of the total strain that are considered to be elastic, plastic collapse, and plastic expansive components of strain in a drained triaxial compression test. Typical observed variations of stress difference $(\sigma_1-\sigma_3)$ and volumetric strain ε_v, with axial strain ε_1, are shown in this figure for a test performed with constant confining pressure σ_3. Both elastic (recoverable) and plastic (irrecoverable) deformations occur from the beginning of loading of a cohesionless soil, the stress-strain relationship is nonlinear, and a decrease in strength may follow peak failure. The volumetric strain is initially compressive and this behavior may be followed by expansion (as shown in Fig. 1(a)) or by continued compression. The plastic strains are initially smaller than the elastic strains, but at higher values of stress difference the plastic strains dominate the elastic strains.

The yield surfaces for the plastic strain components are indicated on the triaxial plane in Fig. 1(b). The conical yield surface may be curved in planes containing the hydrostatic axis, or it may be straight as in the original model (6). In principal stress space the yield and failure surfaces are shaped like asymmetric bullets with their pointed apices at the origin of the stress space as shown in Fig. 2(a). Typical cross-sections of these surfaces are shown in Fig. 2(b). They have been shown to model the experimentally determined three dimensional strengths of sand as well as normally consolidated clay with good accuracy (6,8,9).

The yield surface corresponding to the plastic collapse strains forms a cap on the open end of the conical yield surface as shown in Fig. 1(b). The collapse yield surface is shaped as a sphere with center in the origin of the principal stress space. It should be noted that yielding resulting from outward movement of the cap does not result in eventual failure. Failure is controlled entirely by the conical yield surface.

The plastic expansive strains are calculated from a nonassociated flow rule, whereas the plastic collapse strains are calculated from an associated flow rule. The result of a change in stress is shown in the triaxial plane in Fig. 3. The plastic strain increment vectors are superimposed on the stress space in this diagram. Both plastic collapse and plastic expansive strains are caused by the change in stress from point A to point B, because both yield surfaces are pushed out. The magnitudes of the strain increments are indicated by the lengths of the vectors, and the total plastic strain increment is calculated according to Eq. 1 as the vector sum of the two components. The elastic strain components are further added (not shown in Fig. 3) to obtain the total

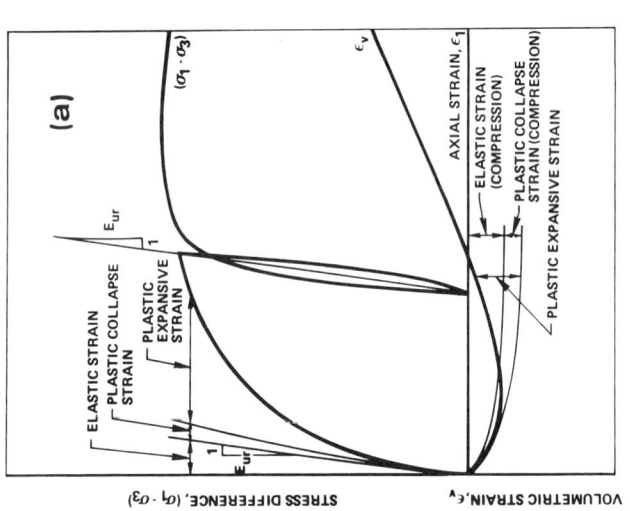

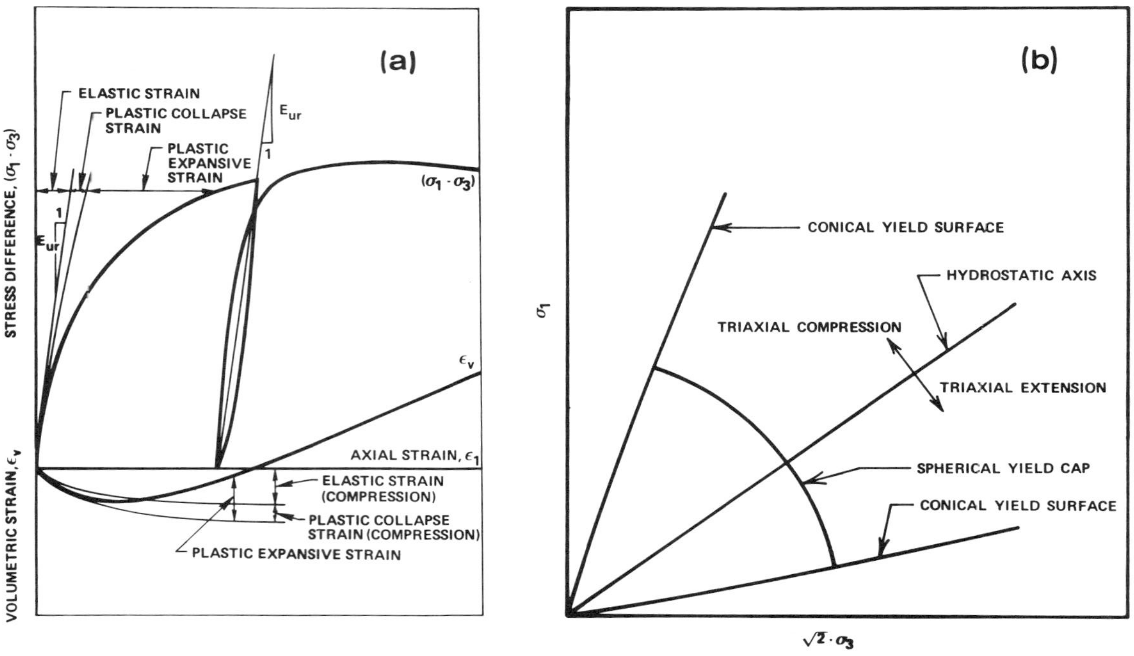

Figure 1. Schematic Illustrations of (a) Elastic, Plastic Collapse, and Plastic Expansive Strain Components in Drained Triaxial Compression Test, and (b) Conical and Spherical Cap Yield Surfaces in Triaxial Plane.

STRESS-STRAIN MODEL

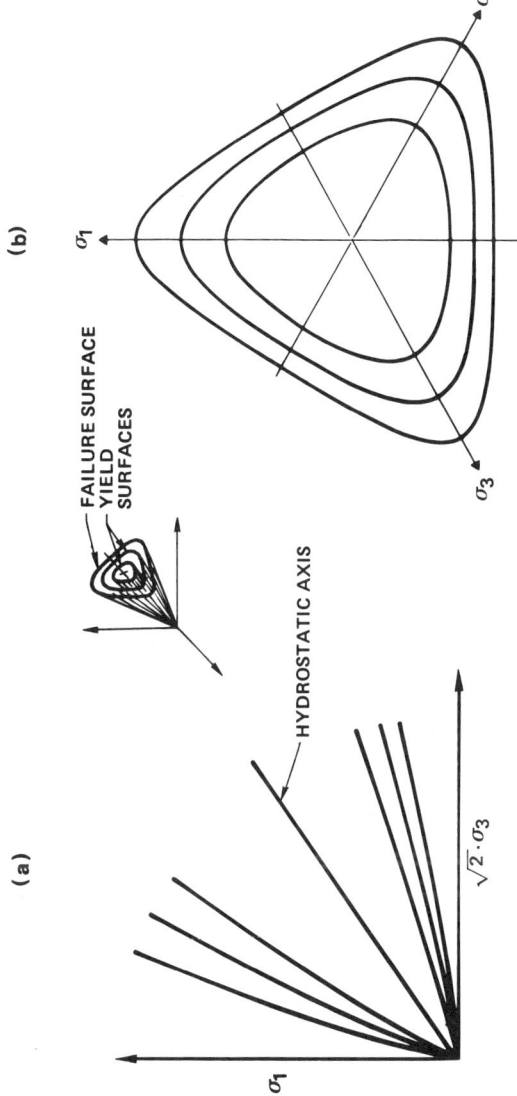

Figure 2. Characteristics of Proposed Failure and Yield Surfaces Shown in Principal Stress Space. (a) Traces of Failure and Yield Surfaces in Triaxial Plane. (b) Traces of Failure and Yield Surfaces in Octahedral Plane.

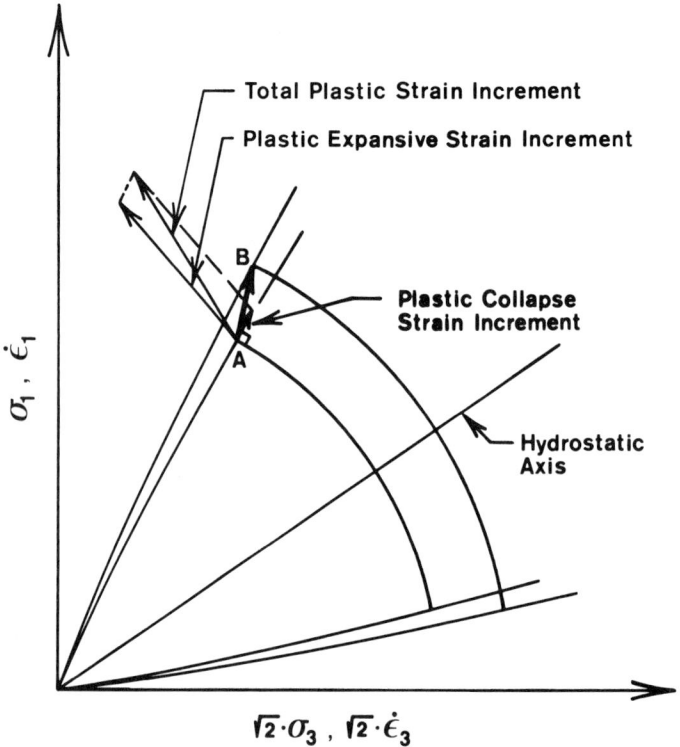

Figure 3. Schematic Diagram of Yielding Process with Plastic Strain Components Superimposed in Triaxial Plane.

strain increment for the stress change from A to B.

The elasto-plastic stress-strain model employed here is applicable to general three-dimensional stress conditions, but the soil parameters required to characterize the soil behavior can be derived entirely from the results of isotropic compression and conventional drained triaxial compression tests. The accuracy of the model has been evaluated by comparing predicted and measured strains for several types of laboratory tests performed on cohesionless soils. Thus, it has been demonstrated that the following important aspects of observed stress-strain and strength behavior are modeled with good accuracy: (1) Nonlinearity; (2) effects of σ_2 and σ_3 including decreasing maximum stress ratio with increasing confining pressure (curved failure envelope); (3) decrease in strength after peak has been reached (strain softening); (4) stress-path dependency including proportional loading and stresses as well as strains occurring under K_o-conditions; (5) shear-dilatancy effect and its variation over a range of confining pressures, i.e. changes in volumetric strain behavior from expansive to compressive with increasing confining pressure; (6) pore pressures and effective stress-paths for undrained conditions over a range of confining pressures, as well as critical confining pressure for given soil density or void ratio; (7) coincidence of strain increment and stress increment axes at low stress levels (elastic behavior) with transition to coincidence of strain increment and stress axes at high stress levels (plastic behavior) (2,3,6,7).

Because the behavior of normally consolidated clay in many respects resembles that of sand (8), it has been possible to use the same model (with one minor modification) for prediction of stress-strain, pore pressure, and strength behavior observed in isotropically consolidated, cubical triaxial tests and for prediction of K_o-compression of normally consolidated, remolded clay (4,9). The necessary soil parameters for this purpose can be derived entirely from the results of isotropic compression and isotropically consolidated, undrained triaxial compression tests (4).

Whereas soil behavior during small to moderate stress reversals may be predicted with reasonable accuracy, the model in its present form cannot accurately predict soil response during large stress reversals and large changes in stress involving unloading and reloading under general three-dimensional conditions. Similarly, the behavior of initially anisotropic soils cannot be accurately predicted by the present model.

Basic Tests on Ottawa Sand

Results of tests performed on dry Ottawa Sand were provided for determination of the necessary soil parameters. The specimens were prepared by aerial pluviation at a dry density of 1.73 g/cm³ and a relative density of 87%. The Ottawa Sand had an average grain size of 0.42 mm and a coefficient of uniformity of 1.3.

All tests were performed in a flexible, fluid cushion, cubical apparatus with stress control. The data base consists of two isotropic compression tests, two triaxial compression tests with constant lateral pressure (σ_3 = 5 and 10 psi), one triaxial extension test with constant minor principal stress (σ_3 = 10 psi), three triaxial compression tests with constant mean normal stress (σ_m = 5, 10, and 20 psi), and three triaxial extension tests with constant mean normal stress (σ_m = 5, 10, and 20 psi). The results of the shear tests in the data base are shown in Figs. 8 and 9 for comparison with the model predictions.

To help designate these tests and those to be predicted, use is made of the stress ratio m defined as

$$m = \frac{\sigma_2 - \sigma_3}{\sigma_1 - \sigma_3} \tag{2}$$

The value of m is zero for triaxial compression tests where $\sigma_2 = \sigma_3$, and it is unity for triaxial extension tests where $\sigma_2 = \sigma_1$. For tests with σ_2-values between σ_3 and σ_1, the value of m is between zero and unity. Each test in the program was performed with constant m-value.

Derivation of Soil Parameters

The mathematical expressions used for modeling the various components of the elasto-plastic stress-strain theory are presented without proof or discussion and used in connection with determination of parameter values for Ottawa Sand. The background and arguments for employing these expressions are contained in the references listed at the end of this paper.

Elastic Strains. -- The elastic strains, which are recoverable upon unloading, are calculated from Hooke's law using the unloading-reloading modulus defined as:

$$E_{ur} = K_{ur} \cdot p_a \cdot \left(\frac{\sigma_3}{p_a}\right)^n \tag{3}$$

The dimensionless, constant values of the modulus number K_{ur} and the exponent n may be determined from unloading-reloading cycles in triaxial compression tests. In Eq. 3 p_a is atmospheric pressure expressed in the same units as E_{ur} and σ_3. Since the unloading-reloading moduli derived from the tests on Ottawa Sand show considerable scatter, the initial elastic moduli were calculated from the initial slopes of the stress-strain curves in the data base and plotted versus the confining pressure as shown on Fig. 4(a). Previous experience indicates that K_{ur} is between one and three times the value of K_i, whereas the exponent n takes the same value for the variation of the initial modulus and the unloading-reloading modulus. For the present purpose K_{ur} is taken as two times K_i as indicated on Fig. 4(a). Thus, the values of the modulus number, K_{ur}, and the exponent, n, for use in Eq. 3 are 3000 and 0.75, respectively. A value of Poisson's ratio of 0.2 is used for calculation of the elastic strains in sand.

Plastic Collapse Strains. -- The plastic collapse strains are calculated from a simple plastic stress-strain theory which incorporates (1) a cap-type spherical yield surface with center at the origin of the principal stress space, (2) an associated flow rule, and (3) a work-hardening relationship which can be determined from an isotropic compression test. The yield criterion, f_c, and the plastic potential function, g_c, have the form:

$$f_c = g_c = I_1^2 + 2 \cdot I_2 \tag{4}$$

where I_1 and I_2 are the first and the second stress invariants:

$$I_1 = \sigma_1 + \sigma_2 + \sigma_3 = \sigma_x + \sigma_y + \sigma_z \tag{5}$$

$$I_2 = -(\sigma_1 \cdot \sigma_2 + \sigma_2 \cdot \sigma_3 + \sigma_3 \cdot \sigma_1) \tag{6}$$

$$= \tau_{xy} \cdot \tau_{yx} + \tau_{yz} \cdot \tau_{zy} + \tau_{zx} \cdot \tau_{xz} - (\sigma_x \cdot \sigma_y + \sigma_y \cdot \sigma_z + \sigma_z \cdot \sigma_x) \tag{7}$$

Performing all the necessary derivations, the final form of the plastic collapse stress-strain relationships become:

$$\begin{Bmatrix} \Delta\varepsilon_x^c \\ \Delta\varepsilon_y^c \\ \Delta\varepsilon_z^c \\ \Delta\varepsilon_{yz}^c \\ \Delta\varepsilon_{zx}^c \\ \Delta\varepsilon_{xy}^c \end{Bmatrix} = \frac{dW_c}{f_c} \cdot \begin{Bmatrix} \sigma_x \\ \sigma_y \\ \sigma_z \\ \tau_{yz} \\ \tau_{zx} \\ \tau_{xy} \end{Bmatrix} \tag{8a-f}$$

where dW_c is an increment in work per unit volume for a given value of f_c and a given increment of the yield function, df_c. The plastic collapse work is calculated from

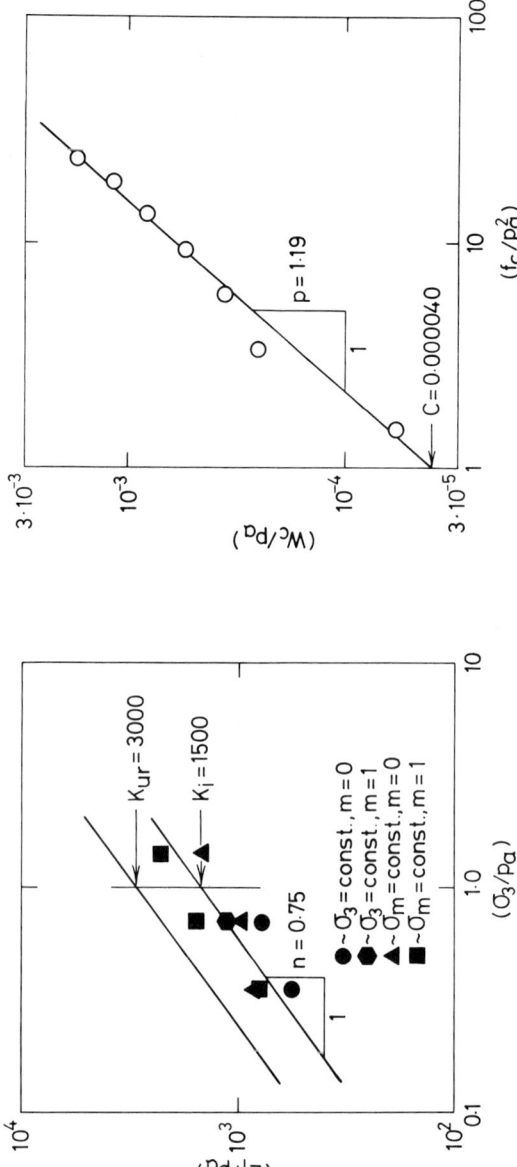

Figure 4a. Determination of the Values of K_{ur} and n Involved in Elastic Modulus Variation for Ottawa Sand.

Figure 4b. Relation Between Plastic Collapse Work, W_c, and the Value of f_c for Ottawa Sand.

$$W_c = \int \{\sigma_{ij}\}^T \{d\varepsilon_{ij}^c\} \tag{9}$$

which for an isotropic compression test reduces to

$$W_c = \int \sigma_3 \cdot d\varepsilon_v^c \tag{10}$$

W_c is plotted as a function of f_c which for isotropic compression can be calculated from:

$$f_c = 3 \cdot \sigma_3^2 \tag{11}$$

The diagram in Fig. 4(b) shows the relationship between W_c and f_c plotted on log-log scales for Ottawa Sand. This relationship can be modeled by the following expression:

$$W_c = C \cdot p_a \cdot \left(\frac{f_c}{p_a^2}\right)^p \tag{12}$$

where the collapse modulus C and the collapse exponent p are determined as shown in Fig. 4(b). On this diagram C is the intercept with $(f_c/p_a^2) = 1$ and p is the slope of the straight line. The increment in plastic collapse work is then determined from:

$$dW_c = C \cdot p \cdot p_a \cdot \left(\frac{p_a^2}{f_c}\right)^{1-p} \cdot d\,(f_c/p_a^2) \tag{13}$$

and used in connection with Eqs. 8a-f for calculation of plastic collapse strains.

Plastic Expansive Strains. -- The curved yield surfaces which best describe the behavior of cohesionless soils are expressed in terms of the first and the third stress invariants:

$$f_p = (I_1^3/I_3 - 27) \cdot (I_1/p_a)^m \tag{14a}$$

$$f_p = \eta_1 \text{ at failure} \tag{14b}$$

where I_1 is given by Eq. 5 and I_3 is given by

$$\begin{aligned}I_3 &= \sigma_1 \cdot \sigma_2 \cdot \sigma_3 \\ &= \sigma_x \cdot \sigma_y \cdot \sigma_z + \tau_{xy} \cdot \tau_{yz} \cdot \tau_{zx} + \tau_{yx} \cdot \tau_{zy} \cdot \tau_{xz} \\ &\quad - (\sigma_x \cdot \tau_{yz} \cdot \tau_{zy} + \sigma_y \cdot \tau_{zx} \cdot \tau_{xz} + \sigma_z \cdot \tau_{xy} \cdot \tau_{yx})\end{aligned} \tag{15}$$

In Eq. 14, the values of η_1 and m are constants to be determined for specific soils at the desired density. For failure conditions the apex angle, indicated on the diagram in Fig. 2(a), increases with the value of η_1, and the curvature of the failure surface increases with the value of m. The values of η_1 and m can be determined by plotting $(I_1^3/I_3 - 27)$ versus (p_a/I_1) at failure on a log-log diagram as shown in Fig. 5(a). On this diagram η_1 is the intercept with $(p_a/I_1) = 1$ and m is the slope of the straight line. Note that the failure conditions for all tests in the data base are plotted on Fig. 5(a). However, the stress-strain curve for the triaxial extension test with constant minor principal stress appears to indicate that peak failure has not been reached in this test. The corresponding failure point (shown in parentheses on Fig. 5(a)) has therefore not been included in the determination of η_1 and m.

The non-associated flow rule employed for the plastic expansive strains is characterized by a plastic potential function, g_p, of a form similar to the yield criterion:

$$g_p = I_1^3 - \left(27 + \eta_2 \cdot \left(\frac{p_a}{I_1}\right)^m\right) \cdot I_3 \tag{16}$$

where η_2 is a constant for given values of f_p and σ_3. The values of η_2 can be determined from the results of triaxial compression tests using the following expression:

$$\eta_2 = \frac{3 \cdot (1+\nu^P) \cdot I_1^2 - 27 \cdot \sigma_3 \cdot (\sigma_1 + \nu^P \cdot \sigma_3)}{\left(\frac{p_a}{I_1}\right)^m \cdot \left[\sigma_3 \cdot (\sigma_1 + \nu^P \cdot \sigma_3) - \frac{I_3}{I_1} \cdot m \cdot (1+\nu^P)\right]} \tag{17}$$

where ν^P is a function of the plastic strain increments determined by subtracting the elastic and plastic collapse strains from the total strains:

$$\nu^P = -\frac{\Delta\varepsilon_3^P}{\Delta\varepsilon_1^P} \tag{18}$$

The variation of η_2 with f_p and σ_3 is shown on Fig. 5(b) for the two triaxial compression tests with constant lateral pressure. This variation can be expressed by a simple equation:

$$\eta_2 = S \cdot f_p + R \cdot \sqrt{\frac{\sigma_3}{p_a}} + t \tag{19}$$

STRESS-STRAIN MODEL 639

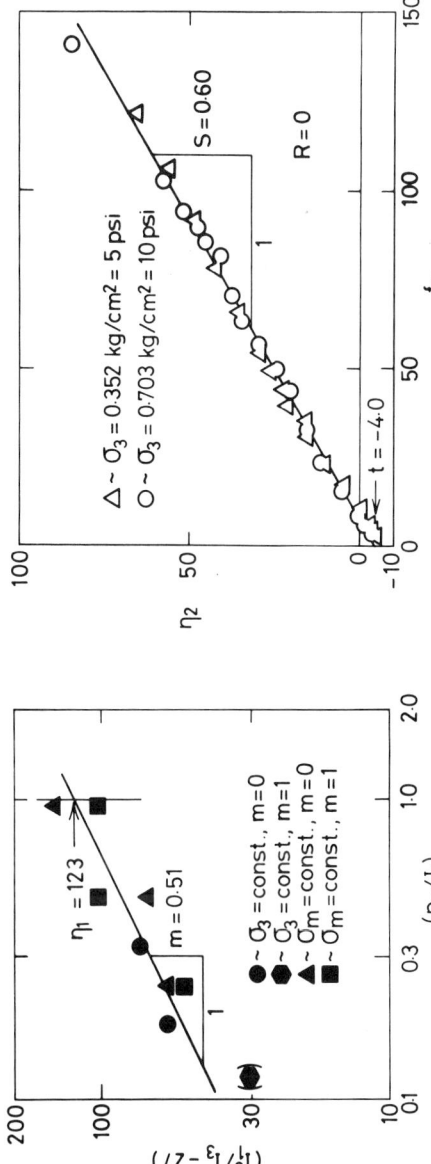

Figure 5a. Determination of the Values of n_1 and m Involved in Failure Criterion for Ottawa Sand.

Figure 5b. Variation of η_2 with f_p for Ottawa Sand.

For the Ottawa Sand the variation of η_2 appears to be independent of the lateral pressure σ_3, and the value of R is therefore zero for this sand. The values of S and t are indicated on Fig. 5(b).

The plastic expansive work is calculated from

$$W_p = \int \{\sigma_{ij}\}^T \{d\epsilon_{ij}^p\} \tag{20}$$

and the variation of W_p with f_p (calculated from Eq. 14a) is shown in Fig. 6 for the two triaxial compression tests with constant lateral pressure. This variation can be approximated by exponential functions for which the following expression is used:

$$f_p = a \cdot e^{-b \cdot W_p} \cdot \left(\frac{W_p}{P_a}\right)^{1/q}, \quad q > 0 \tag{21}$$

where the parameters a, b, and q are constants for a given value of σ_3. The parameter q can be determined for a constant value of the confining pressure according to:

$$q = \frac{\log\left(\frac{W_{ppeak}}{W_{p60}}\right) - \left(1 - \frac{W_{p60}}{W_{ppeak}}\right) \cdot \log e}{\log\left(\frac{\eta_1}{f_{p60}}\right)} \tag{22}$$

where e is the base for natural logarithms, and (W_{ppeak}, η_1) and (W_{p60}, f_{p60}) are two sets of corresponding values from the relation between the work input W_p and stress level f_p. These two points correspond to the peak point and the point at 60% of η_1 on the work-hardening part of the W_p-f_p relation, as indicated on Fig. 6. The variation of q with confining pressure σ_3 may be modeled by the following simple expression:

$$q = \alpha + \beta \cdot \frac{\sigma_3}{P_a} \tag{23}$$

where the values of α and β represent the intercept and the slope of the straight line shown in Fig. 7(a). Since the slope of this line is zero for Ottawa Sand, the value of β is zero.

The parameters a and b in Eq. 21 are calculated from:

$$a = \eta_1 \cdot \left(\frac{e \cdot P_a}{W_{ppeak}}\right)^{1/q} \tag{24}$$

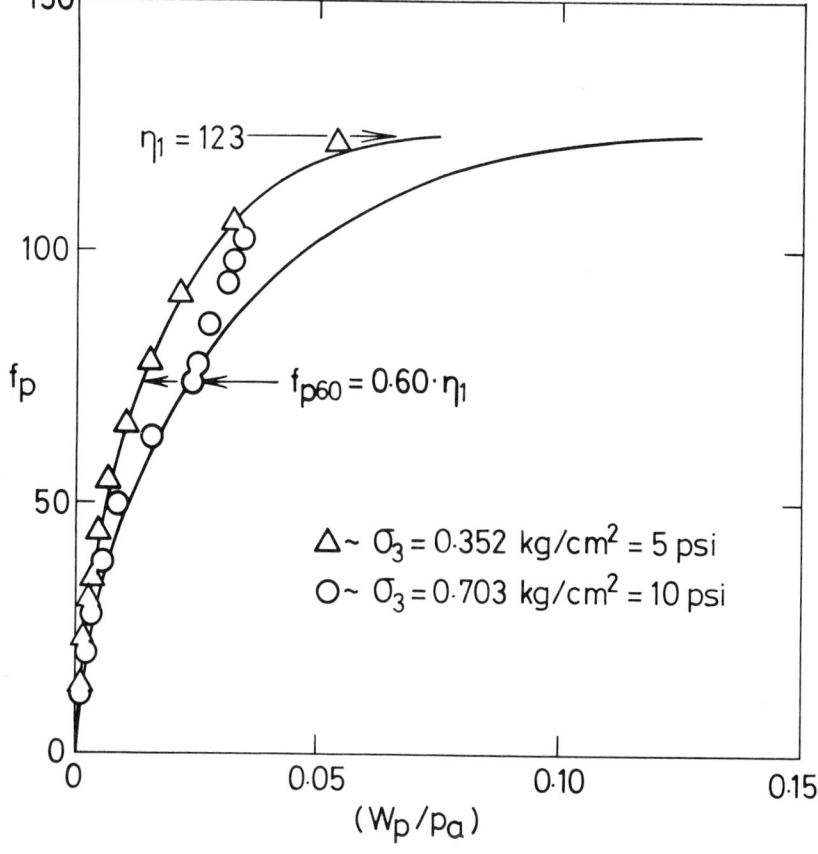

Figure 6. Variation of Total Plastic Work with f_p and σ_3 for Ottawa Sand.

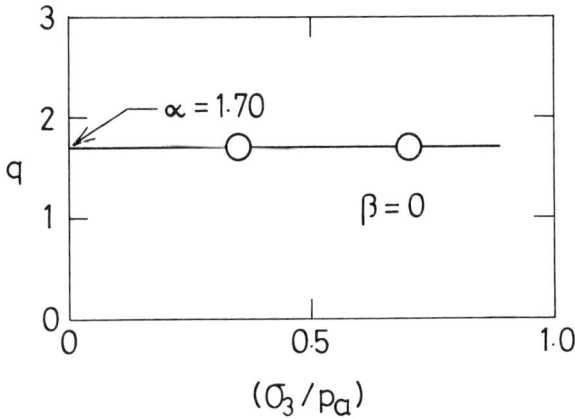

Figure 7a. Variation of q with Confining Pressure σ_3 for Ottawa Sand.

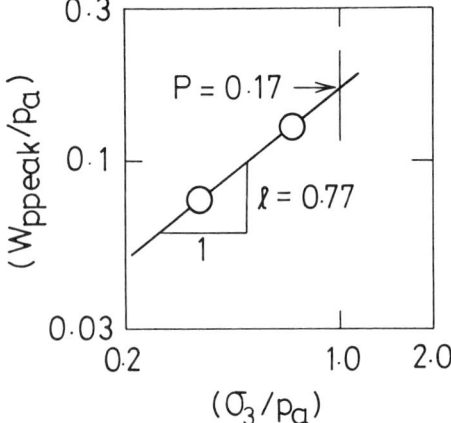

Figure 7b. Variation of W_{ppeak} with Confining Pressure σ_3 for Ottawa Sand.

and

$$b = \frac{1}{q \cdot W_{ppeak}} \quad (25)$$

where q is determined from Eq. 23 and e is the base for natural logarithms. The variation of W_{ppeak} with confining pressure σ_3 can be approximated by the following expression:

$$W_{ppeak} = P \cdot p_a \cdot \left(\frac{\sigma_3}{p_a}\right)^{\ell} \quad (26)$$

where P and ℓ are constants to be determined as shown in Fig. 7(b) for Ottawa Sand. On this diagram P is the intercept with $(\sigma_3/p_a) = 1$ and ℓ is the slope of the straight line.

Using the relation between f_p and W_p in Eq. 21 and the appropriate values of a, b, and q as determined above, the solid lines on Fig. 6 are obtained. These lines are used to represent the experimentally determined work-hardening relationships shown in this figure. It may be seen that the fit between the solid line and the experimental results is not very accurate above f_{p60} for the test with $\sigma_3 = 10$ psi. This is caused by the unloading-reloading cycle performed in this test as shown on Fig. 8(b). A considerable shift in the primary loading stress-strain curve occurs due to this cycle. A similar substantial shift does not occur in any of the other data base tests, as may be seen on Figs. 8 and 9. The solid lines in Fig. 6 are therefore used for the purpose of the following predictions of the behavior of Ottawa Sand.

The form of the stress-strain relationships which account for the plastic expansive behavior of the cohesionless soil can be expressed as:

$$\Delta\varepsilon^p_{ij} = \Delta\lambda_p \cdot \frac{\partial g_p}{\partial \sigma_{ij}} \quad (27)$$

The derivatives of g_p with regard to the normal stresses become:

$$\frac{\partial g_p}{\partial \sigma_x} = 3 \cdot I_1^2 - \left(27 + \eta_2 \cdot \left(\frac{p_a}{I_1}\right)^m\right) \cdot \left(\sigma_y \cdot \sigma_z - \tau_{yz}^2\right)$$
$$+ \frac{I_3}{I_1} \cdot m \cdot \eta_2 \cdot \left(\frac{p_a}{I_1}\right)^m \quad (28a)$$

and similar expressions are obtained for the other normal stresses by interchanging the indices on the stresses. The derivatives of g_p with respect to the shear stresses become:

$$\frac{\partial g_p}{\partial \tau_{yz}} = \left(27 + n_2 \cdot \left(\frac{p_a}{I_1}\right)^m\right) \cdot \left(\sigma_x \cdot \tau_{yz} - \tau_{xy} \cdot \tau_{zx}\right) \tag{28b}$$

and similar expressions can be obtained for the other shear stresses by interchanging the indices on the stresses.

The value of the proportionality constant $\Delta\lambda_p$ in Eq. 27 can be written as:

$$\Delta\lambda_p = \frac{dW_p}{3 \cdot g_p + m \cdot n_2 \cdot \left(\frac{p_a}{I_1}\right)^m \cdot I_3} \tag{29}$$

where g_p is the plastic potential function and dW_p is the increment in plastic work due to an increase in stress level df_p:

$$dW_p = \frac{df_p}{f_p} \cdot \frac{1}{\left(\frac{1}{q \cdot W_p} - b\right)} \tag{30}$$

where f_p is the current value of the stress level.

Summary of Parameter Values for Ottawa Sand

Fourteen soil parameters are incorporated in the elasto-plastic stress-strain model for cohesionless soil. The determination of thirteen of these parameters is illustrated for Ottawa Sand in Figs. 4-7, and Poisson's ratio is 0.2. Only the results of the two triaxial compression tests with constant lateral stress have been employed for determination of parameter values appropriate for the plastic potential function and the work-hardening relationship for the plastic expansive strains. The parameter values for Ottawa Sand are summarized in Table 1. Some of these values are zero for this sand. None of the parameters have dimensions. All dimensions are controlled, where appropriate, by the dimension of the atmospheric pressure, p_a, as e.g. in Eq. 26. The parameters in Table 1 are employed for calculation of strains in Ottawa Sand for any stress-path involving primary loading, neutral loading, unloading and reloading.

Comparison of Measured and Predicted Behavior for Data Base Tests

The results of the data base shear tests are shown in Figs. 8 and 9 together with the predictions of the model indicated by solid lines. Because the triaxial compression tests in Fig. 8 were used for derivation of most of the parameter values, the good agreement was to be expected. However, only the initial slopes and the strengths from the tests in Fig. 9 were employed for determination of some of the parameters, as shown in Figs. 4(a) and 5(a). Thus, the solid lines in Fig. 9 represent, in part, predictions by the model. It may be seen that the comparisons between experimental results and solid lines are reasonably good in most cases. The largest differences occur for the tests where the experimentally obtained strengths deviate most from the assumed strengths, as indicated on Fig. 5(a).

Table 1. Summary of Parameter Values for Ottawa Sand

Parameter	Value	Strain Component
Modulus No., K_{ur}	3000	
Exponent, n	0.75	Elastic
Poisson's Ratio, ν	0.2	
Collapse Modulus, C	0.000040	Plastic
Collapse Exponent, p	1.19	Collapse
Yield Const., η_1	123	
Yield Exponent, m	0.51	
Pl. Potent. Const., R	0.0	
Pl. Potent. Const. S	0.60	Plastic
Pl. Potent. Const., t	-4.0	Expansive
Work-Hard. Const., α	1.70	
Work-Hard. Const., β	0.0	
Work-Hard. Const., P	0.17	
Work-Hard. Exponent, ℓ	0.77	

Appendix I. -- References

1. Lade, P.V., "The Stress-Strain and Strength Characteristics of Cohesionless Soils," thesis presented to the University of California, at Berkeley, Calif., in 1972, in partial fulfillment of the requirements for the degree of Doctor of Philosophy.

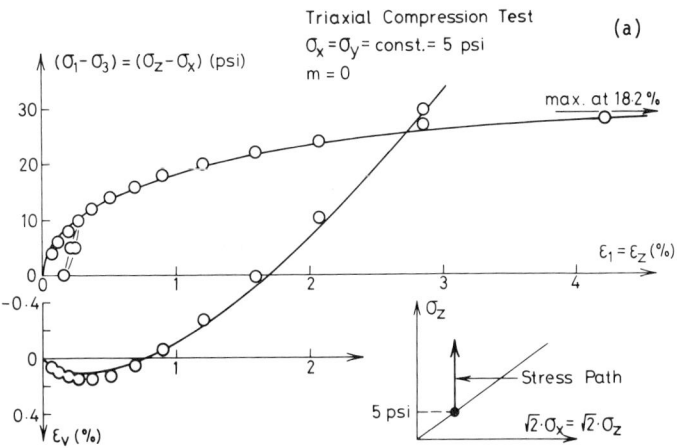

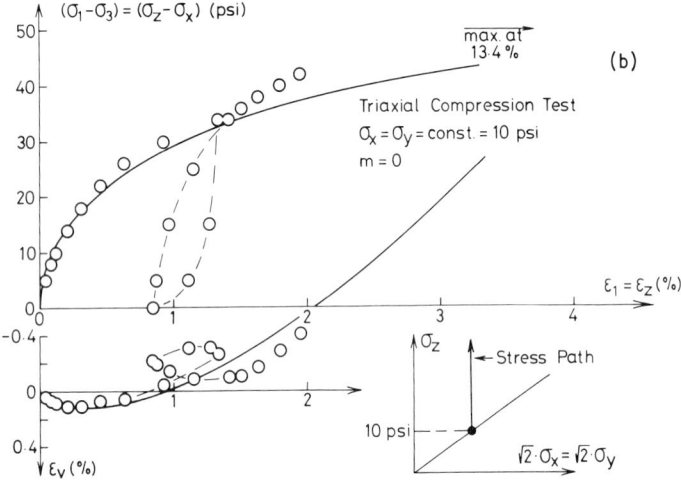

Figure 8. Comparison Between Measured and Predicted Stress-Strain and Volume Change Behavior for Ottawa Sand in Drained Triaxial Compression Tests.

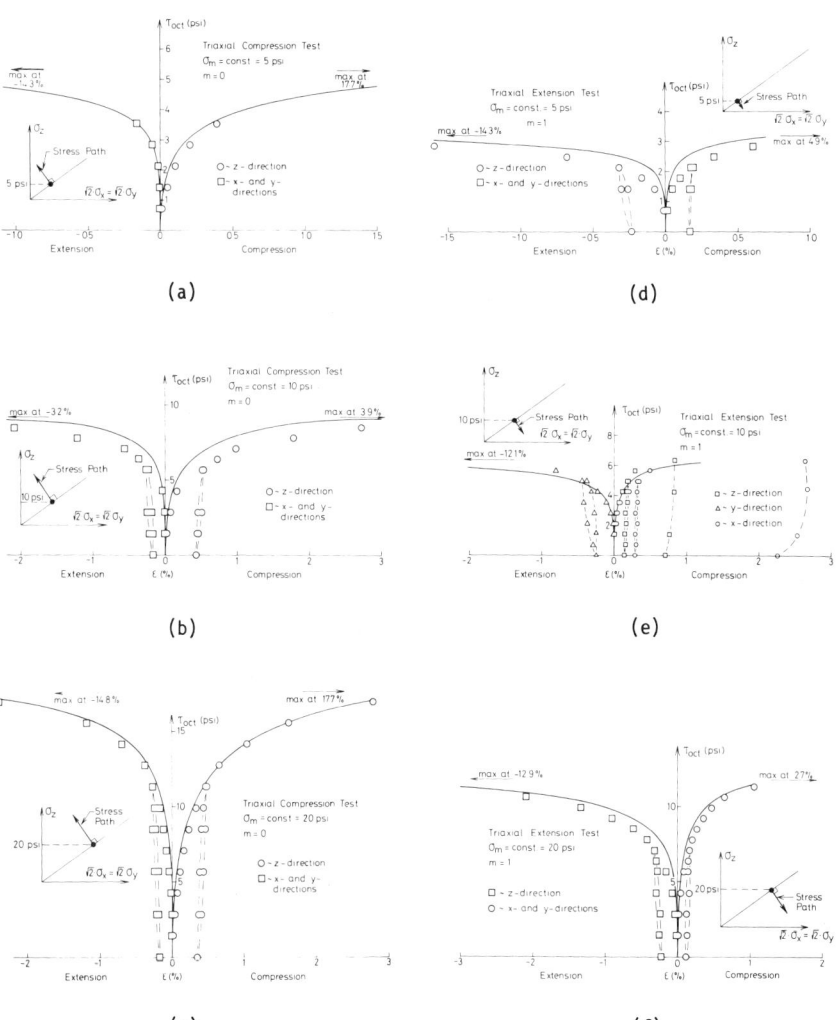

Figure 9. Comparison Between Measured and Predicted Stress-Strain Behavior for Ottawa Sand in Triaxial Compression and Extension Tests with Constant Mean Normal Stress.

2. Lade, P.V., "Elasto-Plastic Stress-Strain Theory for Cohesionless Soil with Curved Yield Surfaces," International Journal of Solids and Structures, Pergamon Press, Vol. 13, 1977, pp. 1019-1035.

3. Lade, P.V., "Prediction of Undrained Behavior of Sand," Journal of the Geotechnical Engineering Division, ASCE, Vol. 104, No. GT6, Proc. Paper 13834, June, 1978, pp. 721-735.

4. Lade, P.V., "Stress-Strain Theory for Normally Consolidated Clay," Proceedings of the Third International Conference on Numerical Methods in Geomechanics, Aachen, West Germany, Vol. IV, 1979, pp. 1325-1337.

5. Lade, P.V., and Duncan, J.M., "Cubical Triaxial Tests on Cohesionless Soil," Journal of the Soil Mechanics and Foundations Division, ASCE, Vol. 99, No. SM10, Proc. Paper 10057, October, 1973, pp. 793-812.

6. Lade, P.V., and Duncan, J.M., "Elastoplastic Stress-Strain Theory for Cohesionless Soil," Journal of the Geotechnical Engineering Division, ASCE, Vol. 101, No. GT10, Proc. Paper 11670, October, 1975, pp. 1037-1053. Closure in Vol. 104, January, 1978, pp. 139-141.

7. Lade, P.V., and Duncan, J.M., "Stress-Path Dependent Behavior of Cohesionless Soil," Journal of the Geotechnical Engineering Division, ASCE, Vol. 102, No. GT1, Proc. Paper 11841, January, 1976, pp. 51-68.

8. Lade, P.V., and Musante, H.M., "Failure Conditions in Sand and Remolded Clay," Proceedings of the 9th International Conference on Soil Mechanics and Foundation Engineering, Tokyo, Japan, Vol. I, July, 1977, pp. 181-186.

9. Lade, P.V., and Musante, H.M., "Three-Dimensional Behavior of Remolded Clay," Journal of the Geotechnical Engineering Division, ASCE, Vol. 104, No. GT2, Proc. Paper 13551, February, 1978, pp. 193-209.

EXAMPLES OF THE USE OF THE CAP MODEL FOR SIMULATING THE STRESS-STRAIN BEHAVIOR OF SOILS

By George Y. Baladi,[1] and Ivan S. Sandler,[2] Members, ASCE

Introduction

The cap model falls within the general framework of the classical incremental theory of plasticity for materials which have time- and temperature-independent properties. It is intended to represent the behavior of soils under a wide range of conditions.

Of course, no single constitutive model can represent any material in all situations. Even water, which is probably the most studied and best understood real material known to man, is never described by a single constitutive law to cover all situations. For example, water can sometimes be considered incompressible and sometimes compressible, sometimes inviscid and sometimes newtonian. For some applications surface tension or cavitation, etc., may be modeled. In short, whenever a constitutive model is selected, only those features of material behavior relevant to the problem at hand should be included.

The cap model has been developed with a number of features which can be included or excluded in specific applications. These involve anisotropy, kinematic hardening, pore pressure effects, limit surfaces, etc. The materials considered in the NSF/NSERC Workshop on Plasticity Theories and Generalized Stress-Strain Modeling of Soils each require different features of the cap model. Therefore, three different versions of the cap model have been utilized. These are presented in the following parts. Part I contains the development of a transversely isotropic cap model and demonstrates its ability to fit the stress-strain behavior of clay X and clay Y (Material A). The development of a kinematically hardening version of the cap model and its ability to fit the stress-strain behavior of kaolinite clay (Material B) is presented in Part II. Part III presents the development of an elastic-plastic isotropic version of the cap model and demonstrates its ability to fit the stress-strain behavior of Ottawa sand (Material C).

---

[1]Research Civ. Engr., Geomechanics Div., Structures Lab., Waterways Experiment Station, Vicksburg, Miss.
[2]Partner of Weidlinger Associates, Consulting Engineers, New York, NY.

PART I: AN ELASTIC-PLASTIC TRANSVERSE-ISOTROPIC CONSTITUTIVE MODEL FOR CLAY
(MATERIALS A)

By George Y. Baladi

The cap model falls within the framework of the classical incremental theory of plasticity (Hill [1]) for materials which have time- and temperature-independent properties and which are capable of undergoing small elastic as well as small plastic strains during each loading increment, i.e., the total strain increment, $d\varepsilon_{ij}$, is assumed to be the sum of the elastic (or recoverable) strain increment, $d\varepsilon_{ij}^E$, and the plastic (or permanent) strain increment, $d\varepsilon_{ij}^P$.

$$d\varepsilon_{ij} = d\varepsilon_{ij}^E + d\varepsilon_{ij}^P \quad \ldots \ldots \ldots \ldots \ldots \ldots \ldots \ldots \ldots \ldots \quad (1)$$

Because natural clays are often anisotropic, it is to be expected that an anisotropic version of the cap model would be appropriate in the current situation. Perusal of the given data in this case confirms this. The required model is constructed through the use of "Pseudo Invariants" of Stress as described below.

Fundamental Definitions of "Pseudo Stress Invariants"

The numerical implementation of a transverse-isotropic model can be greatly simplified if strain can be separated into hydrostatic and deviatoric components. This can be done, as will be shown later in this section, by defining "pseudo stress invariants" as follows (Baladi [2], Sandler and DiMaggio [3]):

$$\phi_1 = \sigma_{11} + \alpha(\sigma_{22} + \sigma_{33}) \quad \ldots \ldots \ldots \ldots \ldots \ldots \ldots \ldots \quad (2)$$

$$\phi_2 = \frac{\beta}{6}\left[(\sigma_{11} - \sigma_{22})^2 + (\sigma_{11} - \sigma_{33})^2\right] + \frac{(\sigma_{22} - \sigma_{33})^2}{6} + \gamma(\sigma_{12}^2 + \sigma_{13}^2)$$

$$+ \frac{\beta + 2}{3}\sigma_{23}^2 \quad \ldots \ldots \ldots \ldots \ldots \ldots \ldots \ldots \ldots \ldots \ldots \quad (3)$$

where σ_{ij} are components of the stress tensor and α, β, and γ are dimensionless material properties (constants). The axis of material symmetry, in this definition, is assumed to be normal to the $\sigma_{22} - \sigma_{33}$ plane (the plane of isotropy). In reality, Eqs. 2 and 3 represent a simplified version of the complete transverse-isotropic elastic, transverse-isotropic plastic model described by Baladi [2]. The complete model contains two descriptions of ϕ_1 and ϕ_2, i.e., one for the elastic portion of the model (ϕ_1^E, ϕ_2^E) and one for the plastic portion (ϕ_1^P, ϕ_2^P). These descriptions require two sets of the parameters α, β, and γ; namely, α^E, β^E, and γ^E and α^P, β^P, and γ^P. For the present discussion, the model has been simplified such that $\alpha^E = \alpha^P = \alpha$, etc. Note that when $\alpha = \beta = \gamma = 1$, ϕ_1 becomes J_1, the familiar first invariant of the stress tensor, and ϕ_2 becomes J_2', the familiar second invariant of the stress deviation tensor. In that case the model reduces to the familiar isotropic cap model (DiMaggio and Sandler [4], Sandler and Rubin [5], and Baladi [6]).

The Plastic Potential Functions

The mathematical basis of the incremental theory of plasticity was established by Drucker [7] when he introduced the concept of material stability, which has the following implications:

1. Yield surface (loading function) should be convex in stress space.
2. Yield surface and plastic potential should coincide (this is to guarantee an associated flow rule).
3. Work "softening" should not occur.

These three conditions can be summarized mathematically by the following inequality:

$$d\sigma_{ij} \, d\varepsilon_{ij}^P \geq 0 \quad \ldots \ldots \ldots \ldots \ldots \ldots \ldots \ldots \quad (4)$$

The above conditions allow considerable flexibility in choosing a loading function, ϕ, for the model which serves as both a yield surface and a plastic potential. For the present model, the loading function consists of two parts: an ultimate failure envelope which effectively serves to limit the maximum shear strength of the material and a strain-hardening surface, Figure 1. The failure envelope portion of the loading function is denoted by

$$\phi = G(\phi_1, \sqrt{\phi_2}) = \sqrt{\phi_2} - f(\phi_1) = 0 \text{ for } \phi_1 < L(\kappa) \quad \ldots \ldots \ldots \quad (5)$$

and the strain-hardening surface by

$$\phi = H(\phi_1, \sqrt{\phi_2}, \kappa) = \sqrt{\phi_2} - F(\phi_1, \kappa) = 0 \text{ for } X(\kappa) \geq \phi_1 \geq L(\kappa) \quad \ldots \ldots \quad (6)$$

Note in Figure 1 that $L(\kappa)$ and $X(\kappa)$ define the intersection of the hardening surface with the failure envelope, $G(\phi_1, \phi_2)$ and the ϕ_1 axis, respectively, and κ is, in general, a function of the history of plastic volumetric strain, ε_{kk}^P. $L(\kappa)$ is chosen so that

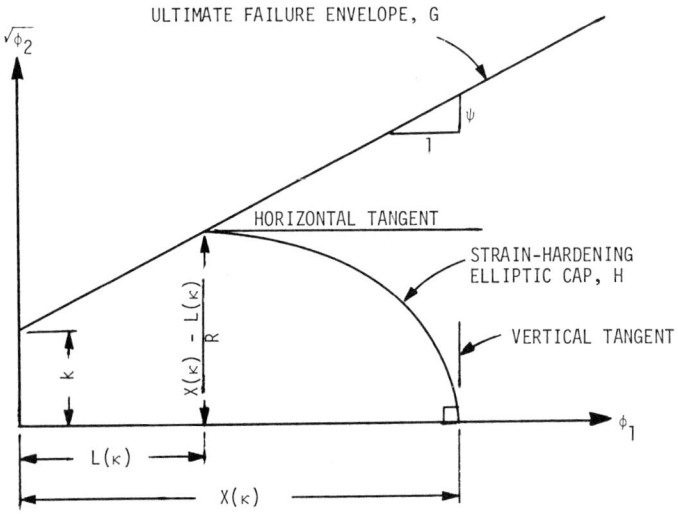

FIG. 1.—Proposed Yield Surface for the Elastic-Plastic Transverse-Isotropic Model

$$L(\kappa) = \begin{cases} \kappa & \text{if } \kappa > \\ 0 & \text{if } \kappa \leq 0 \end{cases} \quad \ldots \ldots \ldots \ldots \ldots \ldots \ldots \ldots \ldots \quad (7)$$

The hardening parameter κ can be one of the following functions

$$\kappa = g_1 (\epsilon^P_{kk}) \quad \ldots \ldots \ldots \ldots \ldots \ldots \ldots \ldots \ldots \ldots \ldots \ldots \quad (8)$$

$$\kappa = g_2 [(\epsilon^P_{kk})_{max}] \quad \ldots \ldots \ldots \ldots \ldots \ldots \ldots \ldots \ldots \ldots \ldots \quad (9)$$

The use of Eq. 8 permits the hardening surface to move back toward the origin (Figure 1) when a point on the failure envelope, G, is reached, thus controlling the dilatancy of the material. The use of Eq. 9, on the other hand, only permits the hardening surface to move away from the origin, thus ensuring dilatancy while still permitting hysteresis in a hydrostatic load-unload cycle.

The plastic loading criteria for the function ϕ are given by

$$\frac{\partial \oint}{\partial \sigma_{ij}} d\sigma_{ij} \begin{cases} > 0 \text{ loading} \\ = 0 \text{ neutral loading} \\ < 0 \text{ unloading} \end{cases} \quad \ldots \ldots \ldots \ldots \ldots \ldots \quad (10)$$

Plastic strains will occur only when $d\oint$ is positive and $\oint = 0$. During unloading or neutral loading, as well as for $\oint < 0$, the material will behave elastically. The prescription that neutral loading produces no plastic strain is called the continuity condition. Its satisfaction leads to coincidence of elastic and plastic constitutive laws during neutral loading (Drucker [7] and Handelman [8]).

Plastic Strain Increment Tensor

Based on Eq. 4, the plastic strain increments in the model are defined through the associated plastic flow rule

$$d\varepsilon^P_{ij} = \begin{cases} d\lambda \frac{\partial \oint}{\partial \sigma_{ij}} & \text{if } \oint = 0 \\ 0 & \oint < 0 \end{cases} \quad \ldots \ldots \ldots \ldots \ldots \ldots \quad (11)$$

where $\oint$ is defined by either Eq. 5 or 6 and $d\lambda$ is a positive function of proportionality which is nonzero only when plastic deformation occurs.

The plastic stress-strain relation can be expressed in terms of the hydrostatic and deviatoric components of strain through the chain rule of differentiation applied to the right-hand side of Eq. 11, thus

$$d\varepsilon^P_{kk} = d\lambda(1 + 2\alpha) \frac{\partial \oint}{\partial \phi_1} \quad \ldots \ldots \ldots \ldots \ldots \ldots \quad (12)$$

and

$$d\bar{e}^P_{ij} = \frac{d\lambda}{2\sqrt{\phi_2}} \frac{\partial \oint}{\partial \sqrt{\phi_2}} n_{ij} \quad \ldots \ldots \ldots \ldots \ldots \ldots \quad (13)$$

in which

$$n_{ij} = \frac{\partial \phi_2}{\partial \sigma_{ij}} \quad \text{and} \quad n_{ii} = 0 \quad \ldots \ldots \ldots \ldots \ldots \ldots \quad (14)$$

where $d\varepsilon^P_{kk}$ is the increment of plastic volumetric strain and $d\bar{e}^P_{ij}$ is the "pseudo plastic strain deviation increment tensor." The plastic strain increment tensor, de^P_{ij}, is related to $d\bar{e}^P_{ij}$ through the following relation

$$d\varepsilon^P_{ij} = d\overline{\varepsilon}^P_{ij} - \left(\frac{\delta_{ij}}{3} - \frac{A_{ij}}{1 + 2\alpha}\right) d\varepsilon^P_{kk} \quad \ldots \ldots \ldots \ldots \ldots \ldots \quad (15)$$

in which δ_{ij} is the Kronecker delta function and

$$[A_{ij}] = \begin{bmatrix} 1 & 0 & 0 \\ 0 & \alpha & 0 \\ 0 & 0 & \alpha \end{bmatrix} \quad \ldots \ldots \ldots \ldots \ldots \ldots \quad (16)$$

Hence, the plastic strain increment tensor becomes

$$d\varepsilon^P_{ij} = d\lambda \left(\frac{\partial f}{\partial \phi_1} A_{ij} + \frac{1}{2\sqrt{\phi_2}} \frac{\partial f}{\partial \sqrt{\phi_2}} \eta_{ij}\right) \quad \ldots \ldots \ldots \ldots \quad (17)$$

In order to use Eqs. 12 through 15 or Eq. 17, the proportionality factor $d\lambda$ must be determined. This can be accomplished in a straightforward manner; however, since the resulting expression for $d\lambda$ contains elastic moduli, the determination of the expression is deferred to a later section.

The Elastic Potential Function

Within the yield surface, f, the material behavior is transverse-isotropic elastic. The complementary energy function for this material is chosen to be

$$d\Omega = \frac{d(\phi_1)^2}{18B} + \frac{d\phi_2}{2S} \quad \ldots \ldots \ldots \ldots \ldots \ldots \quad (18)$$

where B and S are material response functions that will be defined subsequently. The elastic strain increment tensor, $d\varepsilon^E_{ij}$, becomes

$$d\varepsilon^E_{ij} = d\left(\frac{\partial \Omega}{\partial \sigma_{ij}}\right) \quad \ldots \ldots \ldots \ldots \ldots \ldots \quad (19)$$

It can be seen from Eqs. 2, 3, 18, and 19 that the elastic behavior is governed by the three elastic constants (α, β, and γ) and by the two response functions B and S which may take the following form

$$B = B(\phi_1, \kappa) \quad \ldots \ldots \ldots \ldots \ldots \ldots \quad (20)$$

$$S = S(\sqrt{\phi_2}, \kappa) \quad \ldots \ldots \ldots \ldots \ldots \ldots \quad (21)$$

Note that the definition of Eq. 18 and the form of Eqs. 20 and 21 dictate that the model is path-independent during purely elastic deformation (Prager [9] and Sandler, DiMaggio, and Baladi [10].

Certain limitations must be placed on the constants α, β, and γ of Eqs. 2 and 3 and the response functions, B and S. These limitations are discussed in detail by Lekhnitskii [11] and Baladi [2] for a linear elastic transverse isotropic material. In the present model these limitations imply

$$\frac{BS}{3B\beta + S} > 0 \qquad (22)$$

$$\frac{BS}{3B(\beta + 1) + 2S\alpha^2} > 0 \qquad (23)$$

$$\frac{S}{\gamma} > 0 \qquad (24)$$

$$\frac{S}{\beta + 2} > 0 \qquad (25)$$

$$-1 < \frac{1.5B - S\alpha^2}{1.5B(\beta + 1) + S\alpha^2} < 1 \qquad (26)$$

$$\frac{3B\beta + 4S\alpha^2}{6B\beta + 2S} - 2\left[\frac{1.5B\beta - S\alpha}{3B\beta + S}\right]^2 > 0 \qquad (27)$$

Elastic Strain Increment Tensor

The elastic strain increment tensor, $d\varepsilon_{ij}^E$, can be obtained from Eq. 19, with the help of Eqs. 14 and 16; thus

$$d\varepsilon_{ij}^E = \frac{1}{9B} d\phi_1 A_{ij} + \frac{1}{2S} d\eta_{ij} \qquad (28)$$

The hydrostatic and deviatoric components of strain can be easily obtained from Eq. 28 as

$$d\varepsilon_{kk}^E = \frac{1 + 2\alpha}{9B} d\phi_1 \qquad (29)$$

$$d\bar{e}_{ij}^E = \frac{1}{2S} d\eta_{ij} \qquad (30)$$

where $d\varepsilon_{kk}^E$ is the increment of elastic volumetric strain and $d\bar{e}_{ij}^E$ is the "pseudo elastic strain deviation increment tensor." The elastic strain deviation increment tensor, de_{ij}^E, is related to $d\bar{e}_{ij}^E$ through the following relation:

$$de_{ij}^E = \overline{de}_{ij}^E - \left(\frac{\delta_{ij}}{3} - \frac{A_{ij}}{1 + 2\alpha}\right) d\varepsilon_{kk}^E \quad \ldots \ldots \ldots \ldots \ldots \quad (31)$$

Note that when $\alpha = \beta = \gamma = 1$, the above model reduces to the isotropic models described by Sandler, DiMaggio, and Baladi [10] and Baladi [6] and the response functions B and S become the isotropic bulk and shear modulus response functions, respectively.

Flow Rule Proportionality Factor

We return now to the determination of the proportionality factor $d\lambda$. Applying the chain rule of differentiation to the right-hand side of Eq. 10 and using Eqs. 1, 12, 13, 14, 15, 17, 29, 30, and 31 leads to

$$d\lambda = \frac{\left[\dfrac{9B}{(1 + 2\alpha)} \dfrac{\partial f}{\partial \phi_1} - \dfrac{S}{B\sqrt{\phi_2}} \left(\dfrac{1 - \alpha}{1 + 2\alpha}\right) \dfrac{\partial f}{\partial \sqrt{\phi_2}} \eta_{11}\right] d\varepsilon_{kk} + \dfrac{S}{\sqrt{\phi_2}} \dfrac{\partial \phi_2}{\partial \eta_{ij}} \dfrac{\partial f}{\partial \sqrt{\phi_2}} de_{ij}}{9B\left(\dfrac{\partial f}{\partial \phi_1}\right)^2 + S\left(\dfrac{\partial f}{\partial \sqrt{\phi_2}}\right)^2 - (1 + 2\alpha) \dfrac{\partial f}{\partial \phi_1} \dfrac{\partial f}{\partial \kappa}} \quad (32)$$

Total Incremental Strain-Stress Relation

The hydrostatic and deviatoric components of the total strain increment tensor can be obtained by adding Eqs. 12 and 29 to obtain $d\varepsilon_{kk}$, and Eqs. 15 and 31 to obtain de_{ij}; thus

$$d\varepsilon_{kk} = \frac{1 + 2\alpha}{9B} d\phi_1 + d\lambda(1 + 2\alpha) \frac{\partial f}{\partial \phi_1} \quad \ldots \ldots \ldots \ldots \ldots \quad (33)$$

$$de_{ij} = \frac{1}{2S} d\eta_{ij} + d\lambda \frac{\eta_{ij}}{2\sqrt{\phi_2}} \frac{\partial f}{\partial \sqrt{\phi_2}} - \left(\frac{\delta_{ij}}{3} - \frac{A_{ij}}{1 + 2\alpha}\right) (1 + 2\alpha)\left(\frac{d\phi_1}{9B} + d\lambda \frac{\partial f}{\partial \phi_1}\right) \quad (34)$$

Hence, the total strain increment tensor becomes

$$d\varepsilon_{ij} = \frac{1}{9B} d\phi_1 A_{ij} + \frac{1}{2S} d\eta_{ij} + d\lambda\left(\frac{\partial f}{\partial \phi_1} A_{ij} + \frac{1}{2\sqrt{\phi_2}} \frac{\partial f}{\partial \sqrt{\phi_2}} \eta_{ij}\right) \quad \ldots \ldots \quad (35)$$

Similarly, the "pseudo hydrostatic and deviatoric components of the stress increment tensor" can be written as

$$d\phi_1 = \frac{9B}{1 + 2\alpha} d\varepsilon_{kk} - 9B \, d\lambda \frac{\partial f}{\partial \phi_1} \quad \ldots \ldots \ldots \ldots \ldots \quad (36)$$

$$dn_{ij} = 2S\ de_{ij} + 2S\left(\frac{\delta_{ij}}{3} - \frac{A_{ij}}{1+2\alpha}\right) d\varepsilon_{kk} - \frac{S\ d\lambda}{\sqrt{\phi_2}} \frac{\partial \phi}{\partial \sqrt{\phi_2}} n_{ij} \quad \ldots \ldots \ldots \quad (37)$$

Equations 33 and 34, or Eqs. 36 and 37, are the general constitutive equations for a transverse-isotropic, inelastic, work-hardening constitutive model in which the axis of material symmetry is normal to the $\sigma_{22} - \sigma_{33}$ plane. To use these equations it is only necessary to specify the functional forms of B, S, ϕ, and κ and, of course, to determine experimentally the values of the coefficients in these functions as well as the values of α, β, and γ. These are the subjects of the next section.

Basic Features of the Model

Elastic Behavior.--The behavior of the model in the elastic (recoverable) range is governed by the three constants, α, β, and γ (Eqs. 2 and 3), and by the two response functions, B and S (Eqs. 20 and 21). Equation 29 reveals that the parameter α and the response function B are compressibility-related material properties, while Eq. 30 reveals that β and γ and the response function S are shear-related material properties.

The parameter α defines the ratio between the elastic strain in the plane of isotropy, ε_r^E, and the elastic strain normal to the plane of isotropy, ε_z^E, under hydrostatic states of stress; thus

$$\alpha = \frac{\varepsilon_r^E}{\varepsilon_z^E} \quad \ldots \ldots \ldots \ldots \ldots \ldots \ldots \ldots \ldots \ldots \ldots \ldots \quad (38)$$

For the ensuing developments, it will be more convenient to work with a Cartesian coordinate system. Hence, the plane 22, 33 of the Cartesian coordinate system 11, 22, and 33 is designated as the plane of isotropy (Figure 2). Equation 38, therefore, becomes

$$\alpha = \frac{\varepsilon_{22}^E}{\varepsilon_{11}^E} = \frac{\varepsilon_{33}^E}{\varepsilon_{11}^E} \quad \ldots \ldots \ldots \ldots \ldots \ldots \ldots \ldots \ldots \ldots \quad (39)$$

The value of α can be determined experimentally from the slope of the unloading strain path curve obtained from a hydrostatic compression test (Figure 2). It is clear from Figure 2 that for an isotropic material, the value of α becomes

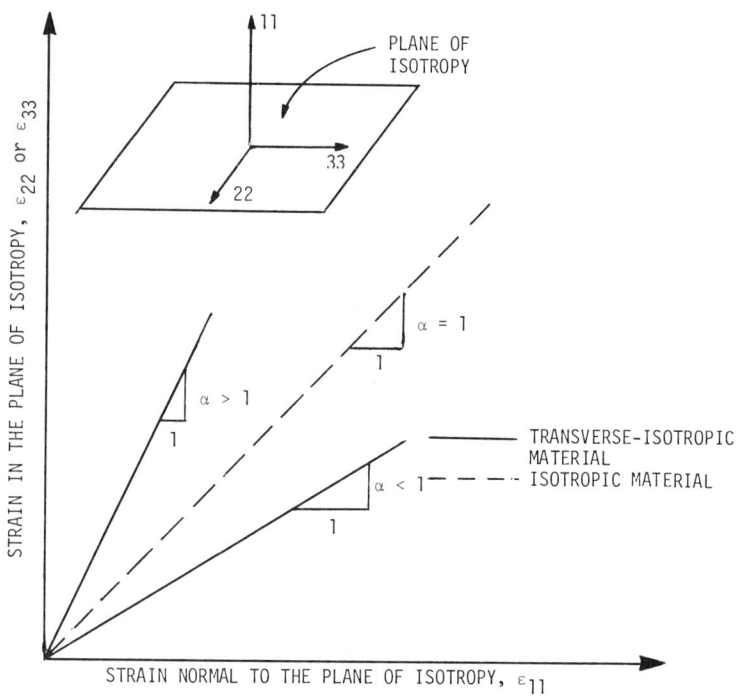

FIG. 2.—Comparison of Strain Paths for Isotropic and Transverse-Isotropic Materials Subjected to Hydrostatic Loading and Unloading

$$\alpha = \frac{\varepsilon_{22}^E}{\varepsilon_{11}^E} = \frac{\varepsilon_{33}^E}{\varepsilon_{11}^E} = 1 \quad \dots \dots \dots \dots \dots \dots \dots \dots \dots \quad (40)$$

The elastic response function B (B is the elastic bulk modulus for an isotropic material) describes the unloading stress-strain response of a hydrostatic compression test. For clay X and Y materials, B can be taken to be a function of the first pseudo invariant of stress ϕ_1; thus

$$B = \frac{B_i}{1 - B_1} \left[1 - B_1 \exp(-B_2 \phi_1) \right] \quad \dots \dots \dots \dots \dots \dots \quad (41)$$

where B_i = initial value of the response function B (Figure 3). Equation 41 together with Eq. 29 indicates that the material constants B_i, B_1, and

B_2 can be readily determined experimentally from the unloading hydrostatic compression test results, as illustrated in Figure 3. Equation 41, therefore, can be written as

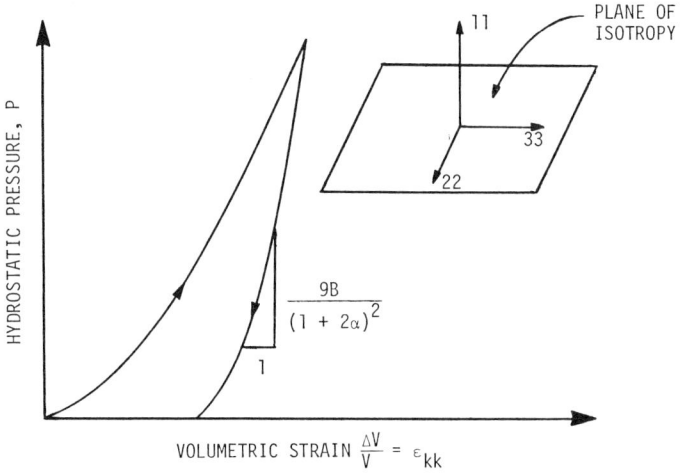

a. HYDROSTATIC COMPRESSION

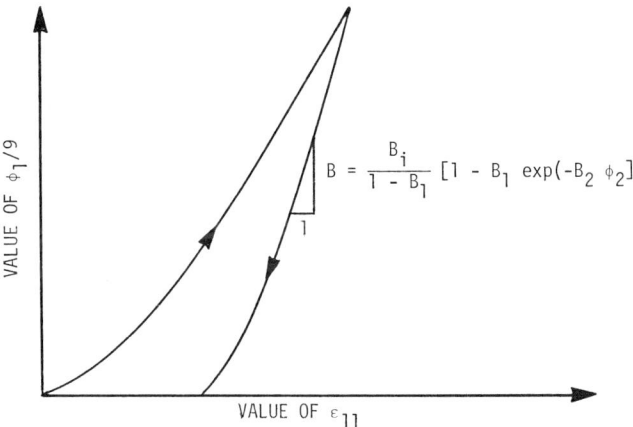

b. METHOD FOR ISOLATING RESPONSE FUNCTION B

FIG. 3.--Elastic Transverse-Isotropic Relationships for Hydrostatic Compression Test

$$B = \frac{B_i}{1 - B_1} [1 - B_1 \exp(-B_2 \phi_1)] = \frac{d\phi_1}{9d\varepsilon_{11}} \quad \ldots \ldots \ldots \ldots \quad (42)$$

The elastic shear response function S accounts for the curvature observed in the stress difference-strain difference results obtained from triaxial compression tests. For clay X and Y materials, S is assumed to be a function of the second pseudo invariant of stress ϕ_2; thus

$$S = \frac{S_i}{1 - S_1} [1 - S_1 \exp(-S_2 \sqrt{\phi_2})] \quad \ldots \ldots \ldots \ldots \ldots \quad (43)$$

where S_i = initial value of the response function S (Figure 4); and S_1 and S_2 = material constants. Equation 30 indicates that the material constants S_i, S_1, and S_2 can be readily determined from the slopes of experimental unloading stress-strain curves obtained from a triaxial test conducted on a cubical specimen, i.e., a 3-D box-type test, in which the axial stress is applied parallel to the 33-axis; i.e., $\sigma_{11} = \sigma_{22}$ = confining pressure (Figure 4). The results of these same tests can be used to determine the value of β from the slope of the unloading strain path shown in Figure 5, thus

$$\beta = \frac{e_{11}^{-E}}{e_{22}^{-E}} \quad \ldots \ldots \ldots \ldots \ldots \ldots \ldots \ldots \ldots \ldots \quad (44)$$

where e_{ij}^{-E} is the elastic pseudo strain deviation tensor defined by Eqs. 30 and 31. There are several ways of determining γ. One way is to measure the unloading shear stress-shear strain response during a direct shear test, or a simple shear test (Figure 6). The slope of the $\sigma_{13} - \varepsilon_{13}$ unloading curve is equal to $2S/\gamma$, from which γ can be easily determined, since the function S is known from Figure 4.

Plastic Behavior.--For the plastic behavior, the loading function $\oint$ (Eqs. 5 and 6) is assumed to consist of two parts (Figure 1): an ultimate failure envelope that effectively limits the maximum shear stress in the material and an elliptically shaped strain-hardening yield surface that produces plastic volumetric and shear strains as it moves. The failure envelope portion of the loading function is mathematically described by

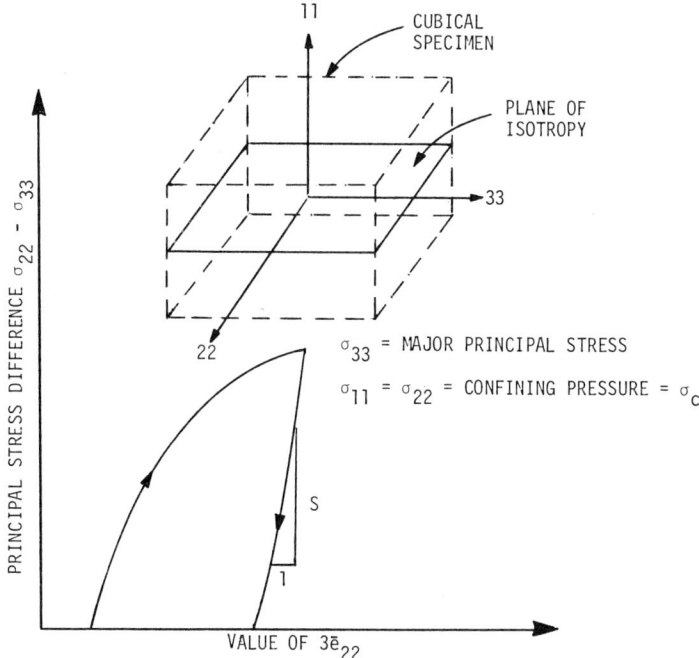

FIG. 4.—Suggested Method for Conducting Triaxial Tests and Plotting Results to Quantitatively Determine the Response Function S

$$G(\phi_1, \sqrt{\phi_2}) = \sqrt{\phi_2} - \psi\, \phi_1 - k \quad \ldots \ldots \ldots \ldots \ldots \ldots \quad (45)$$

and the strain-hardening yield surface by

$$(\phi_1, \sqrt{\phi_2}, \kappa) = [\phi_1 - L(\kappa)]^2 + R^2 \phi_2 - [X(\kappa) - L(\kappa)]^2 = 0 \quad \ldots \ldots \quad (46)$$

where ϕ_1 and ϕ_2 are, respectively, the first and second pseudo invariants of stress, which are defined by Eqs. 2 and 3; and parameters k and ψ are material constants representing pseudo cohesive and frictional strength parameters of the material (Figure 1); R is a parameter which will be defined below; and κ for clay X and Y materials is assumed to be

$$\kappa = X - 3P_c \quad \ldots \ldots \ldots \ldots \ldots \ldots \ldots \ldots \ldots \ldots \ldots \ldots \quad (47)$$

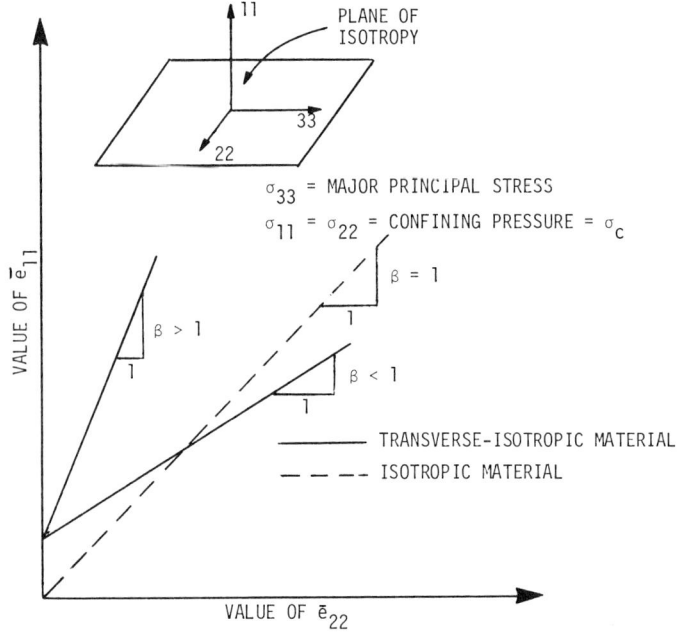

FIG. 5.—Comparison of Lateral Strain Path Results for Isotropic and Transverse-Isotropic Materials Subjected to a Triaxial Test

in which X is related to the plastic volumetric strain in the following relation:

$$\varepsilon_{kk}^P = [1 - \exp(-D\kappa) - AD\,\kappa\,\exp(-D1\kappa)] + W1\,\kappa^2\,\exp(-D2\kappa) - W2\,\kappa^2\,\exp(-D3\kappa) \quad (48)$$

where P_c = preconsolidation pressure; W = maximum plastic volumetric compaction that the material can experience under hydrostatic loading; and D, AD, D1, W1, D2, W2, and D3 = material constants. The material parameter, R, in Eq. 46, is the ratio of the major to the minor axis of the elliptic yield surface (Figure 1). The value of R depends on the state of compaction of the material. For clay X and Y materials, the value of R is assumed to be

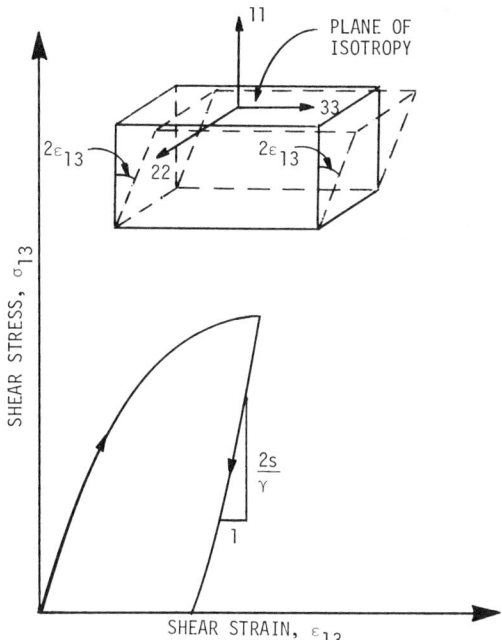

FIG. 6.—Simple Shearing Stress-Strain Response of a Transverse-Isotropic Material

$$R = \frac{R_i}{1 + R_1} \left[1 + R_1 \exp(-R_2 L^2)\right] \quad \ldots \ldots \ldots \ldots \ldots \ldots \quad (49)$$

where R_i, R_1, and R_2 are material constants that can be determined by a trial-and-error process of fitting the model to a uniaxial strain test. Note that the hardening surface (Figure 1) was chosen so that the tangent at its intersection with the failure envelope is horizontal. This condition is guaranteed by the following relationships between L and X:

$$L = \frac{X - Rk}{1 + \psi R} \quad \ldots \ldots \ldots \ldots \ldots \ldots \ldots \ldots \ldots \ldots \ldots \quad (50)$$

The parameters α and β affect the shape of the failure and hardening yield surfaces. This can be seen most clearly in principal stress space (Figure 7). In the octahedral plane, the trace of the failure surface defined

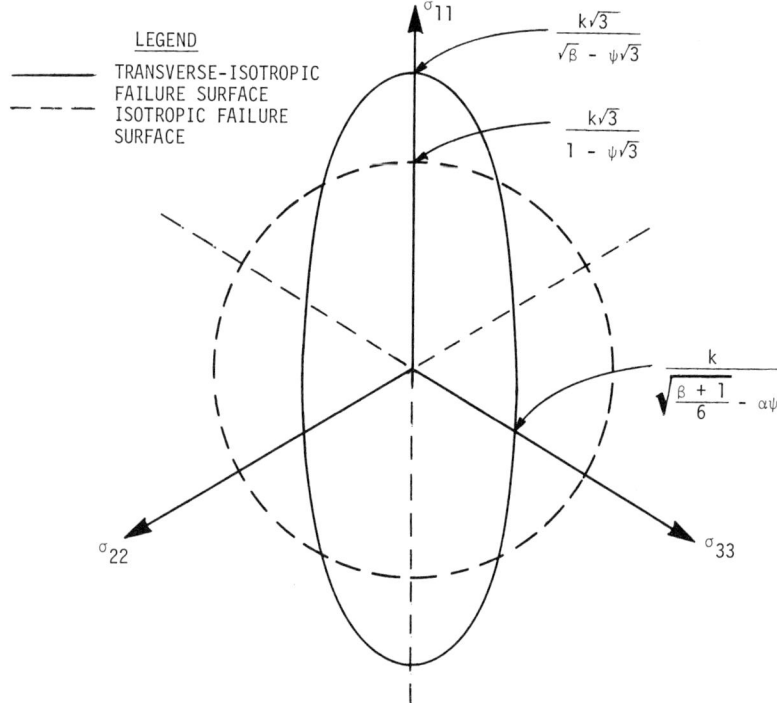

FIG. 7.—Isotropic and Transverse-Isotropic Failure Surfaces in Octahedral Plane

by Eq. 45 is shaped like an ellipse (Figure 7). The dashed circle in Figure 7 represents the trace of a failure surface for which $\alpha = \beta = 1$.

In summary, there are five potential functions (two elastic, three plastic) and three material parameters (α, β, and γ) that describe the complete behavior of the present model. Twenty-two material constants are used in the present model.

Comparisons of Laboratory Test Data with Model Behavior

Experimental Program.—The material used in the experimental program was a natural clay designated as clay X and clay Y. The physical properties of clay X and clay Y are shown in Tables 1 and 2, respectively.

TABLE 1.--Physical Properties of Clay "X"

Soil	Properties
Natural water content (%)	66 ± 2
Liquid Limit (%)	48
Plastic Limit (%)	28
Plasticity Index (%)	20
Liquidity Index	1.8 → 2.0
Clay content (%)	79
Silt content (%)	20
Sand content (%)	1
Activity	0.25
Sensitivity by laboratory vane	50
Specific gravity	2.80

TABLE 2.--Physical Properties of Clay "Y"

Soil	Properties
Natural water content (%)	64 ± 0.5
Apparent preconsolidation pressure	1.6 tsf
Liquid limit (%)	48
Plastic limit (%)	28
Liquidity Index	1.8
Clay content (%)	79
Silt content (%)	20
Sensitivity	80
Specific gravity	2.80

For both soils the triaxial tests were conducted on prismatic samples trimmed from block samples. All samples were 100 percent saturated.

The experimental program for clay X consisted of two isotropic consolidation tests, two unconfined triaxial tests, and a total of six triaxial tests in which the mean normal stress P_c was kept constant. Three different mean normal stresses (P_c - 10, 20, and 30 psi) were used. For each value of P_c two different stress ratios (m = 0 and m = 1) were used.

The experimental program for clay Y consisted of a consolidation test and six conventional triaxial tests in which the minor principal stress was kept constant. Three different confining stresses (σ_c = 2.5, 5, and 10 psi) were used. For each value of σ_c two different stress ratios (m = 0 and m = 1) were used.

During the testing, the measured quantities were axial deformation, lateral deformation in the minor direction, and volume change. Thus, the strain in the intermediate direction is a derived or calculated quantity rather than a direct measurement.

Material Constants.--As indicated above, there are 22 material constants in the present model that must be determined experimentally. The material constants B_i, B_1, B_2, W, D, AD, D1, W1, D2, W2, D3, and α can be determined from the response of the material under isotropic consolidation tests with major and minor principal stress measurements. Unfortunately, such measurements were not made when the isotropic consolidation test was run on clay X and clay Y. Therefore, in obtaining the numerical values of the above constants, several trial-and-error attempts have to be made. The material constants S_i, S_1, S_2, and β can be readily determined from the slopes of experimental unloading stress-strain curves obtained from triaxial tests conducted on a cubical specimen in which the axial stress is applied parallel to the plane of isotropy. Again, the results of such tests were not available.

In summary, the present model requires an elaborate set of experimental data in order to fully describe the material behavior. The numerical values of the material constants for clay X and clay Y are given in Table 3.

Correlation with Test Results.--Figures 8 through 16 compare actual test results with the model fits for clay X. Figure 8 compares effective mean normal stress versus volumetric strain for the isotropic compression test. Unconfined compression stress-strain curves for horizontal and vertical

Table 3.--Numerical Values of Material Constants for Clay X and Clay Y

Name	Notation	Units	Clay X	Clay Y
Failure envelope parameter	ψ	---	0.11	0
	k	psi	5.50	11
Hardening surface parameters	W	---	0.21	0.27
	D	psi^{-1}	0.06	0.02
	AD	---	0.95	1.0
	$D1$	psi^{-1}	0.061	0.021
	$W1$	psi^{-2}	0.0006	0
	$D2$	psi^{-1}	0.1	0
	$W2$	psi^{-2}	-0.000003	0
	$D3$	psi^{-1}	0.002	0
	R_i	---	4.0	0.8
	R_1	---	-0.25	0
	R_2	psi^{-2}	0.05	0
Elastic response functions B and S	B_i	psi	580	105
	B_1	---	0	0.95
	B_2	psi^{-1}	0	0.1
	S_i	psi	318	1500
	S_1	---	0	-2.0
	S_2	psi^{-1}	0	0.5
Elastic parameters	α	---	0.4	0.8
	β	---	0.215	1.5
	γ	---	1.0	1.0

samples are compared in Figures 9 and 10, respectively. Figures 11 through 16 compare consolidated drained triaxial shear test responses for various constant mean normal stress and stress ratios of 0 and 1.

Test results are compared with model behavior for clay Y material in Figures 17 through 23. Figure 22 compares the results of isotropic consolidation tests. Figures 23 through 28 compare consolidated drained triaxial shear test response for various confining pressure levels and stress ratios of 0 and 1. It is clear from Figures 1 through 23 that the proposed constitutive model simulates the stress-strain response of both clay X and clay Y fairly well. The author believes, however, that a better fit could be obtained if all the data required by the model were available.

The model predictions and comparison with actual data are presented in Volume I of the NSF/NSERC Workshop Proceedings.

Appendix.--References

1. Hill, R., The Mathematical Theory of Plasticity, Oxford University Press, London, 1950.

2. Baladi, G. Y., "An Elastic-Plastic Constitutive Relation for Transverse-Isotropic Three-Phase Earth Materials," Miscellaneous Paper S-78-14, August 1978, U. S. Army Engineer Waterways Experiment Station, CE, Vicksburg, Miss.

3. Sandler, I. S. and DiMaggio, F. L., "Anisotropy in Elastic-Plastic Models of Geological Materials," DNA Program Meeting on Material Properties for Ground Motion Calculations, Park City, Utah, November 1973.

4. DiMaggio, F. L. and Sandler, I. S., "Material Model for Granular Soils," J. Engrg. Mech. Div., ASCE, Vol. 97, No. EM3, June 1971.

5. Sandler, I. S. and Rubin, D., "An Algorithm and a Modular Subroutine for the Cap Model," Int'l. J. Num. Anal. Meth. in Geomech., Vol. 3, 1979.

6. Baladi, G. Y., "The Latest Development in the Nonlinear-Elastic-Nonideally Plastic Work Hardening Cap Model," Proceedings of the Symposium on the Role of Plasticity in Soil Mechanics, Cambridge, England, 13-15 September 1973, pp 51-55.

7. Drucker, D. C., "On Uniqueness in the Theory of Plasticity," Q. Applied Mathematics, Vol. 14, 1956.

8. Handelman, G. H., et al, "On the Mechanical Behavior of Metals in the Strain Hardening Range," Q. Applied Mathematics, Vol. 4, 1947, pp 397-407.

9. Prager, W., Introduction to Mechanics of Continua, Ginn, 1961.

10. Sandler, I. S., DiMaggio, F. L., and Baladi, G. Y., "A Generalized Cap Model for Geological Materials," Journal of the Geotechnical Engineering Division, ASCE, Vol. 102, No. GT7, Proc. Paper 12243, July 1976, pp 683-699.

11. Lekhnitskii, S. G., Theory of Elasticity of an Anisotropic Elastic Body, Holden-Day, 1963.

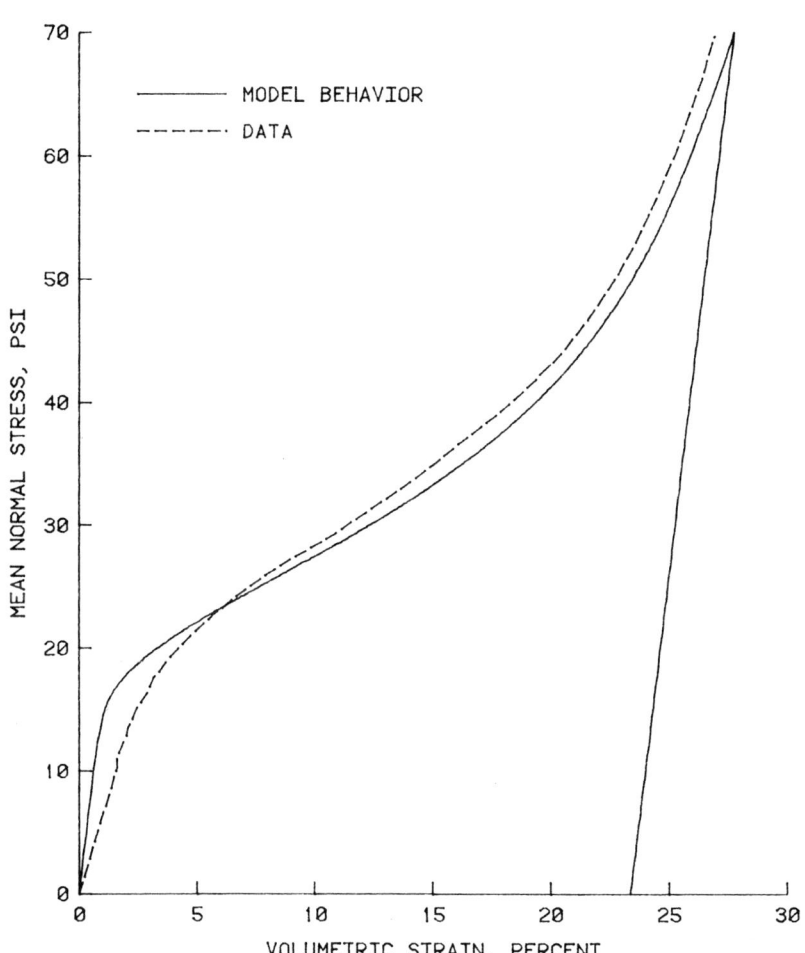

FIG. 8.—Isotropic Consolidation Response, Laboratory Measurements Versus Model Behavior

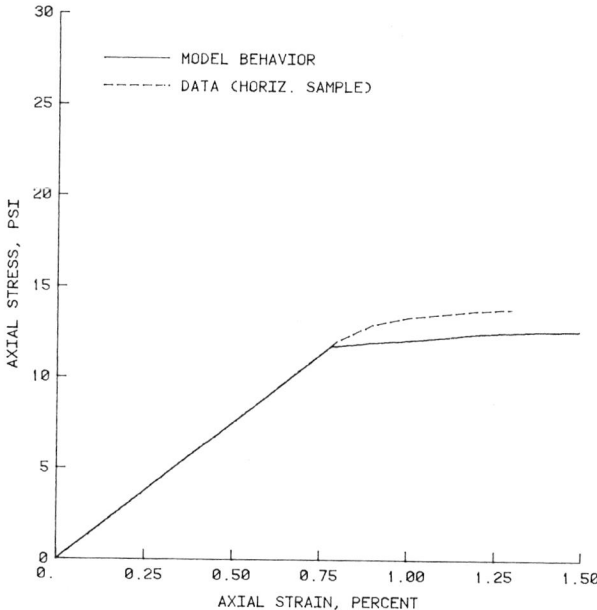

FIG. 9.—Unconfined Compression Stress-Strain Curves, Laboratory Measurements Versus Model Behavior, Clay X

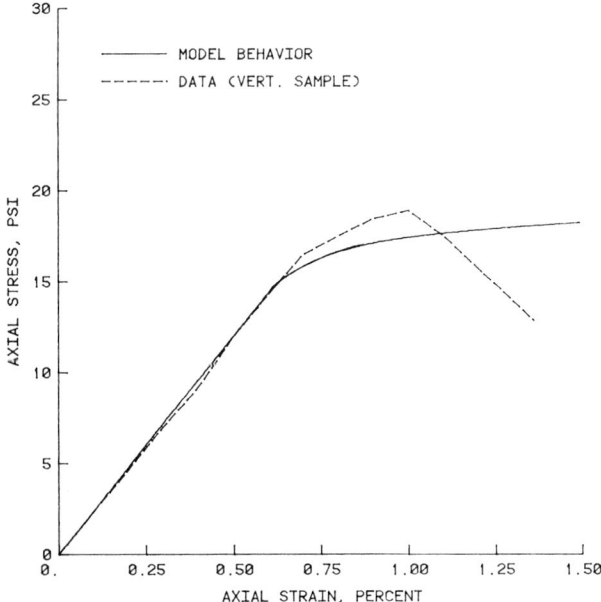

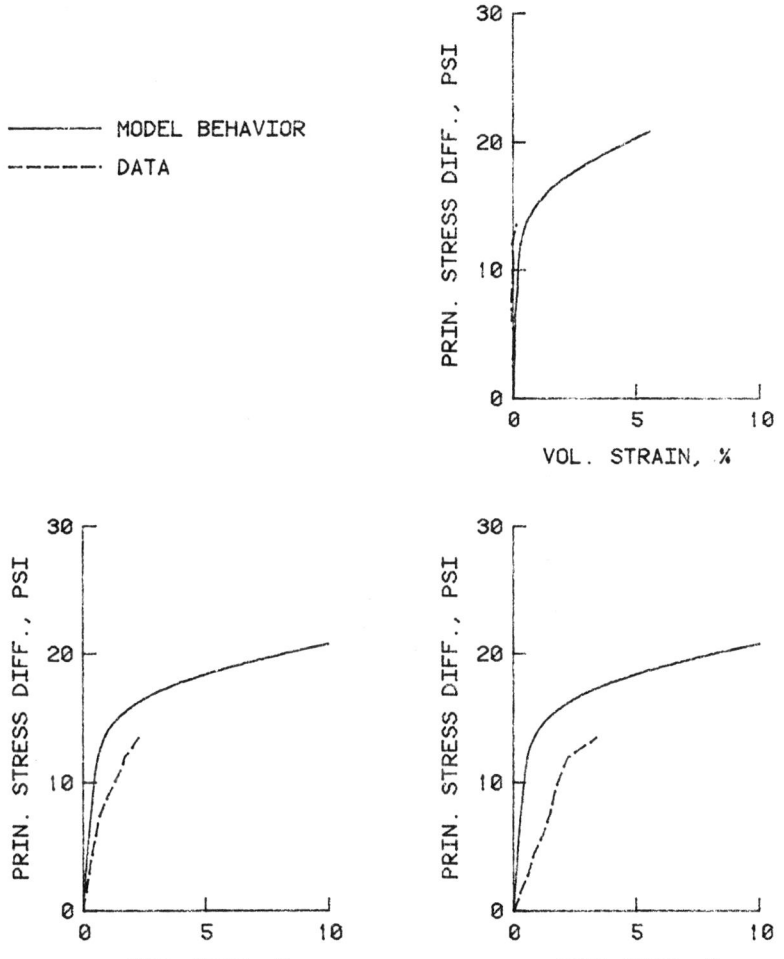

FIG. 11.—Principal Stress Difference Versus Volumetric Strain and Principal Strain Difference, Laboratory Measurements Versus Model Behavior

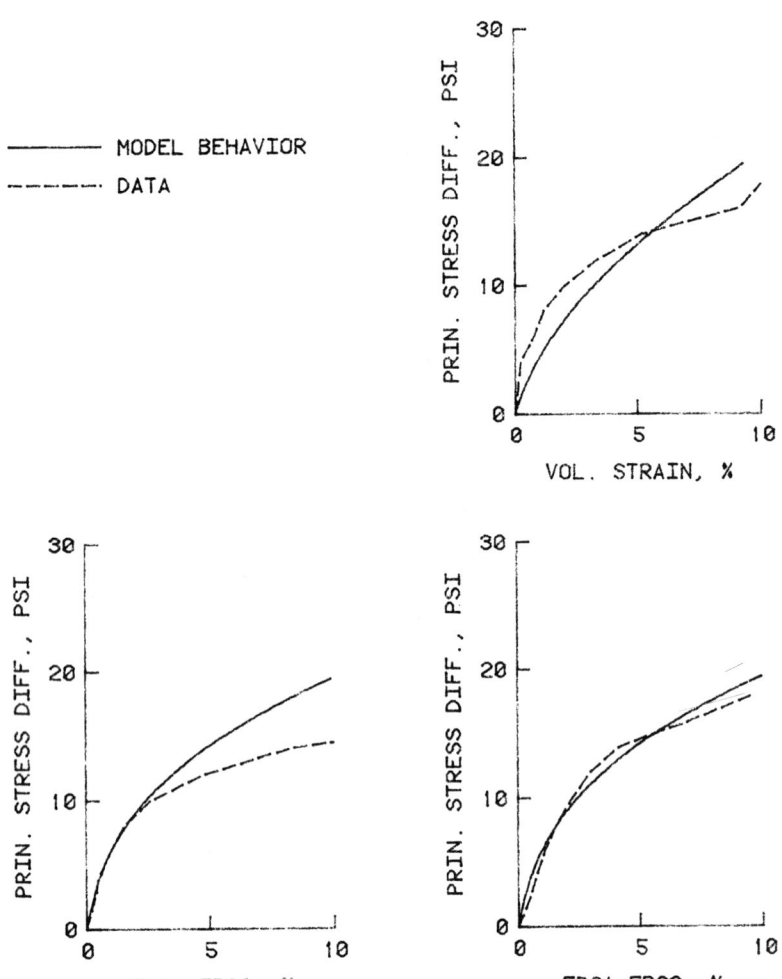

FIG. 12.—Principal Stress Difference Versus Volumetric Strain and Principal Strain Difference, Laboratory Measurements Versus Model Behavior

CONSTANT MEAN STRESS TEST(Pc), CLAY X
Pc = 30.0 PSI; Stress Ratio, m = 0

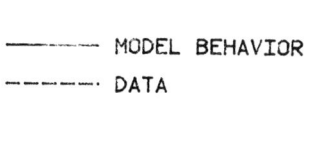
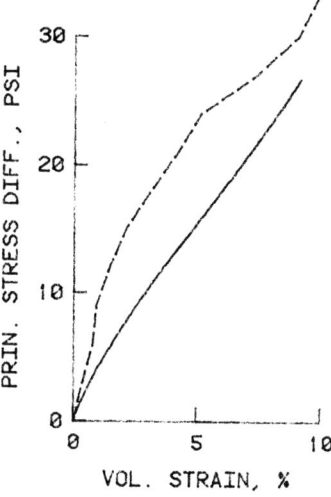
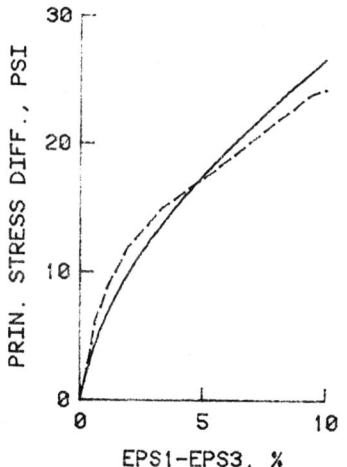
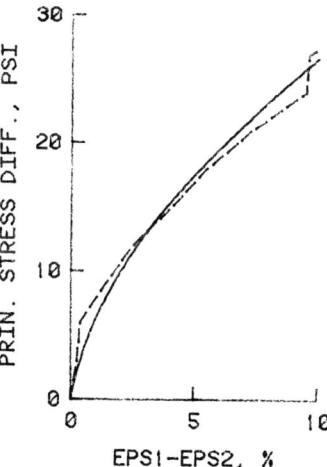

FIG. 13.—Principal Stress Difference Versus Volumetric Strain and Principal Strain Difference, Laboratory Measurements Versus Model Behavior

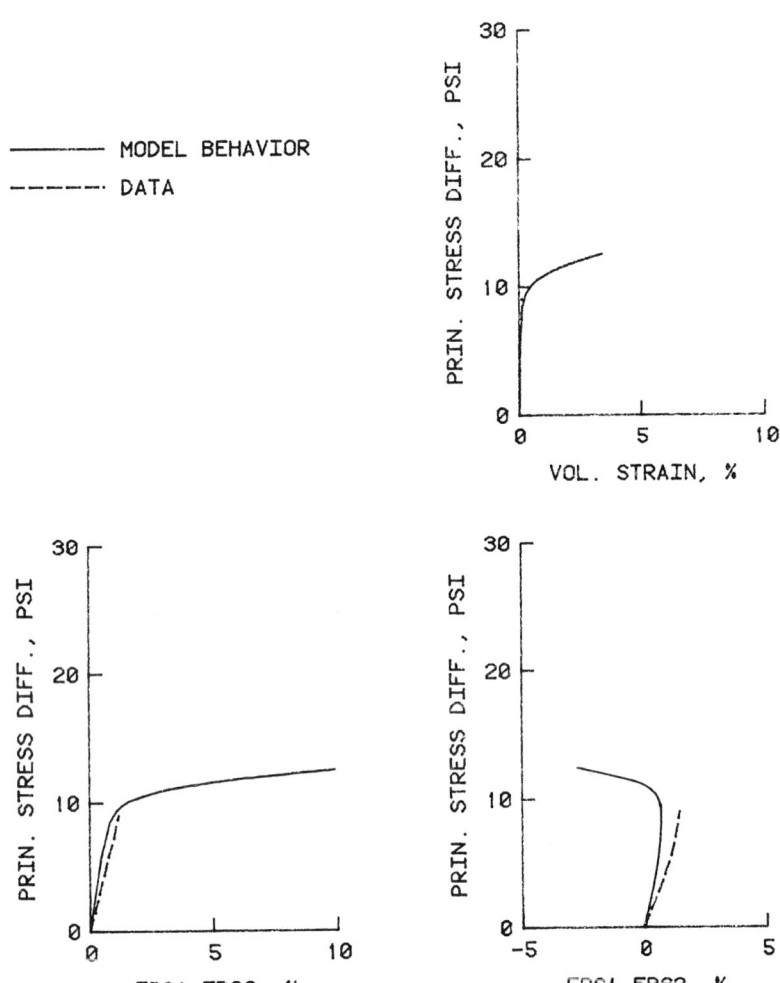

FIG. 14.—Principal Stress Difference Versus Volumetric Strain and Principal Strain Difference, Laboratory Measurements Versus Model Behavior

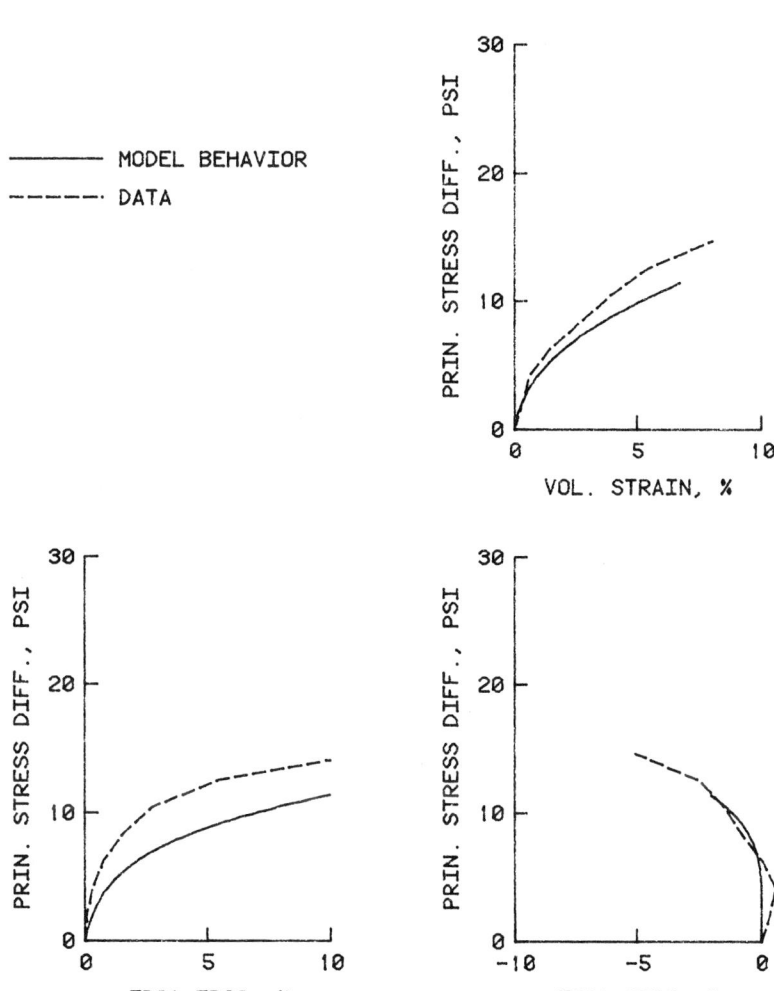

FIG. 15.—Principal Stress Difference Versus Volumetric Strain and Principal Strain Difference, Laboratory Measurements Versus Model Behavior

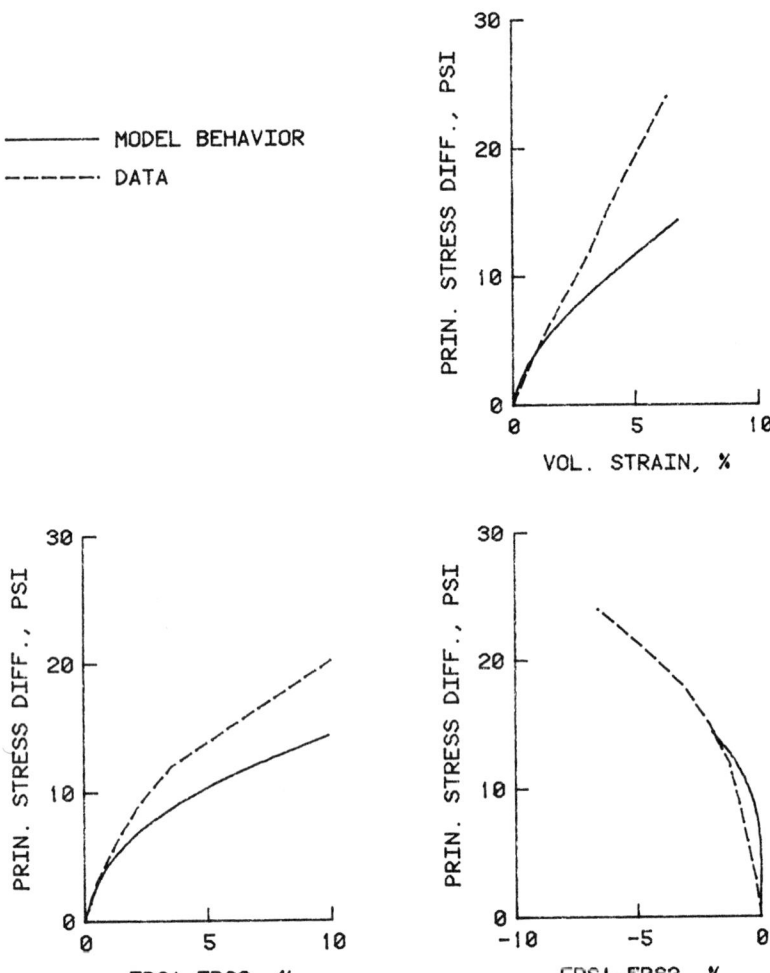

FIG. 16.—Principal Stress Difference Versus Volumetric Strain and Principal Strain Difference, Laboratory Measurements Versus Model Behavior

SIMULATING BEHAVIOR OF SOILS

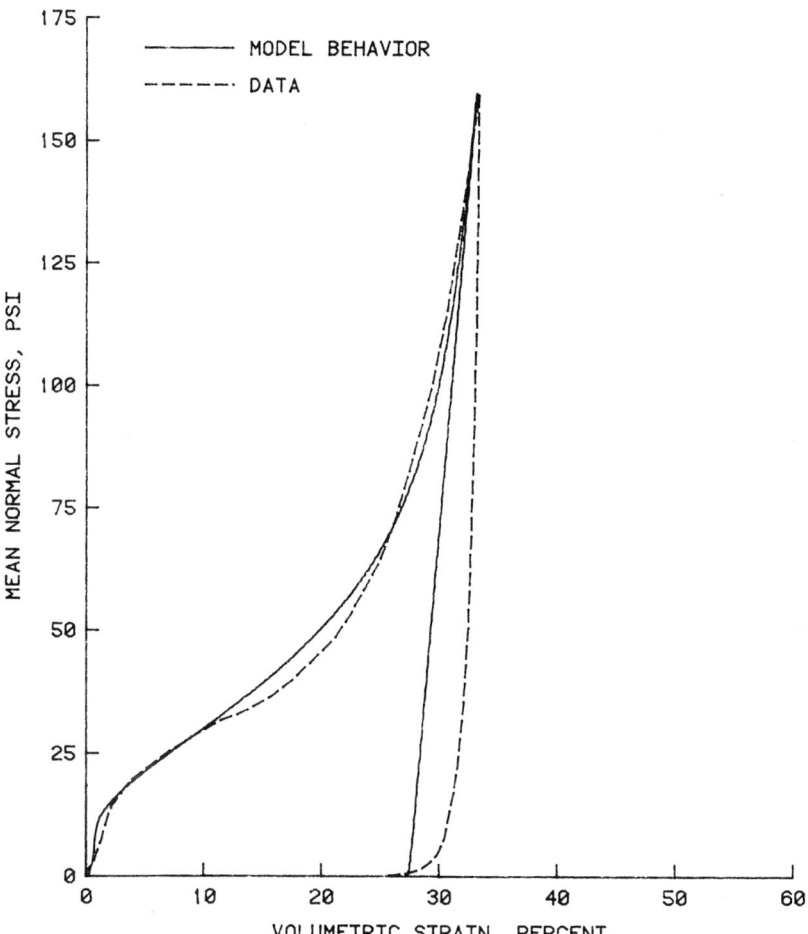

FIG. 17.--Isotropic Consolidation Response, Laboratory Measurements Versus Model Behavior

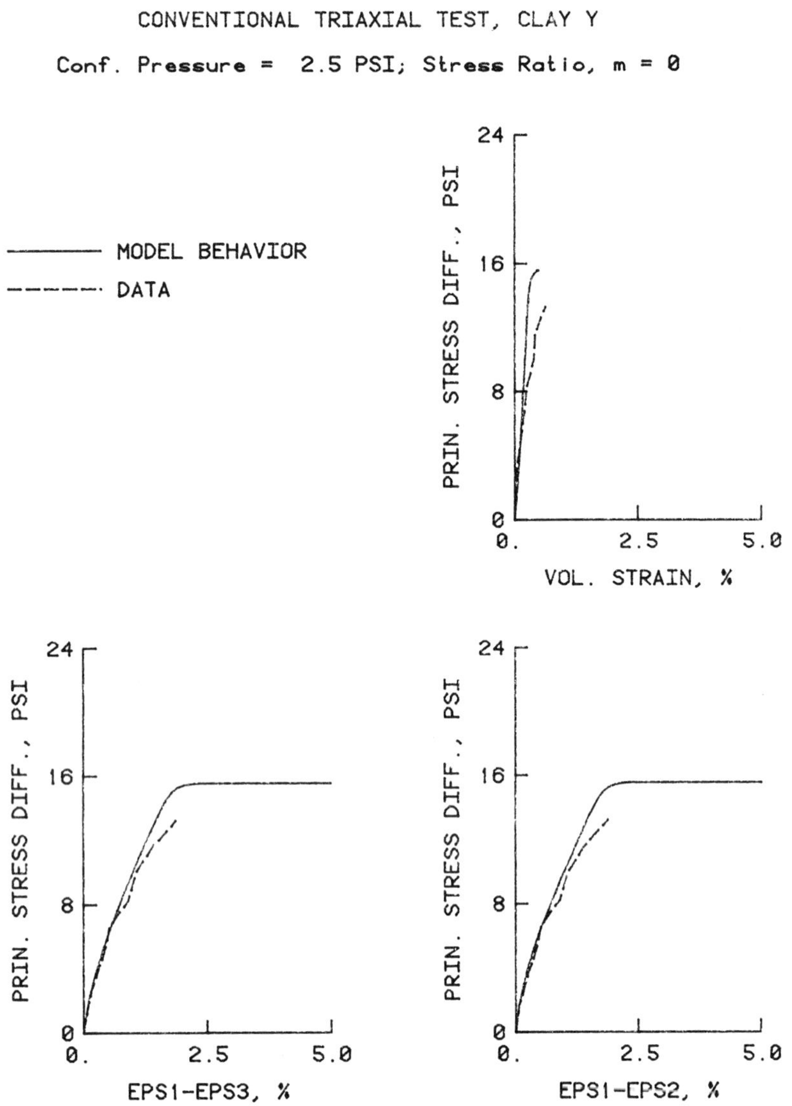

FIG. 18.—Principal Stress Difference Versus Volumetric Strain and Principal Strain Difference, Laboratory Measurements Versus Model Behavior

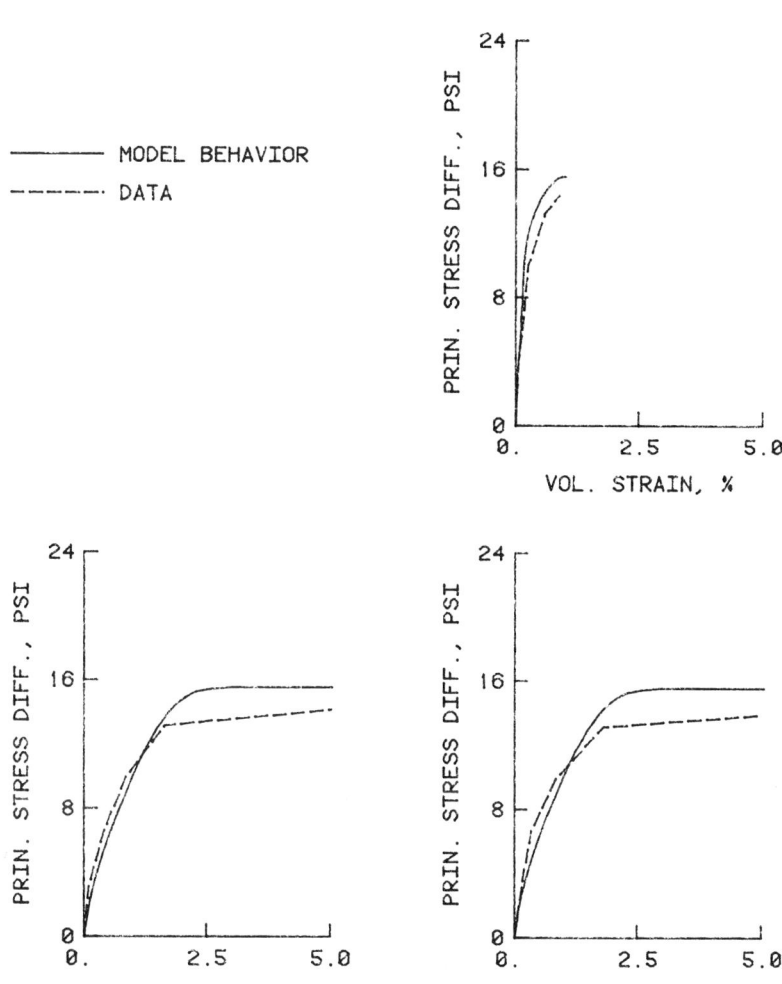

FIG. 19.—Principal Stress Difference Versus Volumetric Strain and Principal Strain Difference, Laboratory Measurements Versus Model Behavior

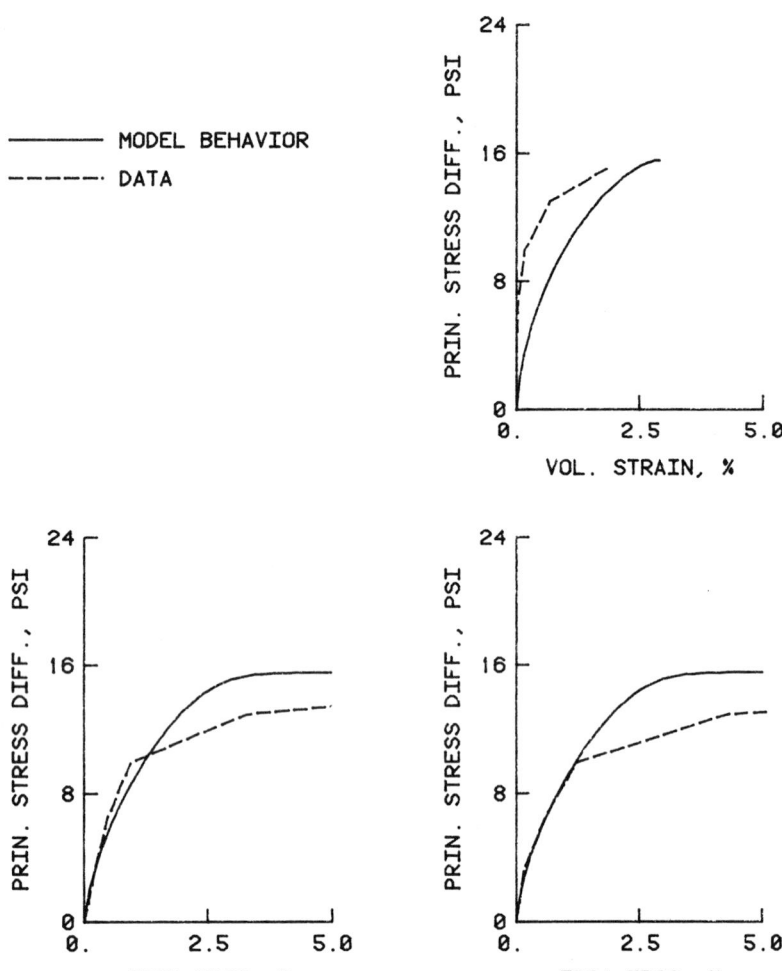

FIG. 20.—Principal Stress Difference Versus Volumetric Strain and Principal Strain Difference, Laboratory Measurements Versus Model Behavior

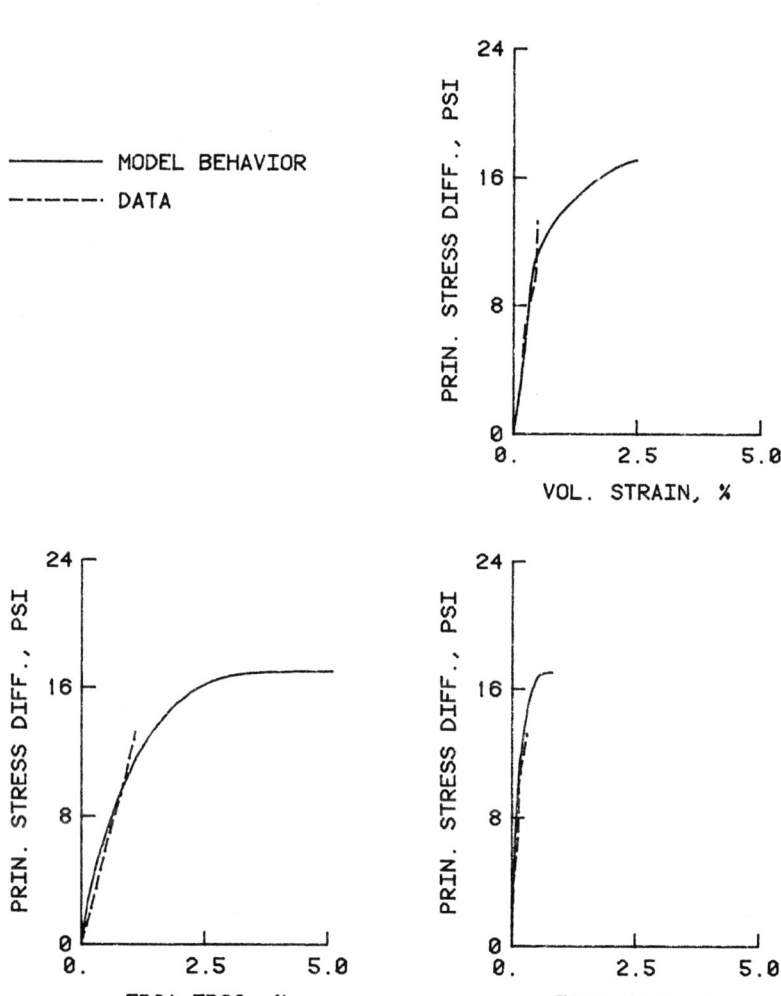

FIG. 21.—Principal Stress Difference Versus Volumetric Strain and Principal Strain Difference, Laboratory Measurements Versus Model Behavior

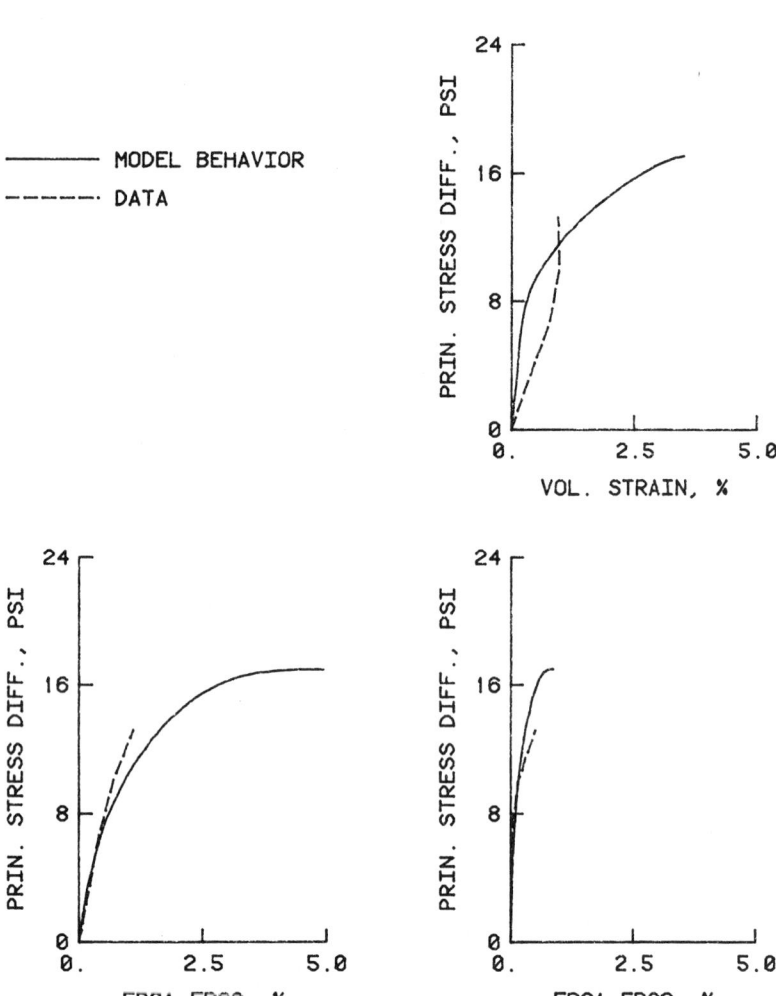

FIG. 22.—Principal Stress Difference Versus Volumetric Strain and Principal Strain Difference, Laboratory Measurements Versus Model Behavior

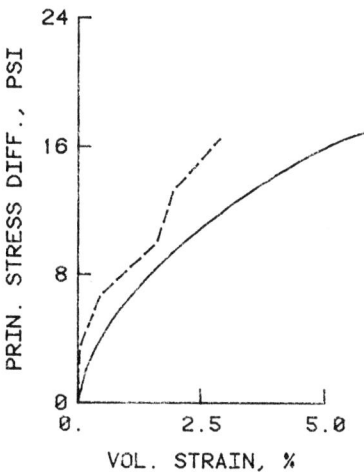
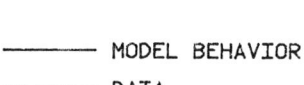
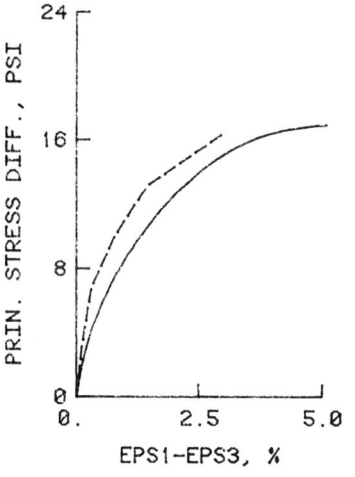
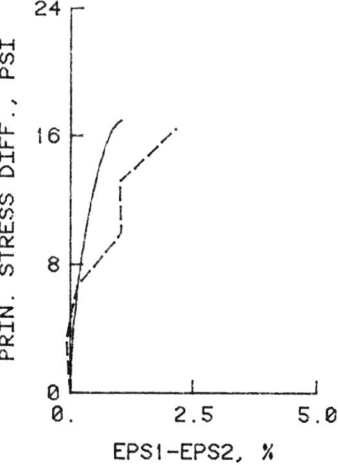

FIG. 23.—Principal Stress Difference Versus Volumetric Strain and Principal Strain Difference, Laboratory Measurements Versus Model Behavior

PART II: A KINEMATICALLY HARDENING CONSTITUTIVE MODEL FOR REMOLDED CLAY (MATERIAL B)

By Ivan S. Sandler

This example deals with a laboratory-prepared kaolinite clay whose behavior for undrained triaxial loading at various angles is required. Because the kaolinite was K_o-consolidated before the triaxial testing, and because the subsequent triaxial tests are known to exhibit substantial anisotropy, a kinematically hardening cap model (Sandler and Baron [1]) where the cap has an initial offset in the direction of vertical compression was assumed. In addition, the material is assumed incompressible, i.e., the elastic and plastic volumetric strains are equal to each other in magnitude but of opposite sign.

To illustrate how the cap model may be simplified substantially if desired, a 9-parameter model was developed which could be run on a hand-held programmable calculator, the Texas Instruments TI-59. All of the model fits and predictions in this section were obtained by hand using only this calculator.

The Model

The behavior of the current model in the elastic range is governed by the bulk modulus K and the shear modulus G. These moduli are assumed constant here.

The plastic behavior of the model is governed by a cap and an ideal failure envelope and by a normal flow rule. The failure envelope is $\sqrt{J_2} = A\, J_1$ (representing a cone in stress space).

The elliptical cap is described by the equation

$$Q_1^2 + R^2 Q_2 = X \quad\quad\quad\quad (1)$$

in which R is the cap shape parameter (assumed constant) and

$$Q_1 = \sigma'_{kk} - \alpha_o \quad\quad\quad\quad (2)$$

$$Q_2 = \frac{1}{2} (s_{ij} - \alpha_{ij})^2 \quad\quad\quad\quad (3)$$

$$X = k + h\, \varepsilon_v^p \quad\quad\quad\quad (4)$$

where σ_{ij}' is the effective stress tensor and s_{ij} its deviatoric part, α_o and α_{ij} are the volumetric and shear hardening parameters defined below, ε_v^p is plastic volumetric strain, and k and h are material constants.

The kinematic hardening of the cap is defined by

$$\alpha_o = \alpha_o^o + \beta \varepsilon_v^p \qquad (5)$$

$$\alpha_{ij} = \alpha_{ij}^o + \beta e_{ij}^p \qquad (6)$$

in which α_o^o and β are constants, e_{ij}^p is the deviatoric part of the plastic strain, and α_{ij}^o is defined through the material constant γ as the tensor

$$\alpha_{ij}^o = \begin{bmatrix} \gamma & 0 & 0 \\ 0 & -\gamma/2 & 0 \\ 0 & 0 & -\gamma/2 \end{bmatrix} \qquad (7)$$

In summary there are nine material constants in this model which must be determined from experimental data. These are K, G, A, R, h, α_o^o, β, and γ.

Comparisons of Laboratory Test Data with Model Predictions

Experimental Program.--The material used in the experimental program was a remolded kaolinite clay for which data for consolidation and rebound as well as a number of undrained triaxial tests were available. Using the model, it is required to predict the stress-strain and pore pressure response for the clay along different stress paths.

Material Constants.--The material constants h and K were estimated from the average loading and unloading bulk moduli in the consolidation tests. G was determined from the low-stress, nearly linear, shear response in the undrained triaxial tests and A was determined from the failure (large strain) states observed in those tests. The values for these parameters are

$$K = 2500 \text{ psi} \qquad G = 1500 \text{ psi}$$
$$h = 2 \times 10^6 \text{ psi} \qquad A = 0.25$$

The remaining five parameters were obtained by trial and error through comparison of the model behavior with the measured results. The parameters thereby determined are

$$R = 4.5 \qquad \alpha_o^o = 95 \text{ psi} \qquad \gamma = 14 \text{ psi}$$
$$k = 8{,}000 \text{ psi} \qquad \beta = 100 \text{ psi}$$

Correlation with Test Results.--The comparisons of the model behavior with the given data are shown in Figures 1 through 5. In these figures Test No. 1 is a conventional triaxial test, Test No. 4 is a compression test with constant mean stress, Test No. 10 is an extension test with constant mean stress, and Test No. 13 is a conventional extension test.

The model predictions and comparisons with actual data are presented in Volume I of the NSF/NSERC Workshop Proceedings.

Appendix.--References

1. Sandler, I. S. and Baron, M. L., "Recent Developments in the Constitutive Modeling of Geological Materials," Proc. 3rd Intl. Conf. Num. Meth. Geomech., Univ. of Aachen, Aachen, W. Germany, 1979.

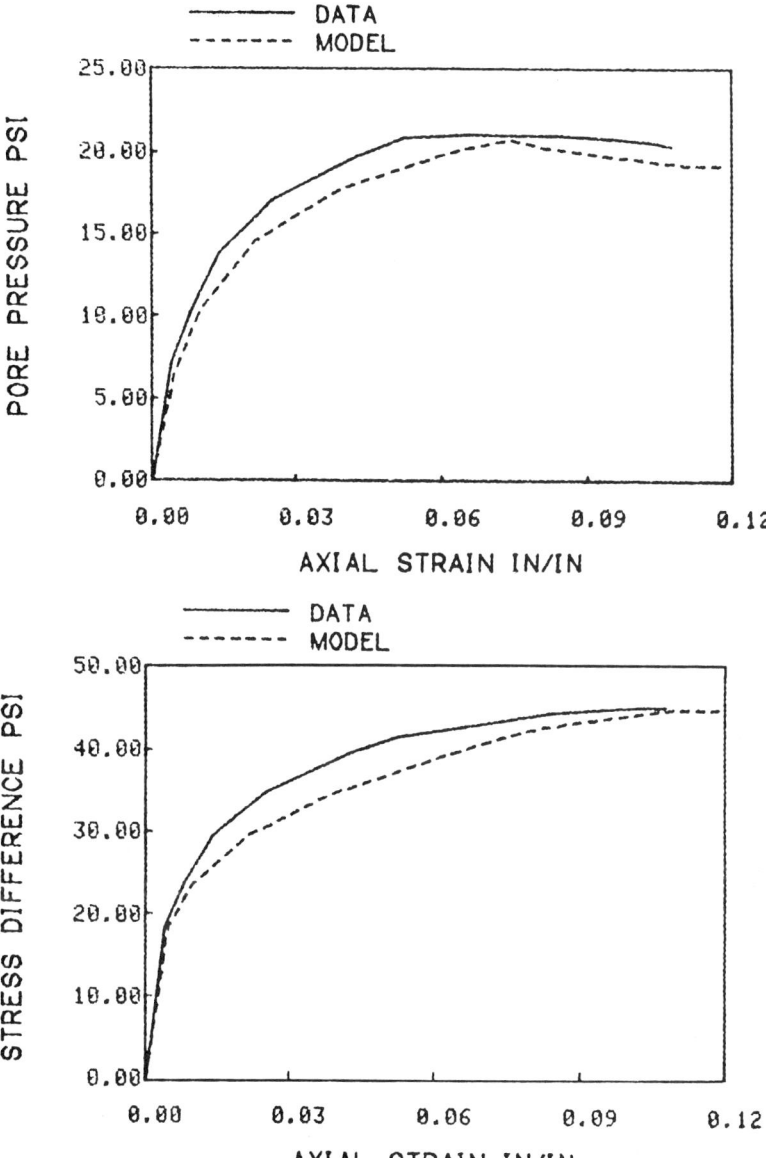

FIG. 1.—Comparison of Model Behavior with Laboratory Data for Material B in a Conventional Triaxial Test on a Horizontal Specimen

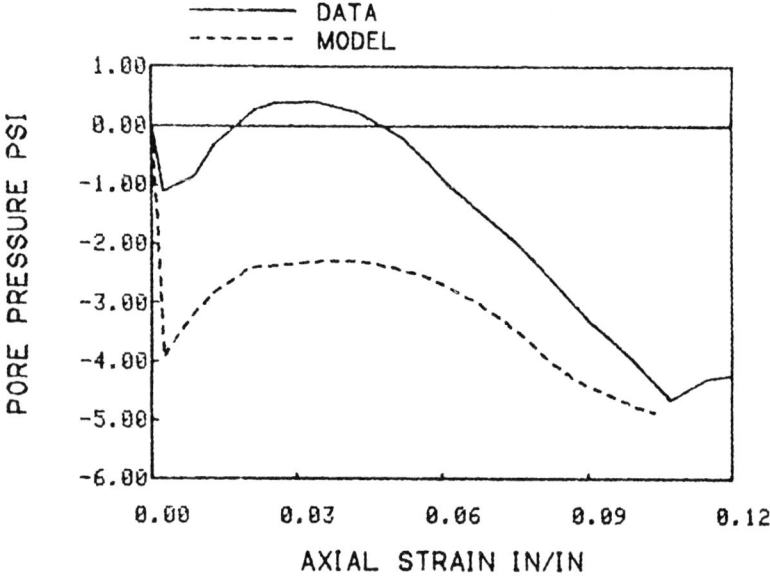

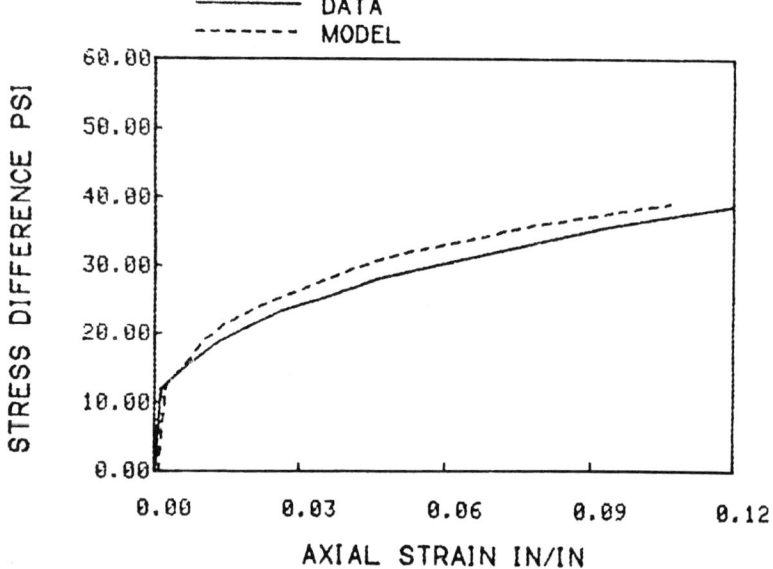

FIG. 2.—Comparison of Model Behavior with Laboratory Data for Material B, Test No. 13

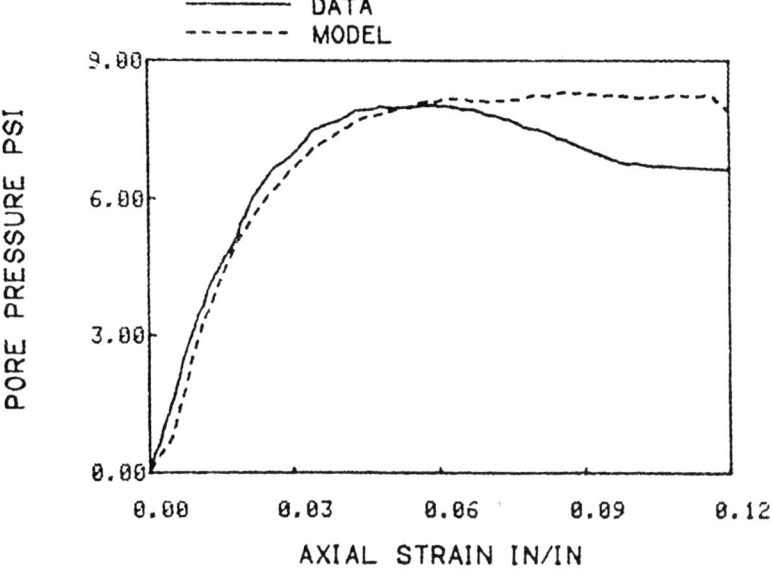
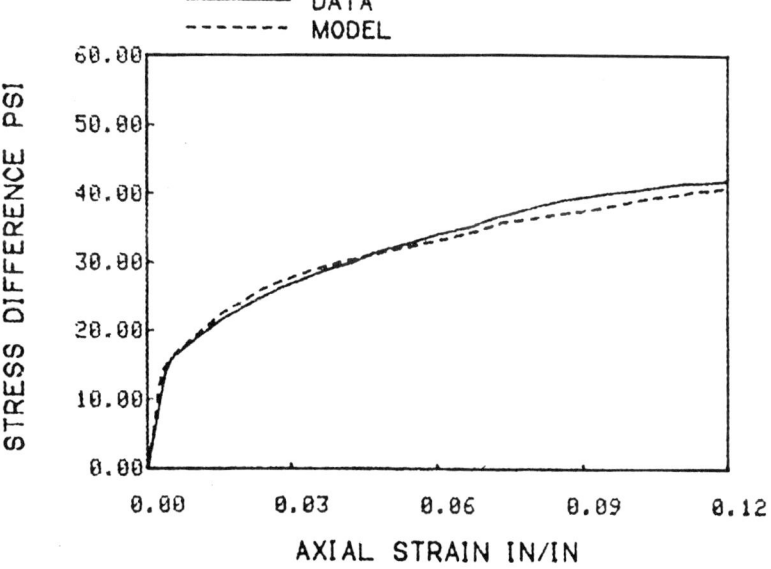

FIG. 3.—Comparison of Model Behavior with Laboratory Data for Material B, Test No. 10

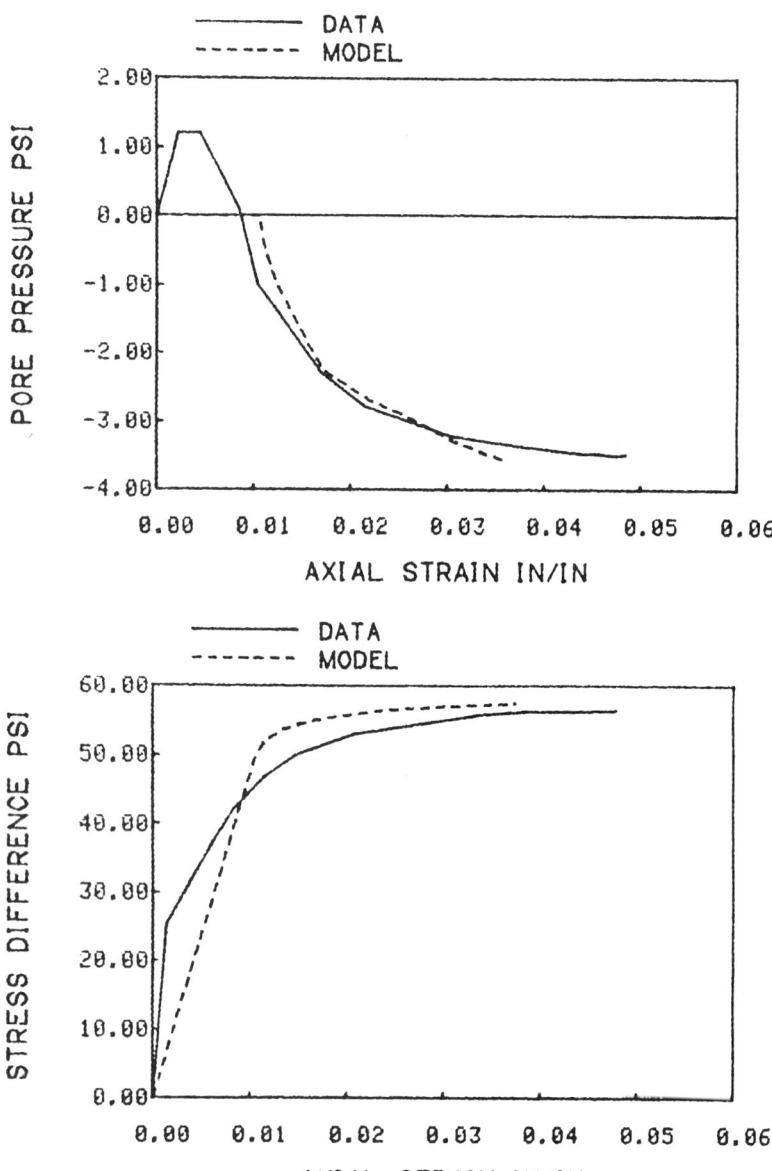

FIG. 4.—Comparison of Model Behavior with Laboratory Data for Material B, Test No. 4

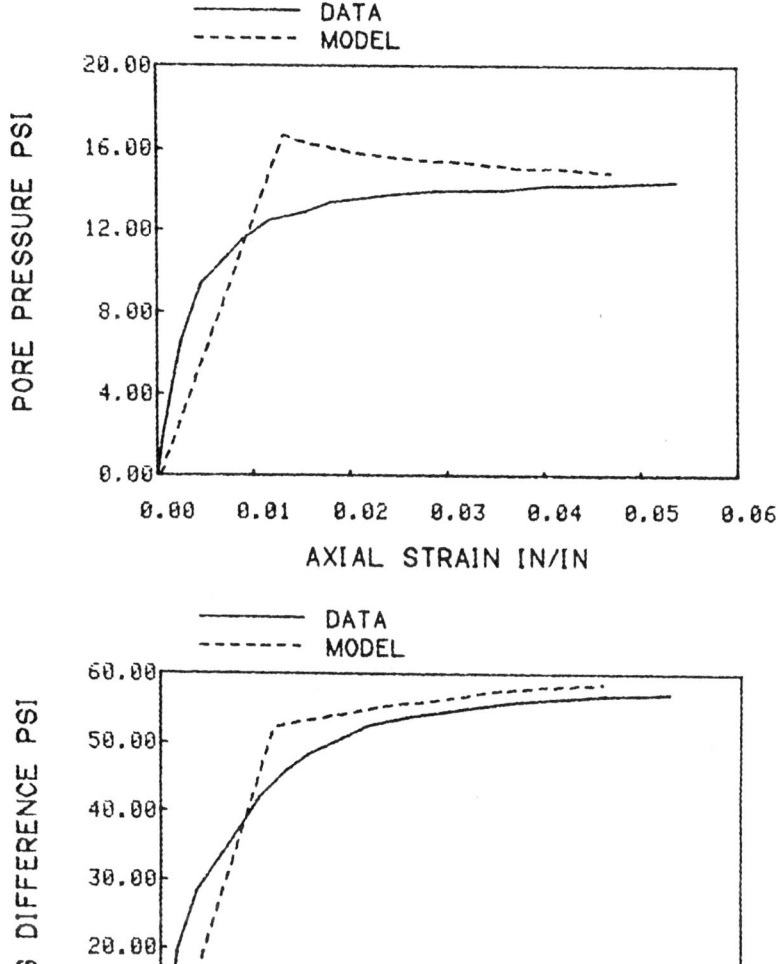

FIG. 5.—Comparison of Model Behavior with Laboratory Data for Material B, Test No. 1

PART III: AN ELASTIC-PLASTIC ISOTROPIC CONSTITUTIVE MODEL FOR SAND (MATERIAL C)
By George Y. Baladi

The material studied in this section is a dry Ottawa sand and would therefore be expected to be most closely modeled by the basic cap model (Sandler, DiMaggio, and Baladi [1], Baladi [2]). Perusal of these data indicates this to be true, with the addition of work hardening to the usually ideally plastic failure envelope. The total strain increment is assumed to be the sum of the elastic and plastic strain increments; i.e.,

$$d\varepsilon_{ij} = d\varepsilon_{ij}^E + d\varepsilon_{ij}^P \quad \quad (1)$$

where $d\varepsilon_{ij}$ = components of the total strain increment tensor; $d\varepsilon_{ij}^E$ = components of the elastic strain increment tensor; and $d\varepsilon_{ij}^P$ = components of the plastic strain increment tensor.

Elastic Behavior.--The behavior of the model in the elastic range is governed by the following incremental equations:

$$dJ_1 = 3K \, d\varepsilon_{kk}^E \quad \quad (2)$$

$$dS_{ij} = 2G \, de_{ij}^E \quad \quad (3)$$

in which dJ_1 = the first invariant of the stress increment tensor; $d\varepsilon_{kk}^E$ = the elastic volumetric strain increment; dS_{ij} and de_{ij} = the stress and strain deviation increment tensors, respectively; and K and G = the elastic bulk and shear moduli, respectively.

The elastic bulk modulus describes the unloading portion of the pressure-volumetric strain response of the material. The following expression is proposed for the elastic bulk modulus:

$$K = \frac{K_i}{1 - K_1} \left[1 - K_1 \exp(-K_2 J_1) \right] \quad \quad (4)$$

where K_i = initial elastic bulk modulus and K_1 and K_2 = material constants.

The elastic shear modulus describes the unloading portion of the deviatoric stress-strain response of the material. The following expression is proposed for the elastic shear modulus:

$$G = G_i + G_1 [1 - \exp(-G_2 \, W^{Pf})] \quad \quad (5)$$

where G_i = initial elastic shear modulus; G_1 and G_2 = material constants; and W^{Pf} = the plastic work associated with the yield envelope.

Plastic Behavior.—The behavior of the model in the plastic range is governed by the flow rule

$$d\varepsilon_{ij}^P = d\lambda \frac{\partial \mathcal{f}}{\partial \sigma_{ij}} \quad \quad (6)$$

in which $d\varepsilon_{ij}^P$ and σ_{ij} = the plastic strain increment and the stress tensor, respectively; $\mathcal{f}$ = the loading function; and $d\lambda$ = a positive scalar factor of proportionality, which is nonzero only when plastic deformation occurs and is dependent on the particular form of the loading function $\mathcal{f}$.

The loading function $\mathcal{f}$ is assumed to be isotropic and to consist of two parts: an elliptically shaped strain-hardening yield surface described by

$$F(J_1, \sqrt{J_2}, \kappa) = (J_1 + C)^2 + 6\bar{J}_2 - \left[X(\kappa) + C\right]^2 = 0 \quad \quad (7)$$

and a Prager-Drucker type yield envelope defined by

$$f(J_1, \sqrt{J_2}, \alpha) = \frac{\sqrt{\bar{J}_2}}{J_1 + C} - \alpha = 0 \quad \quad (8)$$

where $\bar{J}_2$ = the second invariant of the stress deviation tensor; C is a material constant representing the location of the center of the strain-hardening yield surface F and coincides with the intersection of the yield envelope f and the J_1 axis; κ is a hardening parameter which is generally a function of the history of plastic volumetric strain; $X(\kappa)$ represents the intersection of the strain-hardening surface with the J_1 axis; and α is a work-hardening and -softening parameter which allows the yield envelope f to expand and contract isotropically (Figure 1).

The hardening parameter κ in this model was chosen to be

$$\kappa = \varepsilon_{kk}^{PH} = A[1 - \exp(-DX)] \quad \quad (9)$$

where D and A = material constants and ε_{kk}^{PH} is the plastic volumetric strain associated with the strain-hardening yield surface F. The parameter A defines the maximum plastic volumetric compaction that the material can experience under hydrostatic loading. Note that the strain-hardening surface F is chosen so that the plastic strain increment vector associated with it is always parallel to the yield envelope f.

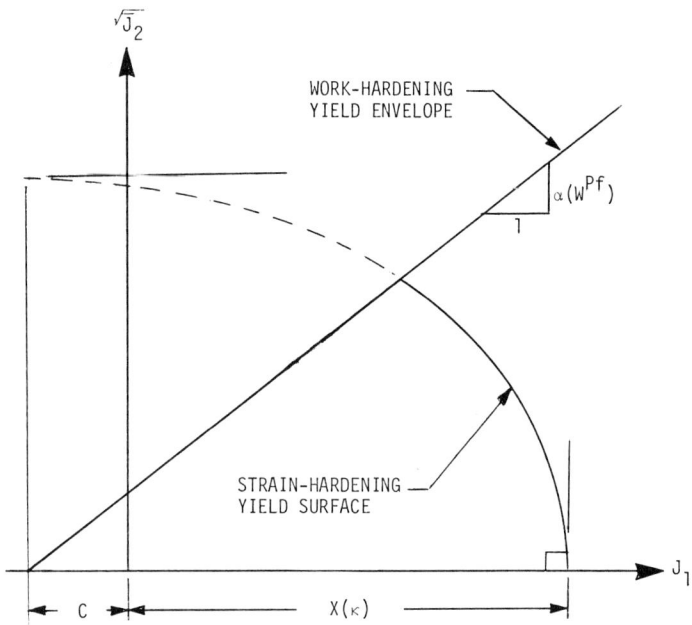

FIG. 1.—Proposed Yield Surfaces for Strain-hardening and -softening Model

The parameter α in Eq. 8 is assumed to be a function of plastic work W^{Pf} associated with the yield envelope f through the following expression:

$$\alpha = C_1[1 - \exp(-C_2 \, W^{Pf})] + C_3[1 - \exp(-C_4 \, W^{Pf})]$$
$$+ C_5[1 - \exp(-C_6 \, W^{Pf})] + \alpha_1 \quad \ldots \ldots \ldots \ldots \ldots \quad (10)$$

where C_2, C_4, and C_6 = material constants, and α_1 = initial value of α, and C_1, C_3, and C_5 are functions of stress ratio through the following expressions:

$$C_1 = (1 - m) \, C_{1max} + m C_{1min} \quad \ldots \ldots \ldots \ldots \ldots \ldots \quad (11)$$

$$C_3 = (1 - m) \, C_{3max} + m C_{3min} \quad \ldots \ldots \ldots \ldots \ldots \ldots \quad (12)$$

$$C_5 = (1 - m) C_{5max} + mC_{5min} \quad \ldots \ldots \ldots \ldots \ldots \ldots \ldots \quad (13)$$

$$\alpha_1 = (1 - m) \alpha_{1max} + m\alpha_{1min} \quad \ldots \ldots \ldots \ldots \ldots \ldots \ldots \quad (14)$$

where C_{1max}, C_{1min}, C_{3max}, C_{3min}, C_{5max}, C_{5min}, α_{1max}, and α_{1min} = material constants. The proportionality factor $d\lambda$ (Eq. 6) can be shown to be (Baladi and Rohani [3])

$$d\lambda = \frac{3K \frac{\partial F}{\partial J_1} (d\varepsilon_{kk} - d\varepsilon_{kk}^{Pf}) + \frac{G}{\sqrt{J_2}} \frac{\partial F}{\partial \sqrt{J_2}} S_{ij}(d\varepsilon_{ij} - d\varepsilon_{ij}^{Pf})}{9K \frac{\partial F}{\partial J_1}^2 + G \left(\frac{\partial F}{\partial \sqrt{J_2}}\right)^2 - 3 \frac{\partial F}{\partial J_1} \frac{\partial F}{\partial \varepsilon_{kk}^{PH}}} \quad \ldots \ldots \ldots \quad (15)$$

for the strain-hardening surface. In the case of the yield envelope f (Eq. 8), due to the hardening parameter α (Eq. 10), it is more convenient to express $d\lambda$ in terms of plastic work. Thus

$$d\lambda = \frac{dW^{Pf}}{\frac{\partial f}{\partial \sigma_{ij}} \sigma_{ij}} \quad \ldots \ldots \ldots \ldots \ldots \ldots \ldots \ldots \ldots \quad (16)$$

for the yield envelope f, where $d\varepsilon_{kk}^{Pf}$ and $d\varepsilon_{ij}^{Pf}$ are, respectively, the volumetric and deviatoric components of plastic strain increment tensor associated with f. The total strain increment tensor given by Eq. 1 can, therefore, be written as

$$d\varepsilon_{ij} = d\varepsilon_{ij}^{E} + d\varepsilon_{ij}^{PH} + d\varepsilon_{ij}^{Pf} \quad \ldots \ldots \ldots \ldots \ldots \ldots \ldots \quad (17)$$

where $d\varepsilon_{ij}$, $d\varepsilon_{ij}^{E}$, $d\varepsilon_{ij}^{PH}$, and $d\varepsilon_{ij}^{Pf}$ = the total strain, the elastic strain, the plastic strain associated with the hardening function, and plastic strain associated with the yield condition, respectively.

In summary, there are 20 material constants that must be determined experimentally. The material constants K_i, K_1, K_2, A, and D describe the complete hydrostatic response of the material (loading, unloading, and reloading). The material constants G_i, G_1, and G_2 describe the shearing response of the material in the elastic range (i.e., during unloading). The material constants C_{1max}, C_{1min}, C_2, C_{3max}, C_{3min}, C_4, C_{5max}, C_{5min}, C_6, α_{1max}, α_{1min}, and C describe the yield condition of the material and the rate of its expansion with plastic work. These constants can be readily determined from standard triaxial compression (with radial strain measurements) and isotropic consolidation tests.

Comparisons of Laboratory Test Data with Model Behavior

Experimental Program.--The material used in the experimental program was an Ottawa sand whose gradation is given in Figure 2. The sand was compacted by aerial pluviation to 1.73 gm/cc (relative density of 87 percent) and tested in the dry state three-dimensionally for its stress-strain properties. The stress paths shown in Figure 3 were used to generate the data base for formulating and fitting the constitutive model. Using the model, it is required to make predictions for the stress-strain response of the same soil when tested in the same test equipment along stress paths different from those used in the model formulation. The stress paths for prediction are shown in Figure 4.

The test equipment used is a flexible, fluid cushion, cubical device with stress control. The vertical axis, z, and the horizontal axes, x and y, are principal stress axes.

Material Constants.--As indicated above, there are 20 material constants in the present model that must be determined experimentally. The material constants K_i, K_1, and K_2 were determined from the unloading portion of the isotropic consolidation test. The material constants W and D were determined from the loading portion of the pressure-volumetric strain response of the material using K_i, K_1, and K_2. The material constants G_i, G_1, and G_2 were determined from the unloading portion of the deviatoric stress-strain response of the material. The remaining constants were determined from the yield condition of the material and the rate of its expansion. The numerical values of these material constants are given in Table 1.

Correlation with Test Results.--Figures 5 through 14 compare actual test results with model fits for Ottawa sand. Figure 5 compares effective mean normal stress versus volumetric strain for the isotropic compression test. Conventional triaxial compression test results are shown in Figures 6 and 7 for constant lateral stresses of 5 and 10 psi and stress ratio of 0. Figure 8 shows model behavior versus test results for a conventional triaxial extension test. Triaxial compression test results for constant mean normal stresses of 5, 10, and 20 psi and for stress ratios of 0 and 1 are shown in Figures 9 through 14.

The model predictions and comparisons with actual data are presented in Volume 1 of the NSF/NSERC Workshop Proceedings.

Appendix.—References

1. Sandler, I. S., DiMaggio, F. L., and Baladi, G. Y., "A Generalized Cap Model for Geological Materials," J. of Geotech. Engrg. Div., ASCE, Vol. 102, No. GT 7, July 1976.

2. Baladi, G. Y., "The Latest Development in the Nonlinear-Elastic-Nonideally Plastic Work Hardening Cap Model," Proc. of the Symp. on the Role of Plasticity in Soil Mech., Cambridge, England, September 1973.

3. Baladi, G. Y. and Rohani, B., "Liquefaction Potential of Dams and Foundation; Development of an Elastic-Plastic Constitutive Relationship for Saturated Sand," Research Report S-76-2, Report 3, 1977, U. S. Army Engineer Waterways Experiment Station, CE, Vicksburg, Miss.

Table 1.—Numerical Values of Material Constants for Ottawa Sand

Name	Notation	Units	Numerical Values
Hardening failure envelope parameter	C	psi	2
	C_{max}	---	0.075
	C_{1min}	---	0.05
	C_2	lb-in.$^{-1}$	4000
	C_{3max}	---	0.1
	C_{3min}	---	0.0667
	C_4	lb-in.$^{-1}$	800
	C_{5max}	---	0.085
	C_{5min}	---	0.0567
	C_6	lb-in.$^{-1}$	125
	α_{1max}	---	0.1
	α_{1min}	---	0.0667
Hardening surface parameter	W	---	0.0024
	D	psi^{-1}	0.002
Bulk modulus function	K_i	psi	37500
	K_1	---	0.75
	K_2	psi^{-1}	0.05
Shear modulus function	G_i	psi	5000
	G_1	psi	-3000
	G_2	lb-in.$^{-1}$	250

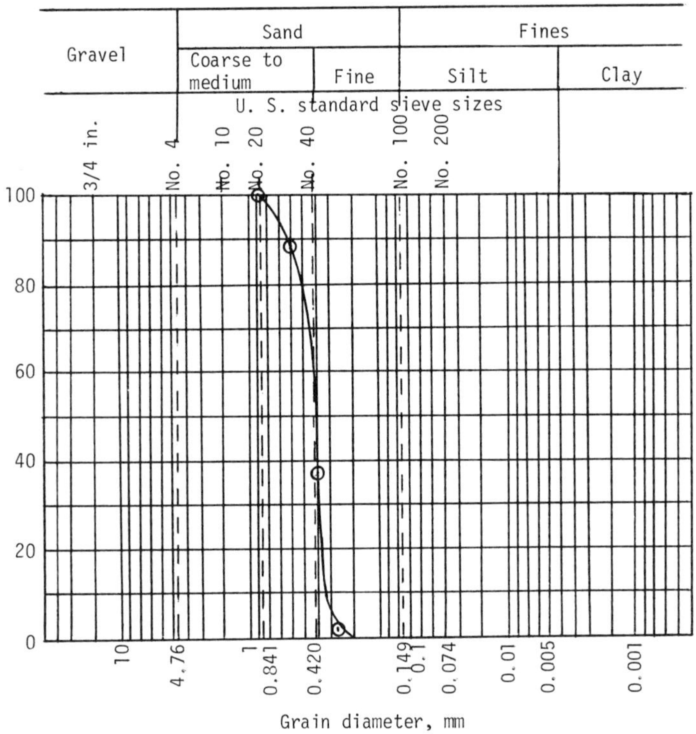

FIG. 2.—Grain Size Distribution for Ottawa Sand

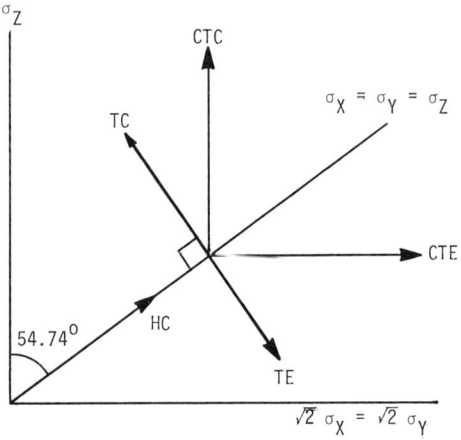

FIG. 3.—Stress Paths Used to Generate Data Base for Modeling

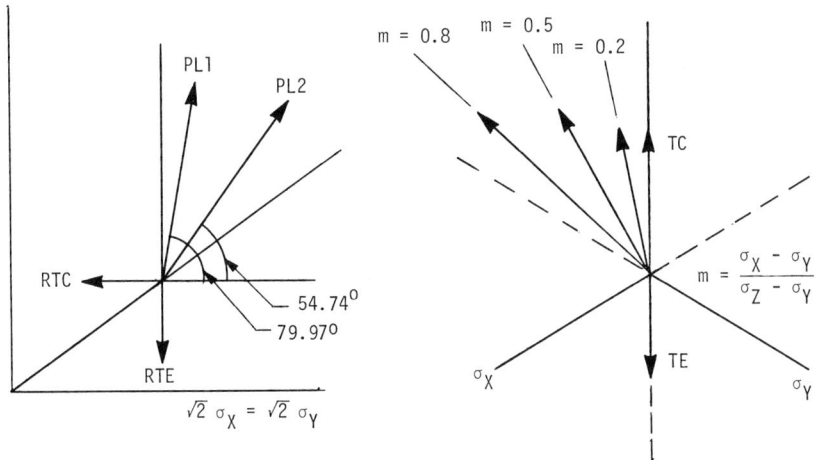
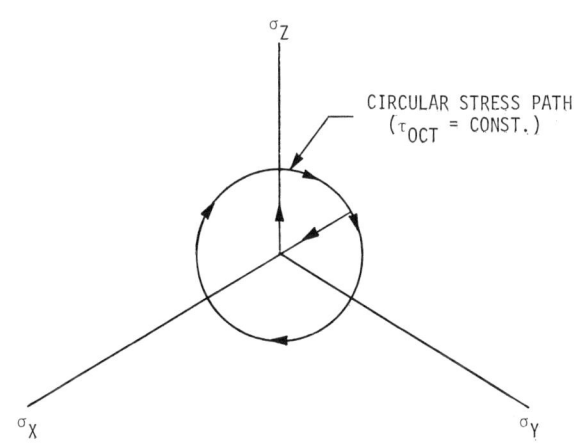

FIG. 4.—Stress Path for Predictions

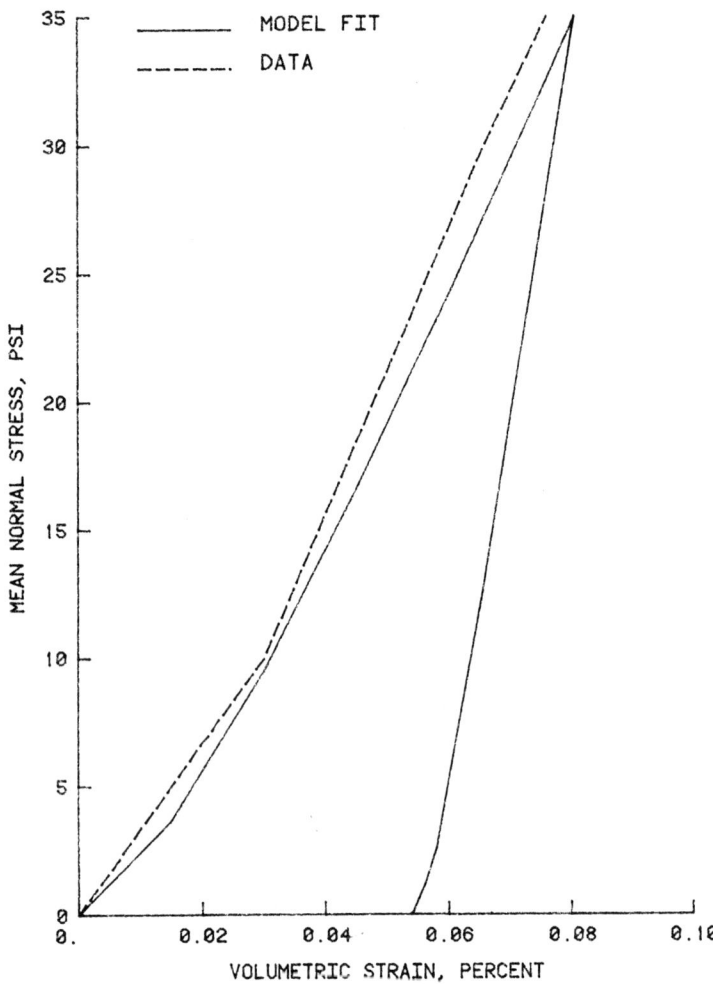

FIG. 5.—Isotropic Consolidation Response, Laboratory Measurements Versus Model Behavior

CONVENTIONAL TRIAXIAL COMPRESSION TEST (CTC) RESULTS (OTTAWA SAND)
Constant Lateral Stress Tests
Lateral Stress = 10 psi; Stress Ratio, m = 0

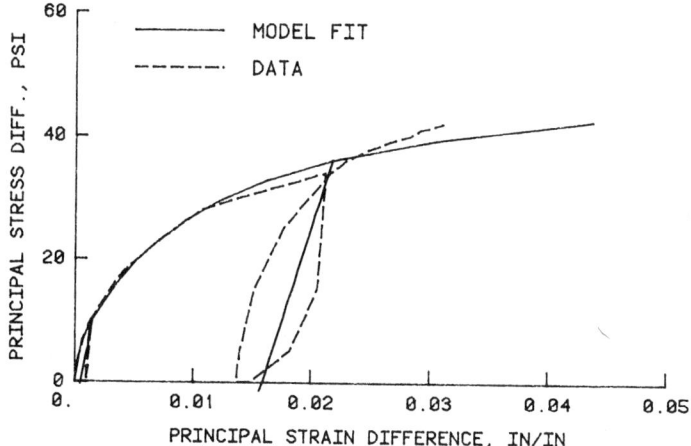

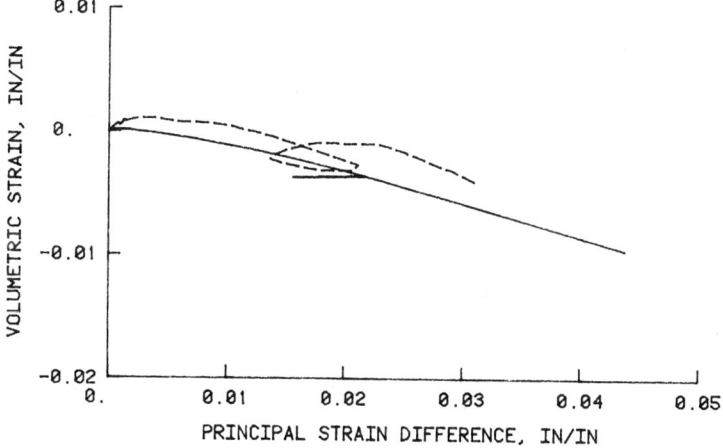

FIG. 6.—Principal Stress Difference and Volumetric Strain Versus Principal Strain Difference, Laboratory Measurements Versus Model Behavior

CONVENTIONAL TRIAXIAL COMPRESSION TEST (CTC) RESULTS (OTTAWA SAND)
Constant Lateral Stress Tests
Lateral Stress = 5 psi;   Stress Ratio, m = 0

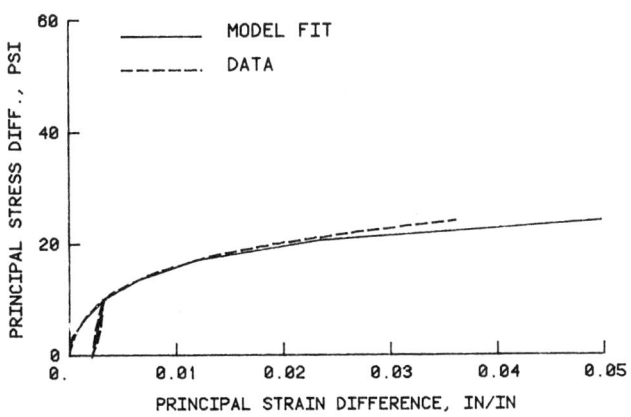

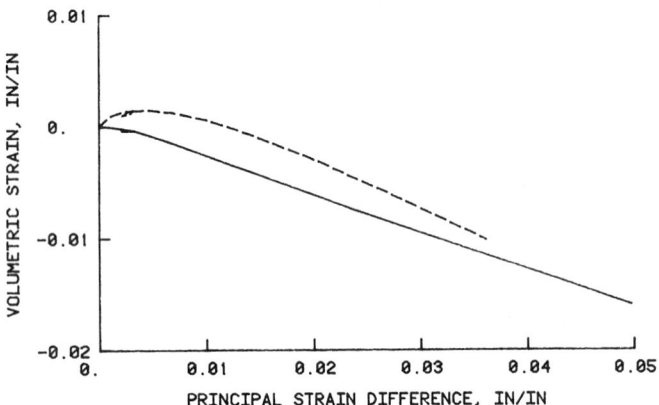

FIG. 7.--Principal Stress Difference and Volumetric Strain Versus Principal Strain Difference, Laboratory Measurements Versus Model Behavior

CONVENTIONAL TRIAXIAL EXTENSION TEST (CTE) RESULTS (OTTAWA SAND)
Stress Ratio, m = 1

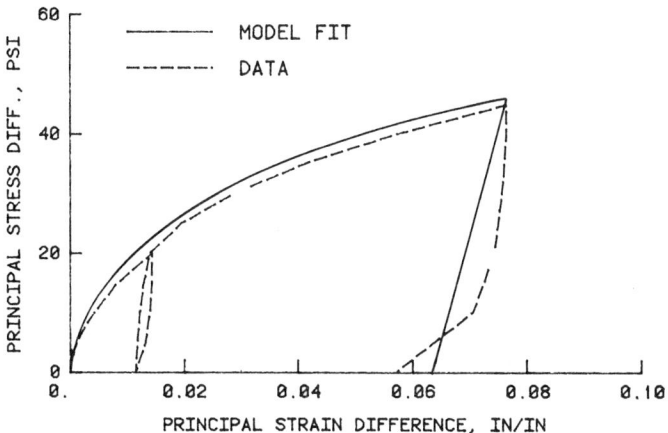

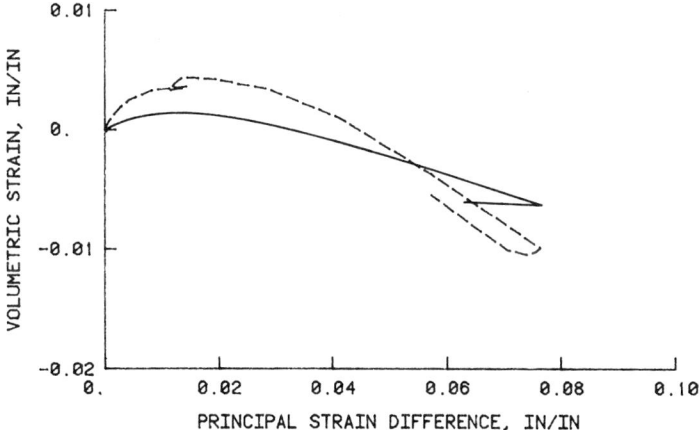

FIG. 8.—Principal Stress Difference and Volumetric Strain Versus Principal Strain Difference, Laboratory Measurements Versus Model Behavior

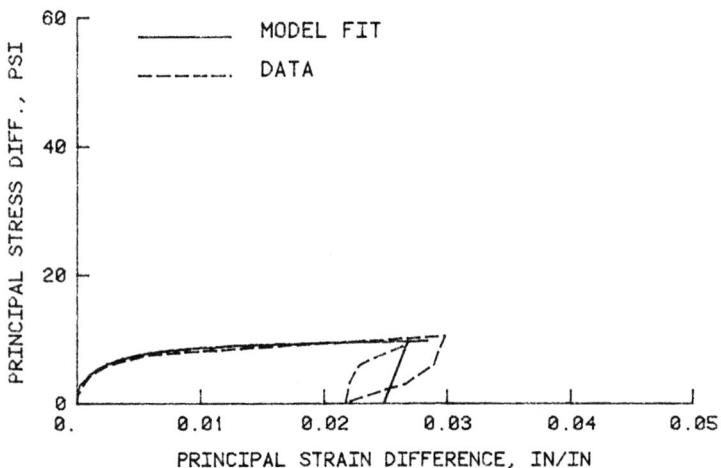
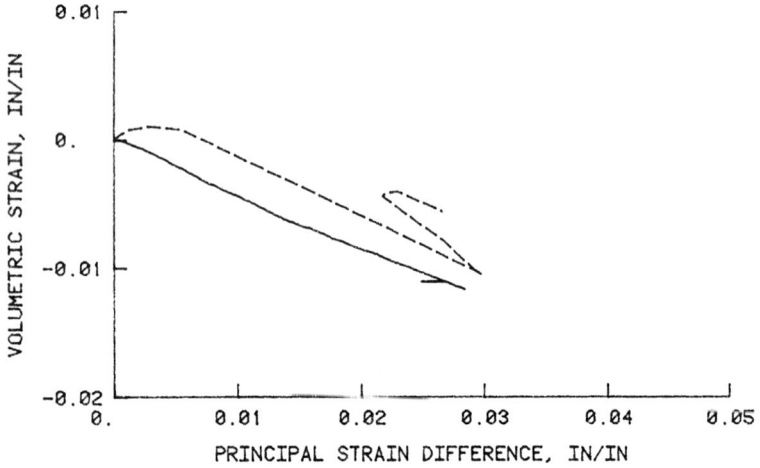

FIG. 9.—Principal Stress Difference and Volumetric Strain Versus Principal Strain Difference, Laboratory Measurements Versus Model Behavior

TRIAXIAL COMPRESSION TEST (TC) RESULTS (OTTAWA SAND)
Constant Mean Stress Test
Mean Stress = 10 psi; Stress Ratio, m = 0

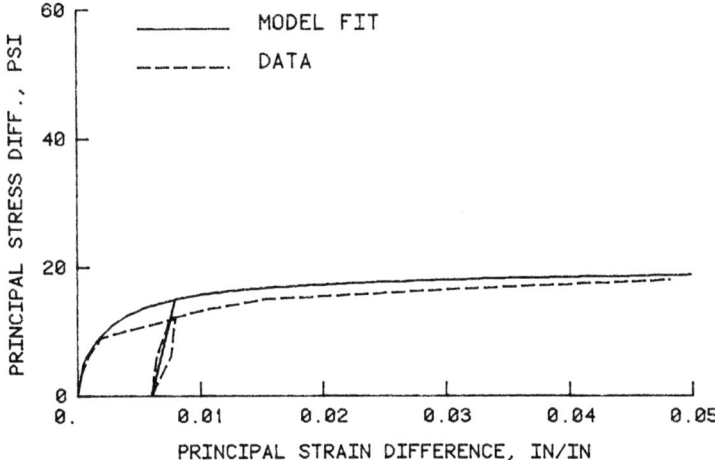

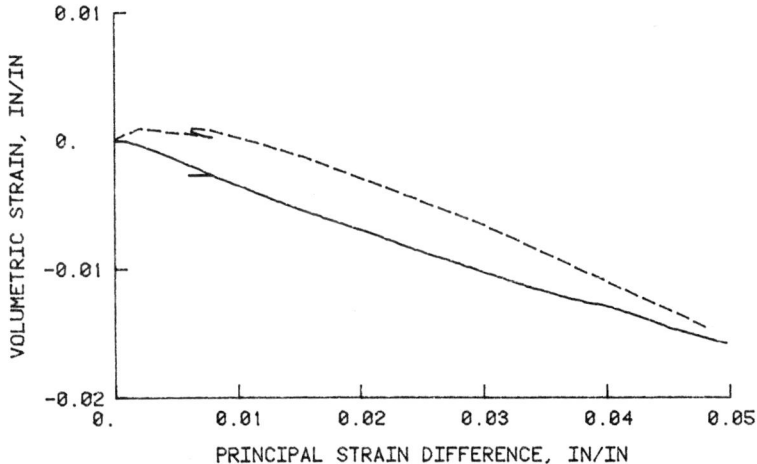

FIG. 10.—Principal Stress Difference and Volumetric Strain Versus Principal Strain Difference, Laboratory Measurements Versus Model Behavior

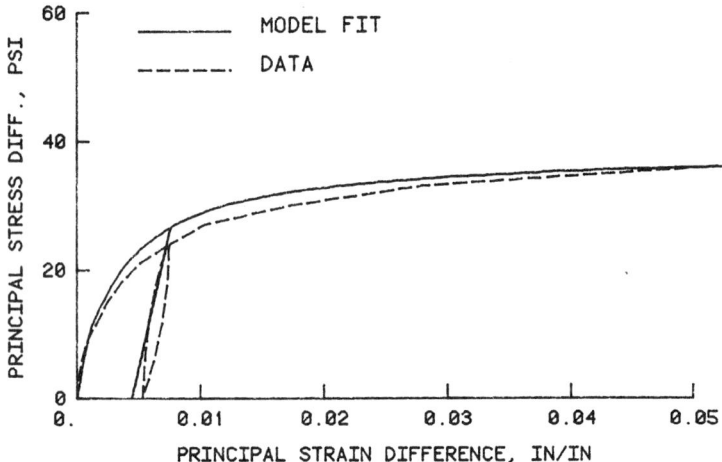
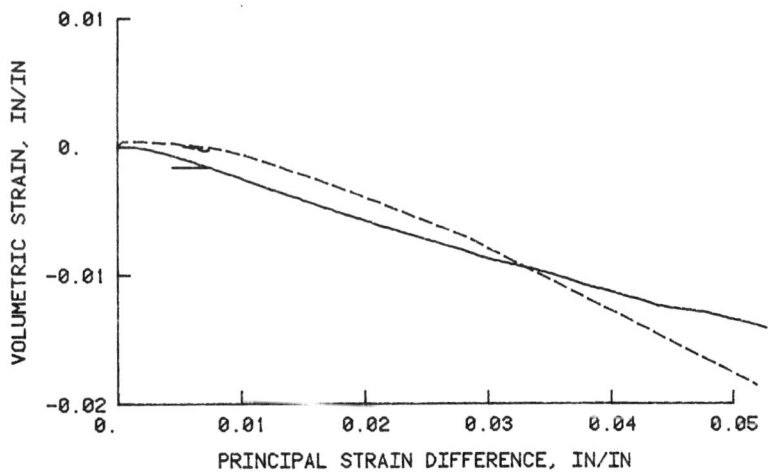

FIG. 11.—Principal Stress Difference and Volumetric Strain Versus Principal Strain Difference, Laboratory Measurements Versus Model Behavior

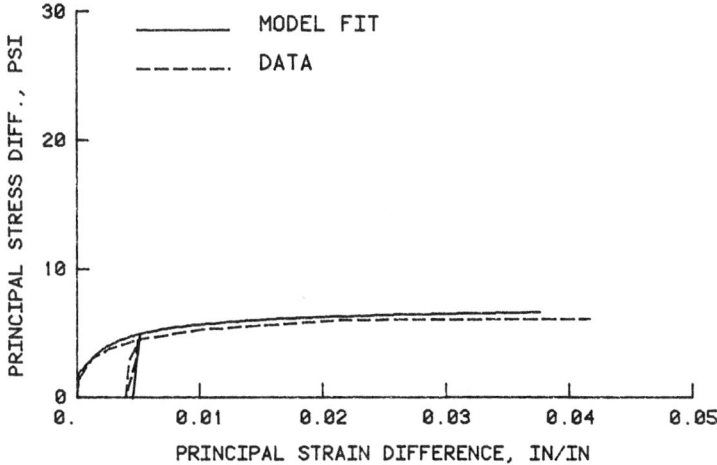
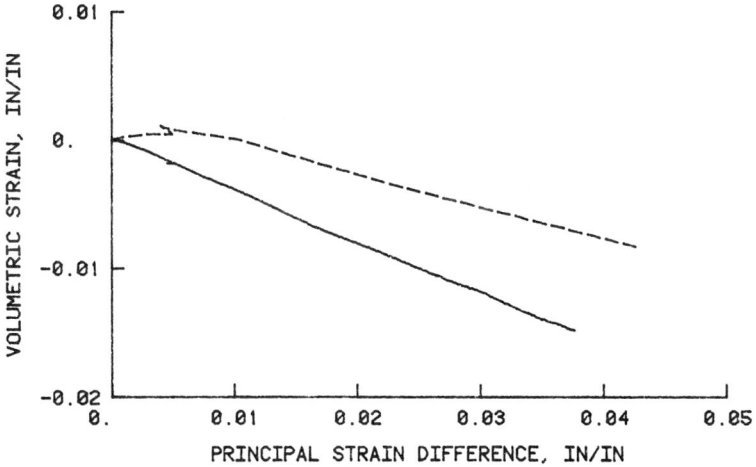

FIG. 12.—Principal Stress Difference and Volumetric Strain Versus Principal Strain Difference, Laboratory Measurements Versus Model Behavior

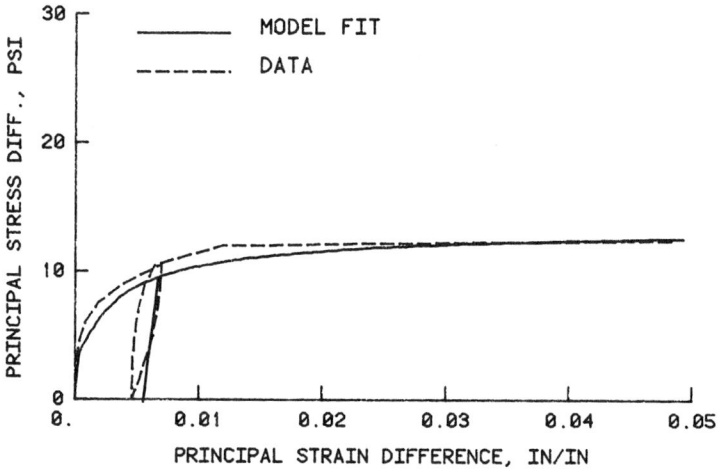
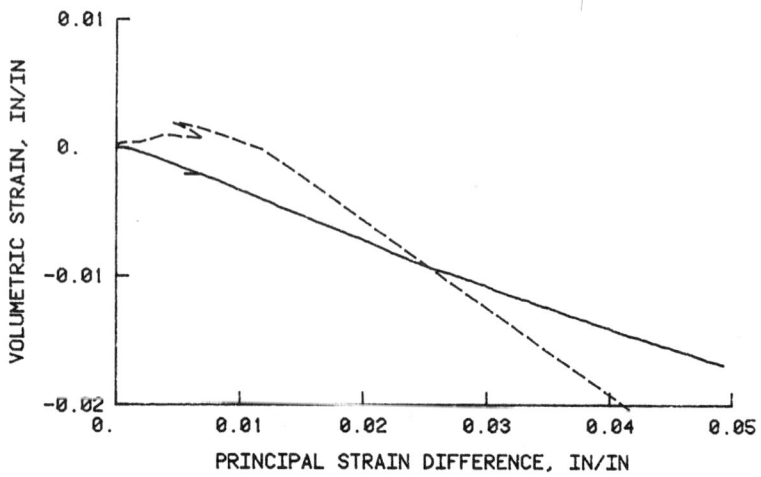

FIG. 13.—Principal Stress Difference and Volumetric Strain Versus Principal Strain Difference, Laboratory Measurements Versus Model Behavior

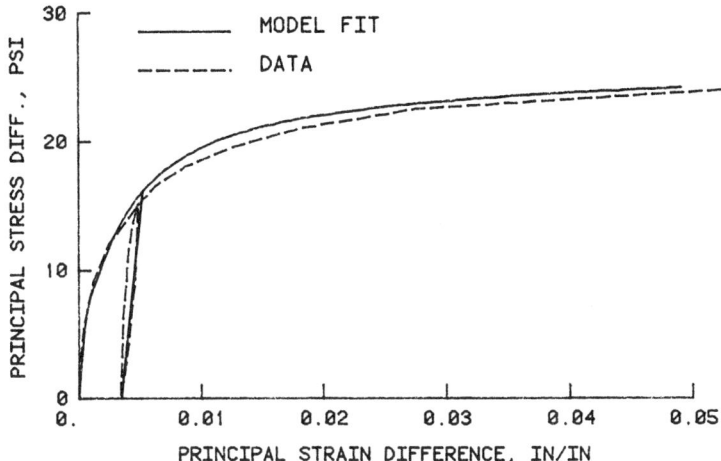
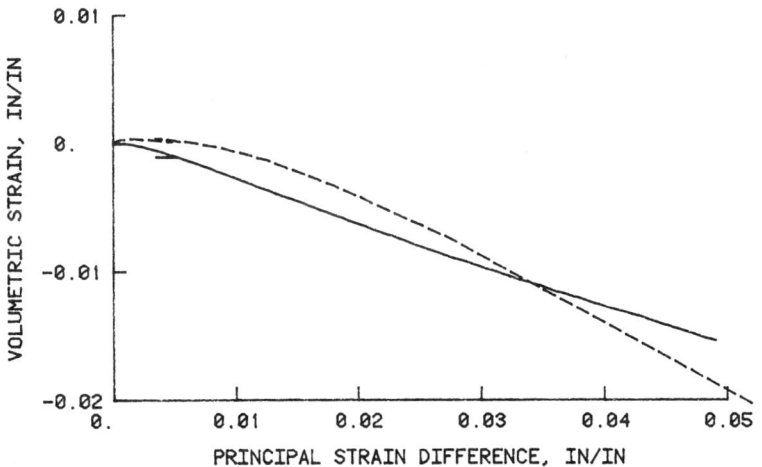

FIG. 14.—Principal Stress Difference and Volumetric Strain Versus Principal Strain Difference, Laboratory Measurements Versus Model Behavior

Conclusion

The cap model is capable of representing the behavior of a wide class of geological materials in a manner consistent with all theoretical continuum mechanics requirements. In this paper several examples are presented in which the model is used to describe the behavior of natural and remolded clays and a dry sand. For this purpose, several features of the general cap model are included. These features involve anisotropic "pseudo-invariants", pore-pressure effects, kinematic hardening and an isotropically hardening "failure envelope". The results illustrate the generality and flexibility of the cap model. One example illustrates the use of a simplified version of the model which can be programmed and exercised on a hand-held calculator.

DESCRIPTION OF NATURAL CLAY BEHAVIOR BY A SIMPLE BOUNDING SURFACE PLASTICITY FORMULATION

By Yannis F. Dafalias,[1] A.M. ASCE, Leonard R. Herrmann,[2] M. ASCE, and Jay Scott DeNatale[3]

INTRODUCTION

In the development of soil mechanics many constitutive laws for soils have been proposed which are appropriate only for particular kinds of loading conditions and initial states such as monotonic, cyclic, drained, undrained, normally consolidated, overconsolidated, etc. Besides the satisfaction of a practical but relatively narrow objective, a law which is good for normally consolidated but not for overconsolidated states, or for monotonic but not for cyclic loading and vice versa is of limited or no value at all for the analysis of soil structures subjected to complex and interchangeable loading conditions. The approach of proposing different stress-strain relations for different loading conditions fails to recognize the fact that the material state parameters are the same in all cases, and under a proper constitutive law for the change of the state the monotonic, cyclic, drained, undrained, etc. behavior should be obtained from one and the same constitutive model.

The aim of this work is to present first a general constitutive model for clays satisfying the above objectives developed by the senior authors (7,14,16) within the framework of critical state soil mechanics employing the new concept of the bounding surface in plasticity theory. Some additional improvements with respect to the previous works are also included. The corresponding general

---

[1] Assoc. Prof., Department of Civil Engineering, Univ. of California, Davis, CA.
[2] Professor, Department of Civil Engineering, Univ. of California, Davis, CA.
[3] Research Asst., Department of Civil Engineering, Univ. of California, Davis, CA.

constitutive relations have been already numerically implemented for use in finite element computations (15). Subsequently, the necessary material constants will be calibrated from the given experimental data on the natural clay samples X and Y. The predictions based on this calibration are presented in the first volume. The reasons for which the present model was not used to predict the response of the Kaolinite clay and the sand are explained subsequently in detail.

The primary motivation which led to the bounding surface concept is the desire to obtain rate independent elastoplastic constitutive relations which can realistically describe the material response under cyclic loading conditions, in addition to monotonic ones. One may observe here that the cyclic loading response is not the focus of the workshop. However, from the point of view of the fundamental character of the constitutive relations a cyclic loading is nothing else but a sequence of monotonic ones. For soils, in particular, this is equivalent to saying that in general a model which can describe realistically the cyclic response is also capable of describing realistically the response of overconsolidated states (in addition to normally consolidated) for the simple reason that cyclic loading brings normally consolidated samples at overconsolidated states. The inverse is also true if one can neglect the anisotropy which may develop during cyclic deviatoric loading. As an example of the above showing the corresponding inadequacies of the classical yield surface soil plasticity, consider the triaxial cyclic deviatoric loading of a normally consolidated clay sample, with fixed deviatoric stress amplitude under undrained conditions. During the first half-cycle a classical yield surface plasticity theory will predict successfully the undrained stress path and the pore-water pressure increase as a direct consequence of interchange between elastic and plastic volumetric strains while the total volumetric strain remains equal to zero, assuming incompressible solid and fluid phases. Subsequently, however, the stress will oscillate within the expanded yield surface at an overconsolidated state, unable to cause any additional plastic strain and as a result no additional pore-water pressure can be predicted contrary to the observed experimental observations. This notion of a purely elastic region implicit in the concept of a yield surface is the major weakness of classical plasticity when applied to soils which do not possess a purely elastic range in the classical sense. The above example shows also the connection between cyclic loading response and response at overconsolidated states. The latter is of great importance for the prediction of the behavior of clay Y which has OCR > 1 for all given confining pressures and for some states of clay X.

It is evident, therefore, that the basic formulation of plasticity must be changed if such important phenomena should be predicted. New concepts are necessary which either present a totally novel approach or modify the classical plasticity theory. Among the former the endochronic concept (27) has been successfully applied to soils (4,28), and among the later the field of work-hardening moduli concept (21) aquired recognition in providing comprehensive constitutive models in soil mechanics (22,24). Without discussing any further the endochronic models here, it should be observed that the field of work-hardening moduli models may be capable of predicting many aspects of soil behavior which cannot be described by the classical yield surface theories, but the necessity to define, up-date and keep in memory the nested surfaces involved in the concept renders them cumbersome and expensive for the numerical analysis of large systems.

A much simpler alternative is offered by the concept of the bounding surface in stress space, first introduced by Dafalias (5), Dafalias and Popov (10-13), and independently by Krieg (20) for metal plasticity. The salient feature of the new concept is that it allows plastic deformation to occur for stress states within the bounding surface on the basis of a very simple idea. For any stress state within (or on) the bounding surface a corresponding "image" point on the surface is specified by an appropriate mapping rule which becomes the identity mapping if the stress state is on the surface. The normal to the bounding surface at this "image" point defines the loading-unloading direction at the stress point. Associated or non-associated flow rules can be employed. At the "image" point a "bounding" plastic modulus is defined by means of the consistency condition for the bounding surface, but the actual plastic modulus is a function of this "bounding" modulus and the distance in stress space between the actual stress point and its "image". When the distance is zero the stress point lies on the bounding surface coinciding with its "image" until further unloading, and the two moduli are identical. Any usual yield surface can be straightforwardly considered as a bounding surface and the only new requirements are the proper definition of the mapping rule associating actual and "image" stress points and the functional dependence of the plastic modulus on the distance between them.

The bounding surface was originally used for metals (10-13) and more recently for soils in conjunction with an enclosed yield surface of vanishing (6) or finite (23) size. There are many possible versions of a bounding surface formulation appropriate for different materials since the concept lends itself to a versatile interpretation and use. The simultaneous use of the two surfaces offers one of the many possible mapping rules mentioned above, and in the case of vanishing elastic range the stress rate direction (but not the magnitude) participates in defining the "image" point (6,11,13). The vanishing elastic range formulation for soils (6) seems promising but needs more investigation before it can be presented with confidence.

Here a very simple version will be presented introducing a quasi-elastic domain enclosed by the bounding surface. This can be achieved following Dafalias (7) who introduced a very simple "radial" mapping rule without the necessity to explicitly define a yield surface: assuming that the origin is always within a convex bounding surface, the "image" point was defined as the intersection of the surface with the straight line connecting the origin with the current stress state. The corresponding bounding surface formulation was found to combine the attractive features of simplicity and predictive capabilities for applications to clays as shown in two papers (14,16), for the triaxial and the two stress invariants stress space within the framework of critical state soil mechanics. A more general development should also include the third stress invariant in order to account for differences in compression and extension and it will be presented in (17). A limited dependence on the third stress invariant is introduced here by rendering the plastic modulus a function of it for stress states within the bounding surface. This improvement was not used for clays X and Y for reasons explained subsequently.

The present formulation of the model is for isotropic clays. A new feature accounting for initial and developing (increasing or decreasing) anisotropy, customarily called induced anisotropy, can be incorporated in the present model with minimum modifications of the basic equations by means of a varying non-associated flow rule (9). It must be emphasized that the description of increasing or decreasing anisotropy can be as important as the inclusion of the initial one for some materials. A detailed presentation of this subject can be found in (8) for orthotropic symmetries. For "soft" materials like clays, an initial anisotropy which cannot be "erased" from the material memory under subsequent isotropic consolidation, for example, may be as disadvantageous as not being included at all from the very beginning. The calibration of the anisotropic parameters associated with the above anisotropic modification of the model, requires a set of experimental data not available for the clays X, Y and the Kaolinite clay. Since the description of developing anisotropy is not an easy task, one can at least try to include initial anisotropy only. We will show that the given experimental data <u>cannot</u> be used towards a rigorous determination of initial anisotropy of clays X,Y and the Kaolinite clay for a number of different reasons, the most important of which is the following.

By its very definition, anisotropy associated with a reference configuration can only be characterized by a proper set of experimental data (stress-strain curves) obtained by imposing <u>the same loading history</u> with respect to different orientations relatively to the reference configuration. None of the given experimental data satisfies this fundamental precondition for the characterization of anisotropy.

In order to be more specific, let us begin with the laboratory prepared Kaolinite clay. First, the unambiguous definition of triaxial compression and extension is that the intermediate principal stress is equal to the minor one for compression and to the major for extension. The "unequal" principal stress (major for compression and minor for extension) can be called the axial stress irrespective of whether or not is in the vertical direction (for the classical triaxial apparatus it is along the vertical or axial direction, hence, the name axial stress). For the two given compression tests, 1 and 4, the axial stress is identified as the σ_1 (effective) which one can plausibly assume is along the K_o vertical consolidation direction. For the two given extension tests, 10 and 13, the axial stress is identified as the σ_3 but now one is left to wonder if the axial stress direction has been rotated or not by 90 degrees around the σ_2 axis before these tests were performed. Even if this is known, the important point is that one <u>cannot</u> characterize anisotropy with or without rotation of the axial stress direction. Indeed, if one assumes that rotation took place, which seems to be the most probable judging from the given sketch for the predictions, then σ_3 becomes parallel to the direction of K_o consolidation (which was occupied by σ_1 during compression) and the extension tests represent loading histories different from the compression tests, applied with respect to the same orientation of the axial stress (defined as the "unequal" stress). Thus, these may be the proper tests to characterize the difference of behavior in compression-extension, but they are exactly the opposite of what is required to characterize anisotropy, i.e., same loading histories for different orientations. If now one assumes that σ_3 was not rotated, then the corresponding data of the extension tests would differ from the compression data due to the combined effect of both anisotropy and difference in compression-extension behavior, rendering impossible to characterize either one. The proper experiments for the characterization of anisotropy should have been only identical compression or only identical extension tests with different orientations of the axial stress, so that response differences could be attributed to anisotropy alone. Since the emphasis of the predictions for the Kaolinite clay is placed mainly on anisotropy, and since it was shown according to the above that such anisotropy cannot be rigorously characterized, it was decided not to use the anisotropic version of the bounding surface model (9) for the predictions due to insufficient data.

Let us now consider the clays X and Y. The situation is similar to the previous case. The given tests are compression for a stress ratio m=0 and extension for m=1, with the axial stress direction (defined earlier as the "unequal" principal stress) being correspondingly the σ_1 (vertical for m=0) and the σ_3 (horizontal for m=1). Again, the effects of anisotropy and different response in compression-extension are mixed making impossible to characterize uniquely either one. There is also one additional difficulty: from the given experimental data, one can detect anisotropy but it is impossible to specify it due to the non-systematic response of the different samples. For example, consider the given data for m=0 which implies $\sigma_2 = \sigma_3$ for clay Y at a confining pressure σ_c = 2.5 psi. It is given that $\varepsilon_2 = \varepsilon_3$ as σ_1 is increased which suggests transversly isotropic symmetries with σ_1 being, for this sample, the axis of rotational symmetries. This is not true, however, for any other test of m=0 for clay Y and the difference in the values of ε_2 and ε_3 are such that cannot be justified by experimental error. The plausible explanation for this discrepancy is that the samples were tested at random principal stress orientations with respect to the vertical direction in the field, which is most probably the direction of the transverse isotropy, and it so happened that for the sample with σ_c = 2.5 psi the triaxial σ_1 direction was identical to the vertical field direction showing the transversly isotropic effects. One such test, however, is insufficient to characterize anisotropy as explained before. Similar arguments apply for clay X. Therefore, during the sampling and testing of natural clays it is very important to keep track of the sample orientation if one wants to characterize unambiguously anisotropy. Another explanation could be that it is possible for a natural clay to have an undetermined anisotropy or strong variations of it. This also may have happened here. In that case the best procedure is to treat the natural clay as isotropic.

The important feature, however, of the different tests on clays X and Y was mainly the effect of overconsolidation. Since the anisotropy and the compression-extension difference effects cannot be isolated and characterized from the given data, it was decided to treat clays X and Y as isotropic with similar response in compression-extension. Thus, a simple isotropic bounding surface model with two stress invariants is used. For the above reasons, the limited possibility to account for differences in compression-extension by the incorporated feature of the plastic modulus dependence on the third stress invariant is not used.

As a closing comment it should be emphasized that the present bounding surface model is appropriate for clays but not for sands, and this is the reason why the sand predictions were not given. The general concept of the bounding surface can certainly be applied to sand behavior but the specific model will be different from the one for clays for the same reason that classical yield surface clay plasticity models are not appropriate for sands and vice-versa. The mechanisms of deformation for clays and sands are different and it would be indeed questionable if one claims that one basic model can describe the behavior of both by changing the parameters only. This is different than saying that a general concept, i.e., endochronic or bounding surface etc., can be properly adjusted in order to account for the unique behavioral features of each material during deformation. That is how the general concept of the bounding surface initially suggested for metals has been extended for application to clays, and it is believed that extension to sands, concrete, rocks, etc., is within the capabilities of the concept.

GENERAL RATE INDEPENDENT ELASTOPLASTICITY

The state of the material is defined in terms of the effective stress σ_{ij} and a set of plastic internal variables q_n which account for the past loading history. Such q_n can be the plastic work, the plastic strains, etc. If u denotes the pore-water pressure the total stress σ^t_{ij} is related to σ_{ij} and u by

$$\sigma^t_{ij} = \sigma_{ij} + \delta_{ij} u \tag{1}$$

with δ_{ij} the Kronecker delta. If now ε_{ij}, ε'_{ij}, ε''_{ij} are the total, elastic and plastic strains respectively and a dot indicates the rate, the relation

$$\dot{\varepsilon}_{ij} = \dot{\varepsilon}'_{ij} + \dot{\varepsilon}''_{ij} \tag{2}$$

is assumed. Small strain formulation is presented henceforth, but the extension to large deformations is straightforward if a proper stress measure and a conjugate rate of deformation measure for which the decomposition (2) holds true are employed.

The elastic incremental constitutive relations are given by

$$\dot{\varepsilon}'_{ij} = C_{ijk\ell} \dot{\sigma}_{k\ell}, \qquad \dot{\sigma}_{ij} = E_{ijk\ell} \dot{\varepsilon}'_{k\ell} \tag{3}$$

where $C_{ijk\ell}$, $E_{ijk\ell}$ are the tensors of elastic compliance and moduli respectively being in general functions of the state with $E_{ijk\ell} C_{k\ell pq} = (\delta_{ip} \delta_{jq} + \delta_{iq} \delta_{jp})/2$.

The plastic constitutive relations require the definition of the direction (or vector) of plastic loading L_{ij} and the plastic modulus K_p, both functions of the state, which in turn determine the loading function L as:

$$L = \frac{1}{K_p} L_{ij} \dot{\sigma}_{ij} \tag{4}$$

Plastic loading, unloading and neutral loading occurs when $L > 0$, $L < 0$ and $L = 0$ respectively. The inclusion of K_p in L allows for the description of unstable behavior when both scalar quantities L_{ij} $\dot{\sigma}_{ij}$ and K_p are negative but $L > 0$. Assuming now linear dependence of $\dot{q}_n$ on $\dot{\sigma}_{ij}$ for rate independence (homogeneity of order one would guarantee rate independence for a more general development), considering the $\dot{\varepsilon}_{ij}^{\prime\prime}$ as one of the q_n but distinguishing it for emphasis, and imposing the requirement of continuous material response with respect to a changing direction of $\dot{\sigma}_{ij}$ across neutral loading (5,12), the constitutive relations are given by

$$\dot{\varepsilon}_{ij}^{\prime\prime} = <L> R_{ij} \qquad (5a)$$

$$\dot{q}_n = <L> r_n \qquad (5b)$$

where the brackets $< >$ define the operation $<z> = zh^*(z)$, h^* being the heavyside step function, and R_{ij}, r_n are functions of the state. The R_{ij} is usually assumed to be the gradient of a plastic potential. In classical plasticity the L_{ij} is defined as the gradient of the yield surface $f = 0$ and K_p is obtained by means of the consistency conditions $\dot{f} = 0$. Inverted constitutive relations can be obtained straightforwardly (16) and the corresponding relations for undrained loading follow from the internal constraint $\dot{\varepsilon}_{kk} = 0$.

THE ROLE OF THE BOUNDING SURFACE

The following sections on the basic equations are a resume of the development presented in (16). The role of the bounding surface becomes instrumental in defining the key quantities L_{ij} and K_p. The previous plastic loading history expressed quantitatively by means of the values of q_n, determines a "Bounding Surface" in stress space, Fig. 1, analytically described by

$$F(\bar{\sigma}_{ij}, q_n) = 0 \qquad (6)$$

where a bar over stress quantities indicates points on $F = 0$. The actual stress point σ_{ij} lies always within or on the bounding surface. To each σ_{ij}, a unique "image" point $\bar{\sigma}_{ij}$ on $F = 0$ is defined according to the simple "radial" rule mentioned in the introduction: with the origin 0 being always within a convex bounding surface, $\bar{\sigma}_{ij}$ is obtained as the intersection of $F = 0$ and the straight line connecting the origin with σ_{ij}, Fig. 1. Analytically, this can be expressed by

$$\bar{\sigma}_{ij} = \beta(\sigma_{k\ell}, q_n) \sigma_{ij} \qquad (7)$$

NATURAL CLAY BEHAVIOR 719

with the radial factor $\beta \geq 1$ obtained from $F(\beta\sigma_{ij}, q_n) = 0$. Only for the origin $\sigma_{ij} = 0$ and $\bar{\sigma}_{ij}$ is undefined without any further consequence. It is important to emphasize that a general bounding surface formulation can utilize different rules associating the "image" to the actual stress point, one of them employing the concept of an enclosed yield surface (12,23).

The direction of plastic loading L_{ij} at σ_{ij} is defined as the gradient of F at the "image" point $\bar{\sigma}_{ij}$, i.e.,

$$L_{ij} = \frac{\partial F}{\partial \bar{\sigma}_{ij}} \tag{8}$$

For any stress rate $\dot{\sigma}_{ij}$ causing plastic loading, a corresponding "image" stress rate $\dot{\bar{\sigma}}_{ij}$ occurs through the hardening of $F = 0$ by means of the $\dot{q}_n$. Thus, the following three key equations complete the bounding surface formulation:

1. The loading function L, eq. (4), is defined in terms of L_{ij}, eq. (8), the two stress rates $\dot{\sigma}_{ij}, \dot{\bar{\sigma}}_{ij}$ and two plastic moduli: the actual one K_p associated with $\dot{\sigma}_{ij}$ and a "bounding" plastic modulus $\bar{K}_p$ associated with $\dot{\bar{\sigma}}_{ij}$ as follows:

$$L = \frac{1}{K_p} \frac{\partial F}{\partial \bar{\sigma}_{ij}} \dot{\sigma}_{ij} = \frac{1}{\bar{K}_p} \frac{\partial F}{\partial \bar{\sigma}_{ij}} \dot{\bar{\sigma}}_{ij} \tag{9}$$

2. The "bounding" plastic modulus $\bar{K}_p$ is obtained from the consistency condition $\dot{\bar{F}} = 0$. Using eqs. (5b), (6) and the second part of (9) one has

$$\bar{K}_p = - \frac{\partial F}{\partial q_n} r_n \tag{10}$$

3. A state dependent relation $\hat{K}_p$ between K_p and $\bar{K}_p$ is established as a function of the Euclidean metric or distance $\delta = [\mathrm{tr}(\bar{\sigma}_{ij} - \sigma_{ij})^2]^{1/2} = (\beta - 1)(\sigma_{ij} \sigma_{ij})^{1/2}$ between the current stress state and its "image", i.e.,

$$K_p = \hat{K}_p (\bar{K}_p, \delta, \sigma_{ij}, q_n) \tag{11}$$

such that $\hat{K}_p$ is an increasing function of δ with $\hat{K}_p > \bar{K}_p$ for $\delta > 0$ (the stress point inside the bounding surface) and $\hat{K}_p = \bar{K}_p = K_p$ for $\delta = 0$ (the stress point on the bounding surface, identical with its "image"). It follows from eqs. (6), (7), (10) and the definition of δ that both $\bar{K}_p$ and δ are functions of the state σ_{ij}, q_n, thus K_p is function of the state.

Eq. (11) embodies the meaning of the bounding surface concept. It allows for plastic deformation to occur for points within the surface at a progressive rate which depends on δ. The closer is the stress point to the surface, the smaller is the K_p approaching the corresponding $\bar{K}_p$, and the greater is the plastic strain rate for a given stress rate. The stress σ_{ij} may eventually reach the bounding surface in the course of plastic loading as it can be seen from eq. (9) where the projection of $\dot{\sigma}_{ij}$ on the gradient of $F = 0$ is greater than the corresponding projection of $\dot{\bar{\sigma}}_{ij}$ since $K_p \geq \bar{K}_p$. The stress point remains on $F = 0$ if loading continues, and upon unloading it detaches from $F = 0$ moving inwards and so forth.

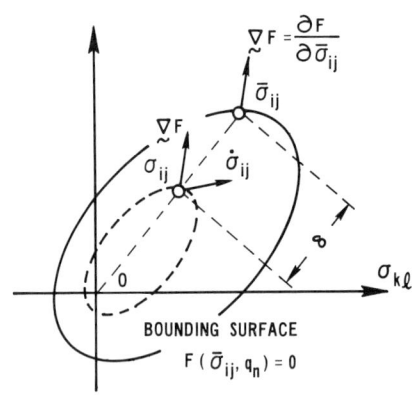

FIG. 1. - Schematic illustration of the bounding surface and the "radial" mapping rule in a general stress space.

As a result of the definition of L_{ij} and L, it follows that at each point σ_{ij} a surface homeothetic to the bounding surface with respect to the origin is indirectly defined, shown by the dashed curve in Fig. 1, which determines all the paths of neutral loading emanating from σ_{ij}. This surface defines a quasi-elastic domain but it is not a yield surface since the stress point may move first elastically inwards and then cause plastic loading before it reaches this surface again. It is closer to the concept of a loading surface (18) but not entirely equivalent since no associated consistency condition is required. As a matter of fact, it never enters the present formulation explicitly.

Finally, it is worth mentioning that a classical yield surface formulation can be obtained easily in the limit if the functional dependence of $\hat{K}_p$, eq. (11), on δ is such that $K_p = \hat{K}_p \simeq \infty$ for $\delta > 0$ and $K_p = \hat{K}_p + \bar{K}_p$ as $w \to 0$. Then the bounding surface becomes a yield surface in the classical sense. This leads to the inverse important conclusion that any classical yield surface formulation can be transformed easily into a bounding surface formulation by identifying the yield surface as the corresponding bounding surface and using the set of the three key equations (9), (10), (11) together with a proper association of an "image" stress point to any actual stress point within or on the surface.

STRESS INVARIANT FORMULATION FOR ISOTROPIC SOILS

The formulation so far can be applied to any material and it is only the concept of the effective stress under undrained conditions which makes it appropriate for soils in particular. Any type of material symmetries can be incorporated by properly defining the elastic moduli and rendering F a function of proper invariant quantities of $\bar{\sigma}_{ij}$ and q_n. Further consideration will be restricted to isotropic soils. Assuming that elastic isotropy is not altered by plastic deformation, the elastic moduli are given by

$$E_{ijk\ell} = (K - \tfrac{2}{3}G)\,\delta_{ij}\delta_{k\ell} + G(\delta_{ik}\delta_{j\ell} + \delta_{i\ell}\delta_{jk}) \tag{12}$$

where the bulk modulus K and the shear modulus G are independent of q_n but can depend on the isotropic stress invariants. Eq. (3) can now be written

$$\dot{\bar{\sigma}}_{kk} = 3K\,\dot{\varepsilon}'_{ii}, \qquad \dot{s}_{ij} = 2G\,\dot{e}'_{ij} \tag{13}$$

with s_{ij}, e_{ij} denoting deviatoric components of stress and strains respectively. Subsequently, compressive stresses and strains are considered positive.

Initial isotropy requires that $F = 0$ must be a function of the basic and mixed isotropic invariants of $\bar{\sigma}_{ij}$ and q_n. This does not exclude, however, the possibility to describe developed plastic anisotropy with respect to subsequent reference states by a proper choice of q_n. For example, if the bounding surface undergoes general kinematic hardening some of the q_n are the coordinates of its center providing a built-in feature which generates anisotropy in the course of plastic deformation (12,23,24). Here an isotropic model is presented.

The stress dependence of $F = 0$ will be restricted to the first stress invariant I and the square root J of the second deviatoric stress invariant defined together with their derivatives with respect to the stress by

$$I = \sigma_{kk}, \qquad \frac{\partial I}{\partial \sigma_{ij}} = \delta_{ij} \tag{14a}$$

$$J = (\frac{1}{2} s_{ij} s_{ij})^{\frac{1}{2}}, \qquad \frac{\partial J}{\partial \sigma_{ij}} = \frac{s_{ij}}{2J} \tag{14b}$$

It will be further assumed that the bounding surface undergoes isotropic and kinematic hardening along the hydrostatic axis controlled by one single scalar q_n measuring the plastic volumetric strain $\dot{\varepsilon}_{kk}''$. If $\dot{e}$ is the rate of the total void ratio and $\dot{e}'$, $\dot{e}''$ are its elastic and plastic parts respectively related by $\dot{e} = \dot{e}' + \dot{e}''$, the $\dot{e}''$ is chosen as the above q_n and its rate is given by

$$\dot{e}'' = - (1 + e_o) \dot{\varepsilon}_{kk}'' = - <L> (1 + e_o) R_{kk} \tag{15}$$

with e_o the initial void ratio corresponding to the reference configuration with respect to which strains are measured. For natural strains follows that $e_o = e$. Thus, eq. (6) becomes

$$F(\overline{I}, \overline{J}, e'') = 0 \tag{16}$$

Subsequently the associated flow rule $L_{ij} = R_{ij}$ is assumed and the partial derivative with respect to an invariant is denoted with a comma followed by the symbol of the invariant as a subscript. Using now eqs. (7), (14), (16) and the associated observations, a straightforward computation yields according to eqs. (8) and (9):

$$L_{ij} = R_{ij} = F,_{\overline{I}} \delta_{ij} + \frac{F,_{\overline{J}}}{2J} s_{ij} \tag{17}$$

$$L = \frac{1}{K_p} (F,_{\overline{I}} \dot{I} + F,_{\overline{J}} \dot{J}) = \frac{1}{\overline{K}_p} (F,_{\overline{I}} \dot{\overline{I}} + F,_{\overline{J}} \dot{\overline{J}}) \tag{18}$$

The form of L can be interpreted as indicating loading whenever the inner product of the stress invariant rates $\dot{I}$, $\dot{J}$ with the gradient of F in invariant stress space divided by K_p is positive. The plastic strain rate is given from eqs. (5a), (17) and (18). Non-associated flow rules can also be used by defining R_{ij} differently.

According to eqs. (5b), (15) and (17), one has from eq. (10):

$$\bar{K}_p = 3(1 + e_o)\ (\partial F/\partial e^{\prime\prime})\ F_{,\bar{I}} \qquad (19)$$

Observe from eq. (19) that with $\partial F/\partial e^{\prime\prime} > 0$, the bounding plastic modulus $\bar{K}_p$ is positive (consolidation), negative (dilatation) or zero (unrestricted shear flow) according to the value of $F_{,\bar{I}}$. Correspondingly the bounding surface expands, contracts or does not harden. This is a particularly interesting property which allows the easy incorporation of the present formulation into a critical state framework.

The changes of the bounding plastic modulus $\bar{K}_p$ on $F = 0$ reflect onto the values of the actual plastic modulus K_p by means of eq. (11). Here the following form of this equation will be assumed

$$K_p = \bar{K}_p + H(I, J, \alpha, e^{\prime\prime})\left[\frac{\delta}{\delta_o - \delta}\right] \qquad (20)$$

where δ is the distance between actual and "image" stress points in the invariant space, δ_o is a properly chosen and possibly varying with the state reference stress or distance such that $\delta_o - \delta \geq 0$, α is the "Lode" angle (19) defined in terms of the second and third deviatoric stress invariants and H is a positive "shape" hardening function of the stress invariants and $e^{\prime\prime}$. Observe that a limited dependence of the constitutive relations on the third stress invariant is introduced by means of α in H and this is an improvement of the previous works (14,16). The exact definition of H will require the identification and experimental determination of certain material constants. The H and the associated constants constitute the "novel" elements of the present formulation with regard to classical yield surface formulations, and are intimately related to the soil response for states within $F = 0$ (overconsolidation). For $H \to \infty$, observe that $K_p \to \infty$ and $K_p = \bar{K}_p$ only for $\delta = 0$. In this case, the bounding surface behaves as a yield surface. It is possible to have $\bar{K}_p < 0$ and $K_p > 0$ if δ is large enough which allows the description of an initially rising stress-strain curve as the stress point approaches the contracting bounding surface ($\bar{K}_p < 0$) during dilatation, and the subsequent unstable falling curve behavior when eventually δ becomes small enough to have both $\bar{K}_p$, $K_p < 0$ as in the case of heavily overconsolidated clays.

From $\dot{\varepsilon}_{kk} = 0$ for undrained conditions and the assumption of incompressible fluid and solid phases the following relations can be derived as shown in (16):

$$L = \frac{F_{,\bar{J}}\ \dot{J}}{K_p + 9\ K\ F_{,\bar{I}}^2} \qquad (21)$$

$$\dot{u} = 3 K F_{,\bar{I}} <L> + \frac{1}{3} \dot{I}^t \tag{22}$$

$$\left[F_{,\bar{I}} + \frac{\bar{K}_p}{9 K F_{,\bar{I}}} \right] \dot{I} + F_{,\bar{J}} \dot{J} = 0 \tag{23}$$

where the first total stress invariant rate $\dot{I}^t$ is indeterminate due to incompressibility and eq. (23) is the differential equation of the undrained stress path in the space of the effective stress invariants I and J.

For any given increment $\dot{\sigma}_{ij} dt$ or $\dot{\epsilon}_{ij} dt$ the soil response is fully determined from the above incremental relations. Recall that the plastic strain rate is always given by eq. (5a). Observe that the material state expressed in terms of the seven quantities σ_{ij}, $e^{\prime\prime}$ suffices to define completely these relations. The triaxial case is obtained as a special case of the general development.

SPECIFIC FORMS OF THE BOUNDING SURFACE

Specific analytical expressions for the bounding surface will be presented here which can be used for the computation of the quantities $F_{,\bar{I}}$, $F_{,\bar{J}}$ and $\partial F / \partial e^{\prime\prime}$ entering the constitutive relations (5a), (17), (18), and eq. (19) which defines $\bar{K}_p$.

Extending the ideas of the critical state soil mechanics from the triaxial to the invariant space a meridional section of the bounding surface is eloquently shown in Fig. 2. The quantity N can be identified as the slope of the projection CSL of the critical state line in invariant stress space which intersects the bounding surface at C where $F_{,\bar{I}} = 0$. For triaxial conditions the N is related to the triaxial CSL slope M by $M = 3/3 N$. The N, and by consequence $F = 0$, can be made functions of the third stress invariant by means of the "Lode" angle α (19) and this is indicated in Fig. 2. Such dependence on α is not considered in the present analytical development but it will be presented elsewhere (17). The radial rule associating the "image" stress point $\bar{I}, \bar{J}$ to the stress point I, J is illustrated. The projection of C on the I axis is denoted by I_1 and is related to I_o by $I_1 = I_o / R$, with I_o the intersection of $F = 0$ with the hydrostatic axis I.

Let us introduce for later reference the quantities

$$\theta = \frac{J}{I} , \qquad x = \frac{\theta}{N} \tag{24}$$

For $0 \leq \theta \leq N$ the bounding surface is the ellipse 1 with center I_1, defined by the equation

$$F = \bar{I}^2 + (R-1)^2 \left(\frac{\bar{J}}{N}\right)^2 - \frac{2 I_o}{R} \bar{I} + \frac{2-R}{R} I_o^2 = 0 \tag{25}$$

Its definition requires one only material constant R in addition to N. In the past R (or its equivalent concept) has been taken equal to 2 (25) or 2.72 (26) but it is proposed here that R should be considered a material constant which can be determined experimentally for a particular clay. Theoretically, R can vary in the range $1 \leq R \leq \infty$.

It is possible to extend the ellipse 1 in the range $N \leq \theta \leq \infty$ under the restriction $R \geq 2$ (for $R = 2$ the ellipse crosses the origin which must lie within or on the bounding surface). It was found, however, that the prediction of the model was unsatisfactory for the deviatoric loading of heavily overconsolidated clays suggesting a curve with a shape "more" parallel to the critical line OC, which is tangent to the ellipse at C. This would yield the advantage of having the possibility to choose $R < 2$ if necessary, at the expense of one additional parameter for the new curve. A hyperbola is therefore proposed whose apex C is at a distance D from its center G, Fig. 2, and the additional parameter A is defined from $D = AI_o$. The analytical expression of the hyperbola is

$$F = \bar{I}^2 - \left(\frac{\bar{J}}{N}\right)^2 - \frac{2 I_o}{R} \bar{I} + 2 I_o \left[\frac{1}{R} + \frac{A}{N}\right] \frac{\bar{J}}{N} - \frac{2A}{RN} I_o^2 = 0 \tag{26}$$

In order to describe the material behavior in tension (1) for the range $-\infty \leq \theta \leq 0$ it is possible to extend the bounding surface into the $I < 0$ range as a smooth curve tangent to the hyperbola at point B, Fig. 2, and intersecting vertically the I axis at point I_t. The I_t measures the tensile strength of the soil. Such an extension as an ellipse 2 was proposed by Dafalias and Herrmann (14) but other shapes may also be suitable. In view of insufficient evidence for the tensile behavior, further discussion is postponed with the observation that the material response will be described by equations similar to the ones applied to the other cases, with attention given in securing the continuity of the constitutive relations at point B.

The expression $dI_o/de^{\prime\prime}$ necessary for the hardening behavior is sought subsequently. Let κ and λ denote the slopes of the rebound and consolidation lines in the $e - \ln I$ plot (same as in the $e - \ln p$ plot). Assuming that there is a limit value $I_\ell > 0$ such that for $I \leq I_\ell$ the relations between I and the elastic part $e^\prime$ of the void ratio changes continuously form logarithmic to linear in order

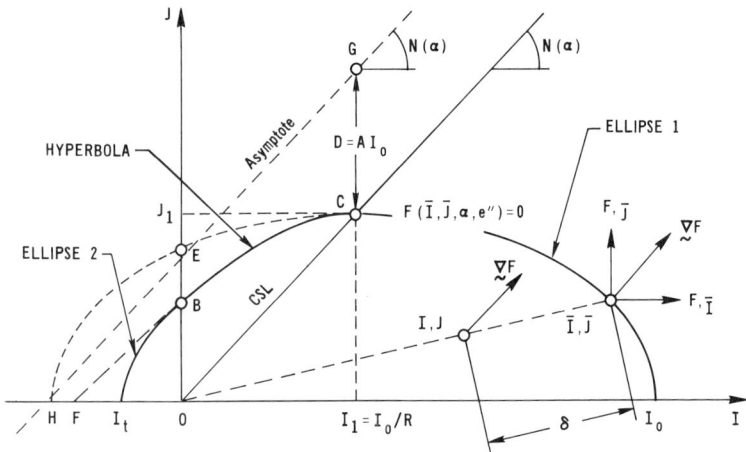

FIG. 2.- Schematic illustration of the bounding surface and the "radial" mapping rule in the invariant stress space.

to prevent excessive softening of the elastic stiffness in the neighborhood of $I = 0$ for cohesive soils, one has

$$\frac{dI}{de^*} = -\frac{\langle I - I_\ell \rangle + I_\ell}{\kappa}, \qquad \frac{dI_o}{de} = -\frac{I_o}{\lambda} \qquad (27)$$

where the first part of eq. (27) applies also for $I = I_o$. It follows immediately from $\dot{e}^{**} = \dot{e} - \dot{e}^*$ and eq. (27) that

$$\frac{dI_o}{de^{**}} = -\frac{I_o}{\lambda - \omega\kappa} \quad \text{with} \quad \omega = \frac{I_o}{\langle I_o - I_\ell \rangle + I_\ell} \qquad (28)$$

The first part of eq. (27) in combination with eq. (13) and $\dot{e}^* = -(1 + e_o) \dot{\epsilon}^*_{kk}$ yields an expression for the bulk modulus K in terms of κ as

$$K = \frac{(1+e_o) \; (\langle I - I_\ell \rangle + I_\ell)}{3\kappa} \qquad (29)$$

It is now straightforward to compute from eqs. (25), (26) and (28) the $F_{,\bar{I}}$, $F_{,\bar{J}}$ and by the chain rule the $(\partial F/\partial e^{\prime\prime}) = (\partial F/\partial I_o)(\partial I_o/\partial e^{\prime\prime})$. Thus, with the exception of K_p all other quantities entering the plastic constitutive relations (5a), (15), (17), and (18) are obtained in closed form including the value of $\bar{K}_p$ from eq. (19). The K_p will be related to $\bar{K}_p$ by a proper form of eq. (20) in the following section.

For undrained deviatoric loading after normal consolidation at I_o the stress point remains in the expanding bounding surface, thus $K_p = \bar{K}_p$ which can be obtained from eqs. (19), (26) and (28). This allows the closed form integration of eq. (23) for the undrained stress path, which in terms of the triaxial variables $p = I/3$, $q = \sqrt{3}J$, $p_o = I_o/3$ and $M = \sqrt{3}N$ is given by

$$\frac{q}{p_o} = \frac{M}{R-1} \left[\frac{2}{R} \left(\frac{p}{p_o}\right)^{\frac{\lambda-2\kappa}{\lambda-\kappa}} + \left[1 - \frac{2}{R}\right]\left(\frac{p}{p_o}\right)^{\frac{-2\kappa}{\lambda-\kappa}} - \left(\frac{p}{p_o}\right)^2 \right]^{\frac{1}{2}} \tag{30}$$

Eq. (30) will be used to determine R.

THE SHAPE HARDENING FUNCTION, IDENTIFICATION AND DETERMINATION OF MATERIAL CONSTANTS

It is only left to specify the most characteristic feature of the bounding surface formulation by defining the shape hardening function H which relates K_p and $\bar{K}_p$ in eq. (20), and subsequently to identify and propose calibration procedures for the material constants entering the constitutive relation on the basis of triaxial experiments. The name "shape hardening" is based on the role of H in defining the shape of the stress-strain curves during plastic hardening (or softening) for points within the bounding surface.

The following form of the key eq. (20) is suggested:

$$K_p = \bar{K}_p + p_a h(\alpha) \left[1 + |x|^{-m}\right] \left[9 F_{,\bar{I}}^2 + \frac{1}{3} F_{,\bar{J}}^2\right] \left[\beta - 1\right] \tag{31}$$

where p_a is the atmospheric pressure providing the proper stress units and recall that $x = \theta/N$, eq. (24). The last bracket expresses the quantity $\delta/(\delta_o - \delta)$ according to the radial mapping rule (7) if $\delta_o = \bar{r}$ with $\bar{r}$ being the distance of the image point $\bar{I}$, $\bar{J}$ from the origin. Clearly, H consists of the product of $p_a h$ with the first two brackets in eq. (31). The introduction of the second bracket is associated with the corresponding "unit normal" triaxial formulation (14,16) and the continuity of the constitutive relations at point C. The exponent m is a positive material constant (its role is explained subsequently) and the shape hardening parameter h is considered a function of the third stress invariant by means of the "Lode" angle α (19) varying according to

$$h(\alpha) = \frac{2\mu}{1 + \mu - (1-\mu) \sin 3\alpha} h_c \tag{32}$$

where $\mu = h_e/h_c$ with h_c, h_e the values of h for triaxial compression and extension respectively. Thus, a limited dependence on α has been introduced through h. For a more general development, α should also enter $F = 0$.

The introduction of $|x|^{-m}$ does not allow plastic deformation to occur within the bounding surface for $\theta = 0$ (zero deviatoric stress) rendering K_p infinite except when $\delta = 0$ for which $K_p = \bar{K}_p$. The absolute value has been introduced for future use in tensile response where $x < 0$. A small value of m suppresses the influence of $|x|^{-m}$ for $\theta > 0$. The shape hardening parameter h bears mainly the responsibility for the description of the material response for states within the surface. Eq. (31) introduces three material constants: m, h_c and μ. All other quantities depend only on the state.

It is now possible to identify and suggest calibration procedures for the material constants entering the formulation. The set of these constants will be divided in two groups, the old and the new.

(i) Old material constants.

This group includes the elastic constants κ (or K) and G (or Poisson's ratio ν), the slope of the consolidation lines λ, the slope of the CSL $N = (1/3\sqrt{3})$ M and l_ℓ, eq. (27). Their determination follows well known methods and some of them can also be evaluated in the process of calibrating the new material constants as it will be shown subsequently. The l_ℓ although introduced as a new parameter is not related to the bounding surface concept and refers to a better description of the elastic response near the origin. It can be taken equal to 1 psi or about 10 kPa, but further investigation is suggested for the elastic response near the origin.

(ii) New material constants.

The new constants are R, A, m, h_c, and μ. The first two refer to the determination of the shape of the bounding surface (R for ellipse 1 and A_c for the hyperbola) while the last three are associated with the change of the plastic modulus for states within the surface. The R, A can be considered as appropriate constants for a classical yield surface formulation improving the surface shape.

The role of m has been explained earlier in connection with the use of $|x|^{-m}$ and does not effect the material response considerably except near $x = 0$.

It was found that m = 0.2 can be used for most clays. For this value the first bracket of eq. (31) varies between 2.58 and 1 as x changes from 0.10 to ∞, a variation which is negligible compared to the values of $\bar{K}_p$ and the other quantities entering eq. (31). The h_c, μ are the most important constants.

The following definite steps are now suggested for the calibration of M, R, G, h_c, μ and A from triaxial experiments once κ, λ have been defined and m = 0.2.

Step 1: Determination of M, R, G

After normal consolidation at p_o, obtain experimentally the undrained stress path in triaxial compression and determine the coordinates p, q of the failure point. From the relation q = Mp, the value of M can be determined and substitution of this relation in eq. (30) eliminates M and yields an equation for R in terms of p and p_o (κ and λ are assumed known). During this process, none of the other new constants are active. With κ, λ known, the corresponding q-ε_1 curve can be used to determine G by numerical curve fitting. If the test is undrained, as in the case of the given data, the q-ε_1 and ε_v-ε_1 curves can be used for the determination of M (failure point), G (initial slope) and R by numerical curve fitting.

Step 2: Determination of h_c, μ

Repeat the same undrained or drained loading as in step 1 but for an overconsolidated sample at an overconsolidation ratio between 1 and R, preferably at OCR = 3R/(2R + 1). Obtain the experimental curves q-p, q-ε_1, u-ε_1, or ε_v-ε_1 for both compression and extension. Determine h_c, h_e and μ = h_e/h_c by curve fitting the experimental data using the developed incremental relations. This can be done by a trial-and-error process observing that increasing h_c or h_e implies stiffer response. For this range of OCR the A does not appear in the constitutive equations because it is associated with the hyperbola which becomes active for the case of heavy overconsolidation.

Step 3: Determination of A

Repeat the same experiment as in step 2 for compression and for OCR $\geq$ 5 at least. With h_c known from step 2, A is determined by a similar trial-and-error numerical process. Increasing A implies a "flatter" hyperbola with a stiffer response and reduced dilatation, while a very small A brings the hyperbola down close to the CSL for materials with small "true" cohesion.

It is interesting to observe that the response to cyclic loading is obtained on the basis of the above state dependent formulation as a sequence of monotonic loading/unloading events without introducing any additional cyclic empirical parameter.

MODEL CALIBRATION FOR CLAYS X AND Y

Before the calibration of the model for clays X and Y, some general comments on the calibration procedures are presented. The steps for the determination of the new and some of the old material constants were explained in the previous section. They basically require a sequence of 3 conventional triaxial experiments in compression and 3 in extension (drained or undrained) at three different overconsolidation ratios: normally consolidated, slightly overconsolidated and heavily overconsolidated. If the difference in compression-extension is not considered, the extension tests can be omitted. It is an important advantage for the calibration of the model the fact that during the series of the above mentioned experiments the new material constants associated with the bounding surface formulation become active one after the other in sequence, permitting their successive evaluation. If the above experimental sequence is not fully available, as in the case of both clays X (slight or no overconsolidation only) and Y (large overconsolidation only), one must rely on past experience for similar clays and on a more extensive trial-and-error effort. This is necessary since more than one constant may be active for such an incomplete set of experimental data. In this case, an overall readjustment of the constants is often desirable in order to achieve a best fit over a large spectrum of given data. Recalling that the model assumes isotropic clays and no difference in compression-extension, a deviation from the above may require such a readjustment even in the case of the availability of the full experimental sequence. For clays X and Y the above reasons necessitate a slight readjustment which relies somewhat on the subjective evaluation of the user in the absence of any formalized optimization or regression analysis. Nevertheless, the suggested steps provide a really good set of initial values for the material constants. The reliability of these steps for the calibration when all the required data are available and isotropy is present has been successfully demonstrated in (14,16) where the material constants were easily evaluated by the above procedure using the experimental data reported by Banerjee and Stipho (2,3).

Initial State of the Material

The initial state requires knowledge of the initial void ratio, the corresponding confining pressure and the preconsolidation pressure. They can be determined from the given e-$\ln p$ curves that were provided for clays X and Y. The initial confining pressure and the preconsolidation pressure yield the overconsolidation ratio (OCR) which is an important quantity for the calibration since it defines the position of the initial stress point within or on the bounding surface (on the hydrostatic axis). A misinterpretation, however, of the preconsolidation pressure on the e-ℓnp curves resulted in an overestimation of the OCR as follows. When the curves were provided as part of the given data, it was not specified exactly what p represented. It could be the vertical stress σ_1 of a K_o consolidation or the average $(1/2)(\sigma_1 + \sigma_3)$ or the mean effective pressure $p = (1/3)(\sigma_1 + 2\sigma_3)$. Since no value of K_o was supplied, it was concluded that the preconsolidation pressure represented the latter while actually was the σ_1. Therefore, for clay X and a preconsolidation pressure of 23.25 psi the subsequent OCR for the confining pressures (in psi) σ_c = 10, 20, 30 and 40 were evaluated as 2.33, 1,17, 1 and 1, respectively. With the proper interpretation of the 23.25 psi as the σ_1 and an assumed K_o = 0.50, the "correct" OCR are 1.55, 1, 1 and 1, respectively ("correct" is put between " since K_o = 0.50 is simply a reasonable assumption). Similarly, for clay Y and a preconsolidation of 22.25 psi, the OCR for σ_c(psi) = 2.5, 5, and 10 were overestimated as 8.90, 4.45 and 2.23 respectively while the correct values are 7.92, 3.96 and 1.48, again for an assumed K_o = 0.50. This affected the calibration of certain material constants and the subsequent predictions to a small degree. Subsequently, and in the first volume, the calibration and predictions for both correct and overestimated OCR will be presented.

Determination of κ, λ

For the case of clay X, the given virgin compression curve yields a value of κ = 0.025. It should be noted that a swell-recompression curve has not been provided. For the case of clay Y, the virgin compression curve and swell-recompression curve yield values of κ = 0.020 and κ = 0.048, respectively. Hence, a value of κ = 0.03 was chosen for both clays X and Y, assuming that if a swell-recompression curve has been given for clay X, it would have increased slightly the value of κ obtained from the virgin compression curve similarly to clay Y.

The determination of λ from the given e-ℓnp curves is more difficult because it varies with pressure. It was estimated that for clay X: 0.25 ≤ λ ≤ 0.92 and for clay Y: 0.23 ≤ λ ≤ 0.63. Alternatively, use of the given index properties in the empirical relations λ ≃ 0.92 (PL-0.09), λ ≃ 0.36 (LL-0.09) and λ ≃ 0.585 (PI) suggested by Schofield and Wroth (26) where PL, LL and PI stand for the plastic limit, liquid limit and plasticity index, yield for both clays X and Y (same index properties) the values λ = 0.175, 0.140 and 0.117 respectively, with an index average value λ = 0.14. Although there is a difference between natural and remoulded clays, the above relations can still provide a reasonable measure. In view of the above variation the following procedure was followed.

For clay X which is normally consolidated or slightly overconsolidated, the value of λ was obtained as the average among the extreme values from the e-$\ln p$ curve on the basis of the average operating pressure, and the index average value 0.14 in order to account for the index properties. This yields a $\lambda = 0.35$ if the extreme values are averaged first, and a $\lambda = 0.45$ if the extreme values and the index average value are averaged together. The $\lambda = 0.35$ was used in the case of the overestimated OCR and the $\lambda = 0.45$ was found to give better results for the correct OCR which seems more logical because it is closed to the average value of the operating pressure. Clay Y has large OCR which implies that the corresponding operating pressure is associated with the lower extreme value $\lambda = 0.23$ for the e-$\ln p$ curve. Thus, λ was chosen as the average between 0.23 and the index average 0.14, i.e., $\lambda = 0.19$ for both correct and overestimated OCR.

It is evident from the above that the assumption of a constant λ for all pressures is not very accurate for natural clays and the problem was circumvented here by averaging. It is possible, however, to have a variable λ with $\ln p$ which would effect the relations (27), (28), but not the general development, and further research is suggested towards this direction.

Determination of M, R, G, h_c and A for Clay X

Assuming that the given last values of the stress for m=0 and m=1 (m is the stress ratio $(\sigma_2-\sigma_3)/(\sigma_1-\sigma_3)$ and should not be confused with the material constant m, eq. (31), which is taken equal to 0.20 for both clays X and Y) correspond to critical or near critical failure and plotting the associated stress measures p and q in the triaxial space, it was estimated that $1 \leq M \leq 1.20$. This variation may be attributed to the strength difference in compression (m=0) and extension (m=1) which is not considered by the present model, and/or experimental error. Following the procedure outlined in steps 1, 2, and 3 for drained experimental data, the final values of the above material constants are tabulated in Tables 1 and 2 for the correct and overestimated OCR, respectively. The value of μ was taken equal to 1 since no difference in compression-extension was considered. The experimental data used for the calibration together with the calibration curves for m=0, 1 are shown on Figures 3,5 for the correct OCR and Figures 4,6 for the overestimated OCR. The predictions for σ_c=40 psi and m=0 are also included. The figures for the same m but the different OCR are placed next to each other for easy comparison. The effect of the overestimation of the OCR is evident in Figs. 4,6 where a poor fitting was obtained especially for the ε_v-ε_1 data where the model predicts smaller volumetric strain for $\sigma_c = 20$ psi than $\sigma_c = 30$ psi.

This is expected since $\sigma_c = 20$ psi was considered a "stiffer" overconsolidated state contrary to the real case. The error in OCR effected also considerably the q-ε_1 curves. The improvement of the calibration with the correct OCR is obvious in Figs. 3,5 where also the predictions for $\sigma_c = 40$ psi are good. The correct OCR necessitated the change of the values of two constants from the ones used with the wrong OCR, i.e., M was changed from 1.05 to 1.20 and λ from 0.35 to 0.45. The A = 0.12 was active to a limited degree for $\sigma_c = 10$ psi. In general, A was not an important constant for clay X since no large overconsolidation was considered.

Determination of M, R, G, h_c and A for Clay Y

The full sequence of experimental data was not available for clay Y, mainly for normally consolidated states. This presented some difficulties in obtaining the proper value of R and h_c and the readjustment mentioned earlier was necessary. Following the prescribed procedures similarly to clay X, the values of the above material constants and the experimental data-calibration curves for correct and overestimated OCR and m=0,1 are shown in Tables 1 and 2 and Figures 7,9 (correct OCR) and 8,10 (overestimated OCR). The $\mu = 1$ for the same reasons as for clay X. Again, figures with the same m but the different OCR are juxtaposed for comparison. Observe the necessity to increase the value of G from 900 psi to 1500 psi and of h_c from 90 to 300 when the correct (and smaller) OCR were considered. This is expected since for smaller OCR, the model yields a "softer" response which must be compensated by increasing G and h_c. A small change of M from 1.15 to 1.20 and of R from 2.00 to 1.80 improved the calibration when the correct OCR were considered. The experimental data-calibration curves fit is satisfactory, especially if one considers the lack of the proper experimental sequence for the model calibration.

CONCLUSIONS

This workshop provided the forum for the presentation of the more "active" constitutive models in modern soil mechanics. One can find in the literature a plethora of stress-strain relations which are characterized as "constitutive". One basic condition of continuum mechanics, however, for the use of the work "constitutive" in association with a certain model is that the corresponding relations must be tensorial in character and should relate all components of stress or stress rates with all components of strain or strain rates. The complexity of soil behavior and the more empirical trend which characterized the historical development of soil mechanics coupled with the necessity to solve practical problems early in this development, introduced soil stress-strain laws which do not always satisfy the

TABLE 1.- Values of Material Constants for Correct OCR

Old Constants	Clay X	Clay Y	New Constants	Clay X	Clay Y
κ	0.03	0.03	R	2.00	1.80
λ	0.45	0.19	A	0.12	0.23
M	1.20	1.20	m	0.20	0.20
G(psi)	600	1500	h_c	120	300

TABLE 2.- Values of Material Constants for Overestimated OCR

Old Constants	Clay X	Clay Y	New Constants	Clay X	Clay Y
κ	0.03	0.03	R	2.00	2.00
λ	0.35	0.19	A	0.12	0.23
M	1.05	1.15	m	0.20	0.20
G(psi)	600	900	h_c	120	90

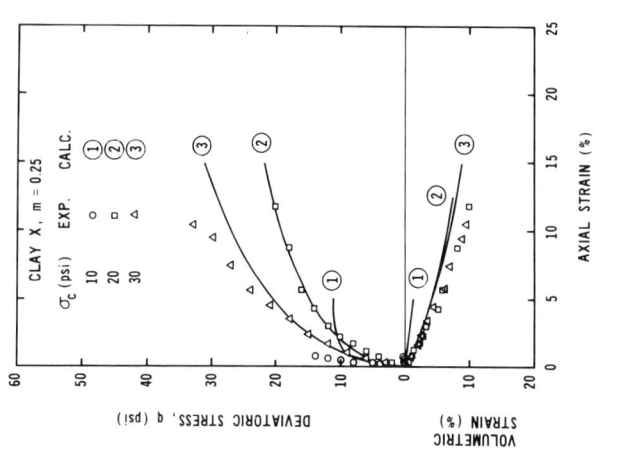

FIG. 3 - Experimental data and prediction curves for correct OCR.

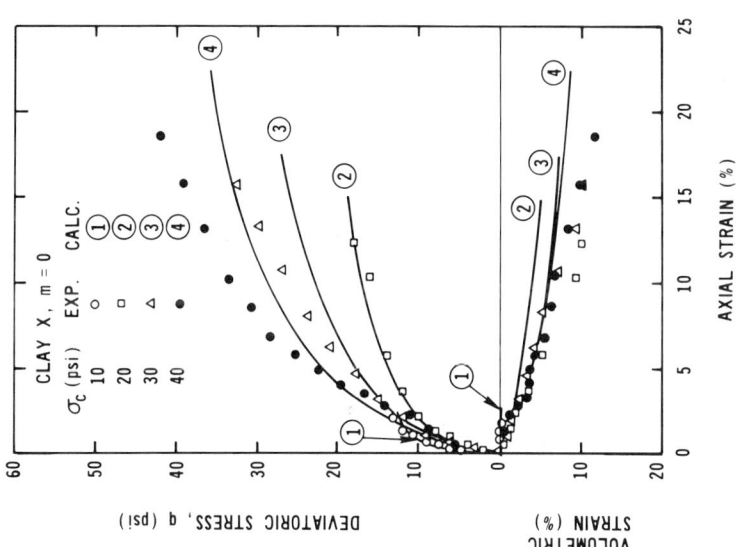

FIG. 4 - Experimental data and calibration curves for overestimated OCR.

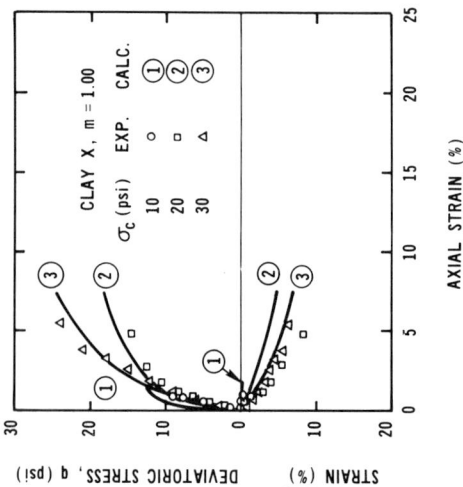

FIG. 6 – Experimental data and calibration curves for overestimated OCR.

FIG. 5 – Experimental data and calibration curves for correct OCR.

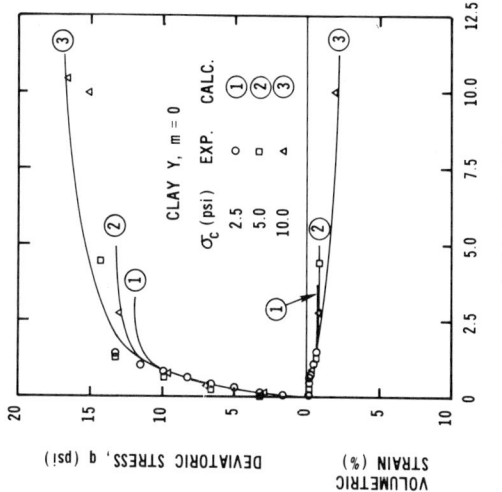

FIG. 8 – Experimental data and calibration curves for overestimated OCR.

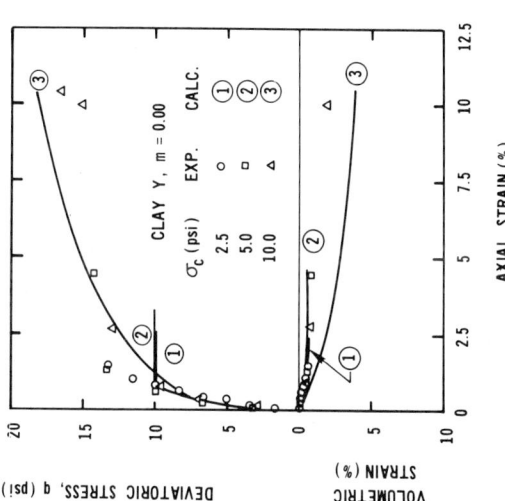

FIG. 7 – Experimental data and calibration curves for correct OCR.

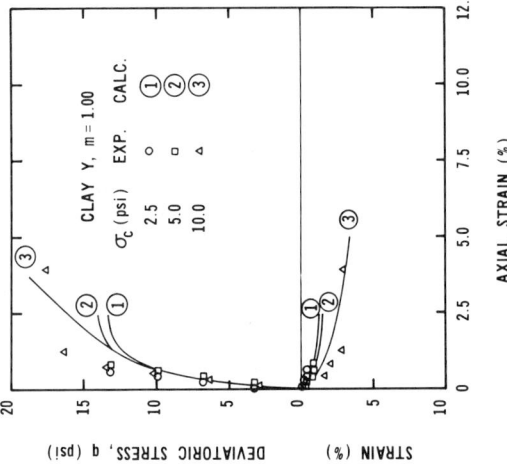

FIG. 10 – Experimental data and calibration curves for overestimated OCR.

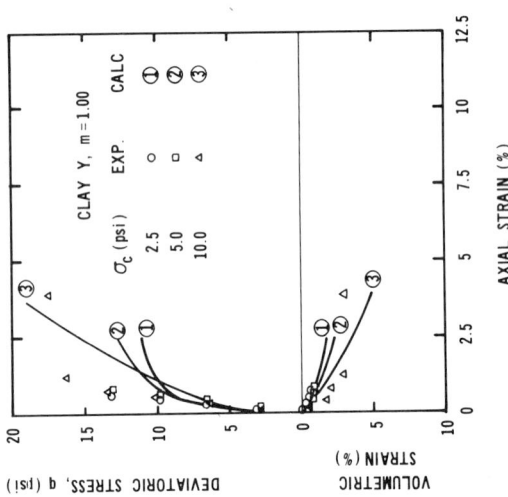

FIG. 9 – Experimental data and calibration curves for correct OCR.

above condition. The loading histories given for calibration and prediction in this workshop were of a general three-dimensional kind, thus only proper constitutive models in the above sense could be rigorously used.

Among the different models, one large class belongs to the general framework of elastoplasticity and the rest do not employ the concept of plastic strain distinct from that of an elastic strain, although irreversibility is incorporated. It is believed that since the soil response can be distinctly characterized by two different deformation mechanisms (elastic and plastic) on the basis of particulate mechanics, the elastoplastic concept is closer to the physical reality and has fundamental advantages over formulations which do no clearly reflect this distinction. The bounding surface model belongs to the elastoplastic class. The bounding surface concept, however, is not applicable only to soils and is not associated just with the specific model presented here, but provides a general framework within which classical elastoplasticity can become more effective for the description of the response of a variety of materials. Broadly speaking, the bounding surface concept reflects a general and simple trend of scientific thinking, that is the proper description of the "bounds" of a process, here the stress-strain laws, allows an easy and "safe" determination of the response at intermediate states.

There is no need to go over the specific features of the present bounding surface plasticity formulation. These can be found in the introduction and the main text of this paper. It will be more interesting to point out what is believed to be a set of important features that a constitutive model must possess and comment on the bounding surface model association with these features.

1. The model must be developed within a fundamental and physically acceptable framework for a class of material behavior, should not violate the basic principles of continuum mechanics, and should be general and tensorial in character. The bounding surface model is essentially a plasticity model and satisfies the above conditions, the tensorial and general character being a result of its formulation in terms of stress invariants and tensorial plastic internal variables.

2. The particular concept characterizing a model must be as simple as possible, if not simple enough. The concept of the "bounds" in any process is a very common and simple approach of scientific thinking and constitutes the foundation of the bounding surface concept and formulation.

3. The constitutive model must have the qualitative and quantitative predictive capabilities for a large spectrum of loading conditions. It is true that a "universal" constitutive model does not exist and may never exist for practical

purposes, but this is not the reason to reach the other extreme of predicting only the response under a very specific kind of loading condition. Especially for soils, it is not an excessive demand to require that one single constitutive model should reasonably well predilct the response under such common loading conditions and states as drained, undrained, normally consolidated and overconsolidated. If cyclic response can also be incorporated (being closely associated with overconsolidated states) then this is indeed a good model. It has been demonstrated (6,7,14-16) that the bounding surface model possesses these features. These features are particularly pertinent with respect to the required predictions for clay X (mostly normally consolidated) and clay Y (overconsolidation) and closely associated with the proper definition and incorporation into the model of the material state in terms of stresses and past loading history (here described by plastic volumetric deformation for isotropy). If a model can predict the response of clay X but not of Y or vice-versa, the model is deficient in the above sense. This is because normal consolidation and overconsolidation are nothing but two different states (due to different loading history) of the same material, and the "constitutive" character of a model which cannot predict both is on weak foundation. Inclusion of anisotropy is also a very desirable feature if one can properly identify the kind of anisotropy and can also incorporate the mechanism for its possible elimination. To this extent the experimental data from the same loading history applied to different orientations of a reference configuration are necessary. The present bounding surface model is isotropic since it was found preferable to treat clays X and Y as isotropic due to insufficient data to characterize anisotropy. For the same reasons, the predictions for the Kaolinite clay were not considered (see introduction). Anisotropy can be relatively easily included (9).

4. The necessary material constants must be easily evaluated by simple conventional experiments and must be as few as possible. The material constants should be distinguished from the model parameters examined in the following. Each material constant should be associated, if possible, with a particular loading condition which gives to the constant a more "graspable" physical meaning. A numerical curve fitting for the determination of these constants is a well accepted process, since one cannot expect that each constant can be exactly specified by a single experiment as for example the modulus of elasticity for metals (even then its determination is the simplest form of a straight-line curve fitting). The present bounding surface formulation possesses

this feature for the four old and four new (total of eight) necessary material constants (the number increases to 10 if extension is included). The new constants introduced become active one after the other in a series of three basic triaxial experiments in normal, light and heavy overconsolidation, permitting their successive and independent evaluation instead of an overall best fit optimization process. Some readjustment, however, of their initial values may be necessary.

5. A very important feature finally is the easy numerical implementation of the model for use in finite elements, finite differences, etc. computations. This is directly related to the number of the associated model parameters. These are the internal quantities (distinct from the material constants) which define the state of the model and the material during the loading history, and which must be kept in memory, up-dated and iterrated upon every incremental step of the computation. A large number of such parameters for each material point renders the application of the model difficult, expensive and perhaps impossible for large scale computations even with the present day computational capabilities. The present isotropic bounding surface formulation is particularly suited for such numerical implementation (15), requiring one single parameter: the plastic change of the void ratio (or plastic volumetric strain), or equivalently the corresponding consolidation pressure from the e-$\ln p$ curve.

In summary, one can refer to the following unique clay response characteristics which can be described by the bounding surface model: consolidation of normally consolidated or slightly overconsolidated states, up to critical failure with increasing deviatoric stress; initial consolidation with subsequent dilatation of heavily overconsolidated states up to critical failure with increasing deviatoric stress; corresponding negative or positive pore-water pressure development for normally and overconsolidated states subjected to undrained deviatoric loading; unstable behavior (falling q-ε_1 curve) for heavily overconsolidated states; capability to include tensile response with failure by loss of cohesion; and prediction of strain accumulation (drained) or pore-water pressure built-up (undrained) under cyclic deviatoric loading. All the above apply to both compression and extension, or a general three-dimensional loading.

Finally, one may describe the bounding surface concept best by the following phrase: its salient feature is that it allows for plastic deformation to occur for stress points within the bounding surface (by contrast to the classical yield surface formulation) at a rate which depends on the proximity of this point from a corresponding one on the surface, thus describing one of the most important behavioral characteristics of soils which do not possess a purely elastic range in the classical sense.

ACKNOWLEDGEMENT

The research reported in this paper was conducted, in part, under National Science Foundation Grant NSF-CME-70-10835.

REFERENCES

1. Al-Hussaini, M.M., and Townsend, F.C., "Investigation of Tensile Testing of Compacted Soils", Miscellaneous paper S-74-10, U.S. Army Engineer Waterways Experiment Station, Vicksburg, Mississippi, 1974.

2. Banerjee, P.K., and Stipho, A.S., "Associated and Non-Associated Constitutive Relations for Undrained Behavior of Isotropic Soft Clays", **International Journal for Numerical and Analytical Methods in Geomechanics**, Vol. 2, 1978, pp. 35-56.

3. Banerjee, P.K., and Stipho, A.S., "An Elastoplastic Model for Undrained Behaviour of Heavily Overconsolidated Clays", **International Journal for Numerical and Analytical Methods in Geomechanics** (Short Communication), Vol. 3, 1979, pp. 97-103.

4. Bazant, Z.P., and Krizek, R.J., "Endochronic Constitutive Law For Liquefaction of Sand", **Journal of the Engineering Mechanics Division,** ASCE, Vol. 102, No. EM2, April, 1976, pp. 225-238.

5. Dafalias, Y.F., "On Cyclic and Anisotropic Plasticity: i) A General Model Including Material Behavior under Stress Reversals, ii) Anisotropic Hardening for Initially Orthotropic Materials", thesis presented to the University of California at Berkeley, Ca, in 1975, in partial fulfillment of the requirements for the degree of Doctor of Philosophy.

6. Dafalias, Y.F., "A Model for Soil Behavior under Monotonic and Cyclic Loading Conditions", **Transactions of the 5th International Conference on SMiRT,** Berlin, Germany, Vol. K, No. 1/8, 1979.

7. Dafalias, Y.F., "A Bounding Surface Plasticity Model", **Proceedings of the 7th Canadian Congress of Applied Mechanics,** Sherbrooke, Canada, 1979, pp. 89-90.

8. Dafalias, Y.F., "Anisotropic Hardening of Initially Orthotropic Materials", **ZAMM,** Vol. 59, No. 9, 1979, pp. 437-446.

9. Dafalias, Y.F., "Initial and Changing Plastic Anisotropy for Pressure Sensitive Materials by Means of a Varying Non-associated Flow Rule", Report No. 80-1, Department of Civil Engineering, University of California, Davis, 1980.

10. Dafalias, Y.F., and Popov, E.P., "A Model of Nonlinearly Hardening Materials for Complex Loadings", **Proceedings of the 7th U.S. National Congress of Applied Mecahnics,** Boulder, USA, 1974, p. 149 (Abstract), and **Acta Mechanica,** Vol. 21, 1975, pp. 173-192.

11. Dafalias, Y.F., and Popov, E.P., "A Simple Constitutive Law for Artificial Graphite-like Materials", **Transactions of the 3rd International Conference on SMiRT,** London, U.K., Vol. C, No. 1/5, 1975.

12. Dafalias, Y.F., and Popov, E.P., "Plastic Internal Variables Formalism of Cyclic Plasticity", **Journal of Applied Mechanics,** Vol. 98, No. 4., 1976, pp. 645-650.

13. Dafalias, Y.F., and Popov, E.P., "Cyclic Loading for Materials with a Vanishing Elastic Region", **Nuclear Engineering and Design,** Vol. 41, No. 2, 1977, pp. 293-302.

14. Dafalias, Y.F., and Herrmann, L.R., "A Bounding Surface Soil Plasticity Model", **International Symposium on Soils under Cyclic and Transient Loading,** Swansea, U.K., Vol. 1, 1980, pp. 335-345.

15. Dafalias, Y.F., and Herrmann, L.R., "Development and Numerical Implementation of a Bounding Surface Constitutive Model for Cohesive Soils", Report No. 80-2, Department of Civil Engineering, University of California, Davis, 1980.

16. Dafalias, Y.F., and Herrmann, L.R., "Bounding Surface Formulation of Soil Plasticity", **Soils under Cyclic and Transient Load,** G.N. Pande and O.C. Zienkiewicz eds., John Wiley and Sons, Inc., New York, in press.

17. Dafalias, Y.F., and Herrmann, L.R., "A General Bounding Surface Plasticity Model for Cohesive Soils," in preparation.

18. Eisenberg, M.A. and Phillips, A., "A Theory of Plasticity with Non-Coincident Yield and Loading Surfaces", **Acta Mechanica,** Vol. 11, 1969, pp. 247-260.

19. Gudehus, G., "Elastoplastische Stoffleichungen fur trockenen Sand", **Ingenieur-Archiv,** Vol. 42, 1973.

20. Krieg, R.D., "A Practical Two-Surface Plasticity Theory", **Journal of Applied Mechanics,** Vol. 42, 1975, pp. 641-646.

21. Mroz, Z., "On the Description of Anisotropic Work Hardening", **Journal of the Mechanics and Physics of Solids,** Vol. 15, 1967, pp. 163-175.

22. Mroz, Z., Norris, V.A., and Zienkiewicz, O.C., "An Anisotropic Hardening Model for Soils and its Application to Cyclic Loading", **International Journal for the Numerical and Analytical Methods in Geomechanics,** Vol. 2, 1978, pp. 203-221.

23. Mroz, Z., Norris, V.A., and Zienkiewicz, O.C., "Application of an Anisotropic Hardening Model in the Analysis of Elastoplastic Deformation of Soils", **Geotechnique,** Vol. 29, No. 1, 1979, pp. 1-34.

24. Prevost, J.H., "Plasticity Theory for Soil Stress-Strain Behavior", **Journal of the Engineering Mechanics Division,** ASCE, Vol. 104, No. EM5, 1978, pp. 1177-1196.

25. Roscoe, K.H., and Burland, J.B., "On the Generalized Stress-Strain Behaviour of Wet Clay", **Engineering Plasticity,** J. Heyman and F.A. Leckie eds., Cambridge University Press, Cambridge, U.K., 1968, pp. 535-609.

26. Schofield, A.N. and Wroth, C.P., **Critical State Soil Mechanics,** McGraw-Hill, London, 1968.

27. Valanis, K.C., "A Theory of Viscoplasticity without a Yield Surface, Part I: General Theory, Part II: Application to Mechanical Behavior of Metals", **Archives of Mechanics,** Vol. 23, 1971, pp. 517-551.

28. Valanis, K.C., and Read, H.E., "A New Endochronic Plasticity Model for Soils", Report SSS-R-80-4294, Systems, Science and Software, La Jolla, California, Dec., 1979.

CONSTITUTIVE THEORY FOR SOIL

Jean H. Prevost[1], M. ASCE

INTRODUCTION

Soil consists of an assemblage of particles with different sizes and shapes which form a skeleton whose voids are filled with various fluids. The stresses carried by the soil skeleton are conventionally termed "effective stresses" [9] in the soil mechanics literature, and those in the fluids are called "pore-fluid pressures". It is observed experimentally that the stress-strain behavior of the soil skeleton is strongly nonlinear, anisotropic and hysteretic. In order to relate the changes in effective stresses carried by the soil skeleton to the skeleton rate of deformation, a general analytical model [5] which describes the nonlinear, anisotropic, elastoplastic, stress and strain dependent, stress-strain-strength properties of the soil skeleton when subjected to complicated three-dimensional, and in particular to cyclic loading paths [3] is used. A brief summary of the model's basic principle is included in the following and the constitutive equations are provided. It is shown that the model parameters required to characterize the behavior of any given soil can be derived entirely from the results of conventional soil tests. The model's accuracy is thereafter evaluated by applying it to represent the behavior of both cohesive and cohesionless soils under both drained and undrained loading conditions for various stress paths.

---

[1] Assistant Professor of Civil Engineering, Princeton University, Princeton, NJ 08544.

CONSTITUTIVE EQUATIONS

The constitutive equations for the solid skeleton are written in one of the following forms:

$$C_{abcd}d_{cd} = \begin{cases} \dot{\sigma}'_{ab} & \text{small deformations} \\ \overset{\triangledown}{\sigma}'_{ab} + \sigma'_{ab}v_{c,c} & \text{finite deformations} \end{cases} \quad (1)$$

in which σ' = effective Cauchy stress tensor; $\underset{\sim}{v}$ = (spatial) velocity of solid phase; $\underset{\sim}{d}$ = rate of deformation tensor for the solid phase (= symmetric part of the spatial solid velocity gradient); $\overset{\triangledown}{\underset{\sim}{\sigma}}'$ = Jaumann derivative, viz

$$\overset{\triangledown}{\underset{\sim}{\sigma}}' = \dot{\underset{\sim}{\sigma}}' + \underset{\sim}{\sigma}' \cdot \underset{\sim}{w} - \underset{\sim}{w} \cdot \underset{\sim}{\sigma}' \quad (2)$$

where w = spin tensor (= skew-symmetric part of the spatial solid velocity gradient). In Eq. 1 C_{abcd} is an (objective) tensor valued function of, possibly, σ'_{ab} and the deformation gradients. Many nonlinear material models of interest can be put in the above form (e.g. all nonlinear elastic materials, and many elasto-plastic materials). The finite deformation form of the constitutive equation above was first proposed by Hill [1] in the context of plasticity theory.

For soil media, the form of the $\underset{\sim}{C}$ tensor is given as follows [5],

$$\underset{\sim}{C} = \underset{\sim}{E} - \frac{1}{H' + \underset{\sim}{Q}:\underset{\sim}{E}:\underset{\sim}{P}} (\underset{\sim}{E}:\underset{\sim}{P})(\underset{\sim}{Q}:\underset{\sim}{E}) \quad (3)$$

in which H' is the plastic modulus; $\underset{\sim}{P}$ and $\underset{\sim}{Q}$ are symmetric second-order tensors, such that $\underset{\sim}{P}$ gives the direction of plastic deformations and $\underset{\sim}{Q}$ the outer normal to the active yield surface; and $\underset{\sim}{E}$ is the fourth-order tensor of elastic moduli, assumed isotropic for the particular class of material models implemented. The plastic potential is selected such that, in agreement with experimental observations, the plastic deviatoric rate of deformation vector remains normal to the projection of the yield surface onto the deviatoric stress subspace, i.e.,

$$\underset{\sim}{P} - \frac{1}{3}(\text{trace }\underset{\sim}{P})\underset{\sim}{1} = \underset{\sim}{Q} - \frac{1}{3}(\text{trace }\underset{\sim}{Q})\underset{\sim}{1} = \underset{\sim}{Q}' \quad (4a)$$

$$\text{trace }\underset{\sim}{P} = \text{trace }\underset{\sim}{Q} + A\frac{\text{trace }\underset{\sim}{Q}}{|\text{trace }\underset{\sim}{Q}|}\{\underset{\sim}{Q}':\underset{\sim}{Q}'\}^{\frac{1}{2}} \quad (4b)$$

in which A is a material parameter which measures the departure from an associative flow rule. When $A = 0$, $\underset{\sim}{P} = \underset{\sim}{Q}$ and consequently the $\underset{\sim}{C}$ tensor possesses the major symmetry. The yield function is selected of the following form [5]

$$f = \frac{3}{2}(\underset{\sim}{s}-\underset{\sim}{\alpha}):(\underset{\sim}{s}-\underset{\sim}{\alpha})+C^2(p'-\beta)-k^2 = 0 \quad (5)$$

where $\underset{\sim}{s}$ is the deviatoric stress tensor (i.e. $\underset{\sim}{s} = \underset{\sim}{\sigma}' - p'\underset{\sim}{1}$, $p' = \frac{1}{3}$ trace $\underset{\sim}{\sigma}'$) α and β are the coordinates of the center of the yield surface in the deviatoric stress space and along the hydrostatic stress axis, respectively; k is the size of the yield surface; and C is a material parameter called the yield surface axis ratio. In order to allow for the adjustment of the plastic hardening rule to any kind of experimental data, for example data obtained from axial or simple shear soil tests, a collection of nested yeild surfaces is used. A plastic modulus is associated with each of the yield surfaces, and

$$H' = h' + \frac{\text{trace }\underset{\sim}{Q}}{\{3\underset{\sim}{Q}:\underset{\sim}{Q}\}^{\frac{1}{2}}} B' \quad (6)$$

where h' is the plastic shear modulus and (h' + B') and (h' - B') are the plastic bulk moduli associated with f which are mobilized in consolidation tests upon loading and unloading, respectively. The projections of the yield surfaces onto the deviatoric stress subspace thus define regions of constant plastic shear moduli.

The yield surfaces' initial positions and sizes reflect the past stress-strain history of the soil skeleton, and in particular their initial positions are a direct expression of the material "memory" of its past loading history. Because the α's are not necessarily all equal to zero, the yielding of the material is anisotropic. Direction is therefore of great

importance and the physical reference axes (x,y,z) are fixed with respect
to the material element and specified to coincide with the reference axes
of consolidation. For a soil element whose anisotropy initially exhibits
rotational symmetry about the y-axis, $\alpha_x = \alpha_z = -\alpha_y/2$ and Eq. 5 simplifies
to

$$[(\sigma_y' - \sigma_x') - \alpha]^2 + C^2(p' - \beta)^2 - k^2 = 0 \qquad (7)$$

in which $\alpha = 3\alpha_y/2$. The yield surfaces then plot as ellipses in the axisymmetric stress plane ($\sigma_x' = \sigma_z'$) as shown in Fig. 1. Points C and E on the
outermost yield surface define the critical state conditions (i.e. H' = 0)
for axial compression and extension loading conditions, resepctively [2].
It is assumed that the slopes of the critical state lines OC and OE remain
constant during yielding.

The yield surfaces are allowed to change in size as well as to be
translated by the stress point. Their associated plastic moduli are also
allowed to vary and in general both k and H' are functions of the plastic
strain history. They are conveniently taken as functions of invariant
measures of the amount of plastic volumetric strains and/or plastic shear
distortions, respectively [3].

Complete specification of the model parameters requires the determination of

(i)   the initial positions and sizes of the yield surfaces
      together with their associated plastic moduli;

(ii) their size and/or plastic modulus changes as loading
      proceeds, and finally,

(iii) the elastic shear G and bulk B moduli.

The soil's anisotropy originally develops during its deposition and subsequent consolidation which, in most practial cases, occurs under no lateral
deformations. In the following, the y-axis is vertical and coincides with the
direction of consolidation, the horizontal xz-plane is thus a plane of material's
isotropy and the material's anisotropy initially exhibits rotational symmetry about the vertical y-axis. The model parameters required to charac-

terize the behavior of any given soil can then be derived *entirely* from the results of conventional monotonic axial and cyclic strain-controlled simple shear soil tests [3-6]. This is explained and further discussed in the following.

DETERMINATION OF MODEL PARAMETERS

In order to follow common usage in soil mechanics, compressive stresses and strains are considered positive in the following. All stresses are effective stresses unless otherwise specified.

As explained previously, for a material which initially exhibits cross-anisotropy about the vertical y-axis, the initial position in stress space of the yield surfaces is defined by the sole determination of the two parameters $\alpha^{(m)}$ and $\beta^{(m)}$ (m=1,...,p) and Eq. 7 simplifies to

$$[q - \alpha^{(m)}]^2 + C^2[p' - \beta^{(m)}]^2 - [k^{(m)}]^2 = 0 \tag{8}$$

for axial loading conditions (i.e., $\sigma'_x = \sigma'_z$ and $\tau_{xy} = \tau_{yz} = \tau_{zx} = 0$), where $q = (\sigma'_y - \sigma'_x)$. The yield surfaces then plot as circles in the q versus Cp' plane (referred to as the axial stress plane hereafter) as shown in Fig. 2. When the stress point reaches the yield surface f_m,

$$q = \alpha^{(m)} + k^{(m)} \sin\theta \tag{9a}$$

$$p' = \beta^{(m)} + \frac{k^{(m)}}{C} \cos\theta \tag{9b}$$

where θ is defined in Fig. 2, and Eq. 3 simplifies to:

$$\frac{\dot{\varepsilon}}{\dot{q}} = \frac{1}{2G} + \frac{1}{H'_m} \sin\theta \,(\sin\theta + C\gamma \cos\theta) \tag{10a}$$

$$\frac{\dot{\varepsilon}}{\dot{p}'} = \frac{1}{B} + \frac{1}{H'_m} (2c \cos\theta + A_m \sqrt{6} \cos\theta |\tan\theta|) \frac{1}{3\gamma} (\sin\theta + C\gamma \cos\theta) \tag{10b}$$

in which (from Eq. 6)

$$H'_m = h'_m + B'_m \cos\theta \tag{11}$$

and

$$\dot{\varepsilon}_v = \text{trace } \underline{\dot{d}} = \dot{\varepsilon}_y + 2\dot{\varepsilon}_x \tag{12}$$

$$\dot{\bar{\varepsilon}} = (\dot{\varepsilon}_y - \dot{\varepsilon}_x) \tag{13}$$

$$\gamma = \frac{\dot{p}'}{\dot{q}} \tag{14}$$

G = elastic shear modulus, B = elastic bulk modulus, $C = 3/\sqrt{2}$. The dependence of the model parameters upon the effective mean normal stress and volumetric strain are assumed to be of the following forms:

$$x = x_1 \left(\frac{p'}{p'_1}\right)^n \tag{15}$$

and

$$y = y_1 \exp(\lambda \, \varepsilon_v) \tag{16}$$

respectively, where $x = B$, G, h'_m, B'_m and $y = \alpha^{(m)}$, $\beta^{(m)}$, $k^{(m)}$; and n are experimental parameters ($n = 0.5$ for most cohesionless soils [7], and $n = 1$ for most cohesive soils); p'_1 = initial effective mean normal stress (i.e. at $\varepsilon_v = 0$); $\varepsilon_v = (v_1 - v)/v_1$ where v = current volume and v_1 = initial volume (i.e., at $p' = p'_1$ and $q = q_1$) of the soil specimen.

When the stress point reaches the outermost yield surface f_p, the soil specimen is in a normally consolidated state. It is assumed that the consolidation soil test results then plot as a straight line parallel to the projections of critical state lines on the ε_v versus Ln p'

plane [8]. The parameter λ is then simple determined from the results of K_0-consolidation soil test results, viz

$$\lambda = \frac{1}{P_K'} \frac{\dot{P}_K'}{\dot{\epsilon}_v^K} \tag{17}$$

where the subscript/superscript K refers to K_0-loading conditions.

(i) Interpretation of Monotonic Drained Axial Compression and Extension Soil Test Results

Let θ_C and θ_E denote the values of θ (Eq. 9) when the stress point reaches the yield surface f_m in axial compression and extension loading conditions. Combining Eqs. 9-16, one finds that:

$$\frac{1}{\tan\theta_C} \pm \frac{1}{\tan\theta_E} = \frac{1}{2C} [3 \gamma_C \frac{x_C}{y_C} \pm 3 \gamma_E \frac{x_E}{y_E}] \tag{18}$$

$$\cos\theta_C - \cos\theta_E = R_{CE}(\sin\theta_C - \sin\theta_E) \tag{19}$$

in which

$$R_{CE} = C \frac{P_C' - P_E' \exp[\lambda(\epsilon_v^C - \epsilon_v^E)]}{q_C - q_E \exp[\lambda(\epsilon_v^C - \epsilon_v^E)]} \tag{20}$$

$$\frac{1}{x_C} = (\frac{P_C'}{P_1'})^n \frac{\dot{\epsilon}}{\dot{q}} - \frac{1}{2G_1} \tag{21}$$

$$\frac{1}{y_C} = (\frac{P_C'}{P_1'})^n \frac{\dot{\epsilon}_v}{\dot{p}} - \frac{1}{B_1} \tag{22}$$

and similarly for x_E and y_E, where the subscripts C and E refer to axial compression and extension loading conditions, respectively. In Eq. 18, the plus sign (+) is to be used when $\tan\theta_C \tan\theta_E < 0$, and the minus sign (-), when $\tan_C \tan_E > 0$.

The smooth experimental stress-strain curves obtained in axial tests are now to be approximated by linear segments along which the tangent (or secant) modulus is constant. Evidently, the degree of accuracy achieved by such a representation of the experimental curves is directly dependent upon the number of linear segments used. The model parameters associated with the yield surface f_m are determined by the condition that the slopes $\dot{q}/\dot{\varepsilon}$ are to be the same in axial compression and extension tests when the stress point has reached the yield surface f_m. The corresponding values θ_C and θ_E are determined by combining Eqs. 18 and 19. For that purpose, note that Eq. 19 is more conveniently rewritten as

$$\left(\frac{1}{\tan\theta_C} + \frac{1}{\tan\theta_E}\right) + \frac{2 R_{CE}}{1-R_{CE}^2} \left(\frac{1}{\tan\theta_C} \frac{1}{\tan\theta_E} - 1\right) = 0 \tag{23}$$

Once θ_C and θ_E have been determined, the model parameters associated with f_m are simply obtained by combining Eqs. 9 and 10, viz.

$$B_m' = \frac{x_C \sin\theta_C \, TC - x_E \sin\theta_E \, TE}{\cos\theta_C - \cos\theta_E} \tag{24}$$

$$h_m' = x_C \sin\theta_C \, TC - B_m' \cos\theta_C \tag{25}$$

$$A_m \sqrt{6} = \frac{1}{|\tan\theta_C|} \left[3\gamma_C \frac{x_C}{y_C} \tan\theta_C - 2C \right] \tag{26}$$

$$k_1^{(m)} = \frac{q_C \exp(-\lambda \varepsilon_v^C) - q_E \exp(-\lambda \varepsilon_v^E)}{\sin\theta_C - \sin\theta_E} \tag{27}$$

$$\alpha_1^{(m)} = q_C \exp(-\lambda \varepsilon_v^C) - k_1^{(m)} \sin\theta_C \tag{28}$$

$$\beta_1^{(m)} = p_C' \exp(-\lambda \varepsilon_v^C) - \frac{k_1^{(m)}}{C} \cos\theta_C \qquad (29)$$

where

$$TC = \sin\theta_C + C \gamma_C \cos\theta_C \qquad (30a)$$

$$TE = \sin\theta_E + C \gamma_E \cos\theta_E \qquad (30b)$$

The yield surface f_1 is chosen as a degenerate yield surface of size $k_1^{(1)} = 0$. Further, in order to get a smooth transition from the elastic into the plastic regime $H_1' = \infty$, so that the material behavior inside f_2 is elastic. The elastic shear, G_1, and bulk, B_1, moduli are then determined from the steepest slopes observed at the origin of the plots q versus $\bar{\varepsilon}$, and p' versus ε_v of the axial compression/extension soil test results, respectively.

At the critical state, $x_C = x_E = 0$, and the stress point is on the outermost yield surface f_p. From Eqs. 24 and 25, it is apparent that $h_p' = 0$ in that case.

(ii) Interpretation of Monotonic Undrained Axial Compression and Extension Soil Test Results

In undrained tests, $\dot{\varepsilon}_v = 0$, and (from Eq. 22) $y_C = y_E = -B_1$ in that case. The model parameters associated with the yield surface f_m are again determined by the condition that the slopes $\dot{q}/\dot{\bar{\varepsilon}}$ are to be the same in axial compression and extension tests when the stress point has reached the yield surface f_m. As previously, the corresponding values θ_C an θ_E are determined from Eqs. 18 and 19, in which

$$R_{CE} = C \frac{p_C' - p_E'}{q_C - q_E} \,. \qquad (31)$$

Knowing θ_C and θ_E, the model parameters associated with f_m are computed using Eqs. 24-30, in which $\varepsilon_v^C = \varepsilon_v^E = 0$.

Note that the sole use of undrained axial soil test results for the determination of the model parameters, does not allow the determination of the parameter λ.

(iii) Interpretation of Simple Shear Soil Test Results

In simple shear soil tests, $\dot{\varepsilon}_x = \dot{\varepsilon}_y = \dot{\varepsilon}_z = 0$. The necessary algebra for the determination of the model parameters is considerably simplified in that case if the elastic constributions to the normal strains is neglected. Eqs. 24-29 then yields: $\sigma_x' = \sigma_z'$,

$$\frac{\dot{\gamma}_{xy}}{\dot{\tau}_{xy}} = \frac{1}{G} + \frac{2}{h_m'} \tag{32}$$

$$\alpha^{(m)} = (\sigma_y' - \sigma_x') \tag{33}$$

$$k^{(m)} = \sqrt{3}\,\tau_{xy} \tag{34}$$

$$\beta^{(m)} = p_S' = \sigma_y' - \frac{2}{3}(\sigma_y' - \sigma_x') \tag{35}$$

The model parameters associated with the yield surface f_m are then simply determined from the above equations and a piecewise linear representation of the shear stress-strain curves obtained in a simple shear test. Note that the sole use of simple shear test results does not allow the determination of the parameters λ, B_m' and A_m. On the other hand, it is apparent from Eqs. 32 and 34 that the degradation of the mechanical properties of the material under cyclic shear loading conditions, i.e.

$$k^{(m)}(\bar{e}) \qquad h_m'(\bar{e}) \tag{36}$$

with

$$\bar{e} = \int \{\tfrac{2}{3}\underline{d}':\underline{d}'\} \qquad \underline{d}' = \underline{d} - \tfrac{1}{3}(\text{trace }\underline{d})\underline{\delta} \tag{37}$$

where the integration is carried along the strain path, are most conveniently determined from the results of cyclic strain-controlled simple shear tests ($\bar{e} = \int \tfrac{1}{\sqrt{3}} |\dot{\gamma}|$ in that case). This is explained and further discussed in Ref. [3].

The procedures described above for the determination of the model parameters may be easily coded, and a simple computer code has been written for that purpose. This code is available to interested readers.

MODEL EVALUATION

The soil data to be used in this section are the ones which were collected by the organizing committee of the NSF/NSERC North American Workshop on plasticity and generalized stress-strain applications in soil engineering held May 28-30, 1980 at McGill University, Montreal, Canada. Laboratory test data on three different soild had been transmitted to the author early in January 1980.

The soils involved had been identified as

(1) natural clay
(2) a laboratory-prepared clay, and
(3) a sand.

All the test results had been obtained from carefully controlled laboratory experiments. Predictions about the constitutive behavior of these soils in the same experimental tests but subjected to loading stress paths not identified in the data had been requested by the organizing committee. This section describes the test results, their analysis, and compares the model predictions with observed behavior in the tests.

1. Natural Leda Clay

The experimental tests had been conducted on prismatic samples in a cubical testing device. The experimental data transmitted to the author hinted that the clay might be orthotropic. In order to determine the appropriate model parameters, extensive modifications to the computer program would have been required. Due to time constraints and work force limitations, it was decided not to undertake that venture.

2. Laboratory Prepared Kaolinite Clay

The experimental tests had been conducted on cylindrical samples in a torsional shear testing device. All samples had first been K_o-consolidated with a cell pressure of 58 psi and a backpressure of 18 psi, and then left to rebound to an equal-all-around cell pressure of 58 psi and a back pressure of 18 psi (in other words, the excess axial load necessary for K_o-consolidation was then released). All the tests were stress-controlled and performed under constant volume conditions (i.e. undrained).

Figs. 3 and 4 show in dashed-lines the experimental results obtained in conventional monotonic axial compression/extension soil tests, and in solid-lines the design curves used to determine the model parameters for that clay. Note that some data points close to failure have been ignored when selecting the design curves because these do not seem consistent with the rest of the data. This inconsistency may be due to experimental difficulties in capturing failure states in stress-controlled testing devices.

Table 1 gives the model parameter values for the Kaolinite clay. Tables 2 - 4 give in tabular form the model predictions for shear tests in which the major principal stress is inclined at $\theta = 15°$, $45°$ and $75°$, respectively, to the vertical axis of the soil specimen. Figs. 5 - 25 show a comparison between predicted (solid-lines) and observed (dashed-lines) behavior of the soil in these tests.

Note that except for the predicted pore-water pressure changes when $\theta = 75°$, all the model predictions agree well with the experimental test results.

3. Ottawa Sand

The experimental tests had been conducted on prismatic samples in a cubical testing device under stress-controlled loading conditions. All samples had first been consolidated under an equal-all-around stress before shearing. Three sets of experimental test results starting at 5 psi, 10 psi and 20 psi equal-all-around stress, respectively, had been provided. These included:

(1) At 5 psi: results of a conventional axial compression test and of axial compression/extension tests in which the mean normal stress remained constant.

(2) At 10 psi: results of a conventional axial compression test, of a non-conventional extension test in which the mean normal stress increased, of an axial compression test in which the mean normal stress remained constant, of a shear test in which both the vertical stress and one lateral stress increased and the other lateral stress decreased such that the mean normal stress remained constant.

(3) At 20 psi: results of axial compression/extension tests in which the mean normal stress remained constant.

As explained previously, complete determination of the model parameters requires information about the shearing behavior of the material in both loading compression (i.e., $\dot{q}/\dot{p} > 0$, $\dot{q} > 0$) and unloading extension (i.e., $\dot{q}/\dot{p} > 0$, $\dot{q} < 0$). It is therefore apparent that the set of data provided is incomplete in that respect since it contains no information about the behavior of the material upon unloading (i.e. shearing with decreasing mean normal stress). Another difficulty encountered in using these data is that shearing was not purely monotonic but included cycles of loading-unloading-reloading. Some adjustments had therefore to be made in order to correct the data for such cycles of shearing as shown hereafter.

Figs. 26-28 and 35-37 show in dashed-lines the experimental results obtained in axial compression/extension tests at a constant mean stress of 5 psi and 20 psi, respectively, and in solid-lines the design curves used to determine the model parameters. Tables 5 and 7 give the model parameter values for the Ottawa sand at 5 psi and 20 psi, respectively. Note that the parameter values β and B' associated with each of the yield surfaces could not be determined from the available test data for the reasons spelled out above. The model is therefore incomplete in that form, and can be used to make statements about the behavior of that sand at constant mean normal stress only. Tables 6 and 8 give in tabular form the model predictions for pure shear loading conditions at 5 psi and 20 psi, resptectively. Figs. 29-34 and 38-43 show a comparison between predicted (solid-lines) and observed (dashed-lines) behavior of the sand in these tests. It is seen that good agreement with the experimental test results is achieved in all cases especially at low strain levels.

Figs. 44-50 show in solid-line the experimental results obtained in axial compression/extension tests with increasing mean normal stress starting at 10 psi. It is apparent by inspection that these tests suffered from many anomalies. Further, the value derived from these tests for the initial shear modulus G = 5555.56 psi is not consistent with the values observed at 5 psi (G = 4695 psi) and 20 psi (G = 9375 psi). It was therefore decided to disregard these tests, and no predictions were made at 10 psi.

ACKNOWLEDGEMENTS

Computer time was provided by Princeton University Computer Center.

REFERENCES

[1] Hill, R., "A General Theory of Uniqueness and Stability in Elastic-Plastic Solids," J. Mech. Phys. Solids, Vol. 6, 1958, pp. 236-249.

[2] Hvorslev, M.J., "Uber die Festigkeitseigenschaften gestorler bindiger boden," Ingeniozvidenskabelige Skrifter, A. no. 45, Copenhagen, 1937.

[3] Prevost, J.H., "Mathematical Modeling of Monotonic and Cyclic Undrained Clay Behavior," Int. J. Num. Analyt. Methods in Geomechanics, Vol. 1, No. 2, 1977, pp. 196-216.

[4] Prevost, J.H., "Anisotropic Undrained Stress-Strain Behavior of Clays," J. Geotech. Eng. Div., ASCE, Vol. 104, No. GT8, 1978, pp. 1075-1090.

[5] Prevost, J.H., "Plasticity Theory for Soil Stress-Strain Behavior," J. Eng. Mech. Div., ASCE, Vol. 104, NO. EM5, 1978, pp. 1177-1194.

[6] Prevost, J.H., "Mathematical Modeling of Soil Stress-Strain-Strength Behavior," Proceedings, 3rd. Int. Conf. Num. Methods in Geomechanics, Aachen, Germany, Vol. 1, 1979, pp. 347-361.

[7] Richard R.E., Woods, R.D. and Hall, J.R., Vibrations of Soils and Foundations, Prentice-Hall, Inc., Englewood Cliffs, New Jersey, 1970.

[8] Roscoe, K.H. and J.B. Burland, "On the Generalized Stress-Strain Behavior of 'Wet' Clay," in Engineering Plasticity, Heyman and Leckis, eds., Cambridge University Press, Cambridge, England, 1968, pp. 535-609.

[9] Terzaghi, K., Theoretical Soil Mechanics, Wiley, New York, 1943.

TABLE 1

KAOLINITE CLAY σ'_{vc} = 40 psi   MODEL PARAMETERS

G_1 = 3121 psi; B_1 = 8323 psi;   n = 1.0; λ = 15.6

m	$\alpha^{(m)}$ (psi)	$\beta^{(m)}$ (psi)	$k^{(m)}$ (psi)	h'_m (psi)	B'_m (psi)	A_m
2	3.120	39.11	15.60	6132.0	-1109.0	-0.25
3	6.303	39.01	18.66	2893.0	-556.1	-0.27
4	9.423	39.00	21.72	1712.0	-365.4	-0.24
5	12.530	39.21	24.86	1701.0	-258.0	-0.27
6	15.650	39.61	28.08	867.5	-56.5	-0.32
7	12.870	37.56	38.56	267.0	-66.7	-0.58
8	8.850	38.25	48.67	0.0	--	-0.49

TABLE 2

KAOLINITE CLAY

SHEAR TEST NO. 2 - Θ = 15°

TOTAL PRINCIPAL STRESSES (PSI)			PORE PRESSURE (PSI)	PRINCIPAL STRAINS (%)		
σ_3	σ_2	σ_1	p_w	ε_3	ε_2	ε_1
58.00	58.00	58.00	18.10	0.000	0.000	0.000
56.50	58.00	78.90	24.57	-0.128	-0.104	0.231
55.98	58.00	86.22	26.87	-0.228	-0.179	0.408
55.51	58.00	92.69	28.28	-0.395	-0.295	0.690
55.22	58.00	96.78	29.00	-0.496	-0.365	0.861
54.99	58.00	100.01	29.34	-0.617	-0.444	1.060
54.82	58.00	102.28	29.60	-0.725	-0.513	1.238
54.68	58.00	104.32	30.03	-0.838	-0.586	1.424
54.58	58.00	105.72	30.10	-0.953	-0.660	1.613
54.42	58.00	107.88				
54.30	58.00	109.60				

TABLE 3

KAOLINITE CLAY

SHEAR TEST NO. 7 - θ = 45°

TOTAL PRINCIPAL STRESSES (PSI)			PORE PRESSURE (PSI)	PRINCIPAL STRAINS (PSI)		
σ_3	σ_2	σ_1	p_w	ε_3	ε_2	ε_1
58.00	58.00	58.00	18.00	0.000	0.000	0.000
51.90	58.00	64.10	18.00	-0.098	0.000	0.098
47.90	58.00	68.10	18.05	-0.175	0.002	0.173
45.90	58.00	70.10	18.41	-0.272	0.018	0.255
43.90	58.00	72.10	19.87	-0.426	0.054	0.373
41.90	58.00	74.10	21.32	-0.728	0.135	0.593
38.20	58.00	77.80	22.16	-1.554	0.324	1.229
36.60	58.00	79.40	22.41	-2.067	0.422	1.645
35.00	58.00	81.00	22.44	-2.621	0.517	2.105
34.20	58.00	81.80	22.57	-3.364	0.634	2.730
33.50	58.00	82.50	22.89	-3.972	0.722	3.250
32.60	58.00	83.40	23.31	-4.756	0.822	3.934
31.90	58.00	84.10	23.72	-5.368	0.892	4.476
31.80	58.00	84.20	23.69	-5.456	0.901	4.554
31.60	58.00	84.40	23.79	-5.631	0.919	4.711

TABLE 4

KAOLINITE CLAY

SHEAR TEST NO. 12 - $\theta = 75°$

TOTAL PRINCIPAL STRESSES (PSI)			PORE PRESSURE (PSI)	PRINCIPAL STRAINS (PSI)		
σ_3	σ_2	σ_1	p_w	ε_3	ε_2	ε_1
58.00	58.00	58.00	17.80	0.000	0.000	0.000
46.47	58.00	58.83	14.43	-0.128	0.057	0.070
43.13	58.00	59.07	10.99	-0.346	0.164	0.183
40.44	58.00	59.26	9.67	-0.708	0.340	0.369
38.61	58.00	59.39	9.28	-0.937	0.450	0.487
36.56	58.00	59.54	7.73	-1.353	0.650	0.702
35.38	58.00	59.62	7.84	-1.530	0.735	0.795
33.98	58.00	59.72	7.58	-1.792	0.860	0.932
32.68	58.00	59.82	6.50	-2.182	1.047	1.136
30.64	58.00	59.96	6.21	-2.654	1.271	1.383
28.91	58.00	60.09	5.37	-3.324	1.588	1.736
26.97	58.00	60.23	4.45	-4.374	2.084	2.289
25.47	58.00	60.33	3.70	-5.197	2.471	2.725
23.96	58.00	60.44	2.94	-6.030	2.862	3.168
22.77	58.00	60.53	2.35	-6.684	3.168	3.517
21.91	58.00	60.59	1.92	-7.159	3.389	3.770
21.37	58.00	60.63	1.65	-7.458	3.528	3.930
20.94	58.00	60.66				
20.51	58.00	60.69				

TABLE 5

OTTAWA SAND $\sigma'_{vc} = 5$ psi - MODEL PARAMETERS

$G_1 = 4688$ psi; $B_1 = 12500$ psi; $n = 0.5$

m	$\alpha^{(m)}$ (psi)	$\beta^{(m)}$ (psi)	$k^{(m)}$ (psi)	h'_m (psi)	B'_m (psi)	A_m
2	0.417	5.000	1.917	4795.00	--	1.551
3	1.000	5.000	2.500	2638.00	--	0.926
4	0.833	5.000	3.833	1172.00	--	0.239
5	1.417	5.000	4.417	5936.00	--	0.226
6	1.625	5.000	5.375	3835.00	--	-0.200
7	1.833	5.000	6.333	2093.00	--	-0.334
8	1.767	5.000	7.567	70.70	--	-0.511
9	2.350	5.000	8.150	0.00	--	-0.508

TABLE 6

OTTAWA SAND

SIMPLE SHEAR TEST (SS) WITH CONSTANT MEAN STRESS AT 5 PSI

(Stress Ratio, m = 0.5)

STRESSES (PSI)			STRAINS (%)		
σ_1	σ_2	σ_3	ε_1	ε_2	ε_3
5	5	5	0.000	0.000	0.000
5	6	4	0.000	0.011	-0.011
5	7	3	0.012	0.042	-0.026
5	7.5	2.5	0.022	0.066	-0.039
5	8	2	0.035	0.111	-0.071
5	7	3	0.035	0.100	-0.060
5	6	4	0.037	0.089	-0.047
5	5	5	0.049	0.072	-0.013
5	6	4	0.049	0.083	-0.024
5	7	3	0.050	0.097	-0.035
5	8	2	0.062	0.130	-0.052
5	8.5	1.5	0.077	0.182	-0.094
5	9	1.0	0.098	0.272	-0.178
5	9.25	0.75	0.110	0.320	-0.225
5	9.5	0.50	0.121	0.385	-0.296
5	9.75	0.25	0.133	0.454	-0.376

TABLE 7

OTTAWA SAND σ'_{vc} = 20 psi - MODEL PARAMETERS

B_1 = 9375 psi; $\bar{B}_1$ = 25000 psi;   n = 0.5

m	$\alpha^{(m)}$ (psi)	$\beta^{(m)}$ (psi)	$k^{(m)}$ (psi)	h'_m (psi)	$\bar{B}'_m$ (psi)	A_m
2	-1.089	20.000	4.91	16000.0	--	1.598
3	0.822	20.000	6.82	6843.0	--	0.223
4	1.233	20.000	10.23	4735.0	--	-0.090
5	3.144	20.000	12.14	2918.0	--	-0.054
6	3.556	20.000	15.56	1662.0	--	-0.103
7	3.967	20.000	18.97	1154.0	--	-0.263
8	4.378	20.000	22.38	415.2	--	-0.435
9	4.789	20.000	25.79	187.3	--	-0.495
10	5.500	20.000	28.90	0.0	--	-0.490

TABLE 8

OTTAWA SAND

SIMPLE SHEAR TEST (SS) WITH CONSTANT MEAN STRESS AT 20 PSI
(Stress Ratio, m = 0.5)

STRESSES (PSI)			STRAINS (%)		
σ_1	σ_2	σ_3	ε_1	ε_2	ε_3
20	20	20	0.000	0.000	0.000
20	22	18	0.000	0.011	-0.011
20	24	16	0.000	0.024	-0.024
20	26	14	0.003	0.055	-0.056
20	28	12	0.005	0.096	-0.102
20	29	11	0.007	0.121	-0.132
20	30	10	0.010	0.150	-0.166
20	28	12	0.010	0.139	-0.155
20	26	14	0.010	0.129	-0.145
20	24	16	0.010	0.118	-0.134
20	22	18	0.012	0.101	-0.116
20	20	20	0.014	0.065	-0.083
20	22	18	0.014	0.076	-0.094
20	26	14	0.014	0.097	-0.115
20	28	12	0.016	0.115	-0.132
20	29	11	0.017	0.130	-0.147
20	30	10	0.019	0.149	-0.167
20	31	9	0.023	0.191	-0.216
20	32	8	0.028	0.238	-0.272
20	33	7	0.029	0.302	-0.354
20	35	5	0.036	0.450	-0.554
20	36	4	0.039	0.603	-0.764
20	37	3	0.041		-1.034
20	38	2	0.032		-1.430
20	39	1	0.012		-1.978
20	40	0			

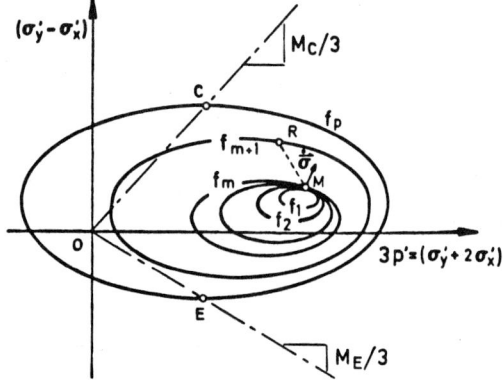

Fig. 1

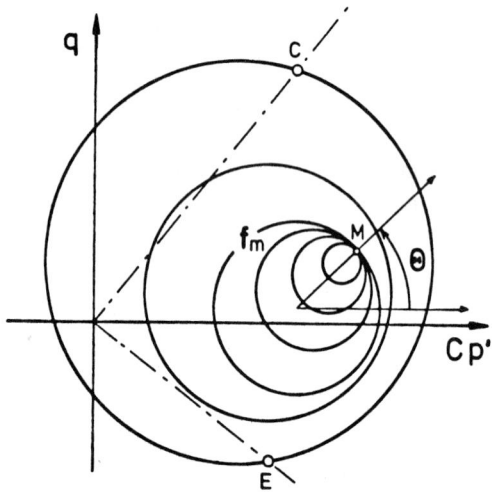

Fig. 2

KAOLINITE CLAY - UNDRAINED AXIAL TEST - INITIAL STRESS = 40 PSI

AXIAL SHEAR PLOT

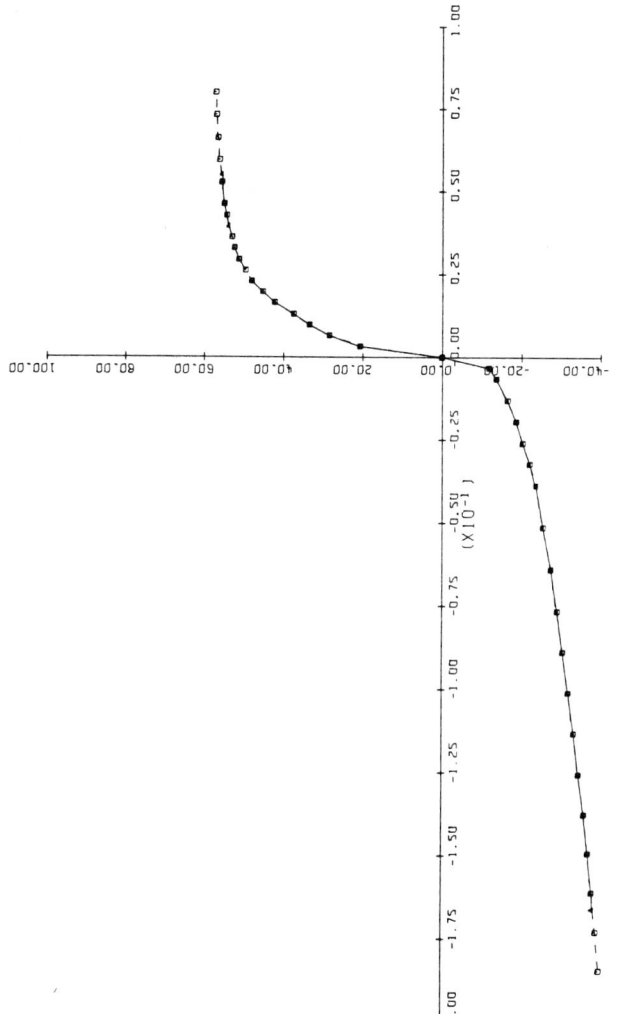

Fig. 3

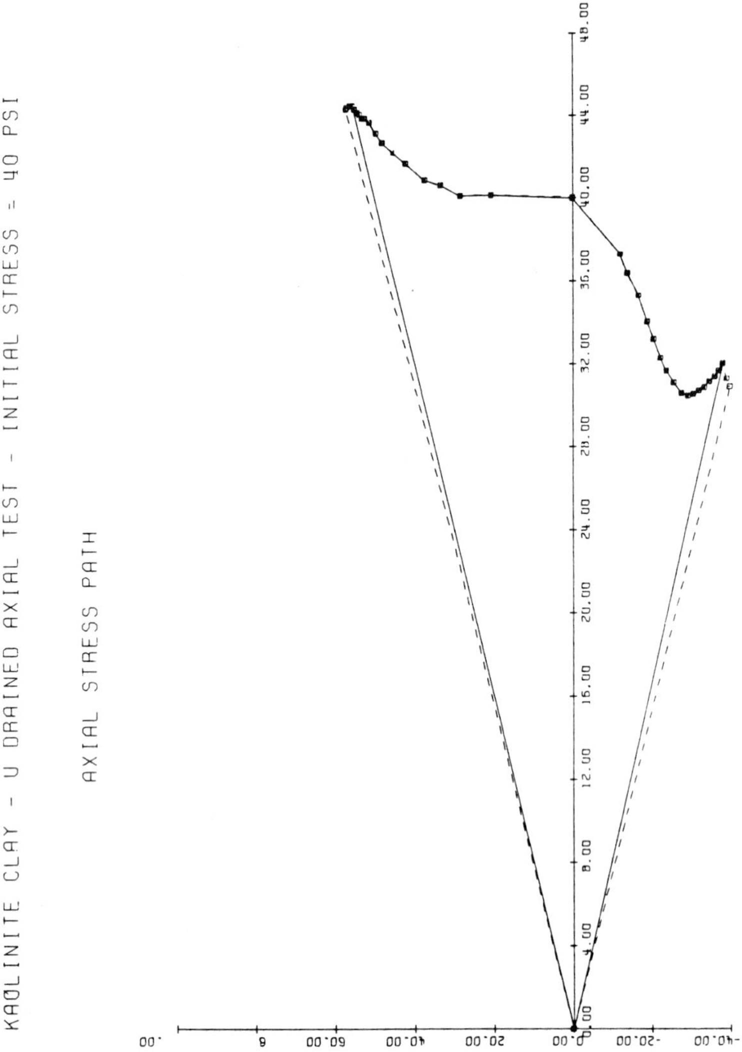

768 STRESS STRAIN FOR SOILS

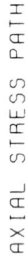

Fig. 4

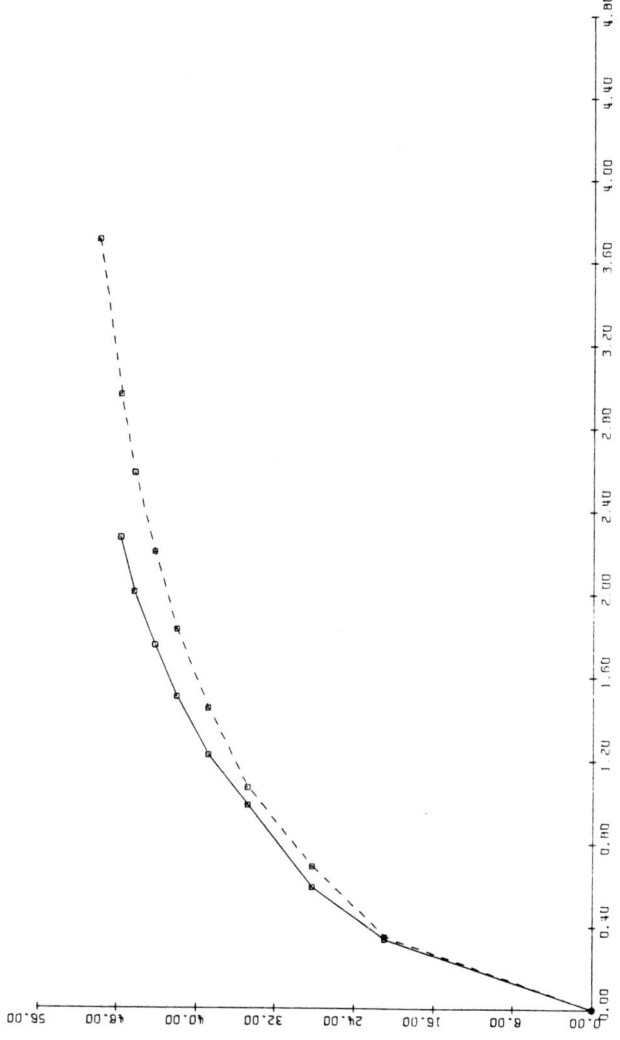

Fig. 5

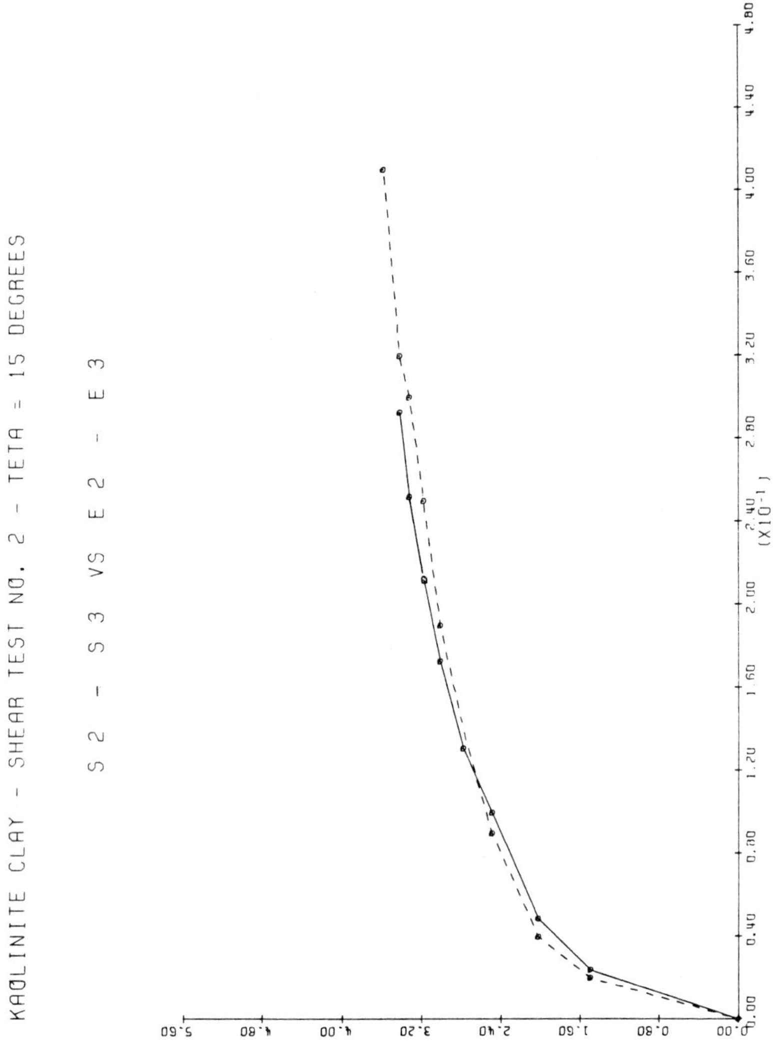

Fig. 6

CONSTITUTIVE THEORY FOR SOIL

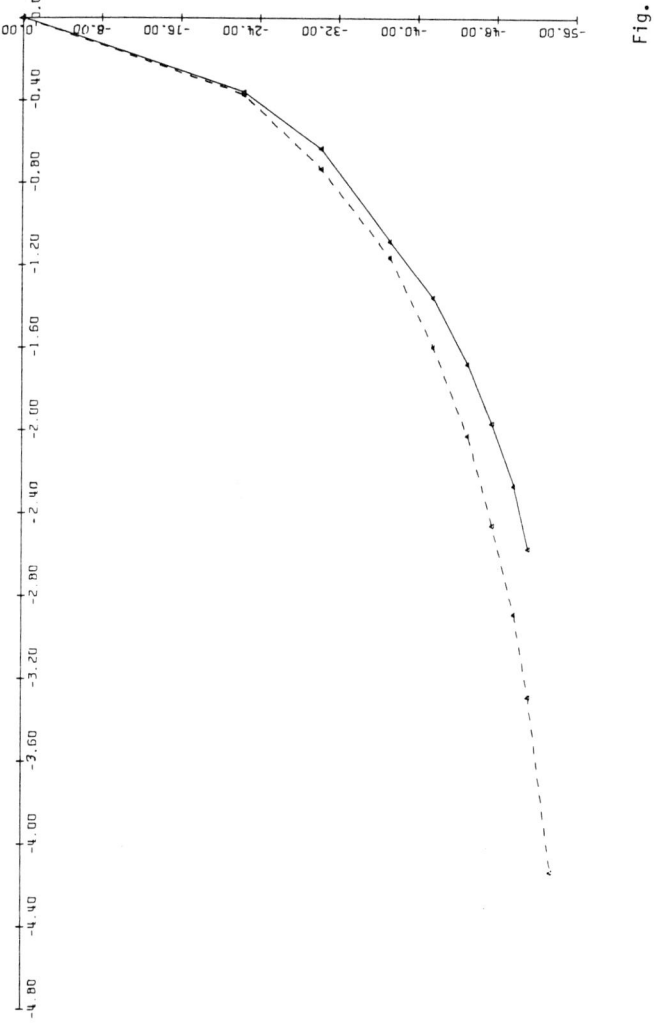

Fig. 7

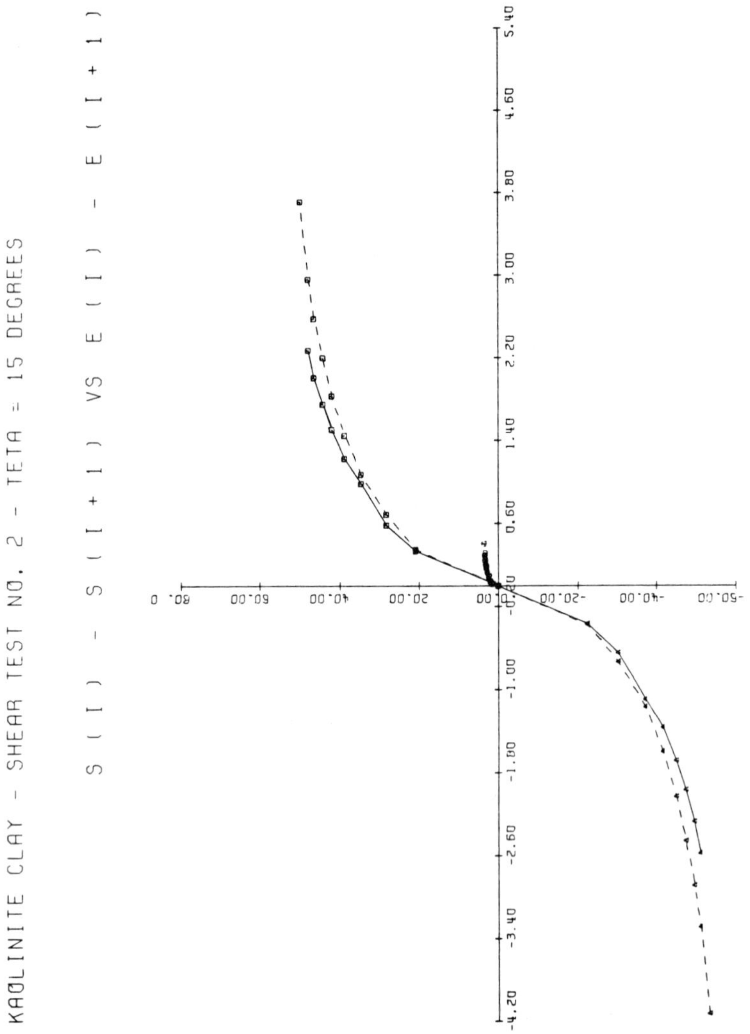
KAOLINITE CLAY - SHEAR TEST NO. 2 - TETA = 15 DEGREES
S (I) - S (I + 1) VS E (I) - E (I + 1)
5.40
4.60
3.80
3.00
2.20
1.40
0.60
-1.00
-1.80
-2.60
-3.40
-4.20
80.0
60.00
40.00
20.00
-20.00
-40.00
-50.00

Fig. 8

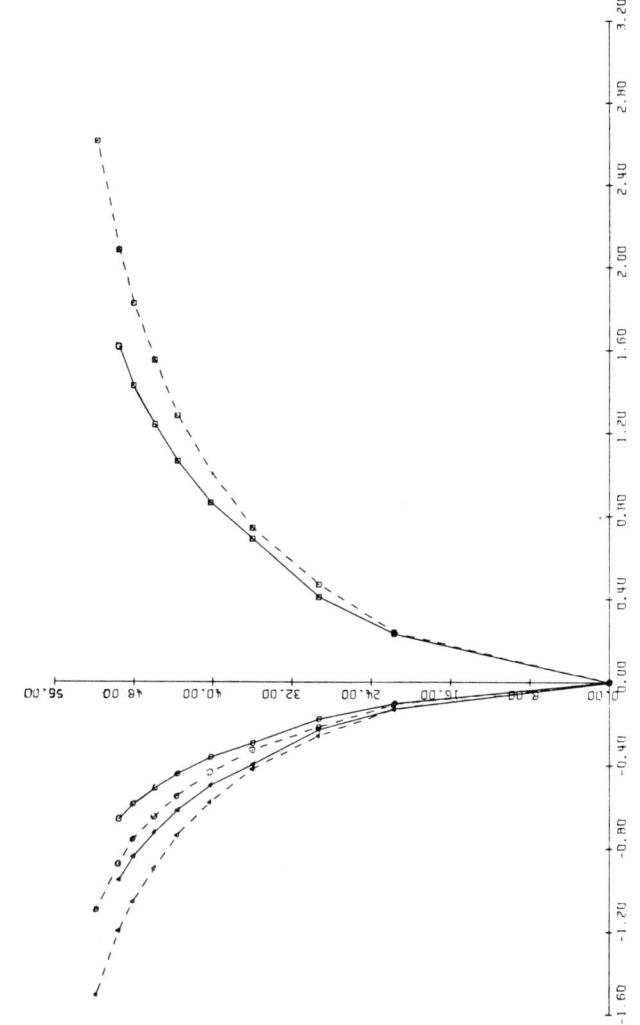

Fig. 9

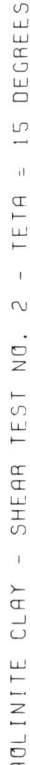

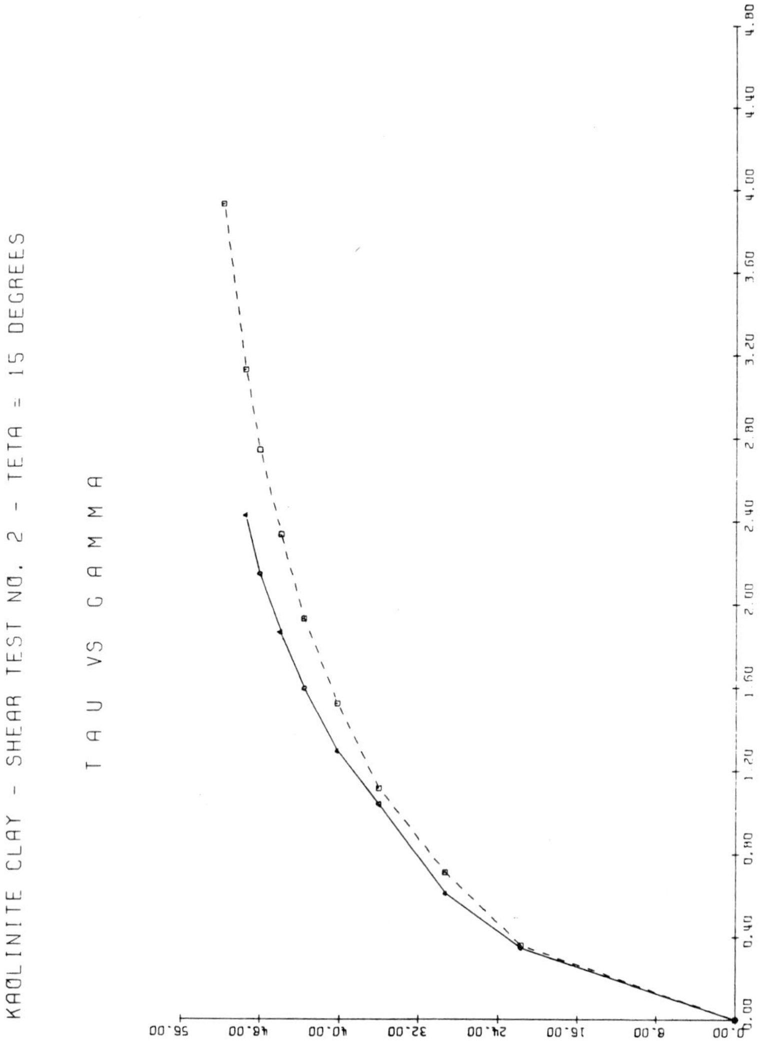
KAOLINITE CLAY - SHEAR TEST NO. 2 - TETA = 15 DEGREES
TAU VS GAMMA
0.00
8.00
16.00
24.00
32.00
40.00
48.00
56.00
0.00
0.40
0.80
1.20
1.60
2.00
2.40
2.80
3.20
3.60
4.00
4.40
4.80

Fig. 10

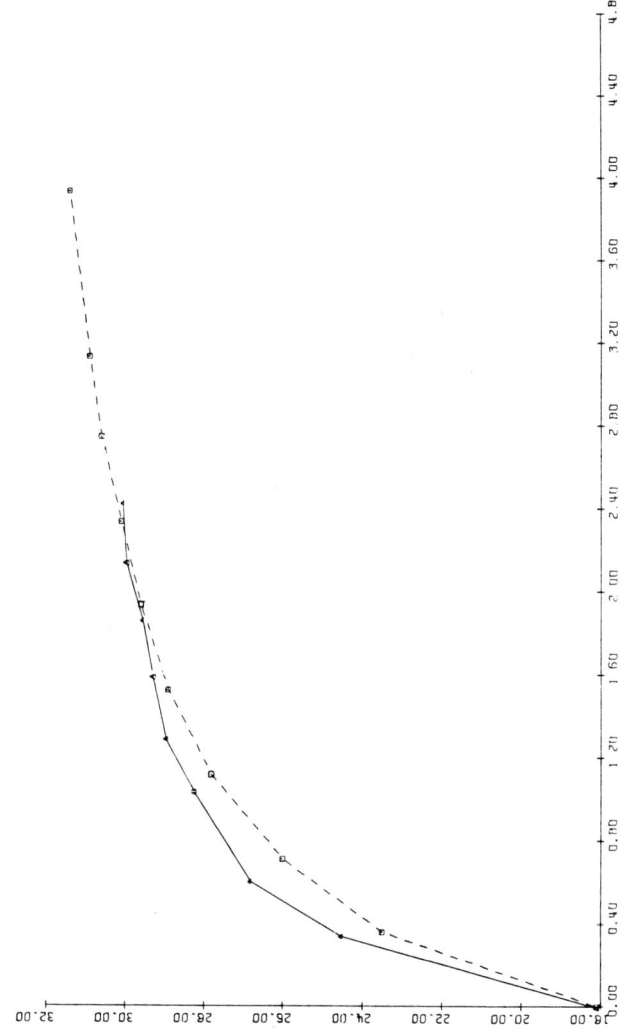

Fig. 11

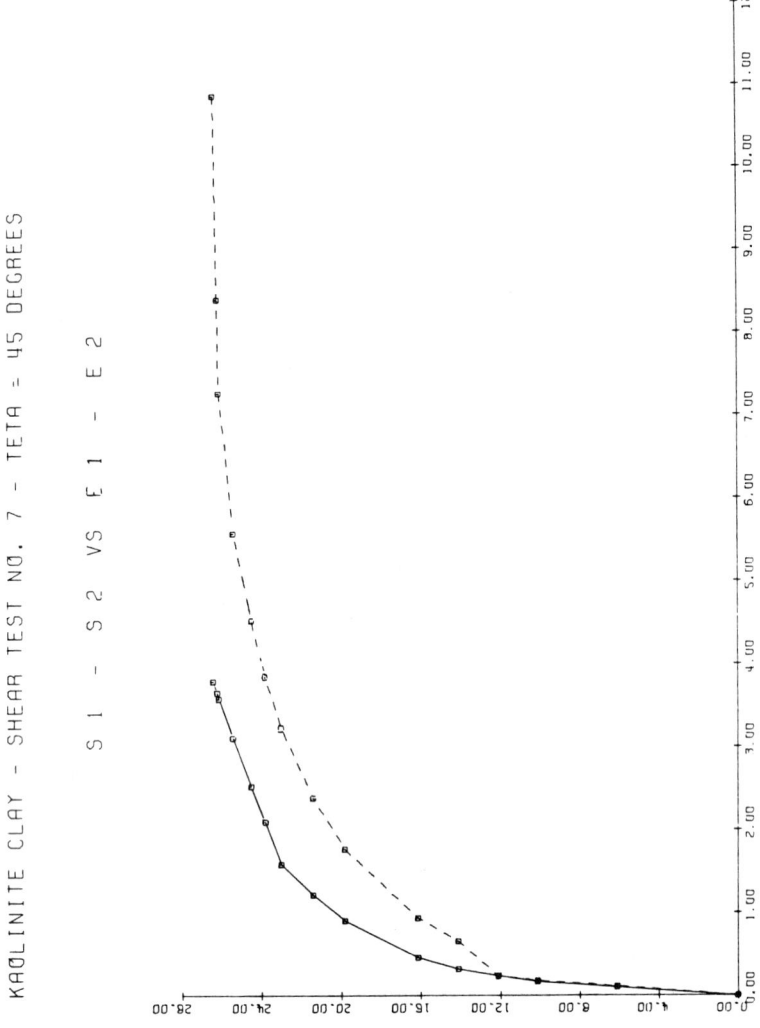

Fig. 12

CONSTITUTIVE THEORY FOR SOIL

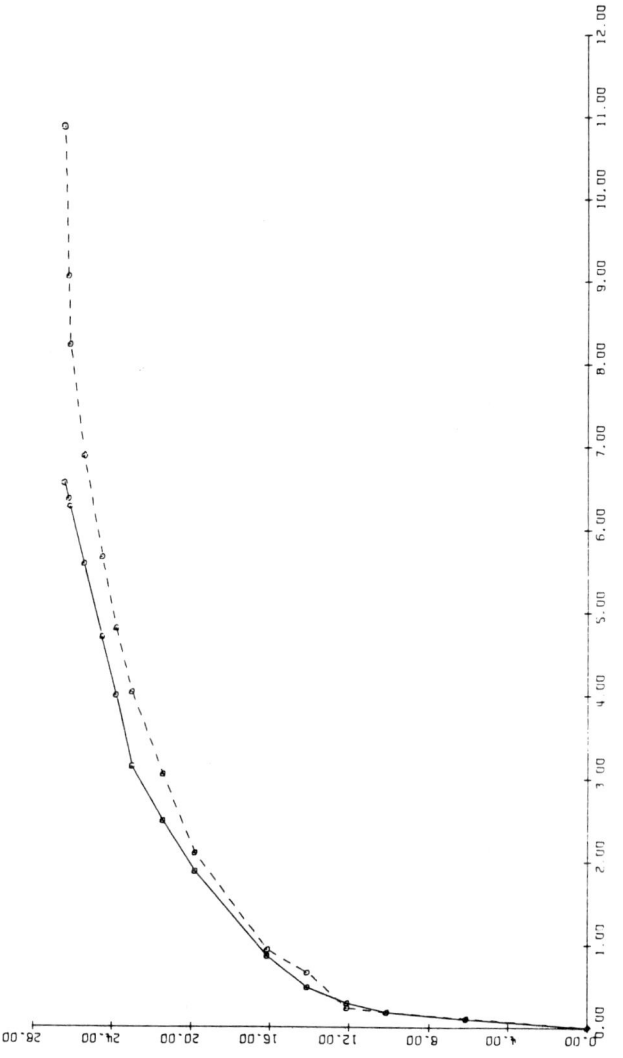

Fig. 13

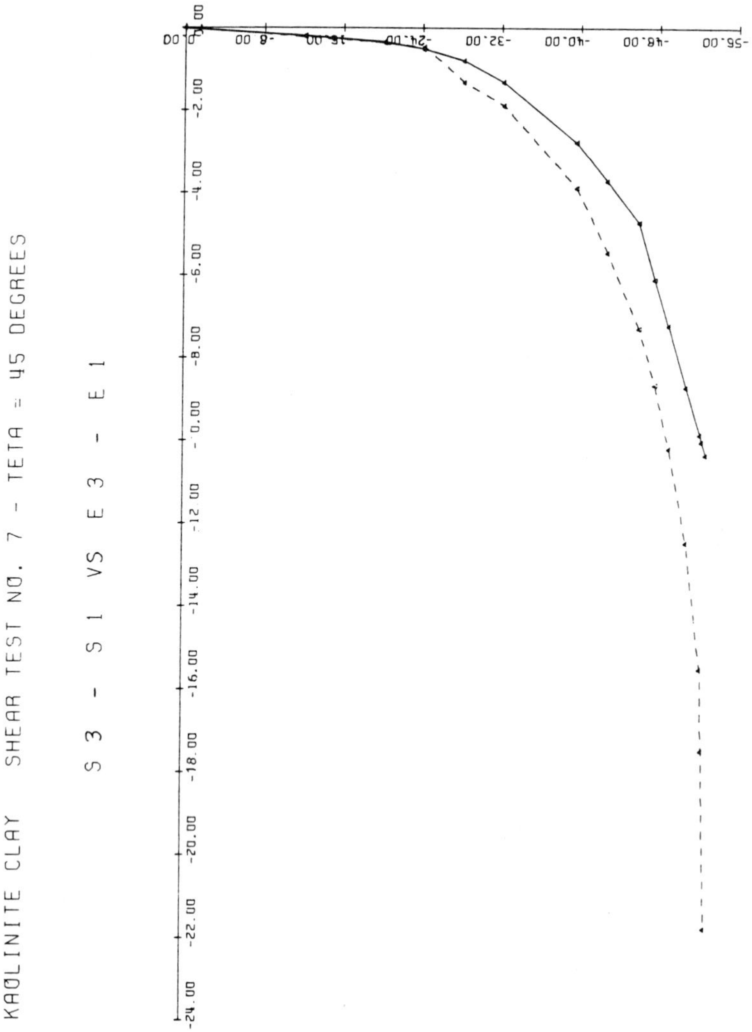

KAOLINITE CLAY SHEAR TEST NO. 7 - TETA = 45 DEGREES

S 3 - S 1 VS E 3 - E 1

Fig. 14

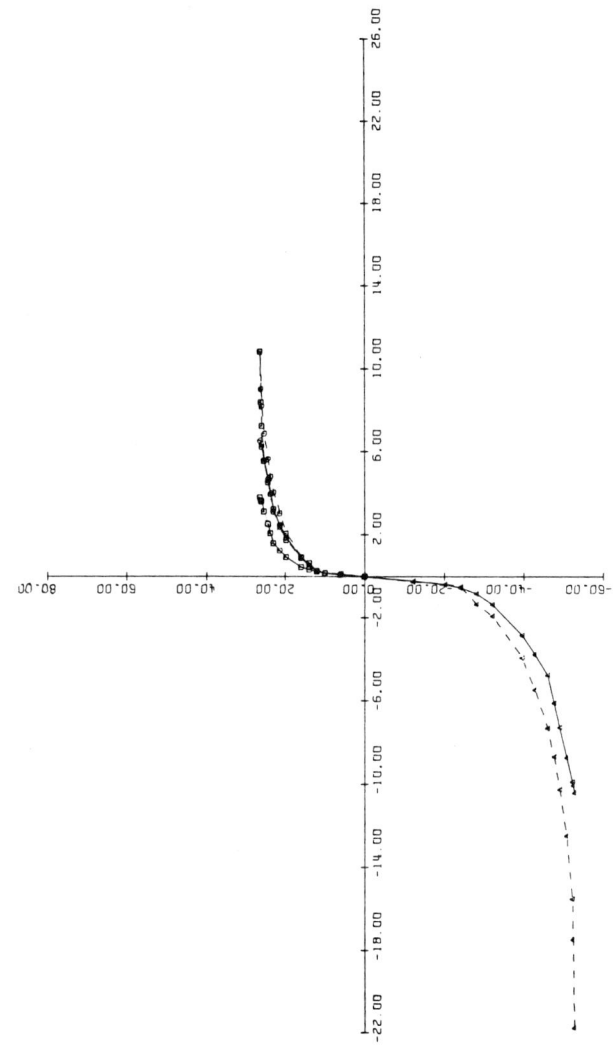

Fig. 15

KAOLINITE CLAY - SHEAR TEST NO. 7 - TETA = 45 DEGREES
TAU VS E(I)
0.00
8.00
16.00
24.00
32.00
40.00
48.00
13.00
11.00
9.00
7.00
5.00
1.00
-1.00
-5.00
-7.00
-9.00
-11.00

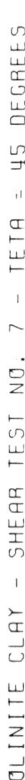

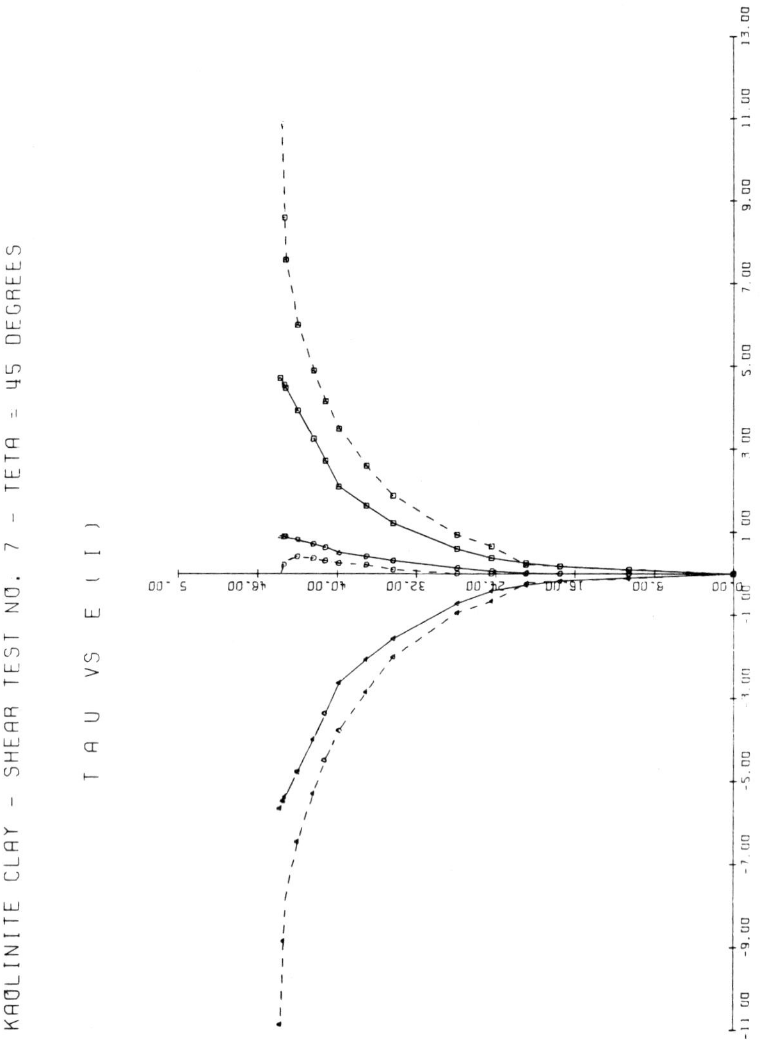

Fig. 16

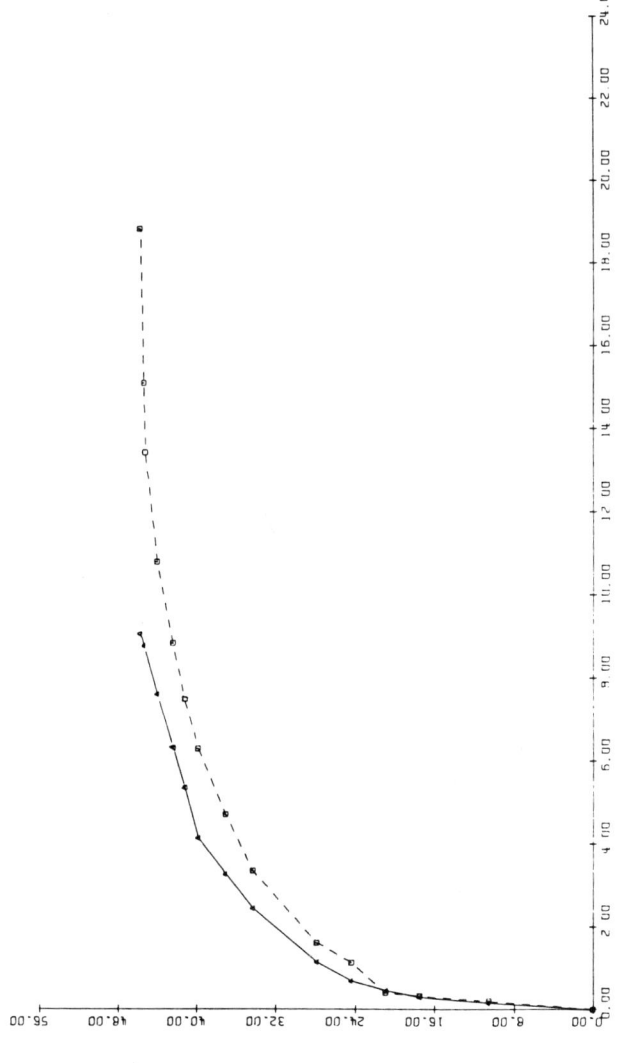

Fig. 17

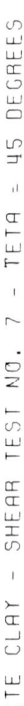

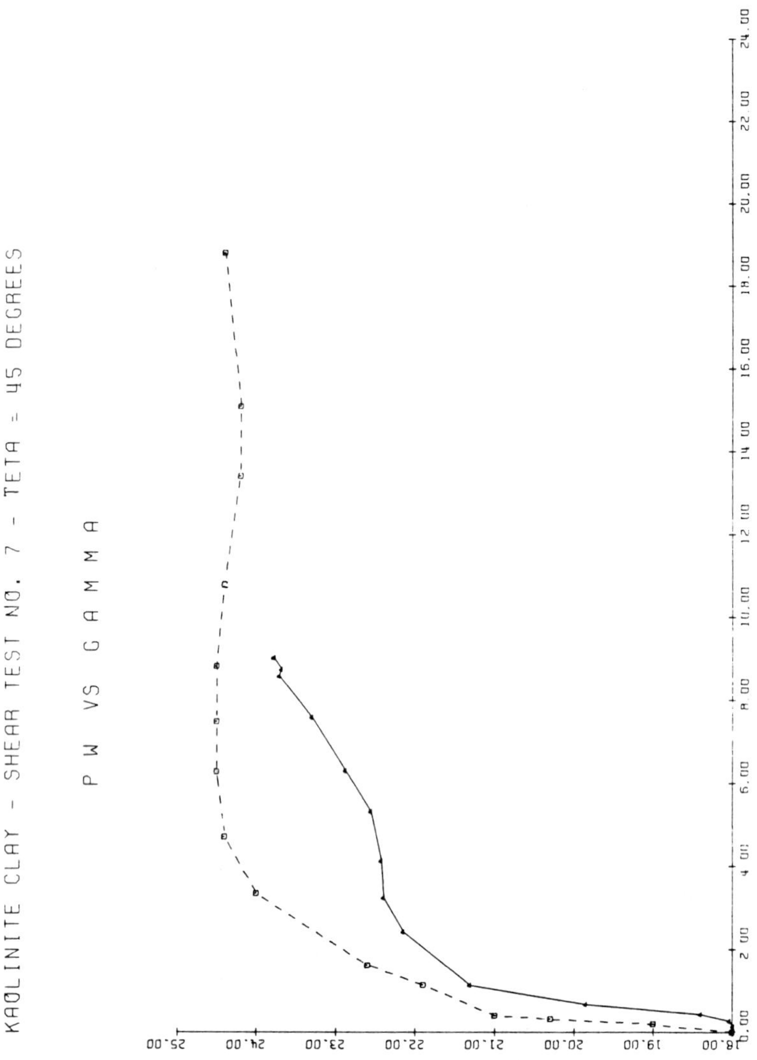
KAOLINITE CLAY - SHEAR TEST NO. 7 - TETA = 45 DEGREES
P W VS G A M M A
25.00
24.00
23.00
22.00
21.00
20.00
19.00
18.00
0.00
2.00
4.00
6.00
8.00
10.00
12.00
14.00
16.00
18.00
20.00
22.00
24.00

Fig. 18

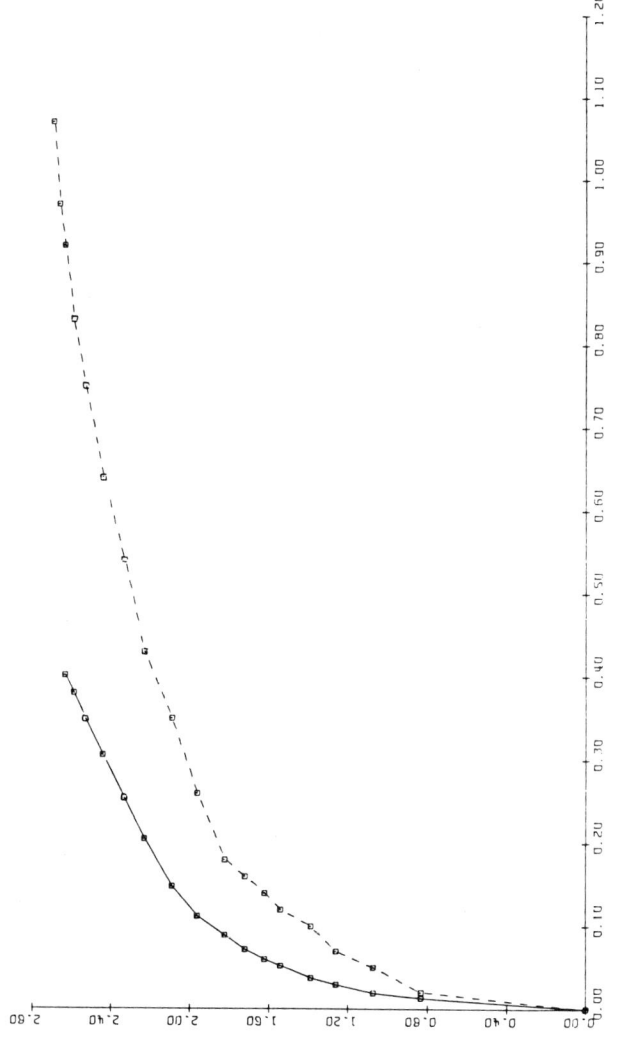

Fig. 19

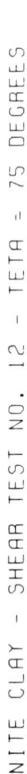

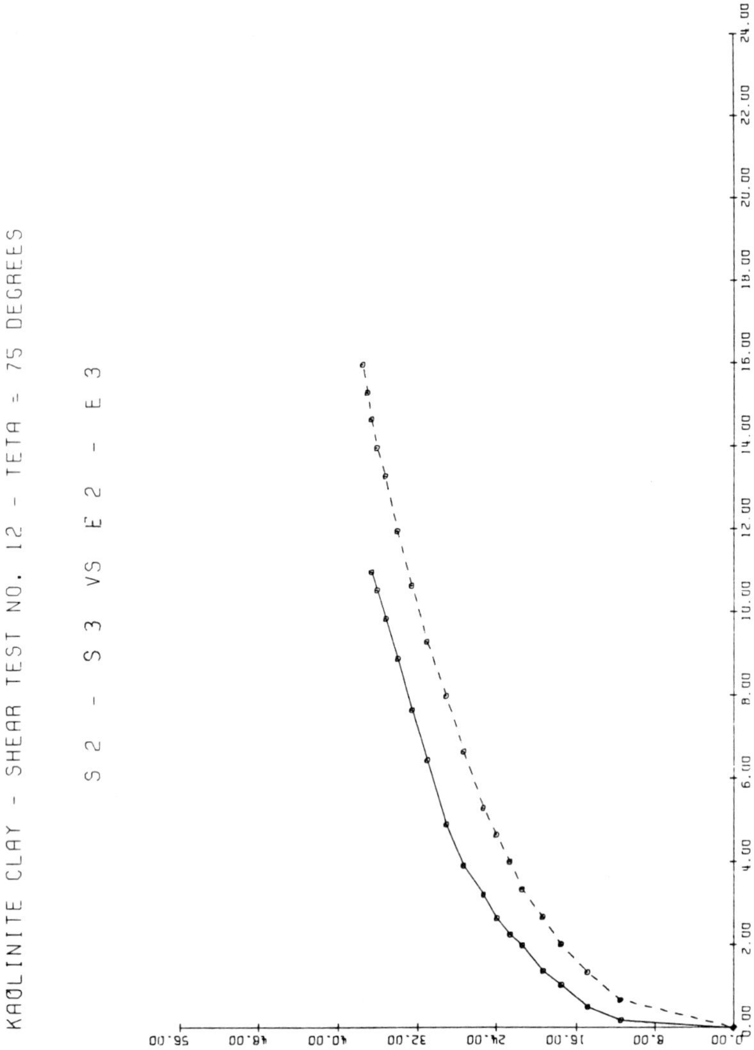
KAOLINITE CLAY - SHEAR TEST NO. 12 - TETA = 75 DEGREES
S 2 - S 3 VS E 2 - E 3
0.00
8.00
16.00
24.00
32.00
40.00
48.00
56.00
0.00
2.00
4.00
6.00
8.00
10.00
12.00
14.00
16.00
18.00
20.00
22.00
24.00

Fig. 20

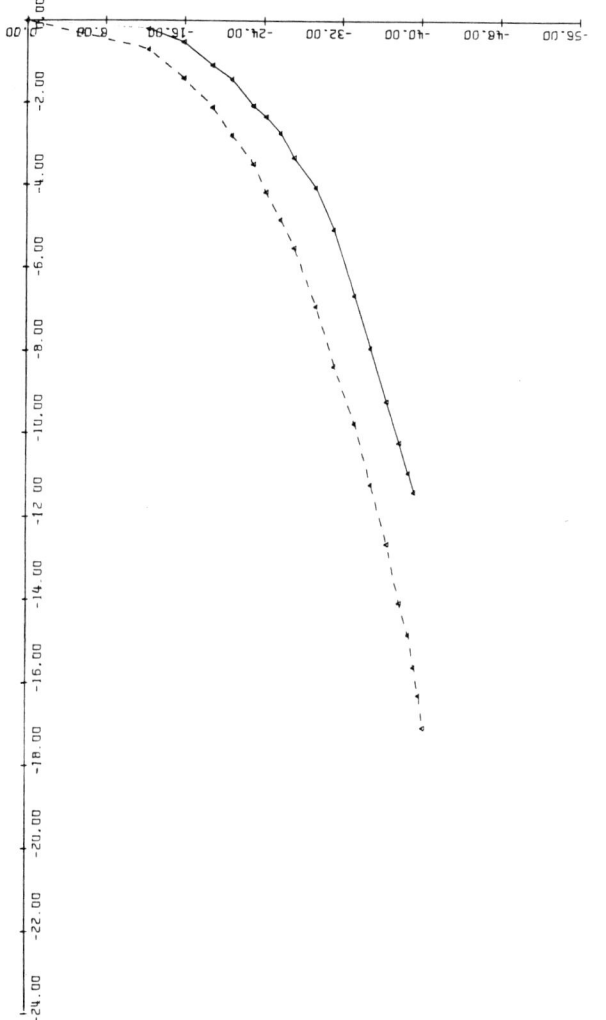

Fig. 21

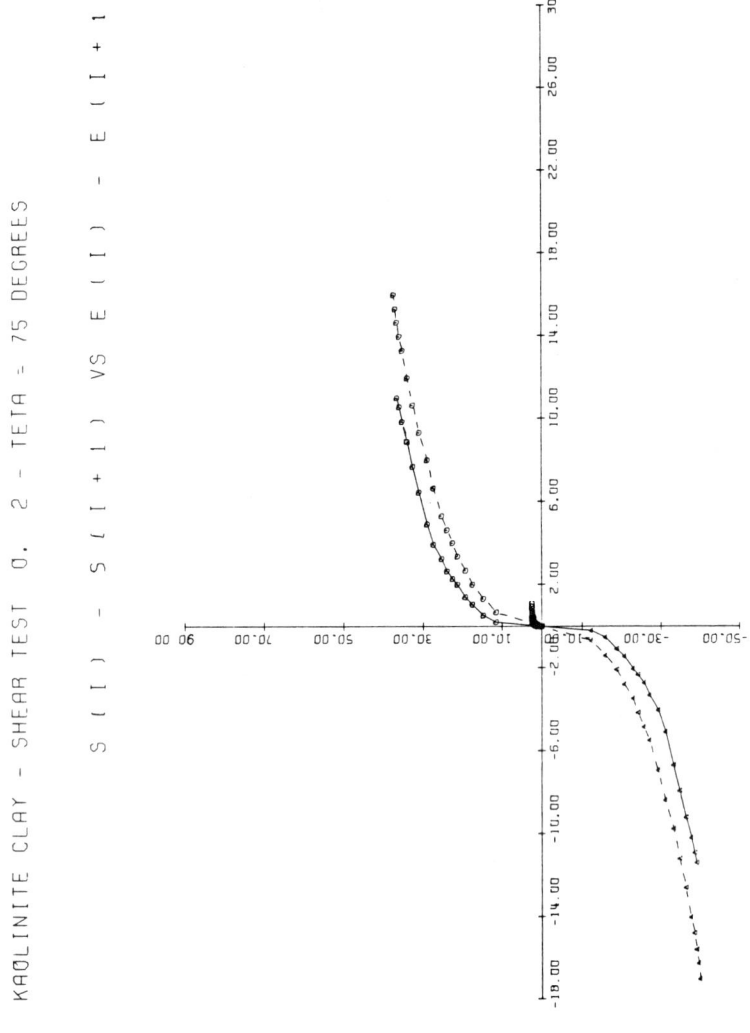

Fig. 22

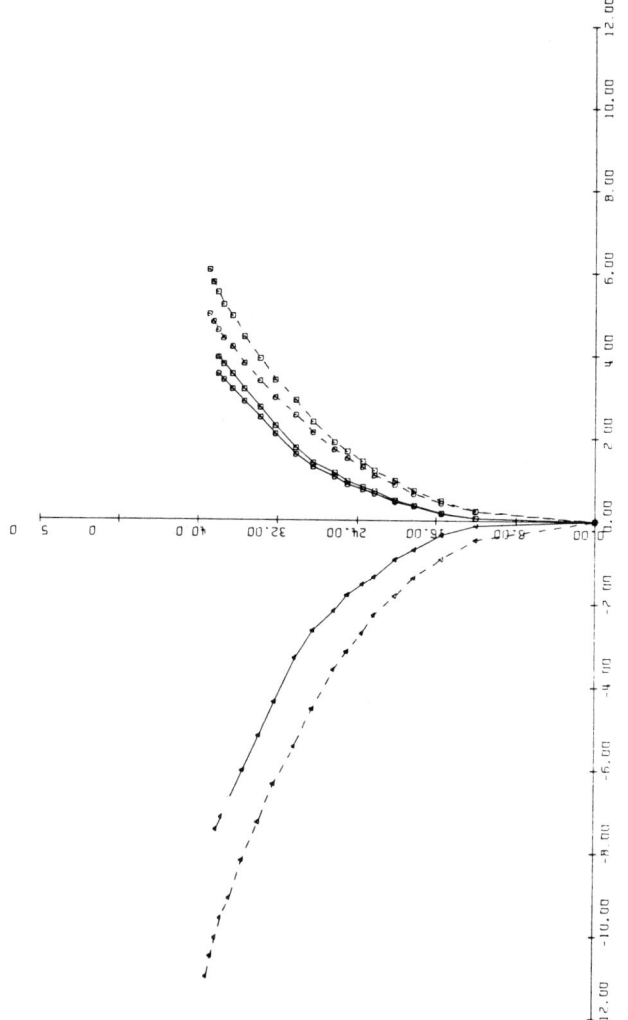

Fig. 23

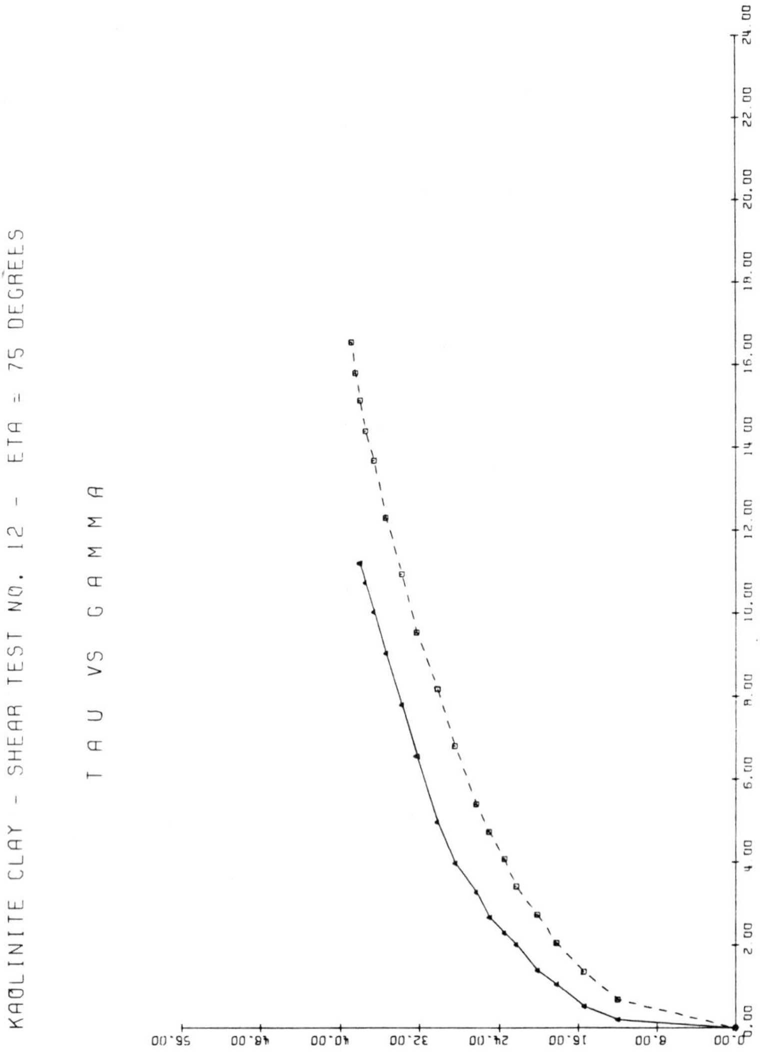
KAOLINITE CLAY - SHEAR TEST NO. 12 - ETA = 75 DEGREES
TAU VS GAMMA
0.00
2.00
4.00
6.00
8.00
10.00
12.00
14.00
16.00
18.00
20.00
22.00
24.00
0.00
8.00
16.00
24.00
32.00
40.00
48.00
56.00

Fig. 24

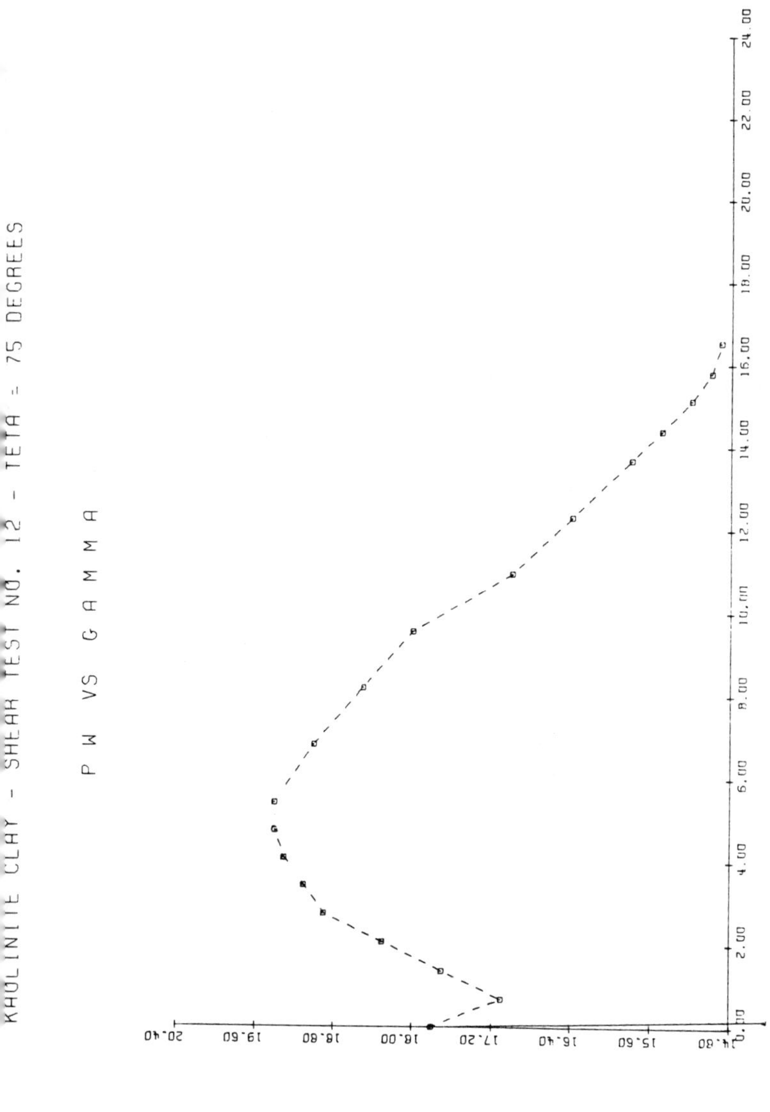

Fig. 25

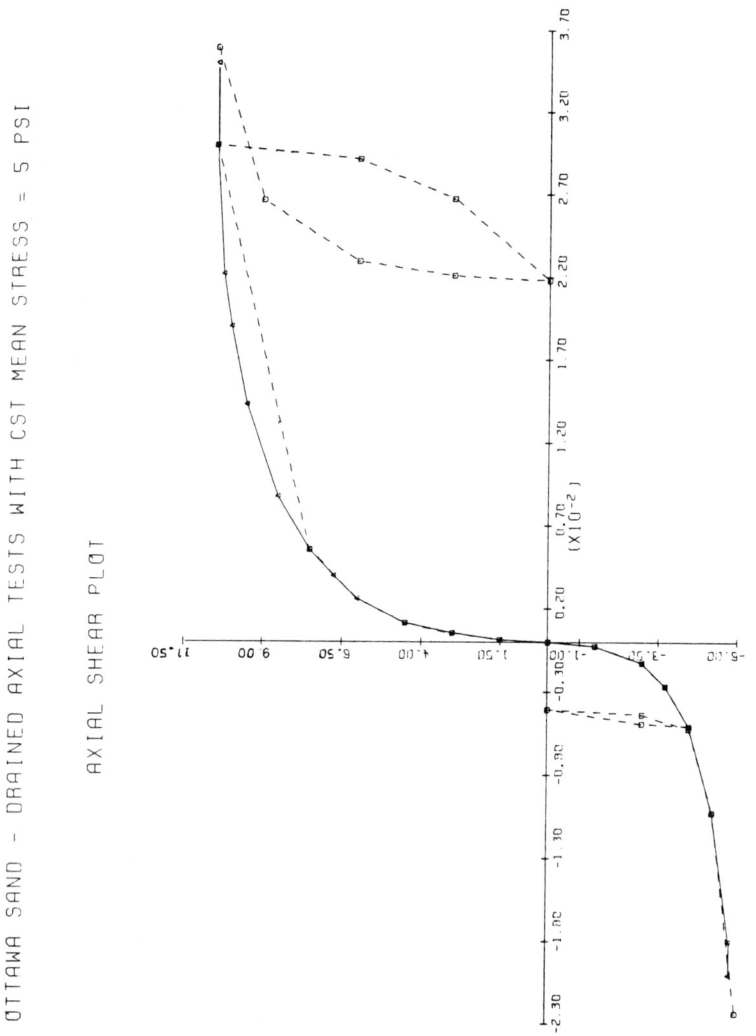

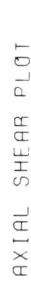

Fig. 26

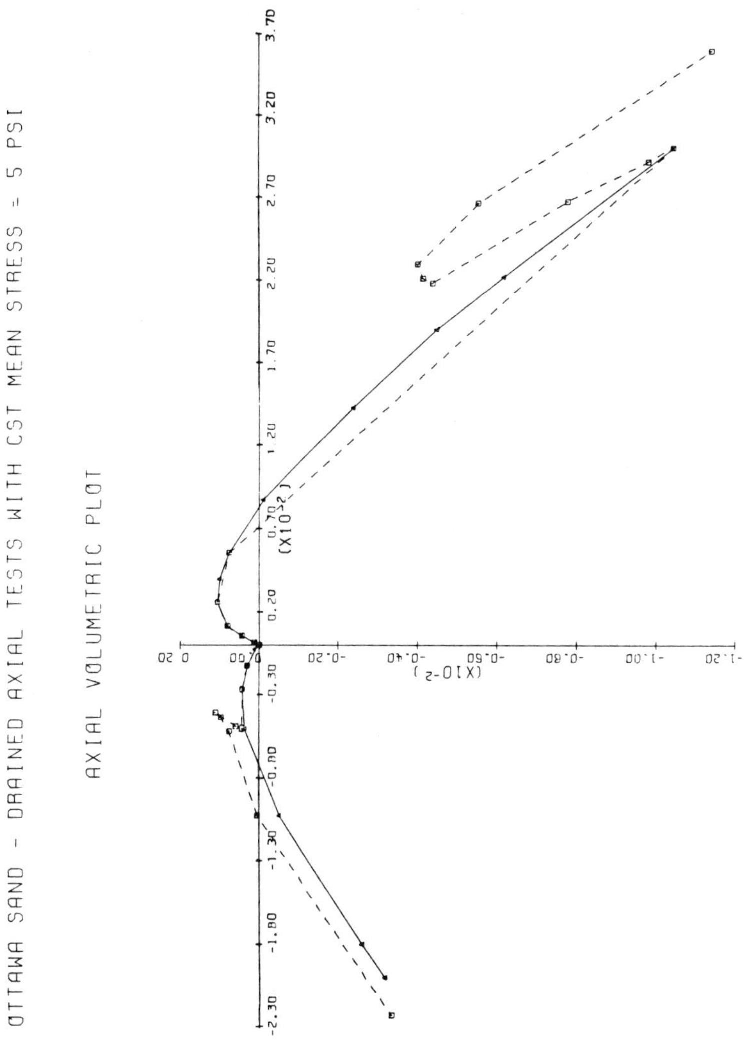

Fig. 27

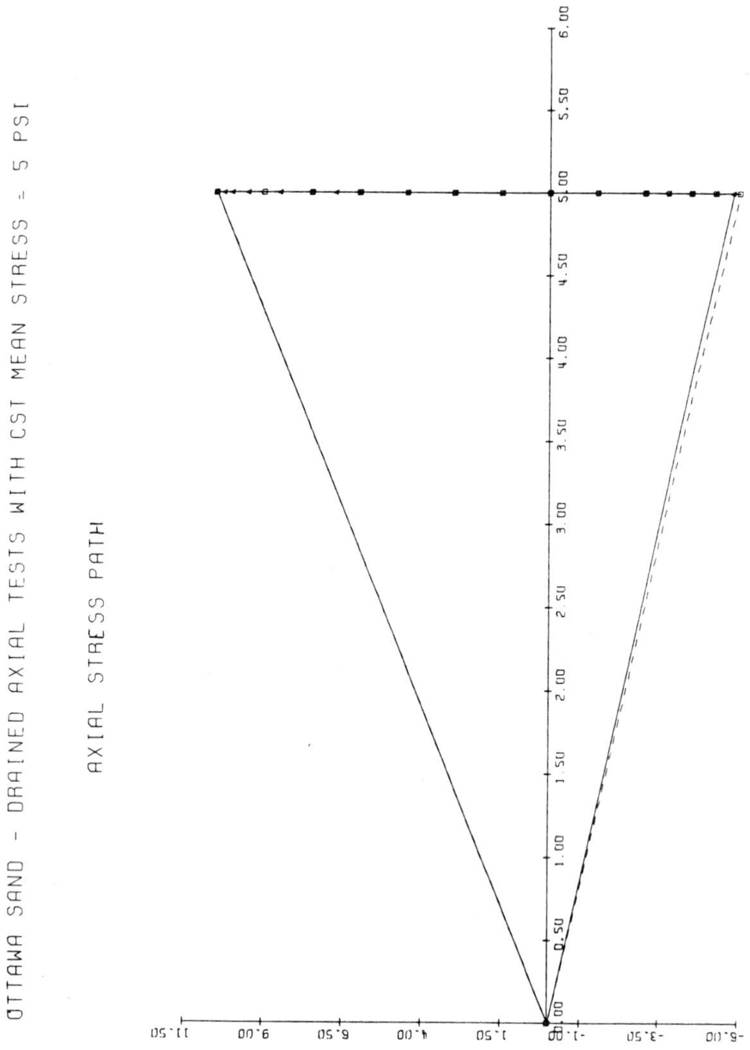

Fig. 28

CONSTITUTIVE THEORY FOR SOIL

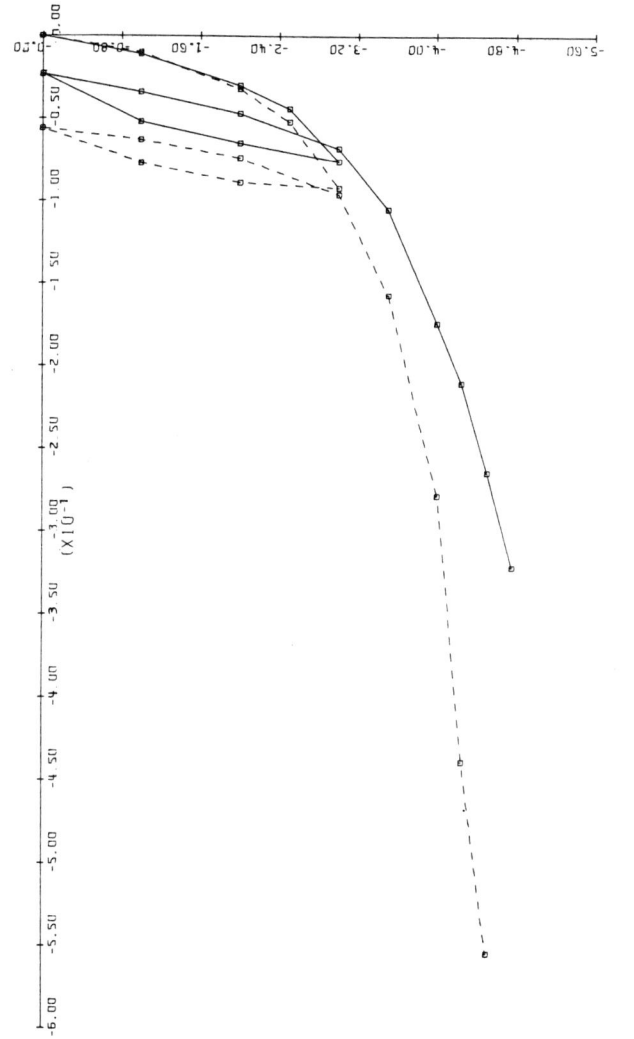

Fig. 29

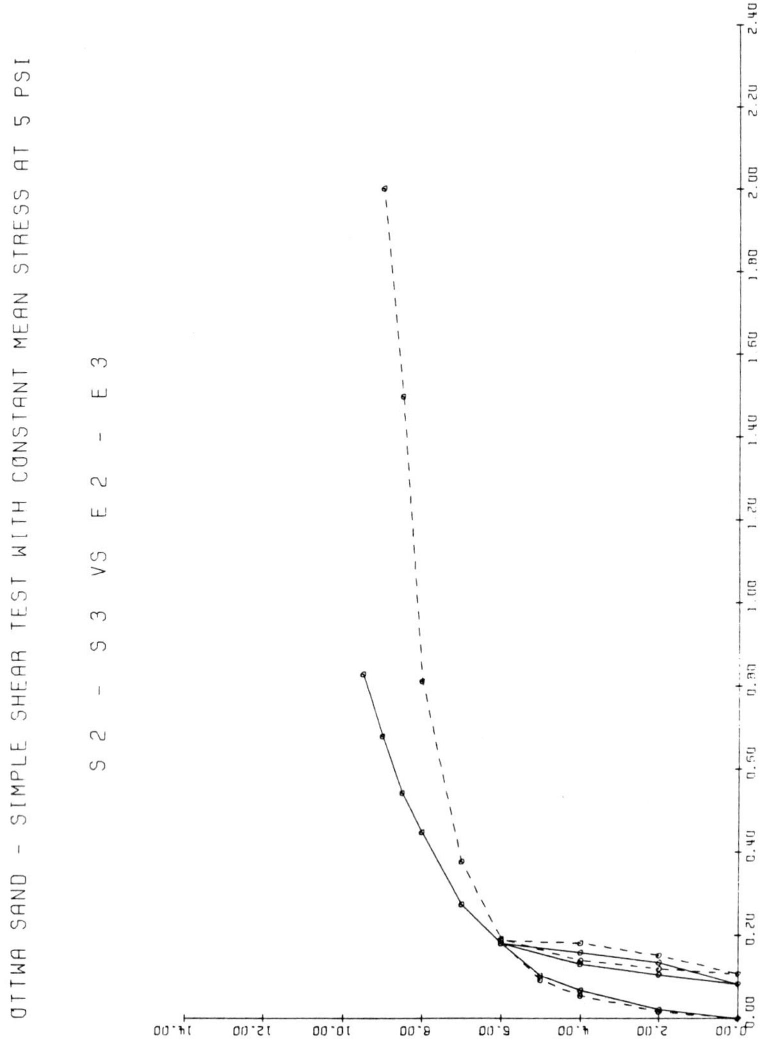
OTTWA SAND - SIMPLE SHEAR TEST WITH CONSTANT MEAN STRESS AT 5 PSI
S 2 - S 3 VS E 2 - E 3
0.00
0.20
0.40
0.60
0.80
1.00
1.20
1.40
1.60
1.80
2.00
2.20
2.40
0.00
2.00
4.00
6.00
8.00
10.00
12.00
14.00

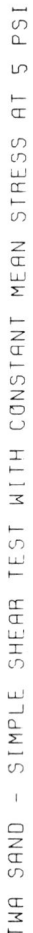

Fig. 30

CONSTITUTIVE THEORY FOR SOIL

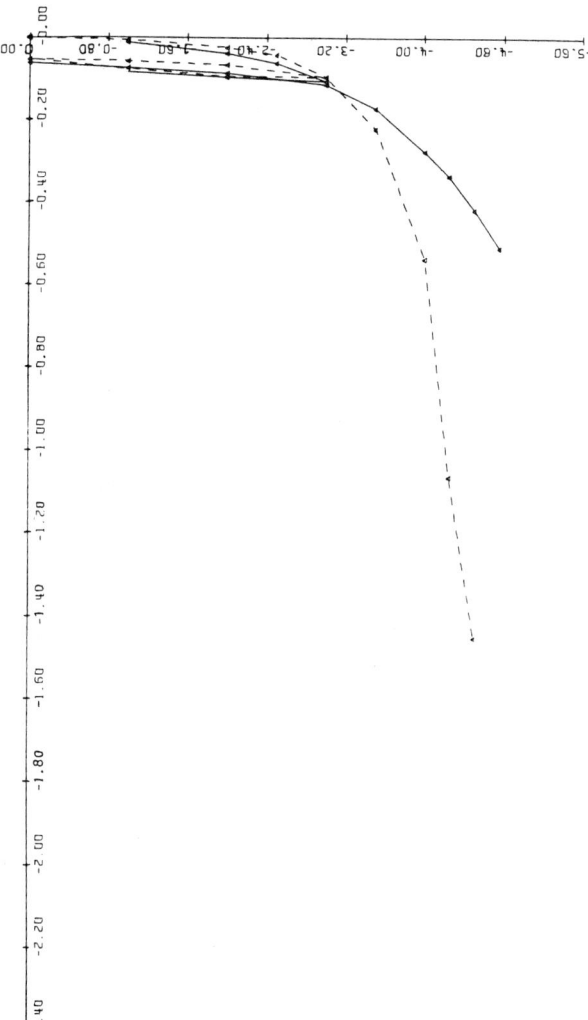

Fig. 31

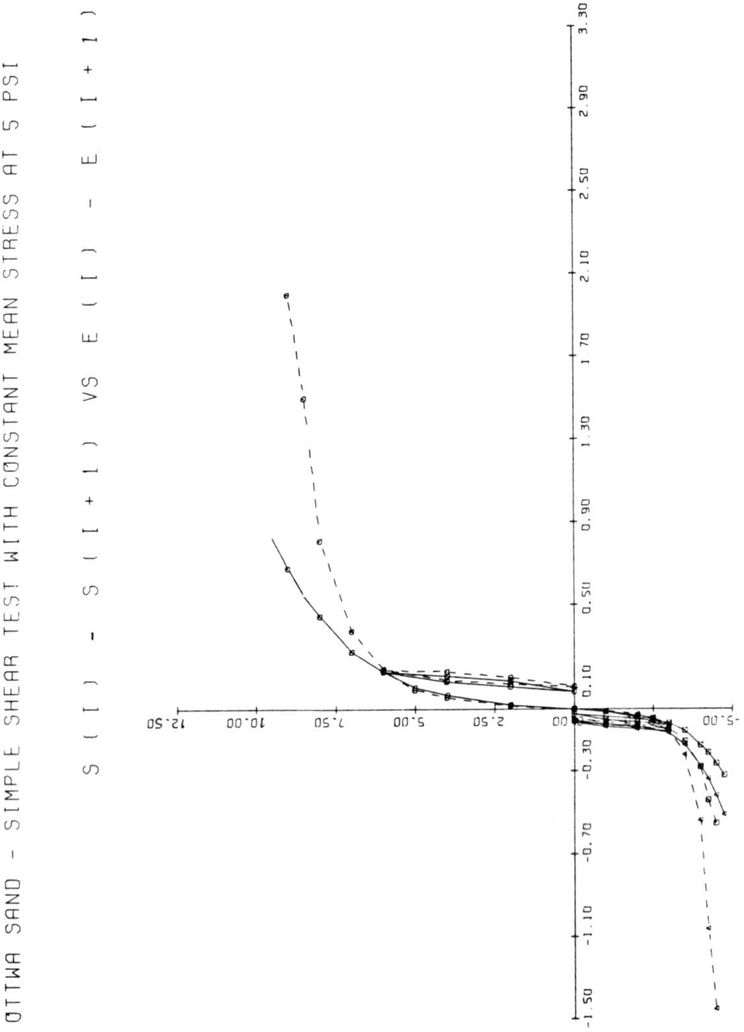
OTTWA SAND - SIMPLE SHEAR TEST WITH CONSTANT MEAN STRESS AT 5 PSI
S (I) - S (I + 1) VS E (I) - E (I + 1)
12.50
10.00
7.50
5.00
2.50
-5.00
3.30
2.90
2.50
2.10
1.70
1.30
0.90
0.50
0.10
-0.30
-0.70
-1.10
-1.50

Fig. 32

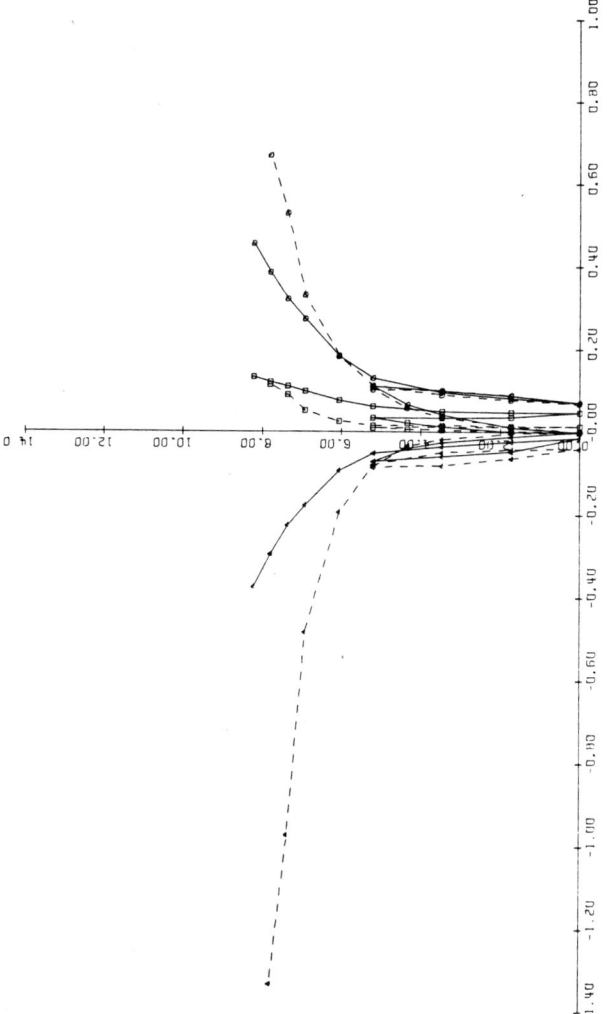

Fig. 33

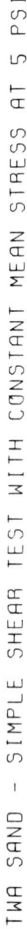

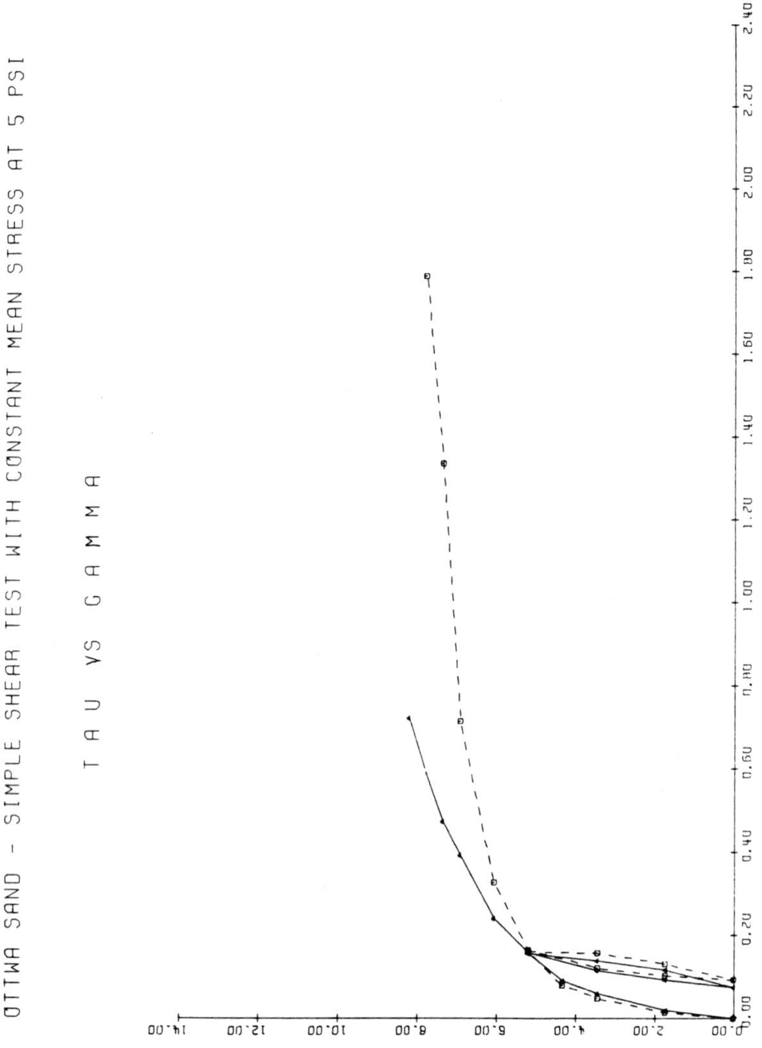

OTTWA SAND - SIMPLE SHEAR TEST WITH CONSTANT MEAN STRESS AT 5 PSI
TAU VS GAMMA
0.00
0.20
0.40
0.60
0.80
1.00
1.20
1.40
1.60
1.80
2.00
2.20
2.40
2.00
4.00
6.00
8.00
10.00
12.00
14.00

Fig. 34

OTTAWA SAND - DRAINED AXIAL TESTS WITH CST MEAN STRESS = 20 PSI

AXIAL SHEAR PLOT

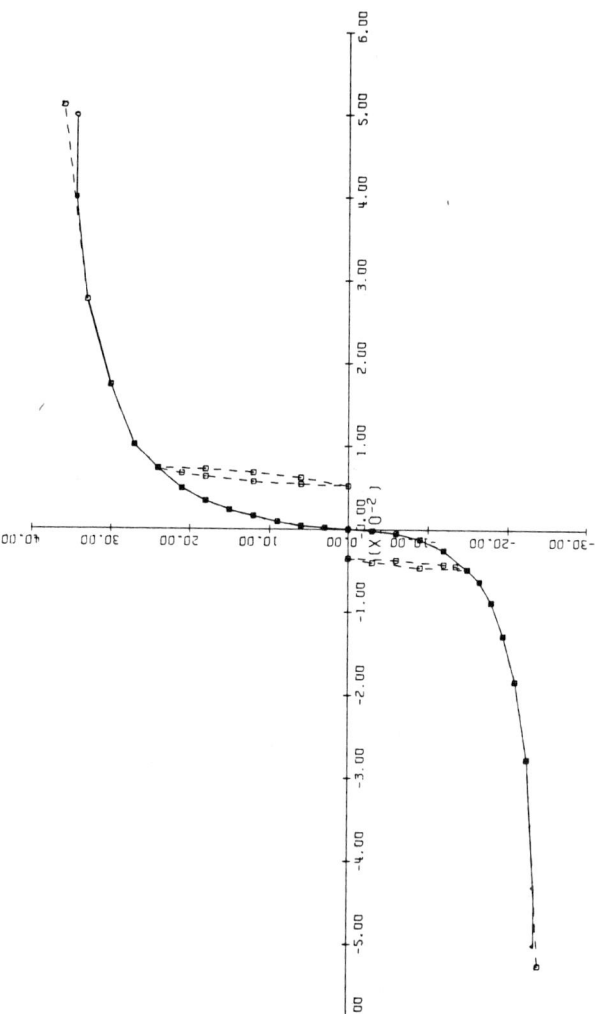

Fig. 35

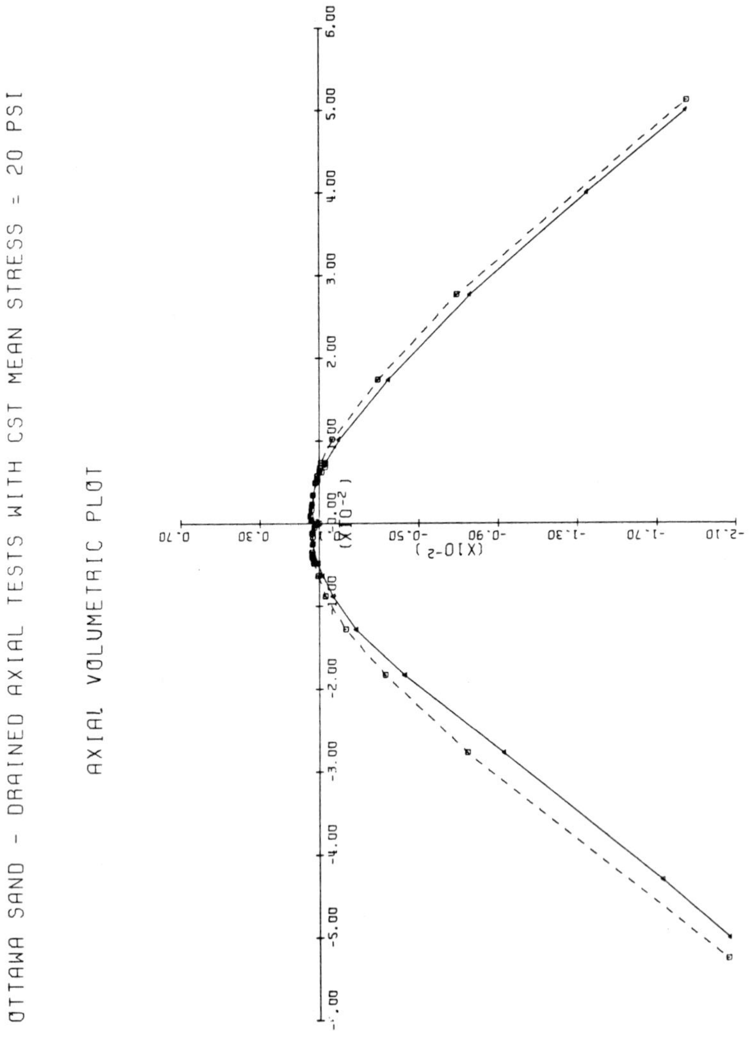

OTTAWA SAND - DRAINED AXIAL TESTS WITH CST MEAN STRESS = 20 PSI

AXIAL VOLUMETRIC PLOT

Fig. 36

CONSTITUTIVE THEORY FOR SOIL

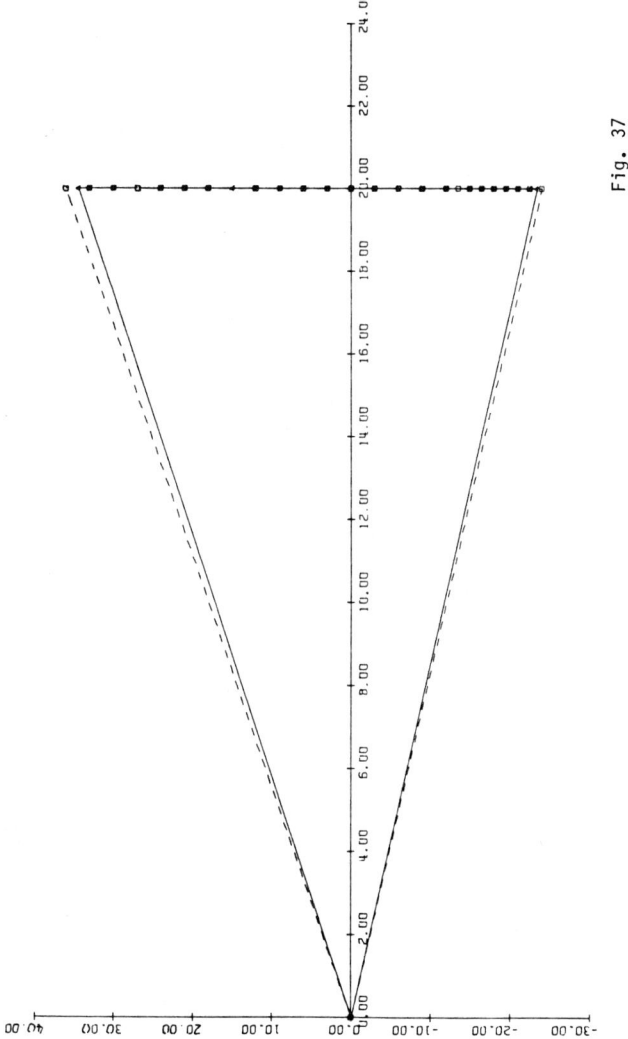

Fig. 37

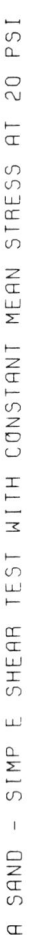

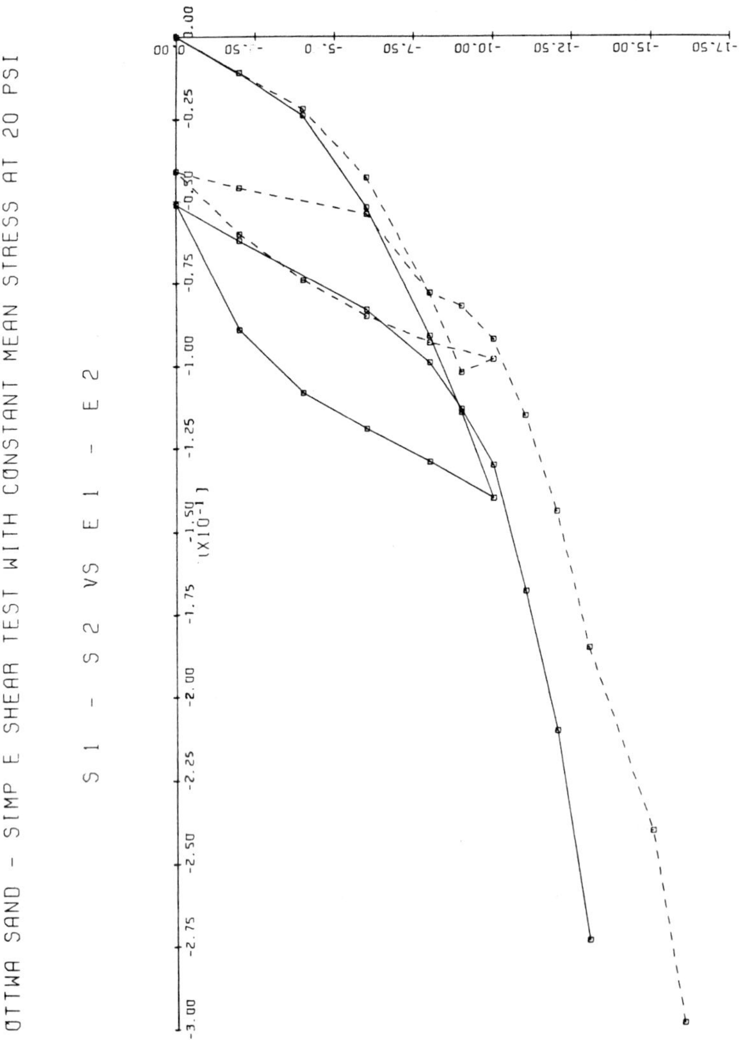

OTTWA SAND - SIMP E SHEAR TEST WITH CONSTANT MEAN STRESS AT 20 PSI
S 1 - S 2 VS E 1 - E 2
0.00
-0.25
-0.50
-0.75
-1.00
-1.25
-1.50
(X10-1)
-1.75
-2.00
-2.25
-2.50
-2.75
-3.00
-5.0
-7.50
-10.00
-12.50
-15.00
-17.50

Fig. 38

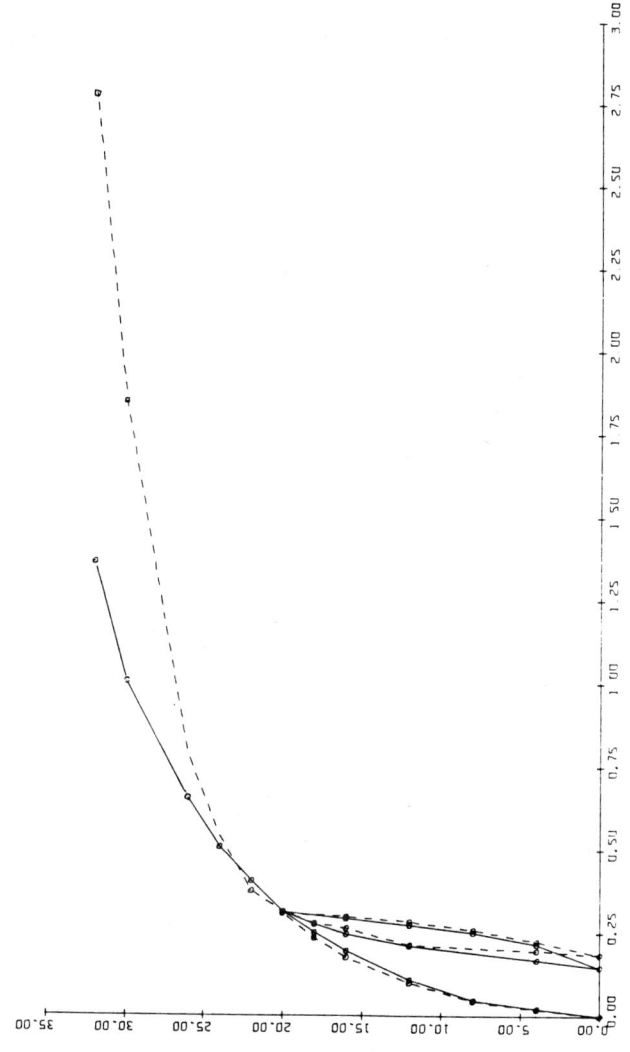

Fig. 39

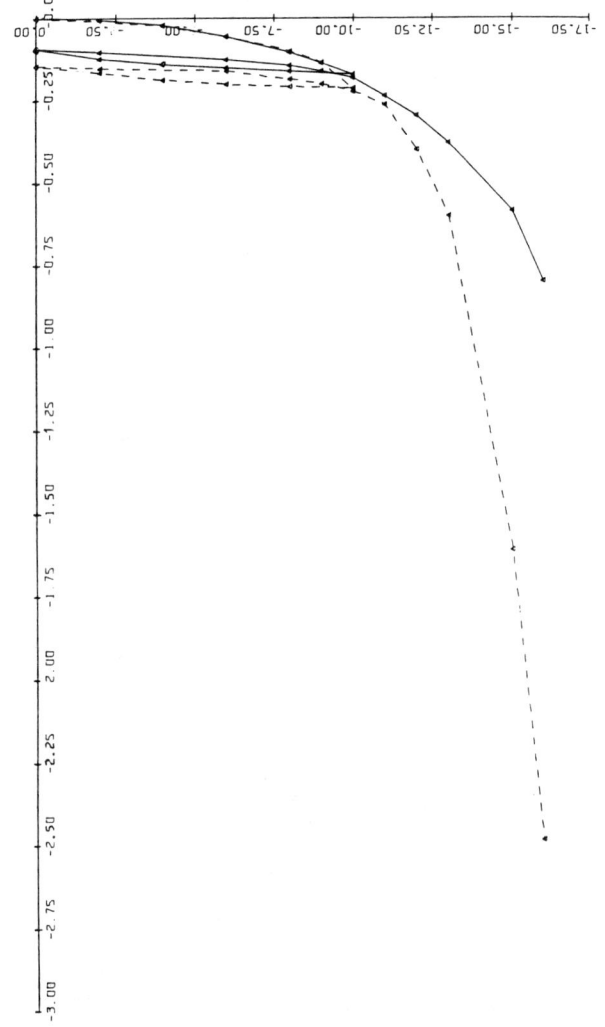

Fig. 40

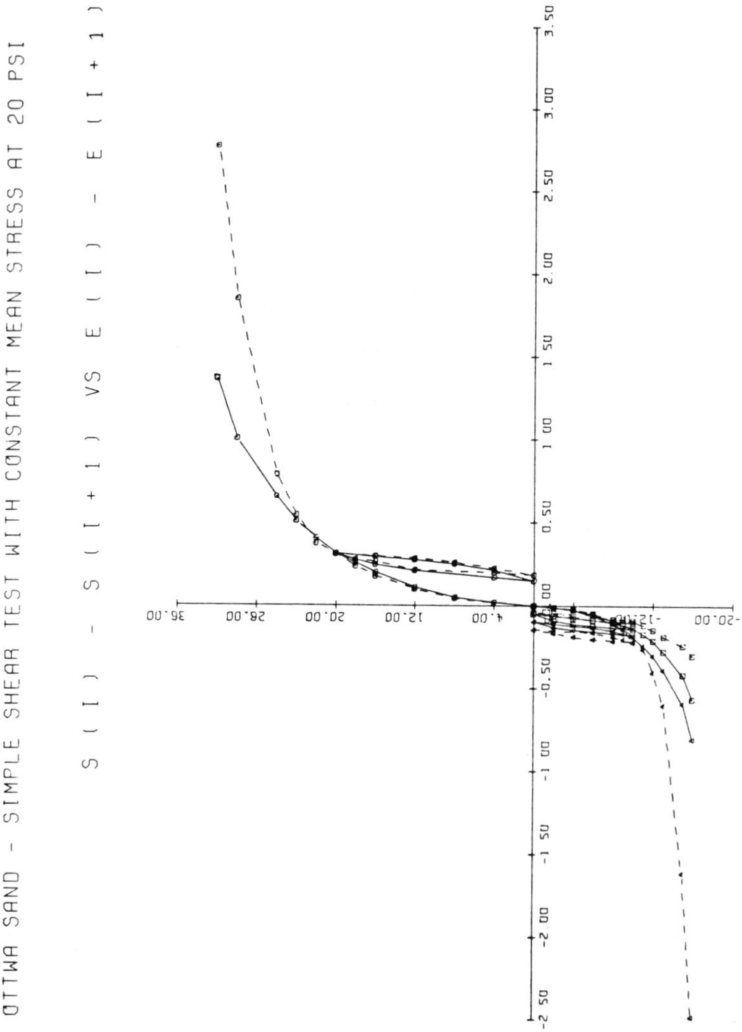

Fig. 41

OTTWA SAND - SIMPLE SHEAR TEST WITH CONSTANT MEAN STRESS AT 20 P I
T A U VS E (I)
28.00
24.00
20.00
16.00
12.00
8.00
4.00
0.00
1.00
0.75
0.50
0.25
0.00
-0.25
-0.50
-0.75
-1.00
-1.25
-1.50
-1.75
-2.00

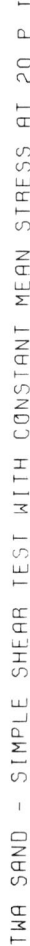

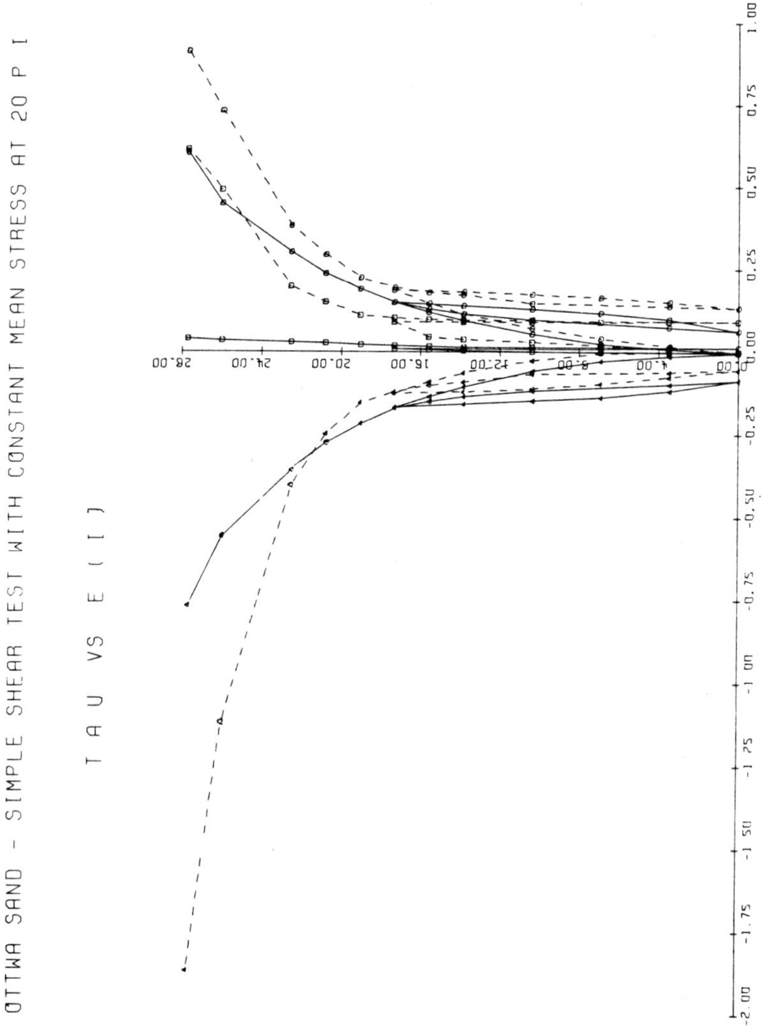

Fig. 42

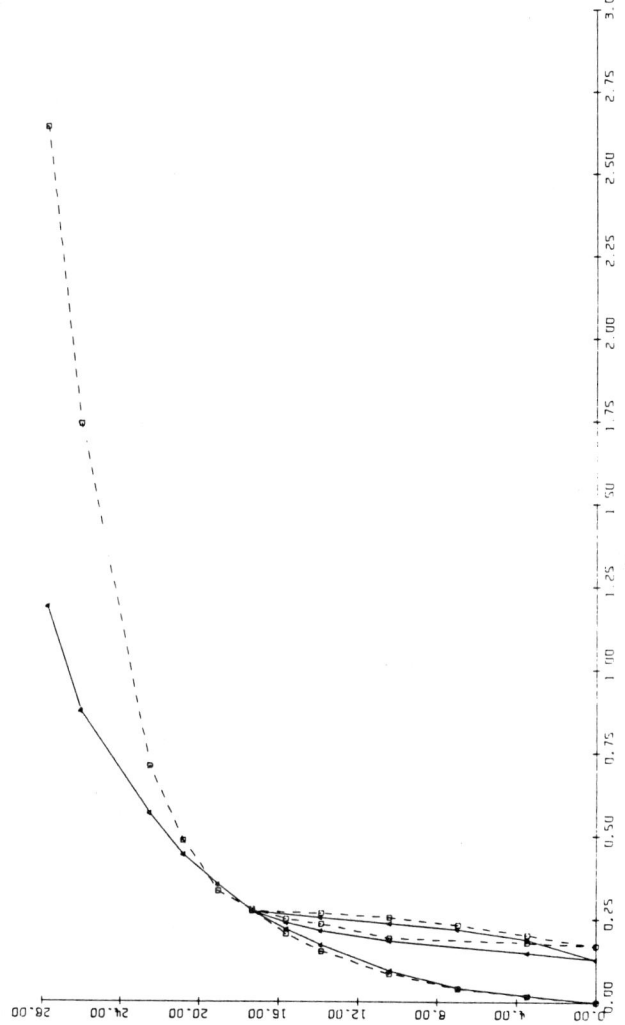

Fig. 43

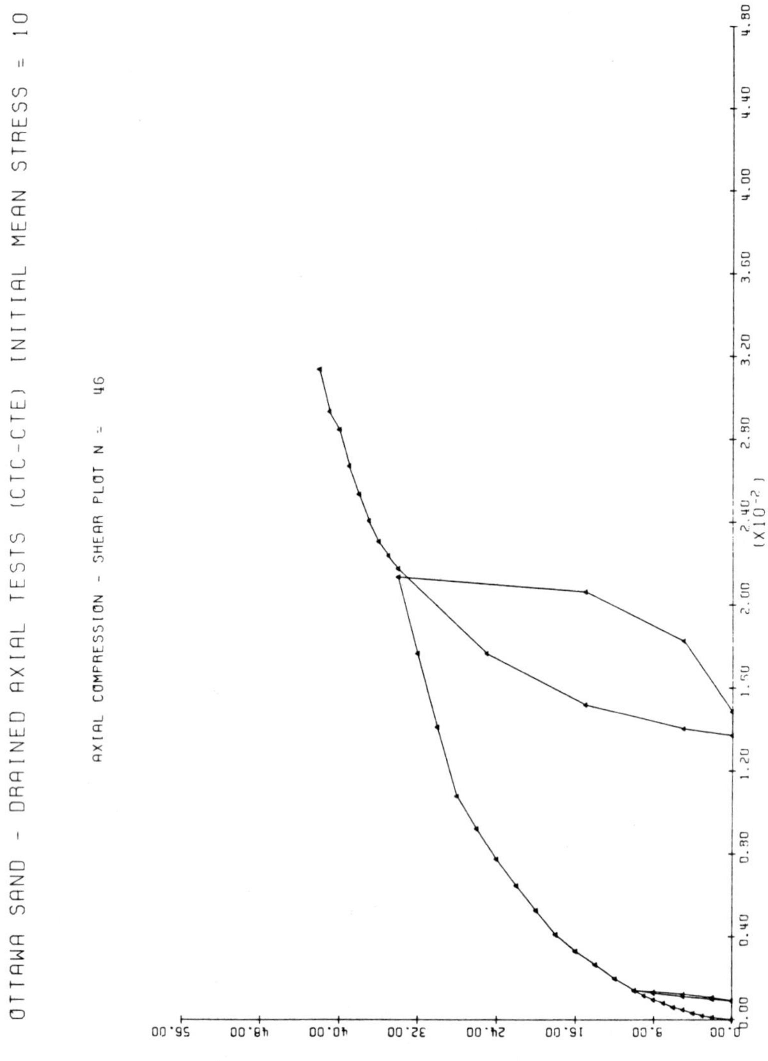

OTTAWA SAND - DRAINED AXIAL TESTS (CTC-CTE) INITIAL MEAN STRESS = 10
AXIAL COMPRESSION - SHEAR PLOT N = 46
0.00
8.00
16.00
24.00
32.00
40.00
48.00
56.00
0.00
0.40
0.80
1.20
1.60
2.00
2.40
(X10⁻²)
2.80
3.20
3.60
4.00
4.40
4.80

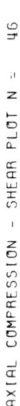

Fig. 44

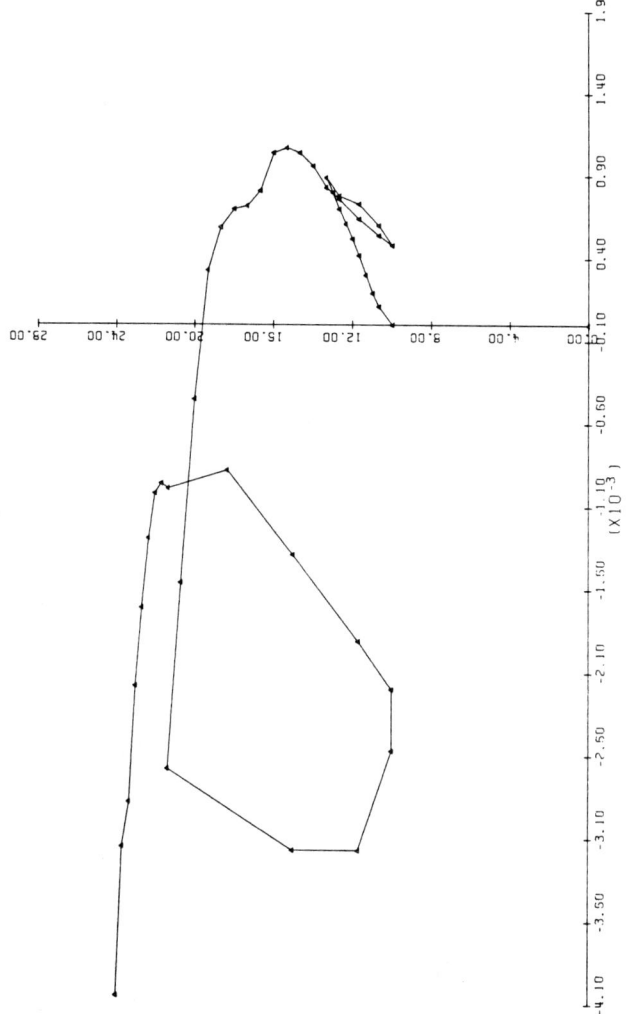

Fig. 45

810

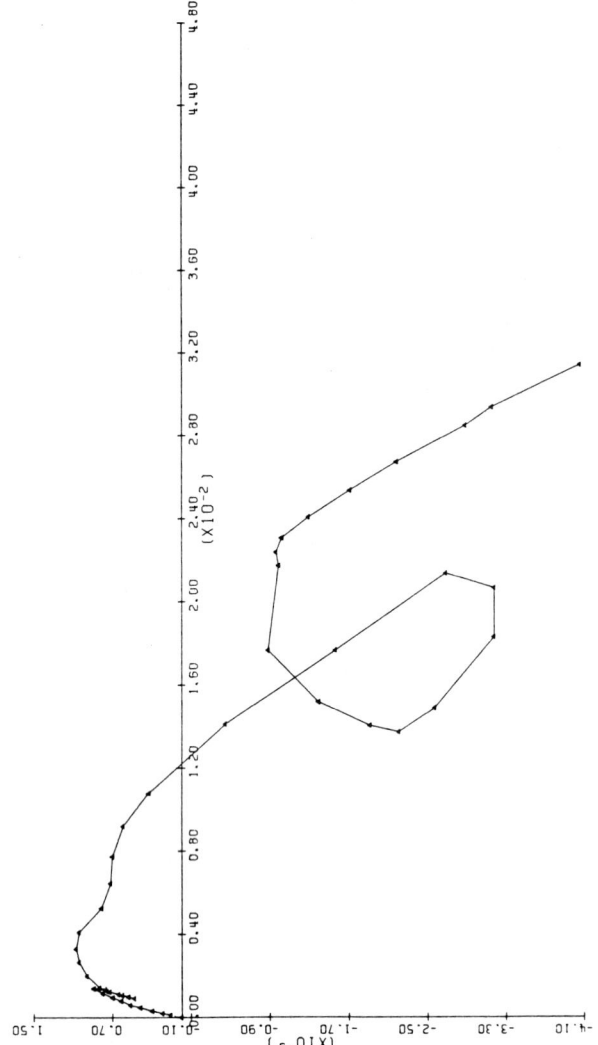

Fig. 46

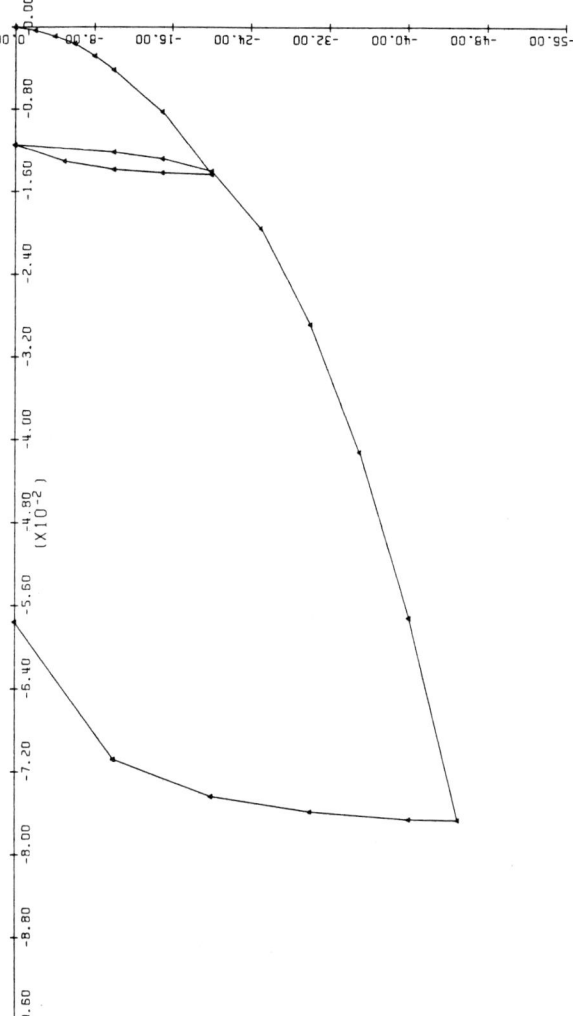

Fig. 47

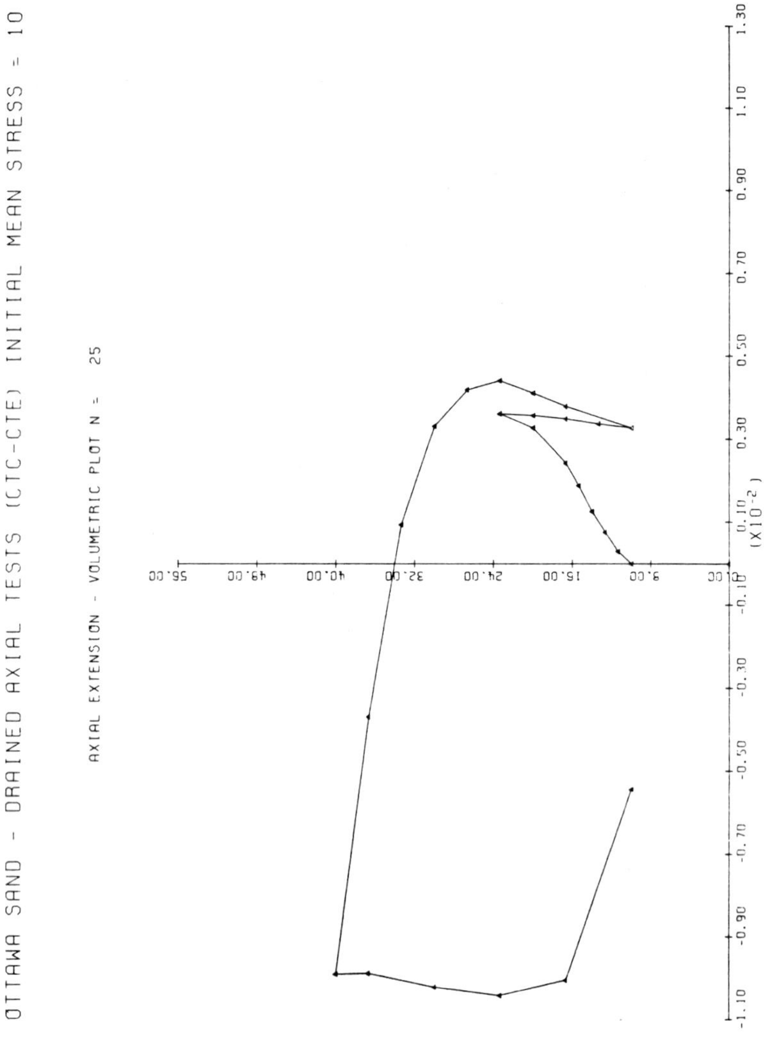
OTTAWA SAND - DRAINED AXIAL TESTS (CTC-CTE) INITIAL MEAN STRESS = 10
AXIAL EXTENSION - VOLUMETRIC PLOT N = 25
0.00
8.00
16.00
24.00
32.00
40.00
48.00
56.00
-1.10
-0.90
-0.70
-0.50
-0.30
-0.10
0.10
0.30
0.50
0.70
0.90
1.10
1.30
(X10⁻²)

Fig. 48

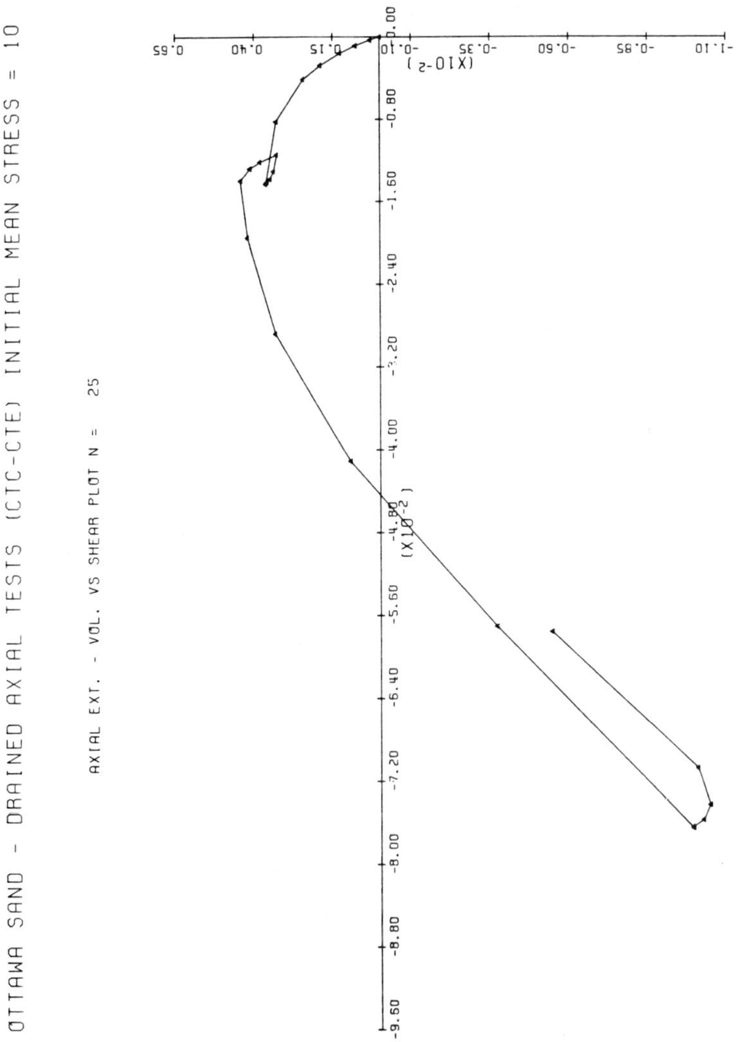
OTTAWA SAND - DRAINED AXIAL TESTS (CTC-CTE) INITIAL MEAN STRESS = 10
AXIAL EXT. - VOL. VS SHEAR PLOT N = 25
0.65
0.40
0.15
-0.10
-0.35
-0.60
-0.85
-1.10
(X10^-2)
0.00
-0.80
-1.60
-2.40
-3.20
-4.00
-4.80
-5.60
-6.40
-7.20
-8.00
-8.80
-9.60
(X10^-2)

Fig. 49

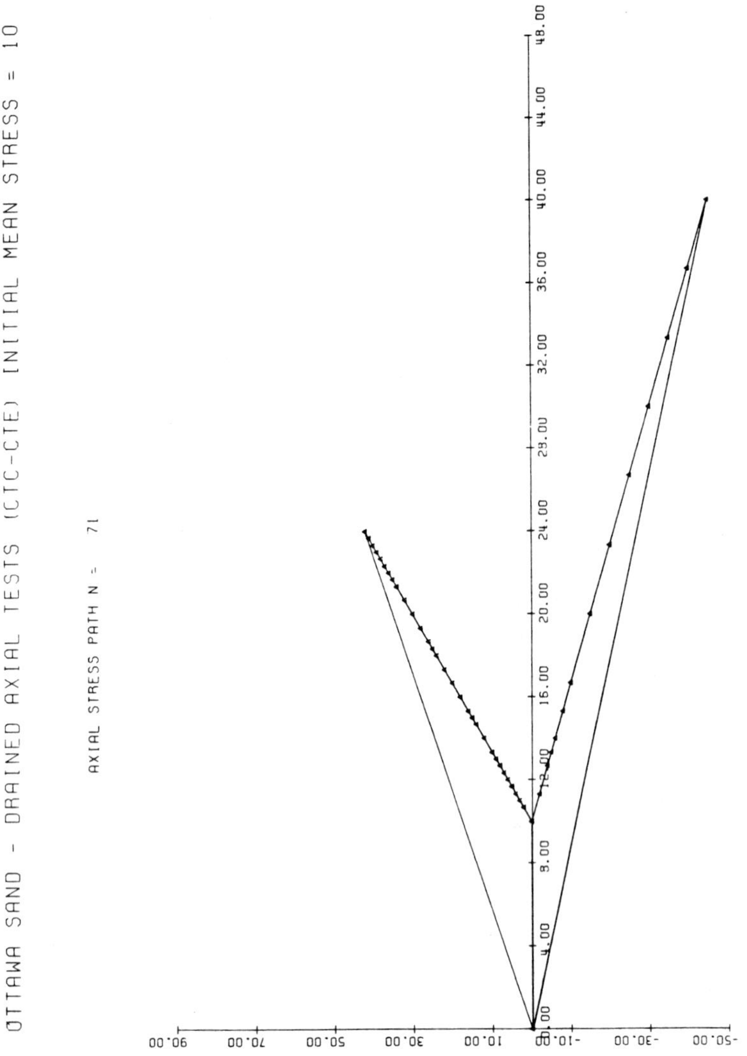
OTTAWA SAND - DRAINED AXIAL TESTS (CTC-CTE) INITIAL MEAN STRESS = 10
AXIAL STRESS PATH N = 71
48.00
44.00
40.00
36.00
32.00
28.00
24.00
20.00
16.00
12.00
8.00
4.00
.00
90.00
70.00
50.00
30.00
10.00
-10.00
-30.00
-50.00

Fig. 50

CONSTITUTIVE MODELS FOR CLAYS AND SANDS

By Koichi Akai[1], M. JSCE, Toshihisa Adachi[2], M. JSCE and Fusao Oka[3], M. JSCE

INTRODUCTION

Two different types of constitutive models have been used to carry out the assignment of this workshop. The first constitutive model is for normally consolidated clays and was derived based on Perzyna's theory of elasto-viscoplasticity (14) so that it could universaly describe such time dependent behaviors as creep, stress relaxation, rate sensitivity and secondary consolidation of the materials. The second model was derived for giving reference to explain the mechanical behaviors of over-consolidated clays and sands specially under cyclic loading processes. The non-associated flow rule in theory of plasticity is applied to the derivation of the second model and thus the time dependent properties of the materials are disregarded.

CONSTITUTIVE EQUATIONS FOR NORMALLY CONSOLIDATED CLAYS

Since normally consolidated clays are regarded as strain-hardening plastic and rate sensitive materials with dilatancy, the constitutive equations must be the one which can describe the behaviors due to those properties. The research concerned with the constitutive equations for the materials may be divided into two categories, namely, (i) the deformation characteristics at the equilibrium state (e.g.,(15)) and (ii) the rate sensitive properties (e.g.,(10)).

In order to construct more general and realistic constitutive equations for fully saturated normally consolidated clays by unifying the results of above two major approaches, Adachi and Okano (2) extended the critical state energy theory (15) for the materials so that it could explain the time dependent behaviors by using Perzyna's theory of elasto-viscoplastic continuum and empirical evidences. They derived the constitutive equations by assuming that clays

---
[1]Prof., Dept. of Transportation Engrg., Kyoto Univ., Kyoto, Japan.
[2]Associate Prof., Disaster Prevention Research Inst., Kyoto Univ., Kyoto, Japan.
[3]Assistant Prof., Dept. of Civil Engrg., Gifu Univ., Gifu, Japan.

attained to their static equilibrium states (states when the volumetric strain rate becomes zero, i.e., $\dot{v} = 0$) at the end of primary consolidation.

Arulanandan et al. (3) made clear, however, that the increase of pore water pressure took place when giving back to undrained condition after the end of primary consolidation. Febre-Cordero and Mesri (7) also pointed out the secondary consolidation occured even in the case of isotropic consolidation. This may correspond to Walker's finding (17), namely that the volumetric strain rate does not depend on the stress ratio, $(\sigma_1-\sigma_3)/\sigma_m'$ in which $(\sigma_1-\sigma_3)$ is stress difference and σ_m' is mean effective stress. These secondary consolidation phenomena are related to Bjerrum's concept of delayed compression (4), though he defined the delayed compression as the reduction in volume at unchanged effective stresses.

Taking into account above experimental evidences, it is natural to assume that normally consolidated clays never reach their static equilibrium state at the end of primary consolidation. In this study we generalize Adachi and Okano's constitutive model so that it can universally explain not only such time dependent behaviors as creep, stress relaxation and strain rate effect, but also as secondary consolidation (or delayed compression). Thus, the derived constitutive equations have a feature to be able to determine the secondary consolidation rate from the results of strain control undrained triaxial compression tests. The constitutive equations have eight material parameters which can be determined by conducting two undrained triaxial compression tests with two different constant strain rate in addition to usual consolidation tests.

Derivation of Constitutive Equations.-It is assumed that clays never reach in their static equilibrium state at the end of primary consolidation even for the case of isotropic consolidation. We define the static equilibrium state as a state at which deviatoric strain rate components $\dot{e}_{ij}$ as well as volumetric strain rate $\dot{v}$ becomes zero. Therefore, any deformation process with definite strain rate is regarded as in non-equilibrium state, namely, in dynamic state.

Perzyna (14) pointed out that the difference of the dynamic and static behaviors of materials occured due to the strain rate sensitivity of the materials and defined this rate sensitive behaviors as viscoplastic. Then, he assumed the existence of the yield function as follows

$$F(\sigma_{ij}, \varepsilon_{ij}^p) = f(\sigma_{ij}, \varepsilon_{ij}^p)/k_s - 1 \tag{1}$$

where the function $f(\sigma_{ij}, \varepsilon_{ij}^p)$ depends on the state of stress σ_{ij} and on the state of inelastic strain ε_{ij}^p, and k_s is the strain-hardening parameter. By using Drucker's postulate (6), Perzyna proposed the following flow rule for viscoplastic deformations in a simple case of infinitesimal strain field.

$$\dot{\varepsilon}_{ij}^p = <\Phi(F)> \frac{\partial f}{\partial \sigma_{ij}} \tag{2}$$

In the equation, the symbol $<\Phi(F)>$ is defined as

$$<\Phi(F)> = \begin{cases} 0 & \text{for } F \leq 0 \\ \Phi(F) & \text{for } F > 0 \end{cases} \tag{3}$$

The functional form of $\Phi(F)$ can be experimentally determined and represents the strain rate effect on the yielding of materials. The following dynamic yield criterion is obtained from Eqs. 1 and 2.

$$f(\sigma_{ij}, \varepsilon_{ij}^p) = k_s \{1 + \Phi^{-1}[(\dot{\varepsilon}_{ij}^p \dot{\varepsilon}_{ij}^p)^{1/2} (\partial f/\partial \sigma_{kl} \cdot \partial f/\partial \sigma_{kl})^{-1/2}]\} \tag{4}$$

This equation shows the change of the dynamic yield function f due to isotropic strain-hardening and inelastic strain rate. Henceforth we indicate the dynamic yield function f by f_d. The dynamic yield condition loses its rate sencitivity only when $F = 0$, i.e., $f = k_s$ and f is denoted by f_s. In this case, inelastic behaviors of materials become inviscid plastic and the following flow rule is satisfied.

$$\dot{\varepsilon}_{ij}^p = \Lambda \frac{\partial f_s}{\partial \sigma_{ij}} \tag{5}$$

in which Λ is obtained from the concept of plastic loading process in theory of plasticity (e.g.,(11)).

In order to construct constitutive equations for clays, we have to give the yield function. In this study, we assume that the mechanical behaviors of clays at their static equilibrium states can be described by the original critical state energy theory (15). According to the theory, the following static yield function is assumed to be valid.

$$f_s = \sqrt{2J_2}^{(s)}/M^*\sigma_m'^{(s)} + \ln \sigma_m'^{(s)} = k_s \tag{6}$$

where $\sqrt{2J_2} = \sqrt{S_{ij}S_{ij}}$ is the second invariant of deviatoric stress S_{ij}, M^* is defined as the value of stress ratio $\sqrt{2J_2}/\sigma_m'$ at the critical state and the superscript (s) denotes the values at the static equilibrium states. The strain hardening parameter k_s is assumed to be given by $\ln \sigma_{my}'$, namely, σ_{my}' represents strain-hardening effect in the change of stress state from $\sigma_m' = 0$ to $\sigma_m' = \sigma_{my}'$ (5). Thus we define

$$k_s = \ln \sigma_{my}'^{(s)} \tag{7}$$

Furthermore, the strain hardening parameter $\ln \sigma_{my}'^{(s)}$ is related to the inelastic volumetric strain v^p through the following equation.

$$dv^P = \frac{\lambda - \kappa}{1 + e} \frac{d\sigma_{my}'^{(s)}}{\sigma_{my}'^{(s)}} \tag{8}$$

Therefore, the constitutive equations for normally consolidated clays at their static equilibrium states are obtained from Eqs. 5, 6, 7 and 8.

Now, let's discuss the case for the dynamic deformation processes. The dynamic yield function f_d should be the same functional form of f_s because of Eq. 1 and we give

$$f_d = \sqrt{2J_2}/M^*\sigma_m' + \ln \sigma_m' = k_d \tag{9}$$

where k_d is the parameter of both strain-hardening and rate effect as seen in Eq. 4. Hereafter, stresses σ_m' and $\sqrt{2J_2}$ without any superscript denote those at the dynamic states except when emphasizing the state in dynamic. In the same way as the static strain-hardening parameter k_s, the parameter k_d in Eq. 9 is given by

$$k_d = \ln \sigma_{my}'^{(d)} \tag{10}$$

It is easily understood in Eqs. 4 and 8 that the parameter $\sigma_{my}'^{(d)}$ represents both effects of strain-hardening due to inelastic volumetric strain v^P and inelastic strain rate $\dot{\varepsilon}_{ij}^P$, namely the dynamic state exceed the static stress state by $k_s \cdot F$ due to inelastic strain rate. Fig. 1 is a schematic diagram of the static and

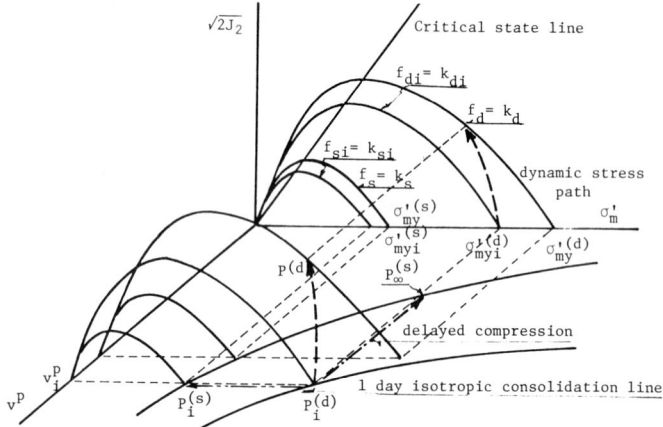

Fig. 1 Schematic diagram of both static and dynamic yield surfaces

dynamic yield surfaces. In the figure, $P_i^{(d)}$ is a dynamic state at the end of one day isotropic consolidation under a prescribed pressure $\sigma_{myi}'^{(d)}$. On the other hand, $P_i^{(s)}$ is the corresponding static state to $P_i^{(d)}$ with the same strain-hardening, namely in the same inelastic volumetric strain state, and lies on the static isotropic consolidation line to which attained by infinite time duration of isotropic consolidation. The state path $P_i^{(d)} \to P^{(d)}$ represents a shear deformation process with inelastic strain rate. The state path $P_i^{(d)} \to P_i^{(s)}$ shows the increase of pore pressure when taking back to undrained condition after the end of one day consolidation and the path $P_i^{(d)} \to P_\infty^{(s)}$ represents the secondary consolidation (delayed compression).

The yield function F can be given by the next relation by reference to Eqs.1, 6, 7, 9 and 10 as well as Fig. 1.

$$F = (\ln \sigma_{my}'^{(d)} - \ln \sigma_{my}'^{(s)})/\ln \sigma_{my}'^{(s)} \qquad (11)$$

Let's find out the relationship between the strain-hardening parameter $\sigma_{my}'^{(s)}$ and the inelastic volumetric strain v^p. Integrating Eq. 8 under the initial condition of $v^p = v_i^p$ and $\sigma_{my}'^{(s)} = \sigma_{myi}'^{(s)}$ gives

$$v^p - v_i^p = \frac{\lambda - \kappa}{1 + e} \ln \frac{\sigma_{my}'^{(s)}}{\sigma_{myi}'^{(s)}} \qquad (12)$$

The line I in Fig. 2 shows this relation and how the strain-hardening parameter $\sigma_{my}'^{(s)}$ increases with the increase of inelastic volumetric strain v^p. It is found that the strain hardening parameter $\sigma_{my}'^{(s)}$, thus the yield function F, can be determined only if the values of v^p and $\sigma_{myi}'^{(s)}$ are given. In principle, it is impossible to get the value of $\sigma_{myi}'^{(s)}$ since no one can obtain the line I which is the infinite time isotropic consolidation line. Fortunately it is, however, not necessary to know the exact value of $\sigma_{mvi}'^{(s)}$ and the reason why will be discussed later on.

Taking into account the elastic stress-strain relations, we obtain the following constitutive equations for normally consolidated clays from Eqs. 2 and 9.

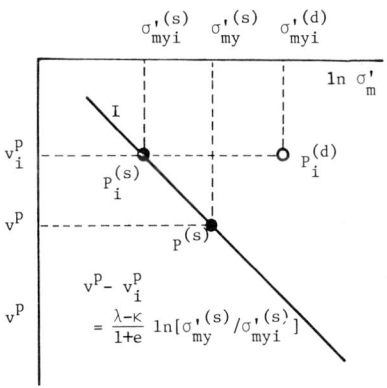

Fig. 2 Relationship between strain-hardening parameter and inelastic volumetric strain

$$\dot{\varepsilon}_{ij} = \frac{1}{2G}\dot{S}_{ij} + \frac{\kappa}{3(1+e)}\frac{\dot{\sigma}_m'}{\sigma_m'}\delta_{ij} + \Phi(F)\frac{\partial f_d}{\partial \sigma_{ij}'}$$

$$= \frac{1}{2G}\dot{S}_{ij} + \frac{\kappa}{3(1+e)}\frac{\dot{\sigma}_m'}{\sigma_m'}\delta_{ij} + \frac{1}{M^\star\sigma_m'}\Phi(F)\frac{S_{ij}}{\sqrt{2J_2}} + \frac{1}{3M^\star\sigma_m'}\Phi(F)[M^\star - \frac{\sqrt{2J_2}}{\sigma_m'}]\delta_{ij} \quad (13)$$

where G is the elastic shear modulus, κ is swelling index, e is void ratio and δ_{ij} is Kronecker delta.

According to the outcome of the previous works (1, 2), the functional form of $\Phi(F)$ in Eq. 13 is assumed to be as follows

$$\Phi(F) = c_o \exp[m' \ln(\sigma_{my}'^{(d)}/\sigma_{my}'^{(s)})] \quad (14)$$

in which c_o and m' are parameters relating to the time dependent properties of the materials. It is, however, noteworthy that the meaning of the parameters, c_o, m' and $\sigma_{my}'^{(s)}$ in Eq. 14, is different from that in the previous works.

There are seven parameters in the constitutive equations, i.e., G, κ, e, $M^\star$, c_o, m' and $\sigma_{my}'^{(s)}$. Among them, the strain-hardening parameter $\sigma_{my}'^{(s)}$ is replaced by λ (consolidation index) and $\sigma_{myi}'^{(s)}$ through Eq. 12.

By substituing both dynamic and static yield functions of Eqs.6 and 9 into Eq.14, $\Phi(F)$ is rewritten as follows

$$\Phi(F) = c_o \exp\{m'[\frac{\sqrt{2J_2}}{M^\star\sigma_m'} + \ln \sigma_m' - \frac{\sqrt{2J_2}^{(s)}}{M^\star\sigma_m'^{(s)}} - \ln \sigma_m'^{(s)}]\} \quad (15)$$

Therefore, altogether 8 parameters are required to be given to complete the constitutive equations. The eight material parameters are summarized as follows

G ; elastic shear modulus.
λ ; consolidation index.
κ ; swelling index.
e ; void ratio (the initial void ratio is used in application).
$M^\star$; the value of stress ratio, $\sqrt{2J_2}/\sigma_m'$, at the critical state.
$\sigma_{myi}'^{(s)}$; initial strain-hardening parameter.
m'; parameter representing time dependent properties.
c_o; parameter representing time dependent properties.

Determination of Parameters.-As already shown, there are 8 parameters in the equations. Let's look at how to determine the parameters. The parameters, λ, κ and e can be obtained from the consolidation and swelling test results. The elastic shear modulus G and the critical state index $M^\star$ can be determined by conducting triaxial compression tests. The determination of these parameters is nothing different from the case in the critical state energy theory. Thus, we will discuss the way to determine the remained parameters, i.e., c_o, m' and $\sigma'^{(s)}_{myi}$.

First of all, the validity of Eq.14 or 15 is examined by the results of undrained triaxial compression tests with constant strain rates. Since the volume change is negligible under undrained condition, i.e., $\dot{\varepsilon}_{kk}= 0$, we obtain the next relationship between mean effective stress σ'_m and inelastic volumetric strain v^p from Eq.13.

$$\dot{\varepsilon}_{kk}= \dot{v} = \frac{\kappa}{1+e}\frac{\dot{\sigma}'_m}{\sigma'_m} + \frac{1}{M^*\sigma'_m} \Phi(F)[\ M^* - \frac{\sqrt{2J_2}}{\sigma'_m}] = 0 \quad (16)$$

Namely,

$$\frac{\kappa}{1+e}\frac{d\sigma'_m}{\sigma'_m} + dv^p = 0 \quad (17)$$

Integrating Eq.17 under the condition of $v^p= 0$ at $\sigma'_m = \sigma'_{me}$ results in

$$v^p = - \frac{\kappa}{1+e} \ln(\sigma'_m/\sigma'_{me}) \quad (18)$$

where σ'_{me} denotes the stress state at the end of primary consolidation.

The meaning of Eq.18 is that the same inelastic volumetric strain v^p takes place at two different stress states having different strain rate, provided two states have the same mean effective stress σ'_m and the same initial condition σ'_{me}. Fig. 3 shows this fact. Namely, the inelastic volumetric strain v^p are same at both stress states represented as P_1 and P_2 lying on two different stress paths which correspond to strain rates $\dot{\varepsilon}^{(1)}$ and $\dot{\varepsilon}^{(2)}$, respectively, as shown in Fig. 3.

Under undrained conditions, the total strain rate component $\dot{\varepsilon}_{ij}$ is equivalent to the deviatoric strain rate component $\dot{e}_{ij}$ because Eq.16 is always satisfied.

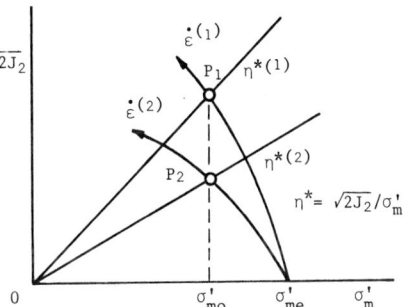

Fig. 3 The same inelastic volumetric strain (same strain-hardening) states on different stress paths

$$\dot{\varepsilon}_{ij} = \dot{e}_{ij} = \frac{\dot{S}_{ij}}{2G} + \frac{1}{M^*\sigma'_m} \Phi(F) \frac{S_{ij}}{\sqrt{2J_2}} \quad (19)$$

We continue to discuss the problem in a simple case of conventional axisymmetric triaxial compression, i.e., $\sigma'_1 > \sigma'_2 = \sigma'_3$. Under this specific condition, the following relations are reduced.

$S_{11}= 2/3(\sigma_1-\sigma_3)$, $\sqrt{2J_2} = \sqrt{2/3}(\sigma_1-\sigma_2)$, $\varepsilon_{11}= e_{11}= 2/3(\varepsilon_1-\varepsilon_2)$

Using these relations in Eqs.19 and 15 results in as follows

$$\dot{\varepsilon}_{11} = \frac{\dot{S}_{ij}}{2G} + \frac{\sqrt{2/3}}{M^*\sigma'_m} \phi(F) \tag{20}$$

$$\phi(F) = c_0 \exp\{m'[\frac{q}{M\sigma'_m} + \ln \sigma'_m - \frac{q^{(s)}}{M\sigma'_m(s)} - \ln \sigma'^{(s)}_m]\} \tag{21}$$

where $q = (\sigma_1 - \sigma_3)$ and $M = \sqrt{3/2} M^*$.

Assuming $\dot{\varepsilon}_{11} = \dot{\varepsilon}^p_{11}$, namely elastic shear strain rate $\dot{e}^E_{11} = \dot{S}_{11}/2G$ to be negligible, the next relation is obtained from Eqs.20 and 21 by comparing the states P_1 and P_2 shown in Fig. 3.

$$\ln\{\dot{\varepsilon}^{(1)}_{11}/\dot{\varepsilon}^{(2)}_{11}\} = \frac{m'}{M^*} \{\sqrt{2J_2}^{(1)}/\sigma'_m - \sqrt{2J_2}^{(2)}/\sigma'_m\}$$

$$= \frac{m'}{M} \{q^{(1)}/\sigma'_m - q^{(2)}/\sigma'_m\} \tag{22}$$

where the superscripts (1) and (2) correspond to the states P_1 and P_2.

Fig. 4 is prepared to evaluate the validity of Eq.22. It is clearly shown in the figure that there exists a linear relationship between the logarithm of strain rate $\dot{\varepsilon}_{11}$ and the stress ratio q/σ'_m as an equi-inelastic volumetric line. Thus, the validity of Eq.15 is evaluated and the parameter m' can be determined from the slope of equi-inelastic volumetric strain line, provided M-value is given.

The next task is how to determine the remained parameters c_0 and $\sigma'^{(s)}_{myi}$. For the purpose, taking into account only the term of inelastic strain rate

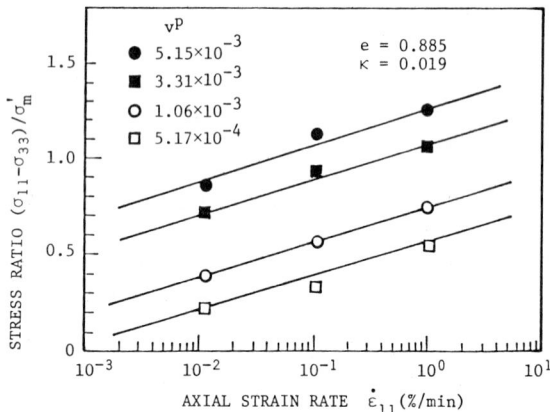

Fig. 4 Relationship between the stress ratio and the logarithm of strain rate.

component $\dot{e}^p_{ij}$ in Eq.13, we have

$$\dot{e}^p_{ij} = \frac{1}{M^*\sigma'_m} \Phi(F) \frac{S_{ij}}{\sqrt{2J_2}}$$

$$= \frac{c_0}{M^*\sigma'_m} \exp\{m'[\frac{\sqrt{2J_2}}{M^*\sigma'_m} + \ln \sigma'_m - \ln \sigma'^{(s)}_{my}]\}\frac{S_{ij}}{\sqrt{2J_2}} \qquad (23)$$

Supposing that only newly developed inelastic volumetric strain v^p, i.e., $v^p - v^p_i$ in Eq.12 is taken into account, the next relation is satisfied.

$$v^p = \frac{\lambda - \kappa}{1 + e} \ln \frac{\sigma'^{(s)}_{my}}{\sigma'^{(s)}_{myi}}$$

or

$$\ln \sigma'^{(s)}_{my} = \ln \sigma'^{(s)}_{myi} + \frac{1 + e}{\lambda - \kappa} v^p \qquad (24)$$

Substituting this relation into Eq.23 results in

$$\dot{e}^p_{ij} = \frac{c_0}{M^*\sigma'_m} \exp\{m'[\frac{\sqrt{2J_2}}{M^*\sigma'_m} + \ln \sigma'_m - \ln \sigma'^{(s)}_{myi} - \frac{1+e}{\lambda-\kappa} v^p]\}\frac{S_{ij}}{\sqrt{2J_2}}$$

$$= \frac{c_0}{M^*\sigma'_m} \exp[-m'\ln \frac{\sigma'^{(s)}_{myi}}{\sigma'_{me}}]\exp\{m'[\frac{\sqrt{2J_2}}{M^*\sigma'_m} + \ln \frac{\sigma'_m}{\sigma'_{me}} - \frac{1+e}{\lambda-\kappa} v^p]\}\frac{S_{ij}}{\sqrt{2J_2}}$$

$$= C \exp\{m'[\frac{\sqrt{2J_2}}{M^*\sigma'_m} + \ln \frac{\sigma'_m}{\sigma'_{me}} - \frac{1+e}{\lambda-\kappa} v^p]\}\frac{S_{ij}}{\sqrt{2J_2}} \qquad (25)$$

where we define C by

$$C = \frac{c_0}{M^*\sigma'_m} \exp[-m'\ln \frac{\sigma'^{(s)}_{myi}}{\sigma'_{me}}] \qquad (26)$$

We again consider the case of undrained axisymmetric triaxial compression and assume that the elastic deviatoric strain rate $\dot{e}^E_{ij}$ is negligible. Under this condition, Eq.25 becomes

$$\dot{e}^p_{11} = \dot{\varepsilon}_{11} = \sqrt{2/3}\, C \exp\{m'[\frac{q}{M^*\sigma'_m} + \ln \frac{\sigma'_m}{\sigma'_{me}} - \frac{1+e}{\lambda-\kappa} v^p]\} \qquad (27)$$

Introducing Eq.18 into Eq.28 gives

$$\dot{\varepsilon}_{11} = \sqrt{2/3}\, C \exp\{m'[\frac{q}{M^*\sigma'_m} - \frac{\lambda(1+e)}{\kappa(\lambda-\kappa)} v^p]\} \qquad (28)$$

When the value of inelastic volumetric strain v^p is prescribed, both values of the strain rate component $\dot{\varepsilon}_{11}$(term in the left hand side of Eq.28) and the stress ratio q/σ'_m (term in the right hand side of Eq.28) are determined from the relation given in Fig. 4. Substituting thus obtained values of v^p, $\dot{\varepsilon}_{11}$ and q/σ'_m into Eq.28, the parameter C is determined. Therefore, both material parameter c_0 and $\sigma'^{(s)}_{myi}$ are expressed through the parameter C. Namely, in order to complete the constitutive equations, it is enough to know the parameter C instead of obtaining individual values of c_0 and $\sigma'^{(s)}_{myi}$.

Thus, all material parameters can be determined from the test results of consolidation, swelling and strain control undrained triaxial compression tests.

Interrelation of Parameter m' and Parameter of Secondary Consolidation α
-The secondary consolidation is well-known time dependent behavior of clay, and we will examine the interrelation between the secondary consolidation rate parameter α and the strain rate parameter m' in the derived constitutive equations. For simplicity, the secondary consolidation is discussed under isotropic stress conditions, i.e., in the case of $\sqrt{2J_2} = 0$. Under the isotropic consolidation, the inelastic volumetric strain rate $\dot{v}^p$ (secondary consolidation rate) is expressed by Eqs. 13 and 15 as follows

$$\dot{v}^p = \dot{\varepsilon}^p_{kk} = \frac{c_0}{\sigma'_m} \exp[m' \ln(\sigma'_m / \sigma'^{(s)}_m)] \tag{29}$$

Since in the case of isotropic consolidation $\sigma'^{(s)}_{my}$ in Eq.8 can be replaced by $\sigma'^{(s)}_m$, Eq.8 may be rewritten as

$$v^p = \frac{\lambda - \kappa}{1 + e} \frac{d\sigma'^{(s)}_m}{\sigma'^{(s)}_m} \tag{30}$$

Supposing that the conditions, i.e., $\sigma'^{(s)}_m = \sigma'_m$ and $v^p = v^p_\infty$, are satisfied at the end of secondary consolidation under the pressure of σ'_m, Eq.30 is integrated with the conditions.

$$v^p_\infty - v^p = \frac{\lambda - \kappa}{1 + e} \ln \frac{\sigma'_m}{\sigma'_m(s)} \tag{31}$$

Substituting Eq.31 into Eq.29 gives

$$\dot{v}^p = \frac{c_0}{\sigma'_m} \exp[\frac{1+e}{\lambda - \kappa} m'(v^p_\infty - v^p)]$$

$$= \frac{c_0}{\sigma'_m} \exp[\frac{1+e}{\lambda - \kappa} m' v^p_\infty] \exp[-\frac{1+e}{\lambda - \kappa} m' v^p]$$

$$= \frac{c_0}{\sigma'_m} \exp[\frac{1}{\alpha} v^p_\infty] \exp[-\frac{1}{\alpha} v^p]$$

$$= C_1 \exp[-\frac{1}{\alpha} v^p] \tag{32}$$

where

$$C_1 = \frac{c_0}{\sigma'_m} \exp[\frac{1+e}{\lambda - \kappa} m' v^p_\infty] = \frac{c_0}{\sigma'_m} \exp[\frac{1}{\alpha} v^p_\infty] \tag{33}$$

$$\alpha = \frac{\lambda - \kappa}{(1 + e) m'} \tag{34}$$

By assuming C_1 to be constant, Eq.32 is integrated under the condition of $v^p = v^p_0$ at $t = t_0$.

$$v^p = \alpha \ln(t/t_0) + v^p_0 \tag{35}$$

where $v_o^p = -\alpha \ln(\alpha/C_1 t_o)$

It is clearly seen in Eq.34 that the parameter α for the secondary consolidation rate can be determined by using the parameter m'.

Constitutive Equations for Anisotropic Consolidated Clays-In order to derive the constitutive equations for anisotropic consolidated clays, both static and dynamic yield functions are assumed to be given by

$$f_s = \bar{\eta}^*/M^* + \ln \sigma_m' = \ln \sigma_{my}'^{(s)} \tag{36}$$

$$f_d = \bar{\eta}^*/M^* + \ln \sigma_m' = \ln \sigma_{my}'^{(d)} \tag{37}$$

$\bar{\eta}^*$ is a stress parameter for representation of anisotropic consolidation introduced by Sekiguchi and Ohta (16), and is defined by

$$\bar{\eta}^* = \{[\eta_{ij}^* - \eta_{ij(o)}^*][\eta_{ij}^* - \eta_{ij(o)}^*]\}^{1/2} \tag{38}$$

in which

$$\eta_{ij}^* = S_{ij}/\sigma_m', \qquad \eta_{ij(o)}^* = S_{ij(o)}/\sigma_{m(o)}' \tag{39}$$

where η_{ij}^* is the (i, j) component of "pressure normalized deviatoric stress tensor" and $\eta_{ij(o)}^*$ is the value of η_{ij}^* at the end of anisotropic consolidation.

Applying the dynamic yield function 37 to Eq.2 gives the next constitutive equations for anisotropic consolidated clays.

$$\dot{\varepsilon}_{ij}^p = \Phi(F)[\frac{1}{M^*}\frac{\partial \bar{\eta}^*}{\partial \sigma_{ij}} + \frac{1}{3\sigma_m'}\delta_{ij}]$$

$$= \Phi(F)[\frac{1}{M^*\sigma_m'\bar{\eta}^*}(\eta_{ij}^* - \eta_{ij(o)}^*) + \frac{1}{3\sigma_m'}\delta_{ij} - \frac{S_{kl}(\eta_{kl}^* - \eta_{kl(o)}^*)}{3M^*\bar{\eta}^*\sigma_m'^2}\delta_{ij}] \tag{40}$$

Generalization of Constitutive Equations to General Loading Conditions-So far it is assumed that the critical state value $M^*(=\sqrt{2J_2}/\sigma_m')$ of a given clay takes a constant value under any loading condition, namely for triaxial compression, triaxial extension or any others. On the other words, the yield surface is same as that of Drucker-Prager's yield criterion. It is, however, well-known that Mohr-Coulomb's criterion is better to be applied to describe the behaviors of geotechnical materials at the critical states.

By assuming that the critical state of the materail is defined by Mohr-Coulomb's criterion, we will modify the constitutive equations so that they can explain the mechanical behaviors under general loading conditions. Introducing so-called Lode angle θ_o, Mohr-Coulomb's yield criterion is expressed as

$$f = -\sqrt{2}\,\sigma'_m \sin\phi' + \sqrt{2J_2}(\cos\theta_o - \sqrt{1/3}\sin\theta_o \sin\phi') - \sqrt{2}\,c'\cos\phi' = 0 \qquad (41)$$

in which

$$-\pi/6 < \theta_o = \frac{1}{3}\sin^{-1}[\frac{3\sqrt{3}}{2}\frac{J_3}{J_2^{3/2}}] < \pi/6$$

and θ_o ; Lode angle.

$J_2 = S_{ij}S_{ji}/2$; second invariant of deviatoric stress tensor.
$J_3 = S_{ij}S_{jk}S_{ki}/3$; third invariant of deviatoric stress tensor.
ϕ' ; internal friction angle.
c' ; cohesive strength

When $c' = 0$, the critical state value M* is obtained from Eq.41 as

$$M^* = [\frac{\sqrt{2J_2}}{\sigma'_m}]_{\text{at critical state}} = \frac{2\sin\phi'}{-\sqrt{1/3}\sin\theta_o \sin\phi' + \cos\theta_o} \qquad (42)$$

Thus for axisymmetric triaxial compression, it results in

$$M = q/\sigma'_m = \frac{6\sin\phi'}{(3 - \sin\phi')} = M_c \qquad (43)$$

and for axisymmetric triaxial extension, it becomes

$$M = q/\sigma'_m = \frac{6\sin\phi'}{(3 + \sin\phi')} = M_e \qquad (44)$$

where $q = (\sigma_1 - \sigma_3)$ and $M_c > M_e$ is always valid.

Substituting M* value given by Eq.42 into the constitutive equations 13 or 40, we will have the constitutive equations which can explain the mechanical behaviors of isotropic normally consolidated clays of anisotropic consolidated clays under general loading conditions.

CONSTITUTIVE EQUATIONS FOR OVER-CONSOLIDATED CLAYS AND SANDS

In this section, we will discuss the derivation of constitutive equations for over-consolidated clays and sands. Recently, various types of constitutive models have been proposed to explain the mechanical behaviors of geotechnical materials under cyclic loading conditions.

Pender [13] considered that whenever the stress ratio $\eta^*_{ij} = S_{ij}/\sigma'_m$ changed even inside of the boundary surface (e.g., critical state energy theory), the plastic yielding occured. In addition, he assumed that the undrained effective stress paths were expressed by parabolic curves and positive dilatancy took place when the stress state was in so-called Dry-side, but negative dilatancy in Wet-side.

Following to Pender's idea and in a way to refine his model, we will derive a constitutive model which can describe the mechanical behaviors of over-consolidated clays and sands. Since there is no elastic region, we use Hill's nonassociated flow rule (8), i.e.,

$$d\varepsilon_{ij}^p = \Lambda \frac{\partial f_p}{\partial \sigma_{ij}} df \tag{45}$$

where f_p is the plastic potential function and f is the plastic yield function. The plastic potential function f_p is assumed to be given by

$$f_p = \bar{\eta}^* + \tilde{M}^* \ln[\sigma_m'/\sigma_{m(n)}'] = 0 \tag{46}$$

In this equation, $\bar{\eta}^*$ is a stress parameter introduced by Sekiguchi and Ohta(16) and in this case we define as follows

$$\bar{\eta}^* = [(\eta_{ij}^* - \eta_{ij(n)}^*)(\eta_{ij}^* - \eta_{ij(n)}^*)]^{1/2} \tag{47}$$

In Eqs.46 and 47, $\sigma_{m(n)}'$ and $\eta_{ij(n)}^*$ are the values of σ_m' and η_{ij}^* at n-th times turning over point of the loading direction. Furthermore, the parameter $\tilde{M}^*$ in Eq.46 is a variable which is defined by

$$\tilde{M}^* = -\frac{\eta^*}{\ln(\sigma_m'/\sigma_{me}')} \tag{48}$$

in which

$$\eta^* = [\eta_{ij}^* \eta_{ij}^*]^{1/2} \tag{49}$$

It is easily understood that $\tilde{M}^*$ is automatically determined once the current stress state (S_{ij}, σ_m') is provided. On the other hand, the yield function is given by

$$f = \eta^* = \sqrt{2J_2}/\sigma_m' \tag{50}$$

Fig. 5 schematically illustrates the plastic potential curve $f_p = 0$ and plastic yield line $f = \eta^*$. In the figure, P_n is the current stress state where naturally both potential curve $f_p = 0$ and yield line $f = \eta^*$ pass through. In addition to the failure line M_f^*, we introduce a constant stress ratio line M_m^* which corresponds to the stress state when the maximum compression of the material takes place. Inside of the state boundary surface, namely in over-consolidated state, we keep $\tilde{M}^* = M_m^*$ after $\tilde{M}^*$ attains to the value of M_m^*. This assumption coincides with the concept of so-called phase transformation line defined by Ishihara et al.(9).

For sands, even in the case of normally consolidated state, we will apply the same constitutive model by always taking $\tilde{M}^* = M_m^*$. In this case, the derived constitutive model coincides with Nishi and Esashi's model for sands (12).

When the loading direction turning over at the point P_n as shown in Fig. 5,

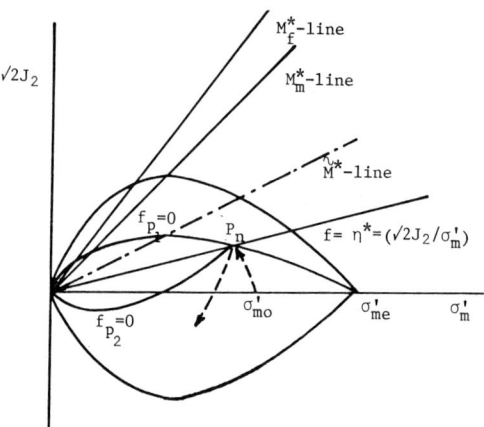

Fig. 5 Schematic diagram of the potential surfaces and yield line.

the plastic potential surface f_{p_2} is given by assuming to be symmetric with respect to the current yield line $f = \eta^*$.

In order to obtain the hardening function, we introduce a strain-hardening parameter $\bar{\gamma}^*$ corresponding to the parameter $\bar{\eta}^*$.

$$\bar{\gamma}^* = [(e^p_{ij} - e^p_{ij(n)})(e^p_{ij} - e^p_{ij(n)})]^{1/2} \tag{51}$$

where e^p_{ij} is plastic deviatoric strain component and $e^p_{ij(n)}$ is the value of e^p_{ij} at the state when the n-th times reverse of loading takes place as shown in Fig. 5. In addition, we assume that the following relationship between the hardening parameter $\bar{\gamma}^*$ and the parameter $\bar{\eta}^*$ is expressed by a hyperbolic curve as given in Fig. 6.

$$\bar{\gamma}^* = \frac{\eta^*(M_f^* + \eta^*_{(n)})}{G'[M_f^* + \eta^*_{(n)} - \bar{\eta}^*]} \tag{52}$$

where G' is the initial tangent modulus, M_f^* is the value of $\bar{\eta}^*$ at the failure state and $\eta^*_{(n)}$ is defined by

$$\eta^*_{(n)} = [\eta^*_{ij(n)} \eta^*_{ij(n)}]^{1/2} \tag{53}$$

This hyperbolic strain-hardening function is a generalization of

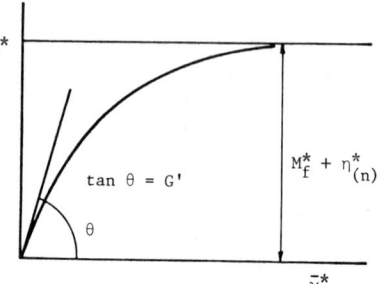

Fig. 6 Hyperbolic representation of strain-hardening parameter

Nishi and Esashi's hardening function (12).

The next task is to determine the parameter Λ in Eq.45. Differentiating Eq.52 with respect to $\bar{\eta}*$ results in

$$\frac{d\bar{\gamma}*}{d\bar{\eta}*} = \frac{[M_f^* + \eta_{(n)}^*]^2}{G'[M_f^* + \eta_{(n)}^* - \bar{\eta}*]^2} \tag{54}$$

On the other hand, differentiating Eq.51 with respect to e_{ij}^p gives

$$\frac{d\bar{\gamma}*}{de_{ij}^p} = \frac{(e_{ij}^p - e_{ij(n)}^p)}{\bar{\gamma}*} \tag{55}$$

Using Eq.45 in Eq.55, we have the following relation.

$$d\bar{\gamma}* = \frac{(e_{ij}^p - e_{ij(n)}^p)}{\bar{\gamma}*} \Lambda \frac{\partial f_p}{\partial s_{ij}} d\bar{\eta}* \tag{56}$$

$\partial f_p/\partial s_{ij}$ is obtained from Eqs.46 and 47 as follows

$$\frac{\partial f_p}{\partial s_{ij}} = \frac{1}{\bar{\eta}*}(\eta_{ij}^* - \eta_{ij(n)}^*) \frac{1}{\sigma_m'} \tag{57}$$

Using this relation in Eq.56 results in

$$\frac{d\bar{\gamma}*}{d\bar{\eta}*} = \frac{(e_{ij}^p - e_{ij(n)}^p)}{\bar{\gamma}*} \Lambda \frac{1}{\bar{\eta}*}(\eta_{ij}^* - \eta_{ij(n)}^*) \frac{1}{\sigma_m'} \tag{58}$$

Comparing Eq.54 with Eq.58, we obtain the parameter Λ as follows

$$\Lambda = \frac{\bar{\gamma}*\bar{\eta}*}{(e_{ij}^p - e_{ij(n)}^p)(\eta_{ij}^* - \eta_{ij(n)}^*)} \cdot \frac{(M_f^* + \eta_{(n)}^*)^2}{G'(M_f^* + \eta_{(n)}^* - \bar{\eta}*)^2} \cdot \sigma_m' \tag{59}$$

Finally, the constitutive equations for over-consolidated clays and sands are obtained from Eqs.45, 46, 47, 50 and 59. Taking into account the elastic component, they can be written by

$$d\varepsilon_{ij} = \frac{dS_{ij}}{2G} + \frac{\kappa}{1+e}\frac{d\sigma_m'}{\sigma_m'} + \Lambda[\frac{\eta_{ij}^* - \eta_{ij(n)}^*}{\bar{\eta}*} \frac{1}{\sigma_m'} + \frac{\tilde{M}*}{3\sigma_m'}\delta_{ij} - \frac{S_{ij}}{2\bar{\eta}*}(\eta_{kl}^* - \eta_{kl(n)}^*)\frac{\delta_{ij}}{\sigma_m'^2}]df \tag{60}$$

There are six material parameters in the constitutive equations, namely G, κ, e, G', M_f^* and $\tilde{M}*$. However, $\tilde{M}*$ is obtained when λ, σ_{me}' (preconsolidation pressure) and the current stress state, and is bounded by the value of M_m^*. Thus, the following seven material parameters have to be given to complete the constitutive equations.

G ; elastic shear modulus.

λ ; consolidation index.

κ ; swelling index.

e ; void ratio (the initial void ratio is used in application).

G'; parameter for strain-hardening parameter $\bar{\gamma}*$.

M_m^*; the value of stress ratio, $\sqrt{2J_2}/\sigma_m'$, at the maximum compression.

M_f^*; the value of stress ratio, $\sqrt{2J_2}/\sigma_m'$, at the failure state.

APPENDIX I - REFERENCES

1. Akai, K. and Oka, F., Thermodynamic theory of inelastic materials and its application to stress wave propagation in cohesive soil, Proc. JSCE, No.253, 1976, pp.109-122.

2. Adachi, T. and Okano, M., A constitutive equations for normally consolidated clay, Soils and Foundations, Vol.14, No.4, 1974, pp.55-73.

3. Arulanandan, K., Shen, C. K. and Young, R. B., Undrained creep behavior of a coastal organic silty clay, Geotechnique, Vol.21, No.4, 1971, pp.359-375.

4. Bjerrum, L., Engineering geology of Norwegian normally consolidated marine clays as related to settlements of Buildings, Geotechnique, Vol.17, No.2, 1967, pp.81-118.

5. Calladine, C. R., The yielding of clay, Correspondence, Geotechnique, Vol.13, No.3, 1963, pp.250-255.

6. Drucker, D. C., A definition of stable inelastic material, J. Appl. Mech., Trans. ASME, Vol.26, 1959, pp.101-106.

7. Febre-Cordero and Mesri,G., Influence of testing conditions on creep behavior of clay, Report No.FRA-ORD & D-75-29, UILU-ENG-74-2031, 1974.

8. Hill,R., The Mathematical Theory of Plasticity, Oxford Univ. Press, 1950.

9. Ishihara, K., Tatsuoka, F. and Yasuda, S., Undrained deformation and liquefaction of sand under cyclic stresses, Soils and Foundations, Vol.15, No.1, pp.29-44.

10. Murayama, S. and Shibata, T., Flow and stress relaxation of clays, IUTAM Rheology and Soil Mechanics Symposium, Grenoble, 1964, pp.99-129.

11. Naghdi, P. M., Stress-strain relations in plasticity and thermoplasticity, Proc. 2nd Symposium on Naval Structural Mechanics, Pergamon Press, 1960, pp.121-169.

12. Nishi, K. and Esashi, Y., Stress-strain relationships of sand based on elasto-plasticity theory, Proc. JSCE, No.280, 1978, pp.111-122.

13. Pender, M. J., A model for the behavior of overconsolidated clay, Geotechnique, Vol.28, No.1, 1978, pp.1-25.

14. Perzyna, P., The constitutive equations for work-hardening and rate sensitive plastic materials, Proc. Vibrational Problems, Warsaw, Vol.4, No.3, 1963, pp.281-290.

15. Roscoe, K. H. and Burland, J. B., On the generalized stress-strain behavior of Wet clay, Engineering Plasticity, Cambridge Univ. Press, 1968, pp.535-609.

16. Sekiguchi, H. and Ohta, H., Induced anisotropic and time dependency in clays, Proc. Specialty Session 9, 9th ICSMFE, Tokyo, 1977, pp.229-238.

APPENDIX II - DETERMINATION OF MATERIAL PARAMETERS

The assigned problems were to predict the mechanical behaviors of two natural clays (Clay-X and Clay-Y), a laboratory prepared Kaolinite clay and Ottawa sand. In this section, we will discuss the constitutive equations used to predict and the determination of the material parameters. Some comparison of previously given test results with their calculated stress-strain relations will be given as examples.

PROBLEM A - *Natural Clay* -

In the test reported and in those to be predicted the stress ratio m is introduced and defined as

$$m = \frac{\sigma_2 - \sigma_3}{\sigma_1 - \sigma_3} \tag{A-1}$$

Clay-X

The preconsolidation pressure σ'_{me} is found to be 23.6psi from the given test results. Since the drained shear tests were carried out under constant mean stress of σ_c = 10, 20, 30 and 40 psi, the behaviors of the clay should be considered in the over-consolidated state as well as normally consolidated state. Thus, the constitutive equations 13 and 60 with taking into account M^*-value given by Eq.42 were applied to the predictions. The required material parameters are λ, κ, G, e, M^*, m' and C for normally consolidated state, and λ, κ, G, e, G', M^*_m and M^*_f for overconsolidated state. Since λ, κ, G and e are common in both cases, total 9 parameters should be given. The material parameters to Clay-X were determined for each confining pressure, i.e., σ_c = 10, 20, 30 and 40 psi, and are listed in Table A-1.

Fig. A-1 shows the comparison of the calculated stress-strain curves with the given test results for the case of m = 0 and 1 under σ_c = 30 psi. In the figure it is found that the stress-strain curve for m = 1 and the volumetric strains are underestimated.

Table A-1 Material parameters for Clay-X

σ_c (psi)	λ	κ	G (psi)	e_o	m'	C	M^*_m	M^*_f	G'
10	0.64	0.01	333	1.76	8	8×10^{-9}	1.23	1.30	1400
20	0.64	0.01	333	1.75	8	8×10^{-9}	0.65	0.73	1400
30	0.64	0.01	333	1.55	8	8×10^{-9}	0.89	1.14	1400
40	0.64	0.01	333	1.30	8	8×10^{-9}	0.89	1.14	1400

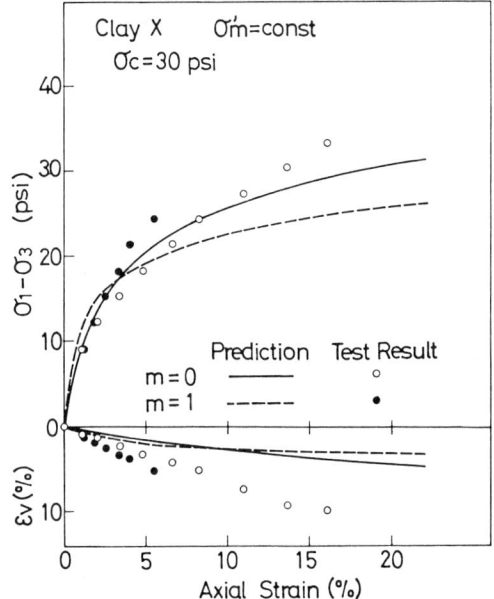

Fig. A-1. Stress-strain relationships of prediction and test results for Clay-X (σ_c = 30 psi, m = 0 and 1)

Clay-Y

Definitely, Clay-Y was in overconsolidated state at the starting state of shear tests. However, we should take into account that it may get into normally consolidated state during the shear process. Thus, total 9 parameters were also determined for every confining pressure, i.e., σ_c = 2.5, 5.0 and 10.0, as same as Clay-X, and are listed in Table A-2.

Fig. A-2 shows the comparison of the calculated stress-strain curves with given test results for the case of m = 0 and 1 under σ_c = 10 psi. The predicted

Table A-2 Material parameters for Clay-X

σ_c (psi)	λ	κ	G (psi)	e_o	m'	C	M_m^*	M_f^*	G'
2.5	0.38	0.03	10^4	1.70	8	8×10^{-8}	1.55	1.63	400
5.0	0.38	0.03	10^4	1.68	8	8×10^{-8}	1.22	1.22	400
10.0	0.38	0.03	10^4	1.66	8	8×10^{-8}	0.82	0.98	400

stress-strain curve and volumetric strain of m = 1 are underestimated, but good agreement is seen in the case of m = 0.

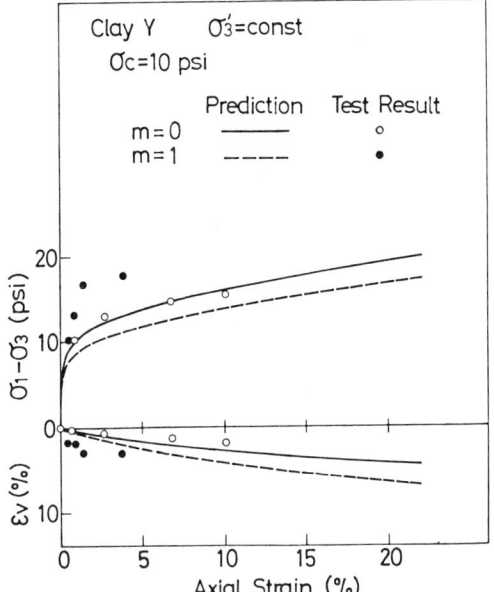

Fig. A-2. Stress-strain relationships of predictions and test results for Clay-Y (σ_c = 10 psi, m = 0 and 1)

PROBLEM B - *Laboratory Prepared Kaolinite Clay* -

This is the problem for anisotropic consolidated clay. In rigorously speaking, our constitutive models can not handle this problem. However, we tried to get the solutions by applying the anisotropic consolidation effect of Eq.38 and the general loading condition of Eq.42 to the constitutive equations 13 and 60. First of all, we considered that the tests from No.1 to No.6 are in so-called compression side, but those from No.7 to No.13 are in extension side. In addition, we assumed that the effect of the rotation of major principal stress could be expressed by the loading parameter m.

Since these tests were conducted under undrained condition, for instance, the test results of test No.1 (triaxial compression, CTC) should be the same of the test No.4 (σ_m- constant compression). According to the above considerations and assumptions, we divided the tests into two groups, i.e., compression side or extension side, as shown in Table A-3. In the table, it is also shown how the effect of the rotation of major principal stress is related to the loading parameter m. From given data, we determined the material parameters and summarized them in Table A-4 for each group.

Table A-3. Assumptions and separation into groups for predictions of Kaolinite Clay

Groups	Angle of principal stress to vertical		Test No.	Given or Predict
Compression side	$0°$	$m=0$	No.1, No.4	given
	$15°$	$m=0.067$	No.2, No.5	predict
	$37.5°$	$m=0.278$	No.3, No.6	predict
Extension side	$45°$	$m=0.5$	No.7	predict
	$58.25°$	$m=0.276$	No.8, No.11	predict
	$75°$	$m=0.067$	No.9, No.12	predict
	$90°$	$m=0$	No.10, No.13	given

Table A-4. Material parameters for Kaolinite Clay

Groups Test No.	λ	κ	G (psi)	e_o	m'	C	M_m^*	M_f^*	G'
Compression No.1-No.6	0.13	0.035	10^4	1.01	18	1.5×10^{-11}	0.84	1.04	400
Extension No.7-No.13	0.13	0.035	10^4	1.01	18	1.5×10^{-11}	0.63	0.73	300

Fig. A-3 shows the comparison of the calculated stress-strain curves with the given test results for the case of the tests No.1, No.4, No.10 and No.13. Both predicted and test results of the effective stress paths for the cases are given in Fig. A-4. It is seen in the figures that the predictions for the tests in extension side, i.e., No.10 and No.13 are underestimated. However, those for the tests in compression side, i.e., No.1 and No.4, are reasonably well.

Fig. A-5 illustrates the stress paths of the tests No.1, No.4, No.10 and No. 13. Clay was anisotropically consolidated at the stress state A and then unloaded up to the isotropic stress state B. After that, clay was sheared in compression or extension side. In the figure, the yield and potential surfaces are also given. It is not in detail, but one may get certain idea about what our model is.

CLAYS AND SANDS

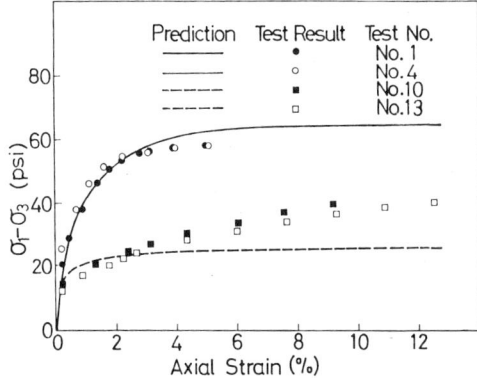

Fig. A-3. Stress-strain relationships of predictions and test results for Kaolinite Clay

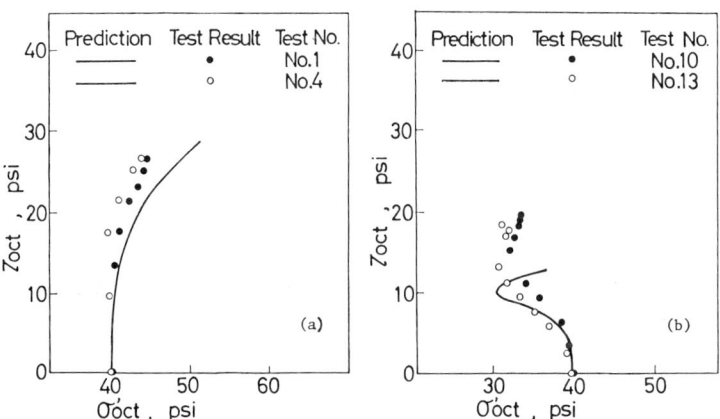

Fig. A-4. Effective stress paths of predictions and test results for Kaolinite Clay, (a) Tests No.1 and No.4, (b) Tests No.10 and No.13.

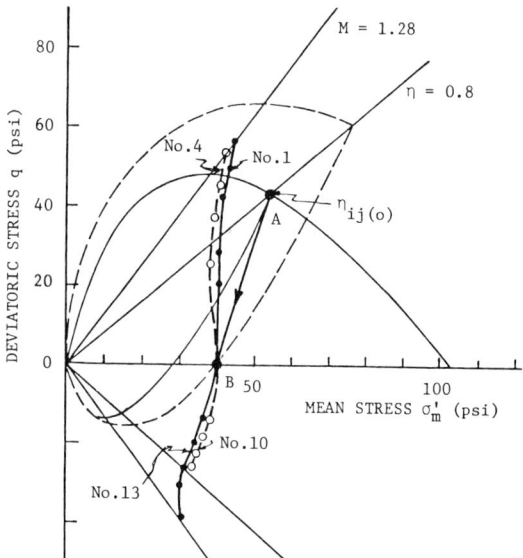

Fig. A-5. Testing conditions, effective stress paths, yield surface and plastic potential surface for Tests No.1, No.4, No.10 and N0.13.

PROBLEM C - *Ottawa Sand* -

This is the problem for dry Ottawa sand. In order to predict the behaviors of this sand, we used the constitutive equations 60 by applying the general loading parameter M^* given by Eq.42. Therefore, the parameters G, λ, κ, e_o, G', M_m^* and M_f^* were required to be determined. In Table A-5, the obtained parameters are summarized for each given test results. Just for convienience, we assigned the test number to each given test as G-1, G-2,······ and also gave the test number to each prediction problem, i.e., P-1, P-2, ······, which are listed in Table A-6 with those testing conditions.

The test number listed in the last column of Table A-5 shows the test whose prediction had been made by using the parameters given on the same line.

Figs. A-6 and A-7 show the comparison of the calculated stress-strain curves with those given by the test results for the test No. G-1 and No.3, respectively. The constitutive model can describe the loading and unloading behaviors as seen in the figures.

Table A-5. Material parameters for Ottawa sand

Test No.	G (psi)	$G'_{(1)}$	$G'_{(2)}$	$M_m (\sqrt{3/2}M_m^*)$	$M_f (\sqrt{3/2}M_f^*)$	Application in Prediction
G-1	8000	944	1600	1.17	1.75	
G-2	7000	960	2740	1.27	1.98	
G-3	7000	330	1330	0.91	1.17	
G-4	5440	1000	2850	1.44	2.29	p-2
G-5	10900	450	900	1.19	2.10	p-1, p-4, p-7, p-8, p-9
G-6	16600	550	1160	1.04	1.87	p-3, p-5
G-7	4710	650	1200	0.88	1.35	
G-8	35400	800	1690	0.91	1.35	
G-9	14100	400	800	0.66	1.24	p-6

as common parameters: $\lambda = 1.15 \times 10^{-3}$, $\kappa = 1.43 \times 10^{-4}$, $e_o = 0.54$

Table A-6. Test number and testing conditions for Ottawa sand

Test No.	Testing Conditions			Test No.	Testing Conditions		
G-1	CTC	σ_3 = 10psi	m = 0	P-1	SS	σ_m = 10psi	m = 0.2
G-2	CTC	σ_3 = 5psi	m = 0	P-2	SS	σ_m = 5psi	m = 0.5
G-3	CTE	σ_3 = 10psi	m = 1	P-3	SS	σ_m = 20psi	m = 0.5
G-4	TC	σ_m = 5psi	m = 0	P-4	SS	σ_m = 10psi	m = 0.8
G-5	TC	σ_m = 10psi	m = 0	P-5	RTC	σ_x = 20psi	m = 0
G-6	TC	σ_m = 20psi	m = 0	P-6	RTE	$\sigma_x = \sigma_y$ = 20psi	m = 1
G-7	TE	σ_m = 5psi	m = 1	P-7	PL1		m = 1
G-8	TE	σ_m = 10psi	m = 1	P-8	PL2		m = 0
G-9	TE	σ_m = 20psi	m = 1	P-9	CSP		

838 STRESS STRAIN FOR SOILS

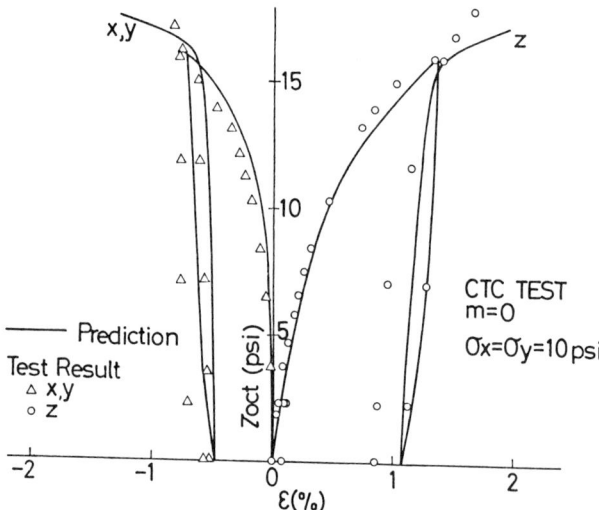

Fig. A-6. Stress-strain relationships of prediction and test results for Ottawa sand (Test G-1)

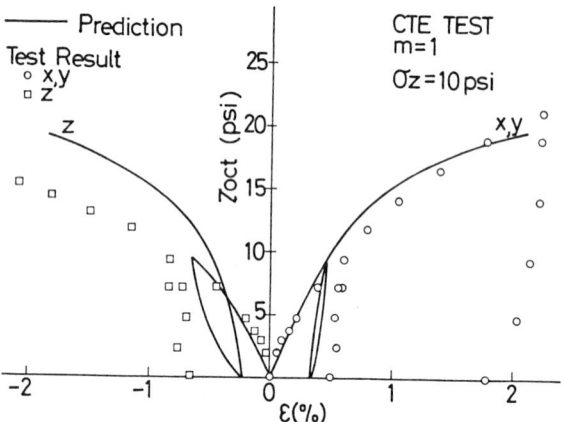

Fig. A-7. Stress-strain relationships of prediction and test results for Ottawa sand (Test G-3)

A CONSTITUTIVE LAW FOR SANDS AND CLAYS -
POSITION, PREDICTION, AND EVALUATION

D. Kolymbas[1] and G. Gudehus[2]

Summary

A theory of soil material behaviour is outlined, critically discussed and evaluated in this paper. This can essentially be done along the philosophical concept of K.R. Popper outlined in the introductory Section 1. The constitutive law is based upon a set of hypotheses; their range of validity is discussed in Section 2.1. Representation theorems and certain approximating functions are proposed in Section 2.2 such that a certain soil is specified by eight dimensionless parameters. The data given to the Predictors are evaluated in Section 3. Determining the material parameters requires a critical discussion of the data set. Contrary to the given format, the predictions are plotted in invariant form. Some concluding remarks in Section 4 refer to the use of the given data set for validation and to recommended further check of the theory.

1. Introduction

A constitutive law is a theoretical model of soil behaviour. The authors have found it useful to follow K.R. Popper's 'Logic of Scientific Discovery' (1977) for judging such models. Any theory serves, according to K.R. Popper, for making _predictions_ of real events. The theory has to be checked in four steps, viz. by
  1) comparing the hypotheses with those of accepted theories, i.e. search for external consistency;
  2) eliminating tautologies which cannot produce any prediction (such as a mere fit);
  3) looking for possible contradictions among the set of hypotheses, i.e. search for internal consistency;
  4) comparing predictions of singular events with observations.

The aim of this process is _falsification_ (in the sense of refutation) of the theory. As one singular observation suffices to

---

[1] Research Associate, Institute of Soil and Rock Mechanics, University of Karlsruhe, W-Germany

[2] Professor of Civil Engineering, same address

falsify a theory the corresponding experiment should be beyond criticism. If a theory stands the critical checking process it is <u>corroborated</u> (or validàted) so far. For reasons of logic a theory can never be verified.

We will follow the concept of Popper in this paper as far as possible. A set of hypotheses is outlined in Section 2.1. This set is checked against established principles of continuum mechanics and soil mechanics. The former imply unit invariance and frame indifference which can be summarized as objectivity. The latter are less well settled and therefore deserve some discussion.

Special representations compatible with this set of hypotheses are outlined in Section 2.2. This specialization implies further hypotheses which cannot, however, as easily be falsified or corroborated as those of Section 2.1. It is also explained how the material parameters of a single material can be determined from a set of test data, and which kind of tests is required. The hypotheses of Sections 2.1 and 2.2 do not yet form a minimal set like a system of axioms; it can only be stated that contradictions have not yet been found.

The submitted test data are evaluated in Sections 3.1 for sand and 3.2 for Kaolin clay. The determination of the material parameters is somewhat arbitrary as the tests do not fully meet the requirements of the theory. The method of prediction for the given stress paths is briefly explained in order to be tractable. An invariant graphical representation different from the given format was chosen.

In Section 4 it will be outlined how far the tests carried out for the Workshop can be used for checking our theory, and what could further be done in that respect. The former discussion is rather fragmentary as no detailed information on testing apparatus and procedure was available, and the latter one is of course only a collection of a few hints.

2. Short representation of the theory

2.1 Defining hypotheses

Any ideal material is defined by a set of assumptions. We briefly outline and discuss as far as possible, in the sense of K.R. Popper, some hypotheses H1, H2 etc. underlying our constitutive law. Each of them has been corroborated within a certain range of validity.

H1 Soils are simple materials of the rate type

$$\overset{\circ}{\underset{\sim}{T}} = \underset{\sim}{f}\,(\underset{\sim}{T},\ \underset{\sim}{T}_e,\ e,\ \underset{\sim}{D},\ \underset{\sim}{D}_2)\ . \tag{2.1}$$

(In this section, we use symbolic tensor notation in order to secure co-ordinate invariance; for the reader's convenience components in the notation of soil mechanics for the case of cuboidal deformations are added.) The symbols in Equation 2.1 have the following meaning:

$\underset{\sim}{T}$, effective Cauchy stress tensor (principal components for cuboidal deformation in soil mechanics notation: $-\sigma_1'$, $-\sigma_2'$, $-\sigma_3'$);

$\overset{\circ}{\underset{\sim}{T}}$, co-rotational effective stress rate (principal cuboidal components $-\dot{\sigma}_1'$, $-\dot{\sigma}_2'$, $-\dot{\sigma}_3'$);

$\underset{\sim}{T}_e$, equivalent stress tensor, to be explained below;

e, void ratio;

$\underset{\sim}{D}$, (first) stretching tensor (principal cuboidal components $-\dot{\varepsilon}_1$, $-\dot{\varepsilon}_2$, $-\dot{\varepsilon}_3$; natural strain rates);

$\underset{\sim}{D}_2$, second stretching tensor (principal cuboidal components $-\ddot{\varepsilon}_1$, $-\ddot{\varepsilon}_2$, $-\ddot{\varepsilon}_3$);

$\underset{\sim}{f}$, tensor-valued response function, to be specified in the following.

Notations and definitions of $\overset{\circ}{\underset{\sim}{T}}$, $\underset{\sim}{T}$, $\underset{\sim}{D}$, $\underset{\sim}{D}_2$ follow modern continuum mechanics. Equation 2.1 deviates from the defining equation of a simple material of the rate type by Truesdell and Noll (1965) in two respects. The right stretch tensor $\underset{\sim}{U}$ as well as the stretching tensors of third and higher orders and stress rate tensors of second and higher orders are omitted. Thus especially relaxation is not covered. On the other hand, $\underset{\sim}{T}_e$ enters as a further variable.

One implication of H1 is that the response function $\underset{\sim}{f}$ can be fully detected in tests with homogeneous deformations. If non-homogeneous deformation tests are used, on the other hand, some information about the material must be given in advance. Strain-controlled cuboidal tests (Goldscheider, 1972) have shown no momentum stresses which is supporting H1. Note that stress-controlled tests do not generally secure homogeneous deformations as various modes of bifurcation are possible (Vardoulakis et al. 1978, Vardoulakis 1979, 1980, Kolymbas 1980).

H2 The response function $\underset{\sim}{f}$ can be decomposed into a rate-independent term not depending on $\underset{\sim}{D}_2$, and a rate-dependent term proportional to $\underset{\sim}{D}_2$, i.e.

$$\underset{\sim}{f} = \underset{\sim}{h}(\underset{\sim}{T}, \underset{\sim}{T}_e, \underset{\sim}{D}, e) + \underset{\sim}{g}(\underset{\sim}{T}, \underset{\sim}{T}_e, \underset{\sim}{D}, e) \underset{\sim}{D}_2. \qquad (2.2)$$

Rate-independence of $\underset{\sim}{h}$ means

$$\underset{\sim}{h}(\lambda \underset{\sim}{D}, \ldots) = \lambda \underset{\sim}{h}(\underset{\sim}{D}, \ldots) \qquad (\lambda > 0), \qquad (2.3)$$

i.e. **positive** homogeneity of first order with respect to $\underset{\sim}{D}$. We only consider representations of $\underset{\sim}{h}$ in the following as no information about rate-effects in the tested materials was conveyed to the Predictors.

We now specialize to cases with $\underset{\sim}{T} = \underset{\sim}{T}_e$. States which are fully determined by stress and void ratio have been introduced as SOM (<u>s</u>wept <u>o</u>ut <u>m</u>emory) states (Gudehus et al. 1977). Deviations from SOM-states (i.e. $\underset{\sim}{T} \neq \underset{\sim}{T}_e$) are possible; they can occur, e.g., during shakedown of sand and with overconsolidated clays. SOM-states are exactly produced by proportional deformation from $\underset{\sim}{T} = \underset{\sim}{0}$ and are always reached asymptotically by proportional deformation. The amount of deformation required to produce SOM states from arbitrary initial states is rather small (well below 1%) for sands; thus for any monotonous strain history and for incremental collapse with sands it suffices to assume SOM-states only.

The amount of deformation to bring overconsolidated clays to SOM-states (i.e. to make them normally consolidated again) is not yet known. We will introduce a simple approach for $\underset{\sim}{T}_e \neq \underset{\sim}{T}$ in order to cover the given Kaolin clay data at Section 3.2.

For $\underset{\sim}{T} = \underset{\sim}{T}_e$, we assume

H3 The rate-independent part $\underset{\sim}{h}$ of the response function is positively homogeneous of first order with respect to stress, i.e.

$$\underset{\sim}{h}\,(\lambda \underset{\sim}{T},\, \underset{\sim}{D},\, e) = \lambda \underset{\sim}{h}\,(\underset{\sim}{T},\, \underset{\sim}{D},\, e) \qquad (\lambda > 0). \qquad (2.4)$$

Several implications of Equation 2.4 help to establish the range of applicability of H3, viz.

a) Material parameters with the unit of stress do not appear. Thus elasticity and strength of the particles cannot be covered, which is only allowable as long as the stress intensity is relatively low. This idealization is widely accepted for clays, but not for sands.

b) For proportional deformations (constant $\underset{\sim}{D}$) starting from $\underset{\sim}{T} = \underset{\sim}{0}$, the stress paths should be proportional (constant $\overset{\circ}{\underset{\sim}{T}}$) too. This has been corroborated repeatedly for sands and unpreloaded clays.

c) For proportional deformations starting from $\underset{\sim}{T} \neq \underset{\sim}{0}$, the stress rates $\overset{\circ}{\underset{\sim}{T}}$ should asymptotically reach the same values as in case (b).

d) For a given stress ratio tensor, defined by $\underset{\sim}{T}/\mathrm{tr}\,\underset{\sim}{T}$, the inviscid part $\underset{\sim}{h}$ of the stress rate is proportional to $\mathrm{tr}\,\underset{\sim}{T}$. This similarity law is widely accepted for unpreloaded clays; it is implied by the Cam clay theory, e.g. It can also be used for sands and is the basis of model laws (Hettler and Gudehus, 1980). Its validity is veiled by other scale effects in model tests and by bifurcation effects in standard cell tests.

Note that implications (b) and (c) hold if $\underset{\sim}{h}$ is positively homogeneous of any order with respect to $\underset{\sim}{T}$.

The next hypothesis refers to dimensionless deviatoric invariants, defined by

$$\cos 3\alpha_h := \sqrt{6}\,\mathrm{tr}(\underset{\sim}{h}^*)^3 / (\mathrm{tr}(\underset{\sim}{h}^{*2}))^{3/2}$$

and

$$\cos 3\alpha_D := \sqrt{6}\,\mathrm{tr}(\underset{\sim}{D}^*)^3 / (\mathrm{tr}(\underset{\sim}{D}^{*2}))^{3/2},$$

wherein * denotes the deviator, as usual defined by

$$\underset{\sim}{h}^* := \underset{\sim}{h} - \frac{1}{3}\underset{\sim}{1}\,\mathrm{tr}\,\underset{\sim}{h} \quad \text{and} \quad \underset{\sim}{D}^* := \underset{\sim}{D} - \frac{1}{3}\underset{\sim}{1}\,\mathrm{tr}\,\underset{\sim}{D}\ .$$

We now can formulate

H4 For a process with $\overset{\circ}{\underset{\sim}{T}}$ = const starting from $\underset{\sim}{T} = \underset{\sim}{0}$, i.e. for proportional stress paths, the quantities $\cos 3\alpha_h$ and $\cos 3\alpha_D$ are uniquely related, i.e.

$$\cos 3\alpha_h = f_\alpha (\cos 3\alpha_D,\ e)\ , \tag{2.5}$$

with a function f_α possibly depending on void ratio e.

For cuboidal deformations the implication of H4 can be represented graphically (Fig. 1). For proportional stress paths one has $\cos 3\alpha_h = \cos 3\alpha_T$. α_T and α_D are angles in the deviator planes $\mathrm{tr}\,\underset{\sim}{T} =$ const and $\mathrm{tr}\,\underset{\sim}{D} =$ const respectively (Fig. 1a). $\alpha = 0$ means cylindrical extension, and $\alpha = 60°$ cylindrical compression. Equation 2.5 is represented in Fig. 1b; in case of $\underset{\sim}{T} = \underset{\sim}{T}_e$ one has $\alpha_T = \alpha_D$ only for $\alpha = 0$ and $\alpha = 60°$.

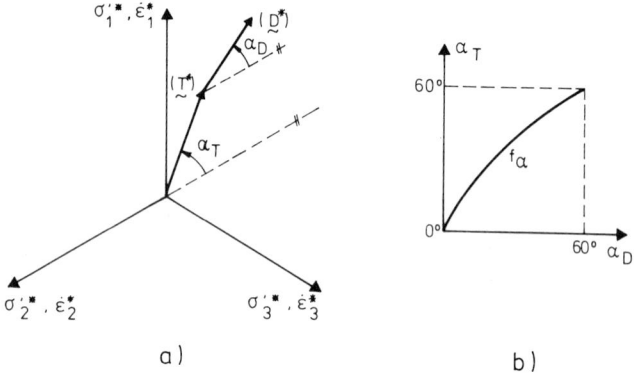

Fig. 1 Geometrical meaning of deviatoric invariants (a) and their mutual dependance (b)

For simple or torsional shear with $\underset{\sim}{T} = \underset{\sim}{T}_e$ and proportional stress paths H4 implies coaxiality of stress rate and strain rate tensors, which is at variance with the hypothesis of de St.Vénant. Simple and torsional shear tests cannot be made with homogeneous deformations and therefore cannot help to clarify this problem. Analyses of bifurcations have disproved the hypothesis of de St.Vénant (Vardoulakis et al., 1978), whereas H4 (together with H3) has been supported for sand by Goldscheider (1976).

We now consider limiting states, which are generally characterized by $\underset{\sim}{h} = \underset{\sim}{0}$, i.e. zero inviscid stress rate response, for certain strain rate directions, i.e. certain values of $\underset{\sim}{D}/\sqrt{\text{tr}\,\underset{\sim}{D}^2}$. For SOM states (i.e. $\underset{\sim}{T} = \underset{\sim}{T}_e$) one can formulate

H5 If $\overset{\circ}{\underset{\sim}{T}} = \underset{\sim}{0}$ holds for a constant stretching tensor, a statical limiting condition

$$f_T(\underset{\sim}{T},\ e) = 0 \qquad (2.6)$$

and a kinematical limiting condition

$$\tan\beta_D := \text{tr}\,\underset{\sim}{D}/\sqrt{3\text{tr}(\underset{\sim}{D}^{*2})} = f_\beta(\cos 3\alpha_T,\ e) \qquad (2.7)$$

are satisfied.

Implications of H5, together with H3, can be represented graphically in case of cuboidal, especially cylindrical deformations. Equation 2.6 is represented by a case in principal stress component space, and by two rays in a Henkel plane (Fig. 2).

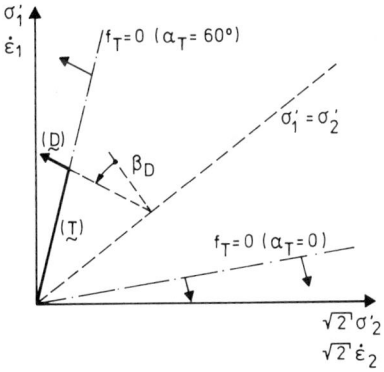

Fig. 2 Representation of statical and kinematical limiting conditions in a Henkel plane

Equation 2.7 can also be represented in a Henkel plane. β_D is a measure of dilatancy in the limiting state which is, due to H3, independent of the first stress invariant $\mathrm{tr}\,\underset{\sim}{T}$. Both Equations 2.6 and 2.7 have been corroborated by cuboidal and cylindrical tests with strictly homogeneous deformation on sands (Goldscheider, 1976) and unpreloaded clays. They are seemingly falsified by tests with nonhomogeneities due to bifurcation (Vardoulakis, 1979) or non-uniform constraints; we will come back to this point in Section 4.

Deviations from SOM states are only considered here for a rather idealized preloaded clay. It is assumed that a certain value $\underset{\sim}{T}_e = \underset{\sim}{T}$ has been produced by preloading and is not diminished afterwards. This assumption implies the neglect of swelling. Furthermore, we assume that $\underset{\sim}{T}_e$ takes the place of $\underset{\sim}{T}$ in the list of arguments of $\underset{\sim}{h}$, and that $\underset{\sim}{T}_e = \underset{\sim}{h}$ holds with the same representation for $\underset{\sim}{h}$ as for the unpreloaded clay. This is a generalization of Hvorslev's equivalent stress concept by taking a tensor $\underset{\sim}{T}_e$ instead of a scalar σ_e.

2.2 Representations

We now consider representations of the tensor valued function $\underset{\sim}{h}$ of the tensor arguments $\underset{\sim}{T}$ (or $\underset{\sim}{T}_e$ in case of preloaded clay as introduced above), $\underset{\sim}{D}$, and e. For reasons of material frame-indifference (Truesdell and Noll, 1965) $\underset{\sim}{h}$ must satisfy the condition

$$\underset{\sim}{h}(\underset{\sim}{Q}\,\underset{\sim}{T}\,\underset{\sim}{Q}^T,\ \underset{\sim}{Q}\,\underset{\sim}{D}\,\underset{\sim}{Q}^T,\ e) = \underset{\sim}{Q}\,\underset{\sim}{h}(\underset{\sim}{T},\ \underset{\sim}{D},\ e)\,\underset{\sim}{Q}^T \qquad (2.8)$$

for any orthogonal tensor $\underset{\sim}{Q}$ (i.e. $\underset{\sim}{Q}\,\underset{\sim}{Q}^T = 1$). Thus $\underset{\sim}{h}$ is a tensor-valued tensor function of two tensor variables; it can therefore always be represented by the expansion (Wang, 1970)

$$\begin{aligned}\underset{\sim}{h} = {} & a_1\,\underset{\sim}{1} + a_2\,\underset{\sim}{T} + a_3\,\underset{\sim}{T}^2 + a_4\,\underset{\sim}{D} + a_5\,\underset{\sim}{D}^2 \\ & + a_6\,(\underset{\sim}{D}\,\underset{\sim}{T} + \underset{\sim}{T}\,\underset{\sim}{D}) + a_7\,(\underset{\sim}{D}\,\underset{\sim}{T}^2 + \underset{\sim}{T}^2\,\underset{\sim}{D}) \\ & + a_8\,(\underset{\sim}{D}^2\,\underset{\sim}{T} + \underset{\sim}{T}\,\underset{\sim}{D}^2) + a_9\,(\underset{\sim}{T}^2\,\underset{\sim}{D}^2 + \underset{\sim}{D}^2\,\underset{\sim}{T}^2),\end{aligned} \qquad (2.9)$$

wherein the scalar coefficients a_1 to a_9 are functions of the nine invariants

$$\text{tr}\,\underset{\sim}{T},\ \text{tr}\,\underset{\sim}{T}^2,\ \text{tr}\,\underset{\sim}{T}^3,\ \text{tr}\,\underset{\sim}{D},\ \text{tr}\,\underset{\sim}{D}^2,\ \text{tr}\,\underset{\sim}{D}^3,$$
$$\text{tr}(\underset{\sim}{T}\,\underset{\sim}{D}),\ \text{tr}(\underset{\sim}{T}^2\,\underset{\sim}{D}),\ \text{tr}(\underset{\sim}{T}\,\underset{\sim}{D}^2),\ \text{tr}(\underset{\sim}{T}^2\,\underset{\sim}{D}^2).$$

We will now reduce this expansion as nine functions of ten scalar variables are scarcely manageable. One of the ten invariants is never independent and can thus be dropped (Rivlin and Ericksen, 1960). For exploiting H4 one can also replace, without loss of generality, $\underset{\sim}{T}$ and $\underset{\sim}{D}$ by the deviators $\underset{\sim}{T}^*$ and $\underset{\sim}{D}^*$ in Equation 2.9. H4 is only satisfied with $a_3 = a_6 = a_7 = a_8 = a_9 = 0$ then, as can be demonstrated by contradiction. H2 and H3 reduce the number of variables by two; the same statement is obtained by means of dimensional analysis from the assumption that no material parameters with the units of stress and time appear. For further reduction we assume

H6 The response function is fully revealed by cuboidal deformations.

This hypothesis is of a pragmatic type: only cuboidal deformations can as yet be realized without non-homogeneities of stress and strain. As $\underset{\sim}{T}$ and $\underset{\sim}{D}$ are co-axial in this case, only six of the ten joint invariants of $\underset{\sim}{T}$ and $\underset{\sim}{D}$ are independent. Without having established an analytical proof yet, we have found that at least two of the mixed invariants, e.g. $\text{tr}(\underset{\sim}{T}\,\underset{\sim}{D})$ and $\text{tr}(\underset{\sim}{T}\,\underset{\sim}{D}^2)$, cannot be dropped.

The following representation was chosen within the frame of the above restrictions:

$$\left.\begin{aligned}
&\underset{\sim}{h} = h\,\underset{\sim}{\hat{h}},\ \text{with} \\
&\underset{\sim}{\hat{h}}^* = \left(C_1\,\tfrac{1}{3}\sqrt{\text{tr}(\underset{\sim}{D}^{*2})}\ +\ \underset{\sim}{D}^*\ -\ C_1\frac{\underset{\sim}{D}^{*2}}{\sqrt{\text{tr}(\underset{\sim}{D}^{*2})}}\right)f_T\ +\ \frac{\underset{\sim}{T}^*}{\text{tr}\,\underset{\sim}{T}}\sqrt{(\text{tr}\,\underset{\sim}{\hat{h}})^2}\ , \\
&\text{tr}\,\underset{\sim}{\hat{h}} = C_3\,\text{tr}\,\underset{\sim}{D}\ +\ C_4\,\frac{\text{tr}(\underset{\sim}{T}^*\underset{\sim}{D}^*)}{\text{tr}\,\underset{\sim}{T}}\ +\ C_5\,\frac{\text{tr}(\underset{\sim}{T}^*\,\underset{\sim}{D}^{*2})}{\text{tr}\,\underset{\sim}{T}\sqrt{\text{tr}(\underset{\sim}{D}^{*2})}}\ , \\
&h = -C_6\,\text{tr}\,\underset{\sim}{T}\left(1\ +\ C_7\,\frac{\text{tr}(\underset{\sim}{T}^*\underset{\sim}{D}^*)}{\text{tr}\,\underset{\sim}{T}\sqrt{\text{tr}(\underset{\sim}{D}^2)}}\ +\ C_8\,\frac{\text{tr}\,\underset{\sim}{D}}{\text{tr}(\underset{\sim}{D}^2)}\right),\quad\text{and} \\
&f_T = C_2\ -\ (\text{tr}\,\underset{\sim}{T})^3\,/\,\det\underset{\sim}{T}\ .
\end{aligned}\right\}\quad(2.10)$$

The eight material constants C_1 to C_8 are dimensionless and depend on void ratio e for sand, but not for unpreloaded clay. By checking this representation one finds that H2, H3 and H4 are satisfied. By some calculation one can also verify that H5 is satisfied and that stress states with $f_T < 0$ cannot be reached by any deformation history. This representation is somewhat different from an earlier version (Gudehus and Kolymbas, 1979).

The following procedure is proposed to determine the material constants. C_1 appears only in the relationship among $\cos 3\alpha_h$ and $\cos 3\alpha_D$. Thus C_1 can be precisely determined in a cuboidal test or in a biaxial test with a device for measuring the intermediate stress σ_2'. However, C_1 seems to be rather invariable for different soils: It may suffice to take $C_1 = 0.6$ for all sands, e.g., which is covering the results of Goldscheider (1976).

The remaining material constants C_2 to C_8 can be determined from cylindrical triaxial test results. To explain this procedure we will make use of the principal components of the stress and stretching tensors. The ratio σ_2'/σ_1' is denoted by K, such that f_T from Equation 2.10e can be expressed as follows:

$$f_T = C_2 - \frac{(1+2K)^3}{K^2} \qquad (2.11)$$

C_2 can now be directly determined from the ϕ'- value for cylindrical compression. For a stress state fulfilling the limiting condition we have:

$$K = K_a := (1-\sin \phi')/(1+\sin \phi') . \qquad (2.12)$$

Since f_T is to vanish for limiting stress states, we finally obtain

$$C_2 = (1+2K_a)^3 / K_a^2 \qquad (2.13)$$

It should be stressed out that only tests with good end plate lubrication and sufficiently low height/diameter ratio to avoid bifurcation should be chosen (Vardoulakis 1979). Otherwise, ϕ' can be grossly underestimated.

In the following the cylindrical compression test is further considered. When a limiting stress state is reached, the kinematical limiting condition is fulfilled, i.e. the angle β_D, defined by Equation 2.7, equals the value β_{Dc} (c for triaxial compression) which can be obtained from the experimental results if the ratio of the radial to the axial strain increments, $\dot{\varepsilon}_2/\dot{\varepsilon}_1$, is registrated (by means of volume change measurements or bandages, e.g.). It can be easily shown that

$$\beta_{De} = \tan^{-1}\left(-\frac{1 + 2\dot{\varepsilon}_2/\dot{\varepsilon}_1}{\sqrt{2} \cdot |1 - \dot{\varepsilon}_2/\dot{\varepsilon}_1|}\right). \qquad (2.14)$$

Now the cylindrical test should be continued in the following way: The sense of deformation is reversed (i.e. an extension test is carried out) with the mean pressure $\sigma'_1 + 2\sigma'_2$ kept constant. This can be achieved by an appropriate control of the cell pressure. Let β_{De} be the angle β_D measured at the beginning (i.e. when $K = K_a$ holds) of this extension test. The knowledge of the two angles β_{Dc} and β_{De} provides the ratios C_4/C_3 and C_5/C_3 via the following procedure: During the compression test the mean stress $\sigma'_1 + 2\sigma'_2$, remains unchanged as soon as a limiting stress state is reached, provided no bifurcation has occured. During the following extension as described above, the mean stress has to be kept constant. Therefore, we require for both compression and extension tr$\underset{\sim}{h}$ = 0. It then follows from Equation 2.10c:

$$\tan\beta_D + \frac{C_4}{C_3} \cdot \frac{\sqrt{2}}{3} \frac{1-K}{1+2K} \cdot \text{sign}(\dot{\varepsilon}_2 - \dot{\varepsilon}_1) + \frac{C_5}{C_3} \cdot \frac{1}{3}\sqrt{\frac{2}{3}} \cdot \frac{1-K}{1+2K} = 0. \qquad (2.15)$$

Since this equation is valid for the compression case at $K = K_a$ and throughout the extension test, it takes the forms

$$\tan\beta_c - \frac{C_4}{C_3} \cdot \frac{\sqrt{2}}{3} \frac{1-K_a}{1+2K_a} + \frac{C_5}{C_3} \cdot \frac{1}{3}\sqrt{\frac{2}{3}} \frac{1-K_a}{1+2K_a} = 0$$

$$\tan\beta_e = \frac{C_4}{C_3} \cdot \frac{\sqrt{2}}{3} \frac{1-K_a}{1+2K_a} + \frac{C_5}{C_3} \cdot \frac{1}{3}\sqrt{\frac{2}{3}} \frac{1-K_a}{1+2K_a} = 0 \qquad (2.16)$$

This is a system of two linear equations with the unknowns C_4/C_3 and C_5/C_3 which now can be determined by solving this system. For the following determination of the material constant C_3 the value of K_o is needed. K_o can be mathematically expressed through the equations: If $K = K_o$ and $\dot{\varepsilon}_2 = 0$, $\dot{\varepsilon}_1 > 0$ then $\dot{\sigma}_2'/\dot{\sigma}_1' = K_o$ must hold. The latter equation $\dot{\sigma}_2' = K_o \cdot \dot{\sigma}_1'$ expressed by means of the constitutive equation for $K = K_o$, $\dot{\varepsilon}_2 = 0$, $\dot{\varepsilon}_1 = 1$ yields an equation for C_3: From $\dot{\sigma}_2' = K_o \cdot \dot{\sigma}_1'$ the equation

$$(K_o + 0,5) \cdot \dot{\sigma}_1'^*/h - \frac{1}{3} \cdot (1-K_o) \cdot \operatorname{tr} \underset{\sim}{h} = 0 \qquad (2.17)$$

can easily be obtained. Into this equation we insert

$$\operatorname{tr} \underset{\sim}{h} = C_3 \cdot (-1 - \frac{2}{3} \cdot \frac{1-K_o}{1+2K_o} \cdot \frac{C_4}{C_3} + \frac{2}{9} \cdot \frac{1-K_o}{1+2K_o} \cdot \frac{C_5}{C_3}) , \qquad (2.18)$$

and

$$\dot{\sigma}_1'^*/h = -(2+C_1 \cdot (2-\sqrt{2/3})) \cdot (C_2 - (1+2K_o)^3/K_o^2)/3 + \frac{2}{3} \frac{1-K_o}{1+2K_o} |\operatorname{tr} \underset{\sim}{h}| \qquad (2.19)$$

and solve the resulting equation obtaining thus C_3. Herewith C_4 and C_5 can also be determined by $C_4 = \frac{C_4}{C_3} \cdot C_3$, $C_5 = \frac{C_5}{C_3} \cdot C_3$.

The coefficients so far determined are controlling the direction of the stress rate related to a given strain rate. The remaining coefficients control the material stiffness and can therefore be obtained by evaluating stress-strain curves. Consider an isotropic compression (i.e. $\operatorname{tr} \underset{\sim}{D} < 0$, $\underset{\sim}{D}^* = \underset{\sim}{0}$) starting from a hydrostatic stress state (i.e. $\underset{\sim}{T}^* = \underset{\sim}{0}$), and an isotropic extension (i.e. $\operatorname{tr} \underset{\sim}{D} > 0$, $\underset{\sim}{D}^* = \underset{\sim}{0}$) starting from the same stress state. Let the corresponding incremental stiffnesses, κ_c and κ_e, defined by $\kappa = \operatorname{tr} \underset{\sim}{\dot{T}}/\operatorname{tr} \underset{\sim}{D}$, be given experimentally. It then follows:

$$\kappa_c = -C_3 \cdot \operatorname{tr} T \cdot C_6 \cdot (1 - \sqrt{3} \cdot C_8) \qquad (2.20)$$

$$\kappa_e = -C_3 \cdot \operatorname{tr} T \cdot C_6 \cdot (1 + \sqrt{3} \cdot C_8) \qquad (2.21)$$

From these equations the values of C_6 and C_8 can be calculated:

$$C_6 = \frac{1}{-C_3 \cdot \text{tr}\underset{\sim}{T}} \cdot \frac{\kappa_e + \kappa_c}{2} \tag{2.24}$$

$$C_8 = \sqrt{3} \cdot \frac{\kappa_e - \kappa_c}{\kappa_e + \kappa_c} \tag{2.25}$$

Alternatively, the initial slope of the stress-strain curve of a cylindrical compression test and either κ_c or κ_e or κ_e/κ_c can be used to obtain C_6 and C_8.

The remaining constant C_7 influences the form (curvature) of the several stress-strain curves calculated by the constitutive law. A convenient way to determine its value is the fit of a calculated to a measured curve by trial and error.

3. Evaluation of the data given to the Predictors

3.1 Ottawa sand

As outlined above, the constant C_1 is supposed to be equal to 0.6. The data given lead to ϕ' values in the range of $43.8°$ to $60.6°$. This great scatter given, the fit of all the constants is problematic. Anyhow, relying upon the triaxial compression test with constant mean stress equal to 20 psi, yielding $\phi' = 43.8°$ [1], the value $C_2 = 76.60$ is obtained. The same test yields $\beta_{Dc} = 20.1°$ (or $18.4°$)[2]. But no information about β_{De}. The other tests give no or contradictory information about β_D at extension (cf compression tests at mean pressure p = 5 psi and at $\sigma_x = \sigma_y = 10$ psi). For lack of reliable information we have assumed $\beta_{Dc} = \beta_{De} = 22°$ such that $C_4/C_3 = 1.44$ and $C_5/C_3 = 0$ follows.

---

[1] The value $\phi' = 45.6°$, which corresponds to the reported stresses $\sigma_x = \sigma_y = 7.5$ psi, $\sigma_z = 45$ psi, has not been considered, since it holds for the extremely high strain of $\varepsilon_z = 17.7\%$, for which the sample deformation was certainly non-homogeneous.

[2] This value is ambiguous as the reported strain-increments in x- and y-directions are not identical.

The given data provide no information about K_o. According to JAKY's formula it is assumed as $K_o = 1 - \sin \phi' = 1 - \sin 43.8° = 0.308$. With $K_o = 0.29$ we obtained: $C_3 = 31.1$ and $C_4 = 44.8$.
The stiffnesses for purely volumetric compression and extension according to the given hydrostatic test results differ for $\sigma = 10$ psi but they are reported to be equal to each other for $\sigma = 4$ psi. The latter case yields $C_8 = 0$.

The remaining constants, C_6 and C_7, are determined by fitting the calculated stress-strain curve to the data measured with the above mentioned triaxial compression test with $\sigma_1' + 2\sigma_2' = 20$ psi. The values so obtained are $C_6 = 1.6$ and $C_7 = 4$.

The obtained fit and predictions are plotted in Figs. 3 to 7.

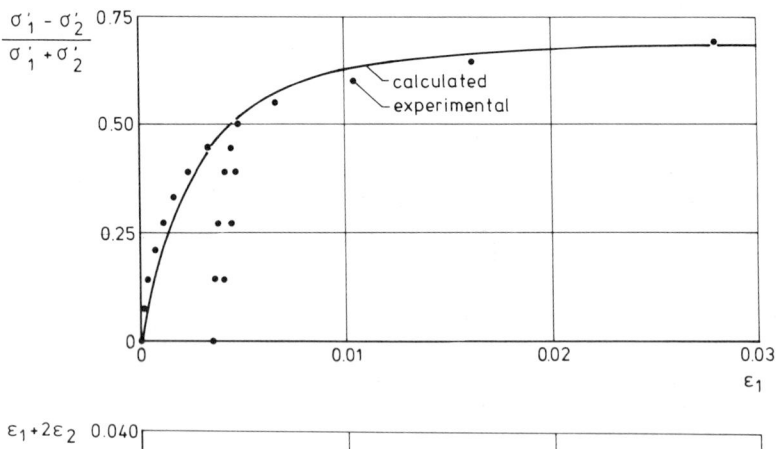

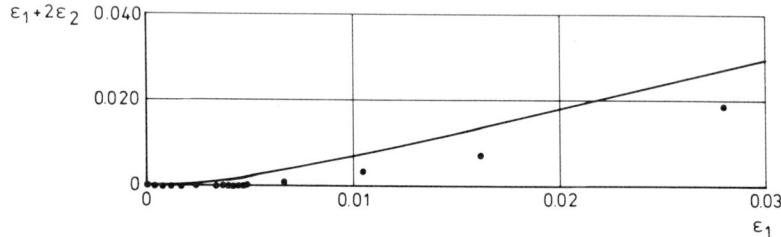

Fig. 3 Fit of the Ottawa sand data for triaxial compression with mean stress $\sigma = 20$ psi.

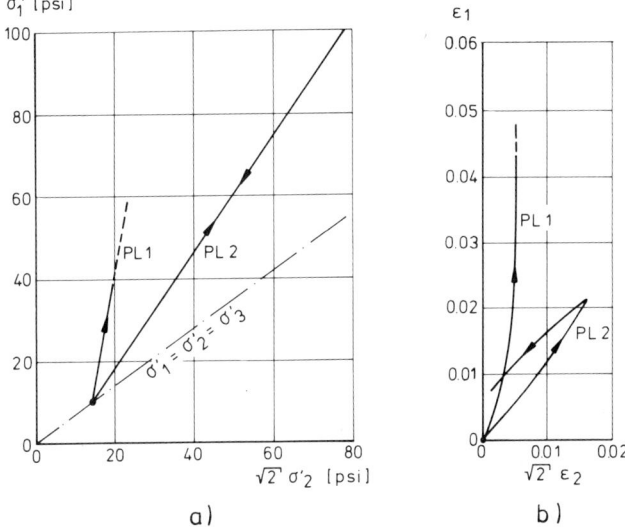

Fig. 4 Predicted strain paths (b) of Ottawa sand for proportional loadings (a)

For the given stress paths Equation 2.10 had to be inverted numerically for every integration step. This comes up to solving a system of non-linear equations. The graphical representations deviate from the formats given to the Predictors for the following reasons: i) Units, as psi e.g., can and should always be avoided by use of dimensionless variables. ii) By use of invariants an arbitrary choice of co-ordinates is avoided. iii) The plots of τ_{oct} vs ε_x, ε_y, ε_z as desired in the format instructions can hardly reveal three-dimensional stress-strain behaviour (note that neither stress nor strain components are necessarily independent variables, and that ordinary σ vs. ε plots can only work for one-dimensional processes).

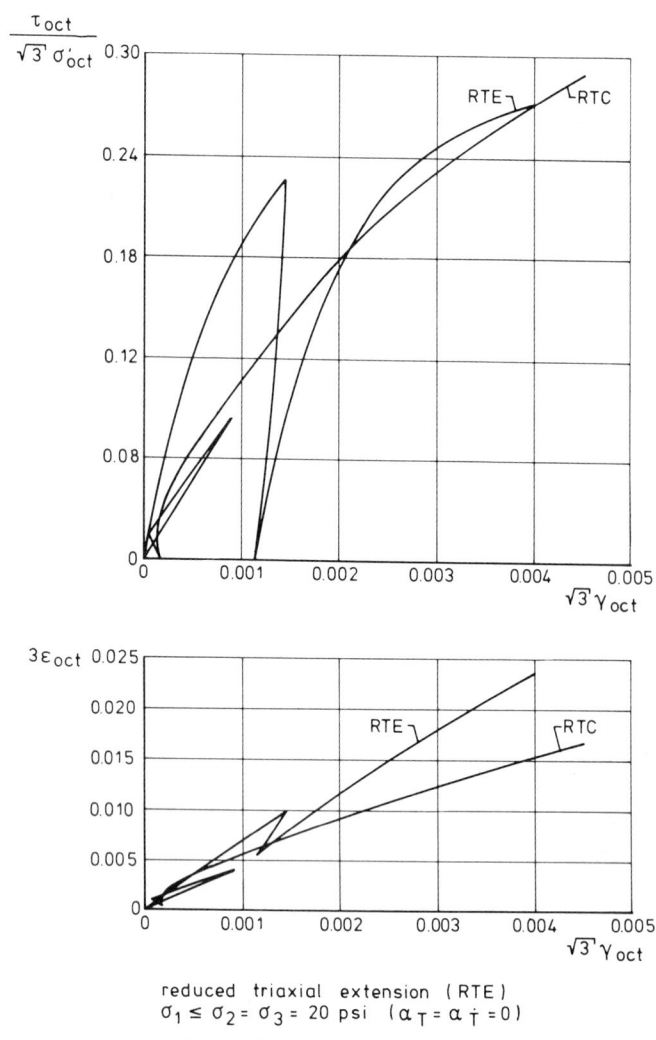

Fig. 5 Ottawa sand: Predicted stress-strain curves and volume changes for reduced triaxial extension (RTE) and reduced triaxial compression (RTC) tests.

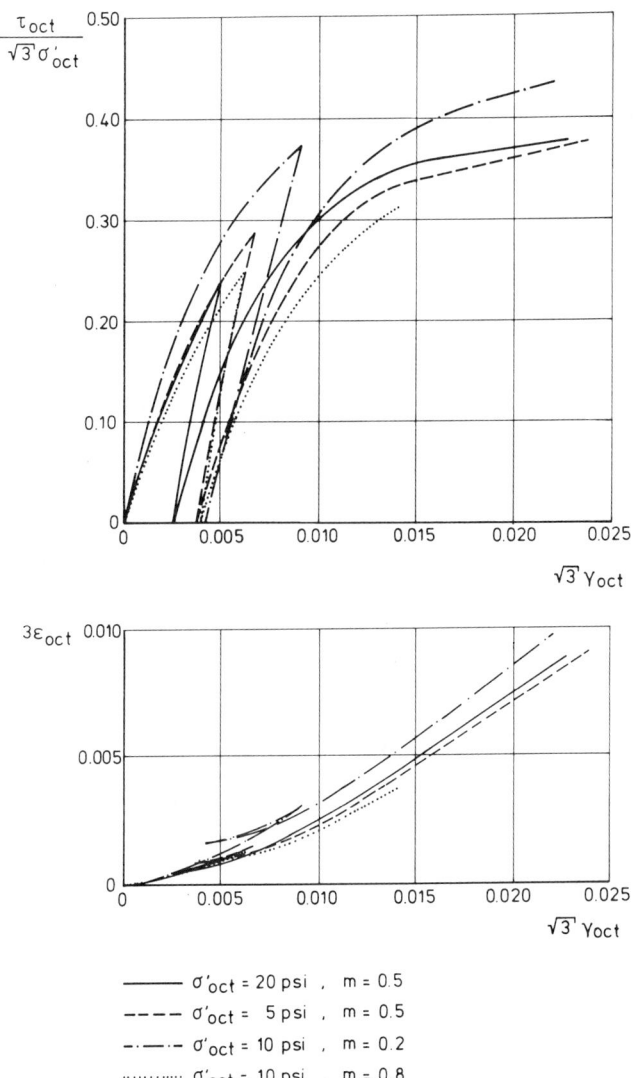

Fig. 6 Ottawa Sand: Predicted stress-strain curves (a) and volume changes (b) for deviatoric stress paths (SS).

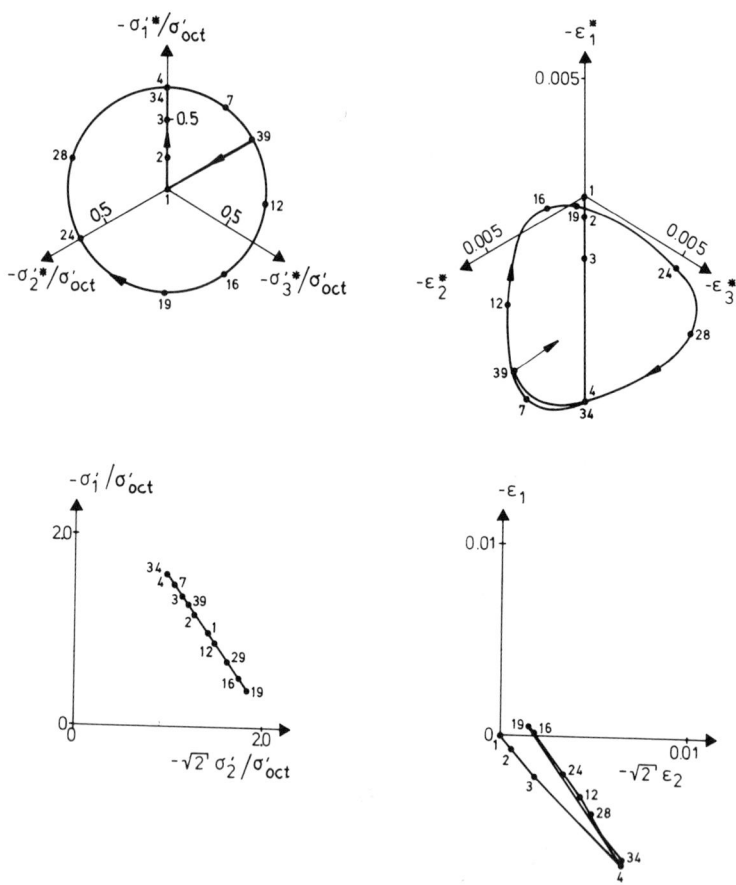

Fig. 7 Ottawa Sand: Predicted strain path for a circular stress path (CSP). Representation with projections onto the deviatoric plane (above) and on the Henkel plane (below).

3.2 Laboratory prepared caolinite clay

Again, $C_1 = 0.6$ has been assumed. Since no test results with strain reversals were available, nothing can be inferred about the constants C_5 and C_8. They have been set to zero. This arbitrarity might have no great influence in the following, since the tests to be predicted contain no strain reversals.

It can be seen from the data that the effective stress path for the compression test is, to a great portion, perpendicular to the line $\sigma_1' = \sigma_2' = \sigma_3'$. This means that $\operatorname{tr}\underset{\sim}{T}$ remains constant for undrained deformation and, as a consequence, $C_4 = 0$, i.e. the considered soil exhibits no dilatancy effects. Admittedly, this is a crude simplification but the data supplied do not allow any more sophisticated fit of our constitutive law to the behaviour of the considered soil. Note that even in a saturated soil with no capability for dilatancy or contractancy pore pressure might be developed in the course of some appropriate undrained deformation due to an increase of total mean stress. As stated in Sec. 2.1 the equivalent stress $\underset{\sim}{T}_e$ has been taken equal to the consolidation stress. According to the data the components of $\underset{\sim}{T}_e$ reads:

$$\sigma_{e2} = \sigma_{e3} = +40 \text{ psi}, \quad \text{and with } K_o = 0.48,$$
$$\sigma_{e1} = +40/0.48 = 83.33 \text{ psi}.$$

To obtain the equivalent stresses in the course of, say, Test No.1, the reported stress increases are added to the initial values given above. For the limiting state we have $\sigma_{e1} = +83.33 +(82.5-40.05) = 125.78$ psi, and $\sigma_{e2} = \sigma_{e3} = 25.20$ psi. It then follows: $K_a = 25.20/125.78 = 0.20$, yielding $C_2 = 68.46$.

The constant C_3 is not needed for the prediction, since only undrained (i.e. $\operatorname{tr}\underset{\sim}{D} = 0$) test results have to be predicted.

The remaining constants C_6 and C_7 are obtained by fitting the calculated stress-strain curve to the experimental one according to the data for compression tests.

Note that for both reported compression tests (i.e. tests No. 1 and No. 4) the soil samples undergo the same deformation and, therefore, the effective stress paths should be identical. It was found that $C_6 = 1.74$ and $C_7 = 2.0$ give a fairly good fit (Fig.8).

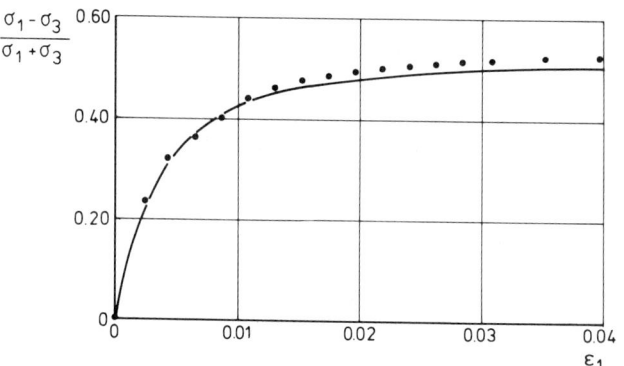

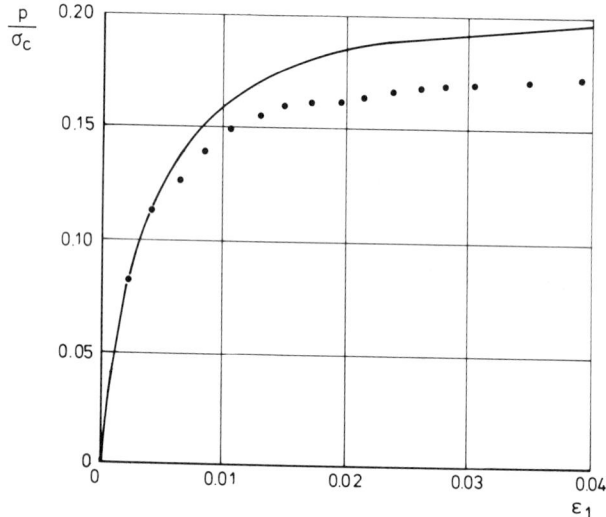

Fig. 8: Kaolinite Clay: Fit of the data (Test No. 1). p denotes pore pressure.

In order to compute the strains for the tests to be predicted, the type of motion underlying these tests must be given. At this point a serious question arises since no statement concerning the type of motion (deformation) of these tests was given to the Predictors. From the enclosed tables to be filled no hint could be gained, since an array of the principal stresses, $\sigma_1, \sigma_2, \sigma_3$, versus the principal strains, $\varepsilon_1, \varepsilon_2, \varepsilon_3$, is conceivable for <u>any</u> deformation, e.g. cuboidal or torsional. By association with the tests with Ottawa sand and natural clays we presumed that rectilinear extensions (i.e. cuboidal deformations or true triaxial tests) had been performed with the clay samples. We interpretated the given information in the following way:

After consolidation of the remolded clay tilted samples were cut and further deformed in some true triaxial apparatus. (Fig. 9).

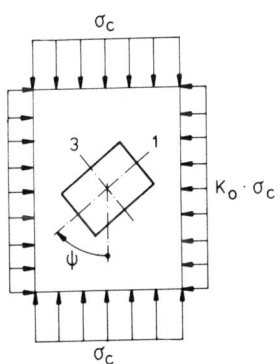

In the cases considered here the principal axes of $\underset{\sim}{T}_e$ do not coincide with the x_1, x_2, x_3-directions which are the principal directions of stress and strain tensors during the presumed cuboidal deformation. These directions are related to the principal directions of the consolidation stress by a rotation leaving the x_2-direction unchanged. The consolidation stress reads, in the rotated system:

Fig. 9: Hypothetical history of Kaolin samples

$$\underset{\sim}{T}_C = \sigma_C \cdot \begin{bmatrix} K_0 \cdot \sin^2\psi + \cos^2\psi & 0 & (1-K_0) \cdot \sin\psi \cdot \cos\psi \\ 0 & K_0 & 0 \\ (1-K_0) \cdot \sin\psi \cdot \cos\psi & 0 & K_0 \cdot \cos^2\psi + \sin^2\psi \end{bmatrix} \quad (3.1)$$

Using these components and the constituve law of section 2 the strains for the given stress paths were calculated. To this purpose our constitutive law, which yields the stress increments as a function of the strain increments, had again to be inverted numerically in every step of the numerical integration. Fig. 10 shows the predicted strains and pore pressures in dimensionless invariant form.

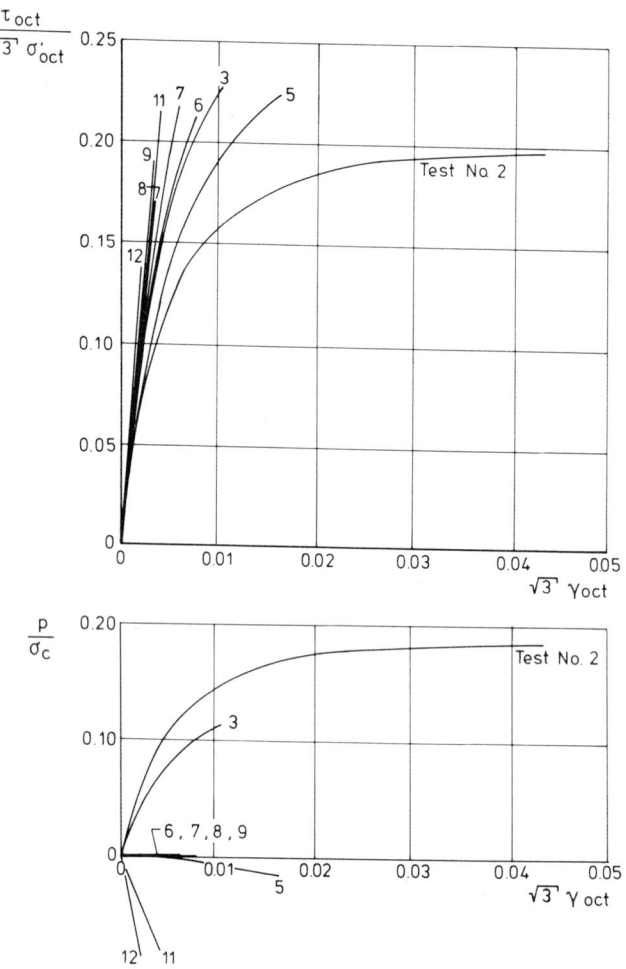

Fig. 10: Kaolinite Clay: Predicted stress-strain curves and pore pressures

4. Concluding remarks

In the preceding Section we have outlined the way we calculated, by means of our constitutive law, the strain paths related to several given stress paths. In order to judge the effectiveness of our law, calculated strain paths should be compared to measured ones. At this point, two questions arise. First, the reliability of the experimental results should be assured, and secondly an appropriate norm should be established in order to value the deviations from the measurements. Usually, the second requirement is realized by a mere visual estimation of the plotted results. Therefore the plots should be as comprehensive as possible. Special difficulties arise if the plots are to represent parametrized three-dimensional curves. For these reasons we didn't follow the proposed formats for the plots but we preferred to plot the relations between some appropriate invariants or to use the projections of the paths onto the deviatoric and Henkel planes. Furthermore, the use of dimensionless variables enabled the plotted results to be independent of any units.

As regards the reliability of the experimental results presented, one has to point out some possible sources of errors. First, the homogeneity of deformation throughout the specimen should always be assured. If not, it makes no sense to mention any strain without specifying the place this strain refers to. Furthermore, the evaluation of a non-homogeneously deformed test requires the stress and strain distribution within the sample to be known. As is well known, no experimental technique permits to fulfil this requirement satisfactorily. Therefore, test results with non-homogeneously deformed samples should be rejected. Consider, in this respect, the triaxial extension test results with mean stress equal to 10 psi: Although the stress histories in x- and z-directions are identical, the final strains are reported to be $\varepsilon_x = 2.230$ % and $\varepsilon_z = 0.707$ %.

Another requirement for test results is that the measured strains and stresses refer to the sample only and not to the experimental device. For example, in experiments with sand the compression due to penetration of sand grains into the rubber should be taken into account if an overestimation of the strains is to be avoided.

Although no experimental details have been enclosed in the data package, we will try to demonstrate the importance of this effect by referring to some hypothetical dimensions of the sample. Consider a hydrostatic test within a cuboidal deformation apparatus (true triaxial test). Let the sample be a cube of 10 x 10 x 10 cm enclosed in a 0.2 mm thick rubber membrane. According to our experience a stress increase from 4 psi to 10 psi, e.g., will cause an average penetration of medium sand grains by 0.1 mm. As a consequence, an average compressive strain of 0.2% would be measured even for an incompressible sand. This value should be compared to the reported value of ε_{avg} = 0.0135% for a hydrostatic stress increase from 4 psi to 10 psi.

Moreover, it should be taken into account that in the course of certain stress-controlled cuboidal deformation tests edge constraints could give rise to erroneous stress registrations. This fact could explain the high amounts and scatter of the reported ϕ'-values.

Consider, for example, the reported results of triaxial extension tests with constant mean stress. The plotted curves of $(\sigma_1' - \sigma_2')/(\sigma_1' + \sigma_2')$ vs. axial strain ε_1 (Fig. 11) do not show

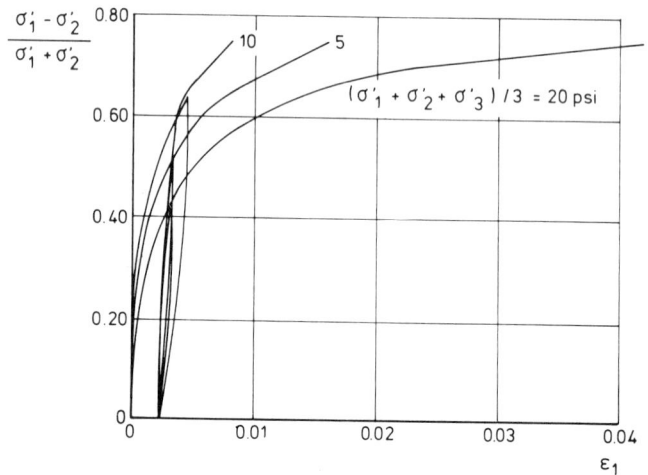

Fig. 11: Some of the stress-strain data given for Ottawa sand (Deviatoric extensions with constant mean stress)

even a constant tendency with increasing mean stress since the highest curve corresponds to the intermediate value of mean stress (10 psi). Consider, further, the "simple shear" test results for Ottawa sand with $\sigma'_{oct} = 20$ psi and $m = 0.5$. It is reported that for $\varepsilon_x = -5.694$ %, $\varepsilon_y = 1.389$ %, $\varepsilon_z = 2.831$ %, $\sigma'_x = 0$, $\sigma'_y = 20$ psi, $\sigma'_z = 40$ psi holds. These values correspond to $\phi' = 90°$, a value which is impossible for dry sand. These data clearly indicate an uncontrolled constraint, probably acting along the edges of the sample.

The predicted behaviour of the laboratory prepared Kaolin clay is considerably "stiffer" than experimentally obtained (Fig.10). This fact is not surprising if the experimental results have been obtained with torsional deformations of hollow cylinders. In this case the motion is entirely different from the motion underlying the predictions. Moreover, no homogeneous shear deformation can be achieved by torsion of a hollow cylinder. This can be easily seen if the components of the stretching tensor $\underset{\sim}{D}$ are expressed in cylindrical co-ordinates r, ψ, z:

$$\underset{\sim}{D} := \begin{bmatrix} \frac{\partial u}{\partial r} & \frac{1}{2}(\frac{1}{r}\frac{\partial u}{\partial \psi} + \frac{\partial v}{\partial r} - \frac{v}{r}) & \frac{1}{2}(\frac{\partial u}{\partial z} + \frac{\partial w}{\partial r}) \\ \frac{1}{2}(\frac{1}{r}\frac{\partial u}{\partial \psi} + \frac{\partial v}{\partial r} - \frac{v}{r}) & \frac{1}{r}\frac{\partial v}{\partial \psi} + \frac{u}{r} & \frac{1}{2}(\frac{\partial v}{\partial z} + \frac{1}{r}\frac{\partial w}{\partial \psi}) \\ \frac{1}{2}(\frac{\partial u}{\partial z} + \frac{\partial w}{\partial r}) & \frac{1}{2}(\frac{\partial v}{\partial z} + \frac{1}{r}\frac{\partial w}{\partial \psi}) & \frac{\partial w}{\partial z} \end{bmatrix}$$

u, v, w are the components of the velocity vector in radial, tangential and axial directions, respectively. Setting the six different components D_{ij} equal to some constants, i.e. D_{ij} = const, $i \leq j$, a system of six partial differential equations is obtained. By taking into account the rotational symmetry condition, $\frac{\partial}{\partial \psi} = 0$, one gets the result that there is no vector field fulfilling these differential equations unless D_{12} and D_{23} vanish. In other words, homogeneity of deformation excludes shear in the direction of the torsional moment.

Due to the reasons pointed out above, we have to state that, unfortunately, the furnished experimental data do not allow to check our theory. It should be kept in mind that our aim is to formulate a theory describing not only certain experimental results, but the material behaviour for a wide variety of stress and strain histories. Checking this theory requires sophisticated experiments such as cuboidal deformation tests with kinematical boundary conditions or improved cylindrical compression tests securing homogeneous deformation and reliable measurement of volume changes. We have to concede that there are certain rather special paths which our constitutive law fails to cover realistically. Such stress paths are the re-loading paths after deviatoric unloading towards a hydrostatic stress state. According to our theory, the stress-strain curves for such paths do not differ from the stress-strain curves for first loadings, whereas the tests show that soil is considerably stiffer under reloading. This kind of hardening is specially covered by elastoplastic constitutive laws, and the stress paths chosen for the predictions are just of the special type mentioned above. It would be much more relevant for checking constitutive laws to realize paths deviating from ordinary reloading. The essential shortcomings of elastoplastic laws arise with marked changes of path directions ("sideward loading"), whereas our constitutive law is just formulated to cover this type of behaviour. As such paths are practically relevant we propose to consider them in more detail when further formulating and checking constitutive laws.

References

GOLDSCHEIDER, M. (1972): Spannungen in Sand bei räumlicher, monotoner Verformung. Dissertation, Universität Karlsruhe

GOLDSCHEIDER, M. (1976): Grenzbedingung und Fließregel von Sand. Mech.Res.Comm. Vol. 3, p. 463-468.

GUDEHUS, G., GOLDSCHEIDER, M. and WINTER, H. (1977): Mechanical Properties of Sand and Clay and Numerical Integration Methods. Ch. 3 of Finite Element in Geomechanics, G. Gudehus (Ed.), Wiley, London.

GUDEHUS, G. (1980): Materialverhalten von Sand: Neuere Erkenntnisse. Der Bauingenieur, Vol. 55, p. 57-67.

GUDEHUS, G., KOLYMBAS, D. (1979): A constitutive law of the rate type for soils. Proceed. 3rd Int. Conf. Num. Meth. Geomechanics Aachen 1979, ed. W. Wittke, Balkema, Rotterdam.

HETTLER, A. and GUDEHUS, G. (1980): Estimation of shakedown displacement in sand bodies with the aid of model tests. Proceed. Int. Symp. Soils under cycl. and Trans. Loading (Swansea) Balkema, Rotterdam.

KOLYMBAS, D. (1980): Bifurcation Analysis for Sand Samples with a non-linear constitutive equation. Accepted for publication in Ingenieur-Archiv.

POPPER, K.R. (1977 and earlier): The Logic of Scientific Discovery Hutchinson, London

RIVLIN, R.S., ERICKSEN, J.L. (1960): Stress-Deformation Relations for Isotropic Materials. Journal of Rational Mech. Anal. Vol. 4, p. 323-425.

TRUESDELL, C. NOLL, W. (1965): The Non-Linear Field Theories of Mechanics. Handbuch der Physik, Vol. III/3, Fluegge (Ed.) Springer, Berlin etc.

VARDOULAKIS, I., GOLDSCHEIDER, M., GUDEHUS, G. (1973): Formation of Shear Bands in Sand Bodies as a Bifurcation Problem. Int. J. Num. Anal. Meth. Geomechanics, Vol. 2, p. 99-128.

VARDOULAKIS, I. (1978): Bifurcation Analysis of the Triaxial Test on Sand Samples. Acta Mechanica Vol. 32, p. 35-54.

VARDOULAKIS, I. (1980): Shear Band Inclination and Shear Moduls of Sand in Biaxial Tests. Int.J.Num.Anal.Methods Geomechanics, Vol. 4, p. 103-119.

WANG, C.C. (1970): A New Representation Theorem for Isotropic Functions. Arch. Rat. Mech. Anal., Vol. 36, p. 203 ff.

LIST OF WORKSHOP PARTICIPANTS

T. Adachi	Kyoto University
A. Ansal	Istanbul Technical University
C.A. Babendreier	National Science Foundation
G.Y. Baladi	U.S. Army Corps of Engineers
M.M. Baligh	Massachusetts Institute of Technology
Z.P Bazant	Northwestern University
R.E. Brown	Law Engineering
C.S. Chang	University of Massachusetts
W.F. Chen	Purdue University
J.T. Christian	Stone & Webster Engineering Corporation
Y.F. Dafalias	University of California at Davis
K.R. Demars	University of Delaware
L. Domaschuk	University of Manitoba
V.P. Drnevich	University of Kentucky
J.M. Duncan	University of California at Berkeley
W.D.L. Finn	University of British Columbia
G. Gudehus	Karlsruhe University
A.M. Hanna	Concordia University
L. Hansen	University of Arizona
B.O. Hardin	University of Kentucky
R.D. Holtz	Purdue University
H.M. Horn	Woodward-Clyde Consultants
G.T. Houlsby	Cambridge University
R.D. Japp	McGill University
E. Kavazanjian	Stanford University
T.C. Kenney	University of Toronto
R.P. Khera	New Jersey Institute of Technology
R.C. Kirby	Woodward-Clyde Consultants
H.Y. Ko	University of Colorado
S. Koh	Purdue University
L.M. Kraft	McClelland Engineers
R.J. Krizek	Northwestern University
C.C. Ladd	Massachusetts Institute of Technology
R.S. Ladd	Woodward-Clyde Consultants
P.V. Lade	University of California at Los Angeles
I. Lam	Fugro, Inc.
K.T. Law	National Research Coundil, Canada
S. Leroueil	Laval University
K.Y. Lo	University of Western Ontario
H. Ludwig	Schnabel Engineering

W.F. Marcuson	U.S. Army Corps of Engineers
P. Mayne	Law Engineering
R.L. McNeill	Sandia Laboratories
J.K. Mitchell	University of California at Berkeley
H.W. Olsen	U.S. Geological Survey
J.C. Olser	McGill University
J.H. Prevost	Princeton University
H. Poorooshasb	Concordia University
S.J. Poulos	Geotechnical Engineers, Inc.
H.E. Read	System, Science and Software
K.R. Rowe	University of Western Ontario
A.S. Saada	Case Western Reserve University
I. Sandler	Weidlinger Associates
R.F. Scott	California Institute of Technology
E.T. Selig	University of Massachusetts
V. Silvestri	L'Ecole Polytechnique de Montreal
R. Singh	Woodward-Clyde Consultants
M. Soulie	L'Ecole Polytechnique de Montreal
K.H. Stokoe	University of Texas at Austin
S. Sture	University of Colorado
M.M. Tabba	Campagnie Nationale de Forage et Sondage
F. Tavenas	Laval University
F.C. Townsend	University of Florida
I. Vardoulakis	University of Minnesota
H.E. Wahls	North Carolina State University
S.G. Wright	University of Texas at Austin
C.P. Wroth	Oxford University
R.N. Yong	McGill University

SUBJECT INDEX

Page numbers refer to first page of paper

Academic user needs, 68
Analytical models, 102, 745

Bounding surface, 711

Cap model, 649
Cap model predictions, 364
Clay behavior, 402, 711, 815
Clay behavior predictions, 592
Clay constitutive models, 711
Clay response predictions, 402
Clay testing, 402, 711
Clays, 151, 176, 245, 254, 265, 328, 364, 416, 443, 461, 592, 649, 815, 839
Computer applications, 48, 151
Consolidated clays, 815
Constitutive law, 839
Constitutive models, 286, 492, 745, 815
Constitutive relations, 151
Constitutive theory, 745
Continuum mechanics, 592
Creep, 815
Cyclic loading, 815

Dams, 443
Deformation, 461
Design, 48

Elastic-plastic materials, 328
Elasticity, 553, 628
Endochronic theory, 286, 539
Excavations, 443

Finite element method, 492

Generalized stress-strain, 1
Geotechnical engineering, 48
Gravels, 443

Hardening, 93
Hyperbolic models, 245, 443
Hyperelasticity, 176, 265

Industry user needs, 68
Industry users, 61

Kaolin, 176
Kaolinite clays, 364

Loading, 166
Loading tests, 166, 539

Mathematical calculations, 592
Mathematical models, 492, 592
Mathematical techniques, 492
Models, 48, 166, 254, 328, 352, 402

Natural clays, 402
Numerical techniques, 48

Ottawa sand, 265, 352, 364, 416, 628, 649
Over-consolidated clays, 815

Phenomenological models, 102, 461
Plastic strain, 553
Plasticity, 1, 93, 102, 132, 151, 176, 328, 628, 649, 711
Plasticity models, 553
Plasticity theory, 815
Porewater pressure, 132
Predictions, 132, 151, 166, 461

Rate sensitivity, 815
Rheological performance, 1
Rock mechanics, 553
Rockfills, 443

Sand behavior, 402, 815
Sand behavior predictions, 628
Sands, 151, 176, 443, 628, 649, 815, 839
Shear tests, 352
Silts, 443
Soil behavior, 48, 93, 102, 151, 265, 286, 461, 628, 649, 745
Soil behavior modeling, 1, 539, 592, 839
Soil behavior predictions, 352, 416
Soil consolidation, 93
Soil constitutive models, 46, 61, 68, 93, 265, 416, 461, 539, 649
Soil constitutive relations, 48, 286
Soil mechanics, 1, 93, 102, 151, 286, 402, 553, 711
Soil modeling, 102, 553
Soil parameters, 628
Soil performance, 166
Soil performance predictions, 176, 328
Soil properties, 48, 254, 492
Soil response, 166, 364, 592
Soil response modeling, 492
Soil response predictions, 286, 539, 839
Soil testing, 254, 286

Soil tests, 151, 166, 245, 265, 416, 745
Soil-structure interactions, 443
Soils, 1, 132, 166, 176, 245, 254, 265, 328, 364, 443, 461, 492, 539, 553, 649, 745
Soils under load, 48
State variables, 151
Strain predictions, 245
Strains, 176, 245, 265, 352, 416
Stress analysis, 443
Stress relaxation, 815
Stress-deformation predictions, 254
Stress-strain, 93, 102, 151
Stress-strain behavior, 352, 492, 649, 745
Stress-strain modeling, 364, 553, 628
Stress-strain models, 132, 245
Stress-strain relationships, 443
Stresses, 176, 265, 416, 461

Theoretical models, 61, 839
Triaxial compression, 352
Triaxial compression tests, 352
Triaxial extension, 352

User needs, 61, 68

AUTHOR INDEX

Page number refers to first page of paper

Adachi, Toshihisa, 416, 815
Akai, Koichi, 416, 815
Ansal, Atilla M., 286, 539

Baladi, George Y., 364, 649
Bazant, Zdenek P., 286, 539
Bonaparte, Rudolph, 254, 461

Chen, W.F., 265, 328, 492, 553
Christian, John T., 61, 93

Dafalias, Yannis F., 402, 711
DeNatale, Jay Scott, 402, 711
Duncan, James M., 245, 443

Finn, W.D. Liam, 132

Gudehus, G., 839

Herrmann, Leonard R., 402, 711
Houlsby, Guy T., 592

Kavazanjian, Edward, Jr., 254, 461
Ko, Hon-Yim, 1, 48, 166, 176
Kolymbas, D., 839
Krizek, Raymond J., 286, 539

Lade, Poul V., 352, 628

Mitchell, James K., 254, 461
Mizuno, E., 328, 553
Mould, J.C., 176

Oka, Fusao, 416, 815

Poorooshasb, H.B., 102
Prevost, Jean H., 745

Saleeb, A.F., 265, 492
Sandler, Ivan S., 364, 649
Scott, R.F., 151
Selig, Ernest T., 68, 102

Wroth, C. Peter, 592

Yong, Raymond N., 1, 48, 166, 176

RAYMOND H. FOGLER LIBRARY
DATE DUE

**BOOKS ARE SUBJECT TO
RECALL AFTER TWO WEEKS**